THE
EIGHTH DAY
OF CREATION

Books by Horace Freeland Judson

❖ ───────────────────────────────

The Techniques of Reading
Heroin Addiction in Britain
The Eighth Day of Creation
The Search for Solutions

THE
EIGHTH DAY
OF CREATION

*Makers of the Revolution
in Biology*

EXPANDED EDITION

HORACE FREELAND JUDSON

Cold Spring Harbor Laboratory Press 1996

THE
EIGHTH DAY
OF CREATION

*Makers of the Revolution
in Biology*

EXPANDED EDITION

Library of Congress Cataloging in Publication Data

Judson, Horace Freeland.
   The eighth day of creation : makers of the revolution in
biology / Horace Freeland Judson. -- Expanded ed.
       p.   cm.
   Includes bibliographical references and index.
   ISBN 0-87969-477-7 (cloth : alk. paper). -- ISBN 0-87969-478-5
(pbk. : alk. paper)
   1. Molecular biology--History.   I. Title.
QH506.J83   1995
574.8'8'09--dc20   95-45151
                                                                CIP

All Cold Spring Harbor Laboratory Press publications may be ordered directly from
Cold Spring Harbor Laboratory Press, 10 Skyline Drive, Plainview, New York 11803-
2500. Phone: 1-800-843-4388 in Continental U.S. and Canada. All other locations: (516)
349-1930. FAX: (516) 349-1946. E-mail: cshpress@cshl.org. For a complete catalog of all
Cold Spring Harbor Laboratory Press publications, visit our World Wide Web Site
http://www.cshl.org/

For Penelope
at the end of the voyage
1979

❖ ❖ ❖

1993
always and always and always
Sunt lacrimae rerum

Contents

I ❖ DNA

Function and Structure:
The elucidation of the structure of deoxyribonucleic acid,
the genetic material

EXHIBITS

INTERLUDE

On the State of Molecular Biology Early in the 1970s

Preface to the Expanded Edition

I am on my second copy of *The Eighth Day of Creation*. Sadly, the first went the way of most valued books—borrowed by a person who no doubt still thinks of me as a friend, but whose name I have forgotten. What I miss are the notes I shamelessly made in the margins. The text, as I found, used to be replaceable; I had not appreciated the danger that it might go permanently out of print. That is why it is such excellent news that the Cold Spring Harbor Laboratory Press is now to reissue the book.

The reasons why Horace Judson's book should remain in print are legion and well-rehearsed, but the promise to write this piece has been a spur to read the book again, from cover to cover, as if picking it up for the first time. What a pleasure! It is, for example, uncanny how accurately Judson's quotations have captured the tones of the voices of the principal actors. Sydney Brenner does indeed pepper his conversation with "OK?", like a teacher anxious that his students should have understood. Francis Crick does indeed have a knack of describing his reactions to past events as if they had been spoken aloud, as in "At the time, we said, 'Isn't that funny, they seem to be arranged in strings' . . ." (about the arrangement of ribosomes in the cell). They were (and luckily are) very real people.

There are three principal episodes in the drama Judson reconstructs: finding the structure of DNA, recognising that messenger RNA is an intermediate between the genetic material and the sites of protein synthesis on ribosomes, and establishing that the genetic code is a triplet code. Now that even schoolchildren know the answers to these questions, Judson's text assumes the importance of an historical document in its own right by its meticulous and vivid account of why such bright people spent so long coming to conclusions that are now self-evident. For what it is worth, the paper in which Crick, Brenner and others showed that three consecutive frameshift mutations will rescue protein synthesis may have been the most elegant paper *Nature* ever published (*Nature* 192, 1227–1232; 1961).

The texture of Judson's account of the antecedents of the important discoveries is similarly thrilling for its richness. T. H. Morgan and H. J. Müller, the grand old men of classical genetics, might well have resented the arrival of the young Turks, but they emerge instead as generous people eager to see their abstract genes made real. Famously, Judson has also done an heroic job in giving the French full credit for their contribution to the understanding of how DNA functions—and not just Monod and Jacob, but the likes of Boris Ephrussi and André Lwoff. (His account of Ephrussi's exasperation with Monod's frustrated career as a conductor of light orchestras is one of the great comic tales of science.) The book is a monumental piece of scholarship that will be read and referred to as long as there is science.

And what fun it all was! That is the lasting message of *The Eighth Day*. How did they, those who organised the intellectual revolution of our times, arrange it to be so? The explanation is all buried here in Judson's text, but perhaps merits dissection and display.

It helped enormously that the founders of molecular biology were well-mannered people. The first sentence of Watson's *The Double Helix* ("I have never seen Francis Crick in a modest mood") may be an accurate record of a close colleague's personal experience, but it gives quite the wrong impression of one whose courtesy, even in disagreement, is legendary. Monod, as sure-footed as they come, was the same.

They were also a gregarious lot, as fond of conferences as of parties. And they worked like dogs. Although the Phage Group and Gamow's RNA Tie Club showed signs of clubbiness, they had a serious purpose as well in allowing the sceptical habits of Delbrück and Crick to provide a kind of built-in quality control. It is remarkable how little rubbish was published during the period of which Judson writes. To tell why, it is only necessary to read his final chapter on how the structure and function of haemoglobin were resolved: these were serious people working at a set of problems not previously defined.

The haemoglobin project was the task to which Max Perutz had set his hand early in the nineteen-fifties, when the idea of obtaining the structure of protein molecules by means of x-ray diffraction seemed an heroic ambition. Perutz was also titularly the head of the Medical Research Council's unit at Cambridge, later to become the Laboratory of Molecular Biology and, in the nineteen-sixties, a mecca for the world's postdocs in the field. Perutz is (and was then) a man whose deep pleasure in discovery is matched only by his modesty; his readiness to offer encouragement to more junior colleagues is legendary. His persistence with the structure of haemoglobin through successive refinements was a Herculean task, brought to fruition in 1961 with the recognition at the Cambridge laboratory that the presence or absence of oxygen in horse haemoglobin is associated with a substantial change of the conformation of the protein chains. Unselfish as always, Perutz told the Paris group what had been learned, from which flowed Monod's theory of allostery and the paper by Monod, Wyman, and Changeux (*J. Mol. Biol.* 12, 88–118; 1965) that remains one of the landmarks in molecular biology.

Sheer ebullience apart, they were also modest people in the sense that matters. Judson makes it plain that even after the structure of DNA had been published, not only the Cambridge group, but those at Harvard, Caltech, Paris and elsewhere accurately appreciated the enormity of the task that lay ahead. That is why what might easily have become a bout of fierce inter-laboratory competition became a kind of collective enterprise without organization; the people acknowledged that the problem they had set themselves was bigger than any one of them. Why else were they forever taking in each others' post-doctoral fellows and carrying out experiments in other people's laboratories? The tale Judson tells of how Brenner, having collaborated at Caltech with Messelson and Jacob on the experiment that first proved that messenger-RNA transfers genetic information from nuclear DNA to ribosomes and thus to proteins, rushed off to Stanford to tell people there what he had found, months before a paper for publication had even been drafted.

No longer is the science like that. Between my two spells as Editor of *Nature*, in 1966–'73 and 1980–'95, the most striking change was the huge increase in competitiveness. In the old days, people would occasionally telephone to say they were sending a paper and to hope that we'd give it a fair wind; by 1980, authors would call to ask whether their manuscript had arrived, whether it had been sent to referees, why we had declined to publish it and why it was *Nature's* consistent practice to rely on referees whose intelligence was below par, whose judgement had been warped by self-interest, whose charitable instincts had been blunted by cynicism and whose parentage must even be in doubt. My colleagues and I could never understand why authors did not appreciate an argument that seemed to us undeniable, that a top-class journal can remain so only by being selective. But these endless telephone conversations from 1980 onwards were a telling sign that good manners had ceased to matter.

Sadly, that is only the tip of an iceberg. By 1980, secretiveness had become commonplace. Authors had taken to sending long lists of names to whom, please, manuscripts should not be sent for review, with the explanation, to authors always self-sufficient, that the listed names were those of people working on the same problem. In that case, why not compare notes just to make sure that the conclusions are correct?

Then there was outright fraud. The first case that came *Nature's* way, in 1980, was that of an ingenious Jordanian in Houston whose technique was to steal manuscripts from other people's mailboxes, retype them, replace the authors' names with those of himself and two or three distinguished people (not at Houston), add a sentence thanking the King of Jordan for "encouragement and support" and send them to obscure journals for publication. Reprints could be had from what turned out to be a non-existent address in Brighton, England. It was a hilarious business, but then the nineteen-eighties turned sour, with a string of cases of misconduct, formal investigations (often on the heels of attempts to hide what had really happened) and careers in ruins. It is probably correct that these pathological happenings are mostly explained by an excess of competitiveness, but personal vanity also plays a part. Latterly, the spate of accusations has been attenuated, but the competitiveness is still there.

Indeed, the competition has even intensified, with the recognition that there are useful things to do with molecular biology, and money to be made as well. None of us can guess what the benefits will be, except that they will be huge. The old innocence has long since gone. Will academics' corporate interests now undermine the integrity of the research university, especially at a time when governments everywhere are ready to shuffle responsibility to any other economic sector willing to pay the bills? There is a sombre undertone in Judson's epilogue for this edition suggesting that he is not sure of the answer.

That is another reason why *Eighth Day* should remain in print. Research at its best is the finding of answers to questions about the world that have not previously been asked. Those of us now alive and older than forty too easily forget that we have lived through a period every bit as revolutionary for science and our view of the world as that between the publication of Newton's *Principia* (in 1687) and the generalisation of it in the eighteen-thirties by William Hamilton, Ireland's prodigy. In the long perspective of history, the structure of DNA and all that has flowed from it will seem the most penetrating of all the insights into the working of the natural world gathered during this remarkable century of discovery. None of us can yet know where it will end. To have the origins of this upheaval so well chronicled, in real time as it were, will be a great boon not just to historians but to everybody with an ambition the better to understand the world.

John Maddox
Editor Emeritus
Nature

April 1996

Foreword to the Expanded Edition

Time and the critics, readers and the scientists themselves have been more than kind to *The Eighth Day of Creation*. Occasionally since its publication I have speculated about the possibility of a new edition. A few errors, typographical or factual—gratifyingly few—had turned up; but these in virtually every case required resetting no more than a single line of type, and we made corrections silently as the book went back for reprinting from time to time. In several places, I would have liked to amplify a passage slightly or perhaps to rebalance an appraisal; although none of these affected more than a paragraph or three, such changes would have required laying out whole chapters anew, and could not be made. The most substantial question was what happened in molecular biology after the early nineteen-seventies, when my account of the science closes. The book deals with the classical era of molecular biology, to the point where the overarching structure was established for the mechanisms of transmission, expression, and regulation of genes—at least, established in outline and for the simplest one-celled creatures. Yet as I interviewed the actors, dug into papers, notebooks, and correspondence, and wrote, I was necessarily following also, even as it happened, the great opening-out of the field. Molecular biologists were shifting to another set of problems—broadly, development and differentiation, or what used to be called embryology. For a hundred years this had proved intractable. Molecular biologists, though, brought to these problems a new view of the science, of the kinds of answers required, and the rudiments of a new set of tools. They had begun to adapt the classical methods of molecular biology and to devise others, in order to break the intractable.

A couple of years ago, when the original American publisher let the softcover edition of *The Eighth Day of Creation* go out of print, several others stepped in wanting to reissue the book. Cold Spring Harbor Laboratory Press was among them. I hesitated. Cold Spring Harbor has been so closely identified with molecular biology since the science's earliest days that I feared I might seem the official historian of the movement. But one aspect of the offer—besides money—was decisive. John Inglis, director of the Press, wanted a new edition, and said he would not buy the original publisher's set type and layout but would have the text scanned into the computer, the changes and new material keyed in, and the book redesigned with new type faces and reset. I'd be able to correct, amplify, rebalance—and append an account of later developments.

The central text of *The Eighth Day of Creation* remains all but identical to what it was. Corrections of fact, though such things are never trivial, were small-scale. I had mistakenly said that Archibald Garrod was Scottish; he was English, and has now been repatriated. Professor Hirohiko Noda, the biochemist who translated the book into what I am told is austere, limpid Japanese, caught a typographical error that had put "nitrate" where "citrate" was meant. A quotation in a notebook of Maurice Wilkins' that he insisted was from Kierkegaard, but which no scholar could verify, was identified by a reader as from Robert Musil's novel *Der Mann ohne Eigenschaften*, in the English translation by Wilkins's sister and a colleague. Max Perutz pointed out a bond to be redrawn in a molecular diagram, and suggested wording to clarify the definition of "cubic lattice" in the structures of simple salts that were among the first molecules that Lawrence Bragg solved by x-ray crystallography. And so on.

Significant amplifications were three. I have added some details to the biographical sketch of Rosalind Franklin. François Jacob had told me, but off the record, about conflicts and strains in his working relationship with Jacques Monod; later, though, he wrote about these himself, so I have put them back in. Most importantly, I have unified and brought forward the discussion of Frederick Sanger's work, adding details, to establish from the start how his sequencing of proteins transformed biologists' understanding. Francis Crick, for a significant example, knew of Sanger's work early, well before publication, and just as he was first sorting out the fundamental questions of biology.

The original edition reprinted the first of Watson and Crick's papers announcing the structure of DNA; here we have added their second paper, about the genetical implications of the structure.

The Afterword of the original edition, the one place that said much about later developments, I have modified considerably, and retitled Conclusion: to discuss the changing nature of the science and to outline those developments, I have added a new Epilogue.

As afterwords, this edition reprints two essays. The first I wrote several years back, retelling the story of the discovery of the structure of DNA but strictly from the point of view of Rosalind Franklin: the essay's aim is to liberate her from those who for polemical purposes have misappropriated her as an example, a prime case, of a woman who had a bad career in science because she was a woman. That use of her is incorrect, anachronistic, irresponsible. We must recognize Franklin's considerable accomplishments: when we do so, we will also understand the motives of those who so demean her. The article appears as "In defense of Rosalind Franklin: the myth of the wronged heroine."

The second essay is about Erwin Chargaff. During the writing of *The Eighth Day of Creation,* living in Cambridge, I renewed an acquaintance with Edward Shils—for decades, the *doyen* of American sociologists—who spent every summer and fall at the University of Chicago, the winter and spring at Peterhouse, oldest and smallest of the colleges of the University of Cambridge. After the book appeared, at Shils' invitation I wrote an extended survey of the other principal published accounts relevant to the history of molecular biology, to that time, omitting only James Watson's *The Double Helix*. Shils published the survey in an eclectic quarterly journal he edited, *Minerva*.* Several of those accounts had been helpful, others trivial; two at least were contemptible, poisoned by misconceptions, raddled with error. The important exception was Chargaff's autobiographical sketch, *Heraclitean Fire*. This is a strange, sad work, written with flair and bitterness; my commentary was extensive, and stands independently of the rest of the article. I reprint it here as "The tragedy of Erwin Chargaff."

❖ ❖ ❖

In the nineteen-eighties and early nineties, the various journals that publish pieces by historians of science threw up a number of studies about one aspect or another of the history of molecular biology; most scant contexts and scope. Several books have appeared. Maclyn McCarty, one of Oswald Avery's colleagues in the crucial paper of 1944 that demonstrated that DNA could transform pneumococcal types, heritably, has published a brief memoir of those long years of research. Two biographies of Linus Pauling have recently been published. Most interesting by far is François Jacob's autobiography, *La statue intérieure*: guarded, as is the man, yet skillfully revealing, too, with a lot to say about the science and about his relationship to André Lwoff and especially to Jacques Monod. In the original French, Jacob rises to elegance and daring. Alas, the English translation limps.

❖ ❖ ❖

For the various corrections and modifications in this edition, and especially to prepare the Epilogue, I have drawn upon conversations over the past decade with scores of scientists; I went back into the transcripts of interviews; and I interviewed a number afresh or for the first time. To all, heartfelt thanks. In particular, I'm grateful to Sidney Altman, Brigitte Askonas, David Baltimore, Paul Berg, Sydney Brenner, Nathaniel Comfort, Jeanne Guillemin, Leroy Hood, Nancy Hopkins, François Jacob, Nicholas Judson, Olivia Judson, Sir John Kendrew, Philip Kitcher, Roger Kornberg, Edwin Lennox, Vittorio Luzzati, Sir John Maddox, Hugh McDevitt, Matthew Meselson, Satoshi Mizutani, Cesar Milstein, Sir Gustav Nossal, Max Perutz, Robert Pollack, Frederick Sanger, Kenneth Schaffner, Tilli Tansey, and Howard Temin.

Baltimore
New Year's Day 1996

*Horace Freeland Judson, "Reflections on the historiography of molecular biology." *Minerva* 18 (Autumn 1980): 369–421.

Foreword to the First Edition

The sciences in our century, to be sure, have been marked almost wherever one looks by momentous discoveries, by extraordinary people, by upheavals of understanding—by a dynamism that deserves to be called permanent revolution. Twice, especially, since 1900, scientists and their ideas have generated a transformation so broad and so deep that it touches everyone's most intimate sense of the nature of things. The first of these transformations was in physics, the second in biology. Between the two, we are most of us spontaneously more interested in the science of life; yet until now it is the history of the transformation in physics that has been told.

The revolution in physics came earlier. It began with quantum theory and the theory of relativity, with Max Planck and Albert Einstein, at the start of the century; it encompassed the interior of the atom and the structure of space and time; it ran through the establishment of the basic modern form of quantum mechanics by about the early nineteen-thirties. Most of what has happened in physics since then, at least until recently, has been the playing out of the great discoveries—and of the great underlying shift of view—of those three decades. The decades, that shift of view, the discoveries, and the men who made them are familiar presences, at least in the background, to most of us: after all, they built the form of the world as we now take it to be. The autobiographies of the major participants, their memoirs and philosophical reflections, have been composed, their biographies written in multiple—and they remain long in print, for these were men of intelligence, originality, and, often, eccentricity. The scientific papers have been scrutinized as historical and literary objects. The letters have been catalogued and published. The collaborations have been disentangled, the conferences reconvened on paper with vivid imaginative sympathy, the encounters, the conversations, even the accidents reconstructed.

The revolution in biology stands in contrast. Beginning in the mid-thirties, its first phase, called molecular biology, came to a kind of conclusion—not an end, but a pause to regroup—by about 1970. A coherent if preliminary outline of the nature of life was put together in those decades. This science appeals to us very differently from physics. It directly informs our understanding of ourselves. Its mysteries once seemed dangerous and forbidden; its consequences promise to be practical, personal, urgent. At the same time, biology has been growing accessible to the general reader as it never was before and as the modern physics never can be. Indeed, part of the plausibility of molecular biology to the scientists themselves is that it is superbly easy to visualize. The nonspecialist can understand this science, at least in outline, as it is—as the scientist imagines it. Yet the decades of these discoveries have hardly been touched by historians before now. *The Eighth Day of Creation* is an historical account of the chief discoveries of molecular biology, of how they came to be made, and of their makers—for these, also, though only two or three are yet widely known, were scientists, often, of intelligence, originality, even eccentricity.

The book took shape from three encounters. I first met Max Perutz in Cambridge, England, in the spring of 1968. On getting to know him and his work with the hemoglobin molecule, I conceived the idea of a book about molecular biology based for the most part on discoveries that had been made, usually to start with in his laboratory, about the structures of large molecules in the cell. That might have been a shorter, quicker book than this. Yet the basic idea persists here that with a large biological molecule, knowledge of the structure, in atomic detail, makes functioning comprehensible.

I first met Jacques Monod in Paris in the fall of 1969. I explained my plan for a book. Monod told me at once, "No. You can't do it that way. You have to do it this way"—and he proceeded to list the discoveries and their makers that comprise the main line of molecular biology. I went away chagrined. Within twenty-four hours I realized that Monod was right, that he had outlined my book, and that the writing would be longer and more arduous than I had supposed. I little understood how long, how arduous—and how unfailingly interesting.

I have known Matthew Meselson since the fall of 1946, but it was only in the spring of 1971, in Lexington, Massachusetts, the research well begun, that I explained my plan to him and recited Monod's list. Meselson told me at once, "It's impossible." Then he said, "Well, there's Jim's book, you can lean on that."* Then he said, "Your essential simplification will be to bring out the way most of these things went by pairs. First an idea, a theoretical insight—then the critical experiment that the insight inspired." In this manner, he paired up many of the items on the list. Then Meselson said, "You realize that after you've done the interviews and written the book, you will have to go back and do the interviews all over again." I have done as he told me.

❖ ❖ ❖

Some things *The Eighth Day of Creation* is not. The book is not a history of biochemistry in its intellectual plenitude but of a younger, more impatient science that borrowed a lot from the biochemistry of the nineteen-thirties and

---

*He meant not *The Double Helix*, James Watson's memoir, but his textbook, *Molecular Biology of the Gene*. In the event, I usually sought instruction from Gunther Stent's strong and elegant, historically minded text, *Molecular Genetics*, and from Max Perutz's smaller, older, but limpid introduction, *Proteins and Nucleic Acids*.

-forties to put to somewhat different purposes, before repaying the loan in the nineteen-sixties and since. Again, the book is not a history of genetics in its full elegance but of a late development in genetics that drove the abstraction of the gene down to the physical reality of the sequence of units built along the fibre of a chemical found in the cell. Similarly, the book is neither a history of bacteriology nor one of x-ray crystallography, except that molecular biologists emerged from these sciences, too. Beyond that, though, *The Eighth Day of Creation* is not a history of scientific ideas in the abstract but of scientists in the process of discovery: its climaxes, if I have succeeded, are those rare occasions when—to three or four, to two, to one alone at first—a great understanding became clear. In the act of discovery, ideas and personal styles fuse.

In *The Eighth Day of Creation* I acknowledge the specialist but have written for the general reader—which I believe is no more than the duty of the historian. I have avoided jargon; essential technical terms I have neither shirked nor taken for granted but have tried to make visual, historically comprehensible, and useful. On the other hand, I have not wanted to re-create the events of discovery in a high-key and illusionary realism. Instead, I have used the voices of the scientists themselves.

For *The Eighth Day of Creation* is built of my conversations with the scientists—over a million words of conversations, some taken in pen and notebook, most on tape (and I transcribed them all myself), with one hundred and eleven scientists in person, another eleven by telephone, and with twelve nonscientists—families and friends—again in person. But more than that, over the eight years since I began the book I have interviewed most of the central figures (thirty-two of that hundred and eleven) not once but repeatedly—indeed, revisiting seven or eight of them so often that they must have come to dread my next appearance. I edited the conversations chiefly by selection and sequence. Quotations selected were edited as lightly as possible, often not at all, though I have elided some repetitions and have silently supplied the occasional antecedent; when somebody said something twice, I chose his more effective or informative phrasing. The conversations have been balanced against each other, and have been carried back to the original events by the letters and unpublished papers and memoranda of several of those who had the ideas and did the experiments. I was particularly lucky to have access to files and unpublished correspondence of Francis Crick, Sydney Brenner, Jacques Monod, Sir Lawrence Bragg, and Matthew Meselson. For crucial passages I have gone to original laboratory notebooks. Then, as each section of the account was put together I gave it to the central figures to read—not only to check the book's accuracy but to cue memories for final re-interviews. This step proved immensely productive with Francis Crick, François Jacob, and Max Perutz, in particular.

❖ ❖ ❖

One overarching characteristic of the revolution in biology I think at the end I may fairly claim as my own perception. Behind the diversity of discoveries moved a unity, a constant direction of change. It was glimpsed only in part by even the most speculative and acute among the scientists: clues and fragments, but no more, appeared in conversations with Jacques Monod, Max Delbrück, Francis Crick, Max Perutz, François Jacob, Sydney Brenner, Roger

Kornberg. In the transformation of biology, the great underlying shift of view was the development of the concept of biological specificity. In the mid-thirties, biologists and biochemists certainly spoke of specificity. They had to do so, for many of the phenomena they dealt with—genes (whatever they were in substance), enzymes and antibodies (known to be protein)—were highly specific in action. Yet specificity was really a term almost empty of meaning. When biochemists attempted to understand proteins, for example, they looked for general chemical rules for their assembly or for repeating physical units in their structure—and reported that they had found them, even though any such rules and units turned the notion of specificity into its opposite. Forty years later, biological specificity is richly stuffed with meaning. How the concept of specificity was filled out and filled in is told in *The Eighth Day of Creation*.

❖ ❖ ❖

My debts are many—though intellectual debts are odd, for they take hard work to accumulate and are a pleasure to acknowledge. First, I could not have written this book without Max Perutz's friendship—beyond valuing—and his imperturbable and always practical enthusiasm. I turned to him for help of many kinds: he tutored me in crystallography; answered interminable strings of questions about his own work and everybody else's; made editorial suggestions from his sound sense for organization and for levels of explanation; made me always welcome in the laboratory of which he was chairman, the Medical Research Council Laboratory of Molecular Biology, Cambridge, England. Francis Crick encouraged the two-step process wherein I interviewed him closely and at length, later presented him with a portion of manuscript that, after reading, he discussed in detail with the tape running again—and then we started over, a stage further along the spiral. We got through two thirds of the book that way, including all that directly related to his work. He read other parts, too, and commented on them by mail. He made the book much better—and more fun (certainly more fun to write). François Jacob read a draft of the chapters that describe the research done at the Institut Pasteur in Paris, and commented on them with devastating honesty. I owe him gratitude and great respect. Sydney Brenner, unpredictable by the moment, in sum was enlightening at every level from the technical to the imaginative. His powers of recall, once switched on, were vivid and seemed total. James Watson tolerated repeated interviews. Later he kindly allowed me to call on the Cold Spring Harbor Laboratory and various members of his staff there for pictures and documents. Matthew Meselson, initial doubt overcome, offered the hospitality—unstinting and unstrained, domestic and intellectual—of an old friend. He read large parts of the manuscript. Often he helped me see more clearly what I wanted to say, and to take it further. He commented on science and scientists with exquisite precision and subtlety.

My gratitude to Jacques Monod is inexpressible and unforgettable, and to Sir Lawrence Bragg.

Many others read parts of the manuscript, from a chapter or two down to the pages that described their own work: to all of you, my thanks—and for the errors that remain despite your generous efforts to set me straight, my apologies.

For interviews and some less formal conversations (and often for other help as well), I have great pleasure in thanking, also, Paul Berg, Max Delbrück, Boris Ephrussi, Walter Gilbert, Aaron Klug, Roger Kornberg, André Lwoff, Linus Pauling, Robert Pollack, Frederick Sanger, Maurice Wilkins, and Paul Zamecnik; and Samuel Beckett, Erwin Chargaff, Manfred Eigen, François Gros, Hugh Huxley, Vernon Ingram, Eugene Jungelson, Sir John Kendrew, Edwin Lennox, Fritz Lipmann, Salvador Luria, Heinrich Matthaei, Barbara McClintock, Zhores Medvedev, Olivier Monod, Philippe Monod, Marshall Nirenberg, Geneviève Noufflard, Mark Ptashne, Alexander Rich, Gunther Stent, and Élie Wollman; and Sidney Altman, Bruce Ames, Ephraim Anderson, David Baltimore, Leslie Barnett, Seymour Benzer, David Blow, Walter Bodmer, Angela Martin Brown, John Cairns, Georges Cohen, Melvin Cohn, Heinrich Cramer, Odile Crick, Cedric Davern, Jerry Donohue, Renato Dulbecco, Vladimir Alexandrovitch Engelhardt, Claudine Escoffier-Lambiotte, Dame Honor Fell, Bertrand Fourcade, Bea Fraenkel-Conrat, Heinz Fraenkel-Conrat, Alan Garen, Raymond Gosling, Marianne Grunberg-Manago, Guido Guidotti, Roger Guillemin, Felix Haurowitz, William Hayes, Mahlon Hoagland, Dorothy Crowfoot Hodgkin, David Hogness, Robert Holley, Bernard Horecker, Rollin Hotchkiss, David Hubel, Sylvia Fitton Jackson, Herman Kalckar, Richard Keynes, Arthur Kornberg, Joshua Lederberg, Herman Lewis, William Lipscomb, Vittorio Luzzati, John Maddox, Janet Mertz, Avrian Mitchison, Severo Ochoa, Arthur Pardee, Peter Pauling, David Perrin, David Phillips, Lincoln Potter, Stanfield Rogers, Andrei Sakharov, Anne Sayre, Howard Schachman, Maxine Singer, S. Jonathan Singer, John Smith, Franklin Stahl, Roger Stanier, Michael Stoker, Alexander Stokes, Hewson Swift, Heinz Georg Terheggen, Charles Thomas, Alfred Tissières, Alexander Lord Todd, Andrew Travers, Agnès Ullman, Klaus Weber, Adrienne Weil, Heinz Winkler, Charles Yanofsky, and Norton Zinder. For answering questions by telephone or letter, I am glad to be able to thank Alice Audureau, Torbjörn Caspersson, René Dubos, Mary Fraser, Sven Furberg, Pauline Cowan Harrison, Leon Heppel, Leonard Lerman, Claude Levy, Ole Maaløe, Werner Maas, Sir Peter Medawar, Marjorie M'Ewen, Margaret Pratt North, Warren Ruderman, Andrew Gabriel Szent-Gyorgi, and Gerard Wyatt.

By the kind invitation of Professor John Edsall, of Harvard University, I have deposited a set of the interview transcripts and tapes—through the Survey of Sources for the History of Biochemistry and Molecular Biology, a joint committee of the American Academy of Arts and Sciences and the American Philosophical Society—in the American Philsophical Society Library, Philadelphia, Pennsylvania. Whitfield Bell, Jr., the librarian there, and David Bearman, secretary to the committee, have been welcoming and helpful. Certain restrictions of access to and quotation from these transcripts will necessarily apply—for example, where passages include statements made in confidence.

Robert Olby is one of the few historians of science to have cast his nets in some of these same seas: I am grateful to him for letting me listen to tape recordings of interviews he conducted with Sir Lawrence Bragg, Francis Crick, and John Griffith. Charles Weiner allowed me to read transcripts of interviews he had gathered with biologists and others involved in the controversy over the potential dangers of the techniques in molecular biology that go by the name of "recombinant DNA."

The staff of the Medical Research Council Laboratory of Molecular Biology in Cambridge have treated me almost as a member of the laboratory for nearly five years; the use of the library there was invaluable. The University Library, Cambridge, was a joy to work in; occasionally I called on the collections of the university's biochemistry and physiology departments, the Scientific Periodicals Library, and the University Postgraduate Medical School, and was always given liberal help. Elsewhere, librarians and archivists at the California Institute of Technology, the Massachusetts Institute of Technology, the Medical Research Council headquarters in London, the Rockefeller University, the Royal Institution, and the Tennessee State Library and Archives have been prompt with help and documents. Many people helped me find pictures; in particular, Susan Gensel, librarian at the Cold Spring Harbor Laboratory, energetically tracked down photographs taken at meetings there, and made prints and negatives available for copying; Karl Maramorosch, who took many of those pictures, searched his collection and lent me negatives; Vittorio Luzzati allowed me to use a photograph of Rosalind Franklin that I believe has not been published before. On the technical side, among others, Thomas Anderson provided prints of some of his historic electron micrographs, Max Perutz and Aaron Klug gave me x-ray-diffraction pictures and photographs of models of proteins, and Maurice Wilkins kindly supplied prints of x-ray-diffraction patterns of DNA and a photo of the space-filling model of the structure. Annette Lenton drew the line drawings—after repeated reassurance that in many cases I really did want them as informal as something a scientist might sketch on a scrap of paper. Glorieux Dougherty put the notes into final order and prepared the index: precision under pressure.

Nobody could be more fortunate than I in my editors—Alice Mayhew and William Shawn—from whom a few words of praise illuminate years of work.

The Master and Fellows of Peterhouse informally, generously, made me their guest for several terms: to participate, even peripherally, in the life and traditions of the college was a high pleasure.

And for encouragement, stimulation, and warm hospitality in the coldest hours, how delightful it is now to thank Edward Shils, Gisela Perutz, Tony Palmer and Annunziata Asquith, Roger Kornberg, Marie Squerciati, Jürgen and Cornelia Rimpau, Jackson and Elena Burgess, Bill McLane, Bryan Robertson, Christopher and Stephanie Porterfield, Peter Newell, Eve and Per Bark, and Virginia and Geoffroy de Vitry.

Westlands, Smith's Parish, Bermuda
2 September 1978

I

DNA

Function and Structure:
The elucidation of the structure
of deoxyribonucleic acid,
the genetic material

1 *"He was a very remarkable fellow. Even more odd then, than later."*

❖ *a* ❖ We were crossing London in one of those commodious taxicabs, Max Perutz and I, on a sun-washed Wednesday afternoon late in May of 1968, heading for Liverpool Street Station and the train to Cambridge, when Perutz, reflecting on the ways men do science, paid James Watson the most exact yet generous compliment I have heard from one scientist about another. At lunch that day, Perutz had talked about his subject, which is molecular biology and, in particular, the hemoglobin molecule. He had talked lucidly but somewhat in the abstract about the science and not the scientists, because, he said, he did not want to deal in personalities. It was enough, for one thing, that Watson's memoir of the discovery of the structure of deoxyribonucleic acid, *The Double Helix*, had just been published and was selling briskly, for that was full of personalities, to the point where many of the scientists Watson had worked with in England as a young man, sixteen years earlier, were now very angry, believing that their privacy had been not only invaded but coined. Francis Crick, I had heard, was one who felt that way. Crick had been Watson's colleague and mentor in the elucidation of the structure of DNA. Watson had begun his memoir, though, with one of those unforgettable first sentences: "I have never seen Francis Crick in a modest mood." Yet, as scientists understand very well (and as Watson had written in his own defense), personality has always been an inseparable part of their styles of enquiry, a potent if unacknowledged factor in their results. Indeed, no art or popular entertainment is so carefully built as is science upon the individual talents, preferences, and habits of its leaders. Perutz himself, though almost unknown outside, is one of these personalities of science: a frail-seeming man, today in his mid-sixties, of gentle courtesy, shy brown eyes, vast information; precise, worrying, abstracted; but behind that with the wiry will to climb for sport the Alps of his native Austria, and to spend forty years working at the cliff face of a problem—the three-dimensional, atomic architecture of the hemoglobin molecule, all ten thousand atoms of it—that was so nearly impregnable it took him the first fifteen years just to discover how to do it, a discovery that itself won him a share in a Nobel prize as it were in passing. Watson is different: impatient, skeptical, forever discontent, swallowing ends of words and ends

3

of thoughts, with a face at fifty only a little less gaunt than in the Cambridge photographs from 1953, with ice-blue, protuberant, even slightly wild eyes and fully modelled lips that retreat in a twitchy preoccupied smile. Watson's field, before the discovery that made him known, had been the genetics of bacteria and the viruses that prey on them, where results in an individual experiment are a matter of hours, or days at most; he has never spent more than a couple of years on any problem. His problem recently has been research into animal cells, their genetics and growth, their surface membranes, and the viruses that transform them into tumors. Study of the mode of action of tumor viruses, he allows, is one approach that may lead to a fundamental understanding of what goes wrong in cancer. But he will say only that these problems are "very, very hard," and that, anyway, he is now an administrator and writes textbooks. For nine years, Watson ran not one but two large research laboratories, the first at Harvard during the week, the other at Cold Spring Harbor, Long Island, on extended weekends and all summer; in 1977 he quit Harvard to direct Cold Spring Harbor full time. Reflecting, though, on Francis Crick and the days of their collaboration, Watson said, a while ago, "It's necessary to be slightly underemployed if you are to do something significant." Did two labs leave him sufficiently underemployed? A bark: "No." A pause. "But I was very underemployed when we solved the structure of DNA." Just so. In large part it had been the barefaced adolescent idleness, the unseriousness, of the long letters Watson wrote from Cambridge home to his mother, in Chicago, and to scientists like Max Delbrück, in Pasadena, and whose mood and prose he later recaptured in *The Double Helix*, that had tickled the general readers of the book and most irritated his former colleagues. It was this, however, that Perutz had seen through, and defended in the taxi on the way to the train that May afternoon. The two men are peers. Perutz had been Watson's immediate superior in Cambridge. Of their first meeting, Watson recalled only that Perutz had assured him one need not be a mathematician to do x-ray studies of molecular structures. Eleven years later, they had stood on the same platform in Stockholm to receive their Nobel prizes, Perutz with John Kendrew, Watson with Crick and Maurice Wilkins. In the photograph of King Gustaf VI Adolf shaking hands with Francis Crick, Watson stands tall, attentive, and haggard in the background, but Perutz has the clasped hands and bashful, proprietary smile of a man pleased that the social occasion is going so well. Now, as we swung past Admiralty Arch, Perutz said, "People sneer at Jim's book, because they say that all he did in Cambridge was play tennis and chase girls. But there was a serious point to that. I sometimes envied Jim. My own problem took thousands of hours of hard work, measurements, calculations. I often thought that there *must* be some way to cut through it—that there must be, if only I could see it, an elegant solution. There wasn't any. For Jim's there was an elegant solution, which is what I admired. He found it partly because he never made the mistake of confusing hard work with hard thinking; he always refused to substitute the one for the other. Of course he had time for tennis and girls."

Perutz and I had met a few weeks earlier, when I went to interview him and others in Cambridge as part of a hurried job of research for a magazine article about molecular biology. I was then a journeyman theatre critic and book reviewer stationed in London, and did science reporting as well. I found that

Perutz and some other scientists possess an unexpected facility in talking about their work to a layman: their explanations opened up a view of the processes of living things that has a coherence, sweep, and clarity of detail I had hardly imagined, and an intellectual daring that even vicariously seems wonderfully bracing. So, as other work has taken me to live in France and then back in the United States, I have kept in touch with several biologists and have sought out and interviewed others when the chance arose—an intermittent seminar that has taught me enough so that I could also read with some understanding the research papers that report the several chief discoveries of the past three decades.

It was a quarter-century ago that Watson and Crick, playing with cardboard cutouts and wire-and-sheet-metal models and sorting out the few controlling facts from a hotchpotch of data, elucidated the molecular architecture of the genetic material itself, the double-railed circular staircase of deoxyribonucleic acid. What has been learned in the years since is full of surprises, full of wit and beauty, full of most gratifying illumination. The culmination is now approaching of the great endeavor in biology that has swept on for a century and a quarter—an achievement of imagination that rivals the parallel, junior enterprise in physics that began with relativity and quantum mechanics. Biologists' pursuit of complete and explicit understanding has begun to list the exact molecular sequences that encode the hereditary message, instruction by instruction; it has tweezed apart the springs and gears by which the message is expressed in the building of the cell, and the ratchets and pawls by which that expression is regulated; it has accustomed men to speak apparently without wonder of the structural transformations by which a single protein molecule, an enzyme, will break or build other proteins, or by which, for example, a molecule of hemoglobin will flex its broad shoulders and bend its knees to pick up oxygen.

To be sure, the discoveries have not produced the great practical payout that has repeatedly been anticipated for them. Biologists have no atomic power stations and no bombs to point to, or at least not yet. No baby has been cured of a congenital deficiency by insertion of a missing gene into its cells. There is no vaccine against human leukemia, not even a cure for hay fever. Though some of the rewards are at last imminent, most scientists have learned that they must speak guardedly and emphasize to laymen the gaps to be filled in.

Yet at certain moments, in talking about these things with scientists or in reading their reports, one realizes just how far they believe they have gone. Watson wrote in a textbook, several years ago, "We already know at least one-fifth, and maybe more than one-third of all the metabolic reactions that will ever be described" in a particular much-studied bacterium, suggesting that "within the next ten to twenty years we shall approach a state in which it will be possible to describe essentially all the metabolic reactions involved"—the total process of life in that creature. Francis Crick said in conversation in Cambridge, England, that the serious problems left for biologists are to understand how growth and differentiation are controlled in higher organisms, how life originated, and how the central nervous system works. Alexander Rich, in Cambridge, Massachusetts, stirred a tray of very early fossils with his finger and said, "Origin-of-life chemistry is about to really explode; in a few years' time we will have a reasonable understanding of all the molecules in the preliving world that were needed for assembling life—in fact, soon we will know so many different possible pathways along which life could have originated that the real problem will be to select the correct one." Seymour Benzer, in Pasadena, explained that he has quit viruses and bacteria to experiment with the genetics of the nervous system in fruit flies: "In particular, I'm working on the relationship between genes and behavior." Sydney Brenner asserted across the high table at King's College Cambridge: "If you say to me, here is a hand, here is an eye, how do you make a hand or an eye, then I must say that it is necessary to know the program; to know it in machine language which is molecular language; to know it so that one could tell a computer to generate a set of procedures for growing a hand, or an eye." Jacques Monod, one of the rare theoreticians in molecular biology, standing in shirt sleeves at the door between his lab and his study at the Institut Pasteur in Paris, answered a question in mild surprise: "The secret of life? But in principle we already know the secret of life."

Many substances with many and subtle interactions have by now been found, and there are more to be found, taking part in the molecular minuet that is each living creature. Although I had gathered some idea of the part taken by DNA—who in this age of popularization has not?—it was a while before I saw clearly that there are several distinct ways in which, for molecular biologists, DNA is primary. Watson in his memoir was hardly shy about the discovery of the structure of DNA: he called it "perhaps the most famous event in biology since Darwin's book." Yet it is not true that that structure is terribly complicated, or was of itself difficult to solve. Protein molecules are much more difficult—and more obviously interesting as well, for they are the agents that carry out the processes of life. The first time I went to see Perutz in Cambridge, I asked him how hard it was to find the structure of a protein, in comparison with DNA. "A protein—an enzyme molecule, for example—is a thousand times more difficult than DNA," he said. "DNA was comparatively simple and could be solved by the method of trial. You have a little information from an x-ray-diffraction photograph; what Crick and Watson had from that was really three limiting measurements: the width of the double helix, the height between the stacked parallel bases—those are the steps up the middle—and the height of one complete turn of the helix. From these, they knew that an essential feature of DNA is its repetitiveness; the same pattern recurs periodically around the fibre axis. Obeying these three parameters, then, they managed to solve the structure by model-building. By trial. Solving the structure of a crystalline protein by this method is impossible because it does not repeat; it does not have a periodic structure. I shall show you some worked-out protein structures presently that will give you an idea of the complexity. To determine such a structure involves determining several thousand parameters from the x-ray photographs."

DNA is primary not just because its structure could be solved first; it comes first conceptually. Proteins are so versatile, so busy, so conspicuous in the economy of the cell and the organism that even at the time Watson arrived in Cambridge, in the fall of 1951, almost all biologists still thought that the hereditary message, the gene, was almost certainly made of protein, too. After all, the variety of proteins is bewildering. Many hormones—for example, insulin—are protein molecules. Respiratory carriers, like hemoglobin, are protein molecules. Antibodies are protein molecules, a different kind to match and neutralize each of the hundreds of thousands of substances that the body may recognize as foreign—and the foreigners themselves are usually protein, at least in part. Enzymes are protein molecules, a different kind to catalyze each chemical event in every metabolic pathway by which cells manufacture energy, structures, wastes, other enzymes, more DNA, and new cells. And in every organism higher than viruses, bacteria, and the simple blue-green algae the chromosomes themselves, where the units of hereditary information reside, are made of protein and DNA intimately bound together.

Biologists looking into cells were like spectators at a building site, peering through a crack in the board fence at the hole in the ground where a new office tower is going up: so much to see, cranes, shovels, power saws, scurrying yellow helmets, welders with dazzling arcs, scaffolding, pneumatic drills, electric cables, riveters, hoists raising rafts of pipe, a huge steel beam swinging precariously overhead—but yes, everything interesting, everything active, seems to be made of metal, even that folding table in the hut over there with a roll of blue paper spread out on it. The structure of DNA, once known, made its function as a blueprint comprehensible and therefore incontrovertible. "Genes are made of DNA—full stop," Perutz said. "The structure of DNA gave to the concept of the gene a physical and chemical meaning by which all its properties can be interpreted. Most important, DNA—right there in the physical facts of its structure—is both autocatalytic and heterocatalytic. That is, genes have the dual function, to dictate the construction of more DNA, identical to themselves, and to dictate the construction of proteins very different from themselves."

Beyond that, deoxyribonucleic acid turned out to be a substance of elegance, even beauty. Structure and those dual functions are united in DNA with such ingenious parsimony that one smiles with the delight of perceiving it. Extreme elegance is almost more convincing to scientists than it should be. Nobody would claim that the first paper Watson and Crick published, "A Structure for Deoxyribose Nucleic Acid," 128 lines in *Nature* on 25 April 1953, had the vast power to reorder men's thinking that marked *On the Origin of Species* in 1859 or the announcement, in 1905, of the special theory of relativity. Yet the discovery of the structure—after a pause for its implications to be absorbed—was brilliantly stimulating. It proved to make possible the analysis in comparable fine detail of the series of events and mechanisms by which the genetic instructions are read off DNA, put to work, and regulated. While these could not be deduced from the structure, they interlock with and flow from it. In sum, the structure demonstrated why and how DNA itself is the master substance in the cell, the point of origin of what happens in the organism; the marvellous unity of structure and function in DNA, once discovered, proved to be the master concept of the next twenty years of biological investigation.

Then there is the undeniable drama of how the discovery was made—a high comedy of intellect, and in action a full-blown thriller. To begin with, the discovery gathered in the threads of several lines of evidence and many experimental antecedents. Equally complex was the web of relationships among the scientists: in this net the purely intellectual details of the process of knowing are tangled inextricably. The discovery was made in a mood of artificially stimulated, frenzied competition. The competition was won by wit, insight, and luck rather than by thoroughness and hard work: victory for the grasshoppers over the ants, and the ants are resentful to this day. Those closest to the discovery were perhaps a dozen in number, comprising six or seven principal actors and the others who provided timely facts or who observed at close range.

THE CHIEF PERSONS OF THE COMEDY

In Cambridge, at the Cavendish Laboratory

- Sir Lawrence *Bragg*, Cavendish Professor of Experimental Physics, originator, forty years before, of the technique of determining the structures of molecules in crystalline substances by x-raying them; an English gentleman in his early sixties and a very senior scientist, with a clear ethical sense, a vigorous competitive instinct, and an unexcelled power of spatial visualization
- Max *Perutz*, chemist, crystallographer, principal discoverer—at just this same period—of the way to apply *Bragg*'s methods to molecules as huge as proteins; head of the unit within the Cavendish where *Crick* and *Watson* worked but in this drama, for the most part, an observer
- Francis *Crick*, English physicist turned to biology after the war's interruption; in his mid-thirties still a Ph.D. candidate; a mind effervescent, stimulating, dry, unmistakable—and at this time not yet matured
- James *Watson*, American biologist on a postdoctoral fellowship, eleven years junior to *Crick*; a young man who had decided that science, whatever else, was the main chance; the equivocal hero who has since by his own sour jest cast himself as maybe the villain
- Jerry *Donohue*, American crystallographer }
  John *Griffith*, Welsh mathematician   } informants and messengers
  Peter *Pauling*, son of Linus *Pauling*   }

In London, at King's College

- Maurice *Wilkins*, another physicist changing to biological substances after the war; one who came early to x-ray studies of DNA and stayed late; friend of *Crick*'s, befriended by *Watson*; quarreling with—
- Rosalind *Franklin*, chemist turned crystallographer, a woman of passionate intelligence and meticulous professional performance; in need of a colleague; supposed by *Wilkins* to have been hired to help him

In New York, at the College of Physicians and Surgeons, Columbia University

- Erwin *Chargaff*, biochemist of the nucleic acids who had found a fundamental ratio but did not see what it signified; by his own description always "an outsider at the inside of science," and yet the mordant defender of the great tradition in biochemistry against such upstarts as *Crick* and *Watson*

At Indiana University and then the University of Illinois

- Salvador *Luria*, geneticist, *Watson*'s doctoral supervisor, friend, and guardian-in-science, a presence off stage

In Pasadena, at the California Institute of Technology

❖ Linus *Pauling,* Professor of Chemistry, the most brilliantly versatile and productive physical chemist of the century; more than a scientist, a force of nature; in his early fifties; *Bragg*'s peer and rival

❖ Max *Delbrück,* physicist turned biologist and a founder of molecular genetics; fastidious aesthetician of science; friend and collaborator of *Luria*'s, preceptor and correspondent of *Watson*'s and so, in this drama, a messenger

Among the protagonists were brilliance, eminence, great ambition, great vanity. Feelings among them were strained, explosive, in one case so unpleasant as to leave a foul taste twenty years later. "DNA, you know, is Midas' gold," Maurice Wilkins told me. "Everybody who touches it goes mad."

Then there is the working out of the intellectual problem of DNA, a story of equal fascination, indeed the driving force within the personal drama. Scientists tell me that it is extremely difficult, when one has found something that makes a difference, to recapture the way one thought before. How individual, or how social, communal, and collective, is the making of a great discovery? Very little of the day-to-day work of science is discovery, so the fact that scientific routine can be carried out by teams, can even be published by teams, says nothing for sure about discovery. In defense of scientific individualism, it must be said that in the process of discovery there comes a unique moment: where great confusion reigned, the shape of the answer springs out —or at least the form of the question. The insight occurs in the prepared mind of some one person. The exaltation of that moment has been compared to the *satori* of the Zen adept. It is the first of the scientist's two ruling rewards for his endeavors—and the second reward, the recognition granted by his peers, depends on the first. Yet from the collective viewpoint it must be said that the insight, however exalting, is not the discovery; it is a moment at the end of a process and the beginning of another. What the insight touches off, even before anything gets published, is the familiar and most characteristic work of the scientific community: criticism, modification, and the development of consequences. It would be neat to say that a discovery is that which gets published—neat, but not quite right. DNA is an outstanding example of how a discovery twists and grows. What is still called the Watson-Crick structure of DNA is different today in factual detail from their first short paper in *Nature*, and very much greater in perceived scientific consequences. The idea now is crammed with meaning. Still, such change is easily followed. To comprehend what went before an insight is much harder. DNA is an outstanding example of this difficulty, too. The preparation of a mind is itself simultaneously communal and individual. The nature and detail of the reigning confusion, against which the insight coalesced, proves particularly difficult to reconstruct. Examine the insight and it shimmers and threatens to dissolve into all its separate points of fact and points of view, available at different moments, and sometimes to several minds. It seems strange that the structure of DNA was not found earlier by others, astonishing that it could be found at all. Once the Humpty-Dumpty of discovery is put together, all the historians and all the sociologists can't really scramble him again—often not even the scientists who were most closely engaged, for their memories are the first to begin to be altered by the persuasiveness of the thing discovered.

Recovery of how the structure of DNA was determined is made easier in some ways, but in others much more difficult, by Watson's *The Double Helix*. The book is great of its kind, exciting, eccentrically outspoken. The book begins in the years just before, and ends in the weeks immediately after, the completion of the paper announcing the structure of DNA; it ends on Watson's twenty-fifth birthday. It was not an exaggeration when reviewers said, as so many inevitably did, that the book is written with an innocent exuberance reminiscent of Pepys' or Boswell's journals; if Watson after all does not have their kindness and charm, he has other qualities. *The Double Helix* is full of delights, high among them, for professional readers as well as laymen, the delight of demystifying science of its more pretentious claims to rationality and orderly procedure, the delight of revealing for once, or seeming to reveal, what scientists "really do"—which turns out to include coffee breaks and gossip. The result is a very beguiling version of the circumstances of Watson and Crick's discovery. The book is irresistible to quote from, dangerous to lean on.

The Double Helix contains more science than most readers notice—and that exactly is its worst flaw. Crick observed recently, "Watson's book was really a fragment of his autobiography. . . . Indeed, it is surprising how many technical arguments he managed to slip in without making his text too heavy." Biologists find the science in the book haphazardly illuminating, mostly at the level of "Why ever did Jim chase after magnesium ions?" When they attempt more serious judgement they find unanswered questions. The historian, trying to see who contributed what to the moment of insight, also needs those missing answers. For laymen, I have always thought, the book scants a clear duty to make the science coherent and significant. Even in the comedy of personalities, *The Double Helix*, while it does not greatly oversimplify, misdirects: watching the frenzied entrances and exits, one must attend closely to what Scaramouche does as well as what he whispers to the audience.

What the book conveys that is probably most valuable is simply how uncertain it can be, when a man is in the black cave of unknowing, groping for the contours of the rock and the slope of the floor, listening for the echo of his steps, brushing away false clues as insistent as cobwebs, to recognize that an important discovery is taking shape. Can it be done at all? Is it as worth doing as we've told ourselves? Why hasn't it been done already? Where's the way in, the vulnerable point? Out of what we think we know, what's unreliable? What's irrelevant? These are the scientific as opposed to the personal circumstances, and they evoke sometimes the mood of the brink of terror, which in good part may explain why such monstrous self-confidence is demanded.

❖ ❖ ❖

At the time, the discovery of the structure of DNA was hard—not intrinsically, but because its importance and uniqueness were not well recognized. The discovery was hard also because the data were scattered, confusing, in some respects meagre, in others overabundant. To begin with, it was not clear what was most relevant in all that was known of the chemical composition of nucleic acids. Neither Watson nor Crick was a biochemist. They were ignorant of a long and erudite scientific tradition, but at least they were not blinded by it.

In the middle of the nineteenth century, the focus of biological investigation had begun to move—in pursuit of the steadily greater resolving power of the microscope and then outrunning it—down from the organ and the tissue to the cell. For a hundred years or more, the cell and its contents were to be the domain of the biochemists, whose interest lay in what they could detect and infer about the subtle flux of materials and energy therein. Biochemists early sorted out the principal distinctive substances they found in living beings into four broad categories—fats (the lipids), sugars and starches (the polysaccharides), proteins, and nucleic acids. The nucleic acids were the last of these to be isolated. In 1868, Johann Friedrich Miescher, a Swiss, twenty-four years old, went to Tübingen to study with Ernst Felix Hoppe-Seyler, an ingenious chemist, the man who gave hemoglobin its present name, and who founded and edited the first journal of biochemistry. Miescher was particularly interested in the chemistry of the cell nucleus. Even with present-day methods, cell nuclei can be hard to separate intact from the cytoplasm and outer membranes surrounding them. Miescher succeeded first with white blood cells from pus, which have big nuclei and not much cytoplasm, and which he got from discarded surgical bandages from the local surgical clinic.

In 1869, he found a new, unexpected compound, acidic, rich in phosphorus, and made up of molecules that were apparently very large. He named it "nuclein." The stuff was so unlike other substances already known in the cell that Hoppe-Seyler repeated the work himself before allowing Miescher to publish in his journal.

What Miescher had discovered, at that point, was a complex of DNA and the protein normally associated with it in higher organisms. In 1870, Miescher returned to his native Basel; there, at the headwaters of the Rhine, he found an excellent and more pleasant source of nuclear material in the sperm of the salmon for which, a century ago, the river was celebrated. The nuclei are large in any sperm cells, remarkably so in the salmon's. From these he first extracted a pure DNA. He and his laboratory went on to characterize these discoveries more precisely; in 1889 a pupil of his, Richard Altmann, introduced the term "nucleic acid."

Miescher's work was technically accomplished, lit by flashes of intuition. The biochemist Erwin Chargaff wrote, several years ago, that "Miescher, much more than his successors, realized the labile character and the macro-molecular behavior of DNA"—although, in life, molecules of DNA are hundreds or even thousands of times larger than the sizes Miescher was able to deduce from his preparations. "Miescher was a sleepwalking kind of scientist," Chargaff said in the course of a conversation. "He never could give reasons for doing what he did." There was no basis for Miescher to guess correctly at the function of the substance he had found. The physical reality of such relatively immense molecules, and even the fact that proteins and nucleic acids occur in long chains, was imperfectly comprehended. The chemistry for handling them had not been invented. The requirements of a science of heredity were still obscure.

Three years before Miescher arrived in Tübingen, Gregor Mendel had presented a paper, "Experiments in Plant-Hybridization," to the Society of Natural Science in Brünn, Moravia, then a quiet corner of the Austrian empire, and in the paper he proposed, as all the world now knows, the idea of a simple, highly regular algebra of heredity in which discrete units—he called them factors; the word "gene" came later—combine and recombine down the generations. But the concept lay entirely dormant until it was rediscovered, along with Mendel's paper, in 1900. Yet at the end of 1892, Miescher pointed out in a letter to his uncle that some of the large molecules encountered in biology, composed of a repetition of a few similar but not identical small chemical pieces, could express all the rich variety of the hereditary message, "just as the words and concepts of all languages can find expression in twenty-four to thirty letters of the alphabet." The history of science is full of speculative asides, and they must not be credited with foresight unduly; Miescher's notion was fatally imprecise. The molecules he offered as examples were albumin and hemoglobin, both proteins. Miescher died in 1895, of tuberculosis, aged fifty-one. "There are people who seem to be born in a vanishing cap. Mendel was one of them . . . and so was also Miescher," Chargaff wrote. "If a glance at the portrait forming the frontispiece to the collection of his papers were not enough to tell us this, his letters show us a man, reticent and intense."

Much of the elementary chemistry of the nucleic acids was settled early, by Miescher and his pupils and in other laboratories as well. Their presence in all cells was as quickly demonstrated. Their function remained unknown. By the beginning of this century the three constituents of nucleic acids had been described. The first of these is a sugar, called ribose, built up from a ring of carbon atoms like all sugars, but in this case a pentagonal ring and five carbon atoms, rather than, say, the linked hexagon-to-pentagon and twelve carbons of common, or table, sucrose. The second constituent is simply a phosphorus atom surrounded by four oxygen atoms, and called a phosphate. The phosphates are responsible both for the acidity and for the richness in phosphorus that were so surprising when nucleic acids were first observed; by labyrinthine series of chemical trials and inferences that took years to perform, it was shown at last that the phosphate groups link and space the sugars, third carbon of one sugar ring to fifth carbon of the ring beyond, over and over in monotonous alternation. The last sort of constituent is called a base, and is built up mostly from nitrogen and carbon atoms—and while the ribose and phosphate constituents are simple, repetitive, and altogether predictable, the bases come in several different kinds and were the central mystery of the nucleic acids. The three-piece subassembly of a base linked to a sugar, with a phosphate also pivoting off the sugar in order to bridge to the next, is called a nucleotide: a homely word, precise, indispensable, and ubiquitous in this science, indeed much like the word "iamb" in poetics, for it expresses not just a particular sort of construction but a unit of length and even a category of significance.

There are five bases that matter, and by the beginning of the century they too had been sorted out: guanine, adenine, cytosine, thymine, and uracil. Some variants of very restricted function were found later. The older generation of biochemists habitually spelled out the names in full; molecular biologists, more familiarly, often use just the initials, G A C T U. In an exuberant moment shortly after the discovery of the structure of DNA, Crick wrote that the relationship among the bases "is likely to be so fundamental for biology that I cannot help wondering whether some day an enthusiastic scientist will christen his newborn twins Adenine and Thymine!"—a fancy that Chargaff did not fail to hold up to scorn. Guanine was first found in 1844, in the excreta of birds, forty years before it was recognized as a nucleic-acid constituent. Crystallized guanine imparts the shine to the scales of fish and reptiles. Adenine was isolated in 1885 from a nucleic acid in beef pancreas. Guanine and adenine are very like each other: in each, the outline of the structure is a flat double ring, a figure 8, a hexagon fused with a pentagon so that the hexagon, of two nitrogen and four carbon atoms, shares two adjoining carbon corners with the pentagon, which has another pair of nitrogen atoms in it. Guanine and adenine differ only in small side groups attached to other corners of the hexagon. These two bases are called purines because of their chemical relationship to uric acid and so to urea—and biochemistry is conventionally said to have begun with the synthesis of urea, by Friedrich Wöhler, in 1828. The three other bases, thymine, cystosine, and uracil, 1893, 1894, and 1900, are called pyrimidines, a longer name of unilluminating origin but a smaller, simpler structure: a single ring, just the same hexagon of two nitrogen and four carbon atoms—with side groups, again, that make the differences.

By the nineteen-twenties, it was realized that there are two kinds of nucleic acid. In one, now called ribonucleic acid, RNA, the bases are adenine and guanine, cytosine and uracil. In the other, the ribose sugar lacks a fringe oxygen atom—hence deoxyribose nucleic acid—and a pyrimidine has been switched, uracil replaced by thymine. Uracil had first been found in yeast, and was known in a species of wheat. Thymine had been discovered in calf thymus gland, whence its name, and was known in every animal cell where it had been looked for. Uracil and thymine are very similar. For a while it was thought, then, that ribonucleic acid, bases G A C *U*, was for plants and deoxyribonucleic acid, G A C *T*, was animal. This idea collapsed in the early thirties under accumulating evidence that both RNA and DNA are universal. By then, too, it was known that the chromosomes are in large part DNA. None the less, DNA was thought to be built up in the simplest way imaginable, with the nucleotides following one another in fixed order in repeated sets of four. This exceedingly elementary picture was called the tetranucleotide hypothesis. It was propounded by Phoebus Aaron Levene, at the Rockefeller Institute for Medical Research, an organic chemist of highest reputation. Accurate measurement of the proportions of the bases in samples of DNA was impossible with the chemical techniques available. And so the belief was held with dogmatic tenacity that the DNA could only be some sort of structural stiffening, the laundry cardboard in the shirt, the wooden stretcher behind the Rembrandt, since the genetic material would have to be protein.

Rigorous proof that the gene is DNA and not protein appeared in 1944, when Oswald Avery and fellow workers at the Rockefeller Institute in New York published a paper in *The Journal of Experimental Medicine* about inheritable transformations that occur in a strain of pneumonia bacteria when they are mixed with DNA extracted from a different strain. Avery's paper is today universally cited as fundamental, always with the reservation that the proof took years to be credited. In February 1944, when the paper appeared,

Crick was working for the British Admiralty as a physicist, designing naval mines, and Watson was a precocious college boy in Chicago, consumed by ornithology the way another might have been absorbed in railway timetables. When they met, seven years later, both knew of Avery's work, though then and for several years more it was still generally believed and widely asserted that genes are protein.

On the point of picking up Avery's paper, I realized that for the moment I was surfeited with abstractions about DNA. I wanted to know what the stuff looks like, how it is prepared, something of how one tells that it is not protein. In the entrance stairwell of Perutz's laboratory building, before a floor-to-ceiling window, I had stopped to look at a molecular model, eight feet tall, of the double helix of DNA, and stopped again, on the landing above, at a model, nearly as large, that claimed it was of the alpha helix, a structure in proteins discovered by Linus Pauling. Surely, I thought, the two models could be told apart if one knew what to look for. The stretch of protein wound up and around, linked back and forth to itself by banisters that were jarringly off the perpendicular, bobbing and weaving upwards in a boxer's shuffling syncopated rhythm, while the DNA, right enough, was double, the strands separated by alternating narrow and wide grooves connected not by balusters but by the horizontal bases, planes laid flat to form a slowly revolving series. One could say the DNA had a calmer, cooler architecture. But really it was preposterous, despite the monumental use to which these models had been put, to think of them as aesthetic objects. The overwhelming impression that each gave was much the same: a visually confusing lacework of thin rods intersecting at knobs colored variously blue, red, black, white. This DNA was still an abstraction. Feeling exceedingly elementary, I asked Sidney Altman, a friend and molecular biologist now at Yale, to show me what DNA really is.

The afternoon I arrived at Altman's laboratory, he produced a large bottle of a gray liquid that looked like dishwater. "We start with this. Bacteria in a culture medium—just water, some salts and a carbon source, and some amino acids they need. They've been growing since I started them last night. That's why it's so cloudy: you're seeing a mist of bacteria—not the bacteria themselves but the turbidity they cause by scattering the light. You realize, the chief thing you're going to learn is how easy things are once you know how to do them. Now we centrifuge them down, to concentrate the bacteria." He poured the gray liquid into four stubby plastic test tubes, which we took over to a large box, like a stainless-steel washing machine, with a top hatch that lifted heavily. Within was a pear-shaped spindle with slanting holes like spokes. Altman put the tubes into the holes and set the machine's speed control at 8,000 rpm. "Come back in half an hour."

When Altman took out the tubes, the liquid had turned water-clear, while at the bottom of each was a small heap, in color the pale yellow-gray of an old nylon shirt. "This is ersatz science," he said. "We're not doing this for any real experimental purpose. Takes the edge off one's precision. The cells are all in those pellets. We can throw out the supernatant." He poured away most of the liquid. Then at his bench, he scraped up the pellets of bacteria with the end of a glass rod and transferred them all to one tube with a little liquid. "They're back in suspension but concentrated a hundredfold." Indeed, the liquid, clear a moment earlier, looked filthy. He turned to an appliance the size and shape of a melon, with a large rubber navel on top that began to shake with silent laughter as he pressed the end of the tube to it. "Breaks up the clumps." In a burlesque of the classic gesture of the movie scientist, Altman held up the tube to the light as he added a liquid from another bottle. "This is EDTA—ethylenediaminetetra-acetate—which takes up the magnesium from the bacterial cell walls. Bacteria have tough cell walls, but this stuff makes them very weak." He searched a shelf, found a plastic jug labelled "10% SDS," added some of that. "Sodium dodecyl sulfate—all it is is a detergent; you could wash dishes with it. We put it in to solubilize the cell walls. The idea is to break the cell walls to get the DNA out." The technical term for rupturing cell walls is "lysis." To lyse cells, biologists use strong chemicals, or even grind the cells in a mortar with sand; animals get the same results with the subtler means of an enzyme, called lysozyme, present in such body fluids as tears, saliva, and intestinal mucus, which has the protective effect of breaking open bacteria that attempt to invade. Altman put a black rubber stopper into the tube. There was about a quarter cupful of liquid in it. "As the cells lyse, the bacteria will vanish and the mixture will begin to clear. At the same time it will get very viscous, because the DNA in each cell is essentially one extremely long molecule, and these are freed. Like a basketball player getting out of a Volkswagen Beetle, only more so." He set the tube half into a fish tank full of running water, where a thermometer said 37°C, blood heat. "Come back in half an hour."

Out of the warm-water bath, the tube of liquid was clear again. "Now we add phenol—what used to be called carbolic acid; our grandmothers used it to disinfect drains. The phenol attacks the protein, which is why it worked for grandmother, but it leaves the DNA alone. And it's heavier than water, so it will sink to the bottom with the protein, while the nucleic acids stay in the aqueous phase at the top." He put the stopper back into the tube, started gently rocking it. The liquid was bubbling slightly and began to look pale gray and thick, like sputum. "Doing this by hand keeps the DNA from breaking so much. Those long fibres, once they're floating free, are exposed to a lot of shearing force. That disgusting glob of white is the protein. In a minute we'll centrifuge them apart." This time he used a small machine standing on his bench, mushroom-shaped, of gray metal. First he weighed the tube, and filled a second one with which to balance the machine. A hand-lettered sign on the centrifuge said *"four buckets are hot—beware."* As he closed the lid, the spin was starting to tilt the tubes up into the horizontal plane. "Come back in half an hour."

As we waited for the spinning to stop, Altman handed me a small brown bottle. "This is what the stuff we're preparing would look like if you made a lot of it and dried it. It's purified DNA, as it happens from calf thymus glands, one of the traditional sources." The bottle was full of small bits of what looked like scraps from an old linen handkerchief, white and obviously fibrous. "They're a lot tougher than lint, though. Those long molecules lying together have a very high tensile strength. No, don't touch, even slight impurities can start breaking the molecules up." With tweezers, he took out the top flake, dropped it into a small vial with a screw lid. "Something to show your friends."

The tube from the centrifuge now had a layer of clear liquid at the bottom, "the phenol," then a layer of white, "the protein—really goopy," and above that another layer of clear liquid. "The DNA is in that top layer. It gets so viscous we'll have trouble getting it out without contaminating it with some of the protein beneath. If this were real, I'd have to be a lot more careful." He packed the tube and assorted glassware into shaved ice in a battered plastic ice bucket. He took a wide-mouthed glass pipette and tried to suck up the layer of liquid at the top of the tube; every time he lifted the pipette slowly away, the liquid simply plopped back into the tube. "It's so viscous it pulls itself out again." He tried other pipettes, at last took a small one over to a bunsen burner, where he held the tip in the flame, and then bent a kink into it. With this he got the liquid up a few drops at a time, and into another, narrower tube. He tilted that. "Watch how it flows. That's *really viscous!* It's good stuff.

"Now, this last step is the spectacular one. I'm going to layer in twice as much ethanol, absolute alcohol." He poured the alcohol slowly down the side of the tube so that it floated, cream on Irish coffee. "Now we stir with this glass rod, gently winding. The alcohol precipitates the DNA, and we pull the fibres out of solution like winding spaghetti out of sauce." As he twisted the glass rod, the tip began to thicken with cobwebs, wet and translucent. As he lifted the rod out, the attached fibre pulled into a long filament. "Amazing. That's a lot of individual molecules lying together, of course. But you could use a fibre like that for x-ray analysis of its structure, the way Rosalind Franklin did."

The principle of DNA extraction is simple: break the cells open, get rid of protein by treatment with phenol or chloroform, precipitate DNA with alcohol. High-school pupils these days learn to extract DNA in science classes, though not, if their teacher is wise, from bacteria; and teacher scrambles to keep ahead with the aid of manuals and source books that tell how to go on to analyze the composition of the DNA by such techniques as chromatography or electrophoresis, in which absurdly small amounts of biological substances can be persuaded to separate themselves for identification and measurement as they migrate, in solution, down a sheet of filter paper or across a hard slab of gelatine, some travelling faster and farther than others because of differences in weight or solubility or electrical charge of the individual molecules.

Methods of such discrimination and finesse—of such chemical resolving power—were only beginning to be available when Avery published his surprising paper. Yet what he accomplished in the early forties is still respected as masterly. His was far from ersatz science. The paper is marked by the probing sensitivity with which he responded to what he did not know. The title is as arid as any in the literature: "Induction of Transformation by a Desoxyribonucleic Acid Fraction Isolated from Pneumococcus Type III," by Oswald T. Avery, Colin M. MacLeod, and Maclyn McCarty. Within, the writing is excellent—supple and taut. Procedures come across vividly. Not just a short run of experiments is reported, but years of work and years of pondering. The argument is tough, clear, close-grained. Scientific papers in our day are written to an artificial, sterilized form; Sir Peter Medawar—an immunologist who shared the Nobel prize for physiology or medicine in 1960—has suggested that they are so deliberately anti-historical as to be a deception, for "They not merely conceal but actively misrepresent the reasoning that goes into the work they describe." Avery's great paper, though, shares with the classics of science of previous centuries at least one quality now grown rare: from the first paragraphs through to the end, one feels an original curiosity working.

Avery was by training a physician, as were his two associates in the paper. As specialisms then went, he was not a geneticist and not exactly a biochemist, but an immunologist and microbiologist. That is, he worked with micro-organisms; and among them microbes rather than viruses, and among microbes the bacteria that cause pneumonia, with the long-term hope of developing sera with which to treat acute cases.

Micro-organisms can go through as many generations in half a week as mankind has had since history began. Each generation for a bacterium is a doubling: many a virus multiplies by the fiftyfold or the hundredfold. The foods such creatures use are so simple that what goes into them can be exactly known, controlled, and compared with what they make of it. Further, it is easy to spot variations. A bacteriologist or a molecular geneticist can routinely select from a billion or more separate cells the single individual that possesses some particular inheritable trait. These and other advantages have made micro-organisms the favorite subjects for many kinds of biology for the middle third of this century, especially for geneticists, though right now a change is taking place, back to fruit flies, mice, and other higher animals, because one-celled creatures are too simple, internally as well as in being one-celled, for the questions molecular biologists have come to. Yet thirty years of intensive scrutiny (as Watson points out to students) mean that next to man himself, the world's most thoroughly understood organism is his small companion through life, the normally benign intestinal bacterium *Escherichia coli*. Determining the results of an experiment with bacteria can be easier than one might suppose. Often enough a microscope is not even needed: just examine where the bugs are growing, on broth or gelatine in one of those fragile low-sided glass dishes, to see the colonies' size, color, texture.

By such simple and visible criteria, the world of Avery's *Streptococcus pneumoniae*, called pneumococcus, is divided into the Rough and the Smooth. S forms are virulent. They kill laboratory animals. A bacterium of S form surrounds itself with a plump, gelatinous capsule, which it builds not of protein but of a complex sugar, a polysaccharide. The capsule partly protects the bacterium from the defenses of the infected animal, and so always goes with virulence in pneumococci. R-form bacteria have lost their ability to make capsules and so to cause infection. (R forms are now understood to be mutants that fail to make the enzyme that knits together the capsular polysaccharide.) To get them, bacteriologists grow pneumococci in a medium made hostile to the ancestral S form. They are called R for no microscopic reason but because, Avery wrote, "On artificial media the colony surface is 'rough' in contrast to the smooth, glistening surface of colonies of encapsulated S cells." R and S forms of pneumococci had first been distinguished in 1923 in London, by Frederick Griffith, a physician doing research in the Pathological Laboratory of the Ministry of Health. Variant virulent S types had also been found, and numbered I, II, III. These differ much less in what can be seen in a glass dish, but can be told apart with certainty by immunological tests. Antibody reactions are among the most exquisitely sensitive detection systems biologists possess. Tested against serum from the blood of rabbits that had survived infection and developed a high degree of immunity, protein from pneumococci betrays its presence, Avery wrote, in dilutions as high as one in fifty thousand, and the capsular sugar in dilutions of one part in six million. Avery himself had discovered the immune reaction to capsular polysaccharide—the first evidence that animals can make antibodies to something not a protein—also in 1923, in the course of the work by which his laboratory established the fixed differences among the S types of pneumococci.

Then, in 1928, Griffith in London had published a startling discovery. He had injected mice with two pneumococcal preparations at the same time: a small amount of a living culture of the R form, derived from Type II and proved to be not virulent by itself, together with a large amount of a dead culture of the S form of Type III—killed by heat, containing no living bacteria, and proved to be not virulent by itself. In short, two different types, one an R, live but not virulent, and the other an S, virulent but killed. Many of the injected mice had died. In the heart's blood of these mice, Griffith had found living, virulent pneumococci, of the S form—and not Type II but Type III.

The change was permanent and inherited. Generations more of culturing had produced nothing but more Smooth, virulent Type III pneumonia germs. Other experiments produced similar transformations of other pneumococcal types. Griffith's discovery suggested doubts about the existence of distinct true-breeding species among bacteria. It opened grave practical problems for epidemiologists and immunologists. It raised clouds of speculative and spurious explanations. All in all, microbiologists found transformation of bacteria about as unsettling as atomic physicists, at that same time, were finding the transmutation of elements by interaction with neutrons and protons. Avery at first found it impossible to credit Griffith's paper. The findings seemed to overthrow his own fundamental demonstration of the fixity of immunological types. But bacterial transformation was confirmed that same year in Berlin and in 1929 was repeated at the Rockefeller Institute.

Two years after that, associates of Avery's found that they could do the same experiment leaving out the mice. They could achieve transformation by growing a culture of R form in a glass dish in the presence of heat-killed pneumococci of the S form. Several months later, James Lionel Alloway, again in Avery's laboratory, took the pursuit of the transforming agent one twist further. Alloway broke open the S-form bacteria to set their contents free, then passed the culture through so fine a filter that the shells, together with any unbroken cells, were removed. When this extract, free of cells, was added to a growing culture of the R form, transformation took place. Further, when he added alcohol to the extract, he got a viscous, "thick syrupy precipitate."

Rollin Hotchkiss, who joined Avery's laboratory in 1935, recalled in a biographical memoir thirty years later that Avery's characteristic question was a quick, insistent "What is the substance responsible?" For the next decade, Avery was increasingly preoccupied with step-by-step purification of the transforming agent and its identification. In the beginning, transformation was an uncertain, delicately balanced phenomenon. "Many are the times we were ready to throw the whole thing out of the window!" Avery said to Hotchkiss. At the last, Avery was able to take a culture of pneumococci of an

R form that had been attenuated from an S of Type II, thirty-six generations back (all the way back to the Crusades, on a human time scale), and add to it what he knew to be a highly purified DNA extracted from an S of Type III—and he got out, in the next generation, fully developed "large, glistening, mucoid colonies" of S Type III. These then remained stable through succeeding generations. Recalcitrant strains of bacteria had been tamed, finicky conditions of culture mastered. Avery's difficulty was no longer the transformation itself but to prove that it was caused by DNA and nothing else, despite the fact that DNA had not been identified in pneumococci before and in defiance of the universal conviction, his own conviction at the start, that DNA was a monotonous molecule and genes were protein. Avery was a small man, a bachelor all his life, smooth-faced and thin; he wore pince-nez. Various friends remember that he would pass his hands across his bald head when perplexed, that he rolled his own cigarettes, that he was fastidious with words and reserved with conclusions, that he was a gentle, versatile, overwhelming monologuist for whom the pneumococcus was the microcosm of biology. Hotchkiss wrote, "My personal notes of 1936 record that in one of his discourses on transformation, Avery outlined to me that the transforming agent could hardly be carbohydrate, did not match very well with protein, and wistfully suggested that it might be a nucleic acid!"

Throughout the paper of 1944, with immaculate caution, Avery, MacLeod, and McCarty speak of their substance as "the transforming principle." To get it, they grew virulent Type III pneumococci at blood heat in twenty-gallon vats of broth made from beef hearts, spun out the bacilli in an iced centrifuge, suspended them in brine, and brought the "thick, creamy suspension of cells" quickly to a temperature hot enough to kill the cells and to inactivate "the intracellular enzyme known to destroy the transforming principle" (an enzyme now called, with brisk inelegance, DNase). They then washed the cooked pneumococci in three changes of brine to remove capsular sugar as well as whatever protein would come away, extracted the bacteria by shaking them for an hour in a solution of bile salt to break the cell walls (and then threw away the cell residue), and reprecipitated the extract with pure grain alcohol.

"The precipitate forms a fibrous mass which floats to the surface of the alcohol and can be removed directly by lifting it out with a spatula," the paper said. This was now washed several times with chloroform to remove protein, and suspended yet again. A digestive enzyme was put in to eat away any remaining capsular sugar. Removal of protein was repeated, "until no further film of protein-chloroform gel is visible at the interface." Pure grain alcohol was added again, "dropwise to the solution with constant stirring." At a concentration where the alcohol nearly equalled the extract, "the active material separates out in the form of fibrous strands that wind themselves around the stirring rod. This precipitate is removed on the rod and washed. . . . The yield of fibrous material obtained by this method varies from ten to twenty-five milligrams per seventy-five liters of culture"—or, at best, just under one hundredth of an ounce from twenty gallons of culture. The method of extraction, before the introduction of detergents and using chloroform rather than phenol, was heroically laborious.

Avery and his colleagues set out to show what their transforming agent was—and what it was not, which was harder. They devised tests with an almost obsessive ingenuity that makes the paper a model of reasoning from and about experiment. The understated iteration takes on rhetorical power. Standard qualitative tests for protein—for example, add a pinch of copper sulphate and see if the solution turns blue-violet—were negative; those for DNA, strongly positive. Chemical analysis found the elements in proportions—particularly the telltale ratio of nitrogen to phosphorus, 1.67 to 1 on average—which agreed closely with what DNA, with its nitrogenous bases and its phosphates, should show but which would have been different if, despite the extraction methods, much protein had remained.

They turned to enzymes. The specificity and speed of enzymes, the power of each to catalyze its own reaction intensively and nothing else, fits them for the burden of proving a biological negative. Pure, crystalline enzymes were just beginning to be available, in great part through the work at a sister unit of the Rockefeller Institute, in Princeton. Other enzymes of proven strength in crude form were obtained from rabbit bones, swine kidneys, and the mucus of dogs' intestines. Of these, enzymes known to digest proteins left the transforming principle intact. Those known to degrade RNA left the transforming principle intact. And those that ignored protein but attacked samples of known DNA destroyed completely the activity of the transforming principle. Avery and his co-workers complicated the enzymatic tests by adding selective chemical inhibitors and by exploring the subtle effects of temperature variations on enzyme activity. Results always agreed, they reported, with what happened in parallel experiments with known DNA.

They went on to immunological tests. These demonstrated that neither pneumococcal protein nor capsular polysaccharide was present in the transforming extract up to the extreme limit of sensitivity of the technique. They spun a sample of the extract on the ultra-high-speed centrifuge, and found that as it sedimented, "the material gave a single and unusually sharp boundary indicating that the substance was homogeneous and that the molecules were uniform in size and very asymmetric"; the result matched with DNA from calf thymus. They tried electrophoresis, and found that as the molecules in a solution of the transforming principle were propelled by a weak electrical current, they stayed together as one substance—and that this moved relatively fast, as nucleic acids do. They found that the transforming principle absorbed ultraviolet light at certain wavelengths to yield the same profile as nucleic acids. They found, and saved for last, that the transforming principle could demonstrate its transforming power in extraordinarily small amounts—down to "a final concentration of the purified substance of 1 part in 600,000,000" of the culture medium containing bacteria of R form.

Avery's concluding discussion is one of those precursors that can sometimes be looked back to in science, or for that matter in philosophy or economic theory or painting, where one seems to see an idea struggling to shake free from a net of previous conceptions. Strikingly,

> the substance evoking the reaction and the capsular substance produced in response to it are chemically distinct, each belonging to a wholly different class of chemical compounds.
>
> The inducing substance, on the basis of its chemical and physical properties, appears to be a highly polymerized and viscous form of sodium desoxyribonucleate. . . . The experimental data presented in this paper strongly suggest that

nucleic acids, at least those of the desoxyribose type, possess different specificities as evidenced by the selective action of the transforming principle.

And those are the attributes of a stuff that is heterocatalytic—that can, as the gene must do, cause the cell to make another specific substance unlike itself. In support of that, Avery also observed,

> Attempts to induce transformation in suspensions of resting cells held under conditions inhibiting growth and multiplication have thus far proved unsuccessful, and it seems probable that transformation occurs only during active reproduction of the cells.

But was the transforming principle autocatalytic as well?

> Once transformation has occurred, the newly acquired characteristics are thereafter transmitted in series through innumerable transfers in artificial media [that is, repeated generations each started in fresh broth] without any further addition of the transforming agent. Moreover, from the transformed cells themselves, a substance of identical activity can be again recovered in amounts far in excess of that originally added to induce the change. It is evident, therefore, that not only is the capsular material reproduced in successive generations but that the primary factor, which controls the occurrence and specificity of capsular development, is also reduplicated in the daughter cells.

Thus Avery circumnavigated the definition of the gene. He was clear and firm about what he had demonstrated; he would not leap.

> Assuming that the sodium desoxyribonucleate and the active principle are one and the same substance, then the transformation described represents a change that is chemically induced and specifically directed by a known chemical compound.

There was an irrepressible doubt—and the call to resolve it by a new order of scientific precision.

> One must still account on a chemical basis for the biological specificity of its action.

So the conclusion checked and stumbled:

> It is, of course, possible that the biological activity of the substance described is not an inherent property of the nucleic acid but is due to minute amounts of some other substance adsorbed to it or so intimately associated with it as to escape detection. . . . If the results of the present study are confirmed, then nucleic acids must be regarded as possessing biological specificity the chemical basis of which is as yet undetermined.

That far, but in public no farther. Privately, Avery did go beyond that. A year before the results were published, he wrote a long letter to his brother, Roy, a bacteriologist then at Vanderbilt University. The letter was meditative, speculative, full of unassuming charm: it defines poignantly the sense of responsibility to science that some acknowledge. Avery first reviewed the years of searching, and then wrote:

> Try to find in that complex mixture, the active principle!! Try to isolate and chemically identify the particular substance that will by itself when brought into contact with the R cell derived from Type II cause it to elaborate Type III capsular polysaccharide, & to acquire all the aristocratic distinctions of the same specific type of

cells as that from which the extract was prepared! Some job—full of headaches & heart breaks. But at last *perhaps* we have it.

He described the experimental tests, and went on:

> In short, the substance is highly reactive & . . . conforms *very* closely to the theoretical values of pure *desoxyribose nucleic acid* (thymus type) Who could have guessed it? . . .
>
> If we are right, & of course that's not yet proven, then it means that nucleic acids are *not* merely structurally important but functionally active substances in determining the biochemical activities and specific characteristics of cells—& that by means of a known chemical substance it is possible to induce *predictable* and *hereditary* changes in cells. This is something that has long been the dream of geneticists. . . . Sounds like a virus—may be a gene. But with mechanisms I am not now concerned—one step at a time. . . . Of course the problem bristles with implications. . . . . It touches genetics, enzyme chemistry, cell metabolism & carbohydrate synthesis—etc. But today it takes a lot of well documented evidence to convince anyone that the sodium salt of desoxyribose nucleic acid, protein free, could possibly be endowed with such biologically active & specific properties & that evidence we are now trying to get. Its lots of fun to blow bubbles,—but it's wiser to prick them yourself before someone else tries to.

Opposition to any identification of DNA as the stuff of the gene was peculiarly concentrated at the Rockefeller Institute. Levene had been there, active until his death in 1940, the world authority on the chemistry of DNA and originator of the tetranucleotide hypothesis, by which repetitive scheme DNA could not possibly specify diversity. Alfred E. Mirsky, working in biochemical genetics there, was convinced and trying to prove that the protein associated with nucleic acids in the chromosomes of higher organisms was the active component. Mirsky argued implacably for many years, both within the institute and in public, that some proteins are resistant to the digestive enzymes used by Avery and his colleagues, so that the DNA must have been contaminated by significant traces of active protein. And every thought and argument there, not least in Avery's own lab, was shadowed by memory of a cautionary triumph of a group at the institute more than a decade earlier—the proof that enzymes are proteins and the humiliation of Richard Willstätter. In Munich, in the early twenties, Willstätter—who was perhaps the foremost organic chemist of the day, and a specialist in enzymes—had claimed that he had gotten enzymatic, catalytic action with preparations that were free of protein. On his evidence, it came to be widely accepted that the biological specificity of solutions containing enzymes was not due to protein. But in 1930 John Howard Northrop at the institute crystallized pepsin and showed that it was protein. That in itself was the second such demonstration, four years after James Sumner had done the same with urease; but Northrop and his associates developed precise techniques for correlating enzyme activity with the quantity of protein present, and showed conclusively that Willstätter's experiments had been contaminated by slight traces of protein. A laboratory colleague of Avery's for many years, René Dubos, when asked about the effect of the Willstätter scandal on Avery, replied, "It was on *everybody's* mind!"

Avery's work, even before the paper came out, was widely though unevenly known, for his laboratory had many visitors. Some papers are great, of course, because they establish, define, settle their issues. This great paper did something else: Avery opened a new space in biologists' minds—a space that his conclusions, so carefully hedged, could not at once fill up. The question was acute: If DNA is the carrier of hereditary specificity, *how?* Two scientists in particular were shocked by that question into the lines—two very different lines—their research took henceforth. Erwin Chargaff, an established biochemist then in his late thirties, on reading Avery's paper switched the work of his laboratory to the study of nucleic acids. Joshua Lederberg, just graduated from Columbia University at the age of nineteen and about to start his doctorate, found the pleasure of reading Avery's paper "excruciating"—so he noted at the time—and its implications "unlimited." He decided that these implications would never be cleared up unless bacterial inheritance could be analyzed by the methods of genetics. But bacteria were generally believed to be asexual, primitive creatures incapable of exchanging genetic information. To do genetics, Lederberg first had to show that their life cycles had a sexual stage—that they mated. On 8 July 1945, he noted down an idea for an experiment to demonstrate genetic recombination in bacteria. The recruiting of Chargaff and the mobilization of Lederberg were among the most important effects of Avery's paper.

Avery's public caution stands in awkward contrast to the self-assurance of Watson and Crick nine years later. The cost may have been great. The Nobel prize selectors had their attention drawn to Avery's work. They waited for the second round of discoveries. Avery was sixty-seven when the paper appeared; it was, Chargaff wrote in tribute, "the ever rarer instance of an old man making a great scientific discovery. It had not been his first. He was a quiet man; and it would have honored the world more, had it honored him more." Avery died in 1955.

I asked Crick one day about boldness and caution. "Some people of course are extremely cautious," he reflected. "Avery, exactly; he only put it in his letter to his brother. Boldness? I would have said that Bragg and Pauling were the people who most influenced me in these matters of style, and both have had that characteristic. Pauling to the point of rashness. I mean, one always knew about Linus that he would probably show an idea even if he realized, even if he knew there was a good chance of being wrong. In fact a lot of his ideas *were* wrong. But the ones that were right were important, and therefore he was forgiven for the fact that his structure of collagen was nonsense, for example, because the alpha helix and the pleated sheet were fine. But what I learned from Bragg was to grasp for the essence of the problem—and then when you've got something, get on with it and by and large publish it reasonably quickly. Though let me tell you that in the past I've often been dilatory, being from time to time of a lazy temperament. But more—from Bragg and Pauling I learned how to see problems, how not to be confused by the details, and that is a sort of boldness; and how to make oversimple hypotheses—you have to, you see, it's the only way you can proceed—and how to test them, and how to discard them without getting too enamored of them. All that is a sort of boldness. Just as important as having ideas is getting rid of them. And you realize that in those days, when we were working on DNA, the pace—things *were* very much quieter then. Now at the moment I've been thinking a lot about this problem of the chromosome structure of higher organisms, and the *rumor* of this has got *around*—the pace has got so much more hectic. I must tell you I prefer the older style—but what can one do! When we started we were living in the woods and now here we are in the middle of a city."

❖ *b* ❖   At Harvard, in the early seventies, Watson several times gave an advanced undergraduate course in the biology of viruses that cause tumors in animals. One September, I went to hear his opening lectures. The course was called Biochemistry 165, and met Tuesdays and Thursdays at eleven o'clock. Watson's office was in the solid, shabby, red brick Biological Laboratories, off Divinity Avenue, where the two magnificent bronze rhinoceroses flank the main door; but the lectures, his secretary said, were in the Science Center, several blocks away. This was a large, newish building, brutalist in style and materials, stepped back abruptly from each story to the next above, a Babylonian temple to the power of science rather than a Palladian reminder of its balance and clarity.

The classroom was crowded, with students standing along the walls. Watson arrived, looked around him, and led us all across the corridor to a large lecture theatre with banks of seats, steeply raked. He was wearing a rumpled suit, in a Prince of Wales check of subdued greenish hue. Standing in the well of the room, he broke a piece of chalk in half and wrote the names of three books on the board: a text, a collection of reprinted research papers, and one called *Biohazards in Biological Research.* As he wrote, his spectacles slipped down to the end of his nose. All three books were published by Cold Spring Harbor Laboratory, of which Watson is the director. He peered at the list, then turned to the announcements with which a course begins—guest lectures, discussion sections, "there'll be no meeting next Tuesday because I have to be in Washington." He stood with hands on the desk, stooping slightly. He spoke to the shoes of the people in the front row. His voice fell away so that many were leaning forward to hear. He made even the routine announcements with an edgy skepticism. "I sort of warn you—anyone here who gets into this course because he thinks it might—make medical school easier, but really doesn't want to do any work, might find himself with an embarrassing grade. I try to distinguish those who care from those who don't. One way to tell is from the term papers. One can read term papers—but only the first five with any interest or enthusiasm; by the time you reach number twenty you're asleep. Awful. But, one thing that always does come through—even if you're asleep—in reading these papers is whether somebody is actually trying to be novel or to think, that's the one thing. . . . The final exam will be open book. Why should you have to take a final? Why should I want to read them?" He looked genuinely perplexed. "I guess, it's a matter of obliging you to get the priorities organized; you have to look back and sort out the important from the unimportant. The questions will be very general, to let you show you can tell the difference between the important and the trivia. . . .

"The origins of tumor virus work have been—many; they have been, sort of—two. One is, people now, over the past sixty years, have tried to find viruses that cause human cancer. And then there's been a second group of people, often for different reasons, who tried to take animal cells and grow them in culture as if they were bacteria. Cell culture. Originally known as tissue culture. . . . This field . . . for some time, but . . . mystique . . . " A voice from the back of the hall shouted, *"Louder!"* Watson looked up, his eyes wide. "Yes, I'll try," he said clearly. "I know I mumble. The only way I don't mumble is if people in the back will be rude. And I don't mind. Call 'louder' and I'll speak louder. For at least five minutes."

What followed, in this and the next couple of lectures, was appropriately technical in detail, and the mumblings and false starts never ceased—but they soon ceased to matter much, because what Watson was talking about was exciting, and his way of seeing it, dubious and depreciative, was exciting. He spoke ad lib and seemed caught up in the subject, thinking it through afresh, so that each lecture grew spontaneously from live and immediate findings, questions, speculations. "It turns out the Salk and the Sabin polio vaccines had SV40 virus, which is an animal tumor virus, in the tissue cultures where they were made; Salk injections carried SV40 virus right in with them. No evidence yet that it causes tumors in people. . . . It's not proved to everybody's satisfaction, but herpes viruses are quite clearly the cause of at least one form of human cancer, probably several. . . . One form of cat leukemia is virus-caused, and -transmitted. To other cats. The open question is, is the cause of leukemia in humans, for example, a cow virus? . . ."

He was funny and scandalous about the shortcomings of scientific ideas and of scientists, though few of these asides are comprehensible out of context. The dry, alert distrust never faltered. He seemed never to miss pointing out doubts, reservations, "murkiness," "holes," the absence of sure knowledge, the facts not worth credence. "I planned today to give a talk on serum factors in cell culture, which I find hard to do because I think a year from now—right now one can report a lot of facts and I have the feeling it's probably not worth your while learning them because someone will find out the truth; but I'll try and go through some of it. The basic fact is that cells in culture require serum in order to grow. That seems absolutely to be solid. And the second fact that to me is a lot less solid, and some people don't believe it at all, is that there's something called contact inhibition. . . ." Later, unexpectedly, "It's my belief that the biochemistry of tumor viruses and cells will show a unity—although that's a hope that now seems almost impossible."

After the second lecture, Watson stayed for a quarter of an hour answering questions from two young men with beards and a young woman with the crinkly golden hair and low forehead of a model for the pre-Raphaelites. Then we walked across to the Faculty Club. His mind was still running on the topic he had announced for the next lecture. His shoes, tan suede with rubber soles, were both untied. "Never bother with them. Only ten years later will one be able to look back and see how terribly we have oversimplified these things." I ventured, as an amateur, that perhaps it was characteristic of modern biology that the great simplifications and discoveries seem soon—as though by the very act of bringing light and order into confusion—to generate evidence that exactly the reverse is also sometimes true. Once the double-stranded structure of DNA was accepted, it was possible to find that in some circumstances DNA occurs in single strands; once the genetic function of DNA had incontrovertibly replaced the notion that DNA is structural, it became possible for those investigating the chromosomes of higher organisms to suggest, as Francis

Crick had begun to do, that much of the DNA in them serves a structural role. There are several more examples. Watson played with the notion of such clarifying opposites for a moment or two, then tossed it away. "That's true and not true. All this extraordinary complication is what you'd really expect, you know. These mechanisms had to accumulate gradually in the course of evolution, so it's obvious that at each point the next variation may not be particularly consistent with what's come before." In a deft appropriation of the nineteenth-century fundamentalist argument against Darwinism, the wonderful, the overflowing, the redundant and self-contradictory complication of the living world becomes the best intuitive support for its piecemeal and undirected evolution.

At lunch, Watson began to talk about why he became director of the laboratories and teaching program at Cold Spring Harbor, and about the research there. "I am fond of the place. It has always for me stood for good science. In the sense that Harvard has always tried to put out, that it stands for good scholarship. Cold Spring Harbor has always had very high standards, and the thought that you might lose an institution with high standards—I think during the summers, well, it is the most interesting place in the world, if you're interested in biology. Only place where you can be stimulated by as much of what is going on; and by the people. In the winters it's small; but the labs are running, you don't go into hibernation for eight and a half months. We have enough people; and it's not clear that if we had twice as many people we'd be any more productive." We ordered lunch. "I think Cold Spring Harbor—well, I'd like it to be a place where advanced science can be done, and also where people can come to talk about it occasionally, talk about it in a formal atmosphere of meetings and advanced courses—and sometimes to make it more permanent by putting it into books."

I asked where he thought research at Cold Spring Harbor should be concentrated, and indeed where the most lively areas were in the whole field. "That's too broad a question. I just wouldn't want to try to answer that. In a small lab, in your own research effort, you can't be too broad. We have a sort of mania—an obsession to understand tumor viruses at the molecular level. And then just—working out a human cell as if it were a Swiss watch, figuring out everything in it. The lab is of a size where—seeing it from a country-wide viewpoint, it would be better to have other groups of about that size and excellence working in the United States rather than to make this one larger. You can't argue that you should be twice as big in order to have a chance of doing something interesting.

"What are the main areas, you ask. I would guess two. This animal-cell culture and its interrelationship with viruses; and neurobiology. Those are the two main fields that really have the momentum; you know they are going to go ahead fairly fast. They are both very difficult—fields where a certain degree of professionalism is necessary. And professionalism generally requires a few years of doing it. Right now cell culture is still largely populated by M.D.s. You can train these people, but the period of time when they are likely to be in active research is relatively short." Wives, children, houses, regular hours are the bane of committed laboratory research, Watson made clear. "The ones who get in by doing a Ph.D. in it I think will have a great advantage. If you get some kid of eighteen or nineteen, begin to get him able to do experiments in this field, then he may have twenty years in which you can count on him being productive. The sorts of breakthroughs that make tumor virus research possible occurred ten years or so ago; not that you then had any illusions it was going to be easy, but you knew you could put a student into it and there'd be something for him to be doing ten and twenty years on. At that level it couldn't fail. The unfortunate thing is the long time when really nobody has moved into this gap.

"Of course, before a field comes into existence you are going to have to use people trained in something else. It's certainly obvious that when molecular—as we now use the term—was starting, Crick and I couldn't have come in through taking a Ph.D. in it; we had to be trained in other things." I said that what was required of students now was incomparably more than then. "It's much harder," Watson said. "I mean, when I was a student there were vast amounts of what you could call classical knowledge of zoology and botany, and I never learned any because I figured it wasn't very relevant. Even lots of genetics I never thought that I needed to know. So I only had to learn a few facts and I got my degree very fast, not because—I was really that bright, but because there was very much less to learn. Now with the knowledge explosion in biology, the mass of facts you should probably know in order to move with some confidence, in any part of cancer research, for example, is enormous. That's one reason why I want to produce books at Cold Spring Harbor." Some scientists had told me, I said, that the professionalism required of students today made them as lopsidedly overtrained and overcompetitive as professional athletes. "If education is too long it'll probably kill you," Watson said. "As a scientist. That's why it's nice if in some way we can put people into a position where they can begin to do science not much after twenty. It's a shame to wait any longer; even then, you see, you are groping, at first. It probably takes a few years before you know what sorts of questions people are trying to answer, but can't. And, you say you have to know all these facts—well, clearly the facts, some of them, that you learn are wrong, so if you take them too seriously you won't discover the truth. You could say that if you become too imbued in the ideas and talk about them too long, maybe your capacity for ever believing they're false would be burned out. Probably what you should learn if you're a graduate student is, not large numbers of facts, especially if they're in books, but what the important problems are, and to sense—which experiments, work that's been done, probably"—he breathed in with a hiss—"probably aren't quite right. And which things you'd like to do yourself if a method came up to do it." I asked how the prospects of the graduate student or the research worker with a new doctorate compared today with what he had faced at the same stage. "Oh, when I was in that position there was no money at all, either. There were a few fellowships, some National Research Council fellowships; they were pretty small. And that was the period when the Rockefeller Foundation had a great influence on science; they were the main supporters of the first work in molecular biology."

He talked some more about the organization of biological research, and its financing. He was going to Washington for the first three days of the next week to attend meetings of the National Cancer Advisory Board. "Washington's a pretty minor aspect of my life. I think a lot of scientists do interact in some way or other with Washington." He mentioned examples.

"But I see myself as an insignificant outsider. The meetings take place in a big room in Building 31 of the National Institutes of Health way out in Maryland. The War on Cancer. This is an annual program review. You'll see a great similarity to a Pentagon briefing on the successful war in Cambodia. No, really—when you see the people leading it, they give you the feeling of the Egyptian army: the parallels are really good, because these people don't expect to win."

We met again on Sunday evening at the airport for the plane to Washington. Watson had spent Friday and Saturday with his wife and children on Martha's Vineyard, at a summer house they had recently bought. "Just had two days' holiday; can't treat today like a Sunday," he said. Settled in the airplane, he began working through a briefcase full of papers. When the ride was bumpy, over New York, he looked out the window for a moment. As we approached Washington and the engines grew quieter, he asked me when I had been at the University of Chicago, and whether we had overlapped there. I had entered in the fall of 1946. "That would have been the year after my sister, Betty," he said.

❖ ❖ ❖

James Dewey Watson entered the University of Chicago in 1943, at the age of fifteen. His schooling had been entirely ordinary, in the Chicago public system, but he was markedly bright and never accustomed to hide the fact; he had been one of the original Quiz Kids in the wartime radio program. The University of Chicago has always had a cross-grained intellectual magnificence. In those years its physics was particularly good, though Watson was not to know what has since become a commonplace, that there beneath the stands of the disused football stadium ticked the world's first atomic pile. That was the decade, too, when Chancellor Robert Hutchins's program for the undergraduate college of the university had its flowering. It seems to be half forgotten now what an extraordinary excitement Hutchins caused in American education—particularly among intelligent adolescents. Chicago would admit them two years before they would normally have graduated from high school, then promised them a four-year degree for which every student was obliged to take all the courses there were, literature through science to mathematics and philosophy; these were taught from original works—the great books in their broader sense, already (we all were aware) the object of considerable condescension but not yet vulgarized and commercialized. Among the great books were great research papers. To be fifteen at the University of Chicago was not prodigious, hardly a matter for remark.

Watson had one stabilizing advantage for an adolescent in that milieu: he lived at home with his family. Many young students, he remembered, had got terribly lost. His family lived a streetcar-ride away. His mother worked in the admissions office of the university; later she moved to the housing office. They had been very poor. There was an uncle, who lived in one of the North Shore suburbs and lectured at the Art Institute; he had been, I gathered, the guardian of the family's cultural values and social standing, but they hardly ever saw him because they had no car.

Graduates from Hutchins's program, having skipped a couple of years in the normal educational sequence, sometimes found their degrees nonnegotiable. Watson graduated in 1946, but stayed on to get a full, normal bachelor's degree by taking a year of zoology. (Thus we had overlapped. We had not met.) Biology at Chicago had been poor and conservative, he thought. "I managed to escape without knowing any genetics or biochemistry." Ornithology had been "a way of playing at science," and as he went further, he said, he had lost his committed interest in birds. Yet, at the time, he told friends that his ambition was to be curator of birds at the American Museum of Natural History, in New York; he still listed the subject on his applications for graduate school; he spent the summer of 1946 taking a course in advanced ornithology at the University of Michigan.

Science is a calling that comes in adolescence. "You get accustomed to doing it yourself in a way that feels good. That sets it for life. By the age of sixteen or seventeen." Teaching at the University of Chicago then was Sewall Wright, a founder of the mathematical discipline called population genetics; Wright has recalled that Watson sat in on a course of his, one in which since 1944 he had discussed the work of Oswald Avery. Watson remembered being interested by population genetics, but has often said that the decisive influence on him was the book *What Is Life?* by the physicist Erwin Schrödinger. This was a short work, published in 1944, that speculated about the physical basis—the atomic and molecular basis—of biological phenomena. In it, Schrödinger, the founder of quantum mechanics in its modern form, an Austrian of Roman Catholic background living in quiet exile in Dublin, popularized some suggestions about the nature of the gene, in the light of the principles of quantum mechanics, which had been made in Berlin in 1935 by another physicist, Max Delbrück. From the moment Watson read Schrödinger's book, he once wrote, "I became polarized towards finding out the secret of the gene." He also became polarized towards Delbrück.

Watson applied to do graduate work at Harvard, which turned him down—fortunately, he later said, because Harvard was not then strong in genetics. He applied to the California Institute of Technology, where genetics was uniquely strong, but which also turned him down. He was accepted by Indiana University, at Bloomington, and was given a research fellowship of nine hundred dollars for the 1947–'48 academic year—along with a crisp note from the dean of the graduate school to say that if he still wanted to study birds he should go somewhere else.

Indiana University, in the late nineteen-forties, had on its staff one of the great founders of classical genetics, Hermann Muller, and younger men of promise. Muller, when Watson arrived, had just won a Nobel prize for his discovery twenty years earlier that x rays cause mutations; mutations obtained one way or another have of course always been the chief tool of experimental genetics. He had begun writing about genes in 1911; 336 publications later, a decade after Watson and Crick found the structure of DNA, Muller was writing about "Genetic Nucleic Acid: the Key Material in the Origin of Life." Muller was a classical geneticist in several of the possible senses, the simplest being that he worked not with micro-organisms but with the fruit fly, *Drosophila melanogaster*, and that he pursued the mapping of its genes and mutations through the endless petty paradoxes of their interactions and their sequences on the chromosomes. Muller's understanding of the nature of the gene was profound, not least when he looked over the shoulders of his colleagues working with bacteria and viruses. In 1921, or almost a quarter century before Avery's paper on the transforming principle, Muller in a brilliant intuitive flash had foreseen how micro-organisms might revolutionize genetics.

Viruses that infect plants and animals had been distinguished only gradually from bacteria—beginning with the first characterization, in 1898, of the viral cause of the mosaic disease of tobacco. Like bacteria, viruses were living, transmissible, multiplying agents. But viruses were far smaller, able to remain infectious after passing through filters of porcelain with holes so fine that all bacteria then known were held back; and viruses were far simpler, able to reproduce only as parasites within living cells. Then, in laboratories in France and England during the first world war, it was observed that bacteria are preyed upon and destroyed by smaller entities that appear to reproduce themselves and multiply. In 1915, a bacteriologist named Frederick Twort, superintendent of a mildly eccentric charitable institution in London, published in *The Lancet* a confused note of a transmissible, filterable agent that destroyed bacteria; he wondered, among several possibilities, whether it might be a virus or an enzyme, or whether a virus is itself some kind of enzyme. Two years later, probably independently, an exactly similar agent was reported by Felix d'Hérelle, a French-Canadian bacteriologist at the Institut Pasteur in Paris. D'Hérelle later wrote that he had first come upon a filterable disease of bacteria in Mexico while studying, of all things, a diarrhoea of locusts. When he returned to the Institut Pasteur, in August of 1915, he was asked to investigate an epidemic of dysentery raging in a cavalry unit stationed near Paris. In cultures of the dysentery bacillus he again found an agent that would get past the finest filter, multiply parasitically within the bacteria, and destroy them. D'Hérelle christened his discovery "bacteriophage," from the Greek root for "eaters" (familiar in English from Othello's tales that captivated Desdemona, of "the cannibals that each other eat, the anthropophagi"). The name bacteriophage has stuck, though these days it is usually shortened to phage. D'Hérelle believed bacteriophages would prove to be a panacea against infectious bacteria; wrong in that, he rightly insisted that they were viruses and went on to characterize their life cycle. In 1921, while many bacteriologists still believed that the Twort-d'Hérelle phenomenon was caused by a bacterial enzyme, not a virus, Muller noted that these entities not only multiply but mutate; and so—speaking at a symposium in Toronto of the American Society of Naturalists—he looked ahead:

> That two distinct kinds of substances—the d'Hérelle substances and the genes—should both possess this remarkable property of heritable variation or "mutability," each working by a totally different mechanism, is quite conceivable . . . yet it would seem a curious coincidence indeed. It would open up the possibility of two totally different kinds of life, working by different mechanisms. On the other hand, if these d'Hérelle bodies were really genes, fundamentally like our chromosome genes, they would give us an utterly new angle from which to attack the gene problem. They are filterable, to some extent isolable, can be handled in test-tubes, and their properties, as shown by their effect on the bacteria, can then be studied after treatment. It would be very rash to call these bodies genes, and yet at present we must confess that there is no distinction known between the genes and them. Hence we cannot categorically deny that perhaps we may be able to grind genes in a mortar and cook them in a beaker after all. Must we geneticists become bacteriologists,

physiological chemists and physicists, simultaneously with being zoologists and botanists? Let us hope so.

Muller's hope of 1921 is justly famous among biologists. Unlike many inspired conjectures in the history of science, it had in it nothing woolly. Twenty years later, to explore that conjecture—that is, to discover and explain the relation "between the genes and them," the bacterial viruses—developed into one of the two great routes that converged on that ambition "to grind genes in a mortar" and so to find in chemistry and physics the foundation of the phenomena of genetics. Then, in 1945, after pondering Avery's work on the transformation of pneumococci, Muller took the leap that Avery had balked at. When the pure DNA was added to the culture, Muller surmised, "There were, in effect, still viable bacterial 'chromosomes' or parts of chromosomes floating free in the medium used. These might, in my opinion, have penetrated the capsuleless bacteria and in part at least have taken root there." He saw great experimental opportunities:

> A method appears to be provided whereby the gene constitution of these forms can be analyzed, much as in the cross-breeding tests on the higher organisms. However, unlike what has so far been possible in higher organisms, viable chromosome threads could also be obtained from these lower forms for *in vitro* observation, chemical analysis, and determination of the genetic effects of treatment.

Joshua Lederberg, in 1961, called Muller's suggestion "an incredibly unconventional interpretation that is now the basic idea in this field."

❖ ❖ ❖

Watson was drawn to Indiana by Muller's presence. Watson's career and his energies and intellect as a scientist were given their shape and direction at Indiana—but not to any large degree by Muller. The great formative influence on Watson was Salvador Luria, who was teaching and doing bacteriophage research in Bloomington, and through Luria his close friend Max Delbrück, and around Delbrück the small, enthusiastic band of researchers, scattered across the country, who thought of themselves as the American phage group. Luria is a short, vibrant North Italian. His voice is husky and deep, with a grin in it. He bursts with ideas whole and in scraps. He trained in medicine in Italy and came to the United States at the beginning of the second world war. He wrote several years ago, with uncharacteristic dryness, "Service in the Italian army in 1936–1937 had given me a chance to study calculus, and a month's wait for the American visa in Marseilles in 1940 had permitted me to read G. N. Lewis's *Physical Chemistry*."

The mass intellectual emigration from continental Europe in the nineteen-thirties, which so stimulated physical science in the United States and England, also had profound consequences for biology, even though the men involved were fewer and younger, with their reputations still to make. They included Perutz, who left Austria for England in 1936, and Chargaff, also an Austrian, who emigrated to the United States in 1934. A less direct influence was the distinguished and passionately intelligent Hungarian physicist Leo Szilard, who in the nineteen-thirties had been the first to envision the possibility that sustained nuclear fission, a chain reaction, could work and cause an explosion, and the first to urge that the United States should try to make an atomic bomb. Szilard wrote the letter about the idea that Einstein signed and that was read to Roosevelt in 1939. Szilard worked on the atomic project at the University of Chicago through the war; afterwards, in reaction against the weapons and against the big-money, big-team physics he had been instrumental in creating, he turned to biology and also to campaigning within the international scientific community for disarmament—for example, through the Pugwash conferences, which he helped to found. In 1947, Hutchins gave Szilard the physicist an appointment as professor of biology and sociology. In the early years of molecular biology, Szilard was an erratic if interesting experimenter and theorist, a cross-pollinator of ideas and an effective critic of others' work, an intellectual and ethical inspiration to younger scientists.

The most important immigrant to biology, however, was Max Delbrück. Delbrück was German, born to the aristocracy of the intellect—his father was the professor of history and his uncle the professor of theology in the University of Berlin—and trained as a quantum physicist. His mind and style had been formed by Niels Bohr, the physicist, philosopher, poet, and incessant Socratic questioner who made Copenhagen one of the capital cities of science between the wars. Delbrück's ideas about the physical properties of the gene, in a youthful paper of 1935, had led Schrödinger to write *What Is Life?* Delbrück was perhaps the earliest of the theoretical physicists who have crossed over to biology; Szilard, Crick, Maurice Wilkins were others, while Linus Pauling, arriving at biology from a different tangent, was a physical chemist whose strength was founded in quantum mechanics. The move from physics has been the intellectual immigration that has mattered most to biology.

Delbrück arrived in the United States in 1937, going first to the California Institute of Technology as a Rockefeller Fellow. He immediately began research in the biology division there, with Emory Ellis, who was working with bacteriophage. Then in 1940, Delbrück moved to Vanderbilt University, in Nashville, where he remained through the war; hired to teach physics, he concentrated his research on phage. By 1926, d'Hérelle had described the life history of a bacterial virus, in three steps: attachment of the phage particle to the susceptible bacterium, multiplication of phage within the bacterial cell, and lysis, or bursting, of the bacterium to set free the progeny virus particles, which can then attach to other susceptible bacteria. D'Hérelle had also invented the simple techniques for detecting and counting viruses that are still the basis of the work. At the time Delbrück arrived at Caltech—"intent on discovering how his background in physical sciences could be productively applied to biological problems," Ellis later wrote of him—nobody had yet seen a phage, for the electron microscope was not introduced until 1939. Ellis showed him some charts of phage growth which demonstrated what d'Hérelle had found, the step-wise cycle of multiplication. Delbrück's immediate response was, "I don't believe it." He went to work with Ellis, repeating and sharpening the step-wise growth experiments and attempting to disentangle the separate stages in the life cycle of bacteriophage.

"One might wonder how the biologist can learn anything about the behavior of organisms so small," Delbrück wrote of those first phage experiments, a few years later while most of the mysteries were still unsolved.

The answer is that bacterial viruses make themselves known by the bacteria they destroy, as a small boy announces his presence when a piece of cake disappears. Much of what we know about the viruses is based on the following experiment, which requires only modest equipment and can be completed in less than a day.

Bacteria first are grown in a test tube of liquid meat broth. Enough viruses of one type are added to the test tube so that at least one virus is attached to each bacterium. After a certain period (between 13 and 40 minutes, depending on the virus, but strictly on the dot for any particular type), the bacterium bursts, liberating large numbers of viruses. At the moment when the bacteria are destroyed, the test tube, which was cloudy while the bacteria were growing, becomes limpid. Observed under the microscope, the bacteria suddenly fade out.

The investigations that Ellis and Delbrück published together in 1939, known in the trade as the one-step growth experiment, proved that lysis of the host bacteria and release of the daughter burst of phage indeed occur strictly on the dot. The beauty of the one-step growth experiment in the eyes of other biologists resides first in the ingenious simplicity by which concentrations of bacteria and phage were manipulated to make sure that the phage began penetrating the bacteria all in synchrony. Then there were the methods Delbrück perfected of assaying, every few moments, the presence and number of free phage particles, by taking a measured sample from the broth of bacteria and phage and spreading it across a plate of nutrient jelly, where each virus starts a colony of bacterial destruction that multiplies so quickly that within a few hours it can be seen by the naked eye, as a bare spot in the bacterial lawn—called a plaque, d'Hérelle's original term—to be counted and characterized for size, texture, outline. Then too, scientists found this paper, like most of Delbrück's, memorable for clarity of argument and laconic style.

The beauty of one-step growth for Delbrück lay rather in the way it raised far more questions than ever it answered. In a lecture in 1946 to the Harvey Society—a medical society in New York City which invites research biologists to speak; the invitations are a signal honor—Delbrück described the fascination in the field of bacteriophage, "a fine playground for serious children who ask ambitious questions." He went on:

> You might wonder how such naive outsiders get to know about the existence of bacterial viruses. Quite by accident, I assure you. Let me illustrate by reference to an imaginary theoretical physicist, who knew little about biology in general, and nothing about bacterial viruses in particular, and who accidentally was brought into contact with this field. Let us assume that this imaginary physicist was a student of Niels Bohr, a teacher deeply familiar with the fundamental problems of biology, through tradition, as it were, he being the son of a distinguished physiologist, Christian Bohr.
>
> Suppose now that our imaginary physicist, the student of Niels Bohr, is shown an experiment in which a virus particle enters a bacterial cell and 20 minutes later the bacterial cell is lysed and 100 virus particles are liberated. He will say: "How come, one particle has become 100 particles of the same kind in 20 minutes? That is very interesting. Let us find out how it happens! How does the particle get in to the bacterium? How does it multiply? Does it multiply like a bacterium, growing and dividing, or does it multiply by an entirely different mechanism? Does it have to be inside the bacterium to do this multiplying, or can we squash the bacterium and have the multiplication go on as before? Is this multiplying a trick of organic

chemistry which the organic chemists have not yet discovered? Let us find out. This is so simple a phenomenon that the answers cannot be hard to find. In a few months we will know. All we have to do is to study how conditions will influence the multiplication. We will do a few experiments at different temperatures, in different media, with different viruses, and we will know. Perhaps we may have to break into the bacteria at intermediate stages between infection and lysis. Anyhow, the experiments only take a few hours each, so the whole problem can not take long to solve."

Perhaps you would like to see this childish young man after eight years, and ask him, just offhand, whether he has solved the riddle of life yet? This will embarrass him, as he has not got anywhere in solving the problem he set out to solve. But being quick to rationalize his failure, this is what he may answer, if he is pressed for an answer: "Well, I made a slight mistake. I could not do it in a few months. Perhaps it will take a few decades, and perhaps it will take the help of a few dozen other people. But listen to what I have found, perhaps you will be interested to join me."

Several scientists had already responded to the charm of the man and the subject. On 28 December 1940, at a meeting of the American Physical Society in Philadelphia during the Christmas vacation, Delbrück and Luria met. Luria had recently arrived in the United States. He was not a physicist, despite his presence at a physicists' convention; yet he knew some physics, was a friend of Enrico Fermi, and had begun phage research of his own while still in Italy. After several hours' conversation, Luria wrote later, they had dinner with Wolfgang Pauli and another European physicist "during which the talk was mostly in German, mostly about theoretical physics, mostly above my head." Afterward, "Delbrück and I adjourned to New York for a 48-hour bout of experimentation in my laboratory at the College of Physicians and Surgeons" of Columbia University. "It was not an experiment to do," Luria said in the fall of 1973. "It was just to see—because, what happened is that Max wanted to work with a particular phage that attacks staphylococcus, a phage named Krueger phage after the man that first wrote it up, and Max had been told by Krueger that this phage should not be assayed by plaques, the way one does *all* the assays, but should be assayed by a very complicated method, which partly turned out later to be the cause why Krueger got very crazy results, results that he interpreted in a very crazy way. And I had gotten the same phage from Krueger, and I was assaying it all the time in the usual way like every other, and I told that to Max, and he said, 'I want to see,' so I said, 'Fine. If you come up on Monday I'll have plates and everything ready and we'll play around.' And he did, and we did."

The two men planned a series of experiments to do together and debated where to do them, whether at Vanderbilt in Nashville or in New York. Three weeks later, Delbrück wrote to Luria that he had been invited to attend the next annual symposium at Cold Spring Harbor and to spend the rest of the summer there; could they work together there? "If that could be arranged satisfactorily at C.S.H., I might overcome my antipathy to the place." In fact, Delbrück got married that summer, and spent the first weeks of his marriage at Cold Spring Harbor. At that symposium, Luria wrote much later, "The whole idea of the nature of the gene—I do not know if it was the first time; it was the first time in my experience—was dealt with in an environment in which geneticists and people interested in molecular structure were interacting freely." Delbrück has returned there to work and teach many summers since.

At the end of January 1943, at Delbrück's invitation, a microbiologist named Alfred Hershey visited Nashville for a few days; he had written papers about phage that caught Luria and Delbrück's attention. In a letter to Luria, along with the draft of a new theoretical idea, Delbrück gave his first impressions of Hershey: "Drinks whiskey but not tea. Simple and to the point. Likes living in a sailboat for three months, likes independence." The three men were the nucleus of the phage group. Delbrück recently characterized this beginning as two enemy aliens and "another misfit in society." The three shared the Nobel prize for physiology or medicine in 1969, seven years later than Watson, Crick, and Wilkins.

❖　❖　❖

The next essential discovery about the relation of genes to bacteriophage was that phages have complex structures. The opposite had been widely assumed: the mystery of viruses in the late nineteen-thirties was that they seemed like chemicals rather than organisms. One stumbles over that assumption in the opening sentence of Delbrück's report of the one-step growth experiment: "Certain large protein molecules (viruses) possess the property of multiplying within living organisms," a process "at once so foreign to chemistry and so fundamental to biology."

The electron microscope was invented in Germany just before the second world war. The first models in the United States were built by RCA in Camden, New Jersey, in 1939. They were, of course, immediately trained on bacteria, revealing structural details never before seen. Within a year, bacteriophage preparations had been found to contain vaguely sperm-shaped particles. Early in November of 1941, Luria came to RCA to ask whether the electron microscope could be used to see how big phages are. Luria talked to Thomas Anderson, who had spent a year in Camden, on a fellowship, taking pictures of specimens of all sorts sent in by biologists of all sorts. Luria told Anderson that with an associate he had just estimated the sizes of several types of phage by the indirect method of dosing them with x rays and counting the proportion left active—like firing a shotgun shell (of known size) into a swarm of bees (of known density) and estimating, from the reduction in stinging power of the swarm, the diameter of the individual bee. Luria wanted to confirm the estimates. Anderson thought the idea was reasonable; Luria said he would come back in a month with some highly concentrated phage stocks. He arrived with samples of three different phages, on Monday, 8 December 1941, the day after Pearl Harbor. They didn't get much done that day, nor even that week, because the phage were not sufficiently concentrated to be spotted for sure in the microscope. So in the first week in March 1942, Luria came back with small vials of phage stocks that assayed as high as ten *trillion* particles per litre, a billion or so in a few drops.

Now, Anderson later wrote, "We could really see the phage as tadpole-shaped particles, whose heads ranged from 600 to 800 Å [angstrom units: named after the nineteenth-century Swedish physicist Anders Ångstrom: one Å is a ten-billionth of a meter; there are two hundred fifty-four million angstrom units to an inch] in diameter, depending on the species." They found that each species had a characteristic structure, and that the sizes were roughly what Luria had expected. They ran some quick controls to see whether they could find anything similar in plain broths, in cultures of bacteria without phage, and so on—they couldn't—and, within a week, sent off a paper and a photo for publication. "There was no doubt in our minds that we had found the phage particles, but certain other workers remained skeptical for years," Anderson wrote. "I remember particularly the reaction of Alfred Hershey's teacher, kindly old Professor J. J. Bronfenbrenner, who had worked on bacteriophages for many years at Washington University in St. Louis. . . . When he first saw our pictures . . . he clapped the palm of his hand to his forehead and exclaimed, '*Mein Gott!* They've got tails!'"

Permutations of the basic phage experiments were endlessly rich. Although through the mid-forties fewer than a dozen investigators were active, since almost every one of them was exploiting his own private system of phages and host bacteria the results grew so complicated—and so difficult to compare—that in 1944, during the summer at Cold Spring Harbor, Delbrück negotiated what came to be called "the phage treaty." He persuaded most phage workers to concentrate on a set of just seven bacteriophages, T (for Type) 1 through T7, of the same bacterial host, a particular strain of *E. coli* (strain B) and its mutants, to be grown in broth at blood heat. In the next fifteen years, some of the most significant discoveries (and eventually a couple of Nobel prizes) were won by departures from the phage treaty, but even so their significance was in large part to be measured against the basis established through the T-set of phages acting on *E. coli* B.

The most fundamental line of research grew directly from the experiments Delbrück and Luria planned at their first meeting, at the end of 1940. Bacteriologists had for some time been puzzling, rather ineffectually, over the phenomenon of resistance—that is, the occasional appearance, in an enormous population of bacteria susceptible to a particular phage, of variant bacteria that the phage can't prey upon. By the wonderful dynamics of microbiology, these resistant individuals can be identified, two or three in a billion, because they alone survive saturation by phage to produce colonies of progeny. Since antibiotics began to be widely used, later in the nineteen-forties, the more familiar form of resistance is the rise of strains of bacteria that are not susceptible to penicillin, tetracycline, and so on.

Today, even the layman thinks of resistant bacteria as originating from mutation—rare but stably heritable change in the genetic makeup—that lets the bacteria survive to reproduce. But when Luria and Delbrück first got together, conventional bacteriologists were by no means clear that microorganisms could be thought about genetically or in terms of natural selection. Many believed that resistance was some kind of adaptation induced, in a few of the bacteria in a culture, by the exposure to the antibacterial agent; the mechanism of induction was never specified, but one might imagine a metabolic change so drastic that it would rarely succeed but when it did would be passed on to the progeny. The idea smacks of the pre-Mendelian pre-Darwinian notion of the inheritance of acquired characters; Luria damned bacteriology as "the last stronghold of Lamarckism."

One Saturday evening in January of 1943, after Luria had moved to Indiana University, he went to a faculty dance held at the Bloomington Country Club. There, watching the fluctuating returns obtained by colleagues gambling on a slot machine, he thought of the experiment that would distinguish between resistance induced in bacteria and resistance resulting from previous spontaneous mutation upon which selection acts. What Luria perceived was that previous spontaneous mutations would pay out jackpots of resistant bacteria that would fluctuate much more widely in size than those paid out by induction. He tried the first experiment on the following morning and wrote off to Delbrück; Delbrück promptly replied that Luria really ought to go to church, but a fortnight later sent along the manuscript of the fully worked-out mathematical basis of the experiments.

The point of the fluctuation test can be seen without the detailed mathematics, however. As a friend explained it to me, "What they did was statistically very simple-minded, but such intelligent and simple-minded things just hadn't been done in this field before." Luria understood that he could begin with a broth seeded with *E. coli;* distribute equal small amounts—hardly a drop—into each of twenty small, separate test tubes and a larger amount—ten milliliters, amounting to two teaspoonfuls—into another, larger tube; and allow the bacteria to multiply overnight, through a dozen or more generations, into rich cultures. Then he could take thirty petri dishes containing jelly saturated with phage T1. Into twenty of these glass plates, on one side, he would distribute the contents of the twenty small tubes. Into the remaining ten plates of T1 he would put equal small amounts all drawn from the bulk culture in the larger tube.

If the production of resistance began only at the moment of exposure of the bacteria to the phage, then it wouldn't matter whether the bacteria came from many individual cultures or from one bulk culture: the twenty plates on one side should show about the same distribution of resistant colonies, the same fluctuation from the average count, as the ten plates on the other side. When Luria performed the experiment, though, the twenty separate cultures showed much wider fluctuations from the average number of resistant colonies, indicating—as Luria had understood when he watched the slot machines—that a few of the individual tubes had contained resistant bacteria from near the beginning of the overnight growth period, while in a few other tubes resistant bacteria had appeared and begun to multiply very late or not at all.

The fluctuation test was the first real evidence that bacteria underwent mutation. And if stable, heritable mutants occurred in bacteria, then genetic analysis of their recombinations began to be possible. Years later, other members of the phage group found more direct ways of proving the same points; none the less, bacterial genetics was born with the publication in 1943 of the paper by Luria and Delbrück reporting the fluctuation test, and the event has been compared to the birth of genetics itself, in 1865, on the appearance of Mendel's paper.

Meanwhile, better electron microscopes took better pictures, which showed that phage have a structure of startling intricacy. Species can differ, but the phage T2, T4, and T6 look like a modern water tower out on the prairie: a tank with a conical top and bottom for a head, below that a narrow sheath as tall as the head, a core within the sheath, and a base plate with six spikes and six long, thin, stiff tail fibres with knees in them—which provide the finishing sinister touch. What struck Delbrück and Luria as strange in Anderson's micrographs was the evidence that few of the phages, perhaps none, appear to enter the host bacteria. Instead, they attach to the outside by the tail fibres and seem to squat down. The analogy of a hypodermic syringe was not thought of until much later. Chemical analyses of great delicacy established that these phage are made up of approximately equal parts of protein and DNA, and that the protein occurs in at least five different types, one for each separate tail structure and a large proportion of head protein.

Then, but not until 1949, Anderson found that phage T4, if grown in salty broth which is suddenly diluted with distilled water, was promptly inactivated. His electron micrographs showed phage ghosts, with empty heads. A colleague, Roger Herriott, demonstrated chemically that the ghosts consist entirely of protein, with none of the DNA normally present. Evidently, osmotic shock bursts the heads, liberating the phage DNA. Some beautiful electron micrographs taken years later show the phage ghost lying in the midst of its looping, crisscrossed DNA threads—or rather, thread, for it can be traced from one free end to the other, and is about 550 times longer than the phage head that contained it. In November 1951, Herriott wrote to Alfred Hershey, with the unmistakable freshness of a new idea, "I've been thinking—and perhaps you have, too—that the virus may act like a little hypodermic needle full of transforming principles; that the virus as such never enters the cell; that only the tail contacts the host and perhaps enzymatically cuts a small hole through the outer membrane and then the nucleic acid of the virus head flows into the cell."

❖ C ❖   On an afternoon in May of 1943, Delbrück was walking across the Vanderbilt campus when he met Roy Avery, who said he had received that day a long and interesting letter from his brother in New York and Delbrück must come to read it. Fourteen pages, handwritten—this was the letter, since become famous, in which Avery described his experiments with the transforming substance and allowed himself to speculate about what they meant. Oswald Avery was not part of the phage group. Sociologists of science have concluded that that was a chief reason why the paper about transformation of pneumococci appears to have had so little consequence for so long. They have backed up the conclusion by counting and mapping the worldwide distribution of copies of *The Journal of Experimental Medicine* in 1944, to find that geneticists might have been unlikely to see the work when it appeared; they have tabulated references at the ends of papers published by leading members of the phage group and have skimmed through texts of reports and discussions at the annual Cold Spring Harbor symposia in the decade from 1944, to find that Avery's work was irregularly and on the whole rarely mentioned. The conclusion from all this irritated Delbrück and amused Luria, because it supplied a false and blindingly trivial explanation for what was, they thought, a much more interesting passage in the history of ideas.

In the summer of 1972, I went out to Cold Spring Harbor Laboratory to interview Delbrück. Cold Spring Harbor is situated along the Sound side of Long Island, hard by Oyster Bay, Roosevelt and Rockefeller country, just past an hour on the railroad from Pennsylvania Station but more Midwest than Manhattan in feeling, the greenery fatly humid, the houses wide-spaced, frame, large, understated, weathered. A sweep of highway leads down a hill to the town; towards the bottom are an inconspicuous sign, on a wooden board, and a road in. The road is to an extent the life of the place—the road and the long lawn down to the water. Buildings are scattered either side of the road, the houses erect, somewhat shabby, Victorian or earlier, some of them offices now, some dormitories, the newer laboratory buildings sitting low, less conspicuous; the lawn slopes off to the right beyond the first group of buildings, down to a volleyball court, more low labs, a boxy, airy, pleasant little library. Across the pretty cove, Cold Spring Harbor itself, lies a boat basin. Meetings, meals, parties, discussions spill across the lawn or up onto the road; groups of swimmers wander down the road and off to the right through woods to the sandspit; a lot of science gets talked out on the road; the administration buildings, and beyond them the tennis courts and then the director's big house, lie to the left along the road; Watson, when he's around, walks the length of the road four or six times a day and probably does more administration there than in his office. I found Delbrück by stopping somebody on the road, and was directed to a gray building nearby. Bulky equipment stood in the corridor. A metal trunk labelled "Controlled Environment Incubator Shaker" chugged and gasped.

Delbrück was at work that morning in a cool room, alone by the far wall, beyond several ranks of lab benches. Surfaces all around were agreeably littered the way a painter's studio might be littered, except here were bottles and beakers, and microscopes covered with crumpled plastic bags. A flask was marked "radioactive waste—liquid." A table held a row of flat glass dishes with glass covers, gray-green growing inside. A handwritten sign above said, "Don't leave open flower pots!" Delbrück sat on a high wooden stool with a spring back, looking down a binocular microscope and working with something on its stage. He wore a short-sleeved tan knit shirt, tan-and-blue shorts, and pale tan sandals. He had muscular calves, gray hair closely cropped, large and graceful hands. A bunsen burner to his left on the bench flared a high almost invisible flame, through which from time to time he flicked the tips of a pair of tweezers. At the other end of the laboratory, two students, a Japanese and a black-bearded European, whistled and chattered in cryptic phrases over their own experiment. Delbrück sneezed vigorously. He had on round, steel-rimmed spectacles with smaller square magnifying lenses added in front. He put a cover on a flower pot and marked it with a felt marker. "Did you bring your bathing suit? High tide is after lunch." We went outside to talk, and sat in the shade of a tree, beside a low wall at the top of the lawn, looking out over the water to the sailboats.

Delbrück refused my first question. "No, let's start with Avery," he said. "Several people have made a big deal out of the fact that Avery was ignored. For a number of years. And have bolstered their assertions with statistics of citations and things like that. But I think that's not proper. First of all, I know that, for instance, Luria and I were very well aware of Avery's work at least since '42." In the tree, a bird sang. "Maybe even '41. Which was the first summer we were together here at Cold Spring Harbor. I know that in two successive summers before Avery's great discovery was announced, we went to visit him, at the Rockefeller. And he was very—attentive to what we were doing and we were very attentive to what he was doing." He talked for a minute about the background to Avery's work with transformation. He spoke at a reflective pace, very precisely, with a quirky fluency that broke up ordered prose with unexpected dashes of out-of-date American slang. He would pause to find a word, then neatly put all the rest of the sentence in.

"So, these methods were all very crude and uncertain at the time, and the main thing that Avery worked on was to make it more reproducible. Even towards the end they were still terribly cumbersome and uncertain in their results. Nevertheless"—Delbrück's voice grew firmer—"it was obvious that he had something very interesting there. And then when he was sure of the discovery—I think I was among the very few, first people who heard about it, because he wrote this long letter to his brother, Roy Avery, at Vanderbilt. Which is where I was." I said I knew of the letter but that it had slipped past me that Roy was at Vanderbilt. "He was at Vanderbilt; and the day he received the letter I met him and he told me about it; and I read it then. And in fact it was I who unearthed the letter years later, because there was some occasion at the National Academy where I was supposed to give a talk, they wanted to emphasize DNA, and I remembered this letter and wrote to Roy Avery and he and his wife spent a whole week looking through old boxes and finally found it." The occasion, though Delbrück didn't say so, was a memorial address after Avery's death. "I go into all this by way of showing that we were very much aware of what Oswald Avery was doing—and also of his discovery, that this transformation worked if you took a preparation which ostensibly contained only DNA."

Ostensibly? "Well." But was there skepticism, then? "Certainly. I mean, everybody who looked at it, and who thought about it, was confronted with this paradox, that on the one hand you seemed to obtain a specific effect with DNA, and on the other hand at that time it was believed that DNA was a *stupid* substance, a tetranucleotide which couldn't do anything specific. So, one of those premises had to be wrong. Either DNA was not a stupid molecule, or—the thing that did the transformation was not the DNA." Delbrück's accent was German, but lightly and attractively so, a matter of rhythms, the letter *r*, and an extra purity of vowels. "So this was very clearly the dilemma, and at that time it was not a matter of right thinking or of profundity or of anything except that you had to find out. Which was right. And Hotchkiss was the one who then joined the Avery group, and in several years of very careful work established that indeed it was the DNA.

"So then the dilemma became—can DNA carry specificity, or are we really not dealing with specificity at all but with a third possibility, some very special case where all the specificity is already there but just needs a stupid kind of molecule to switch it from one kind of production to another. Today we would say that the information is already in the genes; we might say now that you 'de-repress' a gene. At that time we might have said, for example, you can switch a tree from nonflowering to flowering"—he gestured at the tree—"and if you do it by manipulating the daylight-and-dark cycle then you are not putting in a molecule with information but just throwing a physiological switch. Everybody who was interested in these basic questions discussed this. I distinctly remember wading here at low tide with Rollin Hotchkiss, every so often, and he plugging for the idea that DNA might contain enough specificity. For example that the base ratio was not equal parts of the four nucleotides but somewhat off, though he was completely vague how that could get enough specificity into this kind of macromolecule. And even after people began to believe it might be DNA, that wasn't really so fundamentally a new story, because it just meant that genetic specificity was carried by some goddamn other macromolecule, instead of proteins.

"Both proteins and nucleic acids were hopelessly inadequately characterized, in those days. Proteins were characterized a little more, because we knew that there were twenty essential amino acids—or twenty-*odd* amino acids: one didn't, after all, know really how many. But whether proteins were built in a regular repetitive way or in a very specific way was still very unknown in the forties. And similarly, DNA—one knew it was a fibrous molecule, one didn't really know how the nucleotides went in together, at this time, or even whether it had branches in it or not. So, while it was true that Avery's discovery, after several years of refinement, led to the conviction that there had to be enough specificity in the DNA—you really did not know what to do with it.

"So I don't really see, in retrospect, that Avery had been neglected. I think certainly all the people who were seriously interested in what we now call molecular genetics, which at that time didn't have that name—interested in the *nature of the gene*—those who were fully aware of the data and were discussing it, to the extent that it *could* be discussed, they clearly saw the three alternatives. That it was contamination and not DNA at all. Or that the transformation was not really specific but just a special case, a physiological switch. Or that the DNA does carry specificity—but even if it did carry specificity then nobody, absolutely nobody, until the day of the Watson-Crick structure, had thought that the specificity might be carried in this exceedingly simple way, by a sequence, by a code. This dénouement that came then—that the whole business was like a child's toy that you could buy at the dime store"—Delbrück laughed and shook his head—"all built in this wonderful way that you could explain in *Life* magazine so that really a five-year-old can understand what's going on. That there was so simple a trick behind it." He laughed again, the same rueful way. "This was the greatest surprise for everyone."

I asked Delbrück how he meant that the Watson-Crick structure was a dénouement. "Excuse me," he said. "The dénouement at that time was only with respect to the structure and the means of replication, but not with respect to the readout. How you use this Watson-Crick structure to—as we say now, to *code* for anything, to carry specificity, that was still very very obscure." Was it not even evident that the place to look was in the sequence of nucleotides? "In some way, of course, for that was the only degree of freedom left; but how you could use the sequence to code for amino acids was just completely baffling." We talked about that puzzle. "So in retrospect what the dénouement was, was that both the principle of replication and the principle of readout are very simple, and the actual machinery for doing it is immensely complex. That's the way it has turned out."

Therefore, I asked, what the Watson-Crick structure did was define the problems next to be solved? Delbrück said, slowly, "Yes. Yes. Yes—it gave a very marvellous fixed point from which to start on both these problems. Replication and readout. Marvellous in that it was so concrete." In the tree, the bird sang again (and sings still, in my tape recorder). "I mean that it gave the hope that the whole solution will be possible in terms of concrete, three-dimensional chemistry. Stereochemistry. Enzyme chemistry. Which before was not clear."

We talked then about somewhat different things, beginning with his role in Schrödinger's book. What Delbrück's words alone cannot fully convey is his touch: quick, courteous, accessible, subtle, aware, contemptuous of pretense. His bumpy nose and his lantern jaw accentuated the light ironic smile that invited one to join him for a moment to watch the mortal pageant pass. He had the gift—which we know about from, say, Sir Thomas More, because he had it too—of a joyful seriousness. He was inordinately fond, I'd been told, of practical jokes. We wound up talking a little about revolutions in science. I mentioned the suggestion, interesting and fashionable, by Thomas Kuhn, then at Princeton, that science has happened in epochs, each epoch to be known by what Kuhn has termed a paradigm: such a paradigm is not simply a grand theory but more, the epoch's ruling way of conceptualizing theories, so that the end of a scientific epoch is marked by the total breakdown of the paradigm and its replacement by another. The coming of Darwinism is held to have been such a transformation of paradigms. So is the arrival of quantum mechanics. We agreed that "paradigm," as a term, was already almost as thoroughly debauched as "charisma," after which a brand of scent had even been named ("the woman who wears it's *got* it!"). Were there only four or five revolutions in all the history of science big enough to meet the criteria? "I wonder. Whether there aren't hundreds," Delbrück said. "Except we don't know about them. Some few are vaunted. I don't know. I haven't read Kuhn's book." What sort of change was made by the phage group? "Well, the phage group wasn't much of a group," Delbrück said. "I mean, it was a group only in the sense that we all communicated with each other. And that the spirit was—open. This was copied straight from Copenhagen, and the circle around Bohr, so far as I was concerned. In that the first principle had to be openness. That you tell each other what you are doing and thinking. And that you don't care who—who has the priority."

There was a long pause. "So. It's very trivial." But the success of the enterprise has led to a change of temper and a loss of openness, I said, a sharp increase in competitive pressure. "Well, so," Delbrück said. "Different people have different tastes. Different needs. And anxieties. I don't think that's very interesting."

I asked what he had been working on under the microscope. He had been manipulating a small fungus called *Phycomyces*, trying to join two individuals, a red mutant and a white mutant, into a single chimaera, in order to study its genetics. "Better take a look," he said. We got up. Across the harbor, a small sailing skiff, becalmed near some reeds, caught the breeze again. Up on the road a few days before, I had heard a fragment of a passing conversation:

Watson was arranging for the laboratory to buy the marina and the acreage behind it, the whole curve around to the other side of the cove, in order to protect the site and the view from development; he had found a donor for the purchase price and was pushing the measure through his board of directors. At Delbrück's bench, I looked through his microscope and saw four bits of agar, jaggedly broken, and on them two translucent pastel stalks joined at the middle; one pink, the other clear; one slipped into the other at the tip like two soda straws a hundredth of an inch in diameter.

We went to lunch in the cafeteria. Delbrück introduced me to his wife, whose name is Mary Adeline and who is called Manny. Then we walked across the road and up a steep trail through the woods. This led us to a corner of a large flat clearing with, on the left and far sides, a long, low, green wooden building fronted by a veranda and screened doors and windows. Summer camp for biology students, and designated the Motel, though no cars were in the clearing at the moment. The Delbrücks took the same room there each year they came to Cold Spring Harbor. I had been told that even when their children were growing up it was hard to persuade him to use the adjoining room as well. He went inside to take a nap, but first lent me a baggy pair of navy blue swimming trunks. I walked down the drive to the road. After several hundred yards I passed the tennis courts. These had been built recently, paid for by personal donations from Watson and from Manny Delbrück. Farther on, shortly before the path that led to the sandspit and the water, stood the director's house—big, shingled, gabled, deep in its grounds. It was being renovated for Watson. A friend and close collaborator of his at Harvard, Walter Gilbert, told me a while later, "Jim always wanted, or at least partly wanted, just to come back and be the boss of Cold Spring Harbor. It traces back to his first summer here as a student, and seeing that as the ideal scientific environment. Of course he'd always seen the big house on the hill and, like the stories, said 'I'm going to have that house someday.' And by God, he finally has it! I'm not sure it's the original director's house, but it plays the role."

When I met Luria again, at the house of a friend in Lexington, Massachusetts, I asked him about the phage group's reception of Avery's work. "I was the one of us who knew him best, I think," Luria said. "I had a great admiration for him. But he was certainly working in what seemed something very different—but then, I must admit, there was probably a weakness in much of our thinking. It was not because of some hifalutin idea that some scientific ideas or discoveries are ripe and others premature, but because of the fact that we were blocked in biochemical thinking. People like Delbrück and myself, not only were we not thinking biochemically, but we were somehow—and probably partly unconsciously—reacting negatively to biochemistry. And biochemists. As such. As a result, for example, I don't think we attached great importance to whether the gene was protein or nucleic acid. The important thing for us was that the gene had the characteristics that it *had to* have. And that's why Watson and Crick were so tremendously significant to us, as *genetic* thinkers. Because their structure had embedded in it—one saw immediately— the properties of the gene.

"I mean—but let me give my story. I got excited about the original transformation phenomenon, in fact I first discovered its existence, from Dobzhansky's *Genetics and the Origin of Species*, the edition of 1941 or so—and as soon as I met that phenomenon I went to talk to Avery. This was sometime in the spring of 1943, or earlier, before he published. And Avery told me the whole story, how it seemed to be nucleic acids, and so on. Talking with Avery was a marvellous experience. He was a wonderful, short man. Very unpompous, very dignif— He had the dignity of the nondignified people, very simple; and as he was talking he would close his eyes and rub his bald head. And always very precise. And every bit of it a chemist—even though he was an M.D."

I mentioned Delbrück's account of having read Avery's letter to his brother the day it arrived, and the explanation that the real problem had been that nobody knew what to do with the fact that DNA might not be a stupid molecule. "You see that you are saying the same thing as when I said that we were not biochemists," Luria said. "To tell us that the transforming principle was DNA, or was protein, it meant very little, because so what? But I think it is completely wrong to say that we were not aware, that is just typical— I mean, we were *perfectly* aware, every time I was in New York I went to see—" Luria had warmed to the work, and the phrases came tumbling out. "In fact it was in the fall of '45 that I was in the library at the Rockefeller and I found a paper in *Experientia* where André Boivin reported that there was transformation in the Avery manner in *E. coli*, which later nobody was able to confirm until at last recently, but Avery came into the library and I showed the paper to him, and he said, 'Oh, come and have lunch; we will please everybody at the table,' and so we took the paper out, and when Hotchkiss and the others joined us at the table Avery said, 'Well, today we have got Continental company—and Continental support.' I think it is complete nonsense to say that we were not aware."

If the phage group failed to seize on DNA as the genetic substance, Luria said, that was not because Avery was an outsider or his discovery in some abstract way premature, but because the evidence really seemed incongruous. Nothing comparable to bacterial transformation implicated DNA in any of the thousands of experiments that phage workers had been carrying out. "Between 1949 and '51 the results pointed the other way, and I'm the one mainly responsible, having found that when you break open the phage-infected bacteria early, to see what you can see, the earliest signs of any phage that one could discover inside the bacteria were some protein shells. So I put forward the suggestion that maybe this protein was the genetic material of the phage."

Would Luria want to try to explain Delbrück? "Oh, it's very difficult," he said. "Very difficult. I expect the element that lots of people tend to ignore is the enormous influence on his personality of the, what to call it, of the dominant tradition of the German intellectuals. The very strong need—the habit of feeling that the intellectual enterprise is really superior. That's one. There are many features: I always think, of course, of the things in which he and I differ—like for example his enjoyment. He is, I won't say eager for enjoyment in a childish way, but he is outgoing in that, for him, to play tennis or go to the mountains or camp in the desert or to do something that comes to his mind and get pleasure out of it is very important. I was brought up in the puritan tradition of Jewish northern Italians, that if something gives you pleasure there's something wrong with it."

Several people had told me that Delbrück's record for picking right ideas and fruitful lines of research is actually very bad over the past quarter-century. "Yes. That's what I mean," Luria said; and then, in a burst, "Partly *because of that:* because it also had to be fun—and, you know, good research is not fun. I discovered. Good research is not the one which is fun to do, necessarily. You have to be able to drop doing something that you do well, and—it's like nature study. Nature study is the enemy of biology."

Was Delbrück good at picking people? "It's not so much that he's good at picking people, as that he is attracting to people. Because he is terribly intelligent. Because it is so very exciting to work with him. His ideas, the way he thinks, the order— I find Francis Crick, for example, probably the only other person who is equally, who is exciting in that same way. Also because of the way they talk about the things they are working on— Max not to the extent of Francis; nobody talks like Francis; Max is rather silent, but— To spend the days chewing on a problem, and writing and erasing things on the blackboard with him, is terribly exciting. He is unusually cultured by American standards. You know, most American scientists are duds; they never have read a sensible book."

What had Watson been like when he first came to Luria? "In the first term he was at Indiana, he took a course with me," Luria said. "But I didn't know him at all. There were forty students in the class. But at the end of that course he came to talk, and asked if it would be possible to do his thesis with me, even though he was a zoology student. I told him to go check with the dean, but I thought it would be perfectly feasible." It's been asserted, I said, that Luria programmed Watson to become the kind of scientist he was. "God only knows. I haven't programmed anybody," Luria said. "But he was a very remarkable fellow. Even more odd then, than later. But tremendously intelligent, with this mixture of self-assurance and uncertainty of himself that very often bright kids have. It's very difficult to say what we— Mainly what we did was to give him a pleasant environment to work. He did a rather simple problem for his thesis but very beautifully. What struck me from the beginning, and I remember comparing him in that to Muller, he is a person who looks completely disheveled all the time, a mess—except in things that mattered. I have never known anybody whose notebooks, for example, were so perfect, as Jim's notebooks. He had lines and different colors, a system; at any moment he could pick out any experiment— Jim makes the very great distinction. If something is not worth doing, it is not worth doing well."

What gave Watson the ambition to find the secret of life in the physical nature of the gene? "You know, one doesn't talk about these things. It never comes up. I mean, it's clear that we were thinking about nothing but the gene. Self-respecting geneticists who are molecular biologists don't think about anything but the gene. Of course one is interested in experiments, but fundamentally, since 1938, until as far as I'm concerned about 1960, one was interested in nothing but the gene. My interest in other problems only started when I noticed too many people were interested in the gene. It was well taken care of."

❖ ❖ ❖

For fifteen years or so, before it grew too big with success and outran its problems, the group around Delbrück and Luria formed, by all accounts—and there have been quite a number of accounts—one of the rare refuges of the twentieth century, a republic of the mind, a glimpse of Athens, a commonwealth of intellect held together by the subtlest bonds, by the excitement of understanding, the promise of the subject, the authentic freedom of the style.

For Watson, the phage group was more than that: it was home. When Delbrück's friends in science put together, for his sixtieth birthday, in 1966, a volume of reminiscences about working with him, Watson wrote awkwardly, brassily, wistfully about his graduate-student days, and called the piece "Growing Up in the Phage Group." In the fall of 1947, Watson, aged nineteen, in his first term at Indiana, had taken Luria's course in bacteriology and also Muller's in gene mutation; looking back, he wrote:

> It seemed natural that I should work with Muller but I soon saw that Drosophila's better days were over and that many of the best young geneticists, among them [Tracy] Sonneborn and Luria, worked with micro-organisms. The choice among the various research groups was not obvious at first, since the graduate student gossip reflected unqualified praise, if not worship, of Sonneborn. In contrast, many students were afraid of Luria who had the reputation of being arrogant toward people who were wrong. Almost from Luria's first lecture, however, I found myself much more interested in his phages than in the Paramecia of Sonneborn. Also, as the fall term wore on I saw no evidence of the rumored inconsiderateness toward dimwits.

At the beginning of 1948, Watson moved over to work in Luria's laboratory.

Delbrück, that same academic year, had gone from Vanderbilt back to the California Institute of Technology, now as professor of biology. In the spring, he stopped in Bloomington for a day. Watson was introduced to him at Luria's apartment. "His visit excited me, for the prominent role of his ideas in *What Is Life?* made him a legendary figure in my mind. My decision to work under Luria had, in fact, been made so quickly because I knew that he and Delbrück had done phage experiments together and were close friends."

As a dissertation topic, Luria assigned Watson a variation on some research he himself was attempting. Phage exposed to ultraviolet light, and then added sparsely to a culture of host bacteria, produce no progeny and so show no plaques, although the electron microscope reveals that they attach themselves to the bacteria in what appears to be normal fashion. Such inactivation is today completely understood to be a selective structural disruption of the pairing of a few of the phage's DNA nucleotides, hit by the energetic ultraviolet radiation. This physical explanation was inconceivable in 1948. Luria had found, though, that phage inactivated by ultraviolet, when added liberally enough to the culture so that most bacteria were attacked by several phage particles at once, produced a few plaques after all. He called the phenomenon "multiplicity reactivation." He was seeking the reasons why it occurred, and asked Watson to pursue a parallel line using x rays, because they are even more energetic than ultraviolet. It looked as though damaged phages were sometimes able to supply each other's deficiencies within a bacterium, so that normal progeny could be produced. This suggested to Luria that the genetic substance of the phage consisted of different bits, and that inside the bacterium these separated, multiplied independently, and were pooled. Luria was on the trail of a number of discoveries that have since proved highly instructive; but the events by which phages grow within bacteria are more complicated than he then envisaged, so that within a few years his simple version of a gene pool had to be abandoned. Watson's x-ray project turned out to be a cul-de-sac down Luria's detour.

That summer, Watson went to Cold Spring Harbor for the first time. Luria and his wife went, too; the Delbrücks came for the second half. Summers at Cold Spring Harbor then were far simpler and less crowded than they have since become. Courses and meetings were small and leisurely. Experiments were no easier, if the comparison means anything: Columbus had a rougher crossing than Lindbergh, after all, and good scientists find themselves doing things of which the first measure of difficulty is the fact that nobody has done them before. Yet certainly the labor of biological discovery then was less massive, less expensive, far less demanding of fancy equipment. Housing was rudimentary at Cold Spring Harbor, too: patched screens, peeling paint, lumpy mattresses—an American summer in the country in the years before the city, whether Wall Street offices, Broadway theatres, or university laboratories, was air-conditioned. Director of the laboratory since 1941 (and until 1960) was Milislav Demerec. Watson remembered Demerec as the man who went around switching off lights and who refused to make inessential repairs like replacing a broken toilet seat, but he was also a practicing *Drosophila* geneticist who had early seen the point of Delbrück's micro-organisms. Demerec, for example, first collected the seven phages that became the T-set.

In 1945, at Cold Spring Harbor, Max Delbrück had given a summer course in bacteriophage research. Deliberately taught at a level that attracted established scientists from other fields, the phage course became a yearly institution and was an important instrument in the growth of the phage group. Leo Szilard, at the age of forty-nine, took the course in 1947, along with an associate, Aaron Novick, who later called it "a biology that had been made comfortable for people with backgrounds in the physical sciences. In that three-week course we were given a set of clear definitions, a set of experimental techniques and the spirit of trying to clarify and understand. It seemed to us that Delbrück had created, almost single-handedly, an area in which we could work."

Another event every summer was the Cold Spring Harbor Symposium on Quantitative Biology, to which scientists came from all over the world; the series had begun in 1933 and after a three-year wartime lapse had been reestablished in 1946 with a two-week meeting on heredity and variation in micro-organisms. Less formally, the phage group got together there to hear about one another's progress. Geneticists and microbiologists set up laboratories, often bringing their own equipment. Cold Spring Harbor in the summer grew to be the place where one could find people and be in useful prolonged contact with them. When Watson went there it was informal, intimate, exclusive, and even the play—swimming, canoeing, gathering clams or mussels, baseball in the evenings at the foot of the lawn, standing around talking on the road, beer at a place in the village called Neptune's Cave—was saturated and preoccupied with science. Years later, Watson wrote, "As the summer passed on I liked Cold Spring Harbor more and more, both for its intrinsic beauty and for the honest ways in which good and bad science got sorted out." It is striking to compare the reminiscences of members of the phage group with the accounts—also numerous—of scientists who worked with Niels Bohr in Copenhagen in the nineteen-twenties and -thirties.

Delbrück succeeded in creating at Cold Spring Harbor that spirit of ceaseless questioning, dialogue, and open-armed embrace of a life in science which he had learned from Bohr—but with a down-to-earth American character and a good measure of his own high-minded intolerance of shoddy thinking.

In the fall of 1948, on a trip to Chicago with Luria, Watson met Szilard. The encounter was unproductive. "Most conversations with Szilard occurred during meals, which seemingly consumed half of his time awake. . . . Soon I was crushed by his remark that I did not know how to speak clearly." In Bloomington, Watson took a course about proteins and nucleic acids from Felix Haurowitz, who—though Watson had no reason to know it—in Prague, twelve years earlier, had first introduced Max Perutz to the mystery of what happens when the hemoglobin molecule takes up oxygen. In 1949, Haurowitz taught that proteins were so complex that only they themselves could direct their construction; he diagrammed a mechanism whereby a protein gene, stretched out on a rack of the DNA moiety of the chromosome, would collect amino acids, pairing like with like, to build an identical chain. (Delbrück, in a short paper he read at a Cold Spring Harbor symposium, had speculated about the energy requirements of such schemes for proteins' building their own duplicates, but that was in 1941.) "I do not know whether my teaching stimulated Watson to contradict me," Haurowitz said many years later. "When he won the Nobel prize I looked up his mark in my grading book and fortunately found that he had received an A in my course."

Watson's own research reached the point where he knew he had a safe thesis—but a dull one. The summer of 1949, he and others migrated to Pasadena instead of Cold Spring Harbor, because Manny Delbrück was expecting a baby. Watson remembered the seminars several times a week, "dominated by Delbrück's insistence that the results logically fit into some form of pretty hypothesis," camping trips that seemed innumerable, glimpses of Linus Pauling. Delbrück told Watson, dryly enough, that he was lucky his thesis was boring; otherwise he might have to follow it up instead of having time to think and learn. That fall, Watson and his mentors began to wonder where he should go once he got his Ph.D. "Europe seemed the natural place since, in the Luria-Delbrück circle, the constant reference to their early lives left me with the unmistakable feeling that Europe's slower paced traditions were more conducive to the production of first-rate ideas." Herman Kalckar, a Danish biochemist, had taken Delbrück's first phage course at Cold Spring Harbor, in 1945, and was known to be interested in phage reproduction and in nucleotide chemistry; Watson could be placed with Kalckar in Copenhagen, spread the phage gospel, and learn something about nucleoproteins. The matter was arranged for Watson early in November, on a weekend when Szilard had brought the phage people in the Midwest to Chicago and Delbrück and Kalckar himself both happened to be in town. Another young member of the phage group, Gunther Stent, who was a postdoctoral fellow with Delbrück, would also go to Kalckar's lab; a friend of Kalckar's in Copenhagen, Ole Maaløe, had taken the phage course with Delbrück the past summer in Pasadena. Luria helped Watson get a Merck Fellowship from the National Research Council—three thousand dollars for a year and normally renewable. That spring, Watson wrote a draft of his thesis in a month, "but Luria did not like it and took it home for rewriting." He passed his Ph.D. exam—"not surprisingly . . . without fuss"—in May of 1950, at the age of twenty-two. He spent June at Caltech and a final six weeks at Cold Spring Harbor.

The phage group was perceptibly growing: thirty-odd people came to the meeting late in August. Watson's interest was attracted by a report of some experiments with bacteria, infected with phage, which had been fed a diet in which the phosphorus was a radioactive isotope so that, after it was made into phosphate groups of the phage DNA, its fate in the next generation could be traced. The experiments were inconclusive—indeed, somewhat misleading—but they prompted a long succession of attempts to distinguish the functions of proteins and DNA in micro-organisms, and in particular to see what happens to the DNA. Anderson had just published his electron micrographs showing the liberation of DNA from phage inactivated by osmotic shock, which pointed down the same line. Then Watson sailed for Europe.

Watson has always trailed after him a pungent mixture of frankness and reticence. One of the nervous springs of his intelligence is that he cannot help saying out loud the impulsive reactions—to a piece of science, to an instance of bureaucracy, to an intellectual foible of an associate—that other people usually bite back. It makes him uneaseful company and the subject of endless anecdote. Probably as involuntary, less often noticed, is his inability to talk about motive, conviction, and belief except obliquely or in accents of self-derision. Watson has called himself Salvador Luria's "first serious student." When he came to write his first book—*Molecular Biology of the Gene*, an authoritative, extravagantly admired introductory text—he dedicated it to Luria. When he and Crick got the structure of DNA, the first person to whom he wrote about it from Cambridge was Max Delbrück. It's true, as Watson makes sure we realize, that Delbrück was a handy listening, and talking, post for Caltech's most formidable scientist, Linus Pauling, the one who Watson has claimed he believed was in direct competition with him and Crick for the structure. It is also true that through all his accounts of the matter, Watson has given his most spontaneous acts the color of calculation.

2

"DNA, you know, is Midas' gold. Everybody who touches it goes mad."

❖ *a* ❖ The origins of molecular biology, before they fused for the first time, in the early fifties, in the discovery of the structure of DNA, were two distinct, even antithetical approaches to understanding the nature of life. The first approach was through the function of the gene. It was biochemical, in the work of Oswald Avery and Erwin Chargaff, and genetical, in the work of Max Delbrück, Salvador Luria, and the others of the American phage group, who were using the most elementary of living creatures, the viruses called bacteriophage that prey on bacteria, to get at the most elementary facts of biological reproduction. James Watson had grown up in the phage group. In the fall of 1951, at the age of twenty-three, Watson brought that approach to England, to the Cavendish Laboratory, at Cambridge, to Francis Crick. The second approach has been structural, through the physical configurations of the large, long-chain molecules of the cell, to characterize their chemical sequences exactly and to reconstruct their three-dimensional architecture, called their stereochemistry. Watson came to England to get that, found Crick doing that, for it happened that research into structures had been a British invention and a British specialty, though not quite exclusively. The chief American exception in structural research into biological molecules had been, of course, Linus Pauling, who had energy, inventiveness, showmanship, and genius enough for a consortium.

Linus Carl Pauling was born to exemplify the advice of Sir Peter Medawar to his fellow scientists: "Humility is not a state of mind conducive to the advancement of learning." No other American scientist now alive compares with him. He calls himself a physical chemist. In over half a century in science, in over five hundred scientific publications, Pauling stretched that designation to cover everything from crystal structure and quantum mechanics, where he began, to molecular biology, molecular medicine, molecular psychiatry, and the structure of atomic nuclei, the areas where many of his last papers lie. Pauling's original and abiding concern was the structure of molecules and the chemical bond. His definitive book was *The Nature of the Chemical Bond*, first published in 1939. His other classic, *General Chemistry*, which appeared soon after the war, was a textbook that grew out of the years when, even as head of the department of chemistry and chemical engineering at the California Institute of Technology, he taught the introductory undergraduate course. More than anyone else, Pauling has made the structure of molecules the central and most productive question of modern chemistry. Watson, in *The Double Helix*, told the story of the discovery of the structure of DNA as an epic race on converging tracks with Pauling. It is doubtful that Pauling was ever competing for the prize as hotly as Watson imagined (or said that he imagined). Yet the competition was both spur and rationalization for both Watson and Crick. There is no doubt whatever that Pauling's lifetime of work on the structures of molecules provided information, insights, rules, techniques, and intellectual approaches that Watson and especially Crick required, more directly than they built on anything learned from phage or bacterial genetics. The discovery of the structure of DNA by James Watson and Francis Crick was itself a tribute—Crick's tribute—to Linus Pauling.

Pauling retired from Caltech in 1963. He later moved up the coast to Palo Alto, where Stanford University gave him a laboratory, graduate students, an occasional course to teach. The campus there, on the spring noontime when I went to see him, was a vast, green, icy calm of long lawns, eucalyptus trees and palms, wide driveways, and not many people. Pauling came into his office with a silk sweat-scarf at his neck, a white stubble over pink cheeks, and a flat tweed cap. I brought him warm-up greetings from his son Peter, whom I had seen not long before in London, and also from a mutual friend, Matthew Meselson, a former student of his now at Harvard, who had said, "Be sure to tell Linus I'm taking a gram of vitamin C every morning!" Pauling pulled off the cap, liberating a corona of gray hair. Watson, when he first met Crick, wrote to Delbrück about him that "he is no doubt the brightest person I have ever worked with and the nearest approach to Pauling I've ever seen—in fact he looks considerably like Pauling." There is a resemblance, particularly in that Crick and Pauling both have something of the convergence of nose and chin the English call a nutcracker face; but Pauling has a generous mouth, heavy eyebrows and large eyes of startling blue, and a rubbery expressiveness of feature. Crick, in looks and speech as well, could only be English. Pauling when he talks could only be an American of the Far West, of the marvellous generation now passing, classless, idiosyncratic, radical, with a zestful rolling gait to his speech.

Pauling sat down at a desk stacked with books and neat piles of papers. The room was crowded with filing cabinets full, their labels said, of reprints of papers and articles. Shelves around the walls held more papers, and knobs and slices of brilliantly colored plastic that represented atoms—shiny white for hydrogen, scarlet for oxygen, black for carbon, and so on—to be fitted together into models of molecules. These knobs have built into them certain of the physical constraints on the chemical bonds between atoms; as ubiquitous as pocket calculators in chemical laboratories, they were invented by Pauling and his colleague of many years, Robert Corey. Pauling ruminated for a moment about the origin of the term "molecular biology." It has often been credited to the English crystallographer William Thomas Astbury, who had begun to make structural investigations of large biological molecules in the nineteen-thirties, slightly earlier than Pauling had; but Pauling wasn't so sure

Astbury deserved credit. He had read a letter in *Science* in which Warren Weaver, who had been director of the program of grants for research in the natural sciences at the Rockefeller Foundation, had pointed out that he had used the term in the foundation's report for 1938. In the report, Weaver had written of aiding explorations in "those borderline areas in which physics and chemistry merge with biology," and had said that "gradually there is coming into being a new branch of science—molecular biology—which is beginning to uncover many secrets concerning the ultimate units of the living cell." Weaver's letter claiming the term had set Pauling thinking.

"Warren Weaver's usage was quite significant," Pauling said. "The Rockefeller Foundation had started supporting my work about 1932, I believe it was, or 1933. And they made it rather clear that if we were working on biological substances they'd be more interested. This was largely Warren Weaver's idea, that the time had come when a more basic attack ought to be made on the problem of *life*, in the field of biology and medicine. They put up a large amount of money for our work in Pasadena, several million dollars over a period of years, and I think at that time I may have used the term 'molecular biology,' too, or several of us together may have used it. My own first work in this field was on hemoglobin, in 1935—a theoretical job. I asked what the structure of the hemoglobin molecule should be in order to account for the way it takes up oxygen. This is just straight physical chemistry. And the conclusions that I reached were mildly interesting. Oversimplified, of course.

"I became interested in chemistry in 1914, at the age of thirteen, because another student showed me the laboratory he had set up in his bedroom," Pauling said. "I decided then to be a chemist, and to study chemical engineering, which was, I thought, the profession that chemists followed. I studied at Oregon State Agricultural College, and had my first full-time job as a teacher there when I was eighteen to nineteen years old, teaching quantitative analysis. The idea of the chemical bonds holding a molecule together was still rather vague. I taught the hook-and-eye idea—that we had to imagine the sodium atom, say, as having a hook attached to it and the chlorine atom as having an eye, into which the hook could fit to make salt. Then, in 1916, Professor Gilbert Newton Lewis, at Berkeley, said that this hook-eye combination is not a very good picture, and that the chemical bond is really a pair of electrons—two electrons that are held jointly by two atoms. I had my desk in the chemistry library that year when I was teaching full-time, and I read the chemical journals as they came in. And I read about the electron-pair bond, the covalent bond. I knew a great amount of descriptive chemistry, and I could see how the shared pair of electrons could explain what the forces are that hold the atoms together. I could see that the first steps were being taken towards a real, systematic science of structural chemistry. It was then that I developed a strong desire to understand the physical and chemical properties of substances in relation to the atoms and molecules of which they are made up. This interest has largely determined the course of my research for fifty years." Pauling had recently written more about his enduring fascination with the physical basis of chemistry, in an autobiographical note I had read before seeing him, and whose style and breadth of claim I now could hear in his voice:

As I try to remember the state of my development at that time, I am led to believe that this desire was the result of pure intellectual curiosity, and did not have any theological or philosophical basis. I was skeptical of dogmatic religion, and had passed the period when it was a cause of worry; and my understanding of the experiential world was so fragmentary as to be unsatisfactory as the basis for the development of a philosophical system. I was simply entranced by chemical phenomena, by the reactions in which substances disappear and other substances, often with strikingly different properties, appear; and I hoped to learn more and more about this aspect of the world. It has turned out, in fact, that I have worked on this problem year after year, throughout my life; but I have worked also on other problems, some closely related, such as the structure of atomic nuclei and the molecular basis of disease, and others less closely, such as the pollution of the earth with radioactive fallout and carbon 14 from the testing of nuclear weapons, the waste of life and the earth's resources through war and militarism, and the maldistribution of wealth.

Pauling is one of the three people who have received two Nobel prizes. Marie Curie shared the Nobel prize in physics with her husband and Antoine-Henri Becquerel in 1903 and won the prize in chemistry, alone, eight years later. John Bardeen, an American, got a piece of the prize in physics in 1956 and again in 1972. Pauling was awarded the Nobel prize in chemistry in 1954, "for his work on the nature of the chemical bond and its application to the elucidation of the structure of complex substances." That was the year the Nobel prize in literature went to Ernest Hemingway, and *his* citation could with justice have been added to Pauling's: "for the vigorous perfection of a style that has put his contemporaries to school." Pauling's second Nobel was the Peace Prize in 1962, which he got for his work to bring about the treaty banning tests of atomic explosives in the atmosphere.*

Pauling's political stand in the last years of the Truman presidency seems mild enough now, a rather flamboyantly idealistic campaign against the cold war, against atomic weapons and the development of the hydrogen bomb. In those days of the rise of Joseph McCarthy and Richard Nixon, when the national anxieties reverberated into hysteria in Southern California, Pauling's course took courage and principle. His politics had unpleasant consequences at the time. At the end of April 1952, Pauling was supposed to go to London to attend a meeting of the Royal Society on the structure of proteins, but at the last minute was refused a passport. Some claim that had he got to London, he would have seen the newest x-ray-diffraction pictures of DNA from Maurice Wilkins's and Rosalind Franklin's laboratories, and from them learned what he needed to solve the structure. When I asked Pauling about the political responsibilities of scientists, he laughed as if at the foolishness of his enemies. "I have contended that scientists—first that they have the responsibilities of ordinary citizens, but then that they also have a responsibility because of their understanding of science, and of those problems of society in which science is involved closely, to help their fellow citizens to understand, by explaining to them what their own understanding of these problems is. And I have contended that they have the duty also to express their own opinions—if they have opinions. Of course, just after 1945, there was a good response from

---

*In 1980, Frederick Sanger became the fourth scientist to be awarded a second Nobel prize.

scientists. But then in the United States, McCarthy came along and frightened the majority of scientists out of taking any action. And, of course, there's been the really great effort made by the administration, by the powers in control, to convince scientists that they should stay out of things, you know." The brows contracted furiously and the swell of language deepened. "Year after year, I've been told I might know a lot about chemistry but that doesn't mean that I know anything about world affairs or economics or social problems, so I should just keep my mouth shut. I have replied—well, first by saying"—he laughed again, enjoying himself—"that I *refuse to take* your advice, but second by saying that a lot of other people, the lawyers and politicians, are just as ignorant as I am about these matters, and yet no one tells them to keep their mouths shut. After Dean Acheson had been quoted in *Harper's* magazine as saying that I might know a lot about biochemistry but I didn't know anything about world affairs, and should keep my mouth shut—that there was no justification for me to make pronouncements about world affairs—I didn't bother to say that I wasn't a biochemist, but I wrote back saying that if Dean Acheson had studied biochemistry as much as I had studied world affairs, then I would say we *ought to listen* to what he had to say about biochemistry."

Pauling had become entranced with chemical phenomena at the time when quantum mechanics was beginning to be exploited to bring a new, structural order into the immense mass of information that classical chemists had accumulated about the ways atoms combine. Quantum theory dates from a paper by Max Planck in 1900; but in January of 1926, Erwin Schrödinger, then at the University of Zurich, set off the development of the theory in its modern form, distinguished by the term "wave mechanics," and immediately he and many others—among them Werner Heisenberg and Max Born, in Göttingen, Paul A. M. Dirac, in Cambridge, and Niels Bohr and his group, at the Institute for Theoretical Physics, in Copenhagen—were drawn into the most intense, competitive and fruitful multiple collaboration science has ever known. Pauling arrived in Europe on a Guggenheim fellowship just as all this was starting, in March of 1926; he studied for a year in Munich, stayed a few weeks in Copenhagen at Bohr's institute, and then went for five months to Zurich, where Schrödinger was. In the next ten years, Pauling and others working in the domain where physics shades into chemistry—that is, working at a scale of around a billionth of an inch, and concerned with the energy levels, numbers and positions, and spins of the outermost electrons in different kinds of atoms—were learning just how atoms are allowed in nature to behave in each other's intimate company. Pauling was the anthropologist confronted with an imperfectly known language, which turns out to belong to a rich, strange culture whose every interchange is governed by precise rules of hierarchy, status, and payment. To learn the language was made more difficult, of course, by the need to invent new instruments to ask each new kind of question. But the greatest problem was that in the unfamiliar language the very categories of thought turned out to defy preconception. That was where quantum mechanics offered the promise of order.

By far the most powerful consideration Pauling brought into structural chemistry from the new physics was the idea of resonance. It makes intuitive good sense to say that the atoms composing the molecules of a substance tend to arrange themselves in the way that is most stable, and that this conformation will be the one that requires the least energy. After all, once a molecule falls into its least energetic conformation, work is required to push it into any other. Quantum mechanics can demonstrate this sensible notion mathematically for simple molecules; the balance of energy in large molecules of biological interest is as yet beyond exact computation. Resonance takes the idea of a molecule's stability one step further—a step outside the commonsensical, perhaps. The principle of resonance states, for example, that if a molecule can be described as having the bonds among its atoms arranged in either of two ways, then the molecule is to be considered as existing in both arrangements simultaneously.

"I was thinking about this problem in 1928, in relation to the carbon atom," Pauling said. Carbon is a gregarious stuff; the carbon atom has four electrons available for making shared-electron-pair, or covalent, bonds—four hands, so to speak, to clasp its neighbors'—where oxygen, say, has but two and hydrogen only one. Since 1874, chemists had understood that carbon's four bonds tend to take positions directed out from the atom towards the corners of a regular tetrahedron—that is, splayed tripod-wise as far from each other as possible in three-dimensional space, pointing outwards at the four corners of a pyramid that has triangular sides and a triangular base. But physicists knew from atomic theory, confirmed by unmistakable evidence in the lines produced in the spectroscope, that one of the four valence electrons of the carbon atom is in a different orbital state from the three others—so that while three electrons, in the orbital state termed p, do indeed have their charges splayed apart from each other, the fourth, termed s, has its charge distributed spherically around the atom. The contradiction had blocked the formulation of a coherent theory of bonds. Pauling invoked resonance to unite what the physicist knew about electrons with what the chemist knew about bonds. "I had the idea that these electrons, the s electron and the three p electrons, may be all right for the spectroscopist, but for chemical calculations they can be combined together, hybridized together, into four equivalent electrons occupying four tetrahedral orbits." He published a paper that year about the equivalence of the four bonds of carbon. "But it was a short note, because I had not been able to carry out the mathematical calculations," he said in a letter half a century later.

Pauling wrote another paper that year that brought him into rivalry with Lawrence Bragg for the first time. Bragg, with some help from his father, Sir William Bragg, had invented x-ray crystallography, or the method of determining the repeating structures of atoms arrayed together in crystals by analyzing the way they diffract x rays. Pauling had started teaching himself the subject by reading the Braggs' book on it in the summer of 1922, before he went to Caltech as a graduate student; in December 1922 he made his own first x-ray determination of a crystal structure, the simple mineral molybdenite. While Pauling had been in Europe in 1926–'27, Bragg, then at the University of Manchester, had launched a series of papers giving the crystal structures, beginning with the gemstone beryl, of a class of minerals called silicates. These presented the most complicated molecular structures that had yet been attempted. Bragg devised a theory—hardly a theory, a practical rule—for thinking up structures compatible with a substance's chemical composition, to try out against the x-ray data. Pauling, on his way home from Europe in the fall of 1927, stopped for a week or so in England. There he met the elder Bragg, but not the younger. When Pauling got back to Pasadena, he began investigating silicates and similarly difficult minerals himself. He published several structures, including the gemstone topaz. By the summer of 1928, he understood that the various relations among atoms that dictate the stable forms of a crystalline substance could be stated as a set of six principles. He wrote these down. They were published the next spring. To this day, they are called the Pauling rules. The principles were not entirely original with Pauling. The first three, anyway, had been in the air among crystallographers—had, in effect, been used by Bragg. Pauling clarified them, codified them, demonstrated their generality and power.

To put the consequences in a single long sentence, Pauling taught a couple of generations of chemists that the sizes and electrical charges of atoms determine exactly their arrangements in molecules: there must be neutralization of electrical charges, positive or negative, not just on average over the whole molecule but locally, at the spot, while at the same time atoms have to be packed—they fit snugly together. Structures of the molecules of 225 substances were determined in Pauling's laboratory in those years and a great many others in other laboratories. The suggestions from physical theory were pursued by many physical techniques, but most importantly by x-ray diffraction. Pauling's chief rival in the work was Bragg. According to Max Perutz (and others, too), Bragg was always chagrined that Pauling had first published the rules by which physics applies to the structure of minerals.

In the spring and summer of 1930, Pauling spent another six months in Europe, where his first stay was several weeks at Lawrence Bragg's laboratory. "My visit to Manchester was a disappointment to me," Pauling said in our conversation. "I had essentially no contact with Bragg. And they failed to ask me to present a seminar talk, say, on my work, because I had done a great deal of work that bore on what Bragg's lab was doing." He went on to Germany. Once again, ideas of quantum physics, resonance, valency, and hybrid bonds were in the air; they made rules that created coherence out of the experimental data in just the way that the anthropologist, learning the strange language, will suddenly see that by combining categories he had thought were exclusive or even contradictory he can make sentences where nonsense reigned before. When Pauling got back to Pasadena this time, he proceeded to write the grammar book. One evening in December of 1930, he perceived how to calculate the hybrid orbitals that had eluded him two years before. "This idea gives an almost complete theory of the nature of the chemical bond. I worked at my desk nearly all that night, so full of excitement I could hardly write." His great essay "The Nature of the Chemical Bond" was published in April 1931, thirty-four pages in *The Journal of the American Chemical Society*, followed by six more installments in the next two years, and then by the first edition of his great monograph.

"By 1935 I felt that I had an essentially complete understanding of the nature of the chemical bond," Pauling said. Desmond Bernal, who was a student of the elder Bragg—and was the crystallographer who first showed, in 1934, that such giant molecules as proteins can be studied by x-ray-diffraction methods—wrote, many years later, "Pauling was the man who more than anyone else spread the knowledge of quantum theory in the fields of classical chemistry. . . . He had already proved himself in the early years to have such an ingrown sense of the realities of the quantum as applied to chemistry that he did not need to think about detailed derivations but thought automatically in quantum terms."

By the mid-thirties, the chemical bonds that occur in biological substances could be divided into strong ones and weak. The strong ones are the shared-electron-pair bonds, or covalent bonds, whose first approximate explanation, by Gilbert Newton Lewis, Pauling had found so exciting. These can be single or double (or, rarely, triple). If the bond is single, rotation of the atoms (or groups of atoms) at each end is possible around its axis, and the molecular structure that results is to that extent flexible and less predictable. If the bond is double, the bond axis is rigid and greatly restricts the resulting molecular structure. It's like two people holding hands in a country dance, who are less free to swing to different positions if they hold by both hands than if by just one each. Single and double bonds differ also in length, the double bond being shorter, and stronger too, again like two people joining both hands in the dance. Whether two atoms—of the same element or different ones—can form a valence bond at all, and whether the bond can be a double one, depends on whether those sorts of atoms have the electrons available to share. As more atoms get linked to the first pair, changing the molecule to a different chemical substance, they influence the character of each other's bonds. The concept of resonance indeed predicts all this. The ring that the elementary textbook shows thus—

—with C for carbon, N for nitrogen, H for hydrogen, and double lines for double bonds, is also this—

—and the paradoxical insight by which the idea of resonance departs from common sense is that these bonds lie both ways at the same time. Resonance amounts to the whole square dance swinging in motion.

The physicist might state this fact in terms of the probability of finding a given molecule arranged one way or another at a given split instant. The physical chemist, concerned with the stability of the molecule, addresses another facet of the same fact when he speaks of such a bond's having a certain percentage of double-bond character, which will determine its exact length, its rigidity, the angles it can take in relation to other bonds. What Pauling's laboratory did, more than any other, was to get the measure of these lengths and angles in molecules of hundreds of different substances. By the late thirties he knew he had refined them to a couple of hundredths of an angstrom—eighty trillionths of an inch, more or less—and to a couple of degrees of arc. The detail of bonding was essential to understanding protein structures and, in a most surprising way, proved central to DNA. The particular dance of carbon and nitrogen in the diagrams above is the ring on which two of the bases in DNA, the pyrimidines thymine and cytosine, are built. Written out fully, thymine looks like this:

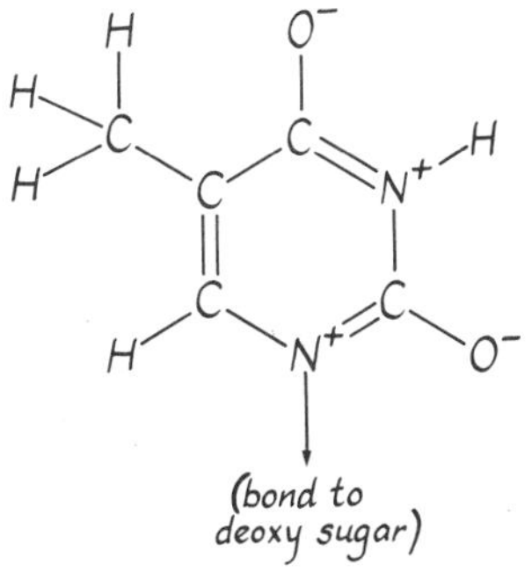

(bond to deoxy sugar)

The purines are double-ringed; adenine looks like this:

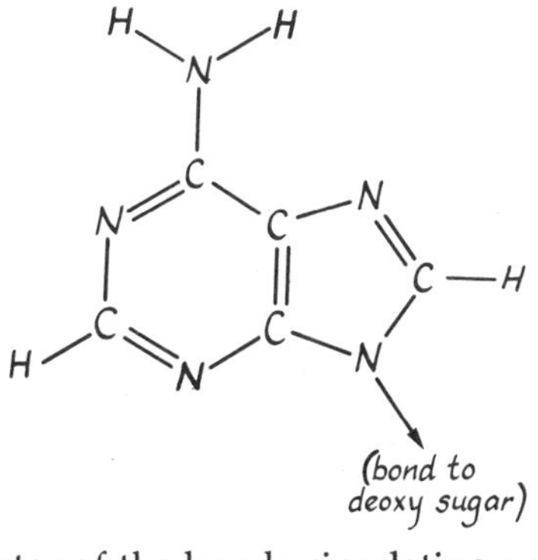

(bond to deoxy sugar)

The partly double character of the bonds circulating around the rings has one general consequence that again makes intuitive sense: the pyrimidines and purines are flat. This fact is of great importance to the structure of DNA—not least because that made it practical for Watson, trying to construct a model for DNA, to cut out the shapes of the four bases with scissors from sheets of cardboard.

Bonds of another sort, called electrostatic, function between atoms of opposite charges but without the sharing of an electron pair; they range from weak to very strong. The weakest of all, not so much a bond as a momentary flirtation, results from the van der Waals forces, named after Johannes Diderik van der Waals, a Dutch physicist, who first described them in 1873. Van der Waals bonds arise because any two groups of atoms as they approach each other closely will induce slight fluctuations in each other's electrical charges. Van der Waals's name is also attached to another aspect of the same phenomenon, the distance at which two atoms begin to get in each other's way. Van der Waals distances, or contacts, are an ever-nagging limitation in any attempt to model the structure of a large biological molecule. In life, the most significant weak bond is the hydrogen bond. This one arises when a hydrogen atom is covalently bound into a molecule and yet, though it has but one electron, has some surplus electrical charge in the outward direction, and so can link lightly with another atom of opposite charge, which in turn is covalently bound elsewhere. The older, more descriptive term for the hydrogen bond was "hydrogen bridge." In forms such as —N—H····O— and —O····H—O— , hydrogen bonds lace together and stabilize the configurations of large molecules. Hydrogen bonds can be formed or broken at living temperatures at a relatively low cost in energy, ranging up to ten per cent of what it takes to make or break a covalent bond. All these weak binding forces work most effectively when molecular surfaces have complementary structures that bring several interacting groups close enough together to form a die-and-coin relation.

In the mid-thirties, Pauling turned from the elementary grammar of bonding to sentences and paragraphs. "While I was doing the hemoglobin studies in 1934 and 1935, I began to speculate more generally about the properties of the large molecules found in living organisms and about the problem of the structure of proteins," he said. Others were beginning to speculate too, notably Astbury and Bernal and their students and, slightly later, Perutz and Lawrence Bragg. The simple first question they all faced was whether protein molecules, known to be made up of very large assortments of a small number of components—the twenty-odd amino acids—were jumbled like paper bags of mixed nuts or were orderly. Bernal's discovery, in 1934, that crystals of the enzyme pepsin, if kept soaking in the liquid out of which they had been precipitated, would yield clear, characteristic x-ray-diffraction patterns, proved that all pepsin molecules possess exactly the same structure—which was implicit, indeed, in the fact that they formed crystals at all. Protein crystals are still mounted for x-ray diffraction as Bernal mounted them, bathed in their mother liquor. Bernal himself understood, from the hour he made his discovery, that it opened the line to the direct and complete determination of the structures of proteins. Yet the molecules were far too complicated to be worked out from those clear diffraction patterns until the decisive next discovery—the way to get controlled variant x-ray patterns for comparative analysis—was made by Perutz nineteen years later.

Bernal had demonstrated that proteins were ordered; not even the most general rules of their order were then certain. One theory said that proteins were long chains of amino acids. This idea had been put forward in the nineteenth century, almost from the moment it began to be understood how large proteins are, but had not been proved to everybody's satisfaction. At the end of the nineteen-thirties, the latest of many alternatives to chains was the cyclol theory, which asserted that a protein molecule was a cage of rings made up of amino acids. In 1939, Pauling and an associate, Carl Niemann—and several other chemists at about the same time—demolished the cages and showed that all the evidence favored the chains. The repeated place in the protein chain where two amino acids link up is called a peptide bond: this is a strong covalent bond, with a lot of double-bond character, between the carbon at the end of one amino acid and the nitrogen that begins the next. The peptide bond is named after the powerful digestive enzyme pepsin, one of the family of enzymes that cleave these bonds and so break up protein chains; "pepsin" is in turn derived from the Greek root *peptos,* meaning cooked. Pepsin was one of the first enzymes to be prepared in pure form, which is why Bernal worked with it. Protein chains are called "polypeptides."

Pauling, once hemoglobin had set him thinking, went after proteins with characteristic energy. "In 1937 I realized that there was very little detailed structural knowledge about amino acids themselves," he said. "Nobody was attacking the problem vigorously, systematically. Bernal had made a preliminary study of some amino acids, without determining any structures. Although there had been several x-ray studies of amino acids reported in the literature, they were all wrong. I knew, of course, what Bill Astbury, at Leeds, had written about the structure of keratin—hair, horn, fingernail, and so on. But I knew that what Astbury said wasn't right."

Astbury, like Bernal, had been trained in crystallography by the elder Bragg. The University of Leeds was set financially as well as geographically in the midst of the great northern English textile industry; Astbury had been hired by the university in 1928 to do research in the physics of textiles. He collected x-ray-diffraction data about organic molecules by working not with crystals but with fibres, which limited him to the scant list of natural materials that grow well-ordered fibres. Even these produced diffraction patterns that were streaks and partial rings rather than the sharp spots produced by crystals like Bernal's. Astbury's abiding interest was wool and the protein keratin, which makes up wool and many related animal substances—hair, horn, fingernail indeed. By the early thirties, Astbury was publishing diffraction patterns that gave tantalizing glimpses of the physical limits any structure proposed for protein would have to meet. His photographs and his vividly plausible interpretations influenced everybody's thinking about large biological molecules. Three indications were most important. Astbury showed a pronounced change in diffraction pattern and thus in structure between unstretched and stretched wool. He called the structures alpha and beta keratin; the style of address persisted, together with the conviction that stretching can produce great and informative changes in fibre structures. The data also suggested, fuzzily, that the polypeptide chains in unstretched keratin—alpha keratin—were coiled helically. Further, a strong, smeared spot in the diffraction patterns told crystallographers that the molecular structure repeated itself every 5.1 angstrom units—every twenty billionths of an inch or so. If the polypeptide chain of unstretched keratin was a helix, 5.1 angstroms should be the height of each turn.

DNA is also a fibrous substance. When Torbjörn Caspersson, a Swedish biochemist working with nucleic acids, sent Astbury a particularly well-prepared sample of DNA from calf thymus, he began a series of diffraction studies. The DNA fibre diagrams, too, offered fascinating suggestions if few certainties. A heavy streak seemed to show a repeating structure every 27 angstroms, and there was evidence from which he concluded that the bases lay flat, stacked, 3.4 angstroms apart. Astbury's diffraction patterns for DNA fibres—like his keratin patterns—were to remain for many years the ones people referred to. But they were not good enough.

Astbury suggested all sorts of structures for keratin and for DNA which he thought were compatible with his data. In 1938, the Symposium at Cold

Spring Harbor was about protein structures. Astbury came and presented with high enthusiasm a matched team of structures that showed, he said, how the spacing of the nucleotides 3.4 angstroms apart along a strand of DNA corresponded almost exactly with the spacing of the amino acids in a polypeptide chain. He believed he had thus explained how DNA and protein are united in the chromosome. But his imagination was sometimes clearer than his data. The information in his fibre diagrams was to prove invaluable and confusing: it was never clinching.

In particular, that 5.1-angstrom repeat in keratin could not be reconciled with Pauling's independent data. "I knew that what Astbury said wasn't right, because our studies of simple molecules had given us enough knowledge about bond lengths and bond angles and hydrogen-bond formation to show that what he said wasn't right," Pauling said. "But I didn't know what *was* right. Although no direct experimental evidence was available about molecular dimensions of simple peptides or closely related substances, I thought the general structure theory should permit predictions to be made. I spent the summer of 1937 in an effort to find a way of coiling a polypeptide chain in three dimensions compatible with the x-ray data reported by Astbury. This effort was unsuccessful, which led me to conclude that I was making some unjustified assumption about the structural properties of the molecules." Twelve years later, Pauling learned that not his structural theory but the interpretation of the x-ray patterns was at fault. The difficulty was a foreshadowing of what was to happen with the structure of DNA.

"Therefore, 1937 was the year when I decided we should determine the structures of amino acids and simple peptides, to see if there wasn't something important that we were overlooking—some structural feature that didn't show up in the more distantly related simple molecules we had been studying." In that year, Pauling was joined at the California Institute of Technology by Robert Corey, who was also interested in the structures of protein. "Well, we carried out that investigation over a period of years," Pauling said. "And by 1948 it was evident that there were no surprises about these molecules, really. We had made our information more precise but hadn't changed our understanding in any qualitative sense. The dimensions I had taken eleven years before for the polypeptide chain were shown to be correct. So in the spring of 1948 I again attacked the problem of coiling the polypeptide chain."

Pauling's method was to build models, sometimes with pencil and paper at first, eventually with precisely scaled physical representations of the atoms—open three-dimensional puzzles in which the individual pieces to be fitted together already carried many of the limitations of angles, lengths, and sizes. These simple toys were one of Pauling's most remarkable contributions to molecular biology: they amounted to a kind of analogue computer that embodied many of the physical rules and restrictions, in order to cut out the endless refiguring of interlocking readjustments. Pauling explained the problem in his lecture to the Swedish Academy on receiving the Nobel prize:

> The requirements are stringent ones. Their application to a proposed hydrogen-
> bonded structure of a polypeptide chain cannot in general be made by the simple

method of drawing a structural formula; instead, extensive numerical calculations must be carried out, or a model must be constructed. For the more complex structures, such as those that are now under consideration for the polypeptide chains of collagen and gelatin, the analytical treatment is so complex as to resist successful execution, and only the model method can be used. In order that the principles of modern structural chemistry may be applied with the power that their reliability justifies, molecular models must be constructed with great accuracy. For example, molecular models on the scale of 2.5 cm = 1 angstrom unit, have to be made with a precision better than 0.01 cm.

Pauling and Corey built with strictest observance of the principles they had learned. One rule that turned out to be crucial was that no rotation is permitted around the peptide bond, because resonance gives it partial double-bond rigidity; it follows that the atoms immediately to each side of the bond should all lie in a flat plane—as indeed they do, to within a tenth of a billionth of an inch. Another rule was that the three atoms of a hydrogen bridge, —O····H—N— , make a straight line. Although it is understood today that neither of these rules is obeyed altogether inflexibly in proteins, as Pauling then supposed, his insistence on reasoning rigorously from principles proved decisive.

From the spring of 1948 through the spring of 1951, well before Watson arrived in England, rivalry sputtered and blazed between Pauling's lab and Bragg's—rivalry over protein. (Bragg had been translated from Manchester to Cambridge in 1938, to be Cavendish Professor of Experimental Physics.) The prize was to propose and verify in nature a general three-dimensional structure for the polypeptide chain. Pauling was working up from the simpler structures of components. In January 1948, he went to Oxford as a visiting professor for two terms, to lecture on the chemical bond and on molecular structure and biological specificity. "In Oxford, it was April, I believe, I caught cold. I went to bed, and read detective stories for a day, and got bored, and thought why don't I have a crack at that problem of alpha keratin." Confined, and still fingering the polypeptide chain in his mind, Pauling called for paper, pencil, and straightedge and attempted to reduce the problem to an almost Euclidean purity. "I took a sheet of paper—I still have this sheet of paper—and drew, rather roughly, the way that I thought a polypeptide chain would look if it were spread out into a plane." The repetitious herringbone of the chain he could stretch across the paper as simply as this—

—putting in lengths and bond angles from memory. He ignored the complex fringe of side chains, one at each point (the alpha carbon) where the remainder (R, for "radical") of that amino-acid residue sticks out from the backbone: though the different side chains give the protein its specific functional characteristics as this or that enzyme, hormone, or whatever, they were irrelevant to Pauling's immediate problem. He knew that the peptide bond, at the carbon-to-nitrogen link, was always rigid:

And this meant that the chain could turn corners only at the alpha carbons. Pauling had come to realize that there was a general argument, much stronger than anything suggested by Astbury's prewar data, for supposing that the structure he wanted would be a helix. "The idea that I had in 1948, and had not had in 1937, is this," he said. "You can show, rigorously, mathematically, that if you have a structure, such as a right hand"—he held his right hand in the air—"and another structure identical with it, but somewhere else in space"—he raised his hand higher, turning it—"the relationship between this structure and *this* one"—he rotated his hand up and down three or four times—"the relation is a rotation around an axis with a translation along the axis. That's the *general* relation between two asymmetric but equivalent objects in space. And if you repeat that, you automatically get a helix. So then I creased the paper in parallel creases through the alpha carbon atoms, so that I could bend it and make the bonds to the alpha carbons, along the chain, have tetrahedral value. And then I looked to see if I could form hydrogen bonds from one part of the chain to the next." He saw that if he folded the strip like a chain of paper dolls into a helix, and if he got the pitch of the screw right, hydrogen bonds could be shown to form, N—H····O—C , three or four knuckles apart along the backbone, holding the helix in shape. After several tries, changing the angle of the parallel creases in order to adjust the pitch of the helix, he found one where the hydrogen bonds would drop into place, connecting the turns, as straight lines of the right length. He had a model.

Worked out, Pauling's paper model met all the criteria. It was similar to the structure he had abandoned eleven years before—and it still failed to account for the strong spot on the x-ray-diffraction pattern that Astbury had obtained with keratin: according to the model, the height of a full turn of the helix should not be 5.1 angstroms, as Astbury's data showed, but 5.4. In May of 1948, Pauling went across from Oxford to Cambridge to give three lectures there. He visited Bragg. Perutz showed him his latest attempts to disentangle the structure of hemoglobin.

"Pauling was very interested," Perutz said later when I asked about the encounter. "I showed him the evidence that the chain in hemoglobin was similar to that in Astbury's alpha keratin. That is to say, there appeared to be a structure which had the same kind of 5.1-angstrom peak along the fibre axis. The neighboring chains seemed to be spaced about ten and a half angstroms apart. But Pauling made essentially no comment."

"You know, I knew about how big the helix should be when I went over to Cambridge," Pauling said. "Perutz showed me the model that he had of hemoglobin that had these ten-angstrom-diameter rods running through, parallel, and I thought, 'Well, those are probably this alpha helix'—which I hadn't named alpha helix yet—'but, you know, I don't know whether this idea is any good or not and I won't say anything about it to Max. When I get back home I'll work through it carefully—more precise calculations.' Of course, that 5.1 against 5.4 stymied me. What I'm telling you now is that I was thinking of the alpha helix in hemoglobin, and I refrained from saying anything to Max, not because I wanted to keep him from having significant information, but because there's no use disturbing people about something unless you feel happy with it yourself." Pauling withheld publication.

A few months later, Bragg got deeply interested in the general structure of the polypeptide chain when he saw a different and what looked like a quick route to the solution, starting with Astbury's old data and adding the latest clues Perutz and John Kendrew seemed to have found in their diffraction studies of hemoglobin and the related protein myoglobin. At Bragg's urging, they tried to exhaust the permutations of structure permitted by the data they had gleaned—they found nearly twenty—and then to light on the likeliest one. Bragg, Kendrew, and Perutz published first—in the spring of 1950, in *Proceedings of the Royal Society.* But Perutz said a while ago that they were always uneasy about the structures they chose: "It was one of those papers you publish mainly because you've done all that work." The paper is certainly odd: long, diffuse, uncharacteristically uncertain, an illustrated mail-order catalogue of the latest polypeptide models. Finally, grudgingly, the authors settled for a flip-flop ribbon structure first proposed by Astbury years before. In the event, they got it wrong, for they fell victim to the 5.1 spot—and they allowed free rotation around the peptide bonds. The error was an intense embarrassment for Bragg, a scientific humiliation. Thirteen years afterwards, he said, "I have always regarded this paper as the most ill-planned and abortive in which I have ever been involved."

Pauling chuckled when he spoke of that in our conversation. "We got the alpha helix, you know, and Bragg and Perutz and Kendrew were trying to do the same thing, and they—ah, failed—because of, really a lack of knowledge of the principles of chemistry, of structural chemistry." He picked up several bright oxygen and nitrogen atoms and rapped them together. "I had even discussed the structures of the amino acids and simple peptides in my book *The Nature of the Chemical Bond,* and I judged that none of them had read this book carefully."

Seymour Jonathan Singer, whom I had met a few days earlier at the La Jolla campus of the University of California, where he was investigating cell membranes, had spent those years in Pauling's laboratory in Pasadena. "To understand Pauling and Watson, you must remember that creativity is an ego drive, as much in science as anywhere else," Singer had said. "I don't think there has ever been anybody doing great science whose ego has not been involved very, very deeply. It's one of the things that make science so competitive. But the competitiveness is just a wart on science. The question is, how deeply is the man involved? If objectivity goes, you get nothing but some kind of personal vision. That's the difference between science and religion. I was there. What Linus did, with his ego drive, was to insist that from his data on crystal structures of simple molecules, he could extrapolate. For example, that the peptide bond had to be planar. So he went ahead and imposed these restrictions. The length of the bond between the carbon and nitrogen atoms was much shorter, according to the x-ray data, than if the bond were single. So it had to have substantial double-bond character. So it had to be flat. Pauling, to begin with, has enormous physical intuition." That, I had learned, is the highest praise. "But what was astonishing about Pauling, and what makes him great, was that he was willing to move to a concept on the basis of certain data whose relevance was not clear to others. Only Linus showed the willingness to take the inductive leap. Then, once he had the idea, he pushed it. The history of science shows that that, in itself, is perfectly justified—yet it's the same egocentricity that led him to collect thousands of signatures of scientists on a petition to ban atomic testing."

I now asked Pauling what he thought about that. He shrugged. "Well, maybe there's something to it." He paused. "I would say that I probably *think more about problems* than other people do."

(Later when I quoted Pauling's remark to Perutz, he laughed and laughed, and then asked, "But didn't Linus also tell you that he has got more imagination than other people? Because, you know, it's true, he has.")

Facts in science are made significant by expectations, a truism patly illustrated by what happened next. The work at Pauling's own laboratory, in 1949 and '50, confirmed and refined what he already knew. He and Corey also devised a second structure for the polypeptide chain, more loosely twisted, that seemed plausible. Meanwhile, however, a research group at Courtaulds, a British corporation that makes artificial fibres for fabrics, had devised a fibrous polypeptide that was altogether synthetic and somewhat different in composition from keratin. When they took a look at its structure, they found that it did not show quite the same diffraction pattern as natural keratin. Reports of their fibre began to appear in 1949. When a preprint of the last of the Courtaulds papers, with the least equivocal x-ray pattern, reached Perutz late in 1950, he was engrossed in a new attack on hemoglobin, by means of the abnormal hemoglobin of sickle-cell anemia. He merely noted that the picture seemed peculiar because it was missing an essential feature of the diffraction patterns from natural keratin. Something was wrong. When Pauling received a preprint of the same paper, he saw at once from the pattern that the long molecules of the artificial polypeptide were packed neatly alongside each other in hexagonal array, like a fistful of pencils or cigarettes; though the men who published the pictures had not recognized it, the individual molecules had to be helices. More important, the artificial polypeptide lacked Astbury's anomalous spot for the 5.1-angstrom repeat which had been holding Pauling back. Something was right.

Pauling and Corey's announcement began with a short letter of the sort that establishes priority, sent to *The Journal of the American Chemical Society* in October 1950 and published in November. They said hardly more than that they had spent fifteen years investigating the detailed atomic arrangement of simple substances related to proteins, that they had used their results in the attempt to canvass all reasonable structures for the protein chain, and that now they had constructed "two hydrogen-bonded spiral configurations" that obeyed the rules and fit the data. That winter, Pauling gave a lecture at

Caltech about protein structures. The theatre was full. Pauling chalked a line of the chemical formula for a polypeptide on the blackboard. He then held up a length of that baby's toy composed of soft plastic bubbles in many colors and shapes that pop together to form a necklace of sausages, and began to demonstrate how amino acids combine. He turned to the molecular models on the bench before him. They followed the stereochemical principles, he said—setting those out in detail with emphatic reference to Bragg, Perutz, Kendrew, and the flat peptide bond. The spiral shapes were held together by hydrogen bonds, and these were straight lines, of correct length.

The first model was the result of Pauling's paper-folding two years before: a tight corkscrew in which each complete turn rose by 5.44 angstrom units—and not 5.1, he said, with emphatic reference to Astbury. This model, sometime that winter, he christened the alpha helix. More specifically, he called it the three-residue spiral: in this model, he said, each nitrogen was hydrogen-bonded to the carbon-oxygen pair that the twist brought into position for it farther along the spiral, three amino-acid residues away. Each full turn of the screw took in about 3.6 amino acids. Here was another lesson in physical chemistry for his rivals. Bragg and Perutz, heeding the conventional crystallographic wisdom, had reasoned from the regularity of the x-ray patterns that the number of amino acids had to come out exactly even along the turns. Pauling had seen that this was nowhere explicitly required. He asserted that this helix would be found in unstretched keratin, in muscle fibres, in hemoglobin.

The second model was the five-residue spiral, similar but more openly wound, with slightly more than five amino acids in a turn, which Pauling presented with equally persuasive detail. I asked him whatever had become of that. "Well, the second one was a nice structure in all respects," he said, with the manner of a veteran teacher who has seen a promising pupil fail his exams. "Except that it had a hole down the middle. I mentioned about that, not in the first paper but later, that the van der Waals instability associated with having that vacant space that's not big enough for a water molecule or anything else to get in, would have made this structure less stable, less likely to occur; and I think it has never shown up." He could shake his head with benignant regret at its fate. The other, the alpha helix, was a rare triumph.

Rumors about Pauling's lecture and models spread quickly even before details of the structures were published. Watson was by then in Europe, on the postdoctoral fellowship that Delbrück and Luria had arranged for him to study nucleic acids with Herman Kalckar. He had found Copenhagen cold, the biochemistry of nucleotides irrelevant, Kalckar incomprehensible—and, after Christmas, preoccupied by a sudden divorce and remarriage. The sort of intense intellectual relationship Watson had had with Luria was precluded. Watson was, as he said later, "very underemployed." He had passed the time by slipping off across town to the laboratory of Ole Maaløe, a Danish member of the phage group, to do experiments on the transfer of radioactive DNA from phage to their progeny, and the two even completed a paper about that. Though the results heightened Watson's suspicion that DNA, not protein, was what genes are made of, they were not conclusive; the research was nothing he could not have done in Pasadena or Cold Spring Harbor. That spring,

Kalckar went to work in Naples for several months, taking Watson with him. Watson's underemployment got worse. A conference on the structures of large molecules, late in May, brought Maurice Wilkins to Naples from King's College London to give a talk on DNA. When Watson attended Wilkins's lecture, he was braced to sharp attention by a slide of a new x-ray-diffraction pattern Wilkins had obtained from what he said was a crystalline form of DNA. Watson had never heard of Wilkins and knew nothing at all about crystallography. But he saw the essential point: if DNA could form crystals, it must have a repeating, regular, orderly structure, and so could be solved.

On the way back to Copenhagen, Watson stopped in Geneva to see Jean Weigle, a Swiss member of the phage group, who had just come back from a winter at Caltech. Weigle told him about Pauling's lecture on the alpha helix. Watson read Pauling's first full report on the two spirals when he got back to Copenhagen, in the April 1951 issue of *Proceedings of the National Academy of Sciences*, and a few days later was astounded, as were biologists everywhere, by the May issue, which carried seven more articles by Pauling and Corey: fifty-one pages that began with more details about the positions of the atoms in the two helices, went on to the structures of hair, feathers, muscle and tendon, silk, horn, quill, gelatine, hemoglobin, and synthetic polypeptides, and included two additional general configurations for protein. One of these was a flat, layered structure called pleated sheet, which Pauling and Corey thought would be found in stretched hair and other fibres that yield similar diffraction patterns. Pleated sheet, while hardly ubiquitous, does occur as a structural element of many proteins, and has entered the lexicon of every biologist. The other general structure was proposed for collagen, which is the tough, flexible protein found in tendon, bone, tusk, and the cornea of the eye; Pauling's model for collagen, a tricky, intertwining three-stranded helix, was entirely wrong.

Watson's remembrance of his reaction to the Pauling bonanza is restless and complicated; he seemed to enjoy displaying himself as shallow. Though he had not been present at Pauling's demonstration in Pasadena, he wrote, in *The Double Helix*, "This show, like all of his dazzling performances, delighted the younger students in the audience. There was no one like Linus in all the world. The combination of his prodigious mind and his infectious grin was unbeatable. Several fellow professors, however, watched this performance with mixed feelings. . . . A number of his colleagues quietly waited for the day when he would fall flat on his face by botching something important." The papers were "above me, and so I could only get a general impression of his argument." He was left in a fever of envy for Pauling's verve: "The language was dazzling and full of rhetorical tricks. One article started with the phrase, 'Collagen is a very interesting protein.' It inspired me to compose opening lines of the paper I would write about DNA, if I solved its structure. A sentence like 'Genes are interesting to geneticists' would distinguish my way of thought from Pauling's." After Wilkins's evidence that DNA had a crystalline structure, the Pauling and Corey papers on protein had the immediate practical effect of convincing Watson that he had to learn x-ray-diffraction techniques if he wanted to find the structure of the gene. This resolve, his boredom with Kalckar, and the luck of an encounter between Luria and

Kendrew at a conference in Ann Arbor that summer, where once more the future was arranged, took Watson within a few months to Cambridge to study under "someone named Max Perutz."

The Cavendish reacted to the alpha helix, too. First with chagrin: the day Bragg saw Pauling's paper, he went across the road to Alexander Todd, the Professor of Organic Chemistry, to ask about the peptide bond, and Lord Todd recently said of their conversation, "I told him if he had asked me at any time in the past ten years, or if he had asked anybody in my lab, we could have told him that the peptide bond had to be planar. Bragg was so horrified that he said that nobody was going to come out with a structure from his lab in the future until he had talked to me to get the okay that it was chemically sound." Then came generous confirmation. Perutz first read Pauling and Corey's seven papers on a Saturday morning at the laboratory. He saw that the alpha helix was obviously right. He saw, further, that if the alpha helix was right, the diffraction pattern of natural unstretched keratin ought to show a spot at a position where nobody had ever reported one, which would confirm the model's uniform but very small rise—about 1.5 angstroms—along the axis of the helix between one amino-acid residue and the next along. Perutz then wondered why Astbury, who had taken hundreds of x-ray-diffraction pictures of fibres containing keratin, had never noticed this spot. But he remembered from visits to Leeds that Astbury's customary laboratory setup for taking diffraction pictures employed too small a photographic plate, and the wrong angle between the fibre and the x-ray beam, to find the 1.5-angstrom spot. So after lunch Perutz took a single horsehair, set it up at the angle he had calculated, placed it down the middle of a cylindrical sheet of film to detect diffraction spots far from the center of the pattern, and took a single picture. There was the predicted spot. Monday morning he showed it to Bragg, saying that the Pauling triumph had made him so angry at their collective stupidity at the Cavendish that he had had to verify the structure at once. Bragg said only, "Perutz, I wish we had made you angry sooner!"

Perutz went on to find the same spot in porcupine quill and in synthetic polypeptides. With Hugh Huxley, a Cavendish colleague, he found it in muscle fibres. Perutz's data on hemoglobin went down only to 2 angstroms, but by taking a new photograph he found the spot once more— "barely visible with the eye of faith," he said recently, but an indication that this functioning, nonfibrous protein contained stretches of alpha helix, too. That summer, Perutz published these findings, the first independent demonstration that the alpha helix was real.

Francis Crick had not known enough structural chemistry to catch the blunder his colleagues had made. A while after Pauling had published, Crick began to wonder about the mathematics needed to predict, from any given model that was helical, what its x-ray-diffraction pattern must be. By the time he met Watson or soon after, Crick got deep into the projective geometry of helices. He also absorbed from Pauling, as he later said, the virtue of boldness in theory-making, boldness as an aspect of rigor. Perhaps most important, he learned the importance of model-building—the lesson that playing with rods and knobs of precise molecular models was the only way to see what theory could be when all the physical limitations were operating at once.

By Watson's account, when the pursuit of the structure of DNA was fairly begun, it was Pauling he imagined himself competing with, Pauling he wrote to friends like Delbrück to ask about, the thought of Pauling he employed to whip himself out of neurasthenic lethargies into frenzies of work. When it came to it, two years later, Pauling and Corey published a structure for DNA some weeks ahead of Watson and Crick—and this time too, in a mirror image of the competition for the alpha helix, the team that published first got it wrong. Indeed, Pauling "botched something important." I asked Pauling about DNA.

His answer was quiet. "We weren't working very hard at it. We had really very little in the way of our own experimental data, a few rather poor x-ray photographs of DNA, not carefully prepared. I wasn't putting in much of my time on determining the structure. I thought I would get it worked out, you know, in a question of time. I didn't know there was competition—that is, I wasn't involved in any race. Corey and I published our paper on a proposed structure, where we had misinterpreted some of the x-ray information— Well, there were two things involved. The value of the density of DNA that we had taken from the literature was wrong. And the x-ray photographs that we were interpreting were really superpositions of the two different crystalline forms of DNA. You know it was Maurice Wilkins's experimental work that put Watson and Crick on the right track. Wilkins made better preparations of DNA, which isolated either one structure or the other structure, and therefore he got x-ray photographs that separated their characteristics."

In a conversation a few weeks earlier at the faculty club of the Massachusetts Institute of Technology, a couple of biologists had speculated whether Pauling, whose recent popular book on the benefits to health and sanity of massive doses of vitamin C was stacked in display near the entrance of the M.I.T. bookstore, was showing signs of what one of the men called "old scientist's disease"—which they defined as what happens to great men when they grow beyond the psychological reach of the salutary system by which scientists blow the whistle on one another's mistakes. What was Pauling now working on? "Well, you see, I'm giving a course on nuclear physics, now; it's called physical chemistry, but it's on the structure of atomic nuclei; I've written a dozen papers in the last six years on this subject. And now I'm working on schizophrenia"—he said it *sheets-oh-frayneea*—"and other diseases. Three years ago, I published a paper called 'Orthomolecular Medicine,' and another, 'Orthomolecular Psychiatry.' I've been working along those lines since. We are trying to analyze people, their metabolisms, accurately enough to determine genetic characteristics that are related to disease."

He went over to a four-drawer filing cabinet, pulled open the bottom drawer, took from a large brown carton a blue-and-white plastic-wrapped bag, stiff and about the size of a thin paperback book. "*Vivonex 100*," the bag read. Standard diet. A chemically defined synthetic low-residue food. Then Pauling took down from a bookshelf a large roll of paper, which he spread out to show a long graph with many jagged peaks on it, each with a number beneath. "See, here is part of a study. 'Linus Pauling, whole urine, night.' And each of these peaks represents a substance, in the urine. We are hooked into a computer which will integrate under the curves and write down the numbers telling us the amounts. We can now calibrate a person, standardize him so that he gives us the same diagram—a kind of urine print—and we can measure more than four hundred substances in his urine. I've been working on this for seventeen years now, and only recently has it got really promising, the main reason being that we now have standardized people with respect to their intake, by giving them a standardized small-molecule diet"—he gestured towards the packet of Vivonex 100—"which is immediately absorbed from the stomach into the bloodstream, no digestion, the contents of the gut quickly disappear and the bacteria too. Perfectly nutritious, I often have a packet in a glass of water for lunch if I don't have time to go home. With this standard diet we avoid two different complications. The first is variations from day to day in the molecular composition of a person because of variation in his diet. And the other, variation due to differences in the bacteria in the gut, because people do have different flora. The bacteria disappear. They are starved to death on this diet. In four days they're gone. After four days, we can measure these four hundred substances with a reproducibility from day to day within about ten per cent, quantitatively. Probably better than anybody else is doing."

Was that accurate enough to begin to demonstrate differences between people? "Oh yes, surely. When we studied mental retardates on this diet, individuals with severe mental retardation, first of course there was a characteristic pattern showing up for the phenylketonurics; but then the others, you know—undiagnosed mental retardates—showed half a dozen different characteristic patterns. We haven't tracked these down yet, but in the course of time we should know just what each abnormality is. There are a lot of genetic abnormalities, of course, that lead to mental retardation. And they often go unrecognized. Patients in mental hospitals haven't been examined enough for these defects to be sorted out and identified. And we find abnormalities among schizophrenics too—with respect, for example, to their vitamin metabolism. Schizophrenics apparently have lower concentrations of some vitamins in their bodies than other people, or a lower vitamin activity, so that smaller amounts of an ingested dose show up in their urines. There are several kinds of schizophrenias, of course, too. But there's nothing surprising in the idea that metabolisms of psychotics might differ. If you look at the population of people with pernicious anemia, they show characteristic psychotic symptoms. Often their psychosis manifests itself months or years before the anemia. It is a severe mental illness. And there is no doubt that it is caused by a low level of vitamin B_{12} in the blood. But if the intrinsic factor in pernicious anemia is provided, or if B_{12} is injected, the psychosis disappears. I think one would—who was it said that the Latin for agnostic is ignoramus?" He laughed. "You can go too far in being skeptical about these matters."

I asked if I could take the packet of Vivonex 100 small-molecule food with me. He said Yes. Later, at my motel, with no time for lunch myself, I read the small print on the packet. "Ingredients: glucose, sodium glycerophosphate, glucono-delta-lactone, L-glutamine, L-arginine hydrochloride, L-aspartic acid . . ." The list of amino acids, all in the left-handed versions the body can use, was followed by vitamins and sugars. Then, in large type, "Orange (mild) flavor," it said. "Also contains glucose oligosaccharides, citric acid, artificial flavor, and artificial color." Reassured that it was fit for consumption, I emptied the packet into a 12-ounce container, added 8 1/2 ounces of water, and stirred vigorously until the powder dissolved. I sniffed, and tasted. Orange (mild) flavor. I drank it down and then, to make sure, ate a couple of grams of vitamin C as well. I'd been captured by that infectious grin and that unquenchable self-confidence, just as Watson had described it.

❖ *b* ❖   "It is also a good rule not to put too much confidence in the observational results that are put forward *until they are confirmed by theory*," Sir Arthur Eddington wrote in 1934: his paradoxical inversion of the inductive system as preached, from Bacon to Russell, has become an epigraph for the latter-day recension of the scientific method as practiced. The biochemistry of nucleic acids in which Watson put so little trust that winter of 1950–'51 in Copenhagen was yielding facts in Paris, New York, and elsewhere that profoundly affected the search for the nature of the gene. No theory sorted out and confirmed the facts. That step was still to come. Several of the facts, all the same, surely looked puzzling enough to be important. More than that, some of the stiffest barriers to theory were at last borne down. DNA as a molecule was growing more intelligent. The new work was directly inspired by Oswald Avery's great paper of 1944, and was being done by biologists very much in Avery's style.

In Copenhagen, Herman Kalckar knew Avery, having met him in New York in 1945. "A delightful old man," Kalckar said in a conversation. "Admittedly, this was the kind of biochemistry that appealed to me more than anything else, since it was exceedingly critical and yet Avery also had a shrewdness. This man was really worthy of two Nobel prizes." Kalckar is famous among biologists for his elliptical, knight's-move manner of speaking. "So that people, I was awfully excited about that work, I must say Max Delbrück too. I told Max, later, after the death of Avery when he gave a memorial talk, gee, remember when you in the old days sailed up to old Avery and said, you know—'I believe you're wasting your time.' He wanted to say that the phage would arrive first. Old Avery smiled, because he could hear the way, you know, Max looked very worried."

In the fall of 1949, after the phage meeting in Chicago when Watson's future had been settled, Kalckar and Alfred Hershey left together for St. Louis by train. On the ride, Kalckar heard all about an experiment Hershey was then planning which might trace, with different radioactive labels, the fates of the phage protein and DNA. Hershey's results came slowly and were not published until 1952; but the train ride, Kalckar said, was one of the reasons he left Watson to get on with his experiments with radioactive phage in Copenhagen. In Avery's own laboratory, the biochemical proof that the transforming principle in pneumococci was nothing but DNA was driven still further. Rollin Hotchkiss there, after many attempts, began to find other inheritable characteristics, such as resistance to penicillin, that could be passed from pneumococcus to pneumococcus by purified transforming principle. Still they remained cautious, Hotchkiss wrote later, about "what biological generalization could responsibly be drawn." In Strasbourg, André Boivin, after claiming to have found the phenomenon of transformation by DNA in *E. coli,* began measuring the amounts of DNA in animal cells. Boivin, with Roger and Colette Vendrely, reported for a variety of species that the germ cells, sperm and ova, which according to Mendelian genetics must have half the genes of the somatic cells, indeed have half as much DNA. Others, at the Institut Pasteur and elsewhere, were finding in DNA the corollary characteristics of the gene: that DNA turns over metabolically at a far slower rate than other substances of the cell, and that the amount of DNA in the nucleus of the somatic cell, in any particular species, is remarkably constant, even when the creature is severely starved. Sometimes these workers were less guarded, reporting, for example, that "the constancy of the deoxyribonucleic acid appears as a natural consequence of the special function now attributed to it, as the depository of the hereditary characters of the species." The change in scientific climate—the sensation that DNA as a genetic material was getting a lot warmer—was perceptible to biochemists and bacteriologists perhaps more than to those who worked with phage. This was a matter of what one read, who visited one's lab, which experimental systems seemed congenial, what colloquia one attended, whom one believed. But Watson has always been extremely sensitive to the heat.

Erwin Chargaff was decisively moved by Avery's work. When he read the paper on the transforming principle, at its publication, he was already running a biochemical laboratory, in the College of Physicians and Surgeons of Columbia University, way uptown in Manhattan at Broadway and 168th Street. The lab was mostly working with lipoproteins, linked fat and protein molecules in the cell, but had done a little with nucleic acids—enough so that Chargaff had read what there was to read about them. Chargaff is a superb critic of science—humane, historical, informed, opinionated, nasty, and funny. Avery's intellectual style, most particularly his fastidious understatement, chimed with Chargaff's own reverence for the complexities that biologists destroy by describing. Chargaff experienced what in a less urbane man might be called a conversion; he himself, several years ago, compared his ensuing quest to that of Cardinal Newman in *The Grammar of Assent.* Avery's discovery "certainly made an impression on a few, not on many, but probably on nobody a more profound one than on me," Chargaff wrote; and he went on:

> I saw before me in dark contours the beginning of a grammar of biology. . . . Avery gave us the first text of a new language, or rather he showed us where to look for it. I resolved to search for this text. Consequently, I decided to relinquish all that we had been working on or to bring it to a quick conclusion. . . . But these biographical bagatelles cannot be of interest to anybody. To the scientist nature is as a mirror that breaks every thirty years; and who cares about the broken glass of past times? I started from the conviction that, if different DNA species exhibited different biological activities, there should also exist chemically demonstrable differences between deoxyribonucleic acids.

The differences would have to reside in the exact proportions and arrangements of the nucleotides.

The chemistry, however, looked insurmountable. Science is helplessly opportunistic; it can pursue only the paths opened by technique. In the event,

Chargaff's chief technique was paper chromatography, which was then brand-new, and which he spent two years adapting for use with nucleic acids. Chromatography is simple, in principle: dissolve the DNA in something that will chop it up into separate components, float the solution down a sheet of wet filter paper and locate the spot where each kind of base comes to rest, cut the spots apart with scissors, wash each fraction off its paper with another solvent and measure it. The problems began, though they hardly ended, with the choice of solvents that would cleave the nucleotides—particularly those containing pyrimidines, which are refractory—without degrading any into something else. In 1949, Chargaff gave a series of lectures about his results for the Chemical Societies of Zurich and Basel, the Society for Biological Chemistry of Paris, and the Universities of Uppsala, Stockholm, and Milan. The lectures summed up four years of published and unpublished research; he revised them into a paper, published in *Experientia* in May 1950. The paper, "Chemical Specificity of Nucleic Acids and Mechanism of Their Enzymic Degradation," was the next high point after Avery's in the biochemical delineation of the gene. Like Avery's, it bears the impress of an individual mind, working.

The minds, despite Chargaff's regard for Avery, are very different: where Avery was careful, direct, and forceful, Chargaff's strength and unmistakable probity stumble sometimes in an excessive self-awareness. The paper, with the work it represented, was the finest thing in Chargaff's scientific career, as well as the culmination of that entire school or line of biochemical reasoning—and, like cultural flowerings of many kinds, it was produced when the impulse and the vision from which it arose were already giving way to a different ideal. The paper has that sort of glow about it, in retrospect, if this is not too fanciful.

It is also scarred by an opportunity missed, a failure to take the inductive leap, which illustrates perfectly why theory is indispensable to confirm fact. What Chargaff found most significant among his own facts, at the time, bore on a theory that was already toothless, enfeebled, and about to be abandoned on the trail. His evidence showed that the four bases in DNA occur in widely varying proportions in yeast, bacteria, ox, sheep, pig, and man. Thus, human thymus nuclei yielded the four bases in the proportion 28 parts of adenine, 19 parts of guanine, the two purines totalling 47 parts; and 16 parts of cytosine, 28 parts of thymine, the two pyrimidines totalling 44 parts.

Such figures buried, at last, the tetranucleotide hypothesis—the idea that the DNA molecule is a monotonous rotation of the four bases one after another. Instead, there exist a vast number of different DNAs. For the first time there was reason to think that nucleic acids are as rich in variety as proteins. Yet within a species, and from organ to organ and tissue to tissue of the same species, the composition of DNA was fixed and typical. Still more, in those few cases where sperm cells had been compared with the nuclei of other cells of the same creature, no chemical differences in DNA were found—while the proteins in these nuclei were *not* the same. And so, Chargaff said, if the long molecules of DNA are to "form an essential part of the hereditary process," the specificity that could be carried by different sequences of nucleotides along the chain "is truly enormous": he mentioned a figure for the number of possible combinations in a reasonable length of DNA that exceeds by many times the number of electrons in the universe.

Chargaff saw one more feature in his tables of data.

> It is, however, noteworthy—whether this is more than accidental, cannot yet be said—that in all deoxypentose nucleic acids examined thus far the molar [that is, molecule-to-molecule] ratios of total purines to total pyrimidines, and also of adenine to thymine and of guanine to cytosine, were not far from 1.

That, in 1950, was all that Chargaff committed to print about the strange uniformity he had come upon in the midst of astounding diversity, the uniformity which sets the nucleic acids off from proteins and all other large molecules, after all—the simple equivalencies among the bases of A to T and of G to C, and so of purines to pyrimidines. Chargaff's brief remark was the first statement of the central feature of DNA: the equivalencies are structural, as Crick and Watson found three years later; they are functional in ways that began to be evident once the structure was known. The equivalencies are now sometimes called the Chargaff ratios. It is not easy to see how, at the time, Chargaff could have understood the significance of the equivalence rule or taken it any further; but it remains that he did not take it further. So the observation he published and abandoned still smolders in his own recollection. Chargaff has become the man of mordant dissent from the Watson and Crick style of science, blind Tiresias wandered into a performance of a comedy, reviling "the degradation of present-day science to a spectator sport," and concluding: "That in our day such pygmies throw such giant shadows only shows how late in the day it has become."

❖ ❖ ❖

Maurice Wilkins shared the Nobel prize with Watson and Crick in 1962 because his laboratory at King's College London produced the x-ray-diffraction pictures of DNA which stated the physical constraints on the structure; by the time the prize was awarded, the scientist who had taken the most important single picture, Rosalind Franklin, was dead. Wilkins has been president of the British Society for Social Responsibility in Science since its founding in 1969. When I first got in touch with him, in June 1971, he said on the phone that a conference on theoretical biology was to meet all the next week in Versailles, with the final session, Saturday, given up to the social responsibility of science, for which he had been asked to come over from London to speak. We arranged to meet there and go on to lunch.

The Hall of the Crusades in the Palace of Versailles was much like any other hotel conference suite, though more florid. Its walls were gloomy with nineteenth-century monumental oils celebrating the conquest of Acre and the lifting of the siege of Rhodes, and its tall columns and cornices were decorated with the heraldic shields of eminent crusaders. Midweek, I stopped in at a session about the origin of life. Then the room was full, the discussion speculative, factual, quick, agreeably tough-minded. Saturday morning's meeting was sparse. When I came in, a heavy-set man was saying from the audience that "Khorana has synthesized the gene; tomorrow we will be able to control the genetics of man." The armorials, high above, suggested a pattern of genetic crosses that related the Count of Flanders to John of France. Wilkins, when his turn came, explained the case made by those who attack science. He warned of "the growth of anti-rational attitudes"; he mentioned the fear of "dehumanization by science, which really amounts to the claim that objective thinking necessarily reduces moral sensibility." None the less, "When those who write against science charge that our curiosity is out of control, I find myself somewhat in sympathy. Kierkegaard said, 'Knowledge is an attitude, a passion. Actually an illicit attitude. For the compulsion to know is a mania, just like dipsomania, erotomania, homicidal mania: it produces a character out of balance. It is not at all true that the scientist goes after truth. *It* goes after *him*. It is something he suffers from.'*

"But the dilemma is there," Wilkins said; "Science, with technology, is the only way we have to avoid starvation, disease, and premature death. The misapplication of science and technology is due to the fact that the politics are wrong. Now my own view is that the politics is indeed wrong; but politics and science are so closely interrelated that they can hardly be separated." He listed several steps, modest enough, that scientists might take. Then, "I think also we need to reject the concept that sets off pure science from all the rest. The general demand for what is called 'relevant' research has got some sense in it." He spoke briefly about "exploring the relation" between art and science. "Today, many artists regard art as merely an inquiry which leads to some statement about man in the universe. . . . I think it may be useful to take the old idea of Tolstoy: we·need to go out to the people. In the ordinary mass of people there is a reserve of wisdom which the specialized thinkers have lost." He spoke of a scientist who had recently returned from China: "He said that in China he saw that every research laboratory has a direct relationship with some particular factory, working with it, interchanging workers, discussing all projects together. What are we doing that resembles that, at all?"

We went out into a light blue-gray rain and to Wilkins's hotel, where he got his overnight bag and made excuses to the organizer, who pressed us to join them all at lunch. As we drove into Paris, he said, "Really, many of the aspects of science and art leave me cold."

Watson, when I had last seen him, had been greatly worried about the possible application with humans of a new technique for growing multiple identical twins from cells taken from a single mature animal. The process is called cloning. The multiple descendants of a single mutant bacterium are a clone, as in Luria and Delbrück's fluctuation test. Many plants produce clones easily, and so can be propagated by cuttings—*klon* is Greek for twig or shoot. The multiple malignant descendants of a single cancer cell that has escaped the surgeon's knife are a clone. In the late nineteen-sixties, John Gurdon, a cell

---

*Several years later I asked Wilkins for the source of that passage. He found it, he said, in a book he borrowed from a London library early in the nineteen-fifties, had copied it out into a notebook with the identification "Kierkegaard" and a page number (page 254) but no further reference; and he had not been able to locate the source since. He showed me the notebook, where the passage sat alone on a left-hand page. I searched for the passage and consulted students of Kierkegaard. The quotation excited the keenest interest. Max Delbrück placed advertisements in a Copenhagen newspaper asking if anyone could identify the source. Years later, a reader of the first edition of this book wrote me that it actually comes from *Der Mann ohne Eigenschaften*, by Robert Musil—or, rather, from volume I of the first English translation, *The Man without Qualities*. This was published in England (by Secker & Warburg) in 1953, I then found—and one of the translators was Eithne Wilkins, Maurice Wilkins's sister. The page reference is correct; Wilkins's transcription contained several inconsequential errors.

biologist then at Oxford, removed the nuclei from fertilized frogs' eggs, replaced them with nuclei taken from cells from the gut of a single tadpole, and grew up a number of frogs with identical genetic constitutions—an animal clone. Other biologists then began trying techniques that might make cloning possible in mammals—much sooner than most scientists thought, Watson warned. The idea of cloning people naturally arises: do we want a dozen Einsteins, Watson asked, a dozen Heifetzes, or "multiple likenesses of Raquel Welch?" Wilkins, though, took Watson's fears with some skepticism when I described them. "There's a lot of hysteria around over the possible effects of science," Wilkins said. "Cloning may be possible, all right, but it's not socially likely. Not on any serious scale. One cannot easily see what are the real social pressures that could lead to any wide attempts at cloning humans."

Wilkins was thin, bespectacled, with a sharp straight nose and a long neck. He had that shade of dark blond hair which will fade imperceptibly into gray. If T. S. Eliot looked like an Anglican bishop, Wilkins looked like a minor canon. He was a fencer in his younger days and looked that too. "But I simply wasn't fast enough." His voice was unexpectedly resonant. He began in science, he said, not as a biologist but in solid-state physics. "It's a field that has become very fashionable just now." As an undergraduate at Cambridge, before the war, Wilkins learned a little crystallography from Bernal. He did his doctoral research with the physicist John Randall, at the University of Birmingham, and when the war began worked first on radar and then on the separation of uranium isotopes, which moved him to Berkeley, California, and the Manhattan Project. The war over, Wilkins felt the moral revulsion that affected many of those who worked on the bomb—and then he too read Schrödinger's *What Is Life?* and got interested in putting physics to work on the complexities of living processes. When his former professor, Randall, invited him to join a new biophysics unit, he accepted. In 1946, Randall's unit was set up in King's College London, which was reconstructing buildings and staff left in ruins by war. The money for the unit came from a governmental agency, the Medical Research Council, set up after the first world war to support research in medicine and in the biology basic to medicine.

Randall's vision of biophysics was from the beginning imperial in scope; the performance turned out diffuse. Wilkins spent a couple of years trying to cause mutations in fruit flies with ultrasonic vibrations, to no effect. He worked on development of reflecting microscopes for spectrographic studies with ultraviolet light, and used the apparatus to study the amounts of DNA in cells. He began structural work, he said, "with a little mucking about with tobacco-mosaic virus"—a recurring favorite of physically oriented biologists. Meanwhile, Randall and several assistants were investigating the sperm of rams with the electron microscope and by x-ray diffraction; Wilkins, by 1950, was looking at sperm heads, too. In the symphony that Randall heard in his mind, these ideas and half a dozen others were but developments of the single theme that he described, in a lecture to the Royal Society on 1 February 1951, as "an experiment in biophysics . . . by study of cells, especially living cells, their components and products." Scientific research is a wasteful process, sure enough. None the less, Wilkins along with the rest of Randall's unit had unusual difficulty in settling on productive lines. Though trained as a physicist,

Wilkins crept up on the structure of DNA from the large-scale end, the cellular end, the biological end of the problem.

❖ ❖ ❖

We got out of the taxi in the rue Monsieur le Prince, near the Odéon, and went around the corner to a small restaurant there. Wilkins had been caught in an excruciating predicament over the solution of DNA. He had started x-ray crystallography with that substance eighteen months before Watson and Crick ever met, and at first had had splendid luck. In May of 1950, Wilkins went to a meeting in London of the Faraday Society, about nucleic acids, where a visitor from Bern, Rudolf Signer, described the greatly improved methods devised in his lab for extracting DNA from calf thymus cells gently enough not to break the long, fragile threads. Signer's recent specimens had preserved the molecules as unbroken strings that were on the order of ten times longer than the DNA molecules then familiar to Erwin Chargaff.

This is perhaps the place to recall that DNA stands for deoxyribonucleic *acid*. The acidity is slight; it is due to the phosphate groups, which are negatively charged. DNA is usually extracted in the form of the sodium salt, which means only that atoms of the ubiquitous metal sodium, which are positively charged, have combined with and neutralized the phosphate groups, just as they will combine with and neutralize negatively charged chlorine atoms to make common salt. The thirst of each charged phosphate group for something to neutralize it, close by, is the sort of phenomenon that Pauling's rules had explained by 1928; it is a fact of intrusive and more than technical importance in the various schemes and fancies for a structure of DNA, right up to the final one. One question, then, had to be whether the substance as extracted really represented the form that occurs in the living cells. A minor curiosity of nomenclature arises here, too: by the old convention that identified the nucleic acids with their most usual sources, Signer's DNA was called sodium thymonucleate by most of the people who eventually did experiments with it.

Anyway, at the meeting of the Faraday Society Signer gave out samples of his best, and Wilkins took some home to King's. A few days later, he was preparing part of this DNA in highly viscous solution for the ultraviolet studies he had been doing. Unexpectedly, he noticed that every time he touched the stuff with the tip of a glass rod and then drew the rod away, "I had spun a very thin fibre of DNA, almost invisible, like a filament of spider web." The fibres seemed so perfect and uniform that the molecules in them must have been neatly aligned alongside one another. Examined under a microscope, in polarized light, the fibres behaved as though they were made up of well-ordered crystals. So Wilkins took the fibres to a graduate student in the unit, Raymond Gosling, who was doing the crystallography with rams' sperm and had borrowed the use of an antiquated x-ray apparatus in a lead-lined basement of the chemistry department. Several dozen of Wilkins's fibres had to be bundled together, in a tiny tungsten-wire frame where they could be tightened like a violin bow, to constitute a large enough specimen for the soft focus of the x-ray tube. Wilkins and Gosling remembered that Bernal, fifteen years before, had got the first good x-ray patterns from protein crystals by keeping them wet. Accordingly, they maintained the DNA fibres at high humidity. They got striking patterns, with many spots, sharp and detailed, far better that ever before seen from DNA, and for the first time unmistakably from a substance in crystalline form. Then—but why, neither Wilkins nor Gosling has every fully explained—Wilkins went on to other projects.

Equipment was one reason. The x-ray tube and camera Gosling had been using gave a diffuse beam of low intensity, not right for fine fibres. That spring, Wilkins arranged for the King's College biophysics lab to be given an x-ray tube of a new model, with a very fine focus, invented and built by Werner Ehrenberg and Walter Eric Spear, two physicists working in Bernal's department at Birkbeck College, London. But Wilkins had collected the new tube by the summer of 1950 or even earlier; Spear wrote to me that he distinctly remembered telling Wilkins at that time about the circuitry, power supplies, and cooling necessary for setting it up. The new equipment was not put to serious use before September 1951—fifteens months or more later—when Rosalind Franklin began to take pictures using it and a new microcamera. At the end of 1950 and the beginning of '51, Wilkins did publish a couple of notes of work on DNA, but these did not emphasize the new x-ray-diffraction pictures and did not illustrate them.

None the less, that series of diffraction pictures taken in May or June of 1950 provided the one Wilkins showed at the meeting in Naples, a year later, that galvanized Watson's enthusiasm for DNA and x-ray crystallography. Wilkins was thus in the position of the inattentive lover who introduces his girlfriend to a chance aquaintance and then must watch her stolen away—and however painful, that position is also ludicrous. There is no easy simile for the other half of Wilkins's problem. "I faced the fact that we would have to go very much further into x-ray-diffraction techniques, in which I was not really qualified. That's why we hired Rosalind Franklin."

We ordered lunch and a glass of the stony wine from the Jura. "We hired her," Wilkins said. "We fell out. It took me a little while to face the fact that our imported x-ray worker was not going to work with us. Collaboration was impossible." Franklin had been brought to King's College, Wilkins said, to get the x-ray data about DNA that were needed. She arrived in January 1951. Randall assigned Gosling to her as a research student. Wilkins turned over to her the rest of the DNA from Signer's laboratory that had yielded the great new diffraction patterns. Then in the summer or fall of 1951—about the time Watson arrived in Cambridge—Franklin refused to collaborate with Wilkins, claiming that she had been given DNA as her own independent project.

The conflict between Maurice Wilkins and Rosalind Franklin, though it seems petty at first glance and is certainly distasteful, ranks as one of the great personal quarrels in the history of science. It was marked, in the way of such quarrels, by a bitterness beyond the personal, for as it went on, Wilkins, Franklin—and Watson observing—realized very well that at issue was the most important discovery then at large in biology. "Basically, I don't think the problem of DNA was so difficult," Wilkins said. "Of course, it's easy to say this in retrospect. But it really was easy to solve. Rosalind Franklin was wrong-headed. Francis Crick said it: she lacked physical intuition. Ah, no." Wilkins took off his spectacles. His eyes were nearsighted, slightly reddened by travel. "She had done some first-class work. It is bad form for me to run her down." Lunch arrived with a bustle and much rearrangement of crockery so that I could keep my notebook by my plate. "Francis Crick, I knew, was in a very difficult position. He had no secure job. He was thirty-five and still working for his Ph.D. He had just got remarried. He wasn't on the best of terms with his boss, Bragg." Wilkins's comments were staccato, somewhat disjointed. It became clear that he was pacing a cage of frustration whose bars should have been fifteen years gone. "Jim Watson was exactly what he says he was: his eye on success and recognition. I knew them both very well. If you read the book—'I'm Jim, I'm smart, most of the time Francis is smart too, the rest are bloody clots.' If you want to make a critical analysis, she was going at it *wrong*."

Linus Pauling had learned that Wilkins had good x-ray diffraction pictures of DNA, and wrote to London—though neither he nor Wilkins, when the point was raised, could remember exactly when—to ask if he could see prints. Wilkins replied that he was not yet ready to show them. An English historian of science, Robert Olby, had since written that if Pauling had seen Wilkins's photos, then he would have had the data for a correct structure of DNA; so I asked Wilkins about that. He disagreed. "Pauling's various contributions on DNA are pretty lamentable. The glib assumption that he could have come up with it—Pauling just didn't *try*. He can't really have spent five minutes on the problem himself. He can't have looked closely at the details of what they did publish on base pairing, in that paper; almost all the details are simply wrong."

What had Wilkins thought when he learned that Watson and Crick were moving in on DNA? "Upset? I was bloody annoyed. But—what do I say about my dear, dead colleague? Do I go along with what Watson wrote about Franklin—and he had it right—or do I play the gent? To assess any scientist is damn difficult to do. How can you be fair? ‑

"I was a bit slow on the base pairing, I must admit. I knew about Chargaff's results, but I didn't cotton on to the fact of a structural implication. Rosalind Franklin put me onto the wrong track. She said the structure couldn't be helical. Rosalind Franklin was doing what I can only, reluctantly, call bad work." Wilkins reached across the table, took my pen, and drew in my notebook a diagram of a structural idea for DNA, joining purines to pyrimidines, that he had been thinking of, he said, in the winter of 1952–'53. "Early in 1953, one evening before they were working on the models, Jim stopped by my lab. The sum total of our conversation was, I said, 'I think Chargaff's data are the key,' and he said, 'I think so too.'

"All I'm saying is—well, I didn't get it. But if people want to speculate about, 'Well, if Watson and Crick hadn't got it, who would?' then I think I could make a good case. As good as anybody's.

"DNA, you know, is Midas' gold. Everybody who touches it goes mad."

❖ ❖ ❖

When Rosalind Franklin died in 1958, she left her scientific notebooks, correspondence, and files to Aaron Klug, a fellow crystallographer. Watson, in his memoir, painted the quarrel between Wilkins and Franklin with the details of purest farce: Rosy the mannish, assertive woman in science, her clothes ridiculous, Wilkins bloodless and dithering, not master in his own lab—Watson dealt an even-handed malice. The civet was not sweetened for everyone by the note about Franklin that Watson appended after Crick and others in England had read his manuscript, a tribute to "her personal honesty and generosity" and to "the struggles that the intelligent woman faces to be accepted by a scientific world which often regards women as mere diversions from serious thinking." In the summer of 1968, after *The Double Helix* appeared, several articles and letters were published in *Nature* to tidy one corner or another of the facts. The most important was by Klug, on behalf of Franklin's role. Klug had been, he wrote, "her last and perhaps closest scientific colleague." After I read his article, I sought him out. We had several conversations, and went through the notebooks Franklin had kept while she was working on DNA. Klug believes, and there is evidence, that Franklin was poised two half-steps from discovery. But the earliest evidence bears not so much on the discovery as on the quarrel; with high punctilio, Klug did not refer to any of this in his article in *Nature*.

Franklin was born in 1920 into a rich, philanthropic, upper-middle-class Jewish family; her ancestors had come to London in the eighteenth century. She went to St. Paul's Girls' School, in London, an academically excellent private school whose graduates were prepared to do well in the professions they took up. At Cambridge University she read chemistry; on her final examinations she achieved not a first—that is, not the highest honors—but a top second. In 1947, she was hired by a French government laboratory in Paris. There she worked with carbon compounds, coal and graphite. She made herself an able and even an original physical chemist. Certain of her papers are among those that founded the science and technology of high-strength carbon fibres. But all her experience in Paris was with amorphous substances like coals and chars, and she had never worked with single crystals, with large molecules, or with those of biological interest.

She gave herself with zest and dedication to research, to laboratory life, to Paris in the late nineteen-forties. A colleague, slightly younger, was the crystallographer Vittorio Luzatti; Franklin, Luzatti, and his wife became close friends. A photograph he took when the three were on a walking trip in the south of France in 1950 or '51 shows Franklin at a table of an evening, in discussion, completing some bit of mending: she had glossy black hair, a high forehead, brilliant eyes and a soft mouth—a face attentive, attractive, feminine, alive with sense and intelligence.

A while ago, Luzatti and I spent a day exploring the Paris Franklin had known. The laboratory where they had worked had small rooms, but light and airy, now converted to offices, strung along a corridor under the eaves of a building off the Quai Henri IV. One apartment where she had lived was up a dim stairway across the grit and pigeons of an unmistakably Parisian courtyard. A lively bistro where members of the laboratory had sometimes lunched and argued was perched deliciously on a bluff at the back of the Latin Quarter. Franklin lived and worked four years in Paris. She had her farewell dinner at the Luzattis'. I asked him why she had left. "Well, she said she was English, and if she wanted a scientific career in England, it was time to go home."

Franklin's gift as a scientist, Klug said, was in dealing with recalcitrant materials. She had the sort of skill that comes from single-minded intelligence.

The techniques she introduced tamed DNA and got much better pictures; but she was self-taught in the art of interpreting them.

Franklin was hired by Randall, at about the time in the spring of 1950 when Wilkins was first beginning to work with Signer's DNA. With Randall's help she was awarded a three-year research fellowship that June, at about the time Wilkins and Gosling were getting their first good x-ray patterns. The fellowship was to begin in September, but in order to finish the experimental part of her work in Paris she delayed her start until January. On December 4, Randall wrote to Franklin in Paris about a thoroughgoing "change in programme" for the work she would soon take up. Randall extended himself to explain the change, saying, in part:

> The real difficulty has been that the x-ray work here is in a somewhat fluid state and the slant on the research has changed rather since you were last yere [sic].
>
> After very careful consideration and discussion with the senior people concerned, it now seems that it would be a good deal more important for you to investigate the structure of certain biological fibres in which we are interested, . . . rather than to continue with the original project of work on solutions as the major one.
>
> Dr. Stokes [Alexander R. Stokes, a physicist in the unit], as I have long inferred, really wishes to concern himself almost entirely with theoretical problems in the future and these will not necessarily be confined to x-ray optics. It will probably involve microscopy in general. This means that as far as the experimental x-ray effort is concerned there will be at the moment only yourself and Gosling, together with the temporary assistance of a graduate from Syracuse, Mrs. Keller. Gosling, working in conjunction with Wilkins, has already found that fibres of desoxyribose nucleic acid derived from material provided by Professor Signer of Bern gives remarkably good fibre diagrams. . . . As you no doubt know, nucleic acid is an extremely important constituent of cells and it seems to us that it would be very valuable if this could be followed up in detail.

There's ambiguity and to spare peeping out from that "consideration and discussion with the senior people concerned" and "followed up in detail." But the letter proves, especially in the light of its timing, that Franklin was not originally brought in to work on DNA in particular, and that when she was assigned the stuff, she had good reason to think—"only yourself and Gosling"—that she headed an independent team. The letter is the only surviving documentary evidence of the terms on which Rosalind Franklin believed she came to King's.

Not long after that, at the turn of the year, Franklin came to a meeting with Randall at King's College, where she met Stokes and Gosling. Wilkins was not there; he once told me that he had been on holiday with his fiancée. In a conversation, Gosling recalled the meeting. "Had Maurice been in the lab, he would have been at that meeting," Gosling said. "All sorts of things might have gone differently. Rosalind Franklin struck me as an intense person. She was medium height, slim, rather dark—with a fairly determined sort of expression. At the meeting she was quiet, slightly nervous, and trying to ask all the right questions—I can see now. At the meeting, I was in no doubt that what Randall was saying was that here was the problem, here was a research student—because Ph.D. students were serfs in those days, and I was definitely being handed over with the problem—here were the x-ray photographs, there are lots of spots on them, get some more, and solve the structure of DNA from the x-ray-diffraction pattern. And that's what Rosalind from then on tried to do."

❖ ❖ ❖

Biology is not the business of the Cavendish Laboratory. For more than a hundred years, the Cavendish Professorship has been the chair of experimental physics in the University of Cambridge. The man in that chair rules the university's research in physics. Indeed, for most of that hundred years the Cavendish Professor was preeminent in British science, with an authority that made him, as it were, the archbishop of physics—an authority variously doctrinal, inspirational, administrative, pastoral, political, or secretly persuasive in the halls of government, all depending, as with Canterbury or York, on the talent and temper of the man himself. In 1974, the Cavendish Laboratory traded the buildings in Cambridge which it had occupied for exactly a century for a new physics factory, functional and uninteresting, out on the fenland a mile west. When Watson came to the Cavendish in September 1951, it was dead central, a heterogeneous heaping-up of Victorian and interwar brickwork, gray and sallow yellow—and, if truth be told, functional and uninteresting. No tablet marks the place where J. J. Thomson discovered the electron. Room 103 in the Austin Wing bears no indication that here, in the southwest corner, stood the original model of DNA. But the archway into Free School Lane is just a minute's walk, through a lacework of lanes, past shops and pubs and glimpses of walled gardens, from the town square where the medieval open-air market still thrives, and equally close to the great colleges along the tame river. On the gravelled paths by the river—but not on the college lawns, mown to the texture of a billiard table, jewel-green, reserved to fellows, not for the rougher feet of foreign visitors or of persons *in statu pupillari*—Watson and Crick used to walk most days after lunch. The Cavendish in the generation before theirs, before the second world war, had grown to be the best laboratory in the world in the physics of atomic particles. The Cavendish Professor who made it so, Ernest Lord Rutherford, a nuclear physicist of genius who predicted the neutron and was first to smash an atom, and a single-minded administrator who ran a dozen brilliant men like boys in school and he the headmaster, was the most powerful man of science in the world. Rutherford died in mid-stride, in 1937. As the question of his successor was being considered, war was obviously coming. But it was not evident that war would transform nuclear physics from a science fit for gentlemen into heavy industry. Most scientists supposed that the chair would go to another nuclear physicist.

The fifth Cavendish Professor, however, was William Lawrence Bragg, who was an adept of what he often described as a specialized branch of optics. Bragg, when Watson was introduced to him by Max Perutz in 1951, was sixty-one, ruddy, tweedy, burly, with a white moustache, and, since 1941, Sir Lawrence: he wore the archetype comfortably, and only his wife called him Willy. Watson thought of him, or wrote that he did, as the establishment administrator of science, ineffectual, largely irrelevant except when something needed to be fixed up over lunch at the Athenaeum. It is hard to believe that

Watson did not know better. Bragg was still working at and publishing research, and was still good at it. He was the only scientist Max Perutz ever knew who could take the notes and data home with him in the evening and come in the next morning with a paper written out in longhand without blot or hesitation, publishable as it stood—"rather like Mozart writing the Overture to *The Marriage of Figaro* in a single night." Although Bragg had liabilities as a leader of science, ones he recognized, these would have been out of Watson's view. Bragg was a highly personal investigator and, it's generally agreed, no great administrator. "His consuming interest was in the nature of things," Perutz once told me. He had worked with Bragg for years and loved him. After the war, Perutz said, "There was a very strong feeling that Bragg was doing nothing to restore the Cavendish as a great center of nuclear physics. Bragg was not a powerful persuader of men, not a person who could sit on committees and get his own way—and get the money. But about nuclear physics I think people were harsh. His hands were tied." The fact was that war had dispersed Rutherford's brilliant Cavendish team, while nuclear physics itself emerged with its organizational scale—and its price—bloated beyond Cambridge's reach.

Bragg had the quality most essential in the leadership of a great laboratory: an instinct for the grand lines of research that were going to pay out next, and for the men to do them. Watson, today, is credited with something of the same gift. Though he has put his name to only one, minor scientific paper since 1962, his Harvard laboratory had an admired record because, observers say, he was an unusual judge of where molecular biology is moving and could tell his graduate students and postdocs what experiments should be profitable. But Bragg had that instinct on a spectacular scale. He fostered at the Cavendish what were to prove the two most important directions in science of the thirty years following the second world war. One was radio astronomy. At Bragg's encouragement, Martin Ryle, a young physicist who had worked on radar during the war and had become curious about operators' reports of signals from the sun and the Milky Way, built the first radio telescope: this led to the discovery of quasars, pulsars, and neutron stars and the accompanying upheaval in cosmology which is far from settling yet. Bragg's other winner was, of course, in recognizing the importance—the very possibility—of structural investigation, by x-ray-diffraction methods, of the large molecules that interest biologists. When Bragg came to Cambridge in 1938, Desmond Bernal moved in turn to Birkbeck College, London, taking with him all that was being done at the Cavendish in crystallography with biological molecules—leaving only Perutz, a diligent young Austrian, and his hemoglobin. Perutz remembered his first meeting with Bragg, and wrote about it thirty-two years later: "I waited from day to day, hoping for Bragg to come round the Crystallographic Laboratory to find out what was going on there. After about six weeks of this I plucked up courage and called on him in Rutherford's Victorian office in Free School Lane. When I showed him my X-ray pictures of hemoglobin his face lit up." Bragg's last paper, published in 1958, was a method for calculating the preliminary but most difficult key coordinates of atoms within the protein molecule, to unlock the mass of diffraction data that Perutz had devised the way to obtain. Perutz was with Bragg when he put the method to the test. "As the correct values . . . emerged from his calculations he realized that the problem of protein structure, that seemingly hopeless venture he had backed against all odds for the past eighteen years, could now be solved, and tears of emotion streamed down his face."

"People thought we were very foolish to dream we could ever do it," Bragg said in a conversation at his London flat, one afternoon late in January of 1971, several months before he died, at the age of eighty-one. "It looked absolutely impossible, when you thought that the most complicated structures ever yet got out were for molecules only about two hundred atoms big. The difficulty goes up, you know, not just as the square but as a high power of the number of atoms in your molecule. With hemoglobin we were trying ten thousand atoms." Photographs of Bragg's youth had shown him dark and startlingly handsome; now his face was slightly out of focus with age, he was wearing carpet slippers, and he would not be able to talk for long. We sat before a gas fire in a generous Victorian drawing-room that looked over a curve of garden to a mossy brownstone church. "Remember, I'm *not* a biologist," Bragg said. "I'm an x-ray crystallographer. My part, my great interest, has been in seeing how one could push x-ray analysis to do more difficult problems of structure. What that structure means biologically, what little I know about it, I only learn through Perutz and Kendrew and people like that. This—art, let's call it, of finding out what the structure is actually like, by purely physical means, optical means, is the thing that I've been interested in, all my life."

I asked the obvious question. "My first x-ray analysis was in 1912," Bragg said. "It was when Max von Laue, in Germany, published his discovery that diffraction effects were produced when a beam of x rays falls on a crystal. Then to explain the symmetrical spots on the photographic plate"—they were looking at the cubic crystal of the mineral zincblende—"Laue brought out a theory of the way this diffraction was produced." X rays had been discovered in 1895; in 1912 it was still not quite settled whether they were particles, like the electron, or waves. Von Laue saw that the diffraction effect showed the wave nature of x rays; but he could explain the complexity of the diffraction spots produced by a simple cubic crystal only by supposing that the x rays were of at least five different wavelengths. Bragg had just graduated from Cambridge with a first in physics; he was to go on as a research student under J. J. Thomson, the then Cavendish Professor. Bragg's father was professor of physics at Leeds and himself a leading experimenter with x rays. "In the long vacation my father showed me Laue's paper, in an obscure German journal." When Bragg returned to Cambridge, "On thinking over Laue's explanation I said, 'This isn't the way they came about; it's really much simpler.'" He visualized the true explanation suddenly one day, walking through the courts of St. John's College. "Laue thought the peculiarities of the pattern were due to peculiarities in the x rays. I said, '*No!* What this is trying to tell you is, the pattern is due to peculiarities in the structure.' And that led on to seeing what the particular structure was, the atomic arrangement in the crystal."

The x rays used in crystallography are shorter than visible light by some four thousand times. Their wavelength is about 1.5 angstroms, which is the same order of size as the spacing between atoms in many solids, and that is why x rays go through substances that are opaque to the eye. Penetrated by a thin beam, Bragg said, the orderly array of atoms in a crystal, layer upon layer, scatters the x rays in an orderly way, causing a repeating series of overlapping circles of waves. At intersection with the flat sheet of film, the troughs and peaks of these waves reinforce each other at some points and cancel each other out elsewhere. The interference pattern that results is characteristic of the structure that produced it, though to figure back to the structure takes mathematics and experience.

"I first worked with Laue's pictures of zincblende. I found the diffraction pattern was explained by the presence of atoms not only at the corners of the cube but also at the centers of the faces. The first pictures I took myself were with sodium chloride and potassium chloride. And there I got the whole structures out. This was the start of the x-ray analysis of molecular structures. And then of course to me it was an enormous challenge, an enormous triumph, to be able to get out the actual molecular arrangement in a biological structure. It looked absolutely impossible. But it so happens that you go on worrying away at a problem in science and it seems to get tired, and lies down and lets you catch it. But my part has been to devise ways of getting the structure out—and that is all a part of optics, of the interference of light, of waves of radiation. And you see, that's why I was interested in Ryle's work, too. Because as you know, radio astronomy is entirely interferometry. It's just a matter of scale."

Bragg was a man who could visualize in space; his sight was as keen and quick as any scientist's since Leonardo. If the prodigious scale of his vision—outwards to the structure of the universe, down and inwards to the structure of life—was not yet altogether evident by the time Watson first met him, one other fact, though Watson fails to mention it, surely bit deep. William Lawrence Bragg was, as he remains, the youngest person ever to win a Nobel prize. The prize was for the invention of x-ray crystallography and the first solutions of crystal structures, beginning with common salt. In 1915, when he received the award, Bragg was twenty-five.

After the war, Perutz, "alone with my hemoglobin crystals," puzzled over the smeared polka dots they produce on x-ray film, in a corner of a laboratory three floors up in one of the newer Cavendish buildings. John Kendrew was the first to join Perutz. Kendrew began by investigating, for his doctoral dissertation, the differences between the hemoglobins of fetal and adult sheep. The substances Perutz's group studied were biological. The money came from the Medical Research Council—an arrangement Bragg had fixed up in 1947, over a lunch at the Athenaeum. The methods and the men were drawn, like Bragg, from the orthodox physical sciences. Perutz was a chemist. Kendrew was a chemist. Francis Crick, when he came to the Cavendish in 1949, was an overage graduate student whose work for a doctorate in physics had been interrupted by the war and seven years in the British Admiralty designing mines, and devices for detecting them, and new mines that would evade the devices.

Francis Harry Compton Crick, in his early sixties, is tall, lean, slightly stooped; he has thinning sandy-gray hair and vigorous reddish eyebrows, and his eyes are sapphire-blue, the irises ringed. His mouth is thin, flexible, amused; his gestures in talk are vivid, and his light voice is full of emphasis and verbal gesture, sometimes almost arch. Those who don't know him, even those outside science, know of his shrill laughter, his loquacity (both characteristics he had even as a schoolboy), and his arrogance in telling his colleagues their scientific business, for these are the qualities that Watson, his friend, made notorious in *The Double Helix* the way a cartoonist will seize a politician by the nose. It should be said, then, that Crick has to a pronounced degree a habit that most scientists cultivate, a careful and constant self-depreciation that distinguishes between what a man knows for sure and what he knows only speculatively or by hearsay. "I do not speak now from secure knowledge," said one, expressing a typical care with a graceful turn; or as Crick sometimes puts it, "I *don't* know this subject, I only know *about* it." The habit has little to do with modesty. It is a hedge against being caught in error. More, it is an intellectual courtesy, a running contribution to evaluation of the weight of statements in truth-seeking talk.

Talk is Crick's life in science as it is to few others, for he has deliberately taken on a singular role in molecular biology: Francis Crick is the theorist. When I have heard him, over the lunch table, explain a new idea to his peers, his reminding them of details they should already know has seemed an almost absent-minded orderliness of exposition, just to save later backtracking. He has not always struck everybody so. "I used to get fiercely irritated at Crick, I'll say that plainly," Bragg said to me. "I realize now I ought to have been far more philosophical about it, not got so annoyed. But the sort of thing was— I remember one occasion when Perutz and I were worrying about the results he was getting on hemoglobin, I came in one morning very excited with an interpretation to suggest to Perutz; I mentioned a certain optical principle, and remember Crick coming in, rather uninvited—because Perutz and I were having a private talk on a point we were very excited about—and listening to us and then saying, 'I must go away and see if you're right.' I went off the deep end. Crick was always— You see, if a man had been sweating away at research for some months, and then might say to himself 'Now I'll have a little rest over the weekend, and I'll come in next week and think what these results mean'—Crick would be very likely to come along on Monday morning and *tell him* what they meant. Like doing someone else's crosswords, you see. Notwithstanding the fact that *of course* he is a great genius. He really is. He reads voraciously."

Crick is to an unusual extent self-educated in biology. He went to a minor English public school, Mill Hill, in northern London; his interest in science was already so single-minded that his family thought him odd. At University College, London, he read physics and had nearly finished his doctorate when the war broke out and a German bomb destroyed his laboratory and gear. When Crick left the Admiralty and physics in 1947, he set out to master the literature of biology, reading with an appetite that has slackened, if at all, only in the last few years. His peers concede without question his astonishing reach. Perutz, whose knowledge is encyclopedic in scope and order: "Francis of course reads more widely than the rest of us." Jacques Monod, the science's other great theorist: "No one man discovered or created molecular biology. But one man dominates intellectually the whole field, because he knows the most and understands the most. Francis Crick."

An important reason Crick changed to biology, he said to me, was that he is an atheist, and was impatient to throw light into the remaining shadowy sanctuaries of vitalistic illusions. "I had read Schrödinger's little book, too. Essentially, if you read that book fairly critically, the main import is very peculiar; for one thing, it's a book written by a physicist who doesn't know any chemistry! But the impact—there's no doubt that Schrödinger wrote it in a compelling style, not like the junk that most people write, and it was imaginative. It suggested that biological problems could be *thought* about, in physical terms—and thus it gave the impression that exciting things in this field were not far off. My own motives I never had any doubt about; I was very clear in my mind. Because when I decided to leave the Admiralty, when I was about thirty, then on the grounds that I knew so little anyway I might just as well go into anything I liked, I looked around for fields which would illuminate this particular point of view, against vitalism. And the two fields I chose were what we would now call molecular biology, though the term wasn't common then, certainly I didn't know it—but I would have said the borderline between the living and the nonliving. That was the phrase I had in my mind, on the one hand. And on the other, the higher nervous system and this problem of consciousness, whatever that may mean. And I had a period of some—weeks, maybe longer, trying to decide between these two. I eventually decided on what we now call molecular biology simply because I thought what I knew, as a physicist, was more relevant! But as you know, in recent years, we're now edging towards the nervous system." In the spring of 1947, in his application to the Medical Research Council for a research student's grant, Crick wrote:

> The particular field which excites my interest is the division between the living and the non-living, as typified by, say, proteins, viruses, bacteria and the structure of chromosomes. The eventual goal, which is somewhat remote, is the description of these activities in terms of their structure, i.e. the spatial distribution of their constituent atoms, in so far as this may prove possible. This might be called the chemical physics of biology.

It might be called a definition of molecular biology.

Crick asked about doing research with Bernal, at Birkbeck, but was turned away. In the fall of 1947, Crick came to Cambridge, but to the Strangeways Laboratory, where he did a piece of cell research that had little structural or any other relevance. More importantly, he soon got to know a wide variety of other Cambridge scientists and their work.

One of these in particular, the biochemist Frederick Sanger, came to have great intellectual importance in Crick's thinking and then to molecular biologists generally as the field developed. Sanger is temperamentally and in scientific style Crick's opposite. Where many scientists, Crick among them, flower at conferences and do a great deal of their science by talking, Sanger is a quiet man—reticent, even shy, a man who worked with his hands, at the bench. He almost never talked to the press, never despite the editor's importuning wrote the big article for *Scientific American*. One might spot him bicycling to work on a spring morning, in a drab brown coat, in the rain. Once I stopped to talk with him in the corridor of the laboratory building, where he was waiting in the queue for his turn at the ultraviolet-light box, in order to illuminate the spots on a sheet of chromatography paper he was holding. Sanger is a Quaker by upbringing, and stayed at Cambridge through the second world war; holding only a junior fellowship in the biochemistry department, and even when the war dried up the usual sources of research funds, with family money he was able to keep on working. In the course of nearly a decade, beginning in the mid-forties, Sanger settled upon the new techniques of chromatography to determine the amino-acid sequences of the two chains of the bovine insulin molecule. He proved that the sequences are unique and always the same, meaning that every molecule of insulin in every cow is exactly like every other. Yet the sequences show no general periodicities: they are not predictable from ordinary chemical rules.

Sanger published fairly rarely. His papers came to be read with heart in mouth by other scientists, for they are technically brilliant. Even as he worked, though, the news slowly spread and the implications sank in. For one thing, his department held a biochemistry tea club where perhaps once a month research that was relatively finished, though not yet submitted for publication, was presented. Brigitte Askonas, later an important figure in immunology in England, came to Sanger's lab as a doctoral student late in 1948, staying on into 1952. "Even then, Fred had only a minor fellowship—and some had wanted to kick him out," she told me once. "When one would ask him how his work was going, he would say very little. 'Oh, I've got another peptide.'" Then at a lab meeting he would bring a stack of cards showing overlapping short sequences, and slowly, diffidently, build up his latest segment of the molecule. "Crick always came to the tea club," Askonas said. "And he always asked awkward questions. *Enfant terrible* questions. And then he would explain, somewhat disingenuously, 'You see, I'm just learning.'" Sanger's general conclusion was forceful by 1949, when he went to the annual symposium on quantitative biology at Cold Spring Harbor (his only such visit). In a paper published on the first of June of that year—the earliest of his magisterial series of papers on insulin appearing every odd-numbered year until 1955—he was already able to say that "there appears to be no principle that defines the nature of the [amino-acid] residue" occupying any particular position in a protein. The conclusion was definitive by 1951. For this work and the methods of sequencing he invented to do it, Sanger was awarded the Nobel prize in chemistry in 1958. (He later turned to the more difficult problem of sequencing nucleic acids, which earned him a share in another Nobel prize, in 1980.) Crick, from his first arrival in Cambridge, knew of Sanger's work step by step, months and even years before new steps were published.

Crick soon learned of the biological work being done by Perutz's small group at the Cavendish. Perutz, thinking back on their first meeting, recalled with amusement that Crick's approach was shy. "At first he sent a messenger, a mutual friend, a mathematician called George Kreisel, who tried to find out whether we might have room for a physicist interested in biology—in a very noncommittal sort of way, not revealing who this person was. And after these preliminary coy inquiries, Crick appeared." What was Crick's reputation as a scientist? "Well, he hadn't got a reputation. He just came and we talked together and John and I liked him." Kreisel had in fact first come to Perutz without Crick's knowledge. Perutz brought Crick to Bragg's attention, invited him to join the hemoglobin group, then got him a staff appointment attached to the Cavendish but paid by the Medical Research Council.

Crick moved over in 1949. He was thirty-three. Perutz's next task was to get him permission, even though he was a member of the Council's research staff, to register as a Ph.D. student. "The first thing, when I went to the

Cavendish, I had to teach myself protein crystallography," Crick told me. "The first thing Francis did was to read everything we had done," Perutz said. "Then he started criticizing."

Crick arrived as Bragg, Perutz, and Kendrew were working up the disastrous paper on protein chains they published the next spring. About the end of his first year, Crick invited his senior colleagues to a seminar which he opened with a twenty-minute monologue on the deficiencies of their methods and the hopeless inadequacy of their hemoglobin molecule. "What Mad Pursuit" he titled the seminar, at Kendrew's suggestion. Many years later, he gave the same title to a slim, sketchy volume of autobiography; recalling that seminar, he wrote, "The subject of my talk . . . broadly speaking, was that they were all wasting their time and that, according to my analysis, almost all the methods they were pursuing had no chance of success. . . . Bragg was furious." Indeed, this was one of the Crick performances that drove Bragg off the deep end—except that Crick was right, and Bragg's own analysis soon confirmed, that the evidence they had already collected was enough to prove that proteins were far more intricate than the geometrically regular models—the "hatbox," the "cigarettes in cigarette box," and other packings of the protein chains—which they had proposed in the past several years. Restless, inquisitive, looking for simplifications rather than the relentless detail of crystallography with large molecules, for more than two years Crick moved from problem to problem, doing other men's crosswords but not finishing his own.

Yet ever nagging at the back of his mind, he remembered, was the question not of DNA in itself but of the relation of gene structure to protein structure, how the one could dictate the other to build the organism. This question was more abstract, yet more precise and more fundamental, than the preoccupations of the rest of the group. Crick kept wanting to look under stones they couldn't lift yet. He made an uneasy fit. "All of us at that time were strictly structure-oriented," Perutz had told me. Crick took up the point. "They were aware of the biological implications," he said. "But they were really thinking about proteins, Bragg especially, as the next biggest interesting unsolved molecule that you could attack by x-ray diffraction. Now, at an early stage, even before I joined them, I came to the conclusion, first, that the self-replication of genes was important; and although I did *not* realize that the genetic material was pure DNA—you remember that in those days this wasn't altogether clear—I conceived secondly the idea of the co-linearity of the genetic material and the amino-acid sequence of the protein. I had already planned, I mean I had conceived, the genetic type of experiments you would have to do. To show that what was in the protein chain, bit by bit, had in some sense to be in the gene. Bit by bit. What I appreciated was that genetics was the key part of biology—and that one had to explain genetics in structural terms."

Crick needed a catalyst. Neither Watson nor Crick remembers for certain their first meeting, sometime in the first week of October 1951, at the Cavendish. "I came home one day," Crick said. "We were living in a little flat in the center of Cambridge in those days, and my wife said to me, 'Oh, Max was round here with a young American, and do you know, *he had no hair!*' What she meant was, he had a *crew cut.* You might say that was the first time I *didn't* meet Jim. And then we must have met, and I don't recall exactly the moment we first met; I remember the chats we had over those first two or three days in a broad sort of way." Mr. Crick was thirty-five, Dr. Watson twenty-three. As Watson had done before with Luria and Delbrück, he was able once more to create almost instantly a mutual intellectual trust with an older scientist of brilliance, which was free of the scathing competitiveness most colleagues his own age have felt. Crick and Watson were soon lunching almost every day on the shepherd's pie or sausages and beans at the Eagle, a more than usually grubby and picturesque pub tucked into a cobblestoned courtyard a crooked block north of the Cavendish. A framed magazine clipping hung until lately in one of the pub's rooms to memorialize those lunches. The crystallography labs were so small, as those who were there recall with some nostalgia, that "we couldn't help falling over each other, always having coffee and tea together, and we all discussed everything all the time that everybody else was doing." At the time Watson appeared, Perutz and Kendrew shared a narrow office, with a single window, the way to the corridor leading through a similar room where various assistants sat. They were one flight up; Bragg had a large office on the floor above, and the x ray machines were in the basement. The walls were brick, painted yellow, with sturdy wooden slats running up them at intervals to which equipment could be bolted. About the time Watson arrived, the unit was given an extra, larger room next door. "Max and John were the senior people," Crick said. "I was just a research student, and Jim was just a visitor. They came in one day and said, 'Well, we've decided what to do about that room, we're going to put you and Jim in the room together so you can talk to each other and not disturb the rest of us!' So clearly we must have got the reputation by that time of rather talking together a lot." They shared the room, cluttered but comfortable, with a succession of third and fourth roommates.

Watson was an exotic addition to the Cavendish. Bragg liked his enthusiasm and its effect on Crick. The two men complemented each other in the kind of many-layered and energizing way that marks great collaborations whether of scientists or of songwriters. Temperament, to begin with. The historian Richard Hofstadter once usefully set out the essential attitudes of the intellectual towards ideas as playfulness and piety: Watson and Crick were both deficient in piety, perhaps, but devoted to the rowdy seriousness of the mind at play. Crick still is. They evidently fell into a kind of intellectual crush on each other, though Watson's mocking self-awareness, fifteen years later, diminishes that: "Finding someone in Max's lab who knew that DNA was more important than proteins was real luck." Watson brought with him a body of knowledge about genetic function, driven down by the phage group to a fine scale, which was almost entirely unfamiliar to the structural analysts. "Watson arrived knowing nothing of our work, as he admitted in his book," Perutz said recently. "But what he fails to say is that we—except maybe Crick, who had read more widely—we knew nothing of *his* work. His arrival at that time was extraordinarily opportune."

Two schools of research that until then had been almost isolated from each other met and seethed in reaction: broadly, the English and the American, the concretely structural and the abstractly functional approaches to biology at the level where chemistry merges into physics. The two approaches had previously come together significantly only at one place: the California Institute of Technology, with Pauling driving the one and Delbrück the other. Yet Crick and Watson were uncharacteristic of the traditions they represented. They were juniors, of the second generation. They were, each in his way, still in transit, not settled in. They were off center, each already prepared to be interested in the approach he found the other brought to their encounter. No other member of the phage group had the structural curiosity or the nagging intuition of the importance of solving DNA that Watson had developed. No other crystallographer was so radically interested in the simple, root questions of the nature of life and so wide-ranging in his pursuit of clues as Crick had become.

Crick was lifted out of himself on meeting Watson; in his own word, "electrified." The two have had their differences since. Crick was, Bragg said, one of those in England who were "really very fiercely against" *The Double Helix* and who urged him to try to stop its publication. Yet Crick's resentment has cooled to humor, and nearly a quarter-century after the meeting he still recalled it with a volatile and generous excitement. "When I met Jim, it was *remarkable*, because we both had the same point of view, but he knew all about phage, which I had only read about in books, you see, and I knew all about x-ray diffraction, which he only knew about secondhand. But it was a remarkable thing to find somebody—you might say that he was the first person from the outside world almost that I, as a new boy, had met who reinforced my own sense of what was important." Was Watson the catalyst, as Perutz and the others had said? "Yes. I think so. But that's because he is very imaginative and he has bright ideas." From those first delighted conversations, they amplified each other's not-quite-certainties. That DNA was pivotal. That solution of its structure would be a sensational coup—and fundamentally explanatory. That somebody, Pauling perhaps or Wilkins, was going to stumble on the solution very soon, but that they could get there first by borrowing Pauling's model-building method.

Watson and Crick made two separate attacks on the structure of DNA. Though the first ended in disaster while the second was a success even more brilliant than they had told themselves it would be, in development the two tries ran parallel. Scientific problems have their unspoken human parameters. Each time Crick and Watson tried it, solving DNA broke down into unknowns to be filled in and outright misconceptions to be overthrown—a gothic landscape of the mind where the solidest boulders might be phantasms and the great citadel loomed now ahead and now behind in the flicker of the light. The uncertainties were compounded by the fact that in the opinion of British scientists, Maurice Wilkins and Rosalind Franklin had, besides their scientific head start, a clear moral priority. "You understand first of all Wilkins was a personal friend, I had got to know Wilkins while I was still in the Admiralty," Crick said. "Not only that, but their lab was a sister unit to ours, of the Medical Research Council." The Council's grants effectively institutionalized Wilkins's priority. But for a pragmatic mind, scruples were features of the shifting landscape.

When Crick and Watson began, they knew very little about DNA for sure, and part of what they were most sure of was wrong. To consider DNA as a physical object, they wanted diameters, lengths, linkages and rotations, screw pitch, density, water content, bonds, and bonds and again bonds. The sport would be to see how little data they could make do with and still get it right:

the less scaffolding visible, the more elegant and astonishing the structure. More than sport was involved. Crick, following Pauling, elevated this penurious elegance into a theoretical principle, the corollary of model-building. "You must remember, we were trying to solve it with the fewest possible assumptions," Crick said. "There's a perfectly sound reason—it isn't just a matter of aesthetics or because we thought it was a nice game—why you should use the *minimum* of experimental data. The fact is, you remember, that we knew that Bragg and Kendrew and Perutz had been *misled* by the experimental data. And therefore every bit of experimental evidence *we* had got at any one time we were prepared to throw *away*, because we said it may be misleading just the way that 5.1 reflection in alpha keratin was misleading." We were in his office in Cambridge; thinking out loud, he got up and began to pace back and forth, with long, loping steps, in the clear lane in front of his desk, speaking in the rhythm of his stride. "They missed the alpha helix because of that reflection! You see. And the fact that they didn't put the peptide bond in right. The point is that evidence can be unreliable, and therefore you should use as little of it as you can. And when we confront problems *today*, we're in exactly the same situation. We have three or four bits of data, we don't know which one is reliable, so we say, now, if we discard that one and assume it's wrong—even though we have no evidence that it's wrong—then we can look at the *rest* of the data and see if we can make sense of *that*. And that's what we do *all the time*. I mean, people don't realize that not only can data be wrong in science, it can be *misleading*. There isn't such a thing as a hard fact when you're trying to discover something. It's only afterwards that the facts become hard."

Among the facts they started with were some x-ray-diffraction patterns. In 1947, Astbury had published one of DNA, previously unseen though it was the best that had been taken before the war, in support of a new, vigorously argued attempt at the structure. But the new picture was no substantial advance, while his interpretation was a giant step backwards into a monotonous repetition of the four nucleotides. The picture still showed the 3.4-angstrom spacing between nucleotides as well as the large structural repeat of some sort at 27 angstroms. The nucleotides, Astbury said, were stacked "immediately on top of one another like a great pile of plates and are not disposed spiralwise round the long axis of the molecule." The sugar components, he thought, lay parallel to the flat bases. Astbury had also figured the density of dry DNA, and found it to be high. This value, if reliable, could be taken together with the dimensions of the DNA molecule to find how many chains of nucleotides the molecule contained.

The number of chains was a crucial question, still unanswered when Watson and Crick began. A mystery only slightly less evident and fully as perplexing was whether in a two-chain or three-chain structure the backbones would be on the inside or the outside. The diffraction patterns that Wilkins and Gosling had got at King's College London, in 1950, were far better for such problems than Astbury's—or at least, they were in the rich promise of information from those crystalline spots. Patience, a better camera, greater skill, some varied preparations of the best DNA—but Watson and Crick had none of these. Rosalind Franklin had them.

Besides the diffraction data there was all that biochemistry. Watson and

Crick had then hardly heard of Chargaff, knew nothing of his ratios. Since the early nineteen-thirties, chemists had been accumulating clues to the links, which atom to which, that tied sugar-phosphate-base together into the single nucleotide and then into a string. About the time Watson came to the Cavendish, the final statement of the chemistry of these linkages in the nucleic acids was being written in another laboratory a block away; this was published early the next year as the tenth in a series of papers by Alexander Todd and various colleagues. Todd settled how the backbone hooks up: third carbon in the sugar ring linked to the phosphate and that in turn linked to the fifth carbon of the next sugar, and so on, with no branching. That known, it was conceivable that DNA could be approached as Pauling had visualized the alpha helix, by ignoring the identities of the side groups—the bases, now— and thinking about the main chain. Move up, give a little twist, move up again, with another little twist—a translation accompanied by a rotation, Pauling had explained to the world, and Crick saw that self-evidently with DNA, too, the result would be a helix of some sort. What sort?

Just this path had been taken several years earlier by Sven Furberg, a young Norwegian crystallographer studying under Bernal at Birkbeck College, London. Furberg, reasoning with marked brilliance and luck from data that were meagre but included his own x-ray studies, got right the absolute three-dimensional configuration of the individual nucleotide: where Astbury had set sugar parallel to base, Furberg, in what he called the standard configuration, set them at right angles. As a structural element, that standard configuration was a powerful help. "Furberg's nucleotide—correcting Astbury's error—was absolutely essential to us," Crick told me. Furberg went on to draw a couple of models of DNA, one of which was a single chain in helical form with the bases sticking out flat and parallel to each other, rising 3.4 angstroms from one to the next, eight nucleotides making one complete turn of the screw in about 27 angstroms. Plausible physically, this helix had too little in it: it failed to account for the density of DNA. Furberg stopped building models and published his results in June of 1949—in his doctoral dissertation. He deposited several copies in the Science Library in London and went home to Oslo.

Over the next three years, Furberg's results appeared piecemeal in a series of papers. From his thesis, his models were well known to Randall's group at King's College. Before Franklin arrived, someone there constructed a nucleotide and a half of a DNA model based on Furberg's helix; Randall discussed and displayed this in his Royal Society lecture on 1 February 1951. Wilkins knew of Furberg's models, too. Franklin owned a copy—made on glossy photographic paper in the days before photocopying was a familiar practice—of the nine pages of Furberg's thesis where he proposed the correct configuration of the nucleotide and put forth his models. Her files also contained reprints of his papers about nucleic acids, and sheets of notes on them. Otherwise, Furberg's models remained almost unnoticed—even by Bernal, who wrote, in 1968, that they had contained "the key to the whole double helix story" and blamed himself for letting "the opportunity slip." Furberg at last got his helical model into print in *Acta Chemica Scandinavica* late in 1952, in time for Watson and Crick to cite it in the notes to their announcement of the successful solution the next spring. Furberg and Franklin did not meet, he recently wrote, until July 1953, in London, where she questioned him closely on his ideas about DNA. At some point, though, he sent her a reprint of a paper of 1950, signed in ink and with a note in pencil: "Hope you have been able to interpret your beautiful fibre diagrams of Na-thymonucleate."

Considering DNA as a physical object, Watson and Crick could not abstract it altogether from the organism. They could never take it for granted that in the wet of the cell, that tiny watery bag of enzymes and energy, DNA in action was much like the stuff that had been extracted and weighed and put under the microscope and examined with ultraviolet light and x-rayed. Was DNA fixed in form at all? Were the x-ray patterns an artifact of the preparation methods? How intimate was the role of protein? Exactly such worry had rationalized Wilkins's long vacillation between further x-ray diffraction of fibres and more work with sperm heads. Considering DNA as a physical object, Watson and Crick had to wonder, overriding all else, how function followed form. Even Pauling's alpha helix had proved sterile and anticlimactic in this way; shortly before Watson left Copenhagen, he had talked to Delbrück, come over from Pasadena for a conference, but Delbrück had turned away Watson's excited questions with boredom, for "the alpha helix, even if correct, had not provided any biological insights." Watson has written since, again and again, of the fear he and Crick felt that the structure of DNA might turn out "very dull" and suggest nothing about how the gene replicates nor about how it controls the chemistry of the cell.

The first assault on the structure began, at least in talk, within days of Watson's settling in to the Cavendish and to the small bedroom that Kendrew and his wife offered him in their house, a block away from the lab. The attempt went through its complete cycle—speculation and preoccupation, a growing sense of competitive pressure, a sudden injection of new data from King's College London, a frenzy of model-building, the conviction of success, an elated effort to persuade others of the model's beauty—in six manic weeks. The model collapsed in ruins before the end of November. The talk that began it must have swarmed like a troop of inquisitive lemurs over every topic in the forest—but what the participants have remembered is the exhilaration, the intensity, not exactly what was said. A couple of events turned the talk serious. Crick invited Wilkins to Cambridge for the weekend of November 9–11. Wilkins came along on the Friday evening to a meeting of the Hardy Club—an elite scientific discussion group that flourished in Cambridge then—where Crick gave a talk on the alpha helix. Later Watson met Wilkins, for the second time. The three men talked about DNA and helices. Wilkins also spoke, doubtless at length, of the trouble he had begun to have with Rosalind Franklin. He had seen little of her that spring and summer. In July, they had both come to Cambridge for a conference Perutz had called to discuss protein structures. Wilkins had taken no new x-ray-diffraction patterns of DNA fibres since the year-old ones he had shown in Naples in May, though he had got some patterns from thin sheets of DNA. These were not richly informative, but they showed, like the fibre pictures, a crossways configuration. Following a conversation with Alexander Stokes, the theoretician in his lab, Wilkins had pondered his way from this configuration back into the molecule and thought it likely to be helical. He said so at Perutz's meeting. While in Naples, Wilkins had obtained samples of the sperm of squid, which occurs in bundles remarkable for the perfection of the parallel array of cells, and at Perutz's meeting he talked about x-ray studies of these that showed, he believed, that DNA in intact cells was helical, too. After the Cambridge meeting, in Wilkins's remembrance, Franklin told him that she resented his continuing to do x-ray work with DNA. Wilkins had spent the end of summer in the United States. Erwin Chargaff had given him a fresh supply of DNA. Wilkins had come back to King's to find that Franklin refused his collaboration, resented his advice or comment, made discussion very awkward. A colloquium on DNA had been called for mid-November. Watson asked if he might go to it.

Ten days earlier, Perutz had got a letter from Vladimir Vand, a crystallographer at the University of Glasgow, who had worked out, he thought, the mathematics by which one could say exactly what x-ray-diffraction pattern will be produced by helical models of any given structure and dimensions. Vand had seen Perutz's note in *Nature* about searching out the 1.5-angstrom spot produced by the alpha helix, and so sent along a draft of a general approach. This was now much needed. Very large molecules in long chains—polypeptides, polynucleotides, polysaccharides—though characteristic of life, are rare elsewhere in nature, and for such chains a helix is a naturally stable configuration. But helical molecules are unlike other substances: despite the intricate geometrical calculations of classical crystallography, at the time Pauling announced the alpha helix there had existed no general way to predict how a helical molecule would betray its presence in an x-ray pattern. Bragg's curiosity had been piqued by this hole in his science, and he and Perutz had asked William Cochran, a young crystallographer at the Cavendish (though not in Perutz's group) to derive the necessary formulas. Cochran hadn't got around to it. Now, as Perutz read Vand's draft, the reasoning didn't seem right. He showed it to Bragg, and to Cochran and Crick. Cochran saw at once that Vand had found the first part of the answer but had then gone astray. Crick puzzled at the equations that morning. After lunch, he went home with a headache, there in his idleness took up the pencil again, and soon saw that he had worked the complete answer. That evening, he went with his wife to a wine-tasting at Matthews', a Cambridge wine merchant. Next morning, when Crick announced that he'd got the solution, Cochran said he had too. They compared. Crick's technique had been clumsy, Cochran's neatly professional, but they agreed on the result. The afternoon headache, the evening wine-tasting, the inelegance of Crick's equations have all become part of the molecular hagiography, like Delbrück's camping trips in the desert.

Independently, several months earlier, Stokes at King's College, with Wilkins's encouragement, had worked out the mathematics of helical diffraction, but had not used it to solve any particular structure, had not even thought it significant enough to write up and publish. Cochran and Crick put their formulation to work at once on the synthetic polypeptides from the Courtaulds laboratory, to demonstrate in more general form than Perutz had done that these fibres were composed of something very like the alpha helix. Crick and Cochran's paper was sent off to *Nature* in mid-December—after a short delay for the rise and fall of a structure for DNA. They took a couple of months longer to finish, with Vand, a complete mathematical presentation of the helical-diffraction theory. The work was Crick's first solid success. Bragg was pleased. Crick said, a while ago, "That was a relatively straightforward bit of crystallography, certainly by today's standards. But of course it did mean that I then had the expertise at my fingertips."

While this was happening, Watson received a letter, forwarded from Copenhagen, from the National Research Council, in Washington. His application to renew his fellowship and transfer to the Cavendish had been turned down. He was unprepared and unqualified to do x-ray crystallography (true enough). The fellowship would be restored if he returned to biochemistry. He was urged to go to the laboratory of Torbjörn Caspersson, in Stockholm, where biochemical research was being done on DNA. Watson wrote to Luria to save him. Luria promptly wrote a persuasive letter to the fellowship board, and a note back to Watson suggesting tactics. Over the next few weeks, with the connivance of half a dozen American and English scientists, a deception was confected: the fellowship board was told that Watson would be working not at the Cavendish, after all, but at a Cambridge biological laboratory, the Molteno Institute, supervised by a biochemist named Roy Markham. Markham's specialty—and Watson's ostensible new subject—was the viruses that infect plants, like tobacco-mosaic virus, and that contain RNA, not DNA. Perutz and Markham between them composed a letter to Washington explaining what Watson would be doing; Perutz got Bragg to write in Watson's support, too. Watson eventually got two-thirds of the fellowship, to run through May.

❖ C ❖   On Wednesday, 21 November 1951, Watson took the train to London for the King's College colloquium. Wilkins had said he would be welcome. There are two very different accounts of that meeting. The first is Watson's, in *The Double Helix*—brief, evasive, and devoted to his failure to take notes, his failure to understand what Franklin was talking about, his failure to remember correctly the little he said he did grasp. Watson's failures were, for his story, all that he found interesting. However, Franklin's version also survives. After her death, Aaron Klug found among her papers a double set of notes in blue ink in her hand: eight sheets headed "Colloquium Nov. 1951" and with them six more pages, much scrappier, which read like preliminary sketching. He also found her laboratory notebooks for the period. The colloquium notes are exactly what one would expect: sensibly, thoroughly organized, somewhat telegraphic in style, much more synoptic by the time she reached the last two pages. They include her stage direction to herself—"[Show photos]." We cannot know how much of the notes she covered at the meeting, though there is no reason to doubt that she presented what she had carefully planned. Stokes and Wilkins also talked about DNA. There was a stranger present: we know Watson stared. "I could not regard her as totally uninteresting. Momentarily I wondered how she would look if she took off her glasses and did something novel with her hair." We are entitled momentarily to wonder what figure Watson, staring, presented to her. Eleven weeks later, she submitted a short report on the first year of her fellowship at King's, and this includes much of the same material in the notes, often in the same language, some things omitted, several things taken further. Notes and report together establish what Franklin thought she knew reliably enough to say outside her lab at the end of 1951.

Franklin, as Maurice Wilkins acknowledged even in the brimstone of his frustration, made one essential discovery about DNA: the fact that its fibres give two distinct types of x-ray diagram. The colloquium took place eighteen months after Wilkins had first found and stopped pursuing the fact that good fibres of DNA, kept wet, yielded diffraction patterns suggesting a crystalline state. Franklin started taking pictures in mid-September. Several weeks before the colloquium, she noticed that when she made fibres even wetter, they stretched somewhat and yielded a new kind of pattern—as she wrote in the colloquium notes, a "complete change in picture towards much simpler." She was not satisfied with the quality of the pictures she was getting from either "Xtalline" or "wet" states, and her notes are full of caution about the need for better fibres and better camera work. Watson heard correctly that she insisted her results were preliminary.

Those early photos of the wet state were of ambiguous promise. The fact that DNA could change to a second form was of itself interesting. Yet even the best of the patterns were cloudy, gray, showing no more than a heavily exposed thick black streak at top and bottom, some vague rings connecting the streaks, and eight or ten smaller blobs, very indistinct, near the center. The contrast with the crisp, complex patterns from the crystalline form demonstrated a considerable breakdown of order within the fibres as they changed from crystalline to wet. Wilkins claimed a quarter-century later that Franklin's early patterns of the wet state "did clearly show helical features," and that when he and Stokes pointed this out to her, she objected to their interference. Her notes for the colloquium say nothing of that: then, her attention centered on the crystalline state, and she didn't make much of the wet state except that it seemed to disrupt the crystalline three-dimensional order and so far gave only a few spots and "smears."

However, she was going ahead on the hypothesis that the structure was helical, in both states. She listed to herself the *"evidence for spiral structure."* First, a straight, untwisted chain would be unstable, unbalanced, "highly improbable." Second, in the x-ray pattern of the crystalline form, the absence of spots or smears on the vertical, up and down from the bull's-eye, suggested a spiral structure. The point was intuitively clear to good crystallographers; exactly that had first led Stokes and Wilkins to think that DNA was helical. Franklin did not know that Crick had just worked out the mathematical explanation; doubtless she should have recognized that Stokes had made the necessary calculations, but in fact there is no evidence that she—or Wilkins, for that matter—put the formal helical-diffraction theory to work until later. Third, the spot that she, like Astbury, found at about 27 angstroms was "much too marked to result merely from diff. betw diff nucleotides, & must mean nucleotides in equivalent positions occur only at intervals of 27 A. Suggests 27 A is length of turn of spiral."

She argued from density measurements, Astbury's and her own, that there were probably twenty-four nucleotides in the amount of DNA along that 27-angstrom turn, and this immediately suggested that more than one chain lay together in each unit of the structure. In the transition from crystalline to wet, the "large length change" meant, she thought, that here the "helix has not got *same* structure as in Xtal—Xtalline form involves some strain of helix"—and she added "cf Pauling."

But despite the example of the alpha helix, she did not consider what bonds held the several chains in each helical unit together. She did ask how the units were bound one to another to form the overall crystal, and for that she thought that hydrogen bonds between bases were "entirely ruled out." Instead, bonds between the charged, acidic phosphate groups were "highly probable." These would be disrupted by water, and so account in part for the loss of three-dimensional order in the wet state. The phosphates would therefore be on the outsides of the helical units, the bases inside. She figured that the crystalline state contained about eight molecules of water per nucleotide. Watson heard this, too, but not correctly. He went away thinking she had said eight molecules of water in each structural unit of the crystal itself, which was far less.

Then, at the end of Franklin's notes for the colloquium, this:

> *Conclusion*
>   Big helix or several chains, phosphates on outside, phosphate-phos-
> phate inter-helical bonds, disrupted by water
>   Phosphate links available to proteins

With the last line she got her head up for a moment from crystallography to biological function. The structure Franklin had begun to imagine is evident, though it was not free of ambiguities. She described it coherently in her fellowship report the following February: the fibre, she thought, was probably "built up of near-cylindrical units," and further:

> The results suggest a helical structure (which must be very closely packed) containing probably 2, 3 or 4 co-axial nucleic acid chains per helical unit, and having the phosphate groups near the outside. It is the phosphate groups which would be capable of absorbing water in large quantities and of forming strong inter-helical bonds [that is, between units, not within a unit] in the presences [*sic*] of considerable quantities of water, thus giving the substance a three-dimensional crystalline structure. These bonds would be disrupted in the presence of excessive quantities of water (leading first to the "wet" structure . . .).

She drew no diagram in the colloquium notes; but in the companion set of rougher sheets that Klug found with the notes there is a quick sketch of two cylinders side by side with a column of dashes for bonds between them, which represents just such a structure.

That November, Franklin possessed one more piece of data that was fully as important as the knowledge that DNA yielded two different x-ray patterns. If she had understood what she knew, that would have been a crucial insight into the nature of DNA. In the language of her trade, she wrote in the colloquium notes that the unit cell of the crystalline state could be indexed as monoclinic, face-centered. She added measurements—three dimensions and an angle—to support the description. But the immediate and necessary consequence of that description was that the coaxial nucleic acid chains she had discerned forming the helical structure could only be running in opposite directions. They had to be upside down to one another. That also meant, taken with the value she had for DNA's density, that there were almost certainly two, not three, just possibly four chains associated. Franklin told Klug, years later, that because she had never done x-ray-diffraction studies with single crystals or with biological substances, she failed for nearly eighteen months to realize the consequence of what she had written down simply as a routine item of crystallographic description. Would Crick have understood it? "Oh, yes. Unquestionably," Klug said. "You're aware that it was Crick who saw the point, the moment he finally did learn of Rosalind's description early in 1953." If at the colloquium Franklin said what she was prepared to say, then Watson's incomprehension and failure to take notes had more ironical meaning than ever was borne in upon him until after *The Double Helix* provoked Klug's reply.

The irony is compounded. Ray Gosling told me recently that before Franklin came to King's College, he had attempted to learn enough crystallography to analyze himself the first x-ray patterns he had obtained with Wilkins—and though his measurements were inaccurate in some details, they established correctly the basic character of the crystals. Whether or not Franklin gave this information at the colloquium—and Gosling was there and remembered that "she talked pretty plainly"—Wilkins already knew the essential facts from Gosling's earlier calculations. But he understood the consequence no better than Franklin did. "And Wilkins and Crick were old friends," Gosling said. "I find it hard to believe that if you're talking about a structure, to somebody like Crick, that what you know about the basic characteristics of the crystal doesn't come up. I find this *very* difficult. I mean, what *else* do you talk about? Surely in a conversation about anything you're interested in, you can romanticize in a pub or strolling the banks of the Cam, but somewhere you've got to introduce the hard data." None the less, it was not until early in 1953 that Crick became consciously aware of the fact that DNA forms what crystallographers call a monoclinic, face-centered unit cell.

❖ ❖ ❖

With helical diffraction, unit cells, space groups, and the rest so central to the elucidation of the structure of DNA, I figured I had better learn more clearly why. Soon after reading Franklin's notes, I asked Max Perutz what these things mean. We were talking in his office at the Medical Research Council laboratory on the south side of Cambridge. He thought for a minute, and then took a blank sheet of paper. The problem was to do what Bragg had first done, to visualize one's way back from the spots, along the paths of the rays, to the object causing the spots. Spots first. Diffraction patterns on x-ray film are usually disc-shaped overall—targets at a molecular shooting gallery. The center of the target is a blank bull's-eye left by a small lead shield that stops the direct, extremely narrow, brilliant x-ray beam. Only the reflections scattered from the beam are interesting. A good diffraction pattern from a protein crystal—Perutz had once shown me some from hemoglobin—will be an array of several hundred small circular spots in many rows, filling the target from edge to edge. The spots range from very faint to very intense in a way that looks random but is not so at all. If it were, structures could not be analyzed.

Next, the object causing the spots. "A perfect crystal consists of a three-dimensional array of atoms or molecules arranged in repeating units of pattern," Perutz said. "In diffraction you get strong intensities from where the atoms lie along repeated planes." These planes are very close together, and the first step in the analysis is always to find the distance between them. Mutual interference between parallel crystal planes blocks off and suppresses the diffracted beams except at certain angles; but at those angles, the reflections from the planes in the set reinforce one another and flash out. The x-ray camera photographs these beams, and its geometry is arranged so that each beam can be identified with the set of planes that must have caused it. Fibres, being disorderly compared with single crystals, produce smears rather than spots, and fewer of them. Some fibrous substances, though, contain many individual crystallites, far too small to separate, lying next to each other but not pointing exactly the same way; such fibres produce diffraction patterns that hint at the clarity obtainable from a perfect single crystal—as though one were looking at a pattern that was the average of many pictures of a crystal taken at different angles, with no easy way to sort out the overlapping aspects. The crystalline pattern of DNA was produced by such fibres.

Helical structures—Perutz drew a multiple zigzag curve across the paper, a helix seen sideways. "Here you must imagine that the x rays are going through this helix perpendicular to the paper." He balanced the pencil for an instant with its point on the zigzag. "The helices then look to the x rays as though the atoms are lying on these repeated planes, this way, that way"—the planes of the zigs and zags. The result, when the rays reach the film, is a striking arrangement of short, horizontal smears that step out along the diagonals from the bull's-eye, in a characteristic X or maltese cross. The zigs cause one arm of the cross, the zags the other. The exact angle of the cross is caused by the angle of the zigs to the zags—that is, by the pitch of the helix.

"Any crystallographer knew that much," Perutz said. "What Cochran and Crick showed, which hadn't been realized, is that if you follow out the layer lines"—step by step along the arm of the cross—"at a certain distance the spots begin to march back together to cross again." The cross doubles at top and bottom of the target and becomes two diamonds. "If you have a setup with the resolving power to show it. Now you must also remember that between the diffraction pattern and the real space back inside the molecule, there is a reciprocal relationship. Big spacings between the repeated planes produce spots close to the center of the target, but as you move out, you are reading planes with finer and finer spacings." I took that reciprocal relation on faith for the moment. "Therefore, in DNA, the Astbury spot at 3.4 angstroms is at the outside edge. The 3.4-angstrom spot happens because the nucleotides—and in particular the heaviest thing in them, the phosphorus atoms—repeat along the helix at that interval. It's the same in principle as the 1.5-angstrom spot I found in the alpha helix, beyond where people had looked before. The fact that in DNA the bases are stacked in the helix parallel to each other at that 3.4-angstrom distance makes the spots more intense."

This was what Crick had discovered about helices by the time of the colloquium in 1951, I said, but Rosalind Franklin had not by then produced any x-ray pictures with striking maltese crosses or diamonds. What did she mean by monoclinic, face-centered? "Do you really want to know what this signifies?" Perutz said. "Crystals can be symmetrical in several ways, or to several degrees. Some are more symmetrical than others. The lowest degree of symmetry is the one we call triclinic, in which the three axes, or planes, of the crystal are all oblique to each other, none of the angles at the corners being right angles. At the other extreme is orthorhombic, where all three planes intersect at right angles. And in between is monoclinic, where two of the three angles are right angles but the third can take any angle. Now, the minimum symmetry that a crystal of *this* sort, monoclinic, may have is twofold, so that if you rotate the crystals by half a revolution they come back into congruence again.

"What Rosalind found was that the crystalline form of DNA *did* have monoclinic symmetry. Furthermore, she found that the axis of symmetry wasn't parallel to the fibres, but perpendicular to them, to the chains. And that had a peculiar geometric consequence, the crucial consequence she failed to recognize." Perutz got out two sharpened pencils and laid them on the table side by side, both points north. "If I rotate this pair together, in the plane of the table, they don't return to symmetry with their original position until they've gone around the full three hundred sixty degrees." He turned the two—west, south, east, north. Then he switched one pencil so that the two now pointed in opposite directions. "Lying head to tail, the two chains return to symmetry after a *half* rotation. You see. They exchange directions. If DNA was monoclinic from its x-ray diagram, and the axis of symmetry was perpendicular to the chains"—he rotated the pair of pencils on the table, this time through a half turn—"then it immediately followed that there must be one chain running up and the other down. That in physical fact there was a dyad, one chain upside down to the other."

What about space groups and face-centered? "But I was avoiding that," Perutz said. "It adds another order of complexity. First, then, the unit cell. We've all often suggested that for this, one should visualize a fancy wallpaper with a repeating pattern. Then the unit cell simply means the smallest repeat of the pattern, identical in size, shape, and contents. Except that in a crystal it's three-dimensional. Now consider the simplest box of atoms, one atom at each of the eight corners." With light, accurate pencil strokes, he drew a box in perspective, dotted lines for the far sides and corner. He thickened the corners into balls. "Or it can be a molecule at each corner. Don't forget that the box repeats and repeats, like the wallpaper pattern, and forms a three-dimensional lattice of atoms. With only the corners occupied, the lattice is called *P*, for primitive, and if it has monoclinic, twofold symmetry it's called a *P2* space group. But in some other substance you could have another atom or molecule at the center of a face of each box, as well." He drew a second box just like the first, then added a ball in the middle of two opposite ends. "This face is called, by convention, the *C* face, and if this box also possesses twofold symmetry then the space group is *C2*, or face-centered monoclinic. Each corner of the unit cell, and each occupied face, is also called a lattice point. But these terms are the abstractions of mathematicians, to analyze the orderly arrangement of the components of the crystal." He drew a box again, but without drawing sides or faces, just balls for atoms. "There are many more geometrical variations. There are two hundred and thirty different—that is to say, geometrically non-equivalent—space groups. They were all worked out in the nineteenth century by classical crystallographers calculating all possible ways to arrange a lattice that will repeat in three dimensions, long before x-ray crystallography began. But it just happened that hemoglobin, in its oxygenated form, also belongs to the face-centered, monoclinic space group C2. So Crick knew immediately, we were *all* very clear about the significance of what Rosalind had found. When we learned it early in 1953.

"Now, her next question had to be, how many nucleotide chains were there in the unit cell? This can be calculated from a very simple equation, one you could derive yourself with a little effort of visualization. But for this calculation she had also to know the density, and as the fibres were wet the density was difficult to measure, and it was not clear how much water and how much salt they contained. On the evidence of the density, the choice was uncertain between two and three chains. But on symmetry grounds—and that is part of the consequence she failed to reach—it seems unlikely that something with twofold symmetry will have an odd number of components per unit cell." Indeed, for DNA it was impossible, because the backbone of the single strand has a polarity, an upness or downness, imposed by the linkages—always phosphate to third carbon of the sugar ring, then fifth carbon to the next phosphate—that Todd and the Cambridge chemists had demonstrated.

❖ ❖ ❖

November 22, the morning after the colloquium, Crick came up to London and Watson met him on Paddington Station, where they took the train to Oxford to spend the weekend. Several levels can be distinguished in what came next. We who know who got the structure sixteen months later have also to remember that several others almost puzzled it out. For at the end of 1951, when nobody was really near, technical details of one or another failed conception of the structure are interesting only as they illuminate the state of knowledge of the substance, the state of the art of model-building, or the state of mind of scientists thinking about the problem. On the train, Crick quizzed Watson about what he remembered of Franklin's talk. "Francis was visibly annoyed by my habit of always trusting to memory and never writing anything on paper," Watson wrote in *The Double Helix*. "Particularly unfortunate was my failure to be able to report exactly the water content of the DNA samples upon which Rosy had done her measurements. The possibility existed that I might be misleading Francis by an order-of-magnitude difference." By Watson's recollection, Crick became convinced before they reached Oxford that "only a small number of formal solutions were compatible both with the Cochran-Crick theory and with Rosy's experimental data. Quickly he began to draw more diagrams to show me how simple the problem was." As jottings and talk progressed over the next several days, Crick evolved not so much a theory about the structure of DNA as a theory about how to construct a theory. By the time, a week later, that they had a model, they also had a memo, eight pages in Crick's handwriting.

Franklin's notes and Crick's memorandum make a rare, miniature case study. They were written within a few weeks of each other, and from the same data—some, to be sure, filtered through Watson. They were worked up, thought out, yet still informal. Franklin prepared a firmly matter-of-fact report. Watson asserted in his memoir, seventeen years later, that she rejected model-building and helical structures as premature. None the less, she unquestionably had the general features of a helical solution in mind, one that would have been more correct than Crick and Watson's. Crick's memo was no report, for he hardly had anything to report; he still refers to it as a plan of attack. It was inconclusive, for all its brave assurance: if Bragg could write an overture in an evening, Crick here produced, in a long weekend, a sketch for an unfinished sonata—Mozart at age nine, style and flair unmistakable, means insufficient, theme not his own—"Sonata on an Air We Heard Someone Whistling." The opening bars:

> Stimulated by the results presented by the workers at King's College, London, at a colloquium given on 21st Nov. 1951, we have attempted to see if we can find any general principles on which the structure of D.N.A. might be based. We have tried, in this approach, to incorporate the *minimum* number of experimental facts, although certain results have suggested ideas to us. Among these we may include the probable helical nature of structure, the dimensions of the unit cell, the number of residues [nucleotides] per lattice point, and the water content. Having arrived at a tentative structure this way, we have generalized what we regard as the important features, and now present these as postulates.

The contrast with Franklin's approach was diametric; the conclusions were inevitably similar in uncertainty and error.

The physical differences among the four kinds of bases were to be ignored, and this would result, following Pauling, in a helix. The helix would not, despite Pauling's example, be held together by hydrogen bonds. The bases were the only serious candidates to form hydrogen bonds. From what Watson and Crick found in the reference books, though, the arrangement of hydrogen atoms fringing the bases was inconstant. Each base was said to occur in either of two forms. In particular, the bases guanine and thymine were supposed to take what are called the enol and the keto configurations. These differed only by the place where a particular hydrogen atom hooked on in the fringe. Thus, in the keto form, the movable hydrogen attached directly to a nitrogen atom in the principal ring of the base; in the enol form, the hydrogen jumped over to an adjacent but more protrusive location, bonded to an oxygen atom that was in its turn attached to the ring.

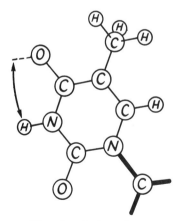

Thymine, in keto form

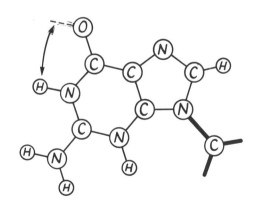

Guanine, in keto form

Such closely similar molecular alternatives are called "tautomers," a piece of international scientific small change coined from the same Greek as "tautology." Truly tautomeric forms of a molecule interchange with each other so readily that they exist side by side in equilibrium in a sample of the substance; the phenomenon was sometimes treated as though the molecule were vibrating rapidly between the two forms in a kind of resonance. Ordinary techniques of chemical analysis won't distinguish between one tautomeric form of a molecule and another. None the less, the diagrams of guanine and thymine as printed in the early nineteen-fifties more frequently showed them in the enol forms. At the outset, Crick could see no way that such capricious configurations could form stable hydrogen bonds. Even if that objection were put aside, it remained true that the configurations of the bases made lumpy pieces in the model-builder's puzzle. Thus the geometry of the tautomeric forms was the basis for Crick's prejudice against hydrogen bonds to knit up the chains, and for Watson's against putting the sugar-phosphate backbones on the outside. These prejudices endured.

Looking for another way to bind the chains, Crick and Watson, like Franklin, came to the phosphate groups of the backbone. Unlike Franklin, they did not link them directly, P—O⋯O—P. Instead, they saw that with so very little water as Watson reported, other atoms present in solution would form electrostatic bridges between the charged phosphates. Crick's memorandum specified sodium to bridge the phosphate groups, since they were working with that salt of DNA; Watson remembered magnesium ions instead, but it's possible that he did not think of them until the next spring. They supposed—though with less conviction—that the backbones would thus be tied together in the center of the molecule, with the bases projecting outwards all around. One tempting advantage of putting the bases on the outside was that the irregularities in their shapes need not be fitted into the regular, crystalline structure. Another was that sequences of different bases would be simply, obviously available to interact with protein. "Starting from these postulates we can now attack the problem of possible structures systematically," Crick wrote. His attack was

> to write down systematically all the possible topological combinations. This is done by writing down all the possible linkage schemes . . . for an infinite plane surface. A strip of this surface [is] then isolated and folded round into an infinite cylindrical sheet. The number of ways of doing this can then be explored systematically.
>
> Having thus obtained all possible linkage schemes, the next step is to build models of them all. . . . The model is then subjected to a series of tests, in which the experimental data from simple structures is increasingly used. For example, . . . the number of residues per 27 Å length of helix may be quite wrong, and so on. In doing this it is particularly important not to use as a basis of rejection rather vague criteria, such as not quite seeing how it fits in with possible preconceived ideas about the X-ray pattern, and not quite seeing how it would stretch. If great care is not taken the right model may be rejected.

Crick hadn't yet heard Pauling tell the story of discovering the alpha helix with paper and scissors, but he was in effect writing a florid program for that procedure. Watson reported none of this, doubtless dismissing the conceptual machinery as overelaborate for the failure that resulted.

Model-building began on Monday. Components had to be improvised out of bits left over from the lab's two-year-old protein chains. Atoms were the wrong sizes. There was no time to order new pieces made. Construction was awkward and hasty; pieces kept slipping out of place. Watson and Crick tried rough models with two intertwined chains, but found they fell short of the density estimates. The structures with three helices, though, Crick wrote, "are more interesting, as they give about 24 residues in 27 Å and have a hole in the middle which is about the right size to accomodate about 8 water molecules." Those, of course, were Franklin's numbers, indifferently remembered. The model took shape by Monday evening. They touched it up on Tuesday morning. At Kendrew's warning that they owed Wilkins the courtesy, Crick telephoned his friend to urge him to take the first train.

Wednesday morning, Wilkins, Franklin, Gosling, and another colleague arrived. Crick preached the glory of helical-diffraction theory, which would allow them to predict from this model the x-ray pattern such a molecule should produce, for ready verification with the King's College pictures. Wilkins told them that Stokes, at King's, had worked out the helical mathematics months earlier; a sentence acknowledging Stokes's independent discovery was added to the forthcoming paper on helical diffraction.

Franklin led the attack on the model itself. By Watson's account seventeen years later, she was adamant that "there was not a shred of evidence that DNA was helical." The notes she had prepared a few days earlier and the fellowship report she submitted a few weeks later bracket this encounter, to give the true range of her opinion: in fact at the time she was not firmly set against helices. What she could correctly have said was that her best x-ray pictures of DNA so far—those of the crystalline form—failed to show the pattern characteristic of a helical structure. It was true also that while her pictures of the second, wet form presented helical features, these were not unequivocal—but she may never have gone into the details of the wet pattern, any more than she had in her notes for the colloquium. The model had two more deficiencies, each crippling. The backbones could not be inside. Finally, a glance told her and a word told them all that the model did not contain enough water—which Crick then had to recognize was not just Watson's lapse of memory but his own lapse of basic chemistry, for those charged sodium or magnesium ions would strongly attract water. "Jim said there was no water there. With little water, the electrostatic forces would be very strong," Crick told me. "That was reasonable given that what Jim said was reasonable, but what I should have known was that what Jim said *wasn't* reasonable. Anybody who knew any chemistry would have known there was a lot of water just from first principles!" With the water put back in, very many other models were once more compatible with the rest of the data. By Wednesday afternoon, less than a week since Franklin's seminar, Watson and Crick's first model for DNA had been flooded out.

From rout to triumph took fifteen months. The rout was moral as well as intellectual, and seemed all but absolute: while successful pirates are sometimes knighted, those who fail get hanged. Randall had a quiet word with Bragg. Bragg, iron showing beneath the tweed, let it be known that Watson and Crick were to leave DNA to the King's lab. Watson recorded that they did not protest. "An open outcry would reveal that our professor was completely in the dark about what the initials DNA stood for," he wrote in *The Double Helix,* but conceded that "we were up the creek with models based on sugar-phosphate cores." The jigs to make the elemental pieces for models were turned over to Wilkins—who, despite Crick's urging and intense frustration, left them idle.

Crick went back to his dissertation. Restless as ever, he was soon wondering about inconsistencies between Pauling's alpha helix and the data. He worked out that the notorious x-ray spot produced by natural alpha keratin at 5.1 angstroms, which had held Pauling back from publication for so long, could be explained by supposing that the alpha helix itself, in that substance, has a larger, slower twist superimposed on it, which he called the "coiled coil." Watson was supposed to be helping Kendrew grow crystals of protein, but his attempts had been desultory and unsuccessful. He now turned to making diffraction pictures of tobacco-mosaic virus, thus combining the methods he had come to the Cavendish to learn with the materials studied by Markham at the Molteno Institute to which the extension of his Merck fellowship was tied. Watson, too, kept worrying away at Pauling's work and methods. Pauling was on lots of minds. Crick pursued the alpha helix with the papers on helical diffraction and the coiled coil. Ruefully, he gave Watson for Christmas a copy of *The Nature of the Chemical Bond.* Surely the gift is the giveaway: for Crick, Pauling was the man to be learned from, the alpha helix the example to be studied. Franklin, too, though she chose crystallographic analysis rather than model-building, felt the force of Pauling's example and cited him in her notes. Only from Watson do we hear that the failed model of DNA was built in a race to anticipate Pauling. The first round had been lost not to Caltech but to King's, and in particular to Franklin.

❖　❖　❖

Over the next months, bits of fact and speculation, curious and not obviously related in their colors or jagged shapes, dropped into the heap; from time to time Watson or Crick would give the kaleidoscope a vigorous shake, but still no intelligible picture appeared. Thinking back, those who took part have been struck by the extreme hazard of the encounters, conversations, and observations that sorted the significant facts together in Crick and Watson's attention. Here in the fine grain of the discovering, scientists themselves are bemused by the attempt to analyze how their minds work.

Watson remained communicant with the phage group, one of their chain of long-winded letter writers. The experiments he and Maaløe had been doing with radioactive phage in Copenhagen the previous year were discussed at Cold Spring Harbor even before publication. Delbrück and others wrote with news of work and ideas. Delbrück fixed up a two-year fellowship from a new source, the National Foundation for Infantile Paralysis, so that Watson could come to Caltech the next fall. That spring, Watson wrote that he'd like more time in Cambridge, so Delbrück got Pauling to say that the facilities were better at the Cavendish at the moment, and got the first year of the fellowship moved to England.

That spring, Luria was to come to Oxford to give a paper at a meeting of the Society for General Microbiology on April 16 and 17. But 1952 was the American presidential election year terrorized by Joseph McCarthy, and Luria was a naturalized immigrant who in his youth, he told me on one occasion, had "passed from a fairly orthodox upbringing to what was no doubt an equally orthodox Marxist philosophy and just little by little evolved in my twenties and early thirties to an existentialist viewpoint." He was refused a passport. Luria's paper for the meeting was the grand summation of the scheme of bacteriophage multiplication he had evolved over several years. In this, the phage protein had the genetic role within the infected bacterium, while the phage DNA was only added at the last step before the burst. The final baking of the protein doughnuts, he had been calling it around the lab, and metaphors can be mind traps. Watson went to Oxford to stand in for his absent mentor. He had just received a long letter from Alfred Hershey reporting a series of experiments he and a colleague, Martha Chase, were about to publish. The results were already shaking the fundamental convictions of the phage group about DNA and were bringing down Luria's scheme in ruins. In Oxford in Luria's place, Watson got up to report Hershey's results, from the letter, and to dilate upon his own work in Copenhagen.

Hershey has since urged the "essential simplicity" of what has come to be known as the Hershey-Chase, or Waring Blender, experiment. The need for it sprang directly from Thomas Anderson and Roger Herriott's discovery that phage had elaborate structures, a proteinaceous head that seemed to encapsulate the phage DNA and a slender protein tail and fine tail fibres that attached to bacterial prey. Looking back in 1966, Anderson found matter for wry wonder in the phage group's long refusal to get the point. In his electron micrographs, he wrote, "One could see empty headed phage ghosts on the bacterial surface. I remember in the summer of 1950 or 1951"—six or seven years after Oswald Avery's paper—"hanging over the slide projector table with Hershey, and possibly Herriott, in Blackford Hall at the Cold Spring Harbor Laboratory, discussing the wildly comical possibility that only the viral DNA finds its way into the host cell, acting there like a transforming principle in altering the synthetic processes of the cell."

Several teams in the phage group besides Watson and Maaløe had then tried to differentiate the fates of the phage protein and DNA by using radioactive markers. Hershey and Chase decided to see if they could strip off the empty phage ghosts from the bacteria and find out what they were and where their contents had gone. DNA contains no sulphur; phage protein has no phosphorus. Accordingly, they began by growing phage in a bacterial culture with a radioactive isotope as the only phosphorus in the soup, which was taken up in all the phosphate groups as the DNA of the phage progeny was assembled, or, in the parallel experiment, by growing phage whose coat protein was labelled with hot sulphur. They used the phage to infect fresh bacteria in broths that were not radioactive, and a few minutes after infection tried to separate the bacteria from the emptied phage coats. "We tried various grinding arrangements, with results that weren't very encouraging," Hershey wrote later. Then they made a technological break-through, in the best Delbrück fashion of homely improvisation. "When Margaret McDonald loaned us her blender the experiments promptly succeeded."

The blender experiment in its simplicity was buried under detail, other experiments, and redundant caution in their report. This paper is tedious and unnecessary labor to read, and the fame of the experiment rests on it only lightly, for Hershey spread the word by letters months before publication and has recalled the procedure succinctly since. "A chilled suspension of bacterial cells recently infected with phage T2 is spun for a few minutes in a Waring

Blendor and afterwards centrifuged briefly at a speed sufficient to throw the bacterial cells to the bottom of the tube," he wrote in 1966. The blender generated strong shearing forces in the liquid, which attacked the fragile connection of viral tail to bacterial wall. After centrifuging, the radioactive labels showed that the sulphur-bearing protein remained in the liquid while most of the DNA went down with the bacteria—which could then be removed and resuspended in more broth to go on with the growth cycle and release new, active phage. The controls to exclude all protein from accompanying the DNA were nowhere so inventive nor so stringent as Avery's had been. It's an accepted fact that the Waring Blender experiment was biochemically sloppy. But Hershey and Chase used a familiar, favorite organic system to produce a pleasingly symmetrical cleavage between the exclusion of phage protein from bacteria and the taking in of phage DNA. They addressed a large audience and a ready one, for where Avery's identification of the transforming principle had seemed an isolated, almost useless result, now the evidence was cumulative. The day I saw Delbrück at Cold Spring Harbor, he brought that up while talking about Avery. "When Hershey and Chase did their experiment showing that in phage, too, it was DNA, by then their discovery was accepted on much poorer evidence," Delbrück said. "But don't make a great spiel out of it, that it was accepted on poorer evidence—because meanwhile the case had been getting stronger anyway. It had been made well for the transforming principle itself, by Hotchkiss in Avery's lab."

The microbiologists in Oxford asked Watson some sensible, responsive questions. One Englishman even proposed a complementary structure for phage rather like a zipper, with phage protein on one side and DNA on the other, to account for their separation when the DNA entered the bacterium and for the assembly of new phage within. Watson answered that he thought the relation of DNA to protein in bacteriophage was more probably "like that of a hat inside a hat box." Watson popped up in the audience later in the meeting to interject more news about phage DNA from Hershey's letter. At one point he remarked that Wilkins had told him "the diameter of a nucleic acid chain is unlikely to be greater than twenty-two angstroms." But he went away convinced that "almost no one in the audience of over four hundred microbiologists seemed interested"—except for three other members of the phage group, André Lwoff, Gunther Stent, and Seymour Benzer, all over from Lwoff's lab in Paris, who understood "that from then on everyone was going to place more emphasis on DNA."

Soon thereafter, on May 1, Watson went with Crick to London for a daylong meeting at the Royal Society on the structure of proteins. It was called to discuss the alpha helix. Astbury, up from Leeds, presided. Linus Pauling and Robert Corey were advertised as participants. But the State Department struck again: this was the notorious occasion when Pauling at the last minute was refused a passport on the suspicion—and despite his firm denials—that he was a Communist. English scientists were incredulous, and wrote letters to *The Times.* "What one remembers is the *scandal* of Pauling's not being there," Crick said. Watson's reaction, reported in *The Double Helix,* was mixed: a careless contempt for American policy and Southern California red-baiters as well as, he suggested, a pleasure that Pauling would not get the chance to see any x-ray pictures at King's College London.

Indeed, the giant had begun to stir: though Watson had no particular reason to know it at the time, Pauling had started thinking about DNA, at least as another large molecule to be determined. And indeed, there was more to be seen at King's than Watson then had any idea. If Pauling had visited, Wilkins and Franklin would have shown him their diffraction photographs from DNA, for though Wilkins turned down a letter from Pauling requesting prints, the courtesies of a personal visit to your laboratory are different and hard to evade. Corey did come to London for the meeting on May 1. He did visit Franklin, though we don't know which day, and she did show him pictures, though we don't know which ones. But Corey did not have Pauling's mercurial intuition for structures.

Whether Pauling himself could have learned enough at King's to be first to solve the structure of DNA is a question that must hang forever in the most intriguing balance. The timing was crucial. Wilkins had the pictures of DNA he had taken two years earlier, which first discovered the crystalline state— pictures Watson had seen and Pauling had not, pictures essential to understanding, yet more stimulating than informative. Shut out by Franklin, Wilkins had now set up his own x-ray equipment. He was able to get some patterns of the wet form, similar to those Franklin had first taken. To his dismay, he found that the samples of DNA brought back from Chargaff in New York could not be made to produce the crystalline pattern; and he had given Franklin the rest of the DNA from Signer in Switzerland that worked. Wilkins's hopes retreated to his varied cellular specimens and his wide travels to obtain them; early in 1952, while on a train ride from Innsbruck to Zurich after a rambling vacation in the mountains, he wrote a long excited letter to Crick about his program of research for the next months. He said, in part:

I am doing a combined holiday & work trip again. I am going to see Signer the n. a. [nucleic acid] man in Bern. . . . [A]lready I have another trip planned on the MRC (actually they didnt pay for my other trips but Unesco & other people). I have got *much* better X ray pictures of the sperm squid which show very nicely a whole series of helical layer lines. . . . These spots will not overlap when a disoriented specimen is used & I want to do it on *living sperm* in glass tubes. I believe your friend in Rome offered me use of an X ray tube (I am sure he did) & what I want you to tell me is *his address* & name so I can confirm this & can go to Naples (you bet) & collect the sperm & go by train to Rome & X ray them. They live for days & it is all very practicable. . . . Also I will try living spermatophore [packets of sperm, as in squid) in glass for orientation. No H_2O humidity fuss. . . . I have found several of your suggestions very valuable but am fairly convinced for many reasons the phosphates must be on the outside. There is just one part of the picture that still puzzles me but if helices are right we must hit on the explanation soon. I am really getting down to the job myself but haven't done anything on models & chemistry as I think the picture holds a key not yet recognized which will then more or less directly give us the model. I now take my own X ray pictures & have made new cameras (hence better sperm pictures). But I believe as you that the key to nucleoproteins lies in the crystalline picture. . . . Franklin barks often but doesn't succeed in biting me. Since I reorganized my time so that I can concentrate on the job, she no longer gets under my skin. I was in a bad way about it all when I last saw you.

When I saw Jim last before Christmas he said you people couldn't build a model along your lines if that is so it simplifies things a bit anyway I am pretty certain the phosphates are on the outside. I wont start making any references to the 'business' between you people & us over n. a. but look forward to discussing all our latest ideas & results with you again. Why don't you come & have lunch with me when you are next in town?

About Naples, we must also separate heads & tails (electrophoretically?). . . . Also I took a nice picture of living rat thymus in a glass tube & think there is a whole new field there . . . to show the pictures of living nuclei or whole cells with negligible cytoplasm. . . . Thus we may be able to prove the helical idea for ordinary cells. . . .

Did you hear I *have* got a Rio invitation? isn't it grand! In some ways I feel I am a very lucky fellow. And there I hope to collect sperm bundles which contain separate chromosomes & do X ray pictures of each of the 2 chromosomes in the cell spiralized & unspiralized etc. This is rather gilding the lily but may be very useful in a year or 2's time to link the X ray business more with *real chromosomes*. I hope I dont go to the divil with all these trips but so far I think they are well worthwhile from the work standpoint.

Franklin had the pictures she had taken the previous fall, which first disclosed the difference in DNA between the crystalline and the wet states. Wilkins of course knew the distinction; Watson had had good opportunity to learn of it. It was unknown to Pauling. Later she began to call these the A and B forms of DNA, and that is how they are designated to this day. Franklin and Gosling, that winter and spring, stuck to the lab and took more pictures, some much richer, which they had not yet shown to anybody else. By the first of May, Franklin's diffraction patterns had not surrendered any more of the essential dimensions of the molecule, whether in A or B form, than had been generally known; but they demonstrated that Astbury's pictures of DNA had mixed the two forms and so were as misleading, in their way, as Pauling had found the patterns of alpha keratin.

Franklin's new pictures from DNA, however, immersed her in a baffling conflict of evidence. To begin with, it had always been obvious to her that the change from crystalline to wet, or the other way, was a change not of a continuous sort—not a steady swelling of the individual molecule with the uptake of water, say, nor some gradual loosening among the many molecules in the fibre—but, rather, a discrete transition, even an abrupt one, between two structural configurations of the molecule. She prepared the DNA specimens herself, drawing the almost invisibly fine fibres out of the viscous solution with the tip of a needle, mounting them singly or in bundles of three or four so as to get the sharpest resolution with the fine-focus x-ray equipment. She told Aaron Klug, some years later, that she had been struck by the change in form when a fibre she was preparing suddenly stretched so much that it fell out of position for the x-ray beam. The transition from A to B was marked by an increase in length of the fibres of fully one-quarter. Early in February, she wrote, in her annual report for her fellowship, "The helix in the wet state is therefore presumably not identical with that of the crystalline state."

In the next few weeks, Franklin's way of thinking about the structure of DNA evolved rapidly. The patterns of the A form were now so good that it appeared possible, given patient measurement and calculation, to work back to the A structure directly, no need for guesswork or model-building, by the quantitative methods of classical crystallography—in particular, by the method called Patterson synthesis, after still another of the Bragg school, Arthur Lindo Patterson. Nobody had ever done a fibre structure this way before; only the unusual, crystalline order of the DNA fibres in the A form made the idea possible at all. The challenge could not help being attractive: a lot of work, a lot of puzzle, and a great intellectual purity. And however difficult the answer, once obtained this way it would be unambiguous: Franklin said to Gosling that with a structure as complicated as DNA's promised to be, even if one could string together a model that might satisfy the x-ray data, how would one show that it was the only model that would do? Figuring a Patterson projection of the A structure took almost all of her attention for the next year.

As Franklin started, that spring, some features of the latest diffraction picture of the A form gave her the idea that that structure could not be a helix after all. Over the Easter weekend, beginning April 10, she took a picture of a single fibre that yielded an unusual variant of the A form. "Double orientation," she wrote in her notebook entry. The spots were all in their right places, but their intensities were asymmetrical in a systematic way that meant that the scattering of x rays was dominated by one or a few large crystals within the fibre. Picture and fibre were unique, fortuitous, impossible to duplicate despite many attempts. Franklin came to set great store by this picture. It gave a partially independent verification of the initial analysis, called the indexing, of the spots—the foundation of all ensuing calculations. The picture's asymmetry seemed stubbornly irreconcilable with a helical structure for the A form. Klug, in his article in *Nature* in 1968, tried in Franklin's defense to reconstruct her state of mind.

> The patterns she was dealing with were fibres or rotation photographs in which the inherent three-dimensional data are to be had only in two-dimensional form, leading to certain possible ambiguities of indexing of the patterns. . . . It was quite natural in the context of the new observations to think that the A structure might not be helical at all and to explore structures that were not helical.

Franklin's notebooks also suggest that an analogy took root in her mind between the two forms of DNA and the ideas that Astbury had advanced about the two forms, stretched and unstretched, of keratin, the fibrous protein of wool; labelling these alpha and beta, Astbury had proposed that one turned into the other not just by a simple change in length of a helical molecule but by its unfolding and rearrangement into something else.

Franklin was in the audience at the discussion of proteins at the Royal Society on May 1, although almost nobody remembers her presence. Wilkins was there, and in Watson's recollection said to him that Franklin was now insisting that "her data told her DNA was *not* a helix."

During the course of that same day, Franklin had the equipment running to take an x-ray picture of a fibre that had previously given her several good patterns of the B form. Next morning, May 2, a Friday, she developed the film. In the notebook she was using—a small, red school-exercise book—she identified the picture as forty-ninth in the series she had started the previous September, and observed, "V good 'wet' photo." She tried another fibre in a short exposure that afternoon. That evening, she set up the good fibre again with the film holder more perfectly centered on the x-ray source. Over the months she had evolved a careful procedure. To eliminate scattering of the x-ray beam by air, the apparatus was enclosed and a steady stream of hydrogen was blown through it—a method Randall had introduced to the lab two years earlier. (The experiment was not without hazard.) To maintain the fibres at a chosen wetness, she bubbled the hydrogen through water—rather, through a saturated solution of ammonium sulphate, a simple salt—so that the gas entered the camera at seventy-five per cent relative humidity. Other humidities, other salt solutions: exact control of water content during long exposures was one of the new techniques by which Franklin kept the A and B forms distinct to get better pictures. At half past six, she switched all this on. An hour later she turned on the x-ray tube to start the exposure. At some time over the weekend there was an interruption of thirty-two hours, but her notes don't say why. At five o'clock Tuesday afternoon, she ended the exposure. She gave the picture the number 51, and figured its total exposure at sixty-two hours.

Compared with all previous B patterns that Franklin had obtained, these two pictures were vivid, No. 51 especially so. The overall pattern was a huge blurry diamond. The top and bottom points of the diamond were capped by heavily exposed, dark arcs. From the bull's-eye, a striking arrangement of short, horizontal smears stepped out along the diagonals in the shape of an X or a maltese cross. The pattern shouted helix.

In a quieter voice it said much more. On each arm, the first three layers out from the center were intensely marked, the fourth layer was almost invisible, the fifth clear again. The memory of Astbury's data, confirmed by a brief calculation from this new diffraction pattern, identified the heavy outer arc, top and bottom, as the 3.4-angstrom spacing between nucleotides. A ruler placed on the print showed Franklin at once—showed anyone who placed it there, showed me twenty years later—that these heavy outside arcs fell on the tenth layer. In the reciprocal relation of real space to diffraction pattern, the innermost layer therefore represented 34 angstroms. (That's a shade less than one hundred and thirty-four billionths of an inch.) And so, inescapably, just this simply, the helix made one full turn in 34 angstroms, and each full turn contained exactly ten nucleotides if the helix were single, ten nucleotide pairs or triplets if there were more than one strand. In a whisper audible to a trained and attentive intelligence, the pattern said more yet. Because the spots making up the arms were so strong and clear, they had to be caused by the heaviest components of the structure—which in DNA were the phosphate groups—lying on the curve of the helix itself. The bases were, accordingly, turned inward. Careful measurement of the angle between the arms of the maltese cross, and another calculation, put the diameter of the helix at about 20 angstroms. How to perform that calculation had been demonstrated in *Nature* three months before by Cochran and Crick in their first paper on helical-diffraction theory.

Ten months later, in the middle of March 1953, Franklin reached all these conclusions about the B structure of DNA, in a thorough analysis of photo No. 51. In the first week of May 1952, she thought briefly and tentatively about No. 49: at the end of a couple of pages of measurements, she saw that the pattern "suggests that there is an integral number (or simple fractional number) of residues per turn of helix (if there is a helix) even in the 'wet' state. In passing from 'crystalline' to 'wet' . . . the fibre-axis period is extended by about twenty-five per cent (27 angstroms to 34 angstroms)." There she stopped, and put the pictures of the B form aside to get on with the solution of the A structure. She turned her back on her own discovery of the B structure of DNA and her own best evidence that that structure, at least, was helical. It was one of these two photos, most likely No. 51, that Watson saw ten months later that gave him the flash of understanding that set him and Crick off on their second attempt to build a model. And so, if Pauling had come to London for the Royal Society discussion on May 1, he would not have seen the most important picture of them all—unless he had stayed a long weekend in England before going around to visit King's College London.

❖ ❖ ❖

"If you want to make a critical analysis, she was going at it *wrong*," Wilkins said to me in Paris. The bitterness he has felt towards Franklin, unslaked after a quarter of a century, has that one scientific aspect. In the spring of 1952, turning away from the B form, Franklin also turned away from helical structures. Over the next months, led by the evidence from the unique picture with spots of asymmetrical intensity that she had taken at Easter, she became firmly convinced that the A form, at least, could not be helical. She said as much. Wilkins allowed himself to be led by her assertions. Indeed, he went further, he wrote recently, and concluded that if A was nonhelical, B could not be helical either. So once again, for several months, he "gave up the frustration of DNA x-ray work." He did not scrutinize Franklin's evidence until Gosling showed it to him after Watson and Crick had got the structure. By the spring of 1952, relations between Franklin and Wilkins had grown so unpleasant that he could not ask her for her data or even for some of the DNA extracted by Signer that alone produced patterns of the A form.

Franklin too was unhappy. That spring, she wrote to Paris to ask if she could go back to the laboratory there where she had worked. She gave up that idea, but then went to see Bernal to ask whether she could move to his crystallographic laboratory at Birkbeck College. On June 19 she wrote to Bernal:

> I have now spoken to Professor Randall about the possibility of moving to your laboratory and applying to the Fellowship Committee for permission to take the remains of my . . . fellowship with me. Professor Randall has no objection to my doing this.
>
> I hope this arrangement is still agreeable to you. If it is, would you suggest a time when I could come to see you to discuss things in more detail?

❖ ❖ ❖

Watson, that spring, was also doing crystallography. His excursion into tobacco-mosaic virus began that winter and stretched into summer. That the virus causing the mosaic blight of tobacco could be precipitated as a crystal out of the juice of infected plants had been found by Wendell Stanley at the Rockefeller Institute in 1935. It was the most portentous and publicized biological discovery of the decade. The idea of a living, self-reproducing entity behaving as a simple chemical, for all the world like salt or sugar, caught the imaginations of laymen and scientists alike: the implications about the secret of life were for those days wildly, provocatively reductionist. The electron microscope, when first turned on the virus a few years later, disclosed featureless rods nearly twenty times as long as they were thick—smaller and much less complex than many bacteriophages but very big compared, say, with protein molecules. As in other viruses that attack plant cells, the nucleic acid in tobacco-mosaic virus was not DNA but RNA; it amounted to six per cent of the particle, while the rest was protein. This was not quite the system of virus and host that Watson was used to, nor the nucleic acid he was interested in.

Although the virus particles were large for x-ray diffraction, Bernal and a colleague had tried some pictures just before the war. The result was one of the most striking pieces of x-ray research of its day: it showed that each tobacco-mosaic virus particle was itself crystalline, made up of regularly repeating identical subunits. Watson, looking at Bernal's diffraction picture with Crick's talks ringing in his ears, thought they suggested something more: that the subunits were put together helically. On the analogy with bacteriophage, the hat would be inside the hatbox, the protein shell wrapped around a nucleic-acid core. To prove it, Watson had to get Crick actually to teach him the mathematics of helical configurations in x-ray diffraction, and persuade others to show him how to grow and mount the crystals and use the x-ray equipment. His first patterns were poor. That spring, a new, fast, much stronger x-ray tube was set up at the Cavendish; with this and practice, Watson's pictures began to improve.

Late one night in June, he came back to the lab to turn off the machine and develop a picture. "The moment I held the still-wet negative against the light box, I knew we had it. The telltale helical markings were unmistakable." Crick confirmed the next morning that Watson had made a solid discovery. It was an early example of what John Kendrew characterized to me as "the Jim Watson type of science—you know, the really clinching experiment which will quickly tell you something dramatic."Watson forthwith dropped any further work on tobacco-mosaic virus. "No more dividends could come quickly," he said in his memoir. The dividend he wasn't reckoning, in the spring of 1952, was one of earned skill: now he, like Crick, could recognize instantly and interpret in detail the x-ray-diffraction patterns produced by helices.

❖ ❖ ❖

Rosalind Franklin at King's was the only person in the world, that spring, working steadily at the problem of the structure of DNA. Watson and Crick at Cambridge fretted at it intermittently but were banned from active pursuit. Wilkins at King's understood the problem's interest but wasn't doing much. Pauling at Caltech, by his own accounts and in the recollections of people close to his work at the time, was still interested chiefly in proteins, particularly in proving that the pleated sheet, as well as the alpha helix, could be found as a live structure.

The question that DNA presented them was extremely simple: How did the strands of the DNA molecule go together? But there are several ways of restating what that meant, so that each time the question was held up to the light, its aspects swam into focus and out again. Seen as an unusual essay in pure crystallography applied to a fibrous structure, as Franklin wanted to solve it, the problem was static: Exactly where was each group of atoms located and how was it aligned? Seen as model-building, by Watson or Pauling, the correct shapes of the parts and at least the broad limitations on the whole had to be sorted out from the work of biochemists and crystallographers, and the puzzle was geometrical: How could the bits all be assembled without breaking any of the rules? Seen as functioning biology, and perhaps Crick understood it this way most clearly, the question became dynamic: How did the structure dictate the assembly of replicas of itself—on itself, from itself—so accurately that DNA could carry the hereditary message? These were all really versions of the same question. The answer to any would be the answer to all.

In still another aspect, what was most difficult to know was what forces held the strands together and in shape. The possibilities were as before: the backbones or the bases. Watson remained obsessed by his guess that small amounts of magnesium were present along with sodium in the salt of DNA—or at least in the living cell—to bridge the phosphates of the backbones. By late spring, Crick had begun to take seriously the second possibility, that the bases strung along one backbone might join to those strung along another. The bases were the variable components in DNA. They had to vary to be free to carry, somehow, the genetic message. Because they varied so, it was hard to conceive that they fit together. If they joined, how? And was there any selectivity about which base joined to which? Astbury had supposed, from the 3.4-angstrom spot in the diffraction pattern, that the bases were stacked one above the other. Crick, in the spring of 1952, was still imagining not that the bases might be paired on the flat like two dominoes in play, but rather that they were interleaved like two decks of cards being shuffled together. Thinking of the tautomeric forms, he still rejected hydrogen bonds between the bases. Curious about other binding forces, he wondered whether the bases were attracted like to like. Yet even to list the sets of alternatives, Crick warned in a recent conversation, would suggest that he had visualized them more clearly at that time than he now can be sure he really did. "I think one must have thought of it in those terms, yes," he said. "But really the question was, was there any pairing at all? Of any sort?"

Crick's interest was desultory. He did not pursue the calculations himself, but casually asked a young mathematician at Cambridge, John Griffith, to work out how the electrical charges on each base might cause adenine to attract adenine, thymine to attract thymine, and so on, to hold the structure together. Such attraction would also explain how a chain could duplicate itself, like to like. Crick put the idea to Griffith over drinks one evening after a talk at the Cavendish by Thomas Gold, a vigorously imaginative theoretical astronomer, on "the perfect cosmological principle." Watson recalled—and among his anecdotes it's one that best captures the effect of Crick in full spate—

> Tommy's facility for making a far-out idea seem plausible set Francis to wondering whether an argument could be made for a "perfect biological principle." Knowing that Griffith was interested in theoretical schemes for gene replication, he popped out with the idea that the perfect biological principle was the self-replication of the gene—that is, the ability of the gene to be exactly copied when the chromosome number doubles during cell division.

What neither Crick nor Watson knew, however, was that Griffith had already been speculating about pairing bases, although not like to like but rather one purine joined to the other (the two large ones, adenine to guanine) and one pyrimidine to the other (the two small ones, cytosine to thymine). Griffith said nothing about his own pairing idea, but agreed to try Crick's.

The basis in quantum mechanics for such calculations, when the units were as large as purines and pyrimidines, was so tentative that Griffith himself regarded his results as highly approximate. None the less, several days later when the two men were standing in line for tea at the Cavendish, Crick asked Griffith whether he had completed the calculations. "Yes," Griffith said. "I find that adenine attracts thymine and guanine attracts cytosine." To which Crick remembered that he answered, "Well, that's all *right,* that's perfectly *okay,* A goes and makes B, B goes and makes A: you can just have complementary replication."

Enter the unquiet spirit of Erwin Chargaff—but softly, at stage right, so that Watson is aware of his presence but Crick not yet. The pairings of adenine with thymine, and guanine with cytosine, are the one-to-one ratios that were submerged in the welter of data that Chargaff had been publishing. Watson's remembrance here differed from Crick's. Watson said in *The Double Helix* that he knew of Chargaff's ratios and had mentioned them to Crick not long before the conversations with Griffith. Crick said recently, though, that if he had been aware of Chargaff's rules he would necessarily have made more of Griffith's results. "Because I did *not,* you see, I am quite sure that I was not conscious of Chargaff's rules at the time. If Jim had told me, I had forgotten them." In the work of a successful collaboration, one can sometimes make out the contrasting approaches that merged into the shared result; in the spring of 1952, one can see in some detail how what Watson knew and what Crick knew, even when they overlapped, had failed to fuse.

What excited Crick in his exchange with Griffith was the idea that replication was complementary. Complementary replication, and the closely related concept that molecules can have surfaces that act as templates on which other molecules are formed, permeate biology today with immense explanatory power. In 1952, these were a pair of theoretical notions—frequently discussed, perhaps, but still in search of their first exemplars. The physical basis for complementary rather than like-with-like replication had been set out with due polemical vigor twelve years earlier in a joint paper by Linus Pauling and Max Delbrück—their one collaboration. I once asked Delbrück how that paper came to be written.

"You haven't heard the story? It's funny," he said. "What happened was that I saw a paper by Jordan, Pascual Jordan, in which he claimed that two macromolecules of identical structure would attract each other. By quantum-mechanical resonance phenomena. And this would explain synapsis of chromosomes—you know, how in higher organisms the chromosomes sort themselves out into correctly matched sets in the cell divisions before sperms or eggs are formed. And how this happens is still a mystery. Anyhow, Jordan claimed this was a direct consequence of quantum mechanics. But I read the paper and was somewhat curious about it.

"That was the summer of 1940. I was already at Vanderbilt, but came back to Pasadena for a few weeks because I had to go to Mexicali, to cross the border, to exchange my student visa with an immigration visa. I met Pauling in the corridor and told him about Jordan's paper, and he went over with me to the library and we looked at it together; and he decided within five minutes that it was baloney." Delbrück smiled with a glint of mischief and said, slowly, "Those were his very words. And I was impressed by the certitude with which, and the *rapidity*"—the bubble of laughter burst—"with which he came to this conclusion, in the face of Pascual Jordan who was one of the founders of quantum mechanics; anyhow he had no doubts.

"A couple of days afterwards, I met Pauling again in the corridor, and he said, 'Oh, Max, I have written a note to *Science* about this Jordan thing and would you like to sign it too?' I was astonished. Well, I succeeded in mitigating some of the rudest comments! I have never written a paper in which I had less part." He laughed again. "Except that I called his attention to the paper by Jordan. The funny thing is that in retrospect, Pauling has forgotten what he did say in the paper, I don't think he has re-read it. And any number of times he has given lectures in which he claims that at that point he"—Delbrück's voice deepened to solemnity—"really foresaw the whole DNA story. But when you read it, there was almost nothing of that in there."

In 1940, Pauling was thinking almost exclusively of proteins. What the paper did contain was a succinct restatement of Pauling's elementary structural doctrines:

> It is our opinion that the processes of synthesis and folding of highly complex molecules in the living cell involve, in addition to covalent-bond formation, only the intermolecular interactions of van der Waals attraction and repulsion, electrostatic interactions, hydrogen-bond formation, etc., which are now rather well understood. These interactions are such as to give stability to a system of two molecules with *complementary* structures in juxtaposition, rather than of two molecules with necessarily identical structures; we accordingly feel that complementariness should be given primary consideration in the discussion of the specific attraction between molecules and the enzymatic synthesis of molecules.
>
> A general argument regarding complementariness may be given. Attractive forces between molecules vary inversely with a power of the distance, and maximum stability of a complex is achieved by bringing the molecules as close together as possible, in such a way that positively charged groups are brought near to negatively charged groups. . . . In order to achieve the maximum stability, the two molecules must have complementary surfaces, like die and coin, and also a complementary distribution of active groups.

Crick had not read the paper; nor had Griffith. To Watson, from the Delbrück circle, the argument was "well worn."

At about that same time, Crick learned from Rosalind Franklin why she considered that the evidence was compelling against a helical structure for DNA, at least in the crystalline, A form. "I met her in the queue at some meeting at the Zoological Lab," Crick said. "I had very little real impression of Rosalind, after all, before the whole business in 1953. After we got the structure, I came to know her better. . . . Beforehand, I'm afraid we always used to adopt—let's say a patronizing attitude towards her. When she told us DNA couldn't be a helix, we said, 'Nonsense.' And when she said but her measurements showed that it couldn't, we said, 'Well, they're wrong.' You see, that was our sort of attitude." I mentioned Franklin's photo of the A form in which the intensities were too asymmetrical to fit a helix. "That's right. That was the one." Crick said that with helical molecules it could be easy, with a small, and unusual, asymmetry in the parallel packing of the molecules in the specimen, to produce the kinds of asymmetries that Franklin had found in the x-ray reflections. "So the result could be spurious, you see. Which Rosalind and

Gosling didn't understand. I never saw the photo at that time or went over it in detail, but she told me about it the time we met and I just said, 'You know, I think it's misleading.'"

As he acknowledged, Crick's manner may have been patronizing; yet Franklin betrayed a grievous slowness of intuitive response by the way she underestimated the man and his advice in that encounter.

Then Chargaff came to Crick's attention inescapably. He arrived in Cambridge one day late in May. John Kendrew arranged a meal at Peterhouse, his college, at which he presented Watson and Crick to Chargaff. A meeting of three men more prone to volatile contempt would be hard to conjure. All three remember the encounter vividly. "He derided my hair and accent, for since I came from Chicago I had no right to act otherwise," Watson wrote in *The Double Helix*. "The high point in Chargaff's scorn came when he led Francis into admitting that he did not remember the chemical differences among the four bases."

Chargaff placed the emphasis differently when I went around to see him. The Columbia Presbyterian Medical Center, where he worked, is a great walled hospital city crammed with activity on the northern cliffs of Manhattan a quarter mile from the George Washington Bridge; Chargaff's laboratories and office were high up in the quiet of the inner citadel. The walls were concrete block, softened by many coats of paint, pastel green. The windows by his desk viewed the river, the New Jersey Palisades, and the bridge. A big Webster's dictionary lay open behind him.

"I take a minority view of science, of course," Chargaff said. "It's a view shared more by philosophers and humanists. The public acclaim given to scientists is exaggerated: scientists are as bright as they happen to be individually. In my opinion, present-day science, especially biological science, is a direct symptom of the decline of the West: all this shameless talk about creating and multiplying will be put down as the barbarism of the twentieth century." He stopped to light a pipe. He was wearing a white lab coat; his white hair was thick and rumpled, his face tanned, with clear, crisply cut features and a flexible mouth. I asked what he remembered about his first meeting with Crick and Watson. The pipe alight, he took it from the center of his mouth.

"Watson's description is fairly correct. Kendrew brought me together with them, not at dinner, as Watson says, but at lunch. They impressed me by their extreme ignorance. Watson makes that clear! I never met two men who knew so little—and aspired to so much. They were going about it in a roguish, jocular manner, very bright young people who didn't know much. They didn't seem to know of my work, not even of the structure and chemistry of the purines and pyrimidines. But they told me they wanted to construct a helix, a polynucleotide to rival Pauling's alpha helix. They talked so much about 'pitch' that I remember I wrote down afterwards, 'Two pitchmen in search of a helix.'

"I explained to them our observations on the regularities in DNA, and told them adenine is complementary to thymine, guanine to cytosine, purines to pyrimidines; that any structure would have to take account of our complementary ratios. It struck me as a typically British intellectual atmosphere, little work and lots of talk. Crick and Watson are very different from each other. Watson is now an able, effective administrator of science. In that respect he represents the American entrepreneurial type very well. Crick is something else—brighter than Watson, but he talks a lot, and so he talks a lot of nonsense."

I said that in reading some of his papers I had formed the impression that he had been cautious, even skeptical, for many years about the Watson-Crick model of the structure of DNA. "I would say we must be cautious still," Chargaff said. "Models in science are necessary as clothes hangers on which ideas and theories can be hung up. But they are usually simplifications. They can be useful in thought; their epistemological value is somewhat limited. Later discoveries have shown that there are many forms of DNA—single-stranded forms, circular forms—and we have been misled, I think, by concentrating so much on micro-organisms. We still know very little of what happens *in vivo*, especially in animal and plant cells. In the chromosomes of higher organisms, the role of nucleoproteins is certainly as indispensable as DNA and evidently much more complex. None the less, I must say that I myself underestimated the role of base pairing, the multitude of functions it performs in the cell."

A few minutes later, Chargaff added, "Pauling in *his* structural model of DNA *failed to take account* of my results. The consequence was that his model did not make sense in the light of the chemical evidence." At the time Chargaff met Watson and Crick, I asked, had he himself understood the implications of the one-to-one ratios he had discovered among the bases? Chargaff answered slowly and distinctly. "Yes and no," he said. "No, I did not construct a double helix."

We talked about Friedrich Miescher and Oswald Avery, and Chargaff modulated from philippic to elegiac. Then he said, "I am against the over-explanation of science, because I think it impedes the flow of scientific imagination and associations. My main objection to molecular biology is that by its claim to be able to explain everything, it actually impedes the flow of free scientific explanation. But there is not a scientist I have met who would share my opinion."

By Crick's account of the meeting, he and Watson put Chargaff "slightly on the defensive." Thinking about structures, "We were saying to him as protein boys, 'What has all this work on nucleic acid led to? It hasn't told us anything we want to know.'" One imagines, perhaps, the conversation of Alaric the Goth with Pope Innocent before the sack of Rome. Chargaff answered, Crick said, "'Well, of course there's the one-to-one ratios.' So I said, 'What's that?' So he said, 'Well, it's *all* published.' Course, I'd never read the literature, so I wouldn't know. So he *told* me. Well, the effect was electric, this is why I remember it. I suddenly realized, by God, if you have complementary replication, you can *expect* to get one-to-one ratios."

But Crick had forgotten which bases Griffith had told him went with which. After meeting Chargaff, he went to Griffith to ask which his pairs were; and by that time he was no longer sure which ones Chargaff had said, and so had to look them up in the literature. To his delight, Griffith matched Chargaff. "So you remember from the history of this thing that *I* realized that the one-to-one ratios could mean complementary replication. It's an interesting question whether *nowadays*, if one had had that idea one would have published it straightaway. Because one could perfectly well write a little note saying one-to-one ratios mean things go together and this in turn means complementary replication. But I should be cautious even now, and in those days it didn't even *occur* to me to publish without some sort of supporting evidence. We could afford to have a beautiful idea like that and just let it sit around for a year. And just talk to our friends, and nobody else would *know*."

Crick was excited enough to think up an experiment that might show whether bases in solution would spontaneously pair in any preferential way, excited enough that he even tried his hand at the laboratory bench himself for a week at the end of July. "The experiments were simply just mixing things in solution and then looking at the way they absorbed ultraviolet light. But one knows now that with the technical means I had available one wouldn't have found it, and I didn't find it. Also I had other things to do. You must remember that I was a research student in my fourth year, and in those days to produce anything at all in protein crystallography, let alone a *thesis*, was really something. Bragg was worried. Jim, on the other hand, had already got his Ph.D. and could afford to be more relaxed."

◆ ◆ ◆

On July 2, Franklin reopened her red notebook, chose a fresh left-hand page, and wrote, "Notes on first cylindrical Patterson. . . . There is *no* indication of a helix of diameter 11 A. . . . *If* a helix, there is only one strand. . . . *If* a helix, it is v far from continuous uniform density. . . ."

A fortnight later, she and Gosling sent around a notice on a small white card, lettered in Franklin's hand, with a neat black border: "It is with great regret that we have to announce the death, on Friday 18th July, 1952, of D.N.A. HELIX (crystalline). Death followed a protracted illness, which an intensive course of Besselized* injections had failed to relieve. A memorial service will be held next Monday or Tuesday. It is hoped that Dr. M.H.F. WILKINS will speak in memory of the late HELIX."

Stokes kept his copy of the card. He said, when I asked about it, "I have very little memory of an actual meeting; the only thing I do remember clearly is a conversation in her room, being shown what must have been a Patterson map, and rather agreeing with her that it couldn't be a helical structure—but rather taking her word for it."

◆ ◆ ◆

Watson took a couple of months off from the hard-edged domain of crystallography and physical chemistry. Towards the end of July, while Crick was experimenting with solutions of bases, Watson went to Paris for a week for the International Congress of Biochemistry, a triennial event. Among the hundreds of papers presented, it is hard to find even three that he might have bothered to listen to. Kalckar gave a talk. Roy Markham, who had started Watson off on tobacco-mosaic virus at Cambridge, spoke about the structure of RNA, but vaguely by Watson's molecular standard. Chargaff was there, but

---

*The word is not written clearly; it refers to Bessel functions, the name of the mathematical tool used in helical-diffraction theory.

his paper offered no data that were not familiar. His name also appeared as co-author on a paper read by a junior colleague, Christoff Tamm. In Chargaff's lab, Tamm had been trying to get at the sequences of bases in DNA. Techniques for this were not devised until nearly a quarter of a century later. Tamm's paper in Paris concluded that the structure of DNA was "a chain in which tracts of pyrimidine polynucleotides alternate with stretches in which purine nucleotides predominate." In other words, the one-to-one ratios Chargaff had described to Watson and Crick a few days before did not necessarily apply at each location along the molecule. Watson did not go to hear either paper. Chargaff and he passed with a nod in a courtyard of the Sorbonne. Pauling appeared on short notice to talk about the alpha helix—the State Department a few days before having issued him a passport good only for France, England, and three months. Wilkins was in Paris, too, on his way to Rio. Delbrück was there, though no friend of biochemistry.

Then Watson and some others moved on to the medieval Abbey of Royaumont, outside Paris, for an International Phage Colloquium, organized by André Lwoff, of the Institut Pasteur. On the train, somebody stole Watson's luggage. At the meeting, Hershey talked about the blender experiment. Watson spoke about tobacco-mosaic virus. Watson remembered later that Pauling asserted at Royaumont that the solution of the structure of DNA would require x-ray studies of its components, like those that he and Corey had carried out on simple peptides before constructing the alpha helix. Lwoff, some years afterwards, set down his memory of Watson:

> It is evening in the solemn drawing room of the Abbaye. In the room is a fifteenth-century oak table, on which there is a bust of Henri IV. A young American scientist, wearing shorts, has climbed on the table and is squatting beside the king. An unforgettable vision!

Watson spent August in the Italian Alps, and a photograph shows him boyish, fit, and vital. Early in September, he went to the Second European Symposium on Microbial Genetics, at Pallanza, on Lake Maggiore. There he learned that several bacteriologists—Joshua and Esther Lederberg in the United States together with Luca Cavalli-Sforza in Italy, and William Hayes independently in England—had found that the increasingly intricate calculus of bacterial genetics made much better sense if one supposed that some bacteria were male, others female.

The idea that bacteria sometimes reproduce not simply by splitting but by conjugating, or mating, was not itself new. Joshua Lederberg, together with Edward Tatum, had demonstrated as much in 1946, for a strain of E. coli and mutants. As Watson noted, Lederberg in 1946 was only twenty. Lederberg had imagined that the mating fused the two bacteria and pooled their entire genetic resources. With zealous ingenuity, he then brought on all the tricks and crosses of the classical geneticist, trying to map the bacterial chromosome much as if it belonged to maize or fruit flies. His work made clear that bacteria had a linear chromosome—one more step in establishing that their genetics were comparable to higher organisms. In six years the map had grown exquisitely paradoxical. Whenever Watson read Lederberg's papers he got stuck, he said, in their "rabbinical complexity."

The revelation that mating bacteria did not fuse after all, but that some donors, promptly called "males," possessed a genetic sex factor that enabled them to transfer genes directly, by contact, to other bacteria called "females," offered the resolution of Lederberg's mapping paradoxes. At the meeting in Pallanza, Watson met Hayes for the first time. Hayes wanted to take the idea of bacterial gender one step further than the Lederbergs were willing to go, by adding the simple, if arbitrary, idea that a male bacterium copies out and donates to the female only a small segment of its genes.

Compressed in these discoveries and arguments, folded tightly like the kind of Japanese paper pellet that on being dropped into water dreamily exfoliates into a many-colored flower, lay an entire other line of development in molecular biology—one that is richly productive to this day, one that was congenial to Watson's training and cast of mind. Watson grew fascinated. He got on well with Hayes, who was a microbiologist working at Hammersmith Hospital, in the west of London, and was three years older than Francis Crick. Watson came back to England after his summer on the Continent, he wrote, with his head full of thoughts about the sex life of bacteria. He soon saw a way to use Hayes's theory in a clever redrafting of Lederberg's map: suppose *E. coli*'s genes were located not on just one but on three chromosomes. The idea "seemed a good one at the time," Hayes later wrote. In *The Double Helix*, Watson said, "As soon as I returned to Cambridge, I beelined out to the library containing the journals to which Joshua had sent his recent work. To my delight I made sense of almost all the previously bewildering genetic crosses. . . . Particularly pleasing was the possibility that Joshua might be so stuck on his classical way of thinking that I would accomplish the unbelievable feat of beating him to the correct interpretation of his own experiments."

Watson began travelling up to London regularly to talk with Hayes. They wrote a paper together. Watson brought Hayes the rest of the way into the phage group, got Delbrück to invite him to talk at the Cold Spring Harbor Symposium the following summer and to work at Caltech on a six-month fellowship the following fall, when Watson would be there too. Hayes's view of bacterial conjugation proved right in essentials, though it was some years before Lederberg conceded as much. Watson's multiple chromosomes, and the joint paper, have been forgotten. And still the energy is so primitive and the giant-killing urge so strong that, for a moment there, one has a glimpse of an alternative future in which the race is to produce the Watson-Hayes model of bacterial genetics and the redoubtable competitor across the sea is Joshua Lederberg.

❖ ❖ ❖

On August 14, Rosalind Franklin reverted to taking x-ray pictures, "to look for double orientation," she wrote in her notebook. This was an attempt to repeat the effect that had produced the one pattern, the previous Easter, that more than anything else had convinced her that structure A was not helical. She took pictures almost every day until September 9, and one more, No. 78, on October 14, all without success.

3 *"Then they ask you, 'What is the significance of DNA for mankind, Dr. Watson?'"*

Politeness, Francis Crick said over the BBC at the time he got the Nobel prize, is the poison of all good collaboration in science. The soul of collaboration is perfect candor, rudeness if need be. Its prerequisite, Crick said, is parity of standing in science, for if one figure is too much senior to the other, that's when the serpent politeness creeps in. A good scientist values criticism almost higher than friendship: no, in science criticism is the height and measure of friendship. The collaborator points out the obvious, with due impatience. He stops the nonsense, Crick said—speaking of James Watson. Rosalind Franklin, in the eighteen months or so from the fall of 1951 to the spring of 1953 when she was working at King's College London on the structure of deoxyribo-nucleic acid, had no such collaborator. It is evident from her notebooks that she needed one, and clear from what we know of her character that she would have worked well—candidly, rudely if need be—with the right one. Franklin died of cancer in April of 1958, at the age of thirty-seven. She was one of the four people closest to the discovery of the structure of DNA.

If it is difficult to form a picture of her from the recollections of those who worked with her, that is chiefly because, unlike the three others, she was without great idiosyncrasy. Firmness of mind and a thoroughgoing scientific professionalism were the dominant impressions Franklin left; she was, above all, determined. She was slim, and walked with an awkward energy. She had clear olive skin, thick, glossy dark hair, good teeth, and brilliant eyes; she dressed sensibly, neatly, tastefully; neither in appearance nor in manner was she masculine; she managed the careful balance that made her sex irrelevant to her intellect. Some found her charming. Some found her feminine. Some found her outgoing. Some found her passionate in opinion and argument. She left these memories vividly—but with only a few people. She was reserved. She made her life so that she would be judged as a scientist—or, at least, it's as a scientist almost exclusively that fellow scientists think of her. She was thirty-one when Watson first met her: eight years older than he, a little younger yet much further along professionally than Crick. How she really affected the extreme and irritable sensitivity with which Watson scanned the human landscape we cannot tell, after all. The "Rosy" in his book is a stage grotesque in prop spectacles and costume blue stockings that Franklin never wore in life, while the nickname he picked up from the King's College lab, where it was used behind her back. ("Because of her rather cold manner," Wilkins wrote to me recently. "I thought the nickname in rather bad taste, but it came into general use." Wilkins used it too.) She had enjoyed the years she had worked with carbons in Paris. The move to King's College London offered some annoyances, of which the worst appears to have been the fact that as a woman she was denied the fellowship of the luncheon club organized by the senior common room. She made no open protest. Although it is true that women were often discriminated against in science in England then, it's also true that Randall's biophysics laboratory at King's College London offered better opportunities to women scientists than most places did. A postwar creation, the lab was notably informal in organization and youthful in staff. Numbers fluctuated, but out of thirty-odd professional scientists there, towards the end of Franklin's second year, eight or nine were women and four of those outranked her.[*] Franklin was profoundly angered, her friends have said she told them, by the way she thought Maurice Wilkins tried to treat her as an assistant rather than a colleague. She got on with the work she believed she had been hired to do.

In 1953, Franklin moved to Desmond Bernal's laboratory at Birkbeck College. She was joined there a year later by Aaron Klug, and the two worked closely together until her death. "She was—well, a firm sort of person," Klug said recently. "She was—I can't say she was like a man, not like that at all; but one didn't think of her being particularly like a woman; she wasn't shy, or self-effacing—but she wasn't blustering, either. She spoke her opinions firmly. I think people were unaccustomed to dealing with that in a woman; I think they expected women to behave—rather differently, quieter. She expected reason to dominate. She was very much a rationalist. I had discussions with her at various times. She could never see the force of non-rational reasons. She was very single-minded. She worked *beautifully*. The kind of single-mindedness that she had made her an absolutely first-class experimental worker. And that sort of skill requires intelligence and determination—it's not just good needlework."

Crick knew Franklin hardly at all in 1952 and 1953. "After we got the structure I came to know her better," Crick said a while ago. "We not only became much more friendly, but she came to consult me quite often on matters of x-ray crystallography. And she and my wife got on well. But, you know, it is very difficult to convey exactly what she was like. I just knew her as another

[*]Franklin's biographer, Anne Sayre, wrote that Randall's laboratory had only one other woman scientist while Franklin was there (*Rosalind Franklin and DNA*; New York: Norton, 1975, p. 99); but Sayre was wrong. A report that Randall prepared for the Medical Research Council in December 1952 listed thirty-one scientific staff. Eight, including Franklin, were women. Not listed was Mary Fraser, who had left the laboratory by then, and Sylvia Fitton Jackson, who was one of the lab technicians—but an unusual technician, for she had already completed and published research at the laboratory and soon afterwards, with Randall's backing, went to Cambridge, where, with no previous degrees, she took a Ph.D. Of these ten women, two have died, Franklin and Dr. Jean Hanson, who was the group's senior biologist. Of the eight living, I have talked with five and corresponded with two others; the eighth was a research student whom I have not been able to trace. The seven agreed in the opinion that as science in England was then practiced, women at their laboratory were well treated.

scientist, really, and as someone who became a sort of family friend. I found her quite a good scientist. She made a great combination with Aaron, because Aaron was a very good, really solid theoretician and Rosalind was really an experimentalist. She was sound, hard-working, knew the right thing to do. And so on. She was always an adequate theoretician—but she wasn't what I would call very imaginative."

Franklin's only close associate when she was working on DNA was Raymond Gosling, the research student who was doing his doctoral dissertation under her direction. "Rosalind had this professional shell; underneath it she could be delightful and relaxed," Gosling said to me. "I think you've got to remember that it was difficult for women in science, much more then than it is now. She was a very intense person, almost eccentrically so. Some people doubtless called her a bluestocking because she was terribly interested in her work. Her interest in music and in art, and so on—though she had great enthusiasms—weren't obvious until you got to know her well. And she was shy, I think—certainly not the person to let her imagination fire up openly about structural ideas to somebody like Maurice. But *we* used to have terrific arguments together. Her great strength was that you could have this *very* frank discussion about the work, and it *never* got personal, it was objective, and it would push along to *reach* somewhere. But she would never get like that with Maurice. She would do that with me. And if at that time she had had someone of her own standing to have those frank and fierce discussions with, it might have helped. And I felt repeatedly that Maurice was trying various ways to stimulate Rosalind into saying something about the structure, but she for her part would say, 'We are not going to speculate, we are going to wait, we are going to let the spots on this photograph tell us what the structure is.' And so since there was nobody here Maurice could talk to who was willing to speculate about the structure, he talked to Crick and Watson about it."

❖ ❖ ❖

In the summer of 1952, Franklin and Gosling began the cumbersome mathematical technique, called a Patterson synthesis, by which they intended to reach the structure of the crystalline, A form of the DNA molecule. "The physical meaning of this so-called Patterson synthesis is one of the most difficult conceptions in crystallography," Max Perutz and John Kendrew had said in an article they wrote together three years before Franklin's attempt. Perutz and Kendrew knew as well as anyone how infuriatingly elusive the method was. They had spent all their ingenuity and countless hours in the effort to apply the method to proteins, with only the glimmering wisps of success sufficient to keep them trying. We need not follow them, and Franklin, very far into that swamp. The mathematician Henri Poincaré, many years ago, gave a lecture about the psychology of mathematical invention to the Psychological Society of Paris. Poincaré warned his audience, "I may tell you that I have discovered the proof of a certain theorem in certain circumstances; the theorem will have a barbarous name, which many of you will never have heard; but that is of no importance: what is interesting, for the psychologist, is not the theorem—it's the circumstances." We can take refuge in Poincaré: Patterson synthesis is one such barbarous theorem.

To be sure, fond crystallographers find this difficult offspring beautiful—an indescribably subtle means of mapping the structure of a molecule without actually having enough information. A Patterson synthesis initially produced what looked like a geologist's contour map, all loops and meandering lines, of a particularly up-and-down square mile of the Dakota Badlands. To reason from this map back to the real structure was twistingly abstruse, and felt like putting one's mind through a sieve. The promise was that in the first step certain features would be revealed—the presence of rods or chains and something about them, or alternatively the presence of a layered structure and something about that, and so on. These clues, if the structure were sufficiently regular, might define its character so that the rest of the data could be interpreted fully in three dimensions. DNA is highly regular. Proteins, as it turned out, are not, except for occasional runs of alpha helix and patches of pleated sheet. Ironically, as Franklin was settling down to the Patterson synthesis of the A structure of DNA, Perutz, Kendrew, and Sir Lawrence Bragg were searching, in something close to despair, for another approach to interpreting the x-ray-diffraction patterns of hemoglobin.

The measurements and calculations that would now be done by machine Franklin and Gosling did for DNA by hand. They chose the most detailed of the photos of the A form and projected them like magic-lantern slides onto white cardboard, enlarged about ten times. They marked the positions of the spots, then estimated the intensity of each by comparison with a brightness scale projected alongside. They quickly found that the eye made the estimates more accurately than any instrument then available.

By the beginning of July, Franklin had confirmed her opinion of the previous fall, at the time of the colloquium Watson attended, that the crystals were monoclinic and face-centered, and she still failed to draw the consequence that the molecule contained two chains running in opposite directions. On July 2, a Wednesday, she sketched in her notebook the main contours of the first Patterson map—four vaguely crescent blobs and a somewhat larger irregular peak in one corner shaped like a sheepskin rug. She measured off three distances. The sketch yielded nothing but perplexity. The feature that caught her attention was "the central banana-shaped peak." This form, she thought, repeated across the map, in the other crescents. It could be made to fit a curve calculated for a single-stranded helix with a diameter of 13.5 angstroms, but that diameter was impossible to reconcile with other dimensions of the crystal.

Puzzling over structures with single helices that she could rule out with ease, Franklin never recalled the idea of "2, 3, or 4 co-axial nucleic acid chains per helical unit" which she had mentioned five months earlier in her fellowship report. "If there is a *flat* banana-like unit in structure, with banana axis parallel to fibre axis," she now wrote, the explanation must be "rather a double-sheet structure"—and she sketched a dozen more tiny bananas viewed from the top and then from the side. The red notebook that had begun as her bench record of x-ray camera work was now the place to which she retreated in the attempt to put her difficult gropings into coherent form. That Wednesday, she fretted about helices, bananas, and double sheets for a couple of pages. By the following Monday the calculations had been extended, and her notes were concerned with peaks that she would expect to observe but that were not to be found. On the fifteenth, in five curt lines, she concluded that "there is no narrow straight chain of high density parallel to the axis" of the fibres. Then not a line more for five weeks.

❖ ❖ ❖

When Watson got back to the Cavendish Laboratory, towards the end of September, from his summer of conferences, he found that he and Crick had two new colleagues. Both were crystallographers from the California Institute of Technology: Linus Pauling's son Peter, in Cambridge as a research student of John Kendrew's, and Jerry Donohue, a former graduate student of Linus's, who was there on a postdoctoral fellowship and had been given a desk in Watson and Crick's office. The room was large and plain, with four desks around the walls—Crick's by the window on the right and Watson's opposite—and a table in the middle. That fall, also, Watson moved into rooms in Clare College, on the river, with some of the most beautiful gardens along the Backs; had stomach trouble; remained convinced that magnesium ions bound DNA together; began taking French lessons; concluded with Crick that RNA most likely played an intermediate role between DNA and the manufacture of proteins and taped to the wall above his desk a piece of paper with a legend to that effect; began a paper on bacterial sex factors and chromosomes with Hayes in London; had dinner several times with Wilkins, on the way home from meetings with Hayes.

❖ ❖ ❖

Wilkins, when he got back to King's College London that fall from lecturing in Rio de Janeiro, began a microscopic study of plant cells—using the pollen of the common spiderwort—in order to measure the total amount of proteins and nucleic acids in the cell, and their relative increase during growth and cell division. The method was a new one, which he had devised himself. That fall, he and Franklin were working completely separately.

❖ ❖ ❖

Linus Pauling, when he got back to Caltech from his conferences in Europe, completed a manuscript on the alpha helix in keratin in which he showed that in these fibres the alpha helices themselves are supercoiled, like ropes laid together into a cable. The geometry of supercoiling at last accounted for the anomalous spot in the diffraction patterns that had misled crystallographers for so long. He mailed the manuscript across the Atlantic to *Nature*, where it arrived within a day or two of Crick's paper on the same subject, the "coiled coil"—a coincidence that was to cause an argument over priority. Pauling then began to read intensively, though more selectively than he realized, through the literature about the physical chemistry of DNA, and to scrutinize x-ray-diffraction pictures of the stuff. One paper he read that had just come out—for scientific journals then, as now, often suffered a backlog of many months—was Crick's full mathematical treatment, with William Cochran and Vladimir Vand, of the x-ray patterns produced by helical molecules generally. Indeed, any advantage in information that Crick and Watson had had at the time of their first try for the structure, a year earlier, had now been dissipated by the appearance of one paper or another. Thus,

Pauling read the paper published that summer of 1952 by the Cambridge biochemists led by Alexander Todd, who had determined chemically the atoms at which the backbone of the DNA molecule link together—phosphate to third carbon of the sugar ring, fifth carbon of the sugar to next phosphate. He recalled that Todd had told him about the linkage during a meeting of the American Chemical Society in New York in 1949. Pauling also examined the latest evidence from the electron microscope—a paper of the previous spring, another still in press—and concluded that the diameter of the DNA molecule was 18 to 20 angstroms. He read Sven Furberg's proposal of a standard three-dimensional configuration for the nucleotides, and of a single-strand helix, which appeared that summer in a Scandinavian chemical journal—three years after Furberg had put them into his doctoral dissertation under Bernal's supervision in London. Pauling also read another, less consequential but more accessible paper by Furberg, which established crystallographically the right-angled structure that one of the bases—cytosine, as it happened—made with the sugar. Of course, Pauling's own understanding of bond lengths and bond angles was unrivalled in the world. Pauling re-read everything William Astbury, at Leeds, had published about DNA before and after the war. He apparently sought out nothing by Erwin Chargaff, though the two men had met. I asked Pauling about that omission. "I knew about Chargaff's one-to-one ratios of adenine and thymine, guanine and cytosine; I had seen Chargaff in December of 1947 and he had told me," Pauling said. "I knew that. But I hadn't carried that over into a physical picture."

On November 25, Pauling went to a seminar where a visiting biologist, Robley Williams, who was working with tobacco-mosaic virus at the University of California at Berkeley, showed an electron-microscope photograph and some x-ray-diffraction pictures of molecules of ribonucleic acid—not DNA—and talked about what their dimensions and density seemed to be. Pauling set to work the next day. "Using the density, I calculated the number of polynucleotide chains per unit to be exactly three," he wrote in 1974.

> This result surprised me, because I had expected the value 2 if the nucleic acid fibres really represented genes. I decided, however, that probably the fibres were artefacts, produced by the process of extraction. . . . I am now astonished that I began work on the triple helix structure, rather than on the double helix.

Pauling also had some diffraction photos that Robert Corey had made, but these were poor. Sometime later that winter he asked Alexander Rich, a postdoctoral fellow, to take some more. But Caltech had neither the fine-focus, high-intensity x-ray equipment nor the carefully prepared DNA fibres that Rosalind Franklin had.

◆ ◆ ◆

By the end of November, Franklin and Gosling completed the first stage of their Patterson synthesis of the A form of DNA. They could not discern the structural regularities they were seeking as clues to the next step. Franklin still believed that the A structure could not be a helix. She was preparing two papers for publication. The first was about the way water content caused DNA fibres reversibly to lengthen and to change from the A to the B structure. The second described the preliminary Patterson map of the atomic relationships within the A structure. At the same time, with no model in mind yet, she and Gosling were beginning to try to convert their data to a full, three-dimensional solution.

At the end of November, Randall, at King's College, asked each scientist or team in his laboratory to write up a few paragraphs describing work in train, so that he could put together a report to a Biophysics Committee of the Medical Research Council. The council was where the money came from, and the Biophysics Committee had been set up five years earlier as a way to improve communications among all the different laboratories in England and Wales financed by the council and doing related research. The committee was made up of the heads of those units and so included, besides Randall, Max Perutz. Randall's report was twelve mimeographed foolscap pages; it had no photographs. Franklin and Gosling's paragraphs said several things about DNA explicitly and precisely. They wrote that the transition from the first to the second structure was marked by a change from 28 angstroms to 34 angstroms—a little less than twenty-four billionths of an inch—in the distance, along the fibre, at which the structure began to repeat itself. They also listed the dimensions of the unit cell of the crystals of the A form—three lengths for the edges, and an angle—and said that the crystals were monoclinic and face-centered. These dimensions were not much different from the preliminary figures she had included in her notes for the colloquium that Watson had attended thirteen months earlier. The Biophysics Committee, including Perutz, visited King's College on December 15, and copies of the report were given to everyone. A fuse was lit.

❖ ❖ ❖

A few days later, Peter Pauling came into Watson and Crick's office to say that he had just received a letter from his father, who had mentioned in passing that he had devised a structure for DNA. In *The Double Helix*, Watson looked back on the day with histrionic despair and determination. "Francis then began pacing up and down the room thinking aloud, hoping that in a great intellectual fervor he could reconstruct what Linus might have done. As long as Linus had not told us the answer, we should get equal credit if we announced it at the same time." No solution emerged. A few days later, Watson saw Wilkins in London and told him the news. "I was hoping that the urgency created by Linus' assault on DNA might make him ask Francis and me for help." Wilkins answered that Franklin had said she was moving to Birkbeck that spring and would not take DNA with her, and as soon as she was gone he would settle himself down to the problem. So Watson went off to Switzerland for a Christmas skiing holiday.

❖ ❖ ❖

A scientist who had worked at Caltech that winter told me that on Christmas Day, Pauling, saying he was depressed about the political difficulties he was having—these included accusations of Communist sympathies made by a witness before a committee of the House of Representatives, and conflict with the administration of Caltech over his campaign against nuclear weapons—asked some colleagues in to see the model he had built of DNA, and after showing it off felt cheered up. On December 31, Pauling and Robert

Corey sent off their manuscript, describing the model and the evidence, to *Proceedings of the National Academy of Sciences.* Pauling that day wrote another letter to Peter in Cambridge, in which he mentioned that the manuscript had been mailed and asked whether Peter would like to see a copy. On 2 January 1953, Pauling and Corey sent a note to *Nature,* twenty-four lines only, announcing the structure, characterizing it, and saying that the detailed description was appearing in the February issue of the *Proceedings.* The letter was printed by *Nature* on February 21.

◆ ◆ ◆

The drums were beating. On January 6, Franklin wrote to Corey to ask for details about the structure. Corey did not reply until April 13.

◆ ◆ ◆

Watson, his skiing holiday over, went to Milan to see Luca Cavalli-Sforza and talk about bacterial genetics some more. He got back to Cambridge towards the middle of January. He urged Peter Pauling to write for a copy of his father's manuscript.

◆ ◆ ◆

On Monday, January 19, in Pasadena, Pauling held a press conference—at which he announced to the world the supercoiling of the alpha helix in fibrous proteins. He did not mention nucleic acid. *The New York Times* gave him a couple of paragraphs the next day and a longer, half-comprehending article the following Sunday, in which the reporter wondered about curing cancer. On January 19, in London, Franklin began a new notebook for the new year, a large thick one with a tan cardboard cover. The previous week, she had written to Bernal, saying, at the end of the note:

> I thought you might like to know, too, that I am giving a colloquium here—together with Gosling—on the nucleic acid work, on Wednesday January 28th. Probably this will not interest you, as there will not be time to go into things in as much detail as has been possible in my private discussions with you. But in case it does, it is to be at 6.30 p.m. in room 27C.

The calculations for the three-dimensional Patterson map of the A form were far enough advanced so that it was time to see what the structure might be. Indeed, at last, it was time to build models. She had abandoned the parallel "double-sheet structure" of "banana-like units" that she had sketched in her notebook the previous summer, but was still rejecting a helix, and she reasoned her way neatly down the page in "search for alternative." By the bottom of the page she thought she had it: *"figure of 8 structure."* Then she wrestled for pages and hours with the worst intricacies of Patterson synthesis, drawing dots and grids, heavy spots and light spots, to try to guide her blind mental fingers shaping the air in the dark box of possibilities. Two pages later she commented, next to a sketch of a seven-nucleotide structural unit, that it was impossible to "reconcile nucleotide sequence with Chargaff's analysis." A few lines lower—and a few days later, perhaps, for her ink changed to a brighter blue—she came back to the problem and wrote, "Nearest to agreement with Chargaff analysis would be 4 purines, 3 pyrimidines, with 2 purines and 2

pyrimidines occupying equivalent positions." Complementary replication—
the notion that there was a structural basis for Chargaff's ratios—flickers there
and dies like a motor that won't turn over that wintry January day. Then she
began to write about model-building. The scale would be a half inch to the
angstrom, the phosphates could be wooden balls, and the sugar ring, she
wrote, would be made of wire and follow Furberg's standard configuration.
Several pages later she arrived at a fully worked out sketch—of the figure-8
flip-flop ribbon, in effect a pseudo-helix, that she was still pursuing.

❖ ❖ ❖

On Wednesday of that week, Pauling mailed a copy of his manuscript to
his son. He sent another copy as a courtesy directly to Sir Lawrence Bragg.

❖ ❖ ❖

Watson had begun writing the paper on bacterial chromosomes that he had
devised with Hayes in London. He was also completing the report of his discovery that tobacco-mosaic virus had a helical structure, as proved by the
telltale pattern on the diffraction photograph he had taken the previous June.

❖ ❖ ❖

On Monday, January 26, two days before her colloquium, Franklin was
writing about the requirements for a "wire model of backbone chain and
sugar rings" and sketching gadgets made of little wire springs that would allow free rotation at bonds that were not rigid. All that week she continued to
draw sugar rings, phosphate groups, bond lengths, bond angles, and figure-8
constructions—head down and doggedly, ingeniously struggling in the
wrong direction.

Probably on Wednesday, the 28th, the copies of Pauling's manuscript
reached the Cavendish. "Peter's face betrayed something important as he entered the door, and my stomach sank in apprehension at learning that all was
lost," Watson wrote, fifteen years later. He remembered that he seized the
manuscript before Crick could ask for it.

❖ ❖ ❖

To read a paper by a great scientist that's all wrong is an odd exercise. To
read it years after the fact prompts a kind of historical meditation that
scientists might try with profit both to their sense of method and to their
vanity—but they almost never do try it, for scientists like to ride the cutting
edge of the subject, close enough to hear the hiss. The great bloopers don't get
reprinted in the collections of classic papers, though some ought to for season-
ing. To read Pauling's paper on "A Proposed Structure for the Nucleic Acids"
the way Watson or Crick would have read it, one would need to be vibrantly
familiar with every subtlety of the subject and yet genuinely not know what to
expect: then the concept of the paper would form in one's mind, and *then*
would come the realization that something may be wrong, must be—*is* wrong.
That experience, when the paper was by a figure as great as Pauling and the
subject as close to one's thinking and ambition as DNA was to Watson's, must
have been rare and intense. Imagine the reflexiveness of it, of finding one's
way to an understanding and then almost instantaneously at the crest of the

wave, another wave, of a second and very different sort of understanding that encompasses and negates the first.

Three things about Pauling's paper were striking. It was uncharacteristically tentative. "We have now formulated a promising structure for the nucleic acids, by making use of the general principles of molecular structure and the available information about the nucleic acids themselves," it said. "The structure is not a vague one. . . . This is the first precisely described structure for the nucleic acids that has been suggested by any investigator. . . . The structure cannot be considered to have been proved to be correct." The paper was also—how strange—entirely synthetic: Pauling offered no new data of his own about DNA. Densities and x-ray-diffraction photographs were those published by Astbury in 1947. "Our own preparations have given photographs somewhat inferior to those." Everything else—dimensions, chemical links, all the rest—came out of Pauling's reading, except for the techniques of model-building, and those were carried over directly from Pauling's work with the alpha helix. But the contrast with the paper twenty-two months earlier announcing the alpha helix was inescapable. The confidence of *that* paper had been based four-square and solid, from the first sentence, on fifteen years of investigation of the detailed atomic structure of simpler substances related to proteins. The new paper was peculiar in a third respect. Although it has ever since been thought of as Pauling's structure for DNA, in fact from its title onwards the paper appeared to claim a structure correct for the nucleic acids generally, meaning RNA as well—a claim that was biochemically bizarre.

After the half-step hesitation of the opening paragraphs, the paper gathered up its authority and strode joyously into battle. "The structure that we propose is a three-chain structure, each chain being a helix," Pauling wrote. Nucleotides would be 3.4 angstroms apart along each helix; the entire screw would make a complete turn in 27.2 angstroms. Both those values, of course, were satisfactorily close to the canonical Astbury numbers. That there were three chains put the density high enough to agree with Astbury's estimate. The phosphate backbones would be intertwined at the axis, the bases turned outwards because they were too irregular to fit inside. As Watson immediately realized, with a lurch of alarm, this was very reminiscent of the three-stranded model he and Crick had abandoned—had they *missed* it, then?—fourteen months before. "The first question to be answered is that as to the nature of the core of the three chain-helical model—the part of the molecule closest to the axis," Pauling wrote. "It is important for stability of the molecule that atoms be well packed together," and that was harder to solve near the axis than farther out, where they were less likely to get in each other's way. To illustrate the need for close packing at the core, Pauling conceded that his second protein helix—the five-residue spiral with the hole down the middle which he had put forward two years before as an alternative to the alpha helix—had never been found in nature. For his closely packed structure for the nucleic acids he gave tables of atomic positions, and six chunky drawings, not easy to understand, not much length or angle data on them, filled with heavy rectangles to show the way the phosphates—a phosphorus atom at the center of four oxygen atoms, one remembers, making a pyramid form—were arranged. The phosphates were tied together, so Pauling wrote, though the pictures don't make that clear, by hydrogen bonds.

Some essential data were wrong, so there was no way the structure could have been right. Pauling said, when I asked him, "We weren't working very hard on it. . . . There were two things involved. The value of the density of DNA that we had taken from the literature was wrong. And Astbury's x-ray photographs that we were interpreting were really superpositions of the two different forms of DNA." But in January 1953, those flaws in the data were as unknown to Watson and Crick as they were to Pauling. Wilkins knew a little better. Rosalind Franklin possessed all the right pieces.

Even on its own terms, Pauling's model had three substantial faults—only one of which the amateur of science, reading the paper today, could be expected to spot. First, behind the clear, firm words about the need for close packing of the atoms at the core lay the fact that Pauling had packed this core so tightly it cried out. He knew that himself. One imagines that, knowing it, he felt much as if his own shoes and gloves and collar were too tight; by the time Watson was reading the manuscript Pauling was already tinkering at the structure to try to ease the pain. Watson and Crick did not immediately notice that failing. But second, the hydrogen bonds that held Pauling's triple helix together at the core made chemical nonsense. They required that the phosphate groups be filled in with extra hydrogen, to connect P—O⋯H—O—P, cancelling the local electrical charges, with the result that "Pauling's nucleic acid in a sense was not an acid at all." Watson could not possibly miss that error, for just here, of course, was the reason why, ever thinking about backbone-inside models, he had been so long preoccupied with the notion that magnesium ions were needed for the bridges joining the helices. He made quick visits to a couple of other Cambridge laboratories to quiz biochemists to be sure he was right. He was right. The appropriate emotion, he wrote, was "pleasure that a giant had forgotten elementary college chemistry."

The third fault of Pauling's model was the most comprehensive (and is most generally comprehensible). The model stood mute. It explained nothing at all. It gave no clue to its own reduplication, nor to its influence over the rest of the cell. In particular, the structure was ignorant of Chargaff's rules. Nor did anything else spring from it to illuminate the secret of the gene. Pauling, the physical chemist who had forgotten the textbook, did seem aware that his model ought to raise a brighter torch. At least, he ended the paper:

> It is interesting to note that the purine and pyrimidine groups, on the periphery of the molecule occupy positions such that their hydrogen-bond forming groups are directed radially. . . . The proposed structure accordingly permits the maximum number of nucleic acids to be constructed, providing the possibility of high specificity [that is, variety of genetic information]. As Astbury has pointed out, the 3.4-Å . . . distance along the axis of the molecule is approximately the length per residue in a nearly extended polypeptide chain, and accordingly the nucleic acids are, with respect to this dimension, well suited to the ordering of amino-acid residues in a protein.

❖ ❖ ❖

On Friday of that week, January 30, Watson went to London to see Hayes. He took the copy of Pauling's manuscript with him, and about teatime that afternoon stopped at King's College. What happened next Watson made the central episode of *The Double Helix*. Finding Wilkins busy, he opened the door of Franklin's lab, saw her bending over a light box examining an x-ray film, and went in to show her Pauling's paper. The two were alone. The only account, of course, is Watson's. Yet there are several things he did not know at the time nor when he wrote. She was ready to build models of the A form of DNA, but her notebooks show her—all that week, possibly that very Friday and again the next Monday—twisting and turning from one cul-de-sac to the next in the effort to visualize what structure to build. She was still intent, at the time of the encounter, on what she called the figure 8. The frustration is palpable in the neat lines of questions and objections, alternatives and more objections, in her notebook. Added to that, she had written to Corey almost a month earlier for information about Pauling's structure, but had received no reply. Now Watson thrust a copy of Pauling's very paper at her and began to talk about helices.

When she looked at it, she could see from the second page or so that the structure could not be right because the x-ray-diffraction photographs on which it was based were Astbury's—and she knew, of course, as Watson of course did not, that those pictures mixed the two forms of DNA. There was a reference in the first pages to Wilkins's better pictures of 1951, which Pauling had not seen, but nothing about hers: her professional courtesy to Corey the previous spring had been doubly fruitless. Watson wrote that he lectured her about helical theory and repeated Crick's assertion that her antihelical evidence was a fluke. So she got mad.

> Suddenly Rosy came from behind the lab bench that separated us and began moving toward me. Fearing that in her hot anger she might strike me, I grabbed up the Pauling manuscript and hastily retreated to the open door.

The scene would have been still more comical if Watson had reminded us that she was short and slim, he over six feet, if scrawny. Wilkins put his head in at the door; Watson sidled out; the two men walked down the corridor in a new brotherhood. Franklin's defenders and even Wilkins have variously claimed that Watson's paragraph about her threatened assault is exaggerated, ridiculous, appropriate to a novel. The more interesting defense is to ask whether she had good reason to be angry.

Wilkins told Watson, as they went down the hall, that Franklin had found that DNA fibres, when kept wet, yielded a different x-ray pattern, suggesting a second structure. Fourteen months after the King's colloquium, despite repeated correspondence and conversations, visits, meals together, between Wilkins and Watson and Crick, the possibility of a second structure was news to Watson, he wrote.

> When I asked what the pattern was like, Maurice went into the adjacent room to pick up a print of the new form they called the "B" structure. The instant I saw the picture my mouth fell open and my pulse began to race. The pattern was unbelievably simpler than those obtained previously ("A" form). Moreover, the black cross of reflections which dominated the picture could arise only from a helical structure. With the A form, the argument for a helix was never straightforward. . . . With the B form, however, mere inspection of its X-ray picture gave several of the vital helical parameters.

The picture that Wilkins showed Watson was Rosalind Franklin's. It was one of the two good pictures of the B form she had taken during the first week in

May, the year before, and was almost surely the better of the two, which she had numbered 51. The helical pattern was inescapable, much clearer than in the x-ray diagram of tobacco-mosaic virus that had elated Watson when he found it the year before.

Despite Watson's palpitations, a first glance told him only that the structure of DNA that gave rise to Franklin's x-ray-diffraction picture was almost certainly a helix. But nobody at the Cavendish had seriously doubted that DNA was helical. To learn more required that somebody carry out the simple measurements that Rosalind Franklin had begun the previous May but had put aside. Watson might have taken the measurements himself, seizing the instant. Wilkins must certainly have already performed them and might have mentioned the results; only a couple of sentences were needed.

Watson knew from Astbury's pictures, as everybody did, that the distance between the parallel bases in DNA was 3.4 angstroms, and a measurement would have confirmed what looked obvious, that that held true for the B structure. This fact was carried not in the black cross at the center of the pattern but in the heavy black arcs, farther out at the very top and the very bottom—and those arcs were so heavy they suggested that something was strongly emphasizing the spacing.

Watson had known for months that the diameter of the DNA molecule was about 20 angstroms. He had first learned that from Wilkins's own research. A measurement of Franklin's photo and a calculation would have confirmed that this too held true for the B structure.

Watson also knew from Astbury that the repeat distance—the height of a single complete turn of the helix—was 28 angstroms or so. A careful look, followed perhaps by a simple measurement, would have found the important and surprising fact, new to Watson, that the B structure had a repeat distance of 34 angstroms, precisely ten times the distance between one nucleotide and the next above it. The ten-times multiple was what reinforced the 3.4-angstrom spacing between nucleotides to make the two arcs so strong.

The three figures—3.4, 20, and 34 angstroms—were the crucial, controlling dimensions to be had from the x-ray-diffraction pattern.

At the end of a conversation with Watson in his Harvard office several years ago, I asked him whether, that afternoon at King's, besides seeing the photograph he had seen any of the calculations worked out by Franklin or Wilkins. "No," Watson said. Did he come away with simply a memory of the picture? "I had a clear memory of it, which could help me with the model because the way the arms of the cross lay in the diagram showed where the main mass of the helix had to be." The point was that when a model was built, the formulas of Crick's helical-diffraction theory could be applied to it to show whether the distribution of the atoms in it—particularly the massive phosphate groups—would produce a pattern generally like the one in the photograph. Watson did not elaborate. I asked, didn't he even have to lay a ruler on the picture? "No," he said. Then he mumbled something I couldn't catch. Then, "If you saw—that's about it. Just knowing it was a helix and it was, ah, a repeat of ten, which I learned on that day, ten, thirty-four angstroms and three point four. That was all." How had he known the repeat was ten? "Oh, Maurice told me, I think, that it was ten. I mean, it was ten. How many layer lines, you see a picture, like that one, you can count."

Mere inspection of the picture did not reveal the dimensions of the unit cell of the DNA crystals, from which followed the number of chains in each molecule. Wilkins knew those dimensions for the A form, but they were not, apparently, among the data brought back to Crick—not this time any more than fourteen months before, after Franklin's colloquium. Watson pressed Wilkins, over the picture and then at dinner in Soho, for measurements of the photograph, density values of samples, and any calculations that might decide the number of chains. He learned a little. Wilkins said the data seemed to favor a three-stranded structure; he knew some of the data about densities and water content that led to that conclusion, and perhaps told Watson the figures; but more important than the information itself was the fact that the argument was tentative. Similarly, Wilkins knew that fibres of DNA, as they got wetter, at the transition from the A to the B structure lengthened abruptly, by about twenty-five per cent. Again the data would have been useful but, as it turned out, were not crucial. Watson also learned that, in Wilkins's belief, "The real problem was the absence of any structural hypothesis which would allow them to pack the bases regularly in the inside of the helix. Of course this presumed that Rosy had hit it right in wanting the bases in the center and the backbone outside."

As Watson rode back to Cambridge that night on a slow, cold train, he sketched what he could remember of the B form's diffraction pattern, and decided that the range of possible densities in the molecule did not, after all, rule two chains out. By the time he had bicycled from the Cambridge station to Clare College and climbed in over the back gate, he had determined to start building models with two backbones—still on the inside.

❖b❖ The arrival of Pauling's paper, the discovery that his structure was impossible, and the new data Watson learned from Wilkins were the explosions that released first Watson, then Crick, for their second attack on the structure. Bragg had read his copy of Pauling's manuscript by the time Watson came to him, on the Saturday morning, to ask that the Cavendish machine shop make up representations of phosphorus atoms and the four bases. To that extent Watson was correct in his insistence, in *The Double Helix,* on the drama of the long-range competition with Caltech. As for King's, "I was never conscious that we were *competing* with Maurice," Crick said when I asked. "We had no intention of doing *experimental* work in the way they were doing. We *found* ourselves in a state of competition. As it were."

How they happened to find themselves in that state was explained by Perutz, who has the honor of having been in the center of events without ever being bloodied by one of Watson's asides. Perutz still heads the unit he started under Bragg at the Cavendish. It is now the Laboratory of Molecular Biology of the Medical Research Council, with about a hundred and fifty research workers in a six-story building on the southern edge of Cambridge. Its reputation has grown to match, as one of the half-dozen biological laboratories in the world so near the top that it's silly to compare them—except that Perutz's is surely the most crowded to work in and the most pleasant. I had come upstairs to the rooftop cafeteria from a seminar Crick was giving on chromosomes, and found Perutz sitting over a late and hurried lunch with his wife, Gisela, a handsome, graying blonde with an oceanic tolerance for the ways of scientists. I asked whether the rivalry with Pauling about DNA wasn't very much exaggerated. Perutz sat for a moment as though he hadn't heard. Then he said, "We none of us had confidence anymore in the group at King's." He spoke quietly and carefully. "Because they were so split among them-selves. And seemed unable to move. In that situation the competition with Pauling was important because Bragg believed it. And so Bragg felt that we could no longer hold back Crick and Watson." The structure in the strict sense as a scientific exercise was embedded in a larger professional and ethical set-ting—and here Watson was the catalyst, and the rivalry with Pauling was in-strumental even though it was in large part symbolic. Bragg's consent was the first rampart breached, and the turning point.

❖   ❖   ❖

Several inner citadels, points of vigorous resistance to insight, had yet to fall. The history of science is liable to a sneaky teleology whereby the future, the truth itself, seems to be drawing the discoverer on: such whig interpreta-tions are harder to purge from this than from other sorts of history because a knowledge of the outcome is often the only redress of the intellectual balance between scientist and historian. Yet if Watson and Crick were drawn into their discovery, they put up a struggle.

When Crick came to work that Saturday, Watson of course told him of all he had learned, and of Bragg's consent to model-building. They would, of course, build a model of the B form. Crick warned, on his usual principle, that they must not rule out three-chain molecules by introducing any assumption that was not forced on them by the evidence, and they did not have the evi-dence. Watson was still resolved to start with two chains, but could do noth-ing without the precisely scaled components from the machine shop. He went back to his manuscript on bacterial genetics.

❖   ❖   ❖

Monday morning, February 2, Rosalind Franklin began a new page of her notebook with the heading *"Objection to figure-eight structure."* Two days later, Linus Pauling wrote a letter to his son admitting that he and Corey were hav-ing to make adjustments to their triple helix because the atoms were packed too tightly.

❖   ❖   ❖

Watson was hampered by the slowness of the shop at the Cavendish in making his toy atoms. He didn't have the simplest pieces until the middle of the week, and the bases would take much longer. He worked at the large table in the middle of the room. He began with the sugar-phosphate backbones, tinkering for hours with wires, supporting rods, and clamps. He went off to play tennis for two hours, three hours, every afternoon. "He had no idea of form, but he had tremendous energy and seemed to reinvent the game every time he played. And he hated to lose," remembered one of his more ac-complished opponents from those weeks. He went off to films and French les-sons. "He was very clever at making a little bit of knowledge go a very long way," said one who taught him. And he returned to prod the model.

Watson spent two days working with backbones inside; the results looked worse, in terms of the packing of the atoms, than the three-chain version they had produced fourteen months before. Crick, not much involved, writing his dissertation, would look up with a comment or suggestion. Watson grumbled at the difficulty of intertwining the chains on the inside. Crick asked, he remembered, "Then, why don't you put them on the outside?" "That would be too easy," Watson said. "Then why don't you do it!" Wednesday night of the first week, over coffee after dinner at the Cricks' place, by Watson's account:

> I admitted that my reluctance to place the bases inside partially arose from the suspicion that it would be possible to build an almost infinite number of models of this type. . . . But the real stumbling block was the bases. As long as they were outside, we did not have to consider them. If they were pushed inside, the frightful problem existed of how to pack together two or more chains with irregular sequences of bases. Here Francis had to admit that he saw not the slightest ray of light.

Next day, Watson did try coiling the backbones as though they were ribbons moving up the outside of a cylinder, barber-pole fashion. They saw quickly that such configuration worked comfortably—as long as the bases were left out.

On Friday morning, Crick got a letter from Wilkins accepting an invitation to come to Cambridge for Sunday lunch. "It's very nice of you to get Paulings paper I will tell you all I can remember & scribble down from Rosie," Wilkins wrote. On the Sunday, almost as soon as Wilkins arrived, Crick started asking questions. Crick and his wife had recently bought a house in a picturesque eighteenth-century lane in the center of Cambridge. The house was narrow, the rooms small, the floors uneven, the proportions pleasing. The dining room was in the basement, and Watson and Peter Pauling were also there. English Sunday lunch is an elastic meal that can reach through tea. Peter Pauling remembered, much later, "We spent most of the afternoon trying to convince Maurice that he must start immediately to build atomic models of DNA. Our chief argument was that if he did not, my father would have another attempt and would get the correct structure." But they learned nothing more than Watson had, nine days earlier. Near the end of the afternoon, Wilkins said he was going to start building models himself, but not until Franklin left for Birkbeck, the middle of March. Crick quickly asked whether Wilkins would mind if *they* started building models. Both Watson and Crick remembered the instant vividly. Wilkins—slowly, clearly—gave his assent. It mattered, to Crick at least. He said to me, "You'll recall that before we started on our second attempt Maurice paid us a visit and we actually did ask his permission first, to work on the thing, which he gave though reluctantly."

❖   ❖   ❖

Tuesday, February 10, Franklin broke away from her monotonous catalogue of rejected impossibilities, and headed her notebook entry, "*Structure B.* Evidence for 2-chain (or 1-chain helix)?" She was looking at photo No. 49, saw that its features were helical, and wrote herself an eight-line review of the mathematics of helical-diffraction theory, trying to distinguish between single and double helices in the data before her. Then, with a couple of vigorous slashes of the pen, she crossed off the last four lines. She turned to some other calculations. She did not come back to structure B for a fortnight.

Pauling and Corey's three-chain model appeared in print that month. Franklin's practice, whenever she read a report that interested her, was to make notes about it on a plain sheet of paper, which she then filed. She demolished Pauling's structure with three laconic observations. She spotted a technical error that, if corrected, would have led him to two chains rather than three. Then she wrote, "Not clear why structure which is so empty in its outer parts wd give by X-rays the outside diameter." Finally, she noted that Pauling, using Astbury's data, had built a model that combined essential dimensions of structure A with others of structure B.

❖ ❖ ❖

At some point, Crick learned—probably from Wilkins—about the report that had been passed out when the Biophysics Committee of the Medical Research Council visited King's College the previous December 15. Crick or Watson asked Perutz for his copy. It's evident that Crick could not have seen the report before Wilkins came to Sunday lunch; very soon after that, though, he got it from Perutz.

Watson wrote in *The Double Helix* that the report contained "Rosy's precise measurements" and so confirmed to Crick "that following my return from King's I had correctly reported to him the essential features of the B pattern." Bragg, in 1971, even told me that the report had contained the photograph itself, though he remembered it wrong, for the report contained no picture of any kind. Franklin had gone no further than to include data about the crystalline geometry of the A form, about the role of water in the transition from A to B, and about the associated change from the 28-angstrom repeat to a 34-angstrom repeat. Watson's breathless account—"Rosy, of course, did not directly give us her data. For that matter, no one at King's realized they were in our hands"—caught the eye of scientist reviewers, and two in particular, André Lwoff, of the Institut Pasteur, in a review in *Scientific American*, and Erwin Chargaff in *Science*, questioned whether Perutz had behaved ethically in giving Randall's report to Crick. So in June of 1969 Perutz published a note in *Science* to print the full text of Wilkins and Franklin's sections of the report, and to show that the committee had been set up—in the words of a letter from the Medical Research Council's archives—expressly "to establish contact between the different groups of people working for the Council in this field," and that the report had not been confidential. Furthermore, nothing was in the report that Watson had not had a chance to learn at the King's colloquium fourteen months before. Perutz's act had been commonplace and not incorrect; Crick and Watson learned nothing from it that Franklin had not previously made public, if quietly. The same issue of *Science* carried a letter from Watson, apologizing to Perutz for creating the wrong impression, in which he maintained, however, "The relevant fact is not that in November 1951 I *could have* copied down Rosalind's seminar data on the unit cell dimensions and symmetry, but that I *did not.*"

As Crick read through Franklin's eleven brief paragraphs, he came upon the following facts—and now it was his turn to experience the sudden leap of recognition.

> The crystalline form of calf thymus DNA is obtained at about 75 per cent RH [relative humidity] and contains about 20 per cent by weight of water. . . . The change from the first to the second structure is accompanied by a change in the fibre-axis repeat period of 28 A to 34 A and a corresponding microscopic length-change of the fibre of about 20 per cent. . . . It was apparent that the crystalline form [structure A] was based on a face-centered monoclinic unit cell with the C-axis parallel to the fibre axis.

She then listed the dimensions of the unit cell of that crystal, three lengths and an angle.

In the long row of candles that had to be lit, one by one, before the structure of DNA emerged out of the darkness and was visible, Crick now set his spark to the one that stood closest to the center of the mystery. Watson had the shock of insight when he recognized the implications of Franklin's photo Wilkins showed him, and that insight was dramatic, easy to visualize and to explain. Crick saw in Franklin's words and numbers something just as important, indeed eventually just as visualizable. But the insight was cerebral and subtle—so subtle that even Watson was not for a long time entirely comfortable in his understanding of the point. There was drama, too: Crick's insight began with an extraordinary coincidence. Crystallographers distinguish 230 different space groups, of which the face-centered monoclinic cell with its curious properties of symmetry is only one—though in biological substances a fairly common one. The principal experimental subject of Crick's dissertation, however, was the x-ray diffraction of the crystals of a protein—the oxygen-carrying hemoglobin from the blood of horses—that was of exactly the same space group as DNA. That was what he was working on the day that Perutz gave him the Medical Research Council report. So Crick saw at once the symmetry that neither Franklin nor Wilkins had comprehended, that Perutz, for that matter, hadn't noticed, that had escaped the theoretical crystallographer in Wilkins' lab, Alexander Stokes—namely, that the molecule of DNA, rotated a half turn, came back to congruence with itself. The structure was dyadic, one half matching the other half in reverse.

"This was the crucial fact," Crick said recently. "Furthermore, the dimensions of the unit cell, which were also in the council report, proved that the dyad had to be perpendicular to the length of the molecule"—switching end for end—"and implied also that the duplication was in fact within the single molecule and not merely between adjacent molecules in the crystal. And so the chains must come in pairs rather than three in a molecule, and one chain must run up and the other down."

What Crick understood ran up and down were the two backbones. The sequence of bonds by which the phosphates and sugars linked up and alternated along one chain was flipped over in the chain running alongside. Discovery of the dyadic symmetry of the molecule settled immediately the spatial arrangement of the backbones as they spiralled up the outside around the core—though Watson at first couldn't follow Crick's reasoning. Watson began his backbone-outside models as though the two chains ran in the same direction. "He was trying to build it with the sugars too close," Crick said when I asked about that. "What I mean is, not moving around the cylinder far enough from one sugar, or one nucleotide, to the next. If the pair of chains ran in the same direction, the way Jim was building it, then each chain *singly* had only to rotate *half* the way around in the height of thirty-four angstroms, for the two-chain structure to repeat itself as Rosalind's x-ray picture showed that it did. Can you visualize that? A lot of people find it difficult. Close your eyes and think about it. But if the two chains ran in opposite directions, as I kept telling Jim they did, then they were not identical and so *both* had to go *all* the way around the cylinder, three hundred and sixty degrees, before the structure made a complete repeat.

"Well, then, if there were ten nucleotides on a chain in the thirty-four angstrom height, that obviously meant that the rotation from each to the next was thirty-six degrees. It was difficult for Jim to grasp that the chains were running in opposite directions, and what that meant. Therefore he was trying to build it to eighteen degrees." Watson went to play tennis one afternoon, telling Crick to try building the model. Crick built one not with an eighteen- but with a thirty-six-degree rotation. Thus, ten bases comprised a complete turn. Crick left a note on the model: "This is it—36° rotation." He told me, "To be fair to Jim, I think it would be ludicrous to say that he didn't understand. But I don't think he liked the type of argument. Therefore I don't think he gave it quite the force that I gave it. If you produce an argument yourself, you like it and give it a lot of force. And after all there are flaws in the argument—it is only suggestive and not conclusive—and I don't think he knew very clearly what the flaws were, he just sensed they were there."

Watson produced at the same time a neat bit of reasoning that confirmed that the molecule had two chains. In his memoir, years later, he tossed off his decision to start with two-chain models: "Francis would have to agree. Even though he was a physicist he knew that important biological objects came in pairs." Crick found the remark irritatingly frivolous. "That's just nonsense! We had a very *good* reason, which Jim's forgotten!" For the B form, Crick explained, the choice between two and three chains was not self-evident: the x-ray data said the nucleotides were 3.4 angstroms apart along the fibres, but the water content of those fibres could not be determined precisely, and so the density—the total number of nucleotides in the 34-angstrom repeat—could not be figured. For the A form, on the other hand, Franklin had worked out the density; but there, the x-ray patterns had not yielded the spacing of the nucleotides. Watson saw, though, that the visible shrinkage that Franklin measured in the transition of fibres from B to A, if applied to the spacing of nucleotides, brought the 3.4-angstrom distance down to about 2.4 angstroms. This figure fit with the density of the A form to settle the number of chains. "It's a nice bit of technical argument, which is always left out of the histories," Crick said. "But, you understand, Rosalind Franklin didn't get that point. *I* didn't do it. It was Jim who did it. If you were reconstructing the history without any evidence you'd have said *I* would have produced an argument like that. Especially as Jim has forgotten."

Crick's argument applied to the backbones of the chains. X-ray crystallography by its nature cannot identify individual bases along the chains, because the bases are not in repetitive array. When Crick settled the problem of coiling, he took DNA almost as far as Pauling had taken the structure of a protein chain with the alpha helix two years before. The differences were two, and crucial.

First, Pauling's alpha helix was held in shape by hydrogen bonds, up and down along the spiral, which were an integral, primary aspect of the structure. Crick and Watson still had no idea how the two backbones of DNA were tied together. Crick still rejected bonds between the flat bases on opposite chains, because the hydrogen atoms fringing the bases that would form such bonds were generally said not to occupy fixed positions but to jump back and forth between alternative spots: this was the problem of the tautomers, or equivalent forms, that Crick had also confronted the first time he and Watson had tried to build a model.

Second, the alpha helix had its backbone at the core, and so, although each of the twenty-odd kinds of amino acids had a residue that was not part of the backbone, those residues pointed outward, and for Pauling's purposes their individual conformations could in principle be ignored. Watson and Crick, with the backbones of DNA on the outside of their model, were left with the problem of how to fit the four kinds of bases into the core. Each sugar had to have its base sticking inwards at the standard position and angle. If the sequence of bases was in some way the vehicle for the genetic information, then the structure surely ought to permit a lot of freedom in which base followed which. Yet the bases were of very different sizes and shapes: some combinations would be too closely packed or would make the sides of the structure bulge; other pairs would leave a gap between them or create a pinched waist in the backbones.

Watson abandoned model-building for the moment and began to soak himself again in the literature about the bases. He browsed through one of the standard monographs, *The Biochemistry of the Nucleic Acids*, by James N. Davidson. There he encountered diagrams of the bases guanine and thymine drawn, as most books drew them, with the hydrogen atoms in place around the fringe of the rings in the tautomeric form called the enol configuration rather than in the alternative form, called keto. Though the difference was a shift of only one hydrogen atom, the enol form of each base was more irregular in shape—a lumpier piece in the puzzle. He reread a series of papers, stretching back many years, by John Masson Gulland and Denis Oswald Jordan, which reported chemical experiments with DNA in various solutions that seemed to show that bundles of several molecules of the stuff were held together by hydrogen bonds between bases. Indeed, such bonds formed very frequently, which suggested that shifts between tautomeric forms of the bases were not the hindrance Crick supposed. Still more, Watson now noticed that the hydrogen bonds between bases were reported to be present even in solutions that contained very little DNA—and this fact hinted, though Gulland and Jordan had made nothing of it, that the bonds linked bases not between but within single molecules of DNA. Was there a rule governing hydrogen bonds between bases?

❖ ❖ ❖

On February 18, Pauling wrote another letter to Peter, in which he said, "I am checking over the nucleic acid structure again, trying to refine the parameters a bit. . . . It is evident that the structure involves a tight squeeze for nearly all the atoms." Then he added, "I heard a rumor that Jim Watson and Crick had formulated this structure already sometime back, but had not done anything about it. Probably the rumor is exaggerated."

❖ ❖ ❖

Watson also read or re-read several reports about x-ray studies of the purine bases, guanine and adenine, each of which had been isolated, purified, allowed to form crystals, and then investigated. One of the reports was a paper that had been published two years earlier by June Broomhead, at the Cavendish, in which she showed that pure guanine made hydrogen bonds from molecule to molecule in a regular, repeated configuration. Adenine, she had found, did the same. At some moment on Thursday, February 19, Watson—by his later memory he was doodling a diagram of adenine at the instant—realized that if he took a pair of adenines hydrogen-bonded, as Broomhead showed them, lying flat side by side, he could revolve them a half-turn and they were back in position. The pair was symmetrical. The hydrogen bonds looked to be of the right length for the pair to be squeezed into the core of the DNA molecule, one base in place on each backbone. He checked from Broomhead's thesis to Davidson's monograph and found that guanine could be shown to form bonds in the same symmetrical way. He copied out the structures of the two pyrimidines, cytosine and thymine, and with a few moments' manipulation spotted ways that they, too, could be bonded to themselves to form symmetrical pairs.

The idea was marvellously simple: that the DNA molecule was made of two chains coiling around each other with identical sequences of bases and held together by hydrogen bonds between every matched pair. Pairing the bases like to like presented obvious difficulties, but perhaps they were not insuperable. The purine pairs were larger and had different shapes from the pyrimidine pairs, so the backbones would have to bulge and pinch. Until the machine shop delivered the pieces, to scale, Watson could not tell just how serious that objection might be. Then there was the problem of mistakes in pairing. From the diagram, Watson could see no reason why guanine should not also bond happily to adenine, while several other mismatches also seemed easy. Frequent pairing errors would make genetic nonsense. None the less, there might be specific enzymes that built the pairs correctly. An explanation could surely be found, for like-to-like pairing made dazzling biological sense: it explained at last the replication of the gene, the preservation of the hereditary message. Replication would occur when the two strands unwound and each became the template on which a new strand was formed, the nucleotides in the growing strand matching themselves one by one to the base sequence on the parent strand. That night as Watson went to sleep, or so he wrote later, visions of adenine pairs danced in his head.

Replication by template, strands unwinding and new ones forming on the old, was an idea that would survive. Watson's like-with-like structure quickly collapsed, but in its collapse brought down the last of the misconceptions that walled off the way of understanding. Coincidence again took part.

Friday morning, February 20, Watson wrote a letter to Max Delbrück to accompany the manuscript of the paper on bacterial genetics that he and Hayes had written. Delbrück was requested to put it in for publication in the *Pro-*

ceedings of the National Academy of Sciences, of which he was a member. Then Watson added:

> I am now extremely busy, largely working on DNA structure. I believe we are close to the solution. We have seen Paulings paper on Nucleic Acid. Have you? It contains several very bad mistakes. In addition we suspect he has chosen the wrong type of model. All in all, however, Paulings paper is at least in the proper mood and the type of approach which the people at Kings College London should be taking instead of being pure crystallographers. I had started on DNA when I first arrived in Cambridge but had stopped because the Kings group did not like competition or cooperation. However since Pauling is now working on it, I believe the field is open to anybody. I thus intend to work on it until the solution is out. Today I am very optimistic since I believe I have a very pretty model, which is so pretty I am surprised no one has thought of it before. When I have the proper coordinates worked out, I shall send a note to Nature since it accounts for the X-ray data, and even if wrong, is a marked improvement on the Pauling model. I shall send you a copy of the note.

The idea of a competition with Pauling served the same function with Delbrück that it had three weeks earlier with Bragg. Watson didn't take time to add any scientific details. He posted the letter and went off to the Cavendish.

There his idea was "torn to shreds"—before noon, by his account, though here he has telescoped into hours the running arguments of a week. When Watson sketched his model for the others in the room, the first reaction came from a figure so far silent, Jerry Donohue. He objected to Watson's using the enol configuration of the bases. Watson was astonished, but he listened, because he knew that Donohue was a physical chemist and crystallographer who had worked for years with Pauling on the structures of small organic molecules.

Donohue said to me, when I visited him at the University of Pennsylvania, "When I went to Cambridge I didn't even know what a nucleic acid was." Perhaps, but he knew as well as anyone in science what a base was. Two years earlier, at Caltech, Donohue—as he explained to Watson—had come across a paper of June Broomhead's about crystals of adenine and guanine. This was one of the things Watson had just been reading. Donohue had based part of a paper of his own on Broomhead's data, and had established for guanine that the positions of the hydrogen atoms were fixed—they did not, after all, jump from place to place. He was convinced that none of the other bases went through tautomeric shifts either.

When Watson showed Donohue the hydrogen bonds he had worked out to pair the other three sorts of bases, Donohue said, "But those are the wrong forms." When Watson showed him the textbook diagrams, Donohue said the diagrams were based on no evidence. The quantum-mechanical arguments, as well as what little evidence there was—he cited the structure of another, similar small molecule that had been determined at Caltech—indicated that the bases were far more likely to take the keto forms. At a stroke, Donohue removed the reason for Crick's long-standing aversion to hydrogen bonding in DNA. At the same stroke, Donohue ruined Watson's pairing of bases like to like, for in the keto forms, though hydrogen bonds could be invented, the difference in size between a pair of purines and a pair of pyrimidines was much greater than before. The pieces could not fit within the cage of backbones.

Crick rejected like-with-like pairing on other grounds. When it came to attaching the pairs to the backbones, Watson's scheme would undo the work of the previous ten days: the two chains would more plausibly be running in the same direction again—abandoning the dyadic symmetry called for by the x-ray evidence. Further, as the argument developed, Crick pointed out that Watson's scheme did not account for the ratios that Chargaff had discovered, whereby however widely the proportions of the bases might vary from one species to another, the amount of adenine seemed always to be about equal to the amount of thymine and guanine equal to cytosine.

They now had everything in mind that they needed for the structure.

❖   ❖   ❖

Several scientists have told me that they wondered how much more Donohue might have said to Watson: in particular, had he suggested any alternative way to pair the bases? The question examines the finely balanced relation between the individual and his community in the moment of discovery. Accordingly, in a conversation in the fall of 1973, I asked Watson whether Donohue had gone further than to say which were the right tautomeric forms. "No, he did not," Watson said. "Jerry really had no interest in DNA. He wasn't thinking about the problem." When I talked to Donohue I asked him the same thing. "I don't recall the conversation exactly, of course —but no, I'm quite sure I went no further than insisting they had to use the keto forms," Donohue said. Had he brought the Chargaff ratios into the discussion, for example? "No," Donohue said. "I probably didn't know who Chargaff was."

❖   ❖   ❖

On Monday, February 23, Franklin headed the next page in her notebook "Structure B: Photograph 51." There is a long day's work in the seven pages of measurements, calculations, and tightly woven reasoning that follow. Using only the mathematical tools of the crystallographer, she was trying to resolve the questions that Watson and Crick at that moment were attacking by model-building. Five pages in, she stopped to sum up. She had the diameter of the molecule right, though she wasn't sure about it. She confirmed her opinion that the backbones were outside. She went on. In her last note of the afternoon, she wrote, "If . . . helix as above is basis of structure B, then structure A is probably similar, with P-P [phosphate-phosphate] distance along fibre axis less than 3.4 angstroms," because of the shortening of the fibres as they changed from the B form back to the A. The next day, Franklin went at it again. She switched repeatedly from A to B, interpreting each by the other. She had spent months puzzling over structure A. Now, in hours, she permitted her analysis of the x-ray pattern from structure B to confute the errors that had been holding her spellbound. She accepted, at last, that structure A would have to be helical too. She concluded that the helix was made up of two chains. She was tormented by the awareness, and showed her torment in line after line, that those two chains bore some strange relation that was eluding her.

Aaron Klug, who lent me Franklin's notebooks, had gone through them to make occasional comments in pencil. Some of these are poignant. On the last page of her notes for February 23, Klug wrote in the margin, "Nearly home."

Farther on, in the notes for the next day, he observed, "R.E.F. is at last making the correct connection between structures A and B."

It is easy to feel great sympathy with Franklin. The fact remains that she never made the inductive leap. Her notes on structure B stopped with February 24. Some days later, still before she learned about the model proposed by Watson and Crick, she and Gosling began to draft a paper to state their conclusions so far about the B structure. That draft also survived among Franklin's papers. It was vigorous, clear, and limited. Klug and I went over it one afternoon in his office. "You see, she was lacking one final step," Klug said. "Well, two steps. What she missed was the presence of the dyad—the symmetry of the molecule. Which Francis had seen immediately from her information about the space group. And even though her notes earlier showed that she knew about Chargaff's ratios, she never got to the base pairing." Klug put the notebooks away in the bottom drawer of a cabinet under the window and put the draft paper back into a file. "She needed a collaborator, and she didn't have one. Somebody to break the pattern of her thinking, to show her what was right in front of her, to push her up and over," Klug said. "You know, in a way, Watson was her collaborator."

❖   ❖   ❖

The Cavendish machine shop still had not provided the representations of the bases, so towards the end of that week Watson drew them to scale on cardboard, in the keto forms, and cut them out. Watson and Crick disagree in their recollections of exactly how the next candle was lit. Crick told me that the insight was first theoretical, a principle revealed, and that it came to them collectively. Crick's mind works that way, of course. It happened late in the week, almost certainly on Friday, February 27. "You know, the way ideas develop is very strange," Crick said. "Base pairing and complementary replication—you'll recall we had that idea the summer before, when Chargaff came through Cambridge, and I talked to John Griffith, and I realized then that one-to-one ratios could mean complementary replication. So the idea long predates the structure. And of course, the paradox of the whole thing is that when we came to build the structure *we did not initially use that idea*. We didn't do it until we were *driven* towards it. The crucial point was after Jerry put us right about the tautomeric forms, because then we could put hydrogen bonds together. And I can remember the moment"—Crick was at his desk; Watson and Donohue were by the blackboard—"when we realized that, by God, we could build a *structure*, in which the bases would be complementary. And explain Chargaff's one-to-one ratios that way. And then next day Jim wrote out and worked out what the base pairs were. I mean, this is how *all* research goes: you *have* an idea, but you can't quite *believe* it."

Watson has not remembered that. He told me, "Francis and I had a rule that we wanted to use as little data, few assumptions, as possible to solve the structure, and we never knew whether Chargaff's rules had some completely extraneous functional reason, and so we didn't put that in. The Chargaff ratios just fell out at the end." Friday night, after cutting out the cardboard bases, still deep in defeat Watson went home and then to the theatre. Saturday morning, February 28, he came in, cleared a place to work, got out his cardboard cutouts.

Though I initially went back to my like-with-like prejudices, I saw all too well that they led nowhere. When Jerry came in I looked up, saw that it was not Francis, and began shifting the bases in and out of various other pairing possibilities. Suddenly I became aware that an adenine-thymine pair held together by two hydrogen bonds was identical in shape to a guanine-cytosine pair held together by at least two hydrogen bonds. All the hydrogen bonds seemed to form naturally; no fudging was required to make the two types of base pairs identical in shape.

Watson stumbled onto his part of the solution visually, from a shape, a representation, and that had happened several times before; that is the way his mind works. No two of the four kinds of bases have the same contour. Watson found that the purine adenine, a fused double ring with other atoms fringing it at several points, could form two hydrogen bonds with the pyrimidine thymine, a single ring, when he placed the two cutouts side by side in the right way. The bonds were the correct length, and were straight lines, N—H···O or N···H—N, as Pauling's model-building precepts required. Guanine and cytosine made hydrogen bonds the same way. The pairing could not be switched, however, for then the various atoms around the fringes got in each other's way. But when an A·T pair was laid on top of a G·C pair, the two compound shapes were exactly congruent. Such pairs could fit inside the backbones without bulges or pinches.

Donohue said these pairs agreed with what he knew. Crick, when he came in, immediately pointed out that the way the bases in these pairs would attach to their sugars meant that the two backbones ran in opposite directions, just as they had to do. Each chain could include both purines and pyrimidines, with pairs flipped over. That satisfied the dyadic symmetry. Chargaff's ratios were satisfied, too. The bases could appear in any order on one chain. Once that order was fixed, though, the base pairing, guanine always with cytosine and adenine with thymine, determined the complementary order on the opposite chain.

That morning, Watson and Crick knew, although still in mind only, the entire structure: it had emerged from the shadow of billions of years, absolute and simple, and was seen and understood for the first time. Twenty angstrom units in diameter, seventy-nine billionths of an inch. Two chains twining coaxially, clockwise, one up the other down, a complete turn of the screw in 34 angstroms. The bases flat in their pairs in the middle, 3.4 angstroms and a tenth of a revolution separating a pair from the one above or below. The chains held by the pairing closer to each other around the circumference one way than the other, by an eighth of a turn, one groove up the outside narrow, the other wide. A melody for the eye of the intellect, with not a note wasted. In itself, physically, the structure carried the means of replication—positive to negative, complementary. As the strands unwound, a double template was there in the base pairing, so that only complementary nucleotides could form bonds and drop into place as the daughter strands grew. Edna St. Vincent Millay wrote a sonnet which is, or so I thought in my youth, as good a poem about science as any since Lucretius'—the sonnet that begins, "Euclid alone has looked on Beauty bare." Perhaps the experience ought to have been like that ("O blinding hour, O holy, terrible day, When first the shaft into his vision shone . . . !"): one doubts, of course, that Crick and Watson altogether realized, that morning, what they had seen. "We have discovered the secret of life," Crick told everyone within earshot over drinks that noon at the Eagle. It was not the entire secret of life, yet truly for the first time at the ultimate biological level structure had become one with function, the antinomy dialectically resolved. The structure of DNA is flawlessly beautiful.

Towards the middle of the next week, when at last the machine shop delivered the flat metal pieces, Watson and Crick began to build the model. These pieces were thin plates of galvanized sheet, cut to the shapes of the purine and pyrimidine rings, with brass rods, three millimeters thick, welded to the corners, sticking out and cut to the right lengths for the bonds. The pieces were joined by brass sleeves slipped over the tips of the rods and fixed by setscrews; they must have been infuriating to build with. The piece in front of me is labelled "Cytosine" on both sides in Crick's hand, and color-coded with a dab of shiny green; the corner where the bond was to be made to the backbone is marked with a spot of black. Crick devised a neat theorem, which he has since forgotten, that allowed them at this stage to erect a section of one strand only, knowing that the geometry of the other was accounted for. Then, when measurements proved that no atoms were squeezed too tight or rattling too loose, Watson began to build the complementary half—and still, at first, had trouble seeing how to make it run in the opposite direction. The model took shape slowly on a table in the middle of the room. The scale was five centimetres to the angstrom, so that one complete turn would be nearly two yards tall. The plates and rods were spidery. The stands and clamps that

Pairing of adenine and thymine and of guanine and cytosine in the double-helical molecule of DNA, as shown by Watson and Crick in 1953. Pauling later showed that guanine and cytosine are joined by a third hydrogen bond. (Drawing from "Genetical Implications of the Structure of Deoxyribonucleic Acid," by J. D. Watson and F. H. C. Crick, *Nature* 171, 30 May 1953.)

held the structure up were visually obtrusive. In a mounting rush of excitement, they completed a full model by Saturday evening, March 7. Crick went home to bed.

That day, Maurice Wilkins wrote a letter to Crick:

> Thank you for your letter on the polypeptides. I think you will be interested to know that our dark lady leaves us next week and much of the 3 dimensional data is already in our hands. I am now reasonably clear of other commitments and have started up a general offensive on Nature's secret strongholds on all fronts: models, theoretical chemistry and interpretation of data, crystalline and comparative. At last the decks are clear and we can put all hands to the pumps!
>
> It wont be long now.
Regards to all,
Yours ever,
M.

Wilkins's letter reached Cambridge the following Monday. "A remarkable thing," Crick said. "I went to my desk, opened the letter from Maurice, you know, dark lady and this sort of thing, and I looked across, and I thought, was it more a question of laughing or—well, you know, sadness almost. You see. There was the model."

Watson and Crick now began to check with plumb line and measuring rod the coordinates of every atom in one of the pairs of nucleotides, so that the structure could be described. Crick began beating Watson to the office in the mornings. Perutz and Kendrew were shown the model several times, with excited lectures by Crick. Bragg was in bed with flu when first told about the model, but when he came in he quickly saw the power of complementary replication; a physical structure with built-in biological consequence stirred his profound interest in the nature of things.

Watson and Crick were riding a surge of interest and enthusiasm among their colleagues. The number of visitors increased daily, and Crick gave his talk over and over. Alexander Todd, called in at Bragg's request, confirmed that the structure was biochemically plausible; there were no obvious oversights. Many of the physicists in the Cavendish first heard of the secret of life one afternoon at tea. G. F. S. Searle, an experimental physicist who had collaborated at the Cavendish with J. J. Thomson before the turn of the century—he was now eighty-eight—asked to be shown the model. He stared at it; then he said that if this was the basis of human heredity, "No wonder we're such a queer lot!" Watson and Crick began drafting a letter to *Nature* to announce the discovery.

Midweek, Watson received a letter from Pauling. A conference about protein structure was being organized to take place in Pasadena the following September, and Pauling had decided to include nucleic acids too, so would Watson be sure to arrive at Caltech to take up his fellowship early enough to attend the conference. Then Pauling wrote that Delbrück had told him Watson had found "a beautiful new structure for the nucleic acids," and he was curious about the details. This, of course, was Watson's ephemeral like-with-like scheme. Pauling added that he had just made modifications in his own structure to increase some of the distances between atoms; he and Corey did not hold "that our structure has been proved to be right, though we incline to think that it is." On Thursday, March 12, Watson wrote a long letter to Delbrück explaining the new model fully, with diagrams, and saying, towards the end:

> The model has been derived almost entirely from stereochemical considerations with the only X-ray consideration being the spacing between the pairs of bases 3.4 A which was originally found by Astbury. It tends to build itself with approximately 10 residues per turn in 34 A. The screw is right handed.
>
> The X-ray pattern approximately agrees with the model, but since the photographs available to us are poor and meager (we have no photographs of our own and like Pauling must use Astburys photographs), this agreement in no way constitutes a proof of our model. We are certainly a long way from proving its correctness. To do this we must obtain collaboration from the group at Kings College London who possess very excellent photographs of a crystalline phase in addition to rather good photographs of a paracrystalline phase. Our model has been made in reference to the paracrystalline form and as yet we have no clear idea as to how these helices can pack together to form the crystalline phase.

In a postscript, Watson asked Delbrück not to mention the letter to Pauling.

About the same time, John Kendrew telephoned Wilkins. By the weekend, a letter to *Nature* had been drafted. Suddenly Watson, to Crick's astonishment, dropped out of the excitement and the revision of the paper and retreated to Paris, to visit friends from the phage group and to do a quick phage experiment.

❖C❖ One summer morning nearly twenty years later, at Cold Spring Harbor, Delbrück and I were sitting on a bench in the shade, looking down the long lawn and out across the water and the small boats moored on the far side. We had been talking about Oswald Avery's first demonstration, in the early nineteen-forties, that DNA carried genetic specificity, and how incomprehensible that discovery had appeared at the time. I asked Delbrück whether he had had a sense of dénouement, in March 1953, when he received Watson's long letter describing the structure. "Oh—I was fascinated," Delbrück said. He fell silent. I asked again, was there a sense of sudden illumination? "I—well, it was obviously right," he said. "There is a funny story connected with that. I had promised Pauling that I would tell him what the structure was they were thinking of, the minute I heard more. Then I got this letter which said at the end, Don't show him. So I was in a dilemma." He wanted to tell others in the laboratory about it, he knew Pauling was eager to hear, and anyway he hated secrecy in scientific matters. "So my first reaction was to call up Pauling and say, I have the news, come on over. And he did, and I just gave him the letter and walked out of the room. And I came back ten minutes later and he laughed and said, 'Did you finish reading the letter? Because at the end it says—' Pauling was convinced, effectively within five minutes. He had this rival structure at the time." A child came up, a young girl, and asked what we were doing, and Delbrück told her in a word or two. We talked some more about the reception of the model.

"But look out," Delbrück said. "The dénouement at that point was only with respect to the structure and the replication mechanism. How you used this Watson-Crick structure to—as we now say, to *code* for anything, to carry genetic specificity, that was still very, very obscure." Was it not even evident that the place to look was in the sequence of nucleotides? "In some way, of course, for that was the only degree of freedom left. But how you could use the sequence to code information for amino acids to build proteins was just completely baffling. It was George Gamow"—a physicist, not a chemist— "who had the boldness a few months later to propose an extremely simple scheme: that the amino acids themselves fitted physically into the DNA," into the slot between the two chains. "Every chemist immediately saw this was utter nonsense. It was manifest it was complete crap, from the stereochemical point of view. And yet as it turned out he was remarkably close to the truth. . . . So in retrospect what the dénouement was, was that both the *principle* of replication and the *principle* of readout are very simple, and the actual machinery for doing it is *immensely* complex."

❖ ❖ ❖

On Tuesday, March 17, Franklin and Gosling finished and dated the rough draft of a paper summarizing their incomplete conclusions about the B structure of DNA. Franklin by then had moved to Birkbeck College; they had not been told of the model built in Cambridge. That day, Wilkins received a copy of the letter that Watson and Crick had drafted for *Nature*. On the 18th, Wilkins wrote cheerfully to Crick; he asked that he be allowed to publish his experimental data simultaneously. Then he added that he'd just learned that Franklin and Gosling wanted to publish *their* data, too. In a postscript, Wilkins suggested a change in the manuscript: "Could you delete the sentence 'It is known that there is much unpublished experimental material' . . . (This reads a bit ironical)."

On the 19th, Crick wrote a letter to his son, Michael, away at school, and began, "Jim Watson and I have probably made a most important discovery. We have built a model for the structure of des-oxy-ribose-nucleic-acid (read it carefully) called D.N.A. for short. . . . Our structure is very beautiful." About the pairing of the bases he said, "It is like a code. If you are given one set of letters you can write down the others. Now we believe that the D.N.A. *is* a code. That is, the order of the bases (the letters) makes one gene different from another gene (just as one page of print is different from another)."

First Wilkins, later Franklin, took the train to Cambridge to see the model. Watson was back from his several days in Paris by the time Franklin visited; he was surprised when she gracefully agreed that the structure must be right, for he did not know how far she herself had come. At the Institut Pasteur, Watson had run into a Canadian biologist, Gerard Wyatt, who had analyzed DNAs from wheat germ, herring sperm, and viruses that attacked various insects and had found in these, too, the one-to-one ratios of adenine to thymine and guanine to cytosine. In particular, Wyatt had discovered that some of his viruses contained an unusual base, 5-methyl-cytosine, which is cytosine with an extra group of atoms attached at one point around the ring. Yet in those viruses, the total of the variant and the normal cytosine was almost exactly equal to the guanine present, and so conformed to the Chargaff ratios. Further, as Watson saw when he talked with Wyatt, the extra group on 5-methyl-cytosine was placed where it did not interfere with hydrogen bonding to guanine. The odd base, which might have upset the structure, was instead a confirmation. Watson and Crick were relieved to have Chargaff's evidence ex-tended, Crick once told me: "The data of Chargaff's, you know, wasn't all that convincing unless you wanted to believe it. And until Wyatt came along and boosted it up, and so forth."

On Saturday, March 21, Watson wrote out a deferential letter to Pauling, which both men signed, enclosing a copy of their manuscript. Bragg, when he got back after his influenza, made clear that he thought three-way simulta-neous publication with Wilkins and Franklin was fair play. But Crick and Watson took longer to put together the final version of the letter to *Nature* than to build the model itself. Its flat wording hardly conveyed the jubilation both men felt, though there were a couple of flourishes Watson could hug to him-self. The letter was dispatched on April 2.

❖ ❖ ❖

Pauling came to Cambridge the first week in April. He saw the model and a print of Franklin's x-ray pattern of the B form, listened to Crick's well-polished explanation, and agreed that the structure looked right. Pauling and Bragg then went to Brussels for the Solvay Conference of chemists, an oc-casional meeting, small and senior in membership and with a distinguished history, named for the Belgian chemist and industrialist Ernest Solvay, who founded them before the first world war. At the Solvay Conference, Bragg talked about Perutz's work with hemoglobin and reported in full detail what Watson and Crick had found—so that was the first public announcement of the structure. Pauling supported Bragg and the structure generously, and said, during the discussion, "Although it is only two months since Professor Corey and I published our proposed structure for nucleic acid, I think that we must admit that it is probably wrong. . . . Although some refinement might be made, I feel that it is very likely that the Watson-Crick structure is essentially cor-rect." He mentioned the method of replication inherent in the model, and said, "I think that the formulation of their structure by Watson and Crick may turn out to be the greatest development in the field of molecular genetics in recent years."

Among the forty-one participants at the conference was the Swedish biochemist Arne Tiselius, who had been a member of the Nobel Committee for Chemistry since 1946 and vice-president of the Nobel Foundation since 1947; Tiselius became president of the foundation in 1960, two years before the prize in chemistry was given to Perutz and Kendrew and the prize in physiol-ogy or medicine to Watson, Crick, and Wilkins.

❖ ❖ ❖

The letter to *Nature* appeared in the April 25th issue. To those of its readers who were close to the questions, and who had not already heard the news, the letter must have gone off like a string of depth charges in a calm sea. "We wish to suggest a structure for the salt of deoxyribose nucleic acid (D.N.A.). This structure has novel features which are of considerable biological inter-est," the letter began; at the end, "It has not escaped our notice that the specific pairing we have postulated immediately suggests a possible copying mechan-ism for the genetic material." That last sentence has been called one of the most coy statements in the literature of science. According to Watson, Crick wrote it. Wilkins's paper followed, signed also by two of his associates at King's College, A. R. Stokes and H. R. Wilson. It was a restatement of helical-diffraction theory, and sprang to life and significance only in the last paragraphs, where Wilkins briefly reported that his x-ray-diffraction studies of intact sperm heads and bacteriophage—both, of course, containing a high proportion of DNA—gave patterns that suggested that DNA in living creatures has a helical structure similar to the model just proposed. The note by Franklin and Gosling came next. It was a revision and extension of their draft from the middle of March, in the light of the model. It presented the crucial diffraction photo of structure B and analyzed that and the other experimental evidence to show—with curt authority—that Franklin's data were compatible with Watson and Crick's structure.

❖ ❖ ❖

"If Watson had been killed by a tennis ball I am reasonably sure I would not have solved the structure alone, but who would?" Crick wrote in *Nature* for the twenty-first anniversary of their first paper.

> Watson and I always thought that Linus Pauling would be bound to have another shot at the structure once he had seen the King's College X-ray data, but he has recently stated that even though he immediately liked our structure it took him a little time to decide finally that his own was wrong. Without our model he might never have done so. Rosalind Franklin was only two steps away from the solution. . . . She was, however, on the point of leaving King's College and DNA, to work instead on TMV [tobacco-mosaic virus] with Bernal. Maurice Wilkins had announced to us, just before he knew of our structure, that he was going to work full time on the problem. . . . I doubt myself whether the discovery of the structure could have been delayed for more than two or three years.
>
> There is a more general argument, however, recently proposed by Gunther Stent and supported by such a sophisticated thinker as [Sir Peter] Medawar. This is that if Watson and I had not discovered the structure, instead of being revealed with a flourish it would have trickled out and that its impact would have been far less. For this sort of reason Stent had argued that a scientific discovery is more akin to a work of art than is generally admitted. Style, he argues, is as important as content.
>
> I am not completely convinced . . . at least in this case. Rather than believe that Watson and Crick made the DNA structure, I would rather stress that the structure made Watson and Crick. After all, I was almost totally unknown at the time and Watson was regarded, in most circles, as too bright to be really sound. But what I think is overlooked in such arguments is the intrinsic beauty of the DNA double helix. It is the molecule which has style, quite as much as the scientists.

❖ ❖ ❖

Discovery, examined closely, I said to Crick, seemed curiously difficult to pin to a moment or to an insight or even to a single person. "No, I don't think that's curious," Crick said. "I think that's the nature of discoveries, many times: that the reason they're difficult to make is that you've got to take a series of steps, three or four steps, which if you don't make them you won't get there, and if you go wrong in any one of them you won't get there. It isn't a matter of one jump—that would be easy. You've got to make several successive jumps. And usually the pennies drop one after another until eventually it all *clicks*. Otherwise it would be too easy!"

But was there a moment when DNA all clicked? "Oh, yes, there was a moment when it clicked as far as *I* was concerned: that was the moment when Jerry Donohue told us about the tautomeric forms. I realized immediately that you could do the thing by base pairing. Though I didn't actually say that we should now build this, that, and the other, I thought it was obvious. I recall it as something I saw very clearly, and I thought they saw very clearly; but I didn't actually spell it out in so many words. But that was the moment when, as it were, the final brick fell into place. It only needed"—he laughed—"From then on you just had to try and see that it worked. But there was no further jump to be made." Which was closer to the beauty of the molecule, the base pairing or the fact that the two chains formed the dyad that ran in opposite directions? "Oh, it's the base pairing. I mean, the two are intimately connected. But the essential feature is the base pairing. And the base pairing of course has turned up *everywhere*, not just in DNA but at every step in getting the message off the DNA to construct proteins. Though it's true that with the two chains running in opposite directions, it's not so *obvious* a structure."

What it came to, I said, was that the difference between Crick's experience of the discovery and Watson's was the moment when the solution was clear. "Yes, well, I think Jim knew very clearly what it was, he experienced the same thing—" But Watson came in the next morning and found the right pairs, before Crick arrived. "That's right, yes, yes. I came in later. Oh, he did it all right, there's no doubt about it. But, but—the feeling we had at that occasion was, it's very curious, that it would be *too good to be true*, if it worked. You see. And if you ask me why I didn't do it myself that evening: I don't know! Jim got there very fast; we could have taken another week."

❖ ❖ ❖

"I solved it, I guess, because nobody else was paying full-time attention to the problem," Watson said. "Sure, you can say that a number of other people could have gotten it in the meantime. Francis wasn't really thinking about it. During that week period or so when I was really worrying about how to put the bases together, I don't think Francis was ever worrying about that. Now *why* he wasn't: 'Well, Jim is thinking about it, and maybe if he doesn't get anywhere I'll think about it.' The thing was not that it was difficult for anyone to think through, but it was getting yourself into a position where it was your only problem to think about. I had nothing else to do. I was totally underemployed. And I was the one who got it." He stopped, and breathed in with a hiss and a grimace. "The only one who might have, would have been Francis, gone back that night, and, aah, done it." Or Donohue, who knew everything Crick and Watson knew? "He could have done it. But he—hadn't come to Cambridge to solve the problem; DNA was just aah—you know. Another big molecule and nothing that unique about it."

❖ ❖ ❖

The strangest character in the pursuit of DNA is the one Watson created, in *The Double Helix,* for himself to hide behind: bumptious brilliance, sure-footed gawkiness, Midwestern American youth in Europe back then before youth fares, growing his hair longer long before long hair. Maurice Wilkins knew Watson well at the time, and reviles Watson's memoir. "I'm Jim, I'm smart. Most of the time Francis is smart too," Wilkins said to me. "The rest are bloody clots." A little calmer, and more perceptively, Wilkins added, "Jim plays himself as the holy fool."

Yes. Yet one sees that Watson was more intuitive than he seemed to want to appear, and surely more hagridden. His working title for *The Double Helix*, when the first draft was circulating in manuscript in the United States and England, was *Honest Jim*. He had conceived of the book, he once told me, as an article for *The New Yorker*—"under the heading 'Annals of Crime'!" His writing is direct, nervous, clear, awkward as though through impatience, more skillful than the self-characterization leads one to expect. Watson's college textbook, published before the memoir, shows qualities that illuminate the later book. *Molecular Biology of the Gene:* the title makes a large claim. The manner is authoritarian. There is urgency, even salt, but reflection is not stayed for, argument not brooked. Thus the intellectual texture of science in process is, I think, stifled, which is a loss of what ought to be a first, if most intangible, concern of great teaching. The text has sold exceedingly well in its own right.

Crick found *The Double Helix* an infuriating invasion of privacy, vulgar, inaccurate, and a gross violation of friendship. After reading one of the drafts, in the spring of 1967, he wrote Watson a six-page, incandescently angry letter, and sent copies to at least ten other people involved, beginning with Bragg and Watson's boss, Nathan Pusey, president of Harvard. "Should you persist in regarding your book as history I should add that it shows such a naive and egotistical view of the subject as to be scarcely credible," Crick wrote. "Anything with any intellectual content, including matters which were of central importance to us at the time, is skipped over or omitted. Your view of history is that found in the lower class of women's magazines." At the heart of Crick's rage was friendship betrayed, but an issue of another kind as well:

> Your book is misleading because it does not in fact accurately convey the atmosphere in which the work was done. Most of the time we were engaged in complicated intellectual discussions concerning points in crystallography and biochemistry. The major motivation was to understand. Science is not done merely by gossiping with other scientists, let alone by quarrelling with them. The most important requirements in theoretical work are a combination of accurate thinking and imaginative ideas.

Crick has got over most of the anger. For a while his lab maintained a list of likelier titles for Crick's counter-memoir. Sydney Brenner suggested *Brighter than a Thousand Jims*. Crick's preference was *The Loose Screw*, and he devised a snappy opening: "Jim was always clumsy with his hands. One had only to see him peel an orange . . . " One point still astonishes him, he told me: Watson's continual references to the lust for a Nobel prize. "Because on no occasion whatsoever do I recall anything of that being said by him to me; it certainly didn't enter *my* thoughts, and it wasn't my impression that it was in *his* mind.

My impression was that we were just, you know, *mad keen to solve the problem,* and all the related problems. So that did surprise me."

Bragg was kindest to Watson and his book. Where Crick had made him "fiercely annoyed," Watson, in his extremity of youth, did not. Bragg wrote a gentle foreword to *The Double Helix*—observing, though, that "those who figure in the book must read it in a very forgiving spirit." Later he told me, "I thought it ought to be published; they—Pauling and Wilkins and Crick—quite rightly thought it wasn't always accurate. That is not the book of a mature man, at all. . . . A rather brash young man coming to Europe for the first time, his violent reactions to it. That's the interest of the book; that's why I thought more or less it ought to be published: as a specimen. But you see, as I suppose young men wouldn't do, he hadn't the responsibility that I think a more mature man would have had, in that he ascribed actions and thoughts to people, not because he knew they had 'em, but because he thought it was the sort of thing they would do. A novelist's interpretation."

❖　❖　❖

"People always ask me, are you doing science, are you doing experiments, and I say No, and then they seem to think that's terrible, and that I should be very unhappy, but I'm really not unhappy at all," Watson said. "The thing that makes one happy is just the appearance of new science, so if you see new science appearing then I'm happy." We were sitting in his office in the Biological Laboratories at Harvard in 1973. The room was narrow and extremely neat, with little obvious sign that science was the kind of scholarship practiced there. A very few pieces of paper were arranged rectilinearly upon his desk, which was standard office issue, not large. Bookshelves covered part of the wall behind his chair. The books included biologists' reference works; several copies of his own books in various translations, though only one of *The Double Helix* in English; and a number of books about theatre, among them *Tynan Right and Left* and Peter Brook's *The Empty Space.*

"Largely, now I guess I'm an administrator," Watson said. His speech was breathy, hesitant, arrhythmic as though always half a step ahead of the thought or behind. Students at Harvard have described his lecture style as talking to his own shirt pocket. He acknowledges these things with the same clarity that he turns on his competitors and his friends. "I've never done much with my hands. I've been trying to get a group together at Cold Spring Harbor to work on mammalian cells; and when you try to create the conditions under which work can be done, that's largely administrative. And I write books now. I've edited together a book on cancer viruses: the lab will punish—not punish, publish it. There's a new edition of *Molecular Biology of the Gene.* It's very hard; you cannot be on top of it all, you oversimplify, you know you oversimplify, and you hope you're not caught—caught in the bad-conceptual sense.

"I think I would hate to be in a position where I didn't have to know facts. To me you're still a scientist if you know the facts. And still try to think about them. So I don't think I've got out of science. I find the question—am I still in science—awful. Some people used to ask me, you know, to explain DNA. They ask you, 'What is the significance of DNA for mankind, Dr. Watson?' It's unanswerable."

❖ ❖ ❖

High achievement in youth must leave a residue of nostalgia. *The Double Helix* ends in the spring of 1953 with Watson in Paris, "looking at the long-haired girls near St. Germain des Prés and knowing they were not for me. I was twenty-five and too old to be unusual." One dismisses that: yet Watson wrote about the structure of DNA, in his textbook, "Before the answer was known, there had always been the mild fear that it would turn out to be dull, and reveal nothing about how genes replicate and function. Fortunately, however, the answer was immensely exciting." Five weeks after Watson and Crick's first paper in *Nature,* their second appeared, in which, after explaining the structure and the evidence all over again, they pursued some of the genetical implications. These flowed from the most novel, most fundamental fact of the model:

> Any sequence of the pairs of the bases can fit into the structure. It follows that in a long molecule many different permutations are possible, and it therefore seems likely that the precise sequence of the bases is the code which carries the genetical information. If the actual order of the bases on one of the pair of chains were given, one could write down the exact order of the bases on the other one, because of the specific pairing.

This immediately suggested, they said, how DNA duplicates itself.

> Previous discussions of self-duplication have usually involved the concept of a template, or mould. Either the template was supposed to copy itself directly or it was to produce a 'negative', which in its turn was to act as a template and produce the original 'positive' once again. In no case has it been explained in detail how it would do this in terms of atoms and molecules.

The elucidation of the structure of DNA called for a new kind of functional explanation.

> Now our model for deoxyribonucleic acid is, in effect, a *pair* of templates, each of which is complementary to the other. We imagine that prior to duplication the hydrogen bonds [connecting the bases in pairs] are broken, and the two chains unwind and separate. Each chain then acts as a template for the formation on to itself of a new companion chain, so that eventually we shall have *two* pairs of chains, where we only had one before. Moreover, the sequence of the pairs of bases will have been duplicated exactly.

Yet perhaps not always exactly: the model, or rather the mistake whose correction by Donohue had cleared the way for the model, suggested for the first time a physical, molecular explanation for the central phenomenon of genetics, namely the occasional, random appearance of mutations. If the sequence of bases carried the information for the organism, then a mutation might be no more than a single change in that sequence. In particular, they wrote, "Spontaneous mutation may be due to a base occasionally occurring in one of its less likely tautomeric forms." For example, though adenine normally paired with thymine, in the rare event that one of its hydrogen atoms shifted to a particular different position at the moment the complementary chain was forming, then the base could bond with the other pyrimidine, cytosine. On the next cycle of replication, the adenine, taking its normal tautomeric form again, would pair as usual with thymine, but the cytosine would pair with guanine and so, on one of the two new double helices, a change in the sequence of bases would have appeared. This was plausible and immensely exciting speculation: proof that a change of a single base pair can cause a mutation was several years away.

They acknowledged the most puzzling objection that had been raised to the model—though they did not answer it:

> Since the two chains in our model are intertwined, it is essential for them to untwist if they are to separate. . . . Whatever the precise structure of the chromosome a considerable amount of uncoiling would be necessary. . . . Although it is difficult at the moment to see how these processes occur without everything getting tangled, we do not feel that this objection will be insuperable.

It has been said that the packing of all sorts of molecules inside the cell is so dense that it makes a rush-hour subway in New York or Tokyo seem uncrowded. A mammalian cell, invisible to the eye, contains—among its enzymes and other proteins, fats, sugars, amino acids and other building blocks, energy-carrying molecules, and the rest—a yard of double-stranded DNA. How uncoiling and recoiling proceeds in that tiny space is still not understood very well. The structure of chromosomes—the arrangement and functioning of the protein and the double helices of DNA in them—is still a problem, too.

In June of 1953, Watson went to Cold Spring Harbor for the annual symposium, which that year was about viruses. Most of the phage group was there. A talk by Watson had been inserted into the program at the last minute. Before the meeting Delbrück distributed, as required reading, copies of Watson and Crick's first paper in *Nature*. They wrote one more paper on DNA, the longest—sixteen pages—and most complete exposition of the arguments they had developed. It was finished that summer and appeared in *Proceedings of the Royal Society* late the following winter. At the end of summer, Watson left Cambridge for his fellowship at Caltech.

From Pasadena that fall, Watson wrote objecting to the idea that a talk of Crick's about the discovery might be broadcast over the BBC:

> I still think a talk on the 3rd [Programme] would be in bad taste. There are still those who think we pirated data and I'm of the belief that a few enemies are worse than a few admirers. Judging it on a monetary basis ($100) seems unfortunate. Basically, however, you are the one to suffer most from your attempts at self publicity. My main concern is not to be dragged into it. . . . If you need the money that bad, go ahead. Needless to say I shall not think any higher of you and shall have good reason to avoid any further collaboration with you.

He then mentioned a plan to study the structure of the other nucleic acid, RNA. He and Crick hoped that its structure would emerge as neatly, but it proved very much tougher. Meanwhile it was left to others to clean up the details of the Watson-Crick model of DNA and resolve some doubts that it inspired.

❖ ❖ ❖

More was involved than simply sweeping up the marble chips: essential points of confirmation were missing. From April through September of 1953,

Watson and Crick had written four papers and given several talks to scientific audiences about their proposed structure; while these had grown in mastery of both the argument and the evidence, all the evidence, even in the paper for the Royal Society, was from other people's work. The model that Watson and Crick had devised, despite its physical presence standing there on a table in the corner of a room at the Cavendish Laboratory, had its only significant status as a theory—a theory in the neutral, nonpejorative sense in which Francis Crick, or Albert Einstein for that matter, was a theoretician. The model of DNA was a proposal about the physical nature of a biological substance. The proposal had consequences: if it were false, certain things should fail to be found by experiment, and the first scientific purpose of the series of papers was to elaborate what the consequences were so that the model, the theory, could be tested. The papers' second purpose, of course, was part of the practical work of science, to bring the structure to general awareness, to begin weaving it ever deeper into the unrolling pattern of thought of the scientific community.

The first independent confirmation of the structure came from Franklin and Gosling. When Franklin had finally arranged her move to Birkbeck College, Randall had ordered her not to take the problem of the structure of DNA with her. Such a prohibition is unheard of in modern science. Crick once described it to me as "outrageous and improper." On 17 April 1953, Randall wrote a letter to Franklin at Birkbeck in which he said, in part:

> Dear Miss Franklin,
>   You will no doubt remember that when we discussed the question of your leaving my laboratory you agreed that it would be better for you to cease to work on the nucleic acid problem and take up something else. I appreciate that it is difficult to stop thinking immediately about a subject on which you have been so deeply engaged, but I should be grateful if you could now clear up, or write up, the work to the appropriate stage. A very real point about which I am a little troubled is that it is obviously not right that Gosling should be supervised by someone not specifically resident in this laboratory.

That spring, she and Gosling finished their crystallographic work on DNA. With the model for structure B in mind, Franklin at last resolved her great difficulties over the Patterson synthesis of structure A. She had the measurements and math in hand, and quickly showed that the A structure, too, fit the sort of model proposed with exactly the changes that the differences in length of the DNA fibres called for. In July, she wrote in *Nature*:

> We suggest that the unit in structure A is, as in structure B, two co-axial helical chains running in opposite directions. In the change from B to A the number of residues [base pairs] per turn increases from ten to eleven and the pitch of the helix decreases from 34 angstroms to 28 angstroms.

The two structures were, after all, so similar that the reversible transition between A and B was fully explained. As the molecules shortened down by twenty per cent to the A structure, the distance between bases shrank from 3.4 angstroms to 2.5, and the base pairs no longer lay flat but tilted at about twenty-five degrees from perpendicular to the axis. The following year,

Franklin and Gosling published their three-dimensional crystallographic solution—the full Patterson synthesis—of structure A. This paper, the last she wrote about DNA, was not her best; in particular, it did not get the spacing of the chains right, because the resolution of the data was not high enough. She told Aaron Klug that she was not very well satisfied with it.

The second kind of confirmation needed was biological. In particular, if DNA multiplied by the two strands' unwinding while each dictated the assembly on itself of a new complementary unit, plucking the uniquely pairing nucleotides out of the soup, then those two original strands were still present in the two daughter molecules, one in each of them, and still present among the four molecules of the next doubling, the eight of the next, and so on. In principle, somewhere in the billions of cells in a fully developed newborn baby, there survived unchanged, though many times unwound and rewound, the original strands of DNA from the father and the mother that had been brought together in a single cell at the moment of conception. This idea is a unique consequence of the theory—of the model. If the original strands could be picked out unchanged, then the method of replication called for by the model would be distinguished from two other possibilities: one where nothing is conserved but all the DNA broken down to fragments and dispersed and new DNA assembled by some means that would be the immediate object of a worldwide hunt; another, totally conservative, where the double strand does not unravel at all but in some other fashion is copied into new double strands.

Gunther Stent was first to set out these alternatives, in a paper he published with Max Delbrück. Perhaps to force the issue, Stent and Delbrück proposed a particular form of dispersive replication. Between the two possibilities lay the one predicted by the Watson-Crick model, which Stent called semiconservative replication. Proof was a matter for phage and bacterial geneticists. A first try to find it was made, soon after Watson and Crick proposed the structure, with the use of micro-organisms and radioactive phosphorus to label the starting generation of DNA. The results were ambiguous, and after that the pursuit became increasingly ingenious.

I first heard of semiconservative replication on New Year's Day, 1958, in Chicago—and a bright, windy, iron-cold morning it was. Seven of us who had been undergraduates together at the University of Chicago (we had overlapped Watson's last year there) were sitting scratchy-eyed over bacon and eggs and coffee when Matthew Meselson, by then a doctoral candidate with Pauling at Caltech, took a photograph from his wallet and passed it around the table. The picture showed a stack of gray stripes, with narrow, dark-gray bands across them—some stripes with one band, some with two or three close together near the middle. The photo was the main result of an experiment that Meselson had devised with a postdoctoral fellow at Caltech, Franklin Stahl.

Their paper was not yet published—not yet written. The work it describes is now recognized as displaying the most rare technical skill, while conceptually its confirmation of the way DNA reproduces itself has become, simply, part of the mainstream. In its place towards the end of the history of the elucidation of the structure and function of DNA, Meselson and Stahl's paper possessed an importance and authority like that of Oswald Avery's an-

nouncement, fourteen years earlier, of the isolation of the transforming principle and its identification as DNA. "Classic" was Watson's epithet for Meselson
and Stahl's paper. Watson's predecessor as director of the Cold Spring Harbor
Laboratory, John Cairns, startled me in conversation when he described
Meselson's central demonstration without qualification as "the most beautiful
experiment in biology."

The experiment was direct, precise; it appeared even easy. What took
several false starts was to work out the way to make extremely fine distinctions between the original DNA, if the strands indeed were separated and conserved, and the much-multiplied progeny. Meselson said something of those
false starts as he explained the photograph that winter morning, and we have
talked about the experiment again since.

He and Stahl first met when Stahl was taking a course at Woods Hole,
Massachusetts, while Meselson was doing research there, in the summer of
1954. In their first conversation about semiconservative replication, they
wondered whether, though others said that radioactive labelling was the obvious technique, they could instead use differences in density to label the
DNA strands. The notion was not unlike radioactive labelling, except that it
would use weight. They thought to grow a crop of micro-organisms in a broth
that provided some element essential to growth in the form of a heavy isotope,
then to grow a single generation in a normal broth, harvest the DNA, and try
to separate heavy molecules from light with a centrifuge. Almost fifteen years
earlier, the kernel of that approach had occurred to the English geneticist John
Burdon Sanderson Haldane; in a vigorously speculative book, *New Paths in
Genetics,* published in 1941, Haldane wondered at one point how the self-
reproduction of the gene could be demonstrated:

> How can one distinguish between model and copy? Perhaps you could use heavy
> nitrogen atoms in the food supplied to your cell, hoping that the "copy" genes
> would contain it while the models did not.

When I came across that passage, I wrote to Meselson about it. "No, I do not
recall ever seeing the astonishing words of J. B. S. Haldane quoted in your letter," he answered. "I got the book from the library to see if anything about it
would jog my memory. Nothing did."

Meselson had first thought of density separations by centrifuge, he said another time, after hearing a lecture at Caltech by Jacques Monod, on a visit a
year after the discovery of the structure of DNA. Monod had talked about experiments that showed that *E. coli* bacteria don't manufacture a certain enzyme that breaks down the sugar lactose until a few minutes after some lactose is added to their broth—which raised the fundamental question how
genes are switched on and off. Meselson thought it might be possible to tell
newly made enzyme molecules from older ones by using heavy hydrogen, the
double-weight isotope of ordinary hydrogen.

At Woods Hole, Meselson and Stahl first talked about using heavy
hydrogen, introduced in the form of heavy water, to label DNA. Then, a year
later, Meselson and Stahl were both at Caltech, where they shared a house in
Pasadena with some other apprentice biologists, just across the street from
their lab. They began trying to label phage DNA with an unusual pyrimidine

called 5-bromouracil, which is similar to the normal base thymine except that it contains an atom of bromine and so is much heavier; some organisms will mistake the fake for the real thymine and incorporate it into their growing DNA. But the labelled phages kept breaking into bits in the centrifuge. "Phages were too tough—that is, too fragile—to handle," Meselson said. "It was a mess."

Meanwhile, they had invented a new way to use the ultra-high-speed centrifuge. The technique is called density-gradient centrifugation, and, despite the opaque term, it is hardly more complicated as an idea than the observation that swimmers don't sink in the Great Salt Lake or the Dead Sea. Indeed, a density gradient is a subtle refinement of the insight upon which Archimedes shouted "Eureka!" Archimedes made his discovery in the bathtub; Meselson began his at the dinner table. "In the house where we lived at Caltech, in the guest room, on the wall, we had a periodic table of the elements," Meselson said. "It was a big chart, made of oilcloth, a beautiful chart. And the first experiment we did was at dinner. We had sugar on the table—I remember the evening very clearly, Frank and I and some other people—and I just put sugar, a *lot* of sugar, in a glass, and filled it with water, and then cut off a piece of fingernail and dropped it in—just to see if you could float, in a solution like that, materials of the density of DNA. I mean, we didn't know exactly the density of DNA—though we sure did later—but fingernail seemed a reasonable analogy. And as I remember, the fingernail sinks, even in the strongest sugar solution. So we needed something denser. We went to the chart in the guest room and said, 'Well, we want something like table salt, sodium chloride, but very dense,' so we read straight down the chart from sodium to the heavier elements that are chemically similar—sodium, potassium, rubidium, and then there's cesium, which is the last naturally occurring element in that group. And we could have used other salts of cesium, the bromide, the iodide, but we knew that the chloride was the most chemically stable and least likely to hurt the DNA. So we chose cesium chloride." Meselson had thus selected a solution of salt in water, like the Great Salt Lake except that the salt was much heavier.

Sure enough, cesium chloride produced a solution whose density was finely tuned to match the densities of the substances that they wanted to float, to sink, or to remain suspended. In the laboratory of Jerome Vinograd, a physical chemist, was a new-model ultracentrifuge that allowed photographs to be taken of a tube as it was spinning. With this machine, they found that if a tube of a solution of cesium chloride is spun for a few hours at about forty-five thousand revolutions per minute, the salt—even though it was in solution, diffused throughout the water—was forced to become more concentrated towards the bottom of the tube.

Here was the crucial and unexpected improvement over Archimedes. The salt concentration formed a gradient of increasing density. When DNA was present in the solution too, and spinning, extreme forces—a hundred forty thousand times the strength of gravity—acted on the individual molecules. Those that started out in the lower part of the salt gradient, where the density soon became greater than their own, rose buoyantly upwards, while the ones near the top were driven down, until they all reached the narrow region of the gradient that matched their own density exactly. The result, when the tube was photographed by passing ultraviolet light through it, was a dark band across the lighter gray of the solution, one band for each DNA of different density.

"We were surprised at how quickly such gradients approach their equilibrium state," Meselson said. After the detour through phage and 5-bromouracil, Meselson and Stahl went back to their original idea of using a heavy isotope for labelling. "And we switched from phage to using pure DNA from bacteria, old *Escherichia coli* itself," Meselson said. "There was a little bit of a thought leap involved in that. Which sounds ridiculous now, but— In fact, it sounds shameful now. But if you work with bacteriophage T4, even though you grow the darned things on *E. coli,* you're a phage worker. To do an experiment on the actual *E. coli* takes a little bit of a leap—it's like going to a mouse. Even if not quite so big a leap. And it was a special kind of leap, in that very sophisticated intelligent people worked with phage T4, but a rather dull lot of people worked with bacteria—or maybe, people who weren't liked so much in our lab. These things were never spoken but they float around at the back of your mind somewhere. But Caltech was Delbrück and the phage group, and people who worked on bacteria were—shall we say, outside the circle. Now *none* of this is sensible. But in the background I think there was this aversion to bacteria. But we took the leap. And then it worked fine."

Meselson and Stahl chose for their label the heavy (and nonradioactive) isotope of nitrogen, an atom of which has a weight equivalent to fifteen hydrogen atoms and is therefore written ^{15}N, whereas ordinary nitrogen has an atomic weight of fourteen and is written ^{14}N. Fortunately, ^{15}N was fairly easy to separate, and was one of the first nonradioactive rare isotopes to be available in high purity; they got theirs from a commercial maker in New Jersey. Meselson and Stahl grew *E. coli* for fourteen generations at blood heat in a broth wherein the only source of nitrogen was ammonium chloride—a pungent white salt known for centuries as sal ammoniac—that had been compounded with the heavy nitrogen. After fifteen hours the bacteria reached a concentration of a billion in every spoonful, and the heavy nitrogen had been built into the bases of all their DNA. Such DNA, Meselson and Stahl reckoned, should have a density almost one per cent greater than normal. They took out a sample of about four billion bacteria—and then, abruptly, switched the rest of the culture to growing on light nitrogen by adding a great excess of ammonium chloride made with the ordinary element.

Now, several times during each of several more generations, they took another sample. Finally, as a reference point, they made up a separate sample of ordinary, unlabelled DNA. Each sample was quickly chilled, and centrifuged lightly for five minutes to drive the bacteria into a pellet. The cells were then broken open, by adding a detergent, to release their DNA; a measured portion of each sample was put into cesium-chloride solution, and each tube spun at 44,770 revolutions per minute for twenty hours.

The DNA from the first sample, harvested before the lighter nitrogen was introduced, formed a crisply defined band towards the denser, lower end of the gradient. The DNA from the unlabelled reference sample formed a band slightly higher up. When the two types were mixed and spun, they separated;

two bands formed with a space between. So much for the controls. In the central experiment, if replication was semiconservative, then precisely at the end of the first generation of bacteria grown with ordinary nitrogen, all the DNA in the sample should be hybrid—each newly grown double helix composed of a heavy and a light strand, so that it would all show up in a single stripe midway between the labelled and unlabelled, the heavy and light positions in the gradient. Then in the second generation, Meselson expected to find DNA molecules of two kinds—some hybrid, others now with unlabelled chains only. (But if replication were totally conservative, then there would always be two bands, labelled and unlabelled, and never a band of mixed density. And if replication were dispersive, not even single strands conserved, then there would be only one band, in the middle position, of average density.)

Early in November of 1957, Meselson wrote a long letter to Watson about the results. "Clean as a whistle!" he said, and gave a table, and went on:

> Who would have imagined that, with all the other great good luck we've had, the DNA molecules would replicate all at the same rate. . . .
> I was all set to send you a collection of verses the overall mood of which is set by the lines
>
> "Now N^{15} by heavy trickery
> Ends the sway of Watson-Crickery . . ."
>
> But now we have WC with a mighty vengeance—or else a diabolical camoflage.

The photograph Meselson showed us that New Year's Day could be read like a chart. It arrayed all the different samples one above another, the bands shifting position and number as one read down. In their paper, Meselson and Stahl said:

> The degree of labeling of a partially labeled species of DNA may be determined directly from the relative position of its band between the band of fully labeled DNA and the band of unlabeled DNA. . . . It may be seen . . . that, until one generation time has elapsed, half-labeled molecules accumulate, while fully-labeled DNA is depleted. One generation time after the addition of N^{14}, these half-labeled or "hybrid" molecules alone are observed. Subsequently, only half-labeled DNA and completely unlabeled DNA are found. When two generation times have elapsed . . . , half-labeled and unlabeled DNA are present in equal amounts.

Sitting around the breakfast table, we could see that in the picture almost without explanation.

> The nitrogen of a DNA molecule is divided equally between two subunits which remain intact through many generations. . . . Following replication, each daughter molecule has received one parental subunit. . . . The results of the present experiment are in exact accord with the expectations of the Watson-Crick model for DNA duplication.

Soon after Meselson got back to Pasadena that winter, Max Delbrück and his wife carried him and Stahl off to the Kerkhoff Marine Station, run by Caltech, on the sea at Corona Del Mar, and locked them into an upstairs room with two sleeping bags and a typewriter until they wrote the paper.

❖ ❖ ❖

Rosalind Franklin, at Birkbeck College, worked with Aaron Klug to get a good second flowering from another of Watson's decapitated rosebushes, the structure of tobacco-mosaic virus. Bernal had taken the first x-ray patterns before the war; Franklin had substantially brought out the details of the virus structure before her death in 1958, at the age of thirty-seven.

Maurice Wilkins, at King's College London, went on refining the structures of the A, the B, and other forms of DNA with more x-ray crystallography, and proved that the B form could be found in the DNA of any species, from viruses to mammals. He and his associates obtained pictures, and from them structures, of DNA in its different forms at higher resolutions, and published a long series of papers. Their improved structure for the B form differed from the original in details of the backbones, most significantly by a shift in the angle of the sugars, which brings the bases in more snugly to the center. The improved model took Wilkins seven years.

❖ ❖ ❖

"You mustn't give the impression that the discovery was all done in a fiercely competitive race," Crick said. I answered that to some extent the impression was inevitable when the relevant events were picked out of the streamy mass and put in order. "I see that; but it wasn't actually *like* that at the time," Crick said. "I mean, people didn't feel as strongly. Life was much more relaxed, you know. Jim—the only person who thought it was a race was Jim, nobody else did. Not only that, but there was more *science* than people realize. I mean, outsiders don't understand the amount of intellectual effort that had to go in—as opposed to just sort of having bright ideas. Although you give the scientific arguments, you have perforce to truncate them. At least that's my impression. Because unless you know what the Fourier transform of a helix *means*—how to manipulate the mathematics of helical-diffraction theory—then you're going to go away with the impression that we could have done it without that. And that's what these young men don't always realize, that you've got to learn a lot of hard thinking in order to have bright ideas, you see. And Jim was quite good, you understand, at picking it up; I mean by the time I'd finished tutoring Jim he understood helical-diffraction theory better than Max or John did at that time, you see. I don't think he'd ever have learnt it by himself; he had to be taught. And then of course, you must remember, I wasn't really working on this problem, I was working on a thesis on proteins; that's why the work was haphazard."

❖ ❖ ❖

A friend who is a scientist asked me one day, "Who was Jim competing with?"

Not with Pauling, I said. That competition was factitious.

"Not with Rosalind Franklin, either," he said. "She didn't really count, after all. Yet Jim's book reeks of competition."

The book laid the competition to Pauling, I said.

"But Jim didn't love Pauling," the friend said. "He loved Francis."

I waited.

"There has to be an extraordinary interaction between two people, before the mind can do what they did. Jim and Francis talked in half sentences. They understood each other almost without words. Modern science is said to be run by teams—but not in this sense at all. Jim and Francis were pretty nearly unique. Perhaps Luria and Delbrück had it at one time. Jacques Monod and François Jacob. I've known it myself briefly. And Jim and Francis. That marvellous resonance between two minds—that high state in which one plus one does not equal two but more like ten. It can't be common. Linus Pauling and Francis are as nearly alike as any two you could imagine, in brilliance, in strength and dominance of mind—even in looks, as Jim recognized at once. But if Jim wasn't competing with Pauling, who was he competing with really? Who are the other candidates?"

Did my friend think Watson was competing with Crick, I asked.

He nodded, with a slight smile. "Yes," he said. "Understand me. The love and the competition are one and the same. You want to have esteem in the other's eyes. Jim was a young Chicagoan with a Midwestern accent, meeting the English for the first time. He brought to Francis that vacuum you carry with you when you come from Chicago. Francis was the one whose regard moved Jim. You want to excel. You want to perform for the other."

❖ ❖ ❖

"I started out in genetics; I'm still interested in genetics," Watson said. It was the first time I had gone to see him at his Harvard laboratory, the spring of 1971. "The work with mammalian cells is no big change. I'm still interested in cells and their organization. One tool which has been very useful in studying bacterial cells has been bacterial viruses. Phage. And now, again, you have viruses which multiply in higher cells, which are very useful tools, and if you study these you can study a tumor, or some kinds of tumor. It's not a big jump. It's a very small jump." He paused, and drew in his breath as though his gums were sore. "You can wonder whether you can just have enough facts in your head sometimes." How serious was that really? "Very serious. One feels that those who are—people like Francis and Sydney Brenner and myself—there will be very few people who will try to be as general as we are."

I remarked that some former colleagues say *The Double Helix* exaggerates the competitiveness of science. Watson's manner changed. He bit the words out sharply. "I probably understated it. It is *the dominant motive* in science. It starts at the beginning: if you publish first, you become a professor first; your future depends on some indication that you can do something by yourself. It's that simple. Competitiveness is very very dominant. The chief emotion in the field. The second is you have to prove to yourself that you can do it—and that's the same thing. You've got to keep doing it: you can't just—once."

Outside on Divinity Avenue, that spring day, ice was still rotting in the shady corners. Seven weeks earlier, in London, crocuses had been in bloom. "I was glad to go to Stockholm when the Nobel was awarded to Jacques Monod," Sir Lawrence Bragg said. "It was the fiftieth anniversary of my own." Wilkins, Bragg said, had been bitterly disappointed. "Because he'd been working at it for *so* long. And then, Perutz and I said, 'Wilkins must be in on this.'

And they published their contributions jointly in *Nature*. And later on when it came to the Nobel prize I put every ounce of weight I could, behind Wilkins getting it along with them. It was just frightfully bad luck, really. It was one of those discoveries where— And the other thing was, that young Watson played an enormous part in it. I don't think Crick would ever have done it, apart from Watson, for a moment. Watson's enthusiasm was so enormous."

Reprinted with permission from
Nature, Vol. 171, No. 4356, April 25, 1953
pp. 737–738.

MOLECULAR STRUCTURE OF NUCLEIC ACIDS

A Structure for Deoxyribose Nucleic Acid

WE wish to suggest a structure for the salt of deoxyribose nucleic acid (D.N.A.). This structure has novel features which are of considerable biological interest.

A structure for nucleic acid has already been proposed by Pauling and Corey[1]. They kindly made their manuscript available to us in advance of publication. Their model consists of three intertwined chains, with the phosphates near the fibre axis, and the bases on the outside. In our opinion, this structure is unsatisfactory for two reasons: (1) We believe that the material which gives the X-ray diagrams is the salt, not the free acid. Without the acidic hydrogen atoms it is not clear what forces would hold the structure together, especially as the negatively charged phosphates near the axis will repel each other. (2) Some of the van der Waals distances appear to be too small.

Another three-chain structure has also been suggested by Fraser (in the press). In his model the phosphates are on the outside and the bases on the inside, linked together by hydrogen bonds. This structure as described is rather ill-defined, and for this reason we shall not comment on it.

This figure is purely diagrammatic. The two ribbons symbolize the two phosphate—sugar chains, and the horizontal rods the pairs of bases holding the chains together. The vertical line marks the fibre axis

We wish to put forward a radically different structure for the salt of deoxyribose nucleic acid. This structure has two helical chains each coiled round the same axis (see diagram). We have made the usual chemical assumptions, namely, that each chain consists of phosphate diester groups joining β-D-deoxyribofuranose residues with 3′,5′ linkages. The two chains (but not their bases) are related by a dyad perpendicular to the fibre axis. Both chains follow right-handed helices, but owing to the dyad the sequences of the atoms in the two chains run in opposite directions. Each chain loosely resembles Furberg's[2] model No. 1; that is, the bases are on the inside of the helix and the phosphates on the outside. The configuration of the sugar and the atoms near it is close to Furberg's 'standard configuration', the sugar being roughly perpendicular to the attached base. There

is a residue on each chain every 3·4 A. in the z-direction. We have assumed an angle of 36° between adjacent residues in the same chain, so that the structure repeats after 10 residues on each chain, that is, after 34 A. The distance of a phosphorus atom from the fibre axis is 10 A. As the phosphates are on the outside, cations have easy access to them.

The structure is an open one, and its water content is rather high. At lower water contents we would expect the bases to tilt so that the structure could become more compact.

The novel feature of the structure is the manner in which the two chains are held together by the purine and pyrimidine bases. The planes of the bases are perpendicular to the fibre axis. They are joined together in pairs, a single base from one chain being hydrogen-bonded to a single base from the other chain, so that the two lie side by side with identical z-co-ordinates. One of the pair must be a purine and the other a pyrimidine for bonding to occur. The hydrogen bonds are made as follows: purine position 1 to pyrimidine position 1; purine position 6 to pyrimidine position 6.

If it is assumed that the bases only occur in the structure in the most plausible tautomeric forms (that is, with the keto rather than the enol configurations) it is found that only specific pairs of bases can bond together. These pairs are: adenine (purine) with thymine (pyrimidine), and guanine (purine) with cytosine (pyrimidine).

In other words, if an adenine forms one member of a pair, on either chain, then on these assumptions the other member must be thymine; similarly for guanine and cytosine. The sequence of bases on a single chain does not appear to be restricted in any way. However, if only specific pairs of bases can be formed, it follows that if the sequence of bases on one chain is given, then the sequence on the other chain is automatically determined.

It has been found experimentally[3,4] that the ratio of the amounts of adenine to thymine, and the ratio of guanine to cytosine, are always very close to unity for deoxyribose nucleic acid.

It is probably impossible to build this structure with a ribose sugar in place of the deoxyribose, as the extra oxygen atom would make too close a van der Waals contact.

The previously published X-ray data[5,6] on deoxyribose nucleic acid are insufficient for a rigorous test of our structure. So far as we can tell, it is roughly compatible with the experimental data, but it must be regarded as unproved until it has been checked against more exact results. Some of these are given in the following communications. We were not aware of the details of the results presented there when we devised our structure, which rests mainly though not entirely on published experimental data and stereochemical arguments.

It has not escaped our notice that the specific pairing we have postulated immediately suggests a possible copying mechanism for the genetic material.

Full details of the structure, including the conditions assumed in building it, together with a set of co-ordinates for the atoms, will be published elsewhere.

We are much indebted to Dr. Jerry Donohue for constant advice and criticism, especially on interatomic distances. We have also been stimulated by a knowledge of the general nature of the unpublished experimental results and ideas of Dr. M. H. F. Wilkins, Dr. R. E. Franklin and their co-workers at

King's College, London. One of us (J. D. W.) has been aided by a fellowship from the National Foundation for Infantile Paralysis.

J. D. WATSON
F. H. C. CRICK
Medical Research Council Unit for the
Study of the Molecular Structure of
Biological Systems,
Cavendish Laboratory, Cambridge.
April 2.

[1] Pauling, L., and Corey, R. B., *Nature*, **171**, 346 (1953); *Proc. U.S. Nat. Acad. Sci.*, **39**, 84 (1953).

[2] Furberg, S., *Acta Chem. Scand.*, **6**, 634 (1952).

[3] Chargaff, E., for references see Zamenhof, S., Brawerman, G. and Chargaff, E., *Biochim. et Biophys. Acta*, **9**, 402 (1952).

[4] Wyatt. G. R., *J. Gen. Physiol.*, **36**, 201 (1952).

[5] Astbury, W. T., Symp. Soc. Exp. Biol. 1, Nucleic Acid, 66 (Camb. Univ. Press, 1947).

[6] Wilkins, M. H. F., and Randall, J. T., *Biochim. et Biophys. Acta*, **10**, 192 (1953).

GENETICAL IMPLICATIONS OF THE STRUCTURE OF DEOXYRIBONUCLEIC ACID

By J. D. WATSON and F. H. C. CRICK

Medical Research Council Unit for the Study of the Molecular Structure of Biological Systems, Cavendish Laboratory, Cambridge

Reprinted with permission from
Nature, Vol. 171, No. 4361, May 30, 1953
pp. 964–967.

THE importance of deoxyribonucleic acid (DNA) within living cells is undisputed. It is found in all dividing cells, largely if not entirely in the nucleus, where it is an essential constituent of the chromosomes. Many lines of evidence indicate that it is the carrier of a part of (if not all) the genetic specificity of the chromosomes and thus of the gene itself.

Fig. 1. Chemical formula of a single chain of deoxyribonucleic acid

Fig. 2. This figure is purely diagrammatic. The two ribbons symbolize the two phosphate-sugar chains, and the horizontal rods the pairs of bases holding the chains together. The vertical line marks the fibre axis

Until now, however, no evidence has been presented to show how it might carry out the essential operation required of a genetic material, that of exact self-duplication.

We have recently proposed a structure[1] for the salt of deoxyribonucleic acid which, if correct, immediately suggests a mechanism for its self-duplication. X-ray evidence obtained by the workers at King's College, London[2], and presented at the same time, gives qualitative support to our structure and is incompatible with all previously proposed structures[3]. Though the structure will not be completely proved until a more extensive comparison has been made with the X-ray data, we now feel sufficient confidence in its general correctness to discuss its genetical implications. In doing so we are assuming that fibres of the salt of deoxyribonucleic acid are not artefacts arising in the method of preparation, since it has been shown by Wilkins and his co-workers that similar X-ray patterns are obtained from both the isolated fibres and certain intact biological materials such as sperm head and bacteriophage particles[2,4].

The chemical formula of deoxyribonucleic acid is now well established. The molecule is a very long chain, the backbone of which consists of a regular alternation of sugar and phosphate groups, as shown in Fig. 1. To each sugar is attached a nitrogenous base, which can be of four different types. (We have considered 5-methyl cytosine to be equivalent to cytosine, since either can fit equally well into our structure.) Two of the possible bases—adenine and guanine—are purines, and the other two—thymine and cytosine—are pyrimidines. So far as is known, the sequence of bases along the chain is irregular. The monomer unit, consisting of phosphate, sugar and base, is known as a nucleotide.

The first feature of our structure which is of biological interest is that it consists not of one chain, but of two. These two chains are both coiled around a common fibre axis, as is shown diagrammatically in Fig. 2. It has often been assumed that since there was only one chain in the chemical formula there would only be one in the structural unit. However, the density, taken with the X-ray evidence[2], suggests very strongly that there are two.

The other biologically important feature is the manner in which the two chains are held together. This is done by hydrogen bonds between the bases, as shown schematically in Fig. 3. The bases are joined together in pairs, a single base from one chain being hydrogen-bonded to a single base from the other. The important point is that only certain pairs of bases will fit into the structure. One member of a pair must be a purine and the other a pyrimidine in order to bridge between the two chains. If a pair consisted of two purines, for example, there would not be room for it.

We believe that the bases will be present almost entirely in their most probable tautomeric forms. If this is true, the conditions for forming hydrogen bonds are more restrictive, and the only pairs of bases possible are :

adenine with thymine ;
guanine with cytosine.

The way in which these are joined together is shown in Figs. 4 and 5. A given pair can be either way round. Adenine, for example, can occur on either chain ; but when it does, its partner on the other chain must always be thymine.

This pairing is strongly supported by the recent analytical results[5], which show that for all sources of deoxyribonucleic acid examined the amount of adenine is close to the amount of thymine, and the amount of guanine close to the amount of cytosine, although the cross-ratio (the ratio of adenine to guanine) can vary from one source to another. Indeed, if the sequence of bases on one chain is irregular, it is difficult to explain these analytical results except by the sort of pairing we have suggested.

The phosphate-sugar backbone of our model is completely regular, but any sequence of the pairs of bases can fit into the structure. It follows that in a long molecule many different permutations are possible, and it therefore seems likely that the precise sequence of the bases is the code which carries the genetical information. If the actual order of the

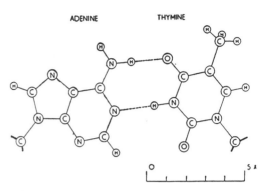

Fig. 4. Pairing of adenine and thymine. Hydrogen bonds are shown dotted. One carbon atom of each sugar is shown

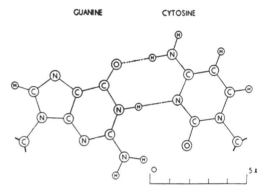

Fig. 5. Pairing of guanine and cytosine. Hydrogen bonds are shown dotted. One carbon atom of each sugar is shown

bases on one of the pair of chains were given, one could write down the exact order of the bases on the other one, because of the specific pairing. Thus one chain is, as it were, the complement of the other, and it is this feature which suggests how the deoxyribonucleic acid molecule might duplicate itself.

Previous discussions of self-duplication have usually involved the concept of a template, or mould. Either the template was supposed to copy itself directly or it was to produce a 'negative', which in its turn was to act as a template and produce the original 'positive' once again. In no case has it been explained in detail how it would do this in terms of atoms and molecules.

Now our model for deoxyribonucleic acid is, in effect, a *pair* of templates, each of which is complementary to the other. We imagine that prior to duplication the hydrogen bonds are broken, and the two chains unwind and separate. Each chain then acts as a template for the formation on to itself of a new companion chain, so that eventually we shall have *two* pairs of chains, where we only had one before. Moreover, the sequence of the pairs of bases will have been duplicated exactly.

A study of our model suggests that this duplication could be done most simply if the single chain (or the relevant portion of it) takes up the helical configuration. We imagine that at this stage in the life of the cell, free nucleotides, strictly polynucleotide precursors, are available in quantity. From time to time the base of a free nucleotide will join up by

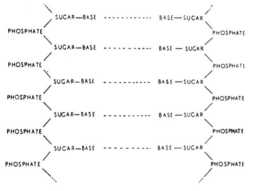

Fig. 3. Chemical formula of a pair of deoxyribonucleic acid chains. The hydrogen bonding is symbolized by dotted lines

hydrogen bonds to one of the bases on the chain already formed. We now postulate that the polymerization of these monomers to form a new chain is only possible if the resulting chain can form the proposed structure. This is plausible, because steric reasons would not allow nucleotides 'crystallized' on to the first chain to approach one another in such a way that they could be joined together into a new chain, unless they were those nucleotides which were necessary to form our structure. Whether a special enzyme is required to carry out the polymerization, or whether the single helical chain already formed acts effectively as an enzyme, remains to be seen.

Since the two chains in our model are intertwined, it is essential for them to untwist if they are to separate. As they make one complete turn around each other in 34 A., there will be about 150 turns per million molecular weight, so that whatever the precise structure of the chromosome a considerable amount of uncoiling would be necessary. It is well known from microscopic observation that much coiling and uncoiling occurs during mitosis, and though this is on a much larger scale it probably reflects similar processes on a molecular level. Although it is difficult at the moment to see how these processes occur without everything getting tangled, we do not feel that this objection will be insuperable.

Our structure, as described[1], is an open one. There is room between the pair of polynucleotide chains (see Fig. 2) for a polypeptide chain to wind around the same helical axis. It may be significant that the distance between adjacent phosphorus atoms, 7·1 A., is close to the repeat of a fully extended polypeptide chain. We think it probable that in the sperm head, and in artificial nucleoproteins, the polypeptide chain occupies this position. The relative weakness of the second layer-line in the published X-ray pictures[3a,4] is crudely compatible with such an idea. The function of the protein might well be to control the coiling and uncoiling, to assist in holding a single poly-nucleotide chain in a helical configuration, or some other non-specific function.

Our model suggests possible explanations for a number of other phenomena. For example, spontaneous mutation may be due to a base occasionally occurring in one of its less likely tautomeric forms. Again, the pairing between homologous chromosomes at meiosis may depend on pairing between specific bases. We shall discuss these ideas in detail elsewhere.

For the moment, the general scheme we have proposed for the reproduction of deoxyribonucleic acid must be regarded as speculative. Even if it is correct, it is clear from what we have said that much remains to be discovered before the picture of genetic duplication can be described in detail. What are the polynucleotide precursors ? What makes the pair of chains unwind and separate ? What is the precise role of the protein ? Is the chromosome one long pair of deoxyribonucleic acid chains, or does it consist of patches of the acid joined together by protein ?

Despite these uncertainties we feel that our proposed structure for deoxyribonucleic acid may help to solve one of the fundamental biological problems—the molecular basis of the template needed for genetic replication. The hypothesis we are suggesting is that the template is the pattern of bases formed by one chain of the deoxyribonucleic acid and that the gene contains a complementary pair of such templates.

One of us (J.D.W.) has been aided by a fellowship from the National Foundation for Infantile Paralysis (U.S.A.).

[1] Watson, J. D., and Crick, F. H. C., *Nature*, **171**, 737 (1953).

[2] Wilkins, M. H. F., Stokes, A. R., and Wilson, H. R., *Nature*, **171**, 738 (1953). Franklin, R. E., and Gosling, R. G., *Nature*, **171**, 740 (1953).

[3] (a) Astbury, W. T., Symp. No. 1 Soc. Exp. Biol., 66 (1947). (b) Furberg, S., *Acta Chem. Scand.*, **6**, 634 (1952). (c) Pauling, L., and Corey, R. B., *Nature*, **171**, 346 (1953) ; *Proc. U.S. Nat. Acad. Sci.*, **39**, 84 (1953). (d) Fraser, R. D. B. (in preparation).

[4] Wilkins, M. H. F., and Randall, J. T., *Biochim. et Biophys. Acta*, **10**, 192 (1953).

[5] Chargaff, E., for references see Zamenhof, S., Brawerman, G., and Chargaff, E., *Biochim. et Biophys. Acta*, **9**, 402 (1952). Wyatt, G. R., *J. Gen. Physiol.*, **36**, 201 (1952).

INTERLUDE

*On the State of Molecular Biology
Early in the 1970s*

4　On T. H. Morgan's Deviation and the Secret of Life

The history of science is dynastic. The most interesting sciences of our day have progressed like some early-modern European state undergoing turbulent overthrow of its ruling house even while the possibilities and the problems, both domestic and foreign, confronting the rulers change only slowly: revolution takes place within a frame of comparatively unyielding continuity. "Molecular biology can be defined as anything that interests molecular biologists," Francis Crick said in *Nature* several years ago. To be sure, he threw the line away in a footnote; but he wasn't just joking. Molecular biology is no one single province, marked off by natural boundaries from the rest of the realm. It is, rather, an intellectual transformation—indeed, a new conceptual dynasty—arisen within the realm. As a dynasty, molecular biology is by no means identical, any longer, with its ancestral heartland, the physical chemistry of the gene-stuff. It has a history now, and, some might claim, a culture of sorts, and, others have charged, an ideology. It is expansionist. I recall Max Perutz remarking once, offhand, about a particular field—it happened to be embryology, but it could have been neurobiology or immunology— "Molecular biologists are going into that, but the science itself has not yet gone molecular." In such fields not yet gone molecular, more than a few scientists will admit to being, in a quiet way, antimolecular biologists.

Molecular biology is a discipline, a level of analysis, a kit of tools—which is to say, it is unified by style as much as by content. The style is unmistakable. The style is bold; it is simplifying; it is unsparing; often it is extremely competitive. The style is also, sometimes, subtle and sophisticated. The style has been set by very few people. A list would have to include Max Delbrück and Linus Pauling, in at the beginning; Jacques Monod; no doubt James Watson; and preeminently Francis Crick. And it would include—quieter men, less known to the public, deeply respected in the community for the beauty and uncanny penetration of their experimental work—François Jacob at the Institut Pasteur, Frederick Sanger in Cambridge, Matthew Meselson at Harvard. These men are stylists of science in the most serious sense. Beyond all that, simply, molecular biology is an expectation. Molecular biologists take it for granted that a certain kind of explanation can be reached, a kind of explanation not even conceivable forty years ago, and they get restless whenever for the moment they must settle for less. The change in the character of acceptable explanations—but it is more than that; it is a change in the ruling way of framing hypotheses in biology—this is the transformation that was culminated and the revolution that has been unleashed by the discovery of the structure-in-function of deoxyribonucleic acid, as the gene. The change can perhaps be characterized in a word: it was the growth, the steadily more focussed clarification, of the idea of biological specificity.

❖ ❖ ❖

In the two decades that followed the elucidation of the structure of DNA, the implications of the discovery were ransacked and the central figure in molecular biology was Crick—not simply as the science's acknowledged theorist, though he was that; nor only for the papers he wrote and the ideas he suggested, though many of these were crucial; but for his generalship. "Where Francis has been all-important has been in his personal contribution, his influence, and in his effect on the movement of information," Jacques Monod said. "No one man created molecular biology. But Francis Crick dominates intellectually the whole field. He knows the most and understands the most." In 1962, Max Perutz's unit moved from the Cavendish Laboratory to a new building built for them by the Medical Research Council on the outskirts of Cambridge. On the door of Crick's office there, a sign invited members of the laboratory, if the door was ajar, to knock and enter; others saw the secretary in the next room. When Watson's *The Double Helix* came out, Crick closed the door firmly for several years to all but fellow scientists. Within was a moderately large, square room, looking north from two flights up over a parking lot. Crick shared the office with Sydney Brenner and a blackboard so that he could always have somebody to talk to. A lane in front of the blackboard between the window wall and Brenner's desk was kept clear, and there Crick, when thinking aloud, paced up and down with a loping, long-legged stride that kept time with the rhythms of his speech. I remarked once that some scientists, with Niels Bohr the greatest example, seemed able to work only by talking, questioning, arguing.

"I do my thinking both ways," Crick said. "I have quite interesting thoughts in the middle of the night. That happened to me for example last week, and there was a wonderful instance when we were working on the genetic code. When I say the middle of the night—you wake up at three in the morning, can't get back to sleep, and you think, By Jove! *that's* the explanation for that! But it *is*, of course, perfectly true that I think talking, and especially to Sydney, it's one of the reasons we share a room—and the blackboard there, you know, that's what we do it on. On the other hand I can get quite a way by myself. Now my colleagues differ very much in this. Jim, I reckon, is a person who is not able to clarify his ideas without discussing them with other people. He's a man whose mind works in jerks, makes imaginative jumps, and he needs others to talk to. This was my impression when I worked with him and I think it's still true. Some others are much more intellectually disciplined in this way. I'm a mixture in between."

"Molecular biology is an ambiguous term," Crick said the first time we met. He was sitting at his desk by the window. Brenner was there, too, writing with his back to us. "The term is used in two rather different ways. First, in a very general sense it can mean almost anything—the attempt to understand any biological problem at the level of atoms and molecules. You could talk about the molecular biology of animal behavior—and that's not so far-fetched as you might imagine; some senior molecular biologists are getting close to that. Second, there is a classical sense of the term, and this is much narrower: classical molecular biology has been concerned with the very large, long-chain biological molecules—the nucleic acids and proteins and their synthesis. Biologically, this means genes and their replication and expression, genes and the gene products. The study of muscle contraction, for example, which has been going on in this laboratory at the level of the molecular structures involved, is molecular biology in the first, broad sense, but not in the second.

"Now, one characteristic of molecular biology in the classical, narrow sense has been that we were studying properties common, or that we thought should be common"—at the beginning they had made some radically simplifying assumptions"—to *all* biological systems. So that one could choose the organism to study that was technically most convenient. Thus for a long time much of the research worked mainly with prokaryotes." Those are the simplest single-celled organisms, the bacteria, with their viruses, plus the blue-green algae. They were convenient for experiments because they multiplied so fast and because their cells had no nuclei and their DNA was not joined with protein in the chromosome. (Some scientific terms, well chewed, are nourishing, but "prokaryote" is a weed with no nutritional value: its Greek roots translate, one for one, into Latin that comes out, comprehensibly, as "prenuclear.") Several ideas underlay the classical research, Crick said. "The most basic idea is that biological information is mainly carried by the sequence of side groups on the regular backbone of a macromolecule. The genetic information is not conveyed and expressed by a large number of intricate symbols—it's not in Chinese—but in two very simple and as it were alphabetic languages. Genetically, the information is carried by nucleic acid, in the sequence of bases; but many such sequences can be translated into the other language—the amino-acid sequences of proteins—by special pieces of biochemical machinery. This machinery is rather elaborate; but the basic biological mechanisms are, nevertheless, in principle, comparatively simple, and they turn out, with minor variations, to be the same throughout nature—just as we had assumed. The simplicity and universality of these mechanisms is, I think, the main reason why molecular biology has been able to advance so rapidly. Because it is impossible to deny that molecular biology, since the discovery of the structure of DNA, has been wildly successful. At least, in this classical stage. For simple cells, many of the things one wants to know *are* known, in outline. Which means that one line that research is taking now, a non-dramatic line but a very important one none the less, is consolidation. Filling in all of the biochemistry so that what we know in outline we also know in detail."

Watson, several years earlier, in his textbook for university undergraduates, had swung hand over hand along that outline of molecular processes to estimate, from the total length of DNA in a bacterial cell, the number of genes, which gave him, next, the number of different kinds of average-size

protein molecules dictated by the genes, and from *that*, the total number of different metabolic steps and intermediate products that the protein molecules
catalyze in the organism—which total he compared with what biochemists
had by then mapped out of the metabolic pathways. He concluded:

> This means that we already know at least one-fifth, and maybe more than one-third
> of all the metabolic reactions that will ever be described in *E. coli.* The conclusion is
> most satisfying, for it strongly suggests that within the next ten to twenty years we
> shall approach a state in which it will be possible to describe essentially all the me-
> tabolic reactions involved in the life of an *E. coli* cell.
>
> Therefore even a cautious chemist, when properly informed, need not look at a
> bacterial cell as a hopelessly complex object. Instead he might easily adopt an al-
> most joyous enthusiasm, for it is clear that he, unlike his nineteenth-century equiv-
> alent, at last possesses the tools to describe completely the essential features of life.

The second current line of research, Crick said, was to work out higher
organisms to the same order of detail that had been achieved for the simplest
forms. The most problematic difference, in a creature whose cells were many
and diverse, was control of the rate and timing of the expression of the genes.
"Because research so far has dealt with prokaryotes, there is the entire question of the eukaryotes—that is to say, everybody else, all the plants and all the
animals. We have an outline idea, in the simplest cell, of how you turn the
genes on and off. But how do we go about asking these fundamental, classical
questions about genes and their products, all over again for higher organisms?"

Brenner by this time had turned on his chair and was listening; he hardly
had any choice. "There are two different sets of problems," he said now. "The
first is to put the knowledge that we've got onto a detailed physical and
chemical basis—most of the research so far has been done at a level that just
opens up the field. Thus, one part of molecular biology is moving to become
almost a branch of physics and chemistry. But then there is the second line,
which is to bring other biological phenomena down to their molecular basis.
These phenomena are very different from each other and understood to very
different degrees. For example, muscle. One would like to know how chemical
energy is released in muscle so that it exerts mechanical force. We want the
detailed molecular interactions. Again, we know that cells can pump selected
sorts of molecules into themselves through their membranes against tremendous electrochemical gradients. This kind of uphill pumping is critically
important. Yet we don't know anything about these pumps: we don't know
their molecular structure; we don't know how they work."

"But at this point one moves beyond classical molecular biology," Crick
said. "As soon as one begins working with multicellular organisms, one faces
the problems of development: embryology, differentiation of the several kinds
of cells, organs, and tissues; the healing of wounds; regeneration; and so on.
This is not one problem but lots of problems. Molecular biologists are getting
into this area now. It is attractive because it appears mysterious—a matter of
scientific temperament. And the other area that's going to grow very fast is the
molecular biology of the nervous system. That too is attractively mysterious,
one of the last real secrets in biology. Or you can combine those two
areas—like Sydney, who is working on how the nerves, in a simple worm he's
chosen, grow to be in the right places."

Brenner was leaning forward over the back of his chair with his arms folded on it. He was barrel-chested, sallow, with a fringe of dark hair, high bony cheekbones, and prominent brow ridges. "There are parts of biology where our basic knowledge is much more diffuse, so that often it has been very difficult to formulate even what the problem is," Brenner said. He spoke in a deep voice, emphatically, between clenched teeth. "Here there is one central field. Development. How the egg turns into the organism. But development ultimately includes *all* of biology: and it will all have to be put on a molecular basis. Molecular biology is what we are beginning to put on that basis; the rest of biology is still waiting there.

"In a funny way, what we have done so far is to work out Morgan's Deviation," Brenner said. Thomas Hunt Morgan was the geneticist identified, more than any other, with the rapid development of classical, Mendelian genetics after 1900. Morgan taught at Columbia University until 1928, when he was invited to set up a biology division at the California Institute of Technology, where he stayed until he died in 1945. Thus he was director of the biology division when Max Delbrück came there in 1937. I asked Brenner what he meant by Morgan's Deviation. "Many of the people who started working in genetics, after the rediscovery of Mendel's laws, were originally interested in development," Brenner said. "Morgan, before 1900, was working on regeneration and on embryology. He was a student of development. He gave that up because, he said, the problems were intractable. He went into the new field of genetics in the hope that it might cast light on the problems of development. They did what they were able to do, which was genetics." Morgan, as a young man, had indeed begun as an experimental embryologist, and though he switched in 1902 he never lost interest in embryology and partly returned to it a quarter of a century later.

Thinking of Gregor Mendel, one thinks first, of course, of garden peas. If one looks at Mendel's paper of 1865 after not having read it since university days, as I have just now done, one is struck by the patience that accumulated and counted those thousands of round seeds and wrinkled seeds, yellow ones and green ones, from tall plants and short, and so on—what did he keep the seeds in?—sorting out the dozens of hybrid combinations that one remembers as being the essence of Mendel: so many AB, so many ABb, so many AaB or AaBb, dominant or recessive, in the ratios that explain, as we all know, how two brown-eyed parents can have a blue-eyed child but not the other way around.

But what was rediscovered in several places simultaneously, about 1900, and then brought recognition to Mendel, was more general: the method. After looking at the work that had been done before him, Mendel put the principles of his approach in a sentence of perfect simplicity. "Among all the [previous] experiments made, not one has been carried out to such an extent and in such a way as to make it possible to determine the number of different forms under which the offspring of hybrids appear, or to arrange these forms with certainty according to their separate generations, or definitely to ascertain their statistical relations." The principles govern genetics to the present day. Mendel's statistical method enabled him to pierce the curtain of the visible, countable characters to a hidden reality of simpler relations, purely abstract, discrete, and elementary.

What Mendel treated only as algebraic units were christened genes in 1909 by the Danish biologist of evolution Wilhelm Johannsen. Genes were the atoms of Mendelism, invisible, indivisible; like the atoms of the physical chemist at much the same time, genes acquired combinatorial rules and otherwise became imaginable as their manifestations were observed. The abstractions lit upon an anatomy in 1903, when, in independent accounts, the American graduate student Walter Sutton and the German zoologist Theodor Boveri pointed out that chromosomes—multiple pairs of threads visible in cell nuclei under the microscope—permuted themselves in cell division, halved their complement in germ-cell formation, and paired again in fertilization, in a physical dance that kept perfect step with Mendel's abstract algebra.

Thinking of T. H. Morgan, one thinks first, or should, of the common vinegar fly, *Drosophila,* whose mutants and hybrids and their multitudinous descendants he examined for red eyes and eosin eyes and white eyes, vestigial wings or wild-type, and so on, and which he kept as best he could in hundreds of milk bottles stoppered with cotton wool. With *Drosophila,* Morgan discovered, for example, the mechanism by which sex is determined, at the instant of the egg's fertilization, by the pairing of the sex chromosomes, either XX or XY, and the consequent phenomenon of sex-linked inheritance that explains, as we all also know, the appearance of disorders like hemophilia among the male descendants of Queen Victoria. And when a student of Morgan's, Alfred Henry Sturtevant, perceived that the statistical evidence for linkage of many genes on one chromosome could be extended to map their relative distance one from another along that chromosome, then the hereditary material became palpably a string of beads, a line of points, each controlling a character of the organism.

The gulf in understanding, between the visible and the hidden, remained. It was clearly recognized, even accepted. Morgan's genetic discoveries won him the Nobel prize in physiology or medicine in 1933, just at the moment when he was finishing a book, *Embryology and Genetics.* In his prize lecture, the following summer, he described the gap that divided the halves of his title:

> Between the characters that are used by the geneticist and the genes that his theory postulates lies the whole field of embryonic development, where the properties implicit in the genes become explicit in the protoplasm of the cells. Here we appear to approach a physiological problem, but one that is new and strange to the classical physiology of the schools.

Good: but Morgan's ideas of the details of embryonic differentiation were inadequate—sometimes wrong, more often peculiarly empty of testable ideas. That lecture must have been nearly the last time that protoplasm was hitched to the plow by a working biologist. In his book Morgan considered but rejected the possibility that "different batteries of genes come into action as development proceeds." More than some of his colleagues, Morgan put aside consideration of the physical nature of the gene as unnecessary or premature. In his Nobel lecture, he also said:

> At the level at which the genetic experiments lie, it does not make the slightest difference whether the gene is a hypothetical unit, or whether the gene is a material particle. In either case the unit is associated with a specific chromosome, and can be localized there by purely genetic analysis.

The algebra remained the same: the formalism was complete and entrancing.

"But to get any deeper than they did, with the system they had, was then technically not possible," Brenner said. "Really there were three requirements to make possible now what was impossible then: good ideas, first; and the choice of biological materials, which meant abandoning *Drosophila* for microorganisms; and then the tremendous technological armamentarium that has grown up."

"Hypotheses and techniques," Crick said. "Molecular biology made progress for several reasons. The experimental techniques have been very powerful. Radioactive tracers, electron microscopy, antibodies as tools to dissect the processes—"

"You can very easily classify them," Brenner said. "Basically, techniques involved in fractionation. Chromatography and electrophoresis, with which you can separate vanishingly small quantities of very similar substances. Then the ultracentrifuge. Secondly, techniques used in chemical detection. Radioactive isotopes. Heavy or light isotopes. Without these labelling methods, almost nothing would have been possible. That can't be overemphasized. The techniques of course get used in combination: what you separate you've got to be able to identify. Then, thirdly, the technology of determining three-dimensional structures. The electron microscope—"

"Although that's not the instrument of choice for getting down to atomic details," Crick said. "It's best of course at the size immediately above the atomic level—one thinks of a lot of the modern work on the structure of viruses, or the demonstrations of the patterns of replication forks in DNA. But then, as you know, at the atomic level x-ray crystallography has turned out to be extremely powerful in determining the three-dimensional structures of macromolecules. Especially now, when combined with methods for measuring the intensities automatically and analyzing the data with very fast computers. The list of techniques is not something static—and they're getting faster all the time. We have a saying in the lab that the difficulty of a project goes from the Nobel-prize level to the master's-thesis level in ten years! Some of the apparatus is now getting expensive—though the cost nowhere approaches the sums required for radiotelescopes or particle accelerators or space research. But the question arises, for the new areas that molecular biology is going into, like development and differentiation or neurobiology, are the techniques adequate? It seems highly probable that new techniques are going to be needed." New sets of experimental organisms and new ways to get them to express biological distinctions were needed too. This area was particularly unsettled, and many different organisms were being tried, sometimes in unnatural combinations—tadpoles, mice, worms, yeasts and fungi, fruit flies once more, human cells, and even such things as mouse cells fused with human cells and made to multiply in glass dishes.

"The other reason for the rapid progress was that at a certain stage a set of hypotheses emerged that were very simple," Crick said. "A well-defined theoretical framework with which we could guide experimentation and from which we could predict, to some extent, what was likely to be discovered. The framework was provided, for the most part, by the middle fifties. The main reason it was possible was the nature of the nucleic-acid molecules, because the functions of these are rather limited. This helps in constructing theories—

because the easiest way to make a theory is to impose a restriction of some sort." The theoretical ideas flowed almost inevitably from the structure of DNA, in a straight functional line beginning with the general nature of the genetic code and the mechanism of translation from DNA, by the intermediary of an RNA messenger, into protein; the means of regulation and control of gene expression; and the nature and functioning of enzymes and other proteins, once constructed.

The restriction essential to theorizing had been perceived in the discussions between Crick and Watson in the winter of 1952–'53, before the structure of DNA was solved. Crick a little later christened this idea the Central Dogma of molecular biology, and invented a variety of simple diagrams with initials and arrows to express it: the Central Dogma asserts that genetic information moves from nucleic acids to protein, and most importantly that no information can get back the other way, from protein into the genetic message. "The hypotheses were simple and right," Crick said. "There was one quite serious error, misidentifying the messenger RNA, that held us up for several years; but then that was cleared up. One is thankful that it wasn't cleared up by someone, as it were, outside the magic circle, because we would all have looked so silly! But now once again the question arises, for the new areas of molecular biology, what are the unifying ideas? It must be said that we don't yet have attractive, unifying hypotheses to test."

"This is really asking whether higher organisms have some unique piece of molecular biology that's unknown to us," Brenner said. "Whether the problems of developmental biology could be solved by one insight like the double helix. In one way, you could say, all the genetic and molecular biological work of the last sixty years could be considered as a long interlude—sixty years of following out Morgan's Deviation into the tractable genetic problems. And now that that program has been completed, we have come full circle—back to the problems that they left behind unsolved. How does a wounded organism regenerate to exactly the same structure it had before? How does the egg form the organism?"

"If you look at the problem in its full horror, it's like Rutherford surveying the atom at the beginning of subatomic physics," Crick said. "We really can't predict—but we find by experience that you can't decide between hypotheses unless you get down to the molecular level."

❖❖❖

Jacques Monod was perhaps the only other person considered by molecular biologists—considered, perhaps, by Crick—to be a theoretician in the same class as Crick. "It's quite ludicrous for Monod to say to you he's not a theoretician," Crick said once, and laughed with pleasure. "First of all he and François Jacob produced the essential saving idea of the messenger. But more particularly, Jack's work on what he calls allostery, for example, where he said—but never mind the mechanism of it, it meant that through the regulatory proteins any metabolic circuit can be connected with any other metabolic circuit, you see. Well, that's an *extremely powerful* theoretical idea, never mind the details."

The Institut Pasteur, where Monod had worked since the war and of which he was director for six years until his death, in May 1976, had a pleasant, slightly musty flavor of the nineteenth century, typical of many Parisian laboratories and hospitals—at once cool and closed in, with high ceilings, yellowing paint, long narrow corridors lined with cupboards full of laboratory glassware, doors with wooden frames and frosted panes. The place evoked an agreeable, old-fashioned, individual kind of science that remembers its vow of poverty. Monod's study, before he became director, was at a corner of the ground floor of one of the oldest buildings, with a laboratory next door; several floors up was the laboratory of his decade-long collaborator François Jacob.

Molecular biology had its origins, as has often been pointed out, in two schools: the one structural and three-dimensional, in effect the British x-ray crystallographers plus Linus Pauling; the other school genetical and one-dimensional, in large part the American phage group. The two came together spectacularly in the structure of DNA, and after that Crick by his choice of problems continued to bestride them both—another way to state his long dominance. But a French school of molecular biology has thrived as well, and from early days. It has been characterized by use of the tools of microbial genetics to probe a different sort of problem: one conceptual step beyond the expression of the gene's information is the interaction and regulation of the gene-determined events. In the late fifties and early sixties, in a series of papers they wrote together and with others, Monod and Jacob produced the essential ideas about the control mechanisms of the cell.

"The aim of molecular biology is to find, in the structures of macro-molecules, interpretations of the fundamentals of life," Monod said, early in the course of the conversation the first time we met. He was short, dark, with massive shoulders and an athlete's poise; his head was large, with strong features; he was self-possessed, and had a smile of disconcerting charm. The September afternoon was warm, and he was in shirt sleeves. His English, pleasantly accented, was perfectly fluent. "What one is interested in, in science, is to arrive at statements that are, if not universal, at least of very wide generality. This has been done for DNA. The essential discoveries about the structure, the base-pairing, and the genetic code—how the gene dictates its orders—have universal application. It is being attempted, at least, for proteins. But here there are complications. Proteins have many and more varied functions. The critical property of proteins is to be Maxwell's Demons, to recognize and sort out small molecules as the servants of the specifications of the DNA. Furthermore, the three-dimensional configurations of proteins—the ways the chains fold up—are intricate and irregular, much more difficult to discover, and not predictable by any rules we know. And the functioning of a protein depends on its three-dimensional structure."

Can molecular biology be defined simply as the attempt to understand living systems by relating changes in structure to function at the molecular scale? "I wouldn't put it just that way," Monod said. "Because this ambition existed long before molecular biology existed. Biologists of the early part of this century—some biologists, the school of Jacques Loeb, for instance—certainly had this ambition, quite decidedly. They were philosophical reductionists and were convinced that the living could be reduced in principle to physics and chemistry, and nothing more." (Loeb was the German-born physiologist who was appointed in 1910 to found the department of experimental biology at the Rockefeller Institute for Medical Research. Two years later, he published *The Mechanistic Conception of Life,* a book that had vast influence; and there he wrote, characteristically, of the "hope that ultimately life, i.e., the sum of all life phenomena, can be unequivocally explained in physico-chemical terms.") "So that's not new," Monod said. "It's of course present in molecular biology, but there's nothing new in it.

"What is new in molecular biology is the recognition that the essential properties of living beings could be interpreted in terms of the structures of their macromolecules. *This,* you see, is much more specific—and in fact it partly contradicts the hope of the physical-chemical school of the beginning of the century. The biologists of that time—and the time extended into the days when I was a student, so I know this quite well—believed, to put it roughly, that the laws of gases would explain the living beings. That is to say, metabolism in the cells would be explained by the general laws of chemistry. This was natural because it was the time when great advances were being made in the understanding of the behaviour of substances in solutions, and of semipermeable membranes, and so on."

Biochemists by then had spent more than half a century, after first opening the bag of the cell, in identifying and trying to interrelate the different kinds of molecules they found within. Considering the unprecedented difficulties, they had made great progress—and had necessarily developed, along with techniques of subtlety and precision, a rigorous, vigorous intellectual tradition. But early biochemists thought small. They knew a lot about the little molecules in the cell, the materials, intermediaries, building blocks. They had begun to analyze the flow of energy as well as matter. With exceptions, they were relatively ignorant of the large molecules, for methods of handling them without breaking them were yet unknown. Into the nineteen-twenties and after, many biochemists doubted the reality of very large molecules, even when they were using the activity of enzymes, unpurified, as practical tools to catalyze small molecular reactions. There was no way biochemists then could discern that the molecules in the cell are all either small or very large, with nothing between—that even small macromolecular chains are forty times or more the size of even the largest, so to speak, micromolecules.

"That was also the time when it was *not* realized that macromolecules have structures that could be attacked and could be defined," Monod said. "It was the time when biologists talked not of macromolecular chemistry but of colloidal chemistry. And colloids were supposed to be things that changed their shapes all the time, that could not be defined, that formed gels and not crystals, that sort of thing." A corollary was the belief that the life of the cell was, as in one widely accepted formula, "the expression of a particular dynamic equilibrium in a polyphasic system." It was as though the imprecision of understanding were displaced, I suggested, and so taken to be an objective property of the systems being investigated.

"But the general point is that if the cell, for regulation, really did have at its disposal only direct chemical interactions, understandable by the general laws of chemistry, then the overall tendency of the system would be to go to equilibrium. Now chemical equilibrium means death," Monod said. "The cell survives only by being very far off equilibrium. What we now understand is quite the reverse: the cell is *entirely* a cybernetic feedback system. The regulation is entirely due to a certain kind of circuitry like an advanced electronic circuit. But it is a chemical circuitry—and yet it transcends chemistry. It is indirect. It enables the cell to gain a degree of liberty from the extreme stringencies of direct, general chemical interactions. And it works virtually without any expenditure of energy. For example, a relay system that operates a modern industrial chemical factory is something that consumes almost no energy at all as compared to the flux of energy that goes through the main chemical transformations that the factory carries out. You have an exact logical equivalence between these two—the factory and the cell. This effect is entirely written in the structure of the proteins, which is itself written in the DNA. And therefore is available to be selected for, and to evolve.

"This is why some of the pre-molecular-biological discoveries of the twenties and thirties were so startling and certainly significant for the whole change of attitude that ensued," Monod said. "I'm thinking now first of all of the crystallization of urease by Sumner in 1926." Urease is the enzyme that catalyzes the breakdown of urea into ammonia and carbon dioxide. Starting with an extract of jack beans, James Sumner had prepared a solution that demonstrated this catalytic activity very strongly; when he let the solution stand overnight in the cold, he found that crystals formed. The crystals were protein. They proved to be pure urease. This was the first pure enzyme ever prepared. It provided the first demonstration that a protein could act catalytically, and confuted the prevalent view, following Richard Willstätter's experiments, that enzymes were not proteins. "Sumner's discovery that one could crystallize an enzyme shocked biologists at the time," Monod said. "In fact the discovery was denied for a long time. And the second discovery—this was of great psychological rather than actual scientific importance—was the crystallization of tobacco-mosaic virus by Wendell Stanley in 1935. Right away there was a lot of stupid discussion about 'Can you crystallize life?' and that sort of thing, which of course is meaningless. But if you could crystallize these biologically active and specific substances, then they had regular structures. With that began the replacement of the colloidal conception of the organization of life by the structural conception.

"Of course, this change owed much of its impetus to the early school of British crystallographers—Bernal, Astbury—who started wondering whether they could get x-ray-diffraction diagrams from these biological crystals," Monod said. I said something about my admiration of the three-dimensional structures of proteins that had recently been worked out in complete detail—hemoglobin by Max Perutz at Cambridge; the first hormone structure, that of insulin, by Dorothy Hodgkin at Oxford; the first complete enzyme structures at the Royal Institution in London, at Harvard, and at Cambridge. The structures of many proteins are so exquisitely precise, it turned out, that the change of a single amino acid—one link—in a chain of a hundred fifty or more can twist the total configuration enough to prove disabling. The best-known instance is the substitution of one amino acid for another at one crucial location in each of a pair of polypeptide chains, among the four chains that make up the molecule of human hemoglobin: the result is sickle-cell anemia. The idea that such a large molecule has a completely specified structure was not unequivocally clear, even to those investigating the structures, until the early nineteen-fifties. The idea that followed, that the secret of the physiological functioning of these molecules should be sought in their structures, epitomizes the molecular transformation of biological explanation. Sir Peter Medawar, himself an immunologist, wrote, in *The Art of the Soluble,* that within the cell:

> There is no dividing line between structures in the molecular and in the anatomical sense; macromolecules have structures in a sense intelligible to the anatomist and small anatomical structures are molecular in a sense intelligible to the chemist. . . . This newer conception represents a genuine upheaval of biological thought.

Not only do proteins have defined structures; some of them—hemoglobin again, enzymes, regulatory molecules—switch back and forth between two shapes as they do their work, in the most elementary manifestations of the property of a living system that can turn chemical energy into specific motion. Max Perutz called the hemoglobin molecule "a molecular lung" and "an organ on a molecular scale." Some enzymes are now known to have slots or jaws that open and shut; they are often described as miniature machine tools.

"I wouldn't like you to believe that I am not paying due respect to the great importance of this school of crystallography," Monod said. "But I want to suggest that the present very beautiful, detailed pictures of protein molecules that are being obtained, by your friends in Cambridge, and elsewhere, are milestones—but not milestones in the development of molecular biology *sensu stricto.*

"By contrast, the first determination of the exact amino-acid sequence of a protein by Sanger"—the sequencing of bovine insulin between 1949 and 1955—"was absolutely essential. One could not even have begun to think seriously about the genetic code until it had been revealed, to begin with, that a protein is beyond the shadow of doubt a polypeptide in which the amino-acid residues really are arranged in a definite, constant, genetically determined sequence—and yet a sequence with no rule by which it determined itself. So that therefore it had to have a code—that is, complete instructions expressed in some manner—to tell it how to exist, you see. Suppose that instead Sanger had found—and that's what many biochemists would have guessed, in those days, that he would find—that there were general rules of assembly, that a polypeptide was made of a repetitive sequence of amino acids, for example lysine, aspartic acid, glutamine, threonine, repeated however many times. Then that would have been a chemical rule"—of the kind that governs the assembly of sugars into monotonous polysaccharides, for example—"and so you didn't need a genetic code. Or you needed only a partial code. But Sanger's discovery, since it revealed a sequence that had no rule, where—" The sequence was full of information, I said, nowhere redundant, no part of it predictable from other parts. "Exactly," Monod said. "And so to explain the presence of all that information in the protein, you absolutely needed the code."

I spoke of Brenner's assertion that molecular biology, in the period now past, had been engaged in completing the program of classical genetics. "Molecular biology is the modern chemical outgrowth of genetics, there's no question about that." Primary, however, had been the work that led to the understanding of DNA as the hereditary substance—"the fundamental biological invariant." This is why Mendel's conception of the gene as the unvarying carrier of the hereditary characters, and Avery's chemical identification of the gene, confirmed by Hershey and Chase, and the elucidation by Watson and Crick of the structural basis of the gene's invariance from generation to generation, comprised unquestionably the most important discoveries that have ever been made in biology—"except only the theory of evolution by natural selection, which itself required these discoveries in its support." These provided the context for the narrower line by which classical genetics, that rather natty and aristocratic caterpillar, emerged from its chrysalis as the drab and ubiquitous moth now usually called molecular genetics.

Classical genetics reached its turning point with the realization, which slowly gained acceptance in the nineteen-forties, that what a gene did was tell the organism to manufacture a particular enzyme. "Beadle and Tatum. One gene, one enzyme. About 1941," Monod said. When a gene was altered or missing, the corresponding enzyme was constructed wrongly or not at all, and the particular step in the metabolic pathway which that enzyme caused to happen was slowed or stopped. *Then* the seeds grew wrinkled instead of smooth, or the eye of a fly was white because the wild red had not been made. Recognition of the connection was not new. Forty years earlier, Archibald Garrod, an English physician, had shown that certain rare, inherited human disorders—albinism, the total lack of pigment in skin, hair, and eyes, was one he identified, and another was alkaptonuria, in which the patient's urine turns terrifyingly but harmlessly black—were caused by the absence of specific enzymes. In 1908 Garrod called such disorders by the term we still use, inborn errors of metabolism. Garrod's inborn errors are often bracketed with Mendel's laws, and indeed with Avery's transforming principle, as cases of scientific prematurity. When the relation of gene to enzyme was recognized again, beginning with a paper presented in the summer of 1940 by George Beadle and Edward Tatum from Caltech, the idea was like the wave that breaks on the beach but is swamped and lost in the much bigger wave that crashes behind. "One gene–one enzyme"—Beadle's slogan exactly captured the conceptual pivot. The one gene was still the classical gene, unitary and algebraic, but its face was now turned away from the next generation and towards its molecular complement and consequence in the present organism. Beadle and Tatum got their result not with flies but with a red mold that commonly grows on bread in tropical climates. Along the corridor, Delbrück was learning about bacteriophage. In Morgan's biology division at Caltech, the dissolution of classical genetics had begun.

The Watson and Crick model of DNA thus towered at the intersection of several historical avenues: biochemistry, molecular structure—from x-ray crystallography and physical chemistry—and classical genetics. What constitutes an acceptable explanation in science? As so often, it's hard to see from the light into the dark: the apparent brilliance of present understanding hides from us what went before.

For Pauling and his colleagues and students at Caltech, and for the Braggs and the several generations of their school in England, the structure of DNA represented not transformation at all, but triumphant proof that the sort of answers they had imagined for nearly twenty years really did offer great explanatory power. This was their answer, reached by their kind of reasoning and by their people. The disappointment was to be that for years DNA remained unique: three-dimensional molecular structure would not be united to biological function again, with remotely comparable generality, until the nineteen-sixties.

For biochemists, the transformation in the nature of acceptable explanations was not one change but several, going on at different depths, each bristling with points of resistance. Nobody mourned the passing of protoplasm: still, when enormous molecules with fixed structures emerged from the colloidal slime and seized the positions of power in the cell, DNA the dictator ordering its protein soldiers, the previous concerns of biochemists were subsumed as intermediate processes. Against that disorientation of conceptual focus was worked out the explicit argument about the chemical substance of the gene—the loosening and then abandonment of the conviction that DNA was a stupid molecule. Because the question was obviously important, and because Avery's demonstration of the genetic function of DNA was impeccable science and yet left a conspicuous trail of doubt and controversy, this aspect of the multiple transformation has since been given almost too much attention. Meanwhile, as understanding of the cell was invaded by new kinds of molecules, biology was invaded by new kinds of scientists, most famously the physicists who were generating the dynamism of the phage group. For biochemists, molecular biology also represented a painful loss of control of the century-old continuity of their tradition—a break which became irreparable as the number of students entering research doubled and doubled again in the fifties.

Geneticists found the transformation less disturbing to established ways of thought than biochemists did, on the whole—once the geneticists' biochemical habits could accommodate the role of DNA. The gene, for Morgan and his contemporaries a point on the line of the chromosome, now came under a lens of far greater resolving power; the gene was not merely pinned down at last to an identified conformation of a particular chemical substance, but was seen to be itself a line made up of hundreds, maybe thousands, of units. Worlds within worlds—the effect was dizzying, with the camera zooming closer, a sudden blur, a shake of the head, and there one was, inside the indivisible gene, regarding phenomena that were a thousandth the size but in other respects familiar. In particular, within months of the publication of the model of DNA, Seymour Benzer—a physicist recruited by Delbrück—began a purely genetic attempt to map the internal sequence of the gene, right down to the mutational events differentiating adjacent nucleotide molecules, using the twenty-minute hundredfold multiplying power of bacteriophage much the way Morgan and his pupils had first mapped genes along the chromosomes of *Drosophila*. Yet, in genetics as in biochemistry, what qualified as a solution to a problem was forever changed. For many biologists, especially the young ones growing into the science, genetical events, like biochemical ones, now really felt as though they were rightly explained only when they could be conceived mechanically, in terms of the pieces and links and angles and local electrical charges that molecules are made of.

By the late nineteen-fifties, the endeavor to determine the molecular machinery by which the commands on the DNA are translated into proteins was sweeping in almost everyone, or so it has seemed to those looking back. A heroic age was fairly begun. Monod named a dozen scientists, listed nearly a score of experiments—a hierarchy of results, obtained by those whom Crick

called "the magic circle" and others "the club," with a few newcomers and
fewer outsiders. They were in and out of one another's labs; small groups
formed for one experiment, fell into other patterns for the next; information
moved ever faster, and the velocity of ideas was most exhilarating of all.
Crick, Brenner, and Jacob, in a conversation in Cambridge on Good Friday of
1960, broke the impasse of the code by conceiving the idea of the messenger—the carrier that brought the gene's instructions to the machinery that
knit together the protein chain. Jacob and Brenner then joined Meselson at
Caltech and found the elusive messenger, while a colleague of Jacob's from
Paris went to Harvard to join Watson and Walter Gilbert—yet another
physicist turned biologist—to pursue the same quarry by a different method.
Monod and Jacob, in the last days of 1960, announced the theory of gene
regulation—that a molecule they called the repressor blocked the DNA to prevent the gene from being read out, until the cell's need of the particular
protein inscribed there caused the repressor to be deactivated and the messenger to be made. Even Crick came into the wet lab and did some experiments with Brenner and bacteriophage to prove that the genetic code was read
in groups of three nucleotides, without overlapping, each triplet specifying
one amino acid in the polypeptide chain. Marshall Nirenberg, an outsider, astonished everyone in the summer of 1961 by announcing the decipherment of
the first code group, a sequence of nucleotides that specified the amino acid
phenylalanine. After several years and the most intense competition among
many laboratories to complete the solution, Crick was able to sit back to consider how it all began and to write, "The genetic code originated at least three
billion years ago, and it may be impossible to reconstruct the sequence of
events that took place at such a remote period. The origin of the code is very
close to the origin of life." Then Gilbert and a colleague at Harvard announced
the isolation of a repressor molecule and Mark Ptashne, a newcomer, also at
Harvard, announced the isolation of another. The first structures of enzymes
were announced; and Max Perutz completed the mapping of hemoglobin at
high resolution—a task of three decades—and at last was able to perceive the
mechanism by which this structure performs its vital functions.

The heroic age of molecular biology lasted perhaps twenty years. As intellectual movements go, in this century, that was a long time. "The secret of
life? But this is in large part known—in principle if not in all details," Monod
said. "For a simple living creature to be synthesized, in my opinion there is no
further principle that would need to be discovered."

I spoke of Brenner's idea that the genetics and molecular biology of the
past sixty years had been the working out of T. H. Morgan's diversion from
embryology. "Yes, quite right," Monod said. He had had a year's fellowship in
Morgan's department at Caltech in 1936. "Morgan *was* an embryologist. To the
extent that we might say he went from embryology to genetics because he felt
the genetic problem had to be solved before we could even begin to think
about the embryological problem, I would say that the genetic problem, by
and large, *is* solved. But the embryological problem remains. Yet at last the
basis to start solving that is there, and that's what Sydney is doing and what
others are trying to do.

"I think another point which is important, which I want to stress, is that
while all thoughtful, hard-headed biologists believed in the neo-Darwinian
theory of evolution, and although it was clear that that was the only one that

gave a rational description of how evolution could have happened, yet it was *still* true that that theory was profoundly incomplete so long as one did not also have a physical theory of heredity. Since the whole Darwinian concept is based on change through the inheritance of new traits, on selection pressing on a somewhat varied population, so long as you could not say exactly how inheritance occurred, physically, and what the generator of variety was, chemically, Darwinism was still up in the air.

"So what molecular biology has done, you see, is to prove beyond any doubt but in a totally new way the complete independence of the genetic information from events occurring outside or even inside the cell—to prove by the very structure of the genetic code and the way it is transcribed that no information from outside, of any kind, can ever penetrate the inheritable genetic message. This was believed but never proved until the structure of DNA and the mechanisms of protein synthesis were understood. This was what Francis called the Central Dogma: no information goes from protein to DNA. And he had to call it a dogma at the time, because at the time so little was yet known." Crick's assertion, since amply borne out, was of fundamental significance: Lamarckism, or the hereditary passing on of characteristics that an organism has picked up or had changed by the action of the environment—whether thought of as the giraffe's neck, the Jew's intellectualism, the winter-hardiness of wheat treated by Stalin's geneticist Trofim Lysenko, or the esoteric collective unconscious postulated by Carl Jung—had been declared dead before, but here at last received its definitive postmortem and was buried, clearing the air. "With that, and the understanding of the random physical basis of mutation that molecular biology has also provided, the mechanism of Darwinism is at last securely founded," Monod said. "And man has to understand that he is a mere accident. Not only is man not the center of creation; he is not even the heir to a sort of predetermined evolution that would have produced either man or something very like him in any case. It is not true that evolution is a law; it is just a phenomenon, which is quite different."

❖ ❖ ❖

Four months after these conversations, I was in Cambridge once more and talked with Sydney Brenner again. He invited me to dinner at King's College, of which he was a fellow. "Come along on Tuesday; it's a wine night." Tuesday was a mild, misty night at the end of January, already several hours past sunset when I crossed King's Parade, leaving the low, nineteenth-century shops behind me, the chapel high and rectilinear and black and bigger than usual off to the right, the porter's lodge first entrance to the left. The senior combination room was three doors farther along. I found a long room, panelled in wood and portraits in oils and bookshelves, warm to the eye and comfortably worn, with scores of places to sit, and decanters and glasses on a small table: the sort of background for conversation that the English set perhaps better than anyone else. One other person was in the room, a white-haired man reading the *Morning Star*. Brenner came in a minute later, rubbing his hands—enjoying the gesture—and then settling the academic gown on his shoulders. He poured two medium-dry sherries, signed the small writing tablet with the pencil, and we went and sat down along the window wall. He spoke in a low voice, almost a confidential whisper.

Science was changing. "You can't grow exponentially all the time. After all, if it continued at the past rate, by 1990 every man, woman, and child would be a scientist. But I think there is a much more insidious change. I think there is going to be much less pure science practiced. Everyone, the man in the street, government, business, students—all distrust pure science. Legislators are suspicious of pure science. They don't think scientists can produce the effects fast enough; many of them don't think they can produce the effects at all. They are suspicious of universities. It's a very complicated business; but they want mission-oriented science. I mean, that's the squeeze from the top. It seems to me there's a big squeeze from below. Which is that the young aren't interested in nature anymore. They're much more interested in not how nature works but how society works and how people work. They don't believe that knowledge is power. This is very strong. The drift from natural science is the modern philistinism."

Was this because science had become so hard? "Partly that, but I think it's deeper than that. The attitude of my generation that all problems can be solved in the next decade, and *should* be solved in the next decade—these expectations are changed. Maybe science should be done better, but more slowly. I think a large number of mediocre people are in science today, and carried along by the system. General concepts are rare. Nobody publishes theory in biology—with few exceptions. Instead they get out the structure of still another protein. I'm not saying it's mindless. But the mind only acts on the day-to-day."

Is there a body of theory? "At least there is a body of slogans. Biologists can only be interested in three questions: How do things work? How are they built? How did they evolve? First, how does it work—how does it run around, how does it breathe and eat, how does it die? But the second question is the deeper question that comes before you can answer the evolutionary question: How are organisms built? We know from molecular biology that organisms not only have something 'like' a computer program, they actually do have a program. But saying that does not, of itself, help me to understand how to make a mouse.

"You see, molecular biology has been terribly mechanism-oriented. Molecular biology has taught, continues to teach biologists to speak the language of chemistry—the biochemists had begun that, but all of biochemistry up to the fifties was concerned with where you get the energy and the materials for cell function. Biochemists only thought about the flux of energy and the flow of matter. Molecular biologists started to talk about the flux of information. Looking back, one can see that the double helix brought the realization that information in biological systems could be studied much in the same way as energy and matter. But because molecular biologists happened to choose a genetic system to work with, they were able to look at the flow of information in terms of molecular hardware. The fact is that it just happened to work out that way—that you had a sequence of bases and a sequence of amino acids and the logic was strong enough and the relation was all one-dimensional. That it was all so simple.

"Look, let me give you an example. If you went to a biologist, twenty years ago, and asked him, How do you make a protein, he would have said, Well, that's a horrible problem, I don't know. But if you were asking a biochemist he would have said, But the important question is where do you get the energy to make the peptide bond. Whereas the molecular biologist would have said, That's not the problem, the important problem is where do you get the instructions to assemble the sequence of amino acids, and to hell with the energy; the energy will look after itself. And I think that's the difference."

The room had filled, behind us and around us, and the noise of conversation was pleasant. "The complete description of the organism is already written in the egg," Brenner said. *Omne vivum ex ovo:* William Harvey had taught that in the seventeenth century. People began getting up and moving toward the door. Dinner had been announced. In the great hall, a step below the dais, the students had eaten and we saw only the last departing backs. The high table was in fact two long, broad oak refectory tables, for King's is rich and populous, though Peterhouse, Perutz's and Kendrew's college, is conceded to have the better cooks. "*Benedic Domine, nobis . . .*" said the Provost; all sat.

"Inside every animal there is an internal description of that animal," Brenner said. Some of the hairs in his eyebrows were three inches long. He was wearing a rusty, dark-green polo-necked sweater beneath the black gown. Along the middle of the table, at intervals, heavy ancient objects of silver shone, their patina deep. "It is an empirical question to find out that description. What is going to be difficult is the immense amount of detail that will have to be subsumed. The most economical language of description is the molecular, genetic description that is already there. We do not yet know, in that language, what the *names* are. What does the organism name *to itself*? We cannot say that an organism has, for example, a name for a finger. There's no guarantee that in making a hand, the explanation can be couched in the terms we use for making a glove." A large tray with fish on it was inserted between us.

"My own conviction is that it will not be enough to know the general mechanism," Brenner said. "Look, let me give you an example. If you were to say to me, Here is a protein, what is its genetic specification?—we could answer in tremendous detail. But if you say to me, Here is a hand, here is an eye, how do you make a hand or an eye, what is *its* genetic specification?—we can't do it. It is necessary to know the exact number and sequence of the genes, how they interact, what they do. We have to know the program, and know it in machine language which is molecular language. We have to know it so that in principle one can generate a set of procedures for making a hand or an eye.

"We are trying to approach it in two ways. Through the whole organism by doing genetic analysis of mutants and so on—the rather classical approach, which depends on the choice of the animal and how deeply you go into it. But the real way, the way one will have to employ in the long run, is actually to work with cell culture, to build organs. Step by step, we will have to make a retina in cell culture.

"What is important is to specify how the computation is performed," Brenner said. I asked whether there didn't have to be levels in the construction of complex beings which would include whole classes of operations. "Look, let me give you an example," Brenner said. "I told you molecular biologists could have said to biochemists, and did say, To hell with the energy; let's find out the order of the amino acids and the energy will take care of itself. Today most molecular biologists would tell you, The important thing to do is to find out

how genes in higher organisms are switched on and off. But you can go farther, to say that the important thing is *which* genes, okay?—and when, and
how they are grouped together, and who is called simultaneously, and who is
sequentially ordered—and never mind the mechanism, that'll look after itself!" We were offered a tray of meat, and dishes of vegetables.

"You see," Brenner said. "I mean, these are all very crude things—but I
think in the next twenty-five years we are going to have to teach biologists another language still. I don't know what it's called yet; nobody knows. But
what one is aiming at, I think, is the fundamental problem of the theory of
elaborate systems. Especially, elaborate systems that arise under conditions of
natural selection. And here there is a grave problem of levels: it may be wrong
to believe that all the logic is at the molecular level. We may need to get
beyond the clock mechanisms. Yet still, if the changes are made at the
nucleotides, then clearly we can't understand anything until we understand
all the levels of computation that connect the change in the gene with the
change in the development, the growth—or in the behavior. That's an *immense*
task. I don't want to say trite things—you know, we all criticized the
biochemists for saying that the cell is a bag of enzymes; well, I mean you
might as well condemn *us* for saying the cell is a bag of repressors and operators.

"There is something to be thought about here that has not yet been formulated successfully. John von Neumann wrote a very interesting essay many
years ago, in which he asked, How does one state a theory of pattern vision?
And he said, maybe the thing is that you can't give a theory of pattern
vision—but all you can do is to give a prescription for making a device that
will *see patterns!* In other words, where a science like physics works in terms of
laws, or a science like molecular biology, to now, is stated in terms of mechanisms, maybe now what one has to begin to think of is algorithms. Recipes. Procedures.

"This has been foresaid by many people. But if your question is, What is a
mouse and how does it come about to be a mouse, it may be that the only way
to give the answer to this is to specify an algorithm for how you could build a
mouse. *In the way the mouse builds itself.* And you must be careful not just to
give a description of the mouse as it exists. Okay? Let me give you a clear-cut
example. You could say, I'll write an algorithm to make a hand. So you'd
write, make a hand, make five fingers, and so on. And you write all that like a
computer program. But if you look at the list of statements, that isn't a program, it's a description of the hand that is there—in curious prose." We indicated, by our choice of medium or small plates, whether we wanted trifle or
cheese and biscuits.

"I believe that in biology, programmatic explanations will be algorithmic
explanations," Brenner said. "You will have to say, Next switch on gene group
number fifty-eight. And then one has that whole lot of molecular biology—what is gene group fifty-eight and what does it do. And one takes for
granted that gene group fifty-eight performs its computation. And then the simultaneous steps, the alternative steps, the sequential steps. In great detail. So.
I feel that this new molecular biology has to go in this direction—to explore
the high-level logical computations, the programs, the algorithms of development, in molecular terms. We've got to try to carry the analysis to that

level—and to carry the molecular biology with it. Because, after all, we're doing the things, asking the questions that people raised in the eighteen-seventies. Not the nineteen-seventies. But of course we're doing it with modern tools and with everything we know. And one would like to be able to fuse the two—to be able to move between the molecular hardware and the *logical* software of how it's all organized, without feeling they are different sciences."

Everybody rose and filed out, and down a corridor to another room, large and low, with a long table down the middle on which were candles burning in silver sticks, glasses, bowls of fruit, and bowls of walnuts with nutcrackers. Everybody sat. From my right, along the polished wood, there arrived a little trolley train on wheels, in the shape of a silver swan, with three parts joined by swivels, and each part bearing a decanter. A card said that the port was Taylor 1955, the white was Raventhaler Herberg Riesling Spätlese Cabinet 1959, and the red was Corton Les Marechaudes 1962. Beyond Brenner sat a gray-haired woman, a computer mathematician from one of the women's colleges. Brenner leaned towards her. "There will be no difficulty in computers' being adapted to biology," he said, with clenched teeth. "There will be Luddites. But they will be buried."

❖ ❖ ❖

"All this shameless talk about creating and multiplying will be put down as the barbarism of the twentieth century," Erwin Chargaff said in a conversation a year later. Many biologists, and biochemists in particular, would like to regard molecular biology as hardly more than patent medicine, putting a new and deceptive label on a familiar bottle—a bottle, namely the cell, that was stolen from them in the first place. Chargaff, sour, literary, and extravagant, found not only the explanations of the new science unacceptable, but the new men and the new style. Chargaff had written that he was a chemist: "a practitioner of the science that has had a monopoly on molecules for such a long time that no one seems yet to have had the potentially lucrative idea of founding the *Journal of Molecular Chemistry.*"

Not long after his first meeting with Watson and Crick, Chargaff elected himself polemicist on behalf of all that has been left out—for the early discoverers, Friedrich Miescher and Oswald Avery; for the role of protein in the chromosome; for complexity and crowding in the cell; for humility and caution in the laboratory. "I am against the over-explanation of science, because I think it impedes the flow of scientific imagination and associations," he said. "My main objection to molecular biology is that by its claim to be able to explain everything it actually hinders the free flow of scientific ideas. But there is not a scientist I have met who would share my opinion."

In *Science*, several years ago, Chargaff wrote, "Today the smallest of the small bacteriophages, tomorrow the brain that conceived *Die Zauberflöte*. But in my laboratory there exists an old house proverb saying 'The first success in an experiment comes from the devil; but then the way drags on.' . . . We really still are very far from an actual grammar of the living cell, not to speak of that of an organ, an organism, or, even more, a thinking organism." In a history of the biochemistry of DNA, he warned his scientific peers:

> In the living tissue many events take place simultaneously; precursors, intermediates, and end products are formed in close propinquity; everything happens on

top of each other, apparently without getting into each other's way; proteins and nucleic acids, lipids and polysaccharides are assembled and deposited where they belong: all presumably under the supervision of the genome [full complement of genes] which at the same time is quite busy reproducing itself. . . . Faith in the unity of nature should not make us forget that the realm of life is at the same time rigidly unified abstractly and immensely diversified phenomenologically. . . .

If DNA, a hundred years ago a humble molecule in Miescher's hands, has been hypostasized into one of the symbols of the ever-increasing divorce from reality that characterizes our living and thinking, this may be taken as one of the signs that the winds of alienation have begun to beat at the doors of what was the most concrete building erected by the Western mind, namely, that of science.

"Eleven years later, they stood on the same platform in Stockholm to receive their Nobel prizes, Perutz with John Kendrew, Watson with Crick and Maurice Wilkins." The Nobel prize ceremonies, 10 December 1962: King Gustaf VI Adolf, right, presenting the prize in medicine or physiology to Francis Crick; background, Max Perutz, John Kendrew, Maurice Wilkins, and James Watson. Perutz and Kendrew shared the prize in chemistry for solving the structures of hemoglobin and myoglobin, the first determinations of the three-dimensional structures of protein molecules. Crick, Watson, and Wilkins shared the prize in medicine or physiology for discovering the structure of deoxyribonucleic acid, DNA, the stuff of which genes are made. *Photograph: Pressens Bild AB/Photoreporters*

"'Perhaps we will be able to grind genes in a mortar and cook them in a beaker after all,' Muller said in 1921." Hermann Muller at the Cold Spring Harbor Symposium in June 1953, where Watson presented the structure of DNA. *Photograph by Karl Maramorosch*

"The relation of gene to enzyme was recognized again, beginning with a paper presented in the summer of 1940 by George Beadle and Edward Tatum from Caltech. . . . 'One gene—one enzyme'—Beadle's slogan exactly captured the conceptual pivot. The one gene was still the classical gene, unitary and algebraic, but its face was now turned away from the next generation and towards its molecular complement and consequence in the present organism." Edward L. Tatum at the Cold Spring Harbor Symposium, June 1953. *From the Norton Zinder photograph collection at Cold Spring Harbor Laboratory Library Archives*

"'This, according to my belief, can be looked upon as the first statement of a correlation between nucleic acids and synthesis of proteins,' Caspersson wrote." Jack Schultz and Torbjörn Caspersson in Caspersson's laboratory in Stockholm, in 1938. *Photograph by Rollin Hotchkiss*

"Delbrück created 'a biology that had been made comfortable for people with backgrounds in the physical sciences.'" Max Delbrück with Wendell Stanley at Cold Spring Harbor Laboratory, about 1946. *Photograph courtesy Cold Spring Harbor Laboratory Library Archives*

"Avery's great paper shares with the classics of science of previous centuries at least one quality now grown rare: from the first paragraphs through to the end, one feels an original curiosity working." Oswald Avery at the Rockefeller Institute, in the late thirties. *Photograph courtesy George Hirst*

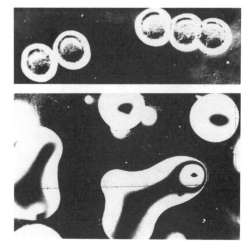

"The world of Avery's *Streptococcus pneumoniae*, called pneumococcus, is divided into the Rough and the Smooth." Abnormal, or rough (above) and normal, smooth, colonies of pneumococcus. *Photograph by Harriett Ephrussi-Taylor*

"'The first determination of the exact amino-acid sequence of a protein, the sequencing of bovine insulin by Sanger, was absolutely essential. It revealed that a protein has a definite, constant, genetically determined sequence—and yet a sequence with no general rule for its assembly. Therefore it had to have a code.'" Frederick Sanger. *Photograph by Arthur Foster, courtesy Cold Spring Harbor Laboratory Library Archives*

"At the time Chargaff met Watson and Crick, had he himself understood the implications of the one-to-one ratios he had discovered among the bases? Chargaff answered slowly and distinctly. 'Yes and no,' he said. 'No, I did not construct a double helix.'" Erwin Chargaff in 1963. *Photograph courtesy American Philosophical Society from the Erwin Chargaff Collection*

"A paper of 1928 brought Pauling into direct rivalry with Lawrence Bragg for the first time. . . . According to Max Perutz (and others, too), Bragg was always chagrined that Pauling had first published the rules by which physics applies to the structure of minerals." Sir Lawrence Bragg, some months before he died, in the summer of 1971, at the age of eighty-one. *Photograph courtesy Max Perutz*

"Pauling chuckled when he spoke of that. 'We got the alpha helix, you know, and Bragg and Perutz and Kendrew were trying to do the same thing, and they—ah, failed—because of, really a lack of knowledge of the principles of chemistry, of structural chemistry." Linus Pauling explaining the alpha helix, in the mid-fifties. *Photograph courtesy Linus Pauling and W. H. Freeman and Co.*

"Delbrück created at Cold Spring Harbor that spirit of ceaseless questioning, dialogue, and open-armed embrace of a life in science that he had learned from Niels Bohr— but with a down-to-earth American character and a good measure of his own high-minded intolerance of shoddy thinking." Max Delbrück and Salvador Luria at Cold Spring Harbor. *Photograph by Karl Maramorosch*

"'*Mein Gott!* They've got tails!'" First electron-microscope photograph of bacteriophage: phage T2, magnified about forty thousand times, taken in March 1942 by Thomas Anderson from phage preparations by Luria. *Photograph courtesy Thomas Anderson*

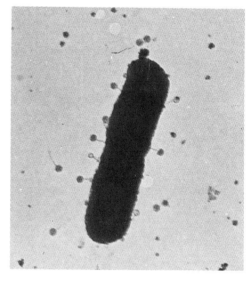

"The analogy of a hypodermic syringe was not thought of until much later." Phage T5 adsorbed on a bacterium of strain *Escherichia coli* B, in an electron micrograph made by Thomas Anderson in 1953. *Photograph courtesy Thomas Anderson*

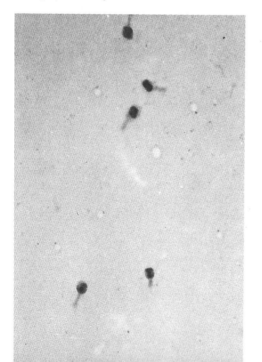

"For fifteen years or so, before it grew too big with success and outran its problems, the group around Delbrück and Luria formed . . . one of the rare refuges of the twentieth century, a republic of the mind." Some of the phage group—Jean Weigle, Ole Maaløe, Élie Wollman, Gunther Stent, Max Delbrück (by window), and Giorgio Soli, a graduate student—in Delbrück's phage laboratory at the California Institute of Technology, late spring of 1949. *Photograph courtesy Cold Spring Harbor Laboratory Library Archives*

"Leo Szilard, at the age of forty-nine, took the phage course in 1947." Szilard and Alfred Hershey at Cold Spring Harbor in the early fifties. *Photograph by Karl Maramorosch*

"To do genetics, Lederberg first had to show that bacterial life cycles had a sexual stage—that they mated." Joshua and Esther Lederberg at the Cold Spring Harbor Symposium in June 1953. *Photograph by Karl Maramorosch*

"Only Hayes interpreted the discovery aright. When bacteria conjugated, he said, one, the donor, or male, passed a copy of its genes to the recipient, or female." William Hayes at the Cold Spring Harbor Symposium in June 1953. *Photograph by Karl Maramorosch*

"'Everybody who was interested in these basic questions discussed this.' Delbrück said. 'I distinctly remember wading here at low tide with Rollin Hotchkiss, every so often, and he plugging for the idea that DNA might contain enough specificity.'" Rollin Hotchkiss and Max Delbrück at the Cold Spring Harbor Symposium in 1953. *Photograph by Karl Maramorosch*

"It's an accepted fact that the Waring Blender experiment was biochemically sloppy. But Hershey and Chase used a familiar, favorite organic system to produce a pleasingly symmetrical cleavage between the exclusion of phage protein from bacteria and the taking in of phage DNA." Martha Chase and Alfred Hershey at the Cold Spring Harbor Symposium in 1953. *Photograph by Karl Maramorosch*

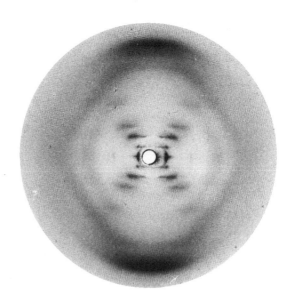

"The picture that Wilkins showed Watson was Rosalind Franklin's." The x-ray-diffraction photograph of the paracrystalline or B form of DNA that Franklin took over the weekend of 1 May 1952 and later published in her paper with Gosling that accompanied Watson and Crick's in *Nature*, 25 April 1953. *Photograph courtesy Maurice Wilkins*

"'All I'm saying is—well, I *didn't* get it. But if people want to speculate about, "Well, if Watson and Crick hadn't got it, who would?" then I think I could make a good case.'" Maurice Wilkins, at King's College London, about 1954, with x-ray equipment of the type also used by Rosalind Franklin and Raymond Gosling. *Photograph courtesy Maurice Wilkins*

"Rosalind Franklin, in the eighteen months or so from the fall of 1951 to the spring of 1953 when she was working at King's College London on the structure of deoxyribonucleic acid, had no such collaborator. It is evident from her notebooks that she needed one, and clear from what we know of her character that she would have worked well—candidly, rudely if need be—with the right one." Rosalind Franklin on a walking trip in France in 1950 or 1951. *Photograph by Vittorio Luzzati*

"'A remarkable thing.' Crick said. 'I went to my desk, opened a letter from Maurice, you know, dark lady and this sort of thing, and I looked across, and I thought, was it more a question of laughing or—well, you know, sadness almost. You see. There was the model.'" Watson and Crick with the first full model of DNA, at the Cavendish Laboratory in the spring of 1953. The slide rule was suggested by the photographer. *Photograph by A.C. Barrington Brown, reproduced with permission of Photo Researchers, Inc.*

"Watson wandered through the week tall and skinny and remote, dressed in loose cotton shorts and short-sleeved shirt, tails flapping. . . . In the lecture hall at Cold Spring Harbor, with the barber-pole schematic drawing of the structure projected on a high screen, Watson pointed to the wide groove that allowed access to the inward-lying base pairs." Watson at the Cold Spring Harbor Symposium in June 1953; in the outdoor picture, Salvador Luria is at left. *Photographs courtesy Cold Spring Harbor Laboratory Library Archives*

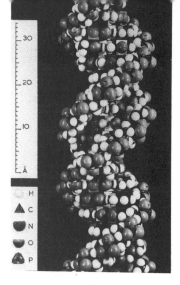

"Maurice Wilkins published a long series of papers culminating in an improved structure for the B form. It differed from the original in details of the backbones, most importantly by a shift in the angles of the sugars, and the shift in turn brought the bases in more snugly to the center." A space-filling atomic model of the structure of DNA as corrected at King's College London. *Photograph courtesy Maurice Wilkins*

"The discovery of the structure—after a pause for its implications to be absorbed—was brilliantly stimulating." Matthew Meselson and Seymour Benzer at the Cold Spring Harbor Symposium in June 1961. *Photograph courtesy Cold Spring Harbor Laboratory Library Archives*

Generations
after transfer

0

0.28

0.71

1.14

1.57

2.00

0 and 2 mixed

N^{14} and N^{15} mixed
for comparison

"'The most beautiful experiment in biology.' If replication was semiconservative, then precisely at the end of the first generation . . . all the DNA in the sample should be hybrid, so that it would all show up in a single stripe. Then in the second generation, Meselson expected to find DNA molecules of two kinds—some hybrid, others now with unlabelled chains only." The results of the density-gradient centrifugation experiment that Meselson and Franklin Stahl carried out in the winter of 1957–'58. *Photograph courtesy Matthew Meselson*

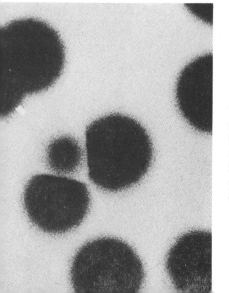

"'My discovery was a gene and a phage and a set of bacterial strains which had the right kind of properties so that one could in a sense split the gene into its internal parts and construct a map of them.' Benzer said." Plaques of bacteriophage T4 growing on a lawn of *E. coli*. The wild type produces small, fuzzy plaques, while an *r*II mutant produces large, smooth-edged ones. *Photograph by Seymour Benzer*

"Gamow's answer came out of the structure of DNA in the most directly physical manner possible: he suggested that permutations of the bases formed holes of different shapes, into which various amino acids fit as specifically as keys into locks." George Gamow, about 1954, with a model of a nucleic acid structure. *Photograph courtesy Igor Gamow*

"When the two men met in 1954, Brenner recognized at once that Benzer had invented what should be the most direct way of all into the relation of gene to protein: map the gene, sequence the corresponding proteins of the mutants, and compare." Sydney Brenner in Cambridge in 1960. *Photograph courtesy Matthew Meselson*

"In the most magnificent achievement of late-classical biochemistry, Fritz Lipmann was finding out how cells make available the energy to drive their manufacturing processes." Fritz Lipmann, at the time he won the Nobel prize, in December 1953. *Photograph: Jacob Maarbjerg, Politikens Presse Foto*

"Berg set out to find some naturally occurring acceptor for the activated amino acid, without knowing what the acceptor might be—and without having heard about Crick's hypothetical adaptor. Later that year, Berg found such an acceptor . . . ; in the presence of the acceptor the amino acid became linked to something. When he purified the something, it was RNA—an unusually small piece." Paul Berg, Alexander Rich, and Howard Dintzis at the Cold Spring Harbor Symposium in June of 1966. *Photograph courtesy Cold Spring Harbor Laboratory Library Archives*

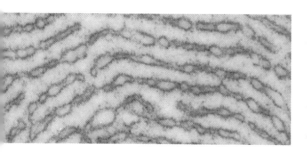

"Palade's electron micrographs showing the endoplasmic reticulum and the enormous numbers of particles attracted intense interest. . . . Paul Zamecnik proved that the particles were the place where the peptide chain was put together." A portion of an electron micrograph of a cell of a guinea pig, taken by George Palade, at the Rockefeller Institute about 1955. *Photograph courtesy George Palade*

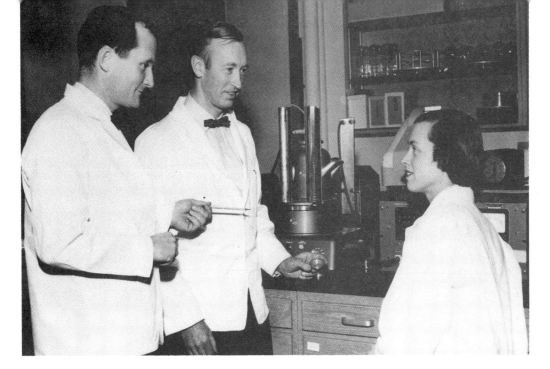

"The cell-free system for protein synthesis was to prove the vehicle of the eventual solution of the coding problem. But even the simplest aim of getting a good and reproducible level of incorporation of amino acids into protein, without complete cells, only began to be possible in Zamecnik's lab in 1953 and '54. At this stage in the work, Zamecnik was joined by Mahlon Hoagland." Mahlon Hoagland, Paul Zamecnik, and Mary Stephenson, in 1954 or '55, in Zamecnik's laboratory at Massachusetts General Hospital. *Photograph courtesy Paul Zamecnik*

"Meanwhile, others—Howard Schachman, Arthur Pardee, and Roger Stanier in Berkeley, Ole Maaløe in Copenhagen—were detecting large numbers of dense spherical particles in bacteria, too." Electron micrograph of microsomal particles from *E. coli*, taken for the Berkeley group by Robley Williams, in 1951. *Photograph courtesy the Virus Laboratory, University of California, Berkeley*

"Lwoff commanded in English as in French a style of sweetest clarity, and his are the only scientific papers I have read that are sometimes intentionally funny. . . . Lwoff was a scientist of great technical finesse, and a perception never bound to a rigid logical line. He was a man of reserve, of severe rectitude, and—the most indelible impression—of balance." Milislav Demerec and André Lwoff at the Cold Spring Harbor Symposium in June of 1953. *Photograph courtesy Cold Spring Harbor Laboratory Library Archives*

"'Boris insisted that there might be some substance in this young guy and that he was really wanting to learn genetics.'" Boris Ephrussi and Jacques Monod on a rooftop at the California Institute of Technology in 1936. *Photograph by Barbara McClintock*

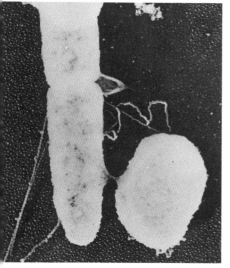

"In 1956, Thomas Anderson visited Paris and took a superb electron micrograph of a long, slender, rod-shaped donor cell lying with a short, rotund recipient." The original picture of conjugating *E. coli*, strain K12. *Photograph courtesy Thomas Anderson*

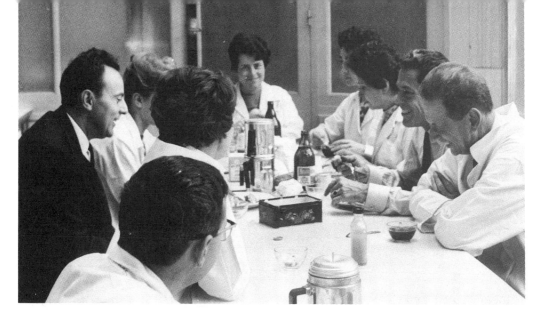

"Lwoff's fabled attic was three flights up, at a back corner, and consisted of a corridor about twelve yards long, high-ceilinged but windowless and crowded with equipment, with five small laboratories opening off it and tucked beneath the slope of the mansard roof. Lwoff's lab was at one end, Monod's diagonally across at the opposite end. . . . 'If I had to tell the story, I would begin with that corridor, the work going on at each end—the phage business of Lwoff and Monod's enzymes,' Jacob said. 'The two things which merged.'" Coffee hour in the Lwoff attic, Institut Pasteur, October 1965. Around the table, left to right: Claude Burstein, a student; Marguerite Lwoff; François Jacob (sitting back); Agnès Ullmann (obscured); Madeleine Brunerie, Monod's secretary (at head of table); Gisèle Houzet, Lwoff and Jacob's secretary; Madeleine Jolit, Monod's technician; Jacques Monod; Andrè Lwoff. *Photograph courtesy Madeleine Brunerie*

"'When I arrived, the lab was divided in two parts.' Jacob said. 'And the great joke was about induction, because Lwoff and Monod were each "inducing" according to his fashion, convinced that the two phenomena had nothing in common except the name." François Jacob and André Lwoff at the Cold Spring Harbor Symposium in 1953. *Photograph by Karl Maramorosch*

"Monod was greatly tempted to become a proselyte to phage." Barbara McClintock and Jacques Monod at Cold Spring Harbor in the summer of 1946. *Photograph courtesy Cold Spring Harbor Library Archives*

"The chief thing that Monod remembered was Szilard's insistent question: Why not suppose that induction could be effected by an anti-repressor?" Jacques Monod and Leo Szilard, Paris, June 1961. *Photograph courtesy Esther Bubley*

"'Simultaneous discoveries have happened to me so often in "important" cases that just once I'd like to come up with something new all alone.'" Arthur Pardee in 1960. *Photograph courtesy Arthur Pardee*

"For twenty days in Pasadena, they could make nothing work at all." From left, François Jacob, Max Delbrück, Matthew Meselson, Ronald Rolfe (partly obscured), Gunther Stent, and Sydney Brenner, in Delbrück's garden in Pasadena in June of 1960, at the time of the messenger experiments. *Photograph courtesy Gunther Stent*

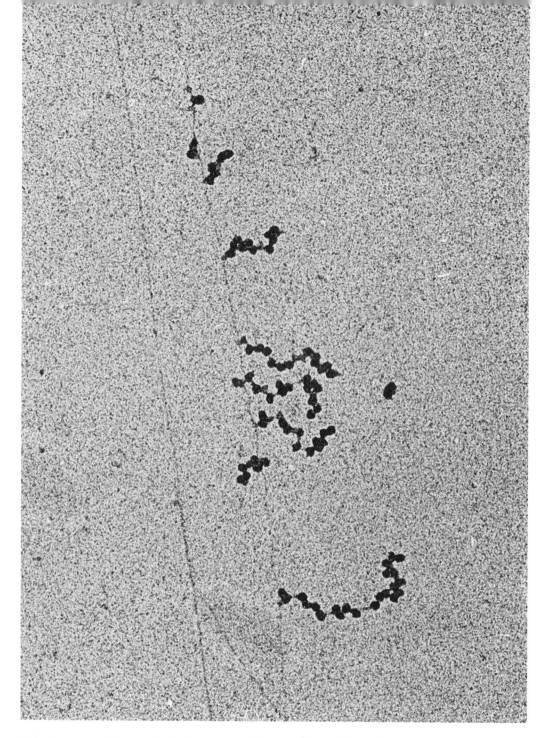

"'That was one of those points when one could see one's own ideas change very abruptly.' Gilbert said. 'You had this nice long messenger; and these ribosomes picked up one end and ran along it.'" A modern electron micrograph of an unidentified, active operon from *E. coli:* the strand at left is an inactive segment of chromosomal DNA; on the strand of DNA at right, several strands of messenger RNA are forming, the newest at the top and just long enough for a single ribosome to attach and begin transcribing the mRNA into protein; at the bottom, the longest polyribosome is on a strand of mRNA near the termination of the gene. Magnification 198,000 times. O. L. Miller, Jr., Barbara A. Hamkalo, and C. A. Thomas, Jr., "Visualization of Bacterial Genes in Action," *Science* 169 (1970): 392–395. *Electron micrograph courtesy Oscar Miller*

"Nirenberg learned of Matthaei's success when he got back from Berkeley." Marshall Nirenberg, at the National Institutes of Health, in the mid-sixties. *Photograph courtesy Marshall Nirenberg*

"Between Nirenberg's technique and Khorana's, the code was almost entirely elucidated by the time of the Cold Spring Harbor Symposium of 1966." Heinrich Matthaei, Har Gobind Khorana (head turned away), and Fritz Lipmann at that Symposium. *Photograph courtesy Cold Spring Harbor Laboratory Library Archives*

"'That night in 1934, Bernal, full of excitement, wandered about the streets of Cambridge, thinking of the future and of how much it might be possible to know about the structure of proteins if the photographs he had just taken could be interpreted in every detail.'" Desmond Bernal and Dorothy Crowfoot in Nottingham in 1937. *Photograph courtesy Dorothy Crowfoot Hodgkin*

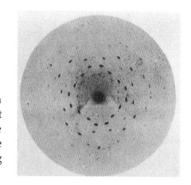

"In 1912, it occurred to a young physicist in Munich, Max von Laue, that the repetitively spaced layers of a crystal would affect x rays just as a diffraction grating affected light, scattering the waves in patterns." Laue's original diffraction pattern from the mineral zincblende, in an x-ray photograph by Lawrence Bragg in 1913. *Proceedings of the Cambridge Philosophical Society 17, 1913*

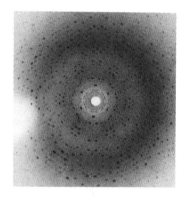

X-ray-diffraction pattern from a crystal of pepsin, the substance with which Bernal got the first such patterns from proteins. *Photograph courtesy Max Perutz*

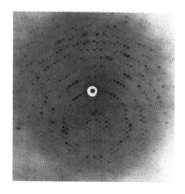

"At the beginning of October 1938, Bragg arrived in Cambridge to succeed Rutherford. . . . Several weeks passed before Perutz gathered his courage and his hemoglobin patterns and went to see Bragg." X-ray diffraction photograph of horse methemoglobin, with the crystal oscillating, taken by Max Perutz in the late thirties. *Photograph courtesy Max Perutz*

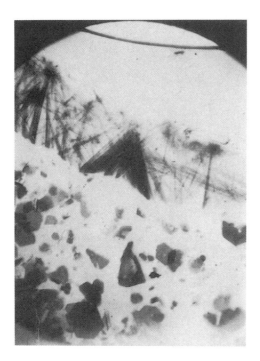

"As Haurowitz looked at the crystals of deoxyhemoglobin, a wave of change moved across the microscope field. Beginning at one side, the hexagonal tablets disappeared, dissolving back into the liquid. Then beginning again at the same side, new crystals formed and quickly grew—the needles of oxyhemoglobin. Air had penetrated from one edge." Photograph by Felix Haurowitz, through the microscope, 1938. *Photograph courtesy Felix Haurowitz*

Felix Haurowitz and Francis Crick at Cold Spring Harbor in 1967. *Photograph courtesy Cold Spring Harbor Laboratory Library Archives*

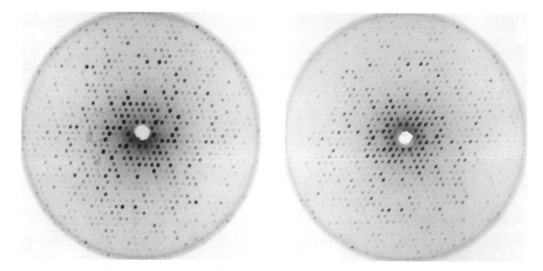

"'When I examined the picture closely, I saw that there were marked changes in the relative intensities of the spots. This perhaps was the most exciting moment in all my research career'—Perutz's voice took on a buttery glow of satisfaction—'because I realized at that moment that now, in principle, the protein problem was solved.'" At left, the x-ray-diffraction pattern from a crystal of native oxyhemoglobin of horse; at right, the pattern from a crystal of oxyhemoglobin of horse carrying heavy atoms, one atom of mercury on each half molecule. The spots are in the same positions but some are markedly different in intensity. *Photographs courtesy Max Perutz*

"On the tables were models. They were surprisingly large, and at first difficult to take in—a jungle of thin metal supporting rods." John Kendrew and Max Perutz with part of a model of the hemoglobin molecule. *Photograph: BBC Copyright*

"Kendrew got out the structure of sperm-whale myoglobin at low resolution in the summer of 1957: that was the first protein to be solved. He constructed a model out of plasticine—the writhing, visceral knot whose photograph I had noticed the first time I met Perutz." John Kendrew with the model of myoglobin at low resolution. The heme group is the disc, gray in the photo and half hidden, at the top center; the long straight portions of the knot turned out to be alpha helices. *Photograph: BBC Copyright*

"The heme groups were set in four separate pockets at the surface of the hemoglobin molecule—and this was the greatest surprise, for they were about as far apart as they could be." Perutz's model of the oxyhemoglobin molecule at low resolution. The layers of plastic sheet he cut to the contours of maximum density in the Fourier maps and then stacked up. The two alpha chains are white, front left and back right; the two beta chains are black, back left and front right; the gray discs show two of the four heme groups. *Photograph courtesy Max Perutz*

"'So you do the experiment, and you stare at it and you say, Now, does it mean anything? What is the world really going to look like? The messenger experiment had that quality,' Gilbert said." Walter Gilbert, at the Biological Laboratories, Harvard, in 1972. *Photograph by William Toby, courtesy Harvard University*

"'And it seemed to me that people who claimed to be trying to isolate the repressor weren't really serious. Weren't really willing to take the kind of risks that were necessary.' What kind of risks were those? 'Well,' Ptashne said. '*Psychic* risks.'" Mark Ptashne, at the Biological Laboratories, Harvard, in 1972. *Photograph by Rick Stafford, courtesy Harvard University*

"Monod and Jacob failed to isolate the substance—the repressor—whose nature and mode of action was their theory's keystone. The failure disappointed Jacob profoundly. . . . In the spring of 1968, the students of Paris rose in insurrection. . . . Monod pleaded for a truce to allow wounded students to be evacuated. His pleas were in vain." After all-night battle between students and police in the Latin Quarter, 10–11 May 1968, Monod leads a girl wounded in the eyes. *Wide World Photos*

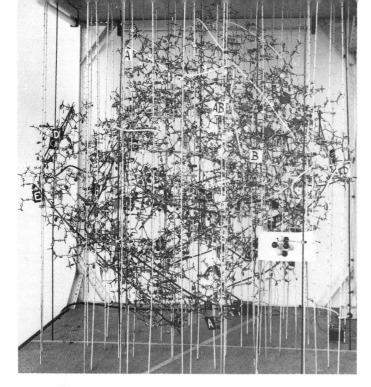

"'People thought we were very foolish to dream we could ever do it.' Bragg said. 'It looked absolutely impossible, when you thought that the most complicated structures ever yet got out were for molecules only about two hundred atoms big. The difficulty goes up not just as the square but as a high power of the number of atoms in your molecule. With hemoglobin we were trying ten thousand atoms.'" The high-resolution atomic model of hemoglobin in the forest of its supporting rods, built by Perutz and Hilary Muirhead in 1967; inset at right is the atomic structure of sodium chloride, to the same scale, which Bragg solved at the founding of x-ray crystallography in 1912. *Photograph courtesy Max Perutz*

Perutz with low-resolution model of hemoglobin at the time of the Nobel prize announcement in 1962. *Reuter*

"'While the artist's communication is linked forever with its original form, that of the scientist is modified, amplified, fused with the ideas and results of others.' Delbrück said." Linus Pauling, Max Delbrück, and Max Perutz, 1976. *Photograph: Floyd Clark, California Institute of Technology*

"Je cherche à comprendre." Jacques Monod, at his house in Cannes two days before his death, on 31 May 1976. *Photograph by Jean Hardy*

II

RNA

*The Functions of the Structure:
The breaking of the genetic code,
the discovery of the messenger*

5 "The number of the beast"

❖a❖   In the fall of 1975, in several long conversations, Francis Crick began to tell me what happened in the years after the structure of DNA was found. I turned with questions, as well, to others who had played a part, and began to read more systematically in the papers written in the mid-fifties. The first impression was of confusion. The discovery of the structure of DNA had been singular in more ways than I had taken into account. One recognized, to be sure, its disproportionate importance for understanding the way things are. But there was also the extreme intensity—the burning focus—of its intellectual and psychological circumstances, and I did not realize how that intensity had influenced the form—how it was itself the form—of the history until I drew breath and asked what came next. The elucidation of the structure of deoxyribonucleic acid had been a concentration of forces, a siege, a conquest. The decade that followed was a story of dispersal and movement, of uncertain direction, wrong direction, multiple lines of work intersecting in fortunate encounters, and a way—found at last through an unexpected break—as different from what went before as the *Odyssey* from the *Iliad*, and yet with continuities of theme, style, character.

After the structure of DNA was solved, biology opened up and opened out. The confusion I perceived was first, trivially, my own, because the science, though not forbiddingly abstruse as some science is, was dauntingly intricate. Beyond that, though, was the confusion of extraordinary growth. In the crudest terms, no other science, not nuclear physics, has ever expanded as biology did in North America and Europe from the mid-fifties to the mid-sixties: new people, new and larger laboratories, more and ever-fuller meetings and journals and books. The dollar was constant and biology flexible, promising, and still cheap. Then there was another kind of opening up that was altogether more interesting: again and again, the smallest, most casual beginnings—a skeptical question asked in a Paris café, some specks seen on an electron micrograph, the idea struck off in a sentence during a drive down from New England—have grown up into specialties that have since engrossed entire lifetimes in science, today command whole teams and laboratories, whose results fill volumes and are not ended.

The web of significance ran horizontally, too, of course: if simple ideas and cheap experiments had great consequence, that was because so much that was done and thought of turned out to be mutually explanatory, instantly suggestive, in relation to so much else. Perhaps this was science as one would expect it to be, in some ideal sense: but it won't do to say that biology after DNA returned to normal, for the norm itself was changing—the sense of what needed explanation and the character of the explanations that would be acceptable. If the psychological fever slowly cooled, the intellectual intensity remained incandescent. Confronted with this living fabric unrolled abundantly upon his worktable, the historian picks up his shears, hesitates, and puts them down again. Whatever he does, he will be like the damned eighteenth-century Dutchman who made Rembrandt's *Night Watch* smaller to fit a smaller room.

Alongside these confusions were still others, of a different order—those of the subject itself as it was worked out in the thoughts of the scientists month by month. It's carminative to reflect that biologists since 1953 have put forward more mistaken ideas than their predecessors did in all the millennia before—and some of the mistakes were beauties. The mistakes are inseparable from the science—not because they do happen but because they must happen. The creative faculty, I think, operates in the same manner when the physicist engages in reasoning and experiment, when the musician composes, or when the historian writes a page, and this is by the argument between invention and disposition, between the voice over one shoulder urging *Try this, then* and the voice over the other shoulder whispering *Not quite right yet.* "Scientific reasoning," Peter Medawar said in a lecture before the American Philosophical Society some years ago, "is a constant interplay or interaction between hypotheses and the logical expectations they give rise to: there is a restless to-and-fro motion of thought, the formulation and rectification of hypotheses, until we arrive at a hypothesis which, to the best of our prevailing knowledge, will satisfactorily meet the case." Thus far, change only the word "hypothesis" and Medawar has described exactly the experience the painter or poet has of his own work. (He knew this, no doubt, but tactfully left the point for his hearers to supply.) "Scientific reasoning is a kind of dialogue between the possible and the actual, between what might be and what is in fact the case," Medawar said a moment later—and there the difference lies. The scientist enjoys that harsher discipline of what is and is not the case. It is he, rather than the painter or poet in this century, who pursues in its stringent form the imitation of nature. The social system of science, from collaboration at the bench or blackboard to formal publication, is first of all a means to enlarge the interplay between imagination and criticism from a private into a public activity. (Perhaps only the playwright actively collaborating with the director and a troupe of actors can be as fortunate as the scientist in the terms of the creative work.)

This is why mistakes are essential to the process of the science. This is why Crick has suggested that his own most elegant disproved theory is a valuable instance to preserve. This is why Jacques Monod, shortly before his death, proposed that in writing a history of biology since the war it would be essential to include the errors and confusion out of which discoveries grew. Besides all that, as was true for the solution of the structure of DNA, only against the ground of the errors and confusions can one measure the difficulty and scale of the achievement.

"The thing is to lodge somewhere at the heart of one's complexity an irresistible *appreciation*," Henry James wrote in the preface to *The Spoils of Poynton*. The one angle of view that commanded a consistent, clear perspective of the next complexities in biology was Crick's. He thought up many of the developments, or stimulated them. Most of what he did not originate he soon became aware of. Besides the talent for theory for which he is known, he had a talent for friendship in science that quickly grew into the role, for this decade, that Max Delbrück's knack for leadership and unification had played before. By brain, wit, vigor of personality, strength of voice, intellectual charm and scorn, a lot of travel, and ceaseless letter-writing, Crick coordinated the research of many other biologists, disciplined their thinking, arbitrated their conflicts, communicated and explained their results. As he went, he sorted the important from the less important with a brisk efficiency that now serves well to distinguish the main line of molecular biology from all else that was happening in biochemistry. Crick's angle of view provides, at the heart of complexity, the irresistible appreciation that is demanded.

By good fortune, the progress of Crick's view of the science can be recovered almost intact. "You better be careful," Crick said. "I must warn you, my memory is now not reliable." Present memory is fortified and set right by the papers he wrote, of course, and particularly by the comprehensive reviews of the state of the subject that he delivered at one meeting or symposium after another. Crick also wrote letters. One fine evening in 1976, at a party in the garden of another Cambridge scientist, a sheep roasting on a spit in a corner by the orchard and a vat of Pimm's Cup with a block of dry ice in it bubbling and smoking on a table beneath a tree, Crick said, "Incidentally, Sydney Brenner has quite a few letters that I wrote to him when he was in South Africa, in 1954 and several years afterwards, which probably give quite a good idea of what we were thinking at the time. Because we had to write because we couldn't discuss. I don't know that that's ever been looked at; I know Sydney has them." The letters, in fact, were frequent, hasty, copious, stuffed with observation and fact. The period has no memoir like James Watson's *The Double Helix*; the correspondence provides a month-by-month account of the science as Crick saw it grow.

❖❖❖

Before I knew of Zen, runs the proverb, *mountains to me were mountains and waters were waters; when I was studying Zen, mountains were no longer mountains and waters no longer waters; then I achieved enlightenment—and mountains were mountains, waters were waters.* So, upon the elucidation of the structure of deoxyribonucleic acid, the remaining problems of biology were transformed—and were still there.

The metaphor must now be musical. The discovery of the structure of DNA had been a comedy of mind and manners, lit by instants, as is fitting in comedy, of poignance and even of beauty. The biology of the next decade took the shape of a rich and most gorgeous fugue. To follow its development we require not a cast list but a set of motifs; to bring these to their resolution demands that we attend contrapuntally, that we think in the special mode in which one keeps multiple, distinct, but interlocking themes all suspended alive in the mind, now one and now another dominating the foreground yet each taking its significance in relation to what was happening in the rest.

That is the kind of enjoyment, it seems to me, that this passage of the history has to offer. Some of the themes are familiar. Even the familiar were transformed. Structure modulated into function. DNA yielded place to RNA—tantalizingly similar, but its function uncertain, its structure amorphous. The central problem in biology became the synthesis of proteins in the living cell. The problem was addressed in three main ways. The first was biochemistry, which grew exquisitely precise as it reached towards the physical chemistry of the transport of energy and the building of individual molecules. The second was genetics, still done with bacteria and bacteriophage but driving its algebraic abstraction so far that it came into direct congruence with the physical gene—and then farther still, into illumination of the means by which genes control genes. The third approach was new, a highly schematized and theoretical entertainment that Crick christened the coding problem. Each of the three depended on the others, for each was but an aspect of that central problem of protein synthesis.

The culmination of those years took place in two meetings—a small, private conference of molecular biologists in Cambridge, England, on the afternoon of Good Friday, 15 April 1960, and a vast congress of biochemists in Moscow, August 1961—and in the intervening sixteen months of intensive work in five laboratories, at Harvard, the California Institute of Technology, the Institut Pasteur, the Cavendish Laboratory, and the National Institutes of Health in Bethesda, Maryland. The dénouement came at the second of those meetings, the one in Moscow. It was altogether unforeseen, even by Crick.

❖❖❖

One morning in the height of the Cambridge summer of 1953, Crick and Watson got a letter from George Gamow, a theoretical physicist in Washington, D.C., whom neither of them had ever met, though they knew of his standing and something of his work. Gamow is celebrated as the proponent of the Big Bang, or primeval fireball—the theory, which has dominated cosmology ever since he elaborated it in 1946, that the universe began in the singular half-hour, billions of years ago, when an inconceivably hot and dense concentration of all energy and all matter was hurled into expansion in an unimaginably powerful explosion. In the spring of 1953, Gamow had been visiting the University of California in Berkeley when a colleague pointed out to him Watson and Crick's paper about DNA in *Nature* of May 30—the second paper published, which had examined the significance of the structure and had suggested that "the precise sequence of the bases is the code which carries the genetical information." He had also seen Frederick Sanger's latest results, published in March, on the amino-acid sequences of insulin. Gamow had clapped the two sides of the problem together and come out with a scheme to explain how the sequential structure of DNA could directly, physically, order the sequential structure of proteins. He explained it in his letter and enclosed the draft of a paper of his own. Crick saw at once that Gamow had made errors at least in detail, and yet that the idea overall was extremely provocative. Gamow's scheme was decisive, Crick has often said since, because it forced him, and soon others, to begin to think hard and from a particular slant—that of the coding problem—about the next stage now that the structure of DNA was known. Crick and Watson carried Gamow's note off to discuss it over lunch at the Eagle.

Fame was already closing in, and although Crick enjoyed it, Watson was self-consciously bemused by the *tristesse* that follows intellectual creation. On March 22, Watson had written to Max Delbrück, his mentor in science still, enclosing the draft of the original announcement in *Nature*; Watson then had just returned from several days in Paris—the skittish flight from writing up their success that had astonished Crick. Mostly Watson's letter rehearsed the evidence for the structure and grumbled about the practical problem of paying his way to the Cold Spring Harbor Symposium—for the subject that year was viruses and Delbrück was organizing it. At the end of the letter, Watson wrote:

> I have a rather strange feeling about our DNA structure. If it is correct, we should obviously follow it up at a rapid rate. On the other hand it will at the same time be difficult to avoid the desire to forget completely about nucleic acid and to concentrate on other aspects of life. The latter mood dominated me completely in Paris, which as expected is by far the most enchanting city I shall ever know.

April 14, Delbrück wrote to Watson. He had been discussing the structure almost nonstop with junior colleagues. "The more I think of it the more I become enamoured with it myself," Delbrück said. Certain questions bothered him, though. For nearly twenty years, from the beginning of his interest in the nature of the gene, Delbrück had bent his entire attention to the problem of biological replication; in the opening lines of most of his major papers, the central concern and the master word had been multiplication. His reservations about the model began with replication. Each thread of the double helix determined completely the sequence of bases in the other, which was, Delbrück acknowledged, what made the manner of replication complementary and brilliantly evident; yet the two strands were wound around each other, not simply laid together so that they could be pulled apart sideways without interlocking. (Delbrück with gentle condescension reminded Watson of two unfamiliar terms, *plectonemic*, from the Greek meaning twisted or braided, opposed to *paranemic*, or side-by-side, for the two modes of coiling.) The two strands of a DNA molecule of typical length would make "about 500 turns around each other," he said.

> These would have to be untwiddled to separate the threads. . . . In any event, one must postulate that the DNA opens up in some manner, both for replication and for doing its business otherwise. In the structure you describe this opening up is opposed both by the two hydrogen bonds per nucleotide, and by the interlocking of the helices, and it becomes a very important consideration to find a way out of this dilemma, or to think of a modification of the structure that does not involve interlocking.

Delbrück had other objections as well. None the less:

> I have a feeling that if your structure is true, and if its suggestions concerning the nature of replication have any validity at all, then all hell will break loose, and theoretical biology will enter into a most tumultuous phase. Only part of this will involve chemistry, analytical and structural. The more important part will consist in attempts to take a fresh view of the many problems of genetics and cytology [study of cells] which came to dead ends during the last 40 years.

That day, Delbrück also wrote a brief letter to Niels Bohr in Copenhagen, *his* mentor, about a symposium on ideas of complementarity in biology and sociology that was planned for the summer of the following year. "Very remarkable things are happening in biology," he said at the end of the letter. "I think that Jim Watson has made a discovery which may rival that of Rutherford in 1911." He said no more; the comparison, addressed to Bohr, reverberated. Ernest Rutherford's announcement in 1911 that atoms have their positive charge (their protons) concentrated in nuclei was the first step to the discovery of atomic structure; it had been the direct inspiration to Bohr himself, then a young scientist in Rutherford's lab at the University of Manchester, to put forward his own first major work, the quantum-theoretical model of the atom that he proposed in 1913.

Late in April or early in May of 1953, a group came across to Cambridge from Oxford to see the model of DNA: Jack Dunitz, a crystallographer who had known Jerry Donohue at Caltech, Leslie Orgel, a biochemist, and with them Sydney Brenner, a South African research student. Brenner's background was unusual. His father had gone to South Africa from Lithuania, his mother from Russia; his father worked as a cobbler. Brenner had won a scholarship, at the age of fifteen, to the University of the Witwatersrand, in Johannesburg, and had got a bachelor's degree in medicine (the M.B., unknown in the United States, is the recognized qualifying degree for physicians in Great Britain, too) as well as a master's in medical biology. That was as far as the South African university system could take him. "It was a very underdeveloped country, scientifically—a provincial country," Brenner said one afternoon. "We didn't have Ph.D.s; facilities for research were quite—primitive. I mean, if you wanted to stain something, you first *synthesized the dye.*" He started independent research in chromosomes, using stains, and in the structures of cells, teaching himself out of books, building his own centrifuge. He began to try to get to a British university, Oxford or Edinburgh.

"I didn't want to go into ordinary biochemistry. I wanted to get into *this* subject—molecular biology—except that I didn't know what it was. So it was suggested I go and work with Hinshelwood"—Sir Cyril Hinshelwood, Professor of Physical Chemistry—"at Oxford, who was a man applying physical chemistry to biology." Hinshelwood accepted Brenner as a doctoral candidate; in an exchange of letters in May of 1952 about the research Brenner might do, Hinshelwood wrote, as his last paragraph, "One matter which I did wonder if you would be interested in, though I am not pressing it, is some work on phage." Brenner arrived in Oxford and threw himself into bacteriophage research with the energy of a man digging a tunnel out of prison. That fall, he heard from Orgel and Dunitz about the Cambridge interest in DNA, and they even speculated about an idea of Brenner's that DNA might be labelled with fluorescent dyes and then crystallized in order to get at its structure. "I was always interested in dyes. Of course, you know, the idea was pretty farfetched."

That winter, Sanger gave a talk in Oxford about the amino-acid sequences of the two chains of the insulin molecule. "Fantastic. After the lecture, one man in the audience made a remark, 'Now we know that proteins have chemical structures.' You see, I had never understood proteins before."

Brenner met Watson and Crick for the first time on the day, next spring, when he came to see the structure. They liked his nimble wit. "I went for a walk with Jim. I told him I was interested in phage, was working on phage then, at Oxford. And doing—well, it doesn't matter, it was pretty idiotic, but I'd learnt the field. And I heard of the Hershey-Chase experiment through Jim."

Watson answered Delbrück's hesitations about the coiling in a letter of May 21. "With regard to your comments on our note (1) biologically we are unhappy with our plectonemic coiling but (2) we believe we should consider the X-ray evidence and stereochemical considerations first and then worry about the biological complications." (About that time, Crick talked to some topologists about how strings link together, but learned nothing that wasn't obvious.) Watson's own worries were of a different order.

> It is all rather embarrassing to me since the Professor (Bragg) is frightfully keen about it and insists upon talking about it everywhere. Until we produced the model Bragg did not know what either DNA or genes were and his reaction to our original Nature Note was "Its all Greek to me." After we had convinced him that DNA might be interesting, he then got out of control and I spend much of my time deemphasizing it since I have not infrequent spells of seriously worrying about whether it is correct or whether it will turn out to be Watsons folly.
>
> Bragg, however, remains cheerful as ever, and has even told the story to the press and so last Fridays "News Chronicle" carried a story on how the secret of life was discovered in Cambridge. This immediately led a reporter of "Time" to Bragg and I am dreadfully afraid that I shall see the story in glaring print when I am in the states.

Three times as many biologists were enrolled at Cold Spring Harbor that summer as had come to Watson's first symposium there, five years earlier. Salvador Luria was due, and Leo Szilard, and most of the other phage workers, and a group from Paris including André Lwoff and François Jacob. Watson arrived on the second of June, to find that Delbrück was giving everybody reprints of the first set of letters to *Nature*—his with Crick, and the two supporting notes from King's College London.

Watson was back in the United States for the first time in nearly three years. His accent was erratically Anglicized, old acquaintances thought, and his conversation sugared with English slang; he wandered through the week tall and skinny and remote, dressed in loose cotton shorts and short-sleeved shirt, tails flapping. The paper he read about DNA, which he and Crick had prepared, was concerned with establishing the central fact of the model—the unique base pairing together with the dyadic, complementary structure—and with rebutting vigorously any suggestion that an alternative structure, one that might somehow avoid intertwined helices, could be fitted to the evidence. The paper was about replication; it was a forestalling of Delbrück. The function of DNA within the cell was acknowledged, but barely:

> A genetic material must . . . duplicate itself, and it must exert a highly specific influence on the cell. Our model for DNA suggests a simple mechanism for the first process, but at the moment we cannot see how it carries out the second one. We believe, however, that its specificity is expressed by the precise sequence of the pairs of bases. The backbone of our model is highly regular, and the sequence is the only feature which can carry the genetical information. It should not be thought that because in our structure the bases are on the "inside," they would be unable to come into contact with other molecules. Owing to the open nature of our structure they are in fact fairly accessible.

In the lecture hall at Cold Spring Harbor, with the barber-pole schematic drawing of the structure projected on a high screen, Watson pointed to the wide groove that allowed access to the inward-lying base pairs. He finished up:

> In any case the evidence for both the model and the suggested replication scheme will be strengthened if it can be shown unambiguously that the genetic specificity is carried by DNA alone, and, on the molecular side, how the structure could exert a specific influence on the cell.

Not all the phage group saw the point of the structure as quickly as Delbrück. Two papers further down the program, Alfred Hershey went methodically over the line of evidence about the infectivity of the DNA in bacteriophage that had culminated in the Waring Blender experiment with Martha Chase. Though that was the experiment that had at last convinced many people of the genetic function of DNA, Hershey didn't mention the proposed structure, and concluded that there was not "a sufficient basis for scientific judgment concerning the genetic function of DNA." He went on:

> The evidence for this statement is that biologists (all of whom, being human, have an opinion) are about equally divided pro and con. My own guess is that DNA will not prove to be a unique determiner of genetic specificity, but that contributions to the question will be made in the near future only by persons willing to entertain the contrary view.

❖ ❖ ❖

"You see, people didn't necessarily *believe* in the code," Crick said one afternoon. "So one used to go around rather as a missionary, in the early days, trying to draw people's attention to the fact that the problem existed. There *were* people who were sufficiently perceptive they didn't have to be told there was a problem, but the majority of biochemists simply weren't thinking along those lines. It was a completely novel idea, and moreover they were inclined to think it was oversimplified. In which they had a modicum of justification—even though they were wrong."

We were in Crick's office, and had been talking, that day and the day before, about the growth of ideas about the function of RNA and the nature of the genetic code. "What I mean is that they thought protein synthesis couldn't be a simple matter of coding from one thing to another; that sounded too much like something a *physicist* had invented. It didn't sound like biochemistry to *them*. They hadn't got a very good idea of what *was* going on; in fact they thought it would be very complicated and involve a lot of enzymes—in which, of course, they were also partly right. So there was a certain resistance to simple ideas like three nucleotides' coding an amino acid; people thought it was rather like cheating—it was *forced*. Unless, of course, you had the sort of background *we* had, when it seemed to be the sensible way to do things. You had to have a good background in genetics and see the problem as a whole—and not just think of protein synthesis as another branch of the biochemistry of intermediate metabolism. That's the real point."

The coding problem and the biochemistry of protein synthesis were pursued in parallel and in considerable competitive tension. From the mid-fifties, these came to be clearly distinguished conceptually, and they enlisted somewhat different groups of biologists. Falling aslant Crick's "simple matter" of moving the genetic information from DNA into protein, the characteristic concern of the biochemists was with "the several steps . . . in the pathway from free amino acid to biologically active protein." That formula—which appeared to ignore altogether the idea of information and even the role of DNA—was advanced as late as May of 1959, in a historical review given before the New York Academy of Sciences by Paul Zamecnik, whose laboratory at Massachusetts General Hospital was the chief place where the several biochemical steps of protein synthesis were discerned. Yet coding and biochemistry were two ways of viewing what was, after all, a single problem. Thus, of course, neither aspect could be solved without constant reference to the other; those groups of biologists overlapped and in any case were small, numbering at first no more than a platoon or two where brigades and armies tramp the field today.

The lines of biochemical inquiry that became relevant are conventionally traced to the late nineteen-thirties; certain papers of these times are always cited—which is proof enough, in a couple of cases, that they are never re-read. In actuality, only when the structure of DNA was solved and the coding problem had emerged, illuminations that cast a new and raking light across the work of twenty years, did the particular biochemical discoveries related to protein synthesis that are now taken as essential stand out clearly. The essential lines were two. One was the function and metabolism of RNA. This, Crick warned, had been particularly vexing. The other was the supply of energy to make chemical bonds.

Ribonucleic acid, one recalls, differs chemically from DNA by an extra oxygen atom attached to the ring of each sugar, along the phosphate-sugar backbone, and by the presence of uracil among the bases rather than thymine. RNA was more difficult to work with than DNA—though for a long time the reasons were not understood. The chief reason is that RNA is generally single-stranded, which means that it lacks the defined structure of DNA and in particular is more susceptible to being chopped up by enzymes. And so, for many years, nobody realized that in life RNA can occur in strands that are far longer than were easily extracted. It was generally thought that RNA was a small molecule. Both structurally and functionally, RNA appeared amorphous even when the structure and function of DNA had been crisply marked off. The obvious convenient source for RNA was always the viruses that attack plants, containing RNA but—outside the plant cells—free of DNA and free also of enzymes; the best studied of these was, of course, the virus of the mosaic disease of tobacco. Technical limits gravely constrained understanding. To work with plant viruses posed the question, not always recognized, whether this RNA had the same function and structure as the RNA found in *E. coli* or the liver cells of laboratory rats.

The earliest research to place RNA somewhere between gene and protein came from the laboratories of Torbjörn Caspersson, at the Karolinska Institutet in Stockholm (where Watson in 1951 was told to go instead of the Cavendish if he wanted his fellowship continued after he left Copenhagen), and of Jean Brachet, at the University of Brussels. Caspersson's lifelong interest was instrumentation. He earned his doctorate by developing a system combining spectroscope and quartz-lensed microscope that could locate precisely the nucleic acids in different parts of an individual living cell, taking advantage of the fact that pyrimidines absorb light very strongly at one particular wavelength—2600 angstroms, in the mid-ultraviolet. The types of nucleic acids could then be differentiated by a staining method called the Feulgen reaction (after the German biochemist Robert Feulgen, who developed it in 1925) that affects deoxyribose sugar—but not ribose, protected by that oxygen atom—so that DNA blooms an intense red.

In the fall of 1937, Caspersson was joined for two years by Jack Schultz, an American geneticist. Schultz had been working under Thomas Hunt Morgan at the California Institute of Technology. With him there, through the year 1936, was a young French graduate student, Jacques Monod. Monod did some experimental work with Schultz, he told me one evening in Paris nearly forty years later. "The subject never came out to anything. But Jack Schultz was an extraordinarily active man, he knew everything," Monod said. Schultz died in 1971. "He was one of the few among those professional *Drosophila* geneticists who really knew biochemistry. And I'll tell you something of which I have no proof but it's very important if true. I remember, from conversations with Jack, that at the time—that is to say, 1936—he was thinking of thymonucleic acid as perhaps important as an intermediate between genes—nobody knew what genes were—and proteins."

Schultz wanted to bridge the chasm between genetics and embryology. His first conjecture about DNA, though, assigned it a function almost perfectly the reverse of what it actually does. He went to Stockholm with the idea that concentrations of DNA along the chromosomes acted to block the expression of the nearest genes, and with the hope that by Caspersson's techniques he could actually locate such blocked genes on abnormal chromosomes of certain mutant fruit flies. Schultz had a professional handicapper's total memory for the intricacies of the *Drosophila* studbook; he travelled by sea from California to Sweden in order to bring with him alive a large collection of flies he had specially selected and bred.

"At that time the whole question of protein synthesis was enveloped in a veil of complete and mysterious darkness," Caspersson said in a letter to me many years afterwards. Schultz's original conjecture did not last. The two men turned Caspersson's microscope system on the chromosomes of mutant flies at different moments in cell division; they went on to the flies' eggs and larval cells in various stages of growth, and then to cells from growing plants—rye germ, onion roots, spinach—and to the eggs of sea urchins. Data mounted up. They published a short note in *Nature,* a paper in a German journal, and then longer papers in *Proceedings of the National Academy of Sciences* of the United States. Caspersson, at the same time, was building an improved spectrographic measuring device. In 1939, Schultz stopped work abruptly to get away before war began.

"We realized the very last few weeks he was here that the first results from the new photoelectric spectrophotometer in early 1939 showed a little too much pentose [that is, ribose and not deoxyribose] nucleic acids in eggs, root-tips and nurse cells to make us feel comfortable," Caspersson wrote. They put the uncomfortable observation in another note to *Nature,* sent off in February. "Of course this, according to my belief, can be looked upon as the first statement of a correlation between nucleic acids and synthesis of proteins." After

Schultz left, Caspersson's new instrument produced more evidence about the distribution of the two types of nucleic acid in different parts of the cell, and about the rates at which nucleic acids were made, in relation to each other and to the synthesis of proteins, at different stages of growth. He analyzed the results in two papers which appeared at length, in German, in *Chromosoma* in 1940 and *Die Naturwissenschaften* early in 1941.

In Belgium in those same years, independently, Jean Brachet had learned how to break open cells to separate nucleus from cytoplasm, gently enough so that their nucleic acids could be characterized chemically. Brachet published the fullest account of *his* findings in a fifty-one-page article in *Archives de Biologie* that came out in Paris in 1942—with a note added at the last minute about the relation of his work to Caspersson's latest, which he had only just read.

From Caspersson's research and Brachet's it emerged that growth in the cell—vigorous protein synthesis—was always associated with plentiful RNA, and that this RNA was located in the cytoplasm, while DNA was confined to the cell nucleus. More sketchily, it seemed from the chromosome studies that what went on genetically in the nucleus influenced the relation of RNA to protein in the cytoplasm. Caspersson also noticed in the microscope that most of the RNA in the cytoplasm was concentrated in minute particles. Brachet had found small, spherical particles in cytoplasm, too, by centrifuging them out, and showed that they contained a good deal of protein bound up with RNA.

These and little more are the observations, in all that bulk, upon which the latter-day biologist seizes: yet in the original papers they were brief, fragmentary, even elusive, for they pointed in a direction different from any the authors thought they were travelling. Thus in 1938, Schultz and Caspersson took it for granted, in their first paper together, that "the structure-forming properties" of DNA were its contribution to the gene, and so concluded, "It may be that the property of a protein which allows it to reproduce itself is its ability to synthesize nucleic acid." Caspersson remained convinced for many years that "the nucleus is . . . organized especially for being the main centre of the cell for the formation of proteins." Brachet, in 1937, and as he said again in 1946, thought it likely that during cell division RNA was broken down and transformed into DNA, so that the total of nucleic acids remained constant. The idea was widely credited.

"You see. Everything is messy," Crick said. "If you look through the literature on the turnover of RNA in cells, you'll find it *exceedingly* complicated and difficult and impossible to understand. And one can see why. Because you had ribosomal RNA, transfer RNA, and messenger RNA. And heteronuclear RNA. And, of course, nobody could distinguish them at all; they were all muddled up together, except that some was in the nucleus and some was in the cytoplasm. It's bad enough *now* to know what goes on in higher organisms anyway. In those days it was impossible. So it isn't surprising that whenever anybody tried to establish something, it always seemed to be *funny.*"

Tiny particles in the cytoplasm, rich in RNA, had also been noticed, and earlier, by Albert Claude, at the Rockefeller Institute. Claude came upon them by methods similar to Brachet's, but in pursuit of a different quarry. In the mid-thirties, he began to try to isolate the infectious agent from transmissible skin-tumors in chickens, by freezing the tumor cells hard to break them up, then spinning them in a centrifuge; the fractions thus separated he injected into the skins of other chickens. He found that the most potently infectious fraction consisted of "small granules" containing protein and ribonucleic acid. But then—in a neat illustration of why control experiments are necessary—he ran comparative studies of normal, uninfected cells from chick embryo and mouse embryo, as well as cells from mouse tumors. In all those, he found, right at the limit of resolution of the microscope, tiny specks of light against a dark field, the same kinds of cytoplasmic granules. Claude reported this in a note in *Science* in January of 1940. He concluded drily that "the demonstration . . . should lead to interesting developments." The developments took twenty years to unravel: no, the details of the granules' functioning are not fully understood even yet.

Work that had been scattered through several languages and erratically circulated, during the war, was brought together and up to date at a symposium on nucleic acids convened in July of 1946, in Cambridge, by the Society for Experimental Biology. Most of the participants were British biochemists, but Caspersson and a colleague came from Stockholm, Brachet from Brussels, Herman Kalckar from Copenhagen, and William Astbury from x-ray crystallography. Nineteen papers were read, discussed, gathered into a book published the following spring; even allowing for the years of war just past and the years of discovery to come, those papers seem today like a flotilla of small boats pointing every which way in the last slack moment before the tide lifts.

Caspersson presented, with earnest insistence, a scheme whereby chromosomes make proteins in the nucleus, which in turn make RNA nucleotides in the cytoplasm, which make cytoplasmic proteins. "The mechanism might look fairly hypothetical. That is, however, not the case."

James N. Davidson (whose book published in 1950 was one of the sources from which Watson later got the wrong tautomeric forms of guanine and thymine) saw as clearly as anyone there that the results obtained by Brachet, Caspersson, and Schultz meant that DNA was confined to the nucleus while RNA was found mainly outside it, and that "a high ribonucleic acid content is characteristic of cells in which protein synthesis is vigorous." (The formula that DNA makes RNA makes protein was first put into print six months later by two bacteriologists in Strasbourg, André Boivin and Roger Vendrely—or rather, by the anonymous editor who compressed their paper in *Experientia* into an English-language summary, less ambiguous than the text.) Davidson had also read Brachet's and Claude's reports that cytoplasmic RNA is concentrated in particles. In his review he listed some of the names these had been given by then—"secretory granules, zymogen granules; small particles, microsomes"—and put the facts about them with perfect clarity:

> It has been stated . . . that in the adult cell nearly all the ribonucleic acid is present in such particulate, and therefore sedimentable, form, while in rapidly proliferating cells such as those of the embryo, the particles account for only 20–40% of the total ribonucleic acid content. The remainder occurs as 'free' ribonucleic acid which does not sediment out in the ultra-centrifuge.

Brachet himself, in a paper lit up by imaginative flashes, took the particles the essential step further:

The results clearly demonstrate that the ribonucleoprotein granules are pre-existing structures in the living cell, where they exert important physiological functions. Another point of interest is the following: when granules are isolated from red blood cells, they are found to contain a small amount of haemoglobin which cannot be eliminated by repeated washings. In the same way, pancreatic granules contain insulin. . . . These facts point towards the following hypothesis: ribonucleoprotein granules might well be the agents of protein synthesis in the cell.

Astbury, the lone crystallographer at the meeting and the one man concerned with three-dimensional molecular structures, delivered an exhortation. He had no new work on nucleic acids to report since the last papers before the war. Indeed, this was the time when he regressed to the notion that DNA was built up, as meaninglessly as a starch, from multiple sets of the four nucleotides. Yet Astbury's language was rousing and certain passages stuck in the mind. Among them:

Right from the beginning of this X-ray study of the nucleic acids we have drawn attention to the fact that the spacing . . . of the nucleotides along the thymonucleic acid column is to all intents and purposes the same as the distance from one side chain to the next along an extended polypeptide, and even at the risk of being accused of dabbling in numerology I should like to say again that I believe this to be no mere coincidence. In a sense the evolution of biological molecules has been one long story of numerology, a sifting and sorting of shapes and sizes and a progressive selection of molecular patterns that has been going on throughout the ages. Biosynthesis is supremely a question of fitting molecules or parts of molecules one against another, and one of the great biological developments of our time is the realization that probably the most fundamental interaction of all is that between the proteins and the nucleic acids. . . . The correspondence between inter-nucleotide spacing and inter-side chain spacing . . . is not an arithmetical accident but a stereochemical correlation of deep significance.

Astbury's stereochemical correlation, so vividly easy to visualize, seemed to almost everyone like the one firm tussock of fact.

That symposium was the first major biological conference in Europe after the war. Errors, insights, enthusiasm and all, it was dense with import and had great influence. "A lot of things, especially during the war, could easily be missed, and papers in obscure places could be missed," Crick said. "I think you can take it that everyone in the field read that symposium." The same summer, Max Delbrück taught the first phage course at Cold Spring Harbor.

❖ ❖ ❖

Meanwhile, separately, in the most magnificent achievement of late classical biochemistry, Fritz Lipmann—in Copenhagen in the late thirties, then at Cornell, and finally at Massachusetts General Hospital through the forties—was finding out how cells make available the energy to drive their manufacturing processes. As so often, the principle was simple; to a greater degree than usual, the chemical network was formidably intricate. The genius lay in holding on to the one through all the twists and interconnections of the other. I met Lipmann in 1971 at Rockefeller University, in New York, where he had worked since 1957. He was then seventy-one years old, a small man, slow-moving, gentle, who spoke in a melancholy rhythm that suggested he habitually shielded his green thoughts from too early exposure to the glare of words. Or perhaps his hesitancy was due merely to the fresh memory of an interview with a newspaper reporter in Houston who had irritated him with her incomprehension. "She had absolutely no concept of what pure science, of what research is. For her, research always has to be application," Lipmann said. We had just sat down in his office. "And I have met other people who have scolded me, about the useful applications of research, and I told them I don't care about that, I do what I do because I want to know it.... I must say it really hurt me, this absolute inability of a lay person—" He let the phrase hang.

He was, he said, "a man who is interested in problems that are soluble in the laboratory and which he chooses only because he thinks that *there* is the edge of knowing. It's not for application. It is where you get the least resistance for what you can work next on. You know approximately where you stand and then, knowing, from this point you have to go further—and you kind of feel out where is the least resistance. But who would understand that when I say it?"

Everyone is familiar with the idea that cells burn food to get building blocks and energy. That is, when a cell splits a complex molecule—when a bacterium, for example, in the most elementary metabolic step of all, uses an enzyme to start breaking up the six-carbon sugar ring called glucose—it gains not only the pieces to recycle or break down further but the energy that was tied up in the chemical bond just split. This is the degradative—the simply downhill—aspect of metabolism. The point is the same, of course, for the cells of a man who has just polished off a plate of oysters and a glass of champagne—perhaps with a salute to the god of fermentation, for this biochemistry began in France in the nineteenth century as research of the most intensely applied kind, responding to the vintners' need to control the steps by which yeast cells change glucose into alcohol and bubbles of carbon dioxide. The practical origin of biochemistry is preserved even in the word "enzyme," whose roots mean "in yeast." Some of the energy that the cell gains by such steps is let go as heat, but some of it is used to put other small molecules together into larger ones. This is the synthetic aspect of metabolism.

That much describes the conflagration—describes the energy of the organism as a bonfire. Lipmann answered the next question: Who's got the matches? That is, when an enzyme builds a complex molecule by catalyzing the joining of a chemical bond, this takes up a certain amount of energy that must be available right there, right then. Right there: within an angstrom or two; right then: within a ten-millionth of a second. "I am a metabolism biochemist," Lipmann said. He came upon the energy problem fortuitously, in 1937. He was looking into a puzzle about one step in the breakdown of glucose by the milk-fermenting bacterium *Lactobacillus delbrueckii* (named, as it happens, after a relation of Delbrück's). It was a sedately traditional sort of question having to do with the step where a small molecule (called pyruvate, but we won't need the name again) was changed by oxidation into acetic acid—the active ingredient in vinegar—and carbon dioxide. He observed, by the accident of a minor change in experimental procedure, that the oxidation could not take place without a supply of phosphate. This is the same simple molecule, a phosphorus atom surrounded by four oxygen atoms, that goes into every nucleotide. When Lipmann put phosphate back in, "*Bumps!*—it goes up," he said: the oxidation took off at a good rate. "I *guessed* that I might be making an intermediate which has a bound phosphate"—that is, a molecule that had a phosphate attached, for which the term is "phosphorylated."

"At this moment I started to become aware of it, that metabolism is not there to oxidize things but to deliver energy into that cell," Lipmann said. "It must generate energy in a form in which it can be used by the cell." When he eventually identified the phosphorylated intermediate, he found that the chemical bond where the phosphate attached was unusual, with an arrangement of groups and electrical charges that made the bond rich in energy compared with other types of bonds. By the biochemical ideas of the day, no such intermediate should have appeared. It was a packet of energy, tipped with phosphorus: the metaphor of a match is apt.

Curiosity aroused, Lipmann remembered that in the nineteen-twenties energy-rich phosphate bonds had overturned received ideas in a neighboring field, the biochemistry of muscle. The immediate source of muscle power had been shown to be the small molecule adenosine triphosphate. Adenosine triphosphate is nothing else than a ribose nucleotide, as in RNA, with the base adenine at one end and two extra phosphate groups strung on at the other. Again, because of the arrangement of groups and charges along the string, in this case four closely spaced negative charges, the bonds between the three phosphates are easily made, very easily broken—and when they are broken, the instantaneous rearrangement of electrons frees an exceptional amount of energy.

Adenosine triphosphate is often abbreviated ATP; the nucleotide itself, in this context, may be described as adenosine monophosphate, AMP; the difference between them, the pair of phosphates joined by an energy-rich bond, is called a pyrophosphate.

Lipmann took an imaginative leap of great boldness. In the maze of biochemical pathways—whose stupefying complexity was only then being fully realized—the small building blocks, we know now, number about sixty different sorts, but are broken down, switched around, and put together in some one thousand to two thousand separate chemical steps. Even as the revelation was dawning that the enzyme that facilitated each step was a huge molecule of extreme specificity, Lipmann proclaimed a counter-revelation: that the energy carriers attending these same steps were a few kinds of small molecules of extreme versatility and one common trait, the energy-rich phosphate bond. ATP, he proposed, "takes out the energy of the metabolic flow" and conducts it to the reactions where needed.

"You see, in every synthesis—but now you have to learn from me a bit of metabolic chemistry," Lipmann said, the gloom lifting slightly. "In every synthetic reaction, we know that you have to prepare the groups that enter into synthesis. You have to hang on to each group the energy which it needs to condense." Thus charged, the group is said to be activated. "Now, in a polymerization reaction a new group is being added to a chain formed of similar groups. It is added by gluing itself to the chain so that it sticks. To do this it needs a certain excess of energy; in order to really stick, it has to throw a little energy away. A very simple example for your genetics business is that nucleic acids are synthesized from the triphosphates."

Any nucleotides, not just the ones with adenine, can form triphosphates with energy-rich bonds. It was established by the mid-fifties that these are precursors of nucleic acids. As a new chain is assembled alongside an existing chain, each link carries in with it its own identification and its own shot of energy. The identification, of course, is the base itself, in the pattern of sites it bears for hydrogen bonding to the complementary base on the existing chain; the energy is at the phosphate end, to spark the covalent bond (snap! and some heat is dissipated and the pyrophosphate pair drifts away) to the sugar just ahead in the new chain.

Peptide bonds to connect up a growing chain of amino acids are now known to be energized in much the same way. "This energy is put on to the group, to the amino acid, before it can take part in the reaction," Lipmann said. He sketched diagrams of molecular parts joining and separating. "That occurs before, and is the step that brings adenosine triphosphate into play, because that's the general carrier of energy." He looked at me doubtfully, paused, then muttered to himself, "Chemistry is something you have to have a passion for, if you really are going to understand."

Lipmann gave the energy-rich phosphate bond its name and invented the mark, ~, called the squiggle, as in —O~P—, by which it is shown. In 1941 he assembled all the evidence then available for its role into a long review, which was profoundly influential. In the fall of 1953, Lipmann was awarded the Nobel prize in physiology or medicine. The synthetic steps he understood, even by that time, were those of the small metabolic building blocks. In his Nobel lecture, that December, he spoke of the way his earliest observations of the energy problem had moved him to propose not only that the phosphate bond provided the general energy-distributing system, "but also to aim from there towards a general concept of transfer of activated groupings by carrier as the fundamental reaction of biosynthesis." In the last sentences of the lecture he spoke "very tentatively" of the first clues that "may foreshadow" extension of his concept to proteins and nucleic acids.

❖❖❖

The other side of the matter, the perception of the biological code as an abstract problem distinguishable from the biochemistry, had few and isolated forerunners—and the fact is surprising. The late forties, after all, were the years when information theory, turned loose from the armorers' workshops, grew with world-conquering confidence and was publicized throughout the sciences as well as to laymen. A fundamental of information theory, unforgettable for example to anyone who read Norbert Wiener's *Cybernetics*—the book came out in 1948 and aroused wide discussion—was the unique status of information as something that could be transferred at almost no cost in energy and yet with hugely multiplied consequences: the instance was the pile driver actuated by the twitch of a rope, or the milling machine under the direction of a punched tape. Another fundamental was the idea of feedback control, whereby the starting up, or the rate, of a complicated process was regulated by the absence, or the amount, of an end product: the instance was the petroleum refinery whose various input valves were adjusted by automatic monitoring of the composition of the output.

But the starting point of the new doctrine was the realization that information was the enemy of chaos. Information was order and specificity. It imposed organization, to hold back or even reverse—though this could be done only temporarily and locally—the inevitable running down of everything into disorder that is described by the second law of thermodynamics and signified by the term "entropy." Information entailed unpredictability, since any part of a message was redundant (like the u after q in English) if it could be predicted from the rest. Information was destroyed by noise, which was the communication engineer's jargon for random change; it could be protected by transmission of a deliberate amount of redundancy from which impaired parts could be restored. These ideas have been fecund, particularly for communications and computers. Their ramifications were expounded with rigor by Claude Shannon, of the Bell Telephone Laboratories, in a pair of articles in *The Bell System Technical Journal*, titled "A Mathematical Theory of Communication," which were published in 1948 just as Wiener's book appeared.

The genetic code, on the face of it, ought to have provoked the most vigorous attempts to apply information theory. Information, for the biologist, was embodied in the extreme specificity of proteins and nucleic acids that was becoming evident in the early fifties. In such terms, what Chargaff and Sanger did was to reduce the chemistry of these molecules to a total dependence on information. Noise was mutation, errors of replication of the gene, errors in getting from gene to protein. Feedback would surely be found in the regulation of cellular processes. And, indeed, these fundamental ideas are almost indispensable in explaining some of the chief discoveries of molecular biology after the structure of DNA. But their usefulness is retrospective, for they played almost no explicit part in the working out of those discoveries.

Shannon was a telephonist. His theory of communication was abstracted—with great power and generality, to be sure—from the problems of transmitting information accurately though fast. He even wrote about coding, in a paper, less known, that was prepared in 1945 but declassified and published only in the fall of 1949; as the paper's history suggests, it was a mathematical analysis of the classical problems of cryptography. Nothing in it was patently applicable to any biological issue, or even suggestive. Wiener, by contrast, was greatly interested in biological applications of cybernetics. He found feedback wherever he looked in the regulation of physiology and behavior. But his suggestions were metaphors rather than worked-out mechanisms, and the ones that excited discussion were about the performance of the nervous system in terms of what had been learned in building the first computers.

Shannon's work and Wiener's were closely related; a third statement, of great potential interest, came at the same time from the mathematician John von Neumann. In December of 1949, von Neumann gave five lectures at the University of Illinois about what he called complicated automata. He wanted to specify the discrete logical functions and connections that self-directing devices would require—in a form general enough to apply equally well to computers fitted with sensory and motor elements, or to organisms. The most complicated automata imaginable were those that could not only direct but reproduce themselves. In the second of the lectures he introduced the strange and glamorous idea that a sufficiently complicated automaton may not permit any complete, written-out description of its functioning that is as simple as the thing itself. His example was a system that would perceive and recognize patterns as subtly as we actually do recognize them. By the last lecture, he had listed the types of logic organs and of logical instructions that a machine would have to contain in order to be able to build another machine like itself by following the instructions—including the instruction to make a copy of the list of instructions.

Today, one sees—as Brenner was first to point out to me—that the absolute distinction between protein and DNA, between the substance that makes up the machinery and the substance that carries the instructions to make the duplicate machine, exactly and fully embodies what von Neumann concluded. But this interesting correspondence could not be recognized until much too late to matter. Von Neumann himself was paying attention—like Wiener and like Shannon—to the comparison between human and mechanical brains. His lectures at Illinois were recorded on tape, but the recording and transcript were defective and never circulated. He came back to the subject in the fall of 1952, to begin the draft of a book that he worked on for a year and abandoned; in Princeton in March of 1953—just when the structure of DNA was found—he gave another set of lectures. Then and later, he was focussed even more on the parallel between the nervous system and computers. His theory of self-reproducing automata was completed from his notes and manuscripts only after his death, and published in 1966.

Information theory made no difference to the course of biological discovery: when the attempt was made to apply it—when a conference was held, a volume published—the mathematical apparatus of the method produced comically little result. The biological papers (and the papers' citations) of the time testify to that by their silence.

The earliest mention of coding that counts was Erwin Schrödinger's, in 1944 in *What Is Life?* Everybody read Schrödinger. The fascination of the book lay in the clarity with which Schrödinger approached the gene not as an algebraic unit but as a physical substance that had to be almost perfectly stable and yet express immense variety. The excitement was concentrated in two ideas in consecutive paragraphs:

THE APERIODIC SOLID. A small molecule might be called 'the germ of a solid'. Starting from such a small solid germ, there seem to be two different ways of building up larger and larger associations. One is the comparatively dull way of repeating the same structure in three directions again and again. That is the way followed in a growing crystal. . . . The other way is that of building up a more and more extended aggregate without the dull device of repetition. That is the case of the more and more complicated organic molecule in which every atom, and every group of atoms, plays an individual role, not entirely equivalent to that of many others (as is the case in a periodic structure). We might quite properly call that an aperiodic crystal or solid and express our hypothesis by saying: we believe the gene—or perhaps the whole chromosome fibre—to be an aperiodic solid.

THE VARIETY OF CONTENTS COMPRESSED IN THE MINIATURE CODE. It has often been asked how this tiny speck of material, the nucleus of the fertilized egg, could contain an elaborate code-script involving all the future development of the organism. A well-ordered association of atoms, endowed with sufficient resistivity to keep its order permanently, appears to be the only conceivable material structure that offers a variety of possible ('isomeric') arrangements, sufficiently large to embody a complicated system of 'determinations' within a small spatial boundary. Indeed, the number of atoms in such a structure need not be very large to produce an almost unlimited number of possible arrangements. For illustration, think of the Morse code. The two different signs of dot and dash in well-ordered groups of not more than four allow of thirty different specifications. Now, if you allowed yourself the use of a third sign, in addition to dot and dash, and used groups of not more than ten, you could form 88,572 different 'letters'; with five signs and groups up to 25, the number is 372,529,029,846,191,405. . . .

Of course, in the actual case, by no means 'every' arrangement of the group of atoms will represent a possible molecule; moreover, it is not a question of a code to be adopted arbitrarily, for the code-script must itself be the operative factor bringing about the development. But . . . what we wish to illustrate is simply that with the molecular picture of the gene it is no longer inconceivable that the miniature code should precisely correspond with a highly complicated and specified plan of development and should somehow contain the means to put it into operation.

They all read Schrödinger: yet what they took from him varied. Crick was aroused by the book's most general qualities, its enthusiasm, its readability, its message that biology could be thought about in a new way and that great discoveries were imminent; in remembrance, at least, the details of Schrödinger's science seemed to him almost embarrassingly gauche. "I wasn't conscious of any influence of what he called the aperiodic crystal—I don't suppose the man had ever heard of a polymer!"

Crick had once told me that even by the time he first came to the Cavendish, even before it was clear that the genetic material was DNA, he had conceived the idea he later called the sequence hypothesis—simply that the information in the gene ran step by step with the sequence of amino acids in proteins. I later reminded him of that. "All that was very private," he said. "It was just something internal; I don't think I put anything on paper. What really wants checking up in the early stages is who actually put things on paper." He has canvassed his memory on several occasions for papers he would have known when he began work on the coding problem: the candidates are no more than four or five.

One that he missed was a proposal in 1947 by Kurt Stern, who ran a group investigating nucleic acids at the Polytechnic Institute of Brooklyn. Stern had read Schrödinger thoughtfully, and Astbury—and from them sprang to the idea that "genes are modulations of a 'neutral' nucleoprotein structure"—to wit, a strand of DNA locked in place between two strands of protein, an arrangement he compared to the groove of a wax phonograph record. He imagined that the "genic modulation" was carried by the alignment of successive bases—no matter which sorts—flipped either to right or to left. What Stern called the "gene code" was not Morse but semaphore. Some read Stern's paper; not Crick.

In 1950, Sir Cyril Hinshelwood, at Oxford, published with some associates a run of papers in *The Journal of the Chemical Society* about the nucleic acids in bacteria, and arrived at a simple theoretical speculation:

> In the synthesis of protein, the nucleic acid, by a process analagous to crystallization, guides the order in which the various amino-acids are laid down; in the formation of nucleic acid the converse holds, the protein molecule governing the order in which the different nucleotide units are arranged.

The first half of that looks fine, but the plausible reciprocity that was the nub of the idea was altogether wrong.

None the less, they understood that what they had to explain was how to order the sequence of amino acids in the chain. The functional properties of the protein, including the way it folded up, would follow from the sequence, they said:

> The specificity of a protein molecule must involve the arrangement of amino-acids in the peptide chain. (It has been ascribed by Pauling [*Endeavour*, 1948, 7, 43] to the folding of the chain, and since this is probably governed by the arrangement of amino-acids in the molecule—in particular by the relative positions of polyfunctional amino-acids capable of forming cross-linkages—these two views are equivalent.)

Consider the setting of that. Pauling, through the nineteen-forties, had insisted that the extreme precision with which an antibody molecule relates to the specific antigen it neutralizes could be explained only by the protein chain's adapting its configuration to fit—an adaptation, therefore, not totally determined by the genes. The alpha helix was still six months away. Sanger was completing the amino-acid sequence of the first chain of insulin. Erwin Chargaff was just publishing the observation on the proportions of the bases in nucleic acids that suggested that *their* sequences could be specific. For Hinshelwood to derive the biological activity of a three-dimensional protein molecule strictly from its one-dimensional sequence was at that time a logical reduction of great power and originality.

Hinshelwood also escaped from Astbury's stereochemical numerology, the similar spacing of nucleotides and amino acids, long enough to state for the first time the genuine problem of correlation—though with one glaring eccentricity of detail:

> In a protein, about 23 different amino-acids may occur, whereas in a nucleic acid only 5 basic units are found—two pyrimidine nucleotides, two purine nucleotides, and ribose phosphate. Clearly there cannot be a one-to-one correspondence between the position of an individual amino-acid in the protein part of a nucleoprotein and the position of an individual nucleotide in the nucleic acid part. If, however, it is assumed that, in the synthesis of a protein at the surface of a nucleic acid polymer, the amino-acid side-chain which is guided into a particular place depends on the nature and relative position of two adjacent nucleotide units, the difficulty can be overcome. Twenty-five different internucleotide arrangements are possible, and this is of the right order to give correspondence with the number of different possibilities in a protein chain.

In a talk that opened the Cold Spring Harbor Symposium in the summer of 1966, which was a vast jamboree—349 participants, eighty-eight papers—that celebrated the complete solution of the genetic code, Crick looked back to

Hinshelwood's proposal as the first that could reasonably be discussed in present-day terms. At the time its influence was negligible.

Early in 1952, Alexander Dounce, at the University of Rochester, put forward a mechanism that addressed both coding and biochemistry. A while later, Dounce led off a round-table discussion of protein synthesis at Oak Ridge National Laboratory, in Tennessee, with a reminiscence. "My interest in templates, and conviction of their necessity, originated from a question asked me on my Ph.D. oral examination by Professor J. B. Sumner," Dounce said. "He inquired how I thought proteins might be synthesized. I gave what seemed the obvious answer, namely, that enzymes must be responsible. Professor Sumner then asked me the chemical nature of the enzymes, and when I answered that enzymes were proteins or contained proteins as essential components, he asked whether these enzyme proteins were synthesized by other enzymes and so on *ad infinitum*. The dilemma remained in my mind, causing me to look for possible solutions that would be acceptable, at least from the standpoint of logic." Dounce's object, he announced at the outset of his paper, was to construct a theory "in sufficient chemical detail to permit of experimental testing. . . . Even if the particular mechanism proposed turns out to be incorrect, it may nevertheless be useful, since the number of imaginable alternative mechanisms is probably quite small."

This was the only early venture in theory whose considerable influence Crick acknowledged, and one sees why he took notice: the speculative strategy and even the tone are very like things he was writing himself at about that time. Dounce had the eye for essentials. He knew that RNA ought to be the template for proteins; the possibility that the RNA originated on the template of a DNA gene seemed obvious. He said, more flatly and clearly than anyone in print before, that the order of amino acids in each specific protein derives from the order of nucleotides in the corresponding nucleic-acid molecule.

Decades later, the sequence hypothesis fits so naturally with the way we have learned to think that it is difficult—once again—to imagine what it was like to advance the idea without a stick of unambiguous evidence. Staying with Astbury's physical lineup of polynucleotide against polypeptide, for that was the only structural idea around, Dounce imagined that each nucleotide would determine an amino acid with the help of its neighbors, one on each side. He counted up that each kind of base could have ten different neighborhoods—adenine between two other adenines, adenine between guanine and adenine, and so on—making forty combinations in all, even if the direction of reading didn't matter. Forty were "more than enough configurations to account for all amino acids known to occur in proteins." Dounce recognized the energy problem as well, and so gave the strand of nucleic acid the extra task of carrying energy-rich phosphate bonds to make the peptide bonds. The rest of his paper elaborated that chemistry—to the point of overemphasis, in Crick's later judgement, so that Dounce's intelligent analysis of the coding problem independent of the chemistry was "somewhat overlooked."

❖❖❖

The same week in June of 1953 that Watson presented the structure of DNA at Cold Spring Harbor, *Nature* carried an article on "Biosynthesis of

Proteins" by two British biochemists, Peter Nelson Campbell and Thomas S. Work, who attempted to review the best evidence and lines of speculation about how cells put protein molecules together. This, they said, was the most formidable and most important unsolved problem in biochemistry. The authors were by no means blind or uninformed. Their five careful pages were soaked in the literature, and to good purpose. It's amusing and unlucky that six weeks after the publication of the structure of DNA, Campbell and Work found themselves in print saying, in a summary written months before, "The conception of the gene is essentially an abstract idea and it may be a mistake to try to clothe this idea in a coat of nucleic acid or protein," and that "if we must have a gene it should have a negative rather than a positive function as far as protein synthesis is concerned." It's telling that their survey could only conclude:

> There are two main streams of thought on protein synthesis: one derived from the study of isolated enzyme systems and suggesting a stepwise coupling of many small peptide units; the other based on the study of genetic inheritance of protein specificity and preferring synthesis on templates, each template being specific for a single protein structure and probably identifiable with a gene. It is impossible with our present knowledge to choose between these two theories.

The same issue of *Nature,* thirty pages on, carried a short note from Charles E. Dalgliesh, at the Postgraduate Medical School of Hammersmith Hospital, London, who challenged the way proteins were imagined to form according to those who argued for templates. Such theories had assumed, Dalgliesh wrote, "that the amino-acids are laid down in their correct order on a template (? consisting of nucleic acid) with subsequent bond formation and separation from the template," or, in other words, "that . . . all amino-acids must be laid down on the template effectively simultaneously." That picture was wrong, he thought, and led to misinterpretation of the results of recent experiments with radioactively labelled amino acids.

> Energy considerations make it most unlikely that a complete protein molecule is formed on a template surface and afterwards separates as a whole from the surface. It is much more probable that amino-acids become attached to the surface of the template, and that as the amino-acid residues link together, the growing peptide chain 'peels off' behind the zone of formation of peptide bonds. . . . In such a case the protein does not finally separate from the surface until it is complete (so that no intermediate steps are observed); but only a small region is attached to the template at any time. Moreover, after a 'wave' of biosynthesis has passed down the template surface, the latter can be reactivated, further replacement by amino-acids can occur, and a further 'wave' of biosynthesis may pass along. Many peptide chains at different stages of growth may thus be attached to the template at the same time.

He included a sketch—a straight line for the template with strings of increasing length erected from it. His point was a clear, small contribution to understanding. Crick read the paper, and even a quarter-century later remembered the sketch.

❖ ❖ ❖

The character of a scientific revolution depends in part on its opposition. The great set pieces of the history of science have been the Copernican, the

Einsteinian, the Darwinian, the quantum-mechanical revolutions. In these, the confrontation was direct and was fought out vigorously. The new men with new ideas met a structure of theory as fully articulated as their own, with a history of success. The last swords of the age of bronze were far better edged, after all, and tougher than the first coarse iron blades: one thinks perhaps of the predictive accuracy of the Ptolemaic system, or of the admiration for Newton that Einstein expressed. The patterns of such classic revolutions have been closely anatomized. It has become—it always was—the naïvest sort of error to suppose that progress has amounted to clearing away the tangled confusions of the past with the scythe of a true idea. It is both the duty and the delight of the historian of science, even more than most, in that matchless phrase "to think the thoughts of the past"—to understand the coherence and explanatory power of the set of ideas now superseded. Nor can rethinking in that special sense ever be said to be finished: witness the present effort to comprehend how Newton's tireless alchemical experiments were of a piece with his mathematics and mechanics.

Not all great scientific revolutions can be fitted to the classic model. To read widely in the biological journals and in the reports of meetings of the nineteen-forties and -fifties—an ambition not so lunatic for then as for now: Campbell and Work in that review in *Nature* in 1953 cited only seventy-four papers and books—amply demonstrates how misleading it can be to survey the past to isolate precursors of present ideas or, more subtly, of present modes of framing and judging hypotheses. There was no dearth of biochemical observation and ingenuity about the making of proteins. Yet it is vain to try to re-create in the historical imagination a coherent or widely accepted early theory of protein synthesis and the making of the cell by its inheritance. Often enough a paragraph will seem prescient—until one finds that paper after paper offered new ideas that were just as reasonable, just as vigorously argued, no more tentatively based on the available evidence, but that came to nothing. The character of this next great movement in biology was formed in part by the lack of articulated opposition. The confusion was densely tangled.

There were practical deficiencies as well. Several of the laboratory techniques that were to prove essential were still new. Only fragments of biological mechanisms or processes were evident. Nor was there much identifiable as a functional anatomy to which the processes might be assigned—and that was crucial.

Ever since the anatomists of the Renaissance, one of the consistently successful strategies of physiology has been to put functions into orderly relation to parts. Molecular biologists in the nineteen-fifties found themselves repeatedly obliged to distinguish new and separate species of RNA, each with its peculiar function, so that—as Crick said—by the end of the decade what had begun as one stuff was split into three or four kinds. Occasionally a great anatomist went deeper, to insist that because, in his understanding, a previously unknown function must take place, an unheard-of structure was required to carry it out. William Harvey did that, writing in 1623, when he said that the blood must find passages, though invisible to the eye, to complete the circulation through the tissues from the finest arterioles to the finest veins. An elect few of the molecular biologists had that gift, extraordinary in any era, for sensing the presence of functional entities and naming them.

❖*b*❖   George Gamow's letter reached Cambridge soon after Watson got back from presenting the model at Cold Spring Harbor. Gamow's was the first specific positive response to Watson and Crick's structure of DNA: where others, notably Delbrück, had reacted for all their excitement by suggesting ways the structure could be tested and might have to be modified, Gamow took it as given and set it to work. The importance of Gamow's idea, Crick later said, "was that it was really an abstract theory of coding, and was not cluttered up with a lot of unnecessary chemical details." Gamow disentangled the problem, stating it for the first time in its modern form.

If genes were DNA, Gamow wrote, and DNA was two chains side by side, formed of only four kinds of nucleotides and joined by the paired bases, "It follows that all hereditary properties of any living organism can be characterized by *a long number . . . written in a four-digital system*, and containing many thousands of consecutive digits." This Gamow called "the number of the beast." He went on:

> The numbers describing two different members of the same species must be very similar to each other, (though not quite identical, unless they belong to a pair of identical twins), whereas the numbers representing the members of two different species must show larger differences. Since the number of all possible arrangements of four elements in sequences of several thousands is incredibly large, we must conclude that all living organisms represent only a negligible fraction of all mathematically possible forms of life. For example, it is extremely unlikely that any organism which ever lived on the surface of the earth was represented by such familiar numbers as π or $\sqrt{3}$ written in the four digital system!

On the other side, as there were twenty amino acids, or perhaps a few more, commonly found in proteins (Gamow thereupon listed twenty-five), it followed that "if one assigns a letter of the alphabet to each amino acid, each protein (and, in particular, each enzyme) can be considered as *a long word based on an alphabet with 20* (or somewhat more) *different letters.*" The question, he said, was how such numbers based on four were "translated" into such words based on twenty.

Gamow's answer came out of the structure of DNA in the most directly physical manner possible: he suggested that permutations of the bases formed holes of different shapes, accessible along the wide groove of the structure, into which various amino acids fit as specifically as keys into locks. The holes were diamond-shaped, each formed by four bases—at the sides a base pair, and at bottom and top the single bases, from the pairs right behind and ahead, that were brought into position by the twist of the helix. The bases at the sides were related by the pairing rules, adenine to thymine and guanine with cytosine, but "the upper and lower corners of each 'diamond' may be occupied by any of the four bases. Thus, the total number of different 'diamonds' is given by the number of different triple combinations of four elements." This was the observation that had fired Gamow's imagination—for the number of combinations was twenty.

"The sequence of bases determines in a unique way the sequence of diamonds," Gamow wrote; "DNA molecules act as highly specialized catalysts, arranging the amino acids from the surrounding medium in well defined sequences, and holding them in that position long enough" for the peptide bonds to form. Neighboring diamonds, of course, shared a side, so

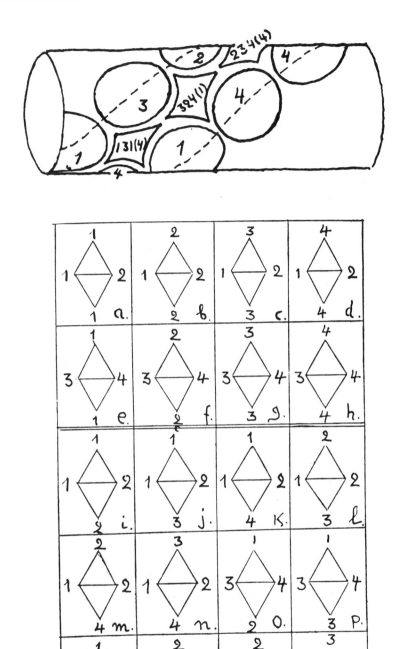

Gamow's diamond-shaped holes in the double helix of DNA, and the coding scheme for protein synthesis he deduced therefrom, in 1953. (Drawing from "Possible Relations Between Deoxyribonucleic Acid and Protein Structures," by G. Gamow, *Nature* 173, 13 February 1954.)

that they had two corners, two nucleotides, in common—and that meant, Gamow pointed out, that there were restrictions on which amino acids could follow which. As more protein sequences besides insulin became known, such regularities would become detectable—or their absence would prove the idea impossible.

Crick and Watson spotted the faults in Gamow's scheme as soon as they began to examine it. His biology was ludicrously patchy. He specified no chemistry, said nothing about energy. He was obviously unfamiliar, at that time, with the involvement of RNA in protein synthesis, or even with the evidence that proteins were made in the cytoplasm. The biochemist Dounce had understood all that thoroughly. Further, Gamow never explained how the amino acids recognized the right cavities, though the implication was clear that they did so by their side chains' slotting in, three-dimensionally; enzymes to assist the process were not mentioned. Pauling, devising the alpha helix, had ignored the differences among amino acids and built his structure on the constant feature of proteins, the planar peptide bond. Sanger had forced the issue of specificity, and now Gamow focussed it on the physical structures of the amino acids. But Crick was a protein crystallographer, and knew no reason to think that Gamow's holes in the helix could provide the variety or precision of shapes necessary to differentiate a score or more of rather similar molecular objects. The three-dimensional structures of valine and leucine, among the simplest amino acids, illustrate the difficulty—while phenylalanine and tyrosine, among the most complex, are even more alike.

valine   leucine   phenylalanine   tyrosine

For that matter, Crick thought, Gamow's list of amino acids was partly wrong. Even the lists given in textbooks and reported by biochemists were wrong. "I remember we got Gamow's paper, and we looked at the list and thought it was a bit screwy, and we were surprised, I think, because we really were sufficiently sold on RNA to believe—as you see, not totally incorrectly—that RNA was the template." So, sitting at lunch in the Eagle that day with Gamow's letter before them, Crick and Watson drew up the correct list.

"For its time, quite audacious and wholly unsupported": so Gunther Stent once wrote of Crick's assumption that there could be a standard set of amino acids that excluded several common and many rarer ones biochemists had found in life. To establish the standard set was a splendid perception of order amidst confusion (in a couple of senses, if one thinks of the babble and crush of an English pub at lunchtime); though Crick and Watson's feat was hardly

THE AMINO ACIDS FROM WHICH
NATURAL PROTEINS ARE SYNTHESIZED

Gamow's most abundant twenty	Crick and Watson's "magic twenty"	Conventional abbreviation
Glycine (6)*	Glycine	Gly
Alanine (10)	Alanine	Ala
Valine (7)	Valine	Val
Leucine (2)	Leucine	Leu
Isoleucine (12)	Isoleucine	Ile
Serine (4)	Serine	Ser
Threonine (11)	Threonine	Thr
Aspartic acid (3)	Aspartic acid	Asp
	Asparagine	Asn
Glutamic acid (1)	Glutamic acid	Glu
	Glutamine	Gln
Lysine (5)	Lysine	Lys
Arginine (9)	Arginine	Arg
	Cysteine	Cys
Cystine (15)		
Cysteic acid (19)		
Methionine (17)	Methionine	Met
Phenylalanine (14)	Phenylalanine	Phe
Tyrosine (13)	Tyrosine	Tyr
Tryptophane (18)	Tryptophan	Trp
Histidine (16)	Histidine	His
Proline (8)	Proline	Pro
Hydroxyproline (20)		

Gamow's remaining five, less frequent: Norvaline (21), Hydroxyglutamic acid (22), Asparagine (23), Glutamine (24), Cannine [*sic*] (25)
Crick and Watson's rejected candidates: Cystine, Hydroxyproline, Hydroxylysine, Phosphoserine, Diaminopimelic acid, Thyroxine and related molecules (that is, the iodinated and brominated tyrosines)

*The sequence here brings structurally similar amino acids together. Numbers in brackets show Gamow's original order of "relative abundance in proteins."

so difficult as Dmitri Mendeleev's, eighty-four years before, in constructing the periodic table of the elements, its clarifying, liberating effects on scientific reasoning were of a similar kind.

Most organisms can make some of the twenty amino acids from others, before using them in protein synthesis. Human adults require eight in the diet, namely, valine, leucine, isoleucine, threonine, lysine, methionine, phenylalanine, and tryptophan. This has practical consequences. For example, wheat is rich in proteins, yet these happen to be low in the amino acid lysine. Plant geneticists strive to breed new varieties that have more lysine to feed the world; until they succeed, the prudent man will put some cheese to his bread—for if a meal lacks an essential amino acid, the organism stops synthesizing protein, and it is no use trying to make up the deficiency a few hours later. Rats require ten amino acids in the diet. Normal *E. coli* can make all twenty—which has important consequences for genetic experiments. While the names and formulas of the twenty amino acids need hardly be committed to memory, a sense of the sorts of differences among them makes it easier to follow biologists when they talk of proteins.

I asked Crick how he and Watson had known they had got the right list. "We didn't know! Just time has shown we guessed correctly!" he said. "Except—let's see what decisions we had to make." Gamow had simply put down twenty-five amino acids in the order of their abundance in nature according to whatever reference he sought out. To fill his diamonds, he drew a line after the first twenty. "It was rather remarkable that Gamow got the number twenty, and gave a list greater than twenty, and when the correct thing was found we put it at twenty—and no fudging."

Crick and Watson had to establish the status of several pairs of closely related amino acids, and then sort out some unusual ones. The first pair was cystine, which Gamow had included, and cysteine, which he had listed in the form of a chemical relation called cysteic acid. Many proteins, including insulin, when their peptide bonds were broken up in the laboratory, could be shown to number cystine among their fragments, which was why Gamow had put it on his list. But cystine is a catamaran among amino acids. It is double, and each of its halves is itself one of the amino acid cysteine (the near identity of names is more confusing than helpful), which contains an atom of sulphur in its side chain. Two cysteines make a cystine not with a peptide bond but by joining at the sulphur atoms with a covalent bond called a disulphide bridge.

cysteine   cystine

"So there, we had to say that the S-S bond was formed afterwards, from the separate S-H groups," once the polypeptide chain was assembled. "All right, so you only have the one to be included, there." In 1955, Sanger published his final paper on the sequencing of insulin, in which he showed how the two chains of the molecule were held together and in shape by disulphide bridges between cysteine residues.

On the other hand, two amino acids not among Gamow's top twenty were near twins of two others that were high on the list, and the chemical difference was of a kind that cast doubt on the frequencies reported. Glutamine and asparagine differ by the presence of an amide group—that is, a molecule of ammonia, NH_2—from glutamic acid, which Gamow had listed in first place, or aspartic acid, in third place. The methods by which chemists usually broke up proteins for analysis, however, split off such amide groups, so that an asparagine molecule showed up as free ammonia and an extra aspartic acid, and similarly a glutamine molecule was read as ammonia and an extra glutamic acid. Such determinations had great authority: Sanger's insulin sequences had originally been done that way. "We said there was no reason for excluding amides; it was just that chemists normally split them off, and counted them as ammonia, and that was an accident of procedure," Crick said. "All right, glutamine and asparagine; so we said we'd put those in. So then the ones we had to be clever about were hydroxyproline, hydroxylysine, things of that sort." Hydroxyproline, in particular, though it differs only slightly from proline, is abundant enough so that it just made Gamow's list; but it is found, as is hydroxylysine, only in one protein, the structural material collagen. They argued that these were formed, like cystine, by modification of the protein after synthesis.

"The general rule was that if an amino acid only occurred in a few odd proteins, we excluded it. But we didn't exclude an amino acid which was in most proteins but wasn't found in *every* protein, because that could be just statistics," Crick said. "So we included tryptophan, even though it wasn't found in insulin, because it was found in lots of other proteins; and we excluded the iodinated and brominated tyrosines, because they were only found

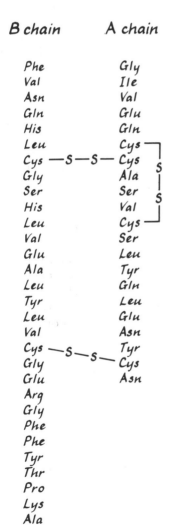

The amino-acid sequence of beef insulin

$$
\begin{array}{ccc}
H & H & O \\
| & | & \| \\
H-N-C-C-OH \\
| \\
CH_2 \\
| \\
C \\
O^{\nearrow} \;\; {}^{\searrow}NH_2
\end{array}
\qquad
\begin{array}{ccc}
H & H & O \\
| & | & \| \\
H-N-C-C-OH \\
| \\
CH_2 \\
| \\
COO^-
\end{array}
$$

asparagine   *aspartic acid*

in a few freak cases." Iodinated tyrosine is the precursor of thyroid hormones and is found only in that gland. Tyrosine with an atom of bromine attached had been found in coral. "We just used our common sense. But the point really is that this job was done in none of the textbooks. The textbooks were just the opposite. They included absolutely everything and made *no* distinctions whatever. In fact it was almost a thing you gained *credit* for—to find a new amino acid in proteins, you see, as *Hopkins* had done; what could you do better than that?" He laughed. Frederick Gowland Hopkins was the most celebrated English biochemist and teacher of biochemists in the first half of the century. In 1901, early in his scientific career, he and a colleague had isolated the amino acid tryptophan, and he had gone on to show that it is an essential dietary component; tryptophan gave Hopkins his first papers in the biochemistry of nutrition, leading him over more than a quarter century to the discovery of vitamins, a Nobel prize, and a knighthood. And indeed, illustrating Crick's point, a paper of 1953 exhibited twenty-eight "uncommon amino acids found in naturally occurring peptides," including several that appeared only in antibiotics and two that had been obtained from a mushroom, the angel of death. But had Crick and Watson had any notion that they must reach exactly twenty? "No. Not at all. There was nothing special about twenty whatsoever. We knew it was about that number, of course—more than ten, less than thirty." Later, however, the number twenty took on for a while the appearance of deep significance.

"The idea of sorting them into the ones which were what you might call freaks, or modifications, and the genuine set, only came up really seriously when Gamow produced his list," Crick said. "Because what had happened before was rather interesting, you see: I knew all that before, but I never bothered to write it down and work it out. I mean I knew that there were freaks and there were ordinary ones. And when Gamow produced his list, we didn't, as it were, have to think out our logical foundations; we merely—" Was that an accident of being in the small protein-crystallography unit at the Cavendish? "Well, I was working with proteins and therefore knew about proteins; but you understand that if one had gone and asked people in *biochemistry* who knew about proteins one would have got the wrong advice. You see. You had to look at it with the eye of the outsider."

Gamow's error in making DNA the template rather than RNA seemed, in itself, trivial; RNA might perfectly well carry the information for the protein sequence in some equivalent fashion. But the fact that there had to be limitations on which amino acids could be neighbors, imposed by the fact that each successive diamond hole shared two nucleotides with its predecessor, was much more interesting. The sequences Sanger had determined totalled no more than fifty amino acids. That was far too meagre, Gamow saw, to allow any amino acids to be identified with particular combinations of nucleotides, for, he wrote, "The number of all possible assignments of 20 amino acids to 20 'diamonds' is given by: 20!"—factorial 20, or 20 times 19 times 18 and so on down to 2, which is 2,432,902,008,176,640,000—"and it would be a tremendous cryptographic task to establish the correlation by the method of trial and error." What Gamow did not quite face, at first, was that Sanger's sequences might be enough to show that the scheme's restrictions on neighbors were not obeyed in insulin. Crick perceived at once that Gamow's idea might be refuted this way, though the details of the disproof were too intricate to work out over lunch. He and Watson wrote back to Gamow with their observations.

Watson, that summer, put Brenner in touch with the phage group, and spoke of him enthusiastically wherever he went. He got Ole Maaløe, in Copenhagen, to send Brenner some serum with which to do immunological tests to identify phage T2; he introduced Brenner to William Hayes at lunch in London; he even prompted an offer to Brenner of a research job in Edinburgh. Then Watson left Cambridge for Caltech. Crick had at last got his degree, and was also on his way to the United States, to spend a postdoctoral year at the Polytechnic Institute of Brooklyn; there had been talk of his then spending a further year at Caltech.

In September 1953, almost the entire cast of the drama of DNA reassembled in Pasadena for Linus Pauling's conference on proteins, as though to take a curtain call. Pauling and Delbrück were there, of course, Watson had taken up residence, Crick came from Brooklyn, and Sir Lawrence Bragg, Max Perutz, and John Kendrew from Cambridge; John Randall and Maurice Wilkins made the trip from London; Astbury and even Lindo Patterson attended. Rosalind Franklin stayed in London. That was a meeting buoyant, even effervescent, with achievement. Pauling still pursued the biological ramifications of the alpha helix—in particular, how it might turn corners—and of pleated sheet in proteins. Watson presented the double helix of DNA. Hugh Huxley, an early member of Perutz's group who was currently at the Massachusetts Institute of Technology and who was applying x-ray crystallography to muscle, described a sliding molecular mechanism for muscle action, fueled by adenosine triphosphate; this was a crucial discovery in that field. Perutz announced that he had found the technique by which a protein molecule could be modified so that two or three variant x-ray diagrams might be obtained and compared in order to locate every piece of the molecule as though stereoscopically and so at last derive its structure.

At Brooklyn Poly, a biochemist named David Harker ran a group studying ribonuclease, and there Crick was to do the crystallography to determine the enzyme's structure. He shared an office with Vittorio Luzzati, another crystallographer, who had worked with Franklin when she was in Paris. "I wrote two very dull papers," Crick said. "And Harker expected me to be in promptly in the morning. To which my response was to leave promptly in the evening! And as we were in a new country, I didn't do as much reading and work out of hours as I would have—as I do now. So it was a year off, in a way." Crick had brought his wife, Odile, their two-year-old daughter, and his teen-aged son by his first marriage, Michael. They rented a house in Fort Hamilton, a section of Brooklyn which was then small-town suburban, next to an army base, a fifty-minute subway ride from midtown Manhattan and almost as far from Crick's lab. "Really, we were miserable," Crick said. "Not only that, but when you go to a country—not now, but in *those* days when one had no money—one arrived with *very little money*. And because you're paid at the end of the month, you've laid out everything ahead. So we were often down to our bottom dollar, and I would go home and hold up a dollar to Odile and say, 'That's our last dollar; I'm not paid for two days!' And I always knew when times were hard because she produced potato pie, which is a delicious thing of potatoes and onions, and that meant that we were getting very hard up. So altogether it wasn't the happiest experience."

Crick met Gamow for the first time that fall. Luzzati was in the room. "I think it was their first discussion," Luzzati said, years later. "About the code, you remember—the triangular holes. That discussion, I remember—they were both shouting! Crick and Gamow. They were in excellent spirits, not quarreling, not at all. Enthusiastic about what they were discussing." The picture is irresistible: two extraverted, opinionated European physicists in a lab in Brooklyn trying to talk each other down about the biological code.

Gamow was a tall man, burly, blond, with a huge head, thick glasses, and exuberant enthusiasms; born in Russia, he was a student in Göttingen in 1928, where he met Max Delbrück, and then like Delbrück became one of the circle around Bohr in Copenhagen, before going to the United States in 1933. Though Gamow was not perhaps one of the central figures in the great decade of quantum theory from the late nineteen-twenties through the late thirties, he made several notable observations, characterized by quirky yet deep-seeing intuition. His career has been called "perhaps the last example of amateurism in scientific work on a grand scale." Again like Delbrück, Gamow had a passion for practical jokes—including that rarity, the scientific joke. He wrote his most important paper on the Big Bang in 1948 with Ralph Alpher, a research student. Then Gamow asked an old friend, the physicist Hans Bethe, to allow *his* name to be added to the list of authors—solely for the fun of being able to cite it as Alpher, Bethe, Gamow, as indeed it is known to this day. By luck, it was published on April 1. Gamow was also a clever, minor popularizer of science; he wrote a series of small books, and illustrated them with his own drawings, in which a fellow called Mr Tompkins escaped the humdrum by exploring, suitably transformed in size, the innards of the atom, the secret of life, and other matters. In the last of these books, Mr Tompkins encounters Max Delbrück on a beach and learns a little about phage. Gamow died in August 1968.

"You never met him?" Crick asked. "Oh. I see. Well, he was extremely jovial; used to drink a bit too much—by the time I knew him, anyway; and was fond of card tricks, which he showed to pretty girls—this sort of thing. He had a marvellous card trick; one of the best amateur card tricks I've ever seen was done by Gamow!" Crick laughed. "And he was what is called good company, was Gamow. I wouldn't quite say a buffoon, but—yes, a bit of that, in the nicest possible way. You always knew, if you were going to spend the evening with Gamow you would have a 'jolly time.' You know. And yet there was something behind it all. But he didn't ever get as good a grasp of molecular biology as for example Leo Szilard did."

By the time the two met, Gamow had worked up the material of the previous summer into a paper he planned to publish in *Proceedings of the National Academy of Sciences,* and he gave Crick a copy: "Protein Synthesis by DNA-Molecules," by G. Gamow and C. G. H. Tompkins. In the paper—Crick preserved the draft—Gamow had figured out the abstract pattern of the nearest-neighbor restrictions imposed by his scheme. The details take up two tables and several paragraphs, and are genuinely unimportant. In principle, by analyzing the overlaps among his four-nucleotide diamonds, he showed that some amino acids would have many fewer chances of combination than others; that all combinations would occur within several defined sets; and—the most crippling restriction—that there would be six amino acids that could never occur twice in a row, while only two could show up in runs of three or more the same. He had not yet applied these rules to the sequence of insulin. Crick urged him to do so.

The problem was like breaking a cryptogram by working out the consequences of inspired guesses. The first step was to find a promising place to start. In the middle of the shorter chain of insulin appeared a sequence that Sanger had identified as:

$$+ \quad \frac{\text{glutamic}}{\text{acid}} \quad + \quad \frac{\text{glutamic}}{\text{acid}} \quad + \quad \frac{\text{cysteic}}{\text{acid}} \quad + \quad \frac{\text{cysteic}}{\text{acid}} \quad +$$

Such a pair of doublets imposed severe limitations; and it occurred within a longer sequence:

$$+ \text{ valine } + \frac{\text{glutamic}}{\text{acid}} + \frac{\text{glutamic}}{\text{acid}} + \frac{\text{cysteic}}{\text{acid}} + \frac{\text{cysteic}}{\text{acid}} + \text{ alanine } +$$

$$\text{serine } + \text{ valine } + \frac{\text{cysteic}}{\text{acid}} + \text{ serine } +$$

There the appearance of valine immediately before one, and then the other, of the types in the doublets tightened the limitations still more.

Working forward and back like that, Gamow found himself forced to demonstrate that there was no possible way that his diamond holes, made by overlapping sets of nucleotides, could allow serine to appear after cysteic acid, as at the right of the longer set above. The fact that cysteic acid is now identified as cysteine makes no difference. The fact that Sanger later showed that one of the glutamic acids in that stretch is actually glutamine, means that the demonstration would have to be carried further than Gamow took it—many pages of analysis further, Crick found—to produce the same result.

Most important was the method of formal argument that Gamow and Crick established: comparing the restraints imposed by any particular coding scheme to the sequences of amino acids so far known in proteins. Simple in this instance, the argument grew subtle and elaborate over the next three years as coding schemes grew more ingenious while Sanger's methods were applied to more proteins. Gamow's proposal had the virtue of being "sufficiently precise to admit disproof," Crick once wrote dryly. "This stimulated a number of workers to show that his suggestion must be incorrect, and in so doing increased somewhat the precision of thinking in this field."

Gamow published three paragraphs and a drawing of the holes in DNA in *Nature,* the following February. The longer paper was turned down by *Proceedings of the National Academy of Sciences* because somebody there objected to Mr Tompkins as coauthor—or so Gamow later told Crick—and eventually appeared, under Gamow's name alone, in the journal of the Royal Danish Academy of Sciences, of which he was a member because of his years in Copenhagen.

❖ ❖ ❖

In Oxford, that fall, Brenner met Milislav Demerec, director of the Cold Spring Harbor Laboratory, who was on a visit to England. Demerec invited Brenner to come to Cold Spring Harbor the next summer. Brenner's friends in Oxford the previous year, Leslie Orgel and Jack Dunitz, had both gone to Caltech with fellowships. Dunitz wrote Brenner a thoughtful letter, full of scientific gossip and advice, and urged him to come to the States. Pasadena, though, was an uneasy place, Dunitz said. "Sometimes I feel that there is something inherently wrong here i.e. the mental domination by L. P. but it's not exactly that. It's more that all the problems being studied are, more or less, his problems." Dunitz wrote from the Delbrücks' house, with Max and others in the next room playing an eighteenth-century trio for flute, recorder, and piano. In a footnote, Dunitz made an astounding announcement: "Max says that he doesn't intend to do any more 'phage work.'"

At Caltech, Watson returned briefly to bacterial genetics. Hayes was there and they had planned to work together. But after the experience of DNA, and with Gamow's twenty-cavity template in mind, Watson soon began trying to get x-ray-diffraction pictures of RNA. He wrote often to Crick. In November, he warned him against coming to Caltech the next year: the smog was nauseating and the scientists boring.

> I thus conclude that Odile will not find Southern California as alluring as she might think. . . . In Biology there is Max,—otherwise fairly dull (from your viewpoint and also mine). In Chemistry—obviously the Great Man—however it is unlikely that you'll eat lunch (et cet.) with him and so I'd suspect you would tell him more than he would tell you. . . . I thus conclude that from the viewpoint of protein crystallography—Cambridge is more attractive. . . . Experimentally I've done almost nothing—RNA interests me more than genetics and so my little effort have been along these lines. I (with Alex Rich) am able to repeat my Cambridge RNA picture—using RNA from calf liver—Hence a unique RNA structure exists! ! ! ! !

The photographs were diffuse; but there were exciting familiar signs:

> The pattern is strongly affected by H_2O content (nothing quantitative as yet)—whether more forms exist, I'm not sure. Naturally I've tried model building and at times (including now) believe I'm learning something. By Christmas we should know more than a little.

Mid-December, he wrote again. Restlessness was growing. He was still finding reasons to think that the structure of RNA could be stormed and sacked and set alight by the same tactics that had conquered DNA. He could find no evidence that RNA could be crystallized; he was still trying to get a first good diffraction pattern. He was coming east for Christmas. "I should like to talk over RNA with you. There is no one here with whom I can profitably talk with as no one here believes himself capable of the Pauling type approach to structures. Crystallographers are basically deadly."

Early in the new year, Odile Crick went back to Cambridge with their daughter, because they had decided not to go to Pasadena after Brooklyn, and she was pregnant and wanted to have the baby in England. Crick, with his son, moved to the apartment that Luzzati was vacating in Brooklyn Heights, whence he could walk to work.

On February 13, Watson wrote an extraordinary letter, beginning "I now understand 1/2 of RNA." Some RNA, that from plant viruses, Watson said, did not show complementary ratios among the bases, but "in RNA from all other sources the ratios *are* complementary. This is not an obvious fact as much of the data is sloppy but the good papers show the ratios and are the ones to be considered." On the other hand, he went on, some RNA was single-stranded and some double; all RNA, single or double, replicated itself like DNA by forming complementary strands. "In any case we now visualize the mysteries of life as follows," Watson wrote, and drew a diagram that looked exactly like a lorgnette: two circles as if for lenses, bridged by an arc, and with a handle hanging down from the right. The left side was labelled DNA, with an arrow circling back on itself to indicate complementary replication. The right-hand circle was RNA, with a similar reflexive arrow. The bridge between them was another arrow, inscribed "chemical transformation? deoribose to ribose." This was a biochemical notion, a throwback to Brachet—offered a fortnight before the first anniversary of the discovery of base pairing. Watson gave it the confident explanation, "This is why we find 2 strands, one to keep code, the other to be transformed to RNA which sneaks to cytoplasm and makes protein." The handle of the lorgnette, hanging downward from RNA, at the right, pointed to the label "Protein (Gamow holes?)"

> All of this is slightly mad but as it is cute I think it is correct. Your comments seriously desired. Have convinced Feynman [Richard Feynman, a physicist at Caltech, now celebrated] and slightly Delbrück. The others are not convinced but are not intuitive creatures and so do not count. I do not believe the base ratios could be so good if they did not imply self replication. . . . Also it would be strange for DNA & RNA to be so similar and yet have different replication schemes.
>
> I think this idea is worth letter to Nature which I am now writing. Do you agree? . . . Idea came 3 days ago after spending 3 days reading all literature on base analysis: Chargaff has failed again!!! Really very funny. The important thing is to ignore data which complicates life.
>
> What do we do after protein specificity is solved. I really do not want to do genetics or watch birds.

The letter was a cruel self-caricature.

"Of course you realize that our ideas on that were totally wrong," Crick said. "We thought that RNA had some structure with twenty cavities—it was that period. Unfortunately, people have forgotten what it is we didn't know at the time." Meanwhile, Crick sat at the bench every day measuring the intensities of spots on diffraction pictures of ribonuclease. "And I produced a totally wrong structure for collagen, for example, which really was a disaster."

Gamow spent the spring of 1954 in the Department of Physics in Berkeley, and early in March wrote to Crick. His letters were as irrepressible as the man, crowding the page with ideas and sketches in several colors, snatches of song and tags of Latin, in a spelling he invented as he went along and a boyish handwriting nearly half an inch high. "Come, listen, my men, while I tell you again, A few unmistakable codels, By which you may know, wheresoever you go, The genuine RNA models," this letter began, and then:

> Dear Francis,
>
> Your letter just arrived forwarded from Washington. . . . I drove to Pasadena a fortnight ago, and spent two days with Jim, Max Delbrück an others. They have a model of RNA, big and nice looking, but they do not believe in it very much themselves. (except of Alex Rich who conceeved it). It has trapezoidal holes formed by two bases, and two different "sugar edges". And there are 20 different holes. *But* I have found (however) that the combination rules do not work at all. According to that model, 10 amino acids can occur only at *even* places in the protein sequence, with the other 10 only in *odd* places, which is certainly not true. In particular, dubbling, like GluGlu or CyCy, is not permited at all! After I came back, I have tried a new code of *triangles* (with three independent bases. . .). . . .
>
> Yours as ever,
> Geo.

That was Gamow's standard signature: he believed unshakably that it was pronounced "Joe," and Joe Gamow he was to his friends.

Rosalind Franklin wrote to Crick in March. She was working on the structure of tobacco-mosaic virus at Birkbeck College, London, and had arranged a visit to the United States for the end of summer. Crick wrote back suggesting where she should go and whom she should see—and most certainly Wendell Stanley, who was now the director of a big virus laboratory recently built at the University of California in Berkeley, where he had several sets of scientists working on tobacco-mosaic virus. (Gunther Stent had joined that lab, too.) Later Crick spoke to Stanley and Pauling on her behalf.

March 30, Dunitz wrote another calm letter to Brenner. Among the news:

> Leslie has become interested in the RNA work of Alex Rich and Jim Watson. This has reached an interesting point, but the evidence is very strange and difficult to understand at present. I won't say anything specific, not because of security reasons, but because the situation seems to change so dramatically from week to week that a comparative outsider like myself gets rather confused by all the rapid series of climax and anti climax. In a nutshell, however, there are 2 theories (1) RNA is just a kind of DNA only slightly different (2) RNA is completely different from DNA. Better experimental evidence is required to settle this interesting point. . . .
>
> How are your plans for coming to the States this summer? Gunther Stent told me that he would be very pleased if you would like to spend some time with him at the Virus Laboratory in Berkeley. Max Delbrück won't be back here until September late. You asked what he was interested in these days. In phototropism—particularly of the fungus phycomyces. Not cancer, as you guessed.

Phycomyces is a primitive fungus that grows on decaying matter, like dung. Its single cells form a slender stalk that is sensitive to at least four distinct stimuli—namely, light, towards which it bends; gravity; stretch; and some unknown stimulus that makes it grow away from nearby solid objects. Delbrück thought this organism might provide an irreducibly simple means of investigating sensory physiology at the molecular level, just the way bacteriophage had opened up molecular genetics. After 1954 he published a few further theoretical papers in the main line of molecular biology, notably on the conundrums of DNA replication and on the code, but his experimentation was thenceforth given over entirely to *Phycomyces*. The work was unflagging, the results—as he said on one occasion when he showed me an experiment under the microscope—scant.

While in Berkeley, Gamow founded an organization he called the RNA Tie Club. Its aims, proclaimed in the charter he sent around to friends, were "to solve the riddle of RNA structure, and to understand the way it builds proteins." Its emblem was to be the "RNA-tie (with green sugar-phosphate chain, and yellow purines and piramidines), as designed by George Gamow, to be produced by an appropriate haberdasher in Oxford, England, through the intermediary of Leslie Orgel." The ties were eventually embroidered in brilliant silk threads on a black ground, in Los Angeles. He proposed sixteen members, Crick and Watson not among them. This later grew to twenty, one for each amino acid; a member was also to wear a club tiepin that carried the three-letter abbreviation of his assigned amino acid. There were to be four honorary members, too—one for each base—who got free ties. The only one who seems actually to have achieved this honor, though, was Fritz Lipmann, alias Cytidine. Indeed, the membership was haphazard, drawn from scientists Gamow happened to know or know about who were interested in the coding problem, but including also several of his physicist friends, notably Edward Teller, of the hydrogen bomb. Later, there was even an RNA Tie Club letter paper, bearing the slogan "Do or die, or don't try," which Delbrück had suggested, and listing the officers, "Geo Gamow, Synthesiser, Jim Watson, Optimist, Francis Crick, Pessimist, Martynas Yčas [a biologist and friend of Gamow's], Archivist, Alek Rich, Lord Privy Seal." Gamow took to saluting members as *Phe* or *Gin* or *Tyr* (*Tyr* was Crick), and so on, and signing himself *Ala*.

The enthusiasm was lunatic even for Gamow. Yet, just as Crick said, there was something behind it. The Tie Club was supposed to encourage discussion and to circulate papers that were more speculative, discursive, and untested than their authors would risk in formal publication. A lot of talk was generated, and half a dozen papers got written. One of them, by Crick, was of first-rate importance, the next great theoretical step in the science.

Late in April, in Washington, the National Academy of Sciences held a symposium on the structure and function of nucleic acids. Pauling was chairman. Alexander Todd talked about the chemistry. Crick presented the complementary structure and reviewed the latest confirming evidence. Watson brought up the attempts he and Rich had been making to relate the structure of RNA to DNA and protein synthesis; but the data conflicted and his confidence was fading fast. A month later, Gamow wrote to Crick again with more coding schemes, saying, in part:

> Now I have comletely shifted to RNA. When Jim was in the east I visited Pasadena, and together with Alek Rich and Leslie Orgel (you must know him, he is from Oxford) we devised a new code in which an a.a. is determined by three consequitive bases, the central one being a "major determinant", with two neighbors "only coloborating."

He drew a diagram, but cancelled it in large letters: "Does not work!"

> I am now trying a new code recently suggested by Edward Teller in which each new a.a. is determined by two bases and the preceeding a.a. (This has a direction!) Do not know yet how good it is.

He drew a funny little sketch of a stack of blocks arranged stepwise to represent the bases, and on each base a man seated, to represent an amino acid, with his back against the base rising behind him and his legs over the shoulders of the next man along. "May work," this label said. Gamow and Watson would be in Woods Hole in August; Crick must come too. "Than we can make a real attack on RNA!"

❖ ❖ ❖

In the early fifties, several people began to turn the electron microscope onto the insides of cells, and especially to look at the small particles of RNA bound up with protein that had first been glimpsed, at the limit of the power of the optical microscope, by Albert Claude at the Rockefeller Institute. What can be done with the electron microscope depends always on the art of preparing the specimen, which must be fixed without damage or distortion and then shadowed with a film of metal atoms so that its features will survive and stand out in the beam of electrons. Biologists learned how to cut and mount extremely thin slices of cells, and sections of the pellets in their centrifuges. With some surprise they saw the clear empty spaces in the chart of the cell, in particular the cytoplasm, filling up with an elaborate anatomy. The dominant feature of this new anatomical realm was a lacy network of filaments, channels, perhaps tubules in section, that Keith Roberts Porter, working with Claude at the Rockefeller Institute, named the "endoplasmic reticulum," meaning literally the network within the cell fluid. For a while it was thought that this was the true material of the intact cell, and that in preparation it broke up, to produce the microsomal particles. But then techniques got much better, and it could be shown that the reticulum was a continuous internal membrane system, a network of tubes with an inside and an outside.

Small, dense particles, about 150 angstrom units in diameter, were clearly to be seen in great numbers attached to the inside of the membranes or scattered nearby. The particles were spherical. They could be separated in the centrifuge. They appeared to contain protein and RNA—indeed, eighty per cent or more of all the RNA in the cytoplasm, though the full accounting took several years. The leader in much of this work, and above all the most skillful at playing off the ultracentrifuge and the electron microscope, was George Palade, also at the Rockefeller. Palade had announced the essential findings by the summer of 1953. He, and others as well, also noticed that even after repeated, prolonged centrifugation at very high speeds, some RNA—about ten per cent of the total—always remained in solution in cytoplasmic fluid.

Meanwhile, others—Howard Schachman, Arthur Pardee, and Roger Stanier in Berkeley, Ole Maaløe in Copenhagen—were detecting large numbers of dense spherical particles in bacteria, too. It turned out that the particles had been seen even in the earliest electron micrographs of the phage group. In the summer of 1942, Luria, Delbrück, and Thomas Anderson had succeeded in capturing individual E. coli at each step of phage infection, including the very moments of lysis when the cell wall, they said, "has burst open and has liberated a flood of material in which several hundred particles of virus ... are visible." Never mind that the deluge had taken place in a drop no more than a few millionths of an inch across; the pictures were dramatic evidence of events that had been inferred but never before seen. But then, the modern Noahs reported, "Along with the virus particles, a granular material has come out from the bacterium." The granules were of very regular size, 100 to 150 angstroms in diameter, which was much smaller than the phage; granules that looked just the same were seen no matter which type of phage was present. The particles were then ignored until the early fifties, when they were found in the ultracentrifuge and compared with Palade's. The ones from bacteria were not so large as those in animals, while nothing in bacteria corresponded to the endoplasmic reticulum.

Palade's electron micrographs showing the membranes and the enormous numbers of particles attracted intense interest. "You should go back and look at the pictures," Crick said. "At the time, we said, 'Isn't that funny, they seem to be arranged in strings.' But you know, it's very easy, if you put a random set of dots, to imagine you see strings. And we didn't think any more about it at the time. And forgot about it. And we didn't do a proper test to see whether the strings were statistical, or not."

By the end of 1953, there could no longer be any doubt that Brachet had been right, seven years earlier, in his surmise about microsomal particles: in higher organisms, anyway, they were the place where the peptide chain was put together. Paul Zamecnik proved that at Massachusetts General Hospital. Zamecnik then had just turned forty. He had been born in Cleveland and trained as a physician. As a youth, he had never met a scientist, he told me in 1977. He was middling tall, with a pleasant voice and a manner that was—above all—deliberate, reflective, careful of statement. He had begun to do research immediately after medical school, he said, because he had failed to get the particular internship he wanted and had a year to wait. He had been interested in protein synthesis since before the war's interruption. At Mass General, in the late forties, he built a small research group; their lab was directly downstairs from Fritz Lipmann's.

A cluster of doctrines about protein synthesis prevailed at that time. One was that amino acids could be exchanged into and out of completed protein molecules. Another was that enzyme proteins were not necessarily built from scratch but were formed by modification of fully formed precursor molecules. Another was that polypeptide chains were joined up by the reverse action of the same sorts of enzymes that also digested them. Zamecnik—as well as others, including Lipmann—questioned those doctrines. His group devised various ways to test them by following the pickup of amino acids into proteins. In one experiment, Zamecnik and a colleague made amino acids containing a radioactive isotope of carbon, ^{14}C. Zamecnik then injected rats with these, and killed the rats at intervals, mashed up each fresh liver, and separated the components of the liver cells—where protein synthesis had been vigorous to that moment. The radioactivity showed up earliest in the microsomal particles. Only after that did it begin to appear in completed proteins; and as it did so, it left the microsomal particles.

Such experiments did not altogether discredit the prevailing ideas of protein synthesis; that was done finally at the Institut Pasteur at about the same time. But they established the microsomal particles as the sites of protein synthesis. Crick found that definitive. And they were part of a series of superb demonstrations to come out of Zamecnik's lab in the next eight years. Zamecnik was an experimentalist rather than a theorist; his results came to be recognized as careful, unhurried, and—above all—trustworthy.

In the early fifties, a number of people were trying to make protein synthesis happen outside the cell, so that the process could be analyzed and played with. Here, Zamecnik's lab led. The idea was to put various combinations of cellular components and juices together, without the presence of living cells, and then to add labelled amino acids to see if the combination would link up some proteins. Such biochemical cocktails for protein synthesis in the test tube were called the cell-free system.

Zamecnik and his colleagues ground up cells with an abrasive, and threw down the grit and cell walls in a centrifuge at low speed. They tried *E. coli*, but after months of work found that they couldn't reliably get rid of whole cells. They turned to fresh rat liver, minced and gently homogenized. After centrifuging that, they were left with a liquid containing all sorts of things—particles, nuclei, fragments of endoplasmic reticulum and other cellular structures, DNA, RNA, many different enzymes, and much else that was unknown or poorly known. This crude mixture could be separated into further fractions by one or repeated runs in the ultracentrifuge at very high speed. Selected fractions could be recombined. Amino acids could be supplied in different combinations. RNA or DNA could be destroyed by adding the appropriate enzymes. Fresh nucleic acids of known source and character could then be introduced. Effects could be followed in great detail when one or another item was labelled with radioactive isotopes. The flexibility and multiplicity of the possible experiments was in principle enormous.

Development, refinement, precise tuning of the cell-free system for protein synthesis was the chief technical advance of straight biochemistry in the decade of the fifties. It was to prove the vehicle of the eventual solution of the coding problem. But even the simplest aim of getting a good and reproducible level of incorporation of amino acids into protein, without complete cells, only began to be possible in Zamecnik's lab in 1953 and '54.

At this stage in the work, Zamecnik was joined by Mahlon Hoagland, who had worked with him once before but had then spent a year in Copenhagen and a year upstairs in Lipmann's laboratory. At that early period, Zamecnik wrote fifteen years later, there were four essential components in the cell-free incorporation system. Amino acids and the microsomal particles had to be present, along with ATP (adenosine triphosphate, the energy carrier). The fourth component was "enzymes from the 105,000 x g supernatant"—which meant that something active remained in the liquid that was left above the pellet in the tube after centrifugation at 105,000 times the force of gravity. Of the four components, Zamecnik wrote, the microsome they then "regarded as the relatively inert marshalling site on which the newly forming chain was built, and from which this nascent protein was subsequently released into the supernatant, soluble protein fraction of the cell." Where the enzymes intervened was unclear. As for the amino acids and ATP, Zamecnik thought it a likely possibility that the two interacted at the very surface of the microsomal particle. "Otherwise, it appeared, there would have to be a separate supernatant enzyme for each amino acid, a possibility that at the time seemed cumbersome"; it was also inconsistent with some simpler experiments in making peptide bonds that Lipmann had begun.

Zamecnik and Hoagland drew themselves schematic diagrams of all this.

ZAMECNIK'S BIOCHEMICAL FLOW FOR PROTEIN SYNTHESIS, 1953

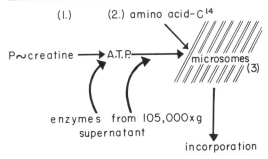

The one that survived is as incorrect and incomplete as the sketch in Watson's letter to Crick at about the same time, but the contrast tells all: Watson dealt with the flux of information, and gave no explicit place to enzymes or to microsomes; Zamecnik was concerned with the flow of energy, materials, and enzymes to the microsomes, and seemed unaware of nucleic acids.

WATSON'S FLOW OF INFORMATION, FEBRUARY 1954

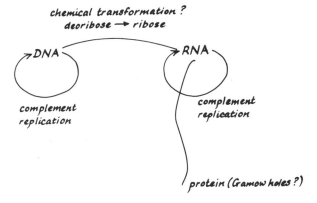

The first collision between the two views was unforgettable, if dubious. Zamecnik wrote:

> In the summer of 1954 we inquired of our scientific neighbor, Paul Doty [a biochemist at Harvard], as to how the RNA of the ribonucleoprotein particle could serve to order the activated amino acids. Dr. Doty mentioned that a Dr. James Watson was visiting him, who with a colleague, Dr. Crick, had recently proposed a model structure for DNA. . . . He would send Dr. Watson over to visit me and I could ask him that question. I looked at Dr. Watson's young face above his white Irish turtle neck sweater, then at his wire model of the double-stranded DNA, and inquired how the message of the DNA made its way into the sequence of protein. Unfortunately, it seemed to me privately, the bases were facing in, rather than out,

as on the biologically appealing earlier Pauling model. Was protein made directly on a DNA template? Probably not, it appeared, because there was no DNA in the ribonucleoprotein particle. Was RNA made on the DNA template? No answer to this question either, although it seemed likely. How did this complicated double helix unwind? There was a gulf between DNA and protein synthesis, Dr. Watson agreed with a diffident smile, as we parted and he took off on a vacation to look at birds.

Watson had come up to Boston from the Marine Biological Laboratory in Woods Hole, Massachusetts. Woods Hole is a summering ground for migratory biologists, a lot like Cold Spring Harbor. Both places support a year-round research program, set quietly by the water, but expand manyfold every summer for scientific meetings and to offer training courses that teach basic subjects but to established scientists at an advanced level and with terrific intensity. Such biologists' summer resorts are a tradition. In Europe, in a more leisured way, a similar function was filled for many decades by the marine biological stations at Roscoff in Brittany and at Naples (where Watson first met Wilkins). Woods Hole is particularly known for its six-week course in general physiology, and Watson went to teach in that. At Woods Hole, that same summer, Matthew Meselson and Frank Stahl first met and talked about demonstrating the semiconservative replication of DNA.

❖C❖   The abstract puzzle of the genetic code was interwoven in a triple fugue with the wet chemistry of RNA metabolism and protein synthesis; the third theme was molecular genetics. This in its own right was elaborated in the early fifties into a science of bewildering complexity at the surface. Its aim, though, was simply stated: to get at the physiology and the chemistry of the gene in action. But that indeed was the eventual point of the coding problem, too. The phage group, in the United States and in ceaseless travels, practiced one kind of molecular genetics. The French at the Institut Pasteur used different organisms to pursue different questions having to do with the expression and regulation in the cell of the functions specified by the genes. Joshua Lederberg was a school of molecular genetics almost single-handed. At the University of Wisconsin and then at Stanford, he investigated what happened when bacteria exchanged genes in a manner analogous to the sexuality of plants and animals, maize and fruit flies. These several lines necessarily intersected. Twice, molecular genetics set the direction that the coding problem took. The later time was in 1960, when the work of Jacques Monod and François Jacob in Paris led Crick and Brenner to see the way out. The earlier time was in 1953 and '54, when Seymour Benzer found the technique by which he could relate mutations to locations on the gene at the scale of the individual nucleotide.

The problem Benzer proceeded to solve was a version of the ultimate classical problem: to map the gene completely and so to drive formal genetic analysis—the algebra that is the essence and glory of Mendelism—down to the level of the chemical gene. His demonstration that that could be done was the necessary logical foundation of a truly molecular genetics. His technique, though laborious, was exciting, for it looked like the direct way to break the genetic code.

Benzer and I talked one afternoon in the spring of 1971, at Caltech, where he had moved six years before. His office was small, bright with daylight, crowded with bookshelves and files all stowed with a mariner's sort of compulsive comfortable neatness. On a shelf was a photograph, enormously enlarged, of nerve connections in the eye of a fly. Benzer was medium dark, medium short, as neat and compact as the room. He was wearing a lightweight tan cardigan over a shirt and tie. The photo, he said, was an electron micrograph: he was presently mapping the genetics of mutations that affected the nervous systems—the behavior—of fruit flies. Half a dozen of the early molecular biologists were then moving into neurobiology; Benzer brought out a cartoon that one of them had sketched, a jokey ancestral tree with the faces of molecular neurobiologists pasted in according to the organisms they were working with.

"It's a new phase," he said. "I feel that, y'know, when I came into molecular biology it was a pioneering science. But when a science becomes a discipline, which is essentially true of molecular biology now, when you can buy a textbook, take a course— There's no question there are many surprises left . . . but a field to work in, to me personally, when it becomes a discipline, becomes less attractive. I find it more fun to be striking out in something which is more on the amorphous side. Which was true of molecular biology when I started. Another thing that becomes unpleasant is the redundancy of effort, a number of people doing the same thing—so that even when you make a discovery, six different guys discover it in the same week. You begin to feel that if it's five guys instead of six guys it doesn't make any difference. But still, my change was not so much to escape from that, as just following my own interests; I've got interested in behavior and I want to look at it."

Benzer was a physicist who turned phage geneticist under the influence—yet another one—of Schrödinger, Luria, and Delbrück. He did his graduate work at Purdue University, in solid-state physics shortly before the appearance of transistors. "I'd always had a latent interest in biology, but it was particularly Schrödinger's book that turned me on. About 1946. One of my ideas initially was to apply solid-state physics to how a nerve works, thinking of a nerve membrane as like a junction between a semiconductor and a metal; but as soon as I got into biology I found that much more exciting things were happening." He met Luria at Indiana University, through a friend. Luria enrolled him in the phage course at Cold Spring Harbor. There he learned about Delbrück's work. "I was an assistant professor of physics at Purdue, but I went on leave of absence to study biology." In 1949, he had the choice of going to Luria's lab or Delbrück's, and asked a student of Luria's then—it was Watson—for advice. Watson told him that Luria "would be likely to ask me every day what I had done, whereas I might not see Delbrück for a week at a time. I chose to join Delbrück." He went to Caltech for two years and then for a year to Lwoff's laboratory at the Institut Pasteur.

By every account, that was a golden time to be there. The central questions in Lwoff's lab were sharply defined and markedly different from those of the phage group. The French worked directly with the genetics of bacteria, using mutations in food requirements to peer into the cell's regulation of biochemical function. They worked with phage, too, but used systems other than the set of T phages grown in the B strain of *E. coli* that Delbrück had established as the phage group's standard. In Paris, Benzer shared a room with Jacob on the attic floor. Stent and others from the phage group were crammed in along the short corridor; Monod was at the far end. *"La belle époque,"* Jacob once called it; the attic, the corridor, and the crowding are part of the legend. The group around Lwoff was close-knit, intellectually vigorous, warmly open to visitors, in a combination that made them unusual in the world, unique in France.

In March 1952, Benzer went with Lwoff and Stent to the meeting of microbiologists at Oxford, where they heard Watson announce the Hershey-Chase experiment. Then came the summer when everyone was on the move: the biochemists' congress in Paris, the phage colloquium at Royaumont—and the news of sexual differentiation in bacteria. This was the distinction between genetic donors and recipients found by the Lederbergs and Luca Cavalli-Sforza, in the United States and Italy, and independently by William Hayes working alone in London; it was the discovery that excited Watson to spend the fall writing the paper with Hayes that claimed, incorrectly, that Joshua Lederberg's published data showed *E. coli* to have three chromosomes.

Benzer came back to Purdue in the fall of 1952. "If I'd stayed any longer I would have been fired." That fall, he was asked to give a genetics seminar, and offered the subject "The Size of the Gene." It was a question decades old, going back indeed to Hermann Muller's speculations in the twenties. Benzer had been reading an article about it just published by Guido Pontecorvo, a geneticist at the University of Glasgow whom he had met that summer. Pontecorvo observed that the classical idea of the gene was ambiguous, because it embraced four or five definitions that were not necessarily equivalent. In particular, he brought together fifteen years of evidence that the gene as the minimum unit of heritable physiological functioning—the unit that determined an enzyme—was not a discrete point but had considerable length along the chromosome. The gene taken another way, as the minimum unit in which mutations can be induced—say, by x rays—had been shown to be much smaller. Thus, mutations probably occurred at different places along a single physiological gene. If these all knocked out the physiological gene altogether, they would obviously be indistinguishable; if they altered the physiological function in slightly different ways without totally crippling it, they might be told apart by genetic experiments—if one had an organism in which extremely rare genetic events could be detected.

Then, the next spring, the announcement of the structure of DNA gave the old question new and very precise meaning: a mutation might be a change in a single base or base pair. In June, Benzer went to the symposium at Cold Spring Harbor, where he heard Watson present the structure. Back at Purdue in the fall of 1953, he thought up an experiment to see if the genes of a phage particle are injected into the host bacteria in a set sequence. He chose a mutant of phage T2 that was convenient to work with because, on the dishes of jelly, it produced large, sharply edged plaques that stood out clearly from the small, fuzzy-edged plaques made by T2 of the wild type. (The term "wild-type" was fixed in the lexicon in the early days of fruit-fly genetics, when one could go out and catch one; now it means the original line of normally functioning individuals.) Benzer's particular mutant phage belonged to a whole set that Hershey had found and tagged *r*II; *r* signified rapid lysis, for these mutants made those large plaques—at least, they did when grown in E. coli strain B—because they burst their bacteria early, so that the circle of infection spread quickly.

What the rapid lysis and difference of plaques actually meant to the phage itself, biochemically, nobody had found out. A practical problem with r mutants, though, was that because the bacteria burst early, they didn't build up high concentrations of new phage particles. But Benzer had other E. coli in the lab—two closely related variants, somewhat unusual, of a strain designated K12 that he had first worked with in Paris. At this point, accident intervened. One day, while he was preparing a demonstration for his students in a phage course, he had the two K12 variants out and so tried plating his mutant phage on them. For comparison, he tried the wild type as well on the K variants— four dishes in all. If, by lucky chance, the r mutant did not burst these bacteria early, its plaques would match the wild type and K12 would be fine for growing high concentrations of phage. And just so, when he looked again, on the two plates of the first K variety he saw small, fuzzy plaques: both the mutant and the wild type looked wild. On the second K strain, the wild phage also made small fuzzy plaques as expected. "But the plate to which r was supposed to have been added had no plaques at all," Benzer said. He first thought that in the rush to prepare his class he had failed to put phage into that dish. But repetition got the same result.

The significance of the result would not have been noticed by anyone else; but Benzer had spent four years handling and pondering the insane complexities of this sort of genetic analysis and these particular organisms, and so he saw at once—there was no way to see it except instantly—that he had been presented with a flawless system for genetic mapping at very high resolution. "My discovery at Purdue was a gene and a phage and a set of bacterial strains which had the right kind of properties so that one could in a sense split the gene into its internal parts and construct a map of them in just the same way as was already done for sequences of genes on a chromosome," Benzer said. He had some rII mutants of phage T2. He had three sorts of bacteria for three different purposes. In E. coli strain B he could find more r mutants, for he would need a great many, by the large plaques they made there. In the first of the variants of strain K12, he could then grow high concentrations of every r mutant he found. The other variety of K12 was the ultrasensitive indicator, for r mutants did not grow there at all and T2 of the wild type grew normally. Therefore to cross two mutants, he mixed them in a culture of the first K variant, putting in enough so that the bacteria were infected by both phages at once. As soon as the culture lysed, he put the viral progeny in their millions on a plate of the second K variety. If, in the first step, within some of the bacteria the strands of phage DNA had exchanged pieces and so produced offspring that made good the mutational defects in the parents, then in the second step these recombinant phage—and only these—would multiply.

"The mutants would not produce plaques, but the recombinants would," Benzer said. "You're familiar with genetic mapping?" I said that I was, in principle, in higher organisms: one did it by crossover frequency, meaning that two sister chromosomes in the cell exchanged parts of themselves so that genes that had been lying on one chromosome were now split up and passed to different descendants—but the closer together the genes, the less likely a split that fell between them. "That's right. Crossover frequency gets smaller and smaller as the two mutations get closer together, and pretty soon you are unable to detect such rare events," Benzer said. But the same thing happened between two different phage DNA strands in a bacterial cell. "So this was a method by which you could detect such events up to the limit of rarity." Down to one quantum of rarity, I said. "Exactly. You could put a hundred million phage particles on a plate; if there was a single recombinant particle, that was the *one* that would produce a beautiful plaque. Calculations showed that you could detect recombinations between two mutants even if their mutations were located only at adjacent nucleotides in the DNA within the same gene."

Benzer began collecting mutants and making crosses. The mapping demanded unbelievable labor and attention: scores, soon hundreds, then thousands of different mutants, tens of thousands of different crosses, hundreds of thousands of plaques examined and counted—and of glass culture dishes to be washed. It was an example, he wrote afterwards, "of what we called 'Hershey Heaven.' This expression comes from a reply that Alfred Hershey gave when Garen [Alan Garen, one of the phage group] once asked him for his idea of scientific happiness: 'To have one experiment that works, and keep doing it all the time.'" After more than five years, Benzer brought the map to such close congruence with the corresponding DNA that he was down to "molecular length (measured in terms of nucleotide units)." Delbrück told him he had "run the genetic map into the ground"—at least, he had done so for a short stretch of about two genes of an enfeebled variety of a small virus that preys on a harmless bacterium.

In the course of that, Benzer went to Cold Spring Harbor to work for the summer of 1954. Sydney Brenner arrived there on July 1. Brenner later described their encounter: "I was carrying around a book on sequence analysis in proteins and Seymour was carrying a map of the *r*II region—consisting of two mutants that mapped in a straight line." The meeting of Benzer and Brenner set up another of the strong partnerships of molecular biology. Their collaboration was episodic, however, limited by accidents of timing and geography—and perhaps limited in small part, as well, by the accident of personal styles. The two men were similar enough, seen from across the long lawn at Cold Spring Harbor, to be taken for brothers. Where Brenner, though, was flamboyant, mordant, quick-tongued, Benzer was controlled in gesture, careful in speech, and, over the next two decades showed himself to be scientifically fastidious, taking his own lone line.

Be that as it may, when the two men met in 1954 Brenner recognized at once that Benzer had invented what should be the most direct way of all into the relation of gene to protein: map the gene, sequence the corresponding proteins of the mutants to find which amino acids were different, and compare. By doing that, they would put the first solid foundation beneath the conjecture upon which the entire speculative palace of molecular biology was being run up: the assumption that the sequences of DNA and protein were, in the lumpy bit of jargon that took hold, "colinear." "That was the period when what was regarded as one of the most exciting problems was to draw the correspondence between the amino-acid sequence of the protein and the nucleotide sequence of the gene," Benzer said. He and Brenner planned the experiments they would have to do. If they could only identify somehow the bases

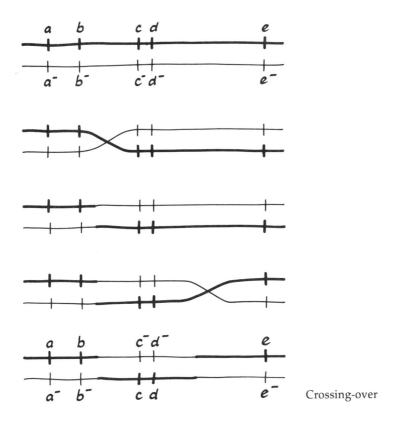

Crossing-over

in the DNA, "it would be possible even to determine the genetic code." The obstacle, to be sure, was that nobody had found the protein that the *r*II gene determined. Benzer intended to put someone to work on that. Meanwhile, he had the map to complete. Brenner, when his visit to the States was done, was committed by the terms of his fellowship to return to South Africa.

❖ ❖ ❖

On 26 July 1954, Watson wrote to Brenner, saying, in part:

> The Physiology Course is frightfully intense and I have never worked so hard in my life. I shall be in Woods Hole till the time of the phage meeting (Aug 25) when I shall drive down to C.S.H. I would be delighted to see you here. Any time in August is convenient provided you let me know a few days in advance. Francis Crick will also be here as will Gamow. Why not come for a week or so.

(That same week, the back-to-college issue of the American edition of *Vogue* appeared, with a photographic essay on "Young Americana Talent"—eighteen people, starlets, set designers, pop composers, and the like, displayed as examples of youthful achievement, and among them "James Dewey Watson . . . a scientist, with the bemused look of a British poet." After Watson's earlier avoidance of publicity, Crick found this intensely irritating.) After Crick had arrived at Woods Hole from Brooklyn and Gamow from Berkeley, Brenner went up for a week or so.

"You were asking me what I was doing in Brooklyn," Crick said to me. "What I did in Brooklyn was to show that Gamow's code was impossible. I invented a neat method of disproving Gamow's code, and I showed that if you used the amino-acid sequence for insulin, and another, beta-cortico-tropin"—a second hormone, whose sequence had been found by then—"you could show that it could not be done by *any* version of Gamow's code."

At Woods Hole, Gamow was staying in a cottage on the grounds of the house of an old friend, the Hungarian biochemist Albert Szent-Györgi, who in 1928 had been the first to isolate vitamin C, and later had been one of the first to show just how adenosine triphosphate energizes the contraction of muscle. Almost every afternoon, Crick and Watson made their way to Gamow's. Brenner, while he was there, came along. "At Woods Hole I got to know Sydney; I had hardly seen him before," Crick said. One day, Gamow came in to give a lecture on the code, though it was sparsely attended; afterwards they adjourned to a bar and kept on arguing.

The arguments began with Crick's disproof. Gamow had wondered during the winter whether his idea could wriggle out of Crick's net with the help of the supposition that only the most vital proteins—namely, the enzymes—were fully determined by the genetic sequence, whereas others such as hormones were put together in some second stage. But the distinction was useless, for the instructions that enzymes would obey to make insulin, say, had to come from somewhere. "I did all possible versions of Gamow's code," Crick said. "I had to be rather ingenious and use very little evidence. In those days there weren't very many sequences. But Gamow's code being a rather restricted family, you didn't *need* much evidence to disprove it."

Crick started, once again, with the restrictions that Gamow's method of coding imposed on which amino acids could be neighbors. He then noted that certain imaginable sequences gave an amino acid the same neighbor on each side, in the form Smith between two Joneses, ABA, PUP, XYX, and so on. Only ten combinations of that form could occur, under the rules of Gamow's scheme. It turned out that no two of the ten could have the same amino acid in the middle position. But the B chain of insulin contained the triplet leucine+tyrosine+leucine, while beta-corticotropin had serine+tyrosine+serine. The two polypeptides could not both be coded by Gamow's scheme.

Crick urged on his companions two other simplifying assumptions of great audacity. The first was not merely that the sequence of nucleotides in DNA dictated the linear sequence of amino acids in the protein, but that the protein needed nothing more than that from the gene—that if the sequence was correct, it would fold into its three-dimensional configuration by itself. The idea seemed reasonable to those who came to the problem from the physical side; Hinshelwood had made the point clearly in the paper of 1950 that Crick, Brenner, and Watson had all read. It was perhaps less obvious to many biochemists. "The important thing Francis did was to take the *strong* line in the sequence hypothesis," Brenner once said. "It was a very ruthless simplification, that all you need is a one-dimensional sequence and the rest looks after itself." Secondly, they assumed, with some apprehension, that the genetic code would be the same for all living things. There was no evidence whatever for this; indeed, the very data in which Chargaff had perceived that the ratios of adenine to thymine and of guanine to cytosine were always unity also demonstrated that, except for those regularities, the nucleotide composition of DNA—that is, the cross ratio adenine *plus* thymine to guanine *plus* cytosine—varied widely from one species to another. Yet universality of the code seemed inevitable for an obvious reason: since a mutation that changed even one word or letter in the code would alter most of a creature's proteins, it looked sure to be lethal.

"The point, the contribution, of Gamow was that he made one realize that there was—or that there *might* be—an abstract problem that was independent of the machinery," Crick said to me. "By thinking of it as an abstract problem of going from one thing to another, you might be able to deduce something about it. As we know, as it turned out—you couldn't. You see. But nevertheless it drew attention to what it was you had to discover—namely what you would now call the genetic code. Which wasn't a thing which most people realized might exist." They went on, as they had been doing all spring, to think up coding ideas that might work where Gamow's did not. The limitations on their fancy were few. Amino-acid sequences had not been established for many proteins yet—though several labs had learned Sanger's techniques and new sequences were announced almost monthly.

As the discussions wore through the days, Crick began to impose analytical order by establishing the terminology. "Code," though Gamow hadn't used it in his papers, was already fixed. "It was only years after that that I found we ought to have used the word 'cipher'," Crick said. I said that Schrödinger had first made the comparison with the Morse code. "Yes, that's right," Crick said. "And of course we all knew about that; so maybe we just took it over—but that's a mistake too: it should be Morse *cipher*." The difference, to a cryptographer, is that a code has a separate arbitrary form assigned to every full word, and so requires its own whole dictionary for encoding and decoding, while a cipher translates letter by letter and so requires only a small key. "You understand, though, I didn't know that difference at the time. 'Code' sounds better, too. 'Genetic cipher' doesn't sound anything like as impressive!"

With four types of bases to encipher twenty amino acids, it had been obvious from the beginning that just two successive nucleotides were not enough to specify a single amino acid, for there were only $4 \times 4 = 16$ combinations. A triplet was required at least—and, although later, in the full delirium of the coding fever, arrangements were imagined that required four, or five, or even six nucleotides to code one amino acid, it was always thought that the triplet codes were most likely. The triplets might be "overlapping," so that three bases coded an amino acid; the second, third, and fourth bases coded the next amino acid; the third, fourth, and fifth bases coded the next, and so on. Gamow's code was overlapping, although because of the structural idea of the diamond-shaped holes with bases at each corner, the overlap was confusing to visualize. Or the code might be partially overlapping, the last of one triplet being the first of the next, or not overlapping at all.

Furthermore, since $4 \times 4 \times 4 = 64$, as many as forty-four of the possible triplets might code for nothing or be available for punctuation marks—for example, to show where on the strand of nucleic acid the sequence for a protein began or ended. Or, possibly, several different base sequences might signify the same amino acid: Crick declared that such codes should be called "degenerate," a usage by which quantum physicists describe multiple states that amount to the same thing, as when a set of electrons can take up several orbital arrangements, all equivalent in energy. "'Degenerate' I certainly invented. Only a physicist would invent it," Crick said. In such terms, he had begun to think of whole classes and families of codes.

Then one morning, Gamow began to get letters from people in Woods Hole thanking him for inviting them to his party. Within a few days he had received scores of acceptances. "The RSVPs kept *rolling* in," Crick said. "But Gamow had never even *heard* of the party. And eventually I had to be peacemaker, because it was Jim and young Szent-Györgi, Andrew, who had done it. And so it was arranged that Gamow supplied the whisky and *they* supplied the beer and the party was on. A few elderly ladies who had heard it was a hoax refused to come! In the end it was quite a good party!"

Sometime that summer, Crick suggested that Brenner ought to come to work with him at the Medical Research Council at the Cavendish Laboratory; Watson decided to visit Cambridge again for the latter half of 1955.

❖❖❖

One day at Woods Hole, Crick had a conversation with the geneticist Boris Ephrussi. Ephrussi had been around at the beginnings of things. He was born in Russia, but was a French citizen by naturalization; he was married to an American, Harriet Ephrussi-Taylor, herself a geneticist who had worked as a postdoc, for three years right after the war, with Oswald Avery at the Rockefeller Institute. In the mid-thirties, Ephrussi had introduced the young Jacques Monod to genetics. Ephrussi and George Beadle, in those years, had exchanged long visits to each other's laboratories in Pasadena and Paris. They had worked together on a tricky genetical and embryological puzzle about the development of eye color in fruit flies—tricky for those days, because they wanted to show that development of the normal, red-brown pigment required a pair of enzymes acting in succession that were determined by two separate genes, and to do so had to transplant unformed eyes from larva to larva. It was shortly after that, when Beadle had switched to the genetics of the bread-mold *Neurospora*, that he and Edward Tatum proposed the formula "one gene–one enzyme." More recently, Watson on his visits to Paris had usually stayed with the Ephrussis; he had told them about the structure of DNA before the first letter to *Nature* had been completed.

Sitting on the grass at Woods Hole, Crick was talking about genes and proteins, in particular about his assumption that they were colinear and Benzer and Brenner's plan to show as much, when Ephrussi took him aback by asking how he knew that amino acids were not put in their primary sequence by something in the cytoplasm, with the DNA in the nucleus merely determining the final, active, folded up configuration of the completed chains. "I don't think Boris necessarily believed it, but it was an idea he thought wasn't impossible," Crick said. "It was because of Ephrussi's remark that I realized that we had to show that a change in a gene made a difference in the amino-acid sequence of a protein."

❖❖❖

Crick also cast his skeptical eye over Watson and Rich's attempts to build models of RNA. "Of course, you realize that our ideas on that were totally wrong. We thought that RNA had some structure with the twenty cavities, it was that period. Mm-hm. Unfortunately people have forgotten what it is we didn't know at the time. Jim and Alex were trying to get x-ray data on RNA, which of course being, as we now know, single-stranded, they weren't having much joy. It was just a mess. And then when one thought about it one realized— You see, we could say, Well, since we can't find the structure of RNA from the data, let's do it the other way around by assuming that there are twenty different cavities and trying to build a structure that *had* twenty different cavities. And as soon as you put it that way, you saw that it was almost impossible to *do*. How would you make a cavity for leucine and a different one for isoleucine, you see?! Well, I suppose nowadays, with modified bases, we might have tried." In certain specialized situations, bases in RNA sometimes have extra atoms fringing the pyrimidine or purine rings. "But we weren't aware of those in those days."

With the problem thus turned around, so that it became obvious there was no way to build RNA or DNA in three-dimensional shapes that would differentiate among amino acids, late that summer Crick took the next great imaginative leap. He invented a new piece of anatomy. He threw out the holes and cavities and went back to the fundamental physical relationship of the two strands of the DNA structure—base pairing, the hydrogen bonds between the complementary sequences of nucleotides. He conceived that there must be a kind of small molecule, never yet detected by biochemists, that had a highly specialized function, and two ends. One end, he thought, would attach to a specific amino acid out of the twenty possible, refusing the nineteen other sorts. Next, the other end would match up to a place on the nucleic acid where the pattern of hydrogen bonds appeared that specified that amino acid. Certain details then sprang to mind. There would have to be a family of at least twenty different sorts of these small molecules. Doubtless there would be twenty different enzymes, too, to match and attach the small molecules to their correct amino acids—exactly the possibility that had occurred to Zamecnik but had then seemed too cumbersome. At that step, the small molecules could also be charged with the energy that would be needed at the next step to make the peptide bonds.

In effect, though Crick never put it this way, he had decided that since he could imagine no nucleic acid that had the shapes necessary to serve as a physical template, he would split off the function of Gamow's holes, which was to pick out amino acids and line them up, from the function of the nucleic acid, which was to specify the order of alignment. The problem of direct matching between amino acids and nucleic acids was shifted to a family of small molecules and enzymes. Perhaps the biochemistry was complicated—though twenty new enzymes were no more than one per cent of all that were known to exist in the cell—but the matter of getting information from DNA to protein was made cleaner, simpler, and far more flexible. These carrier molecules could accommodate almost any coding method.

Crick could not remember with conviction, twenty-one years later, the moment he had had the idea. He said, though, that at the end of August of 1954 he had gone to a Gordon Conference on nucleic acids. The Gordon Research Conferences hold a special place in science in the United States. The American Association for the Advancement of Science sponsors them, every summer, on various topics, in the idle buildings of several New England private boarding schools. The conferences each last a week, are attended by invitation, and have the aim of encouraging informal discussion—so that, for instance, no later quotation of another scientist's remarks is permitted. The one Crick had been invited to was in New Hampton, New Hampshire, from August 30 through September 3. The week before that, Crick, Watson, and Brenner all drove down to Cold Spring Harbor for the annual end-of-summer phage meeting. Among dozens of others, Meselson was there, and Brenner met him for the first time. Brenner presented Benzer's mapping of the *r*II region; Benzer himself was in Amsterdam for a conference (where, coincidentally, he met Delbrück, who read the first draft of Benzer's paper and was scathingly skeptical). Then Crick went back to New England, passing Rosalind Franklin on the road, for she had just finished a Gordon Conference on coal and carbon, the subject in which she had worked for years in Paris. At *his* conference, Crick talked about the structure of DNA. As best he could recall, he first thought of the small carrier molecules while he was being driven back from there to New York.

Then the pot was stirred. Crick sailed from New York on September 8. Gamow returned to George Washington University. Rosalind Franklin spent a couple of weeks at Brooklyn Polytechnic and talked with Crick before he left. She sent an urgent letter back to Desmond Bernal, her chief at Birkbeck College, London, to say that she ought to publish her recent work with tobacco-mosaic virus quickly—even though Bernal was leaving on a trip to Communist China—because she had learnt of three groups working on the virus in the States. Then she went to Pasadena to meet Pauling for the first time, and on up to Berkeley to see Stanley's lab. Watson and Brenner drove to Pasadena, making a holiday of it, stopping to visit Watson's family in Chicago and Luria at the University of Illinois. Alex Rich came East to work at the National Institutes of Health, in the outer suburbs of Washington, and kept in close touch with Gamow.

Brenner spent a fortnight at Caltech, then went up to Berkeley, where he worked with Stent in the virus lab for a month. Benzer wrote to say that his *r*II mutations were now producing "'clusters within clusters' and may be getting near the theoretical limits." Brenner began to think about overlapping codes. Not just Gamow's original code but any that overlapped, since neighbors shared some bases, put restraints on the possible sequences of amino acids. Partially overlapping codes were more flexible, since neighboring amino acids shared one base rather than two; degenerate codes were also flexible, since they offered several ways to indicate the same amino acid. Still, if significant restrictions on neighbors could be shown in the sequences that were known, it might be possible to solve the code from sequences alone, given enough of them. Or, on the contrary, if any amino acid could in fact be the neighbor of any other, that would prove that the code was not overlapping. The pairs of amino acids numbered, of course, a possible $20 \times 20 = 400$. The problem was not quite that big, however: in an overlapping triplet code, any sequence of *four* nucleotides, 1 2 3 4, determined a pair of amino-acid neighbors, 1 2 3 and 2 3 4; the possible sequences of four nucleotides were only $4 \times 4 \times 4 \times 4 = 256$. Still, to find 257 different pairs of amino acids in real life—more than could thus be coded—would take many more protein sequences than had yet been determined.

To demonstrate, however, that the pairing was statistically random would be a powerful lead, if not a rigorous proof—and that demonstration would soon be in reach, as new sequences emerged from the labs. Brenner squared off a twenty-by-twenty grid, listed the amino acids across the top and down the sides, and began entering a reference number in the box for each established pair. The form of the statistical argument was familiar to every phage worker: Brenner was looking for the numbers of citations in the boxes to take a Poisson distribution, which was the same test for fluctuation that Luria and Delbrück had used in 1943 on the numbers of mutant bacteria in test tubes to show that mutations arose at random. By late fall, enough sequences had been published so that the grid looked as if it had been sprinkled with confetti from a great height—and the distribution appeared to be just as random as the simile suggests.

❖ ❖ ❖

On the way home, Brenner stopped in Washington to see Gamow. Rich was there, with another friend of Gamow's, Martynas Yčas, a bacteriologist, and the three were planning several papers together about the coding problem. Brenner showed them his grid and his calculations. In England, Brenner stayed with the Cricks for several days. The two men talked for hours about coding and the ramifications of Crick's idea of small molecules that fitted amino acids to the correct sites on nucleic acids. Brenner made improvements. The hypothetical small molecules needed a name. Brenner suggested calling them "adaptors."

They also talked about getting Brenner a job in England, and Crick introduced him to Perutz and Kendrew. Then Brenner sailed for South Africa and a post as lecturer in the medical school of the University of the Witwatersrand. As soon as he arrived, he began setting up a laboratory to do phage research. He began writing letters. The intellectual isolation was painful, and South African politics oppressive; a fortnight after reaching Johannesburg, he wrote to a friend at Cold Spring Harbor, "It is worse here than I ever imagined in my most terrible nightmares. The whole country just vibrates with tension."

For the next two years, until Brenner was able to move permanently to England, he kept up a vigorous correspondence with Crick, Stent, Benzer, Watson, and several others he had met. One Saturday afternoon in Cambridge, in a book-lined seminar room, Brenner and I were sitting sipping from quarter-liter glass beakers that had been filled with boiling distilled water and a tea bag each, talking about the discoveries of the mid-fifties. Brenner got up and went next door to the office he shared with Crick, and I followed. He took three large cardboard boxes from a cupboard. "The fortunate thing here is, all this is due to my laziness," he said. "This is stuff that I brought with me when I came to Cambridge. And I never filed it. And that's why the girl never destroyed it." Several years earlier, the infamous girl, a secretary, tidying files, had thrown away all of Francis Crick's correspondence from the fifties. "So this is now complete, you see." He handed me the boxes. They contained the unsorted, chronological run of the hundreds of letters he had received in those years. Since his friends were eager to tell him what they were thinking and doing, the collection was gossipy about science and scientists, a fossil record of speculations and inspirations of which only the fittest have surviving offspring.

On 6 December 1954, Stent wrote to Brenner, saying, in part:

Dearest Sydneyboy!

The Lwoffs have just left, and I have a few moments between cocktail-parties to dash off a couple of lines, so that there'll be news from us awaiting you when you make your triumphant return to the Rand. Ta ever so for your Oxonian and Cantabrian letters which both Inga and your serviteur devoured with the greatest of that old gusto. Boy do we miss you here! You simply *must* come back to Berkeley. In future letters, I hope we can discuss in more detail just what we can do to bring you and your gang for an extended stay back to Bagdad-by-the-Bay. . . .

Jim & Leslie were up here over Thanksgiving; that business about how DNA makes RNA is well-nigh forgotten, because now they've got on to an even cleverer idea: DNA switches into an *active* form (the W C structure is only the *resting* form of DNA) by means of divalent ion bridges between the chains, exposing the hydrogen-bonding sites of the bases and creating holes into which amino acids could fit. . . .

So, Wolfgang Amadeus von und zu Witwatersrand, receive thousands of amitiés from your ever-faithful

Inga & Gunther

Crick wrote five sentences on 3 January 1955:

In haste. RNA tie-pin screed enclosed! Hope you had a good journey. I've been spending some time on 'coding' but have got nowhere. Am now writing it up for the RNA Tie Club.

On the twelfth, having received a letter from Brenner, Crick wrote again at greater length:

Sorry to hear about your difficulties. However, everyone is agreed that we must get you back here, and it is just a matter of waiting till you think the time is ripe, so don't take them too much to heart.

I have tried a little more coding, but have become discouraged. I have written a rather negative note for the RNA Tie Club, entitled "On Degenerate Templates and the Adaptor Hypothesis", which is being typed, and I'll send you a copy when it is ready. Within the limits I have explored—which are rather restrictive—I haven't found a code as good as yours. Most codes get defeated by the insulin modifications. I would suggest you write up yours, have it mimeographed and send it to the RNA Tie Club. This seems to be the only way that isolated members like ourselves can keep in touch. I suggest you also write up your table of neighbors, preferably in two forms, the first being restrictive, omitting collagen, silk, etc., and the second the complete one.

I heard from Jim just before Christmas. At that stage he had become fed up with DNA, and was trying to derive models for RNA. I find it very difficult to get a clear idea of what he and Leslie have done. He hadn't mentioned "active" DNA. Today I also heard from Alex, who has also been model building. He thinks that DNA *without* the methyl group of thymine [the carbon atom with three hydrogens filling it out, CH_3, by which thymine differs from uracil] codes for protein, and with it codes for DNA. . . . Again, all details are lacking. However he hopes to come here in the Spring. . . . Jim hopes to be here by early summer.

Shortly after that, Crick's paper was mailed to the membership.

6 "My *mind was, that a dogma was an idea for which there was* no reasonable evidence. *You see?!*"

❖*a*❖ The lineage of Cambridge science—Newton and Harvey, through Darwin, through J. J. Thomson, Russell, Hopkins, Hardy the mathematician, Rutherford, Dirac and Chadwick, Bernal and Bragg, to the biologists and radio astronomers of the present day—has been glorious, and Francis Crick is an ornament to it. Yet his relationship to Cambridge was ambiguous and edgy. Cambridge in its first impact on the young mind is a kind of space: a high Dutch sky and the horizon pitched low on the canvas; buildings and gardens that make it one of the world's most beautiful small cities, preserved from cloying by sudden changes of scale and prospect that I otherwise associate with Venice, and by cold light and a bony austerity. The countryside lies close in. The county people are reticent and thrifty. Cambridge thinks of itself as Roundhead to Oxford's Royalist. Cambridge is a place of overlapping circles and shut doors. It has established intellectual families—the Darwins, the Huxleys, the Keyneses. It has colleges of opulence, of wit and fashion, of crabbed antiquity. Cambridge has the exclusivities and merciless judgements of any university community, of Harvard, Göttingen, or the Sorbonne; any member can name the brilliant men who have left in bitterness. The town has no good restaurant: the good food and great wines are served at certain of the high tables every evening. Trinity College has won more Nobel prizes than France.

Crick was formed by none of these things. His family were provincial merchants, in the shoe business, of limited success, in religion nonconformists. None of his parents' generation went to a university. Crick grew up in North London. He was an undergraduate at University College, London, and began his research as a physicist there before the war. He came to Cambridge only in 1947, aged thirty-one. His college, as a research student, was Gonville and Caius—one of the old, large, rich colleges, with something of a tradition in medicine and the natural sciences, yet with a distinctly plebeian cast. Crick never lived in college. He worked for the Medical Research Council, never the university; he never taught, nor supervised research students. "But I must point out that I never felt in any way excluded from the *scientific* life of Cambridge," Crick said. He was a member of the two physics clubs that flourished

287

then, and of the Hardy Club, which brought biologists and physicists together. A few weeks after meeting Crick, Watson wrote, in a letter to Delbrück, "Francis has attracted around him most of the interesting young scientists in Cambridge and so at tea in his house I'm liable to meet many of the Cambridge characters like the cosmologists [Herman] Bondi, [Thomas] Gold, and [Fred] Hoyle."

Shortly before the discovery of the structure of DNA, Crick bought a narrow house with a high stoop and a stack of bay windows. It was one of a pair in a bend of Portugal Place, a pedestrian lane a minute's walk from the bridge and the round, Saxon church that mark the point of origin of Cambridge. Later he and Odile bought the house adjoining and partly knocked the two together, making a house of precarious steep stairs, irregular levels, and odd angles. He hung a helix, single, painted gold, above the front door. They gave lively parties there. After the Nobel prize, he was granted dining rights by one and another of the colleges. When, in 1960, Churchill College was opened, a new foundation in modern buildings west of the river, Crick was elected a fellow—and when thirty thousand pounds were given to the college for the purpose of building a chapel, Crick protested that a chapel was an offensive anachronism, and quit. In 1975, he was offered—rather, he was quietly asked whether he would accept if elected to—the Mastership of Caius. Odile was even shown around the Master's Lodge, which is one of the most charming in Cambridge. But Crick said no. In the fall of 1976, Crick went to the Salk Institute, in La Jolla, California, just north of San Diego, for a sabbatical year from low British pay and high British taxes; the following spring, he took a permanent job at the Salk and resigned from the Medical Research Council after thirty years and eighty-seven scientific papers.

In the mid-fifties, the standing of Max Perutz's group at the Cavendish grew increasingly anomalous. It was designated, with telltale awkwardness, the Medical Research Council Unit for the Study of the Molecular Structure of Biological Systems—biology forcing its way greenly up between the paving stones of a great physics laboratory. As the group got larger, and attracted visitors, space became cramped. Crick himself was nominally still doing theoretical crystallography. He and those working most closely with him at the Cavendish—as was true, too, for André Lwoff and those under him at the Institut Pasteur—had their important scientific traffic with the United States rather than at home.

❖ ❖ ❖

"On Degenerate Templates and the Adaptor Hypothesis: A Note for the RNA Tie Club," which Crick sent out early in 1955, is the first of his masterworks about protein synthesis and the coding problem. By 1966, he had written two dozen papers related to the subject. Six at least were of great and general importance. Two of those included experiments and were written with collaborators. One more paper, of pleasing ingenuity, happened to be wrong: nature turned out to be less elegant than Crick's imagination. Of the entire run, however, this first was the most unprecedented and original. The paper defined the next questions, and many were new questions. More, it established the way the questions were to be approached, and the terms in which they were to be argued. Most generally, it took for granted that the questions were spatial, physical, logical, easy to apprehend, and therefore tractable and even—in principle—simple. Yet even now, only a few hundred people have ever read the paper—for the surprising reason that it has never been published. It remained a note for the RNA Tie Club: seventeen foolscap pages, typewritten, double-spaced, mimeographed.

The paper is arresting from the moment one begins it: Crick wrote with easy energy, yet lacked nothing in that palpable authority that needs never raise its voice above the conversational. Unexpectedly for science, the tone is cheerful. He took advantage of the Tie Club to put off the academic armor; yet afterwards, when he was writing with full rigor for formal publication, his great papers were always direct, accessible, commanding—and always interesting. I know of no other biologist since the war who has been better at the writing of scientific papers. Crick's was the force of the clear, plain style, rippling with intelligence. He wrote in his own voice, and to be understood. Watson was a plain writer, too: like a jeep ride, efficient, unswerving, and jolting. Crick's writing, like his intellect, was as narrow, flashing, and precisely backed with weight as the blade of an axe. In every way, "Degenerate Templates and the Adaptor Hypothesis" is a document of extraordinary fascination.

At the lower right of the title page was an epigraph:

> "Is there anyone so utterly lost as he
> that seeks a way where there is no way."
>   Kai Kā'ūs ibn Iskandar*

Then Crick began:

> In this note I propose to put on to paper some of the ideas which have been under discussion for the last year or so, if only to subject them to the silent scrutiny of cold print. It is convenient to start with some criticisms of Gamow's paper . . . as they lead naturally to the further points I wish to make.

Crick then analyzed and put right the list of amino acids, and went on to several proofs that Gamow's particular coding idea could not be made to work. Some proofs were new. For instance, Sanger had begun sequencing insulin from animals other than cows, for comparison, and had told Crick of the results. And so:

> Another proof . . . depends on the A chain of two species of insulin. (F. Sanger. Personal communication, and in the press.) The sequences are identical except that one (sheep) has Gly. where the other (bovine) has Ser. The change occurs roughly in the middle of the chain. Both sequences cannot be coded by a Gamow scheme, since changing one pair of bases necessarily alters at least two amino acids, and this cannot be corrected without making further changes in the base sequence. . . . Thus to code both species of insulin A chains is impossible. A third method to disprove Gamow's scheme, given sufficient data, is to count neighbors. This is particularly useful in a scheme which does not distinguish between neighbors-on-the-right and neighbors-on-the-left.

---

*The tag was from a minor medieval Persian writer, Kaikā'ūs ibn Iskandar ibn Qābūs, Amir of Dailam on the Caspian, who had written for his young son a manual of practical advice for princes—don't crack jokes while playing chess; women are best for winter, but try boys in summer; if you must play polo, it's prudent to be goalkeeper; that sort of thing—which had newly been translated into English and published in 1951.

Using the data from the two insulin chains and β-corticotropin one finds 10 amino acids having 8 neighbors or more. Gamow's scheme (see his Table III) allows only 8 amino acids to have more than 7 neighbors. Thus coding would be impossible. . . .

I have set out these at length, not to flog a dead horse, but to illustrate some of the simplest ways of testing a code. It is surprising how quickly, with a little thought, a scheme can be rejected. It is better to use one's head for a few minutes than a computing machine for a few days!

Computers, of course, were still new to science, little known to biology, although five years earlier Kendrew had devised the first computer program for analysis of the spots in an x-ray-diffraction photograph. Crick then went beyond the particulars of Gamow's code to the general limitations on DNA or RNA as a physical template which had concerned him since the previous summer.

The most fundamental objection to Gamow's scheme is that it does not distinguish between the *direction* of a sequence; that is, between Thr. Pro. Lys. Ala. and Ala. Lys. Pro. Thr. . . . There is little doubt that Nature makes this distinction, though it might be claimed that she produces both sequences at random, and that the "wrong" ones—not being able to fold up—are destroyed. This seems to me unlikely.

That observation, made in passing, was the first acknowledgment of a theoretical question that is still unanswered: in general terms, what does the cell do with information it possesses on the DNA—and some organisms possess some DNA sequences in thousands of copies—that it does not use to code for proteins?

This difficulty brings us face-to-face with one of the most puzzling features of the DNA structure—the fact that it is non-polar, due to the dyads at the side; or put another way, that one chain runs up while the other runs down. It is true that this only applies to the backbone, and not to the base sequence, as Delbrück has emphasized to me in correspondence. This may imply that a base sequence read one way makes sense, and read the other way makes nonsense. Another difficulty is that the assumptions made about which diamonds are equivalent are not very plausible. . . . [Gamow's idea] would not be unreasonable if the amino acid could fit on to the template from either side, into cavities which were in a plane, but the structure certainly doesn't look like that. The bonds seem mainly to stick out perpendicular to the axis, and the template is really a surface with knobs on, and presents a radically different aspect on its two sides. . . .

What, then are the novel and useful features of Gamow's ideas? It is obviously not the idea of amino acids fitting on to nucleic acids, nor the idea of the bases sequence of the nucleic acids carrying the information. To my mind Gamow has introduced three ideas of importance:

(1) In Gamow's scheme several *different* base sequences can code for one amino acid. . . . This "degeneracy" seems to be a new idea, and, as discussed later, we can generalise it.

(2) Gamow boldly assumed that code would be of the overlapping type. . . . Watson and I, thinking mainly about coding by hypothetical RNA structures rather than by DNA, did not seriously consider this type of coding.

(3) Gamow's scheme is essentially abstract. It originally paid lip service to structural considerations, but the position was soon reached when "coding" was looked upon as a problem in itself, independent as far as possible of how things

might fit together. . . . Such an approach, though at first sight unnecessarily abstract, is important.

Finally it is obvious to all of us that without our President the whole problem would have been neglected and few of us would have tried to do anything about it.

Crick then confronted that near identity between the distances along the polypeptide and the polynucleotide chains whose significance had entranced everyone since Astbury. But seen in the way imposed by the coding problem—as Gamow had brought them to see it—Astbury's numerology was not a clue but a distraction.

I want to consider two aspects of the DNA structure. Firstly its dimensions; secondly its chemical character.

The dimensional side is soon disposed of. In the "paracrystalline" form of DNA (Structure B) we have one base pair every 3.4 Å in the fibre direction. A fully extended polypeptide chain measured about 3.7 Å from one amino acid to the next. Therefore it is argued that not more than one base pair can, on the average, be matched with an amino acid. If we go up the outside of the helix the position is worse, since the distance per base-pair is now greater, perhaps twice as great.

I want to point out that this argument, though powerful, is not completely water-tight.

Looking for a way out, he wondered whether, for example, DNA in action as a template might have its bases tilted closer together; after all, this was known to happen in the crystalline, A form. "Then, again, we have no evidence to tell us whether the *completed* part of the polypeptide chain stays on the template"—the point first raised by Dalgliesh, eighteen months before.

There follows the moment when the wind drops, the surface of the pool clears and is stilled, and one sees deeply with limpid simplicity into the nature of things.

As regards chemical character, I want to consider not only the DNA structure, but also any conceivable form of RNA structure. Now what I find profoundly disturbing is that I cannot conceive of *any* structure (for either nucleic acid) acting as a direct template for amino acids, or at least as a specific template. In other words, if one considers the physical-chemical nature of the amino acid side chains we do not find complimentary features on the nucleic acid. Where are the knobly hydrophobic [water-repelling] surfaces to distinguish valine from leucine and isoleucine? Where are the charged groups, in specific positions, to go with the acidic and basic amino acids? It is true that a "Teller" scheme, in which the amino acids already condensed act effectively as part of the template, might be a little easier, but a study of [known] sequences from this point of view is not encouraging.

I don't think that anybody looking at DNA or RNA would think of them as templates for amino acids were it not for other, indirect evidence.

What the DNA structure *does* show (and probably RNA will do the same) is a specific pattern of *hydrogen bonds,* and very little else. It seems to me, therefore, that we should widen our thinking to embrace this obvious fact. Two schemes suggest themselves. In the first small molecules (phospholipides? ions chelated on guanine? [he was imagining small structures that might attach]) could condense on the nucleic acid and pad it suitably, and the resulting combination would form the template. I shall not discuss this further here. In the second, each amino acid would combine chemically, at a special enzyme, with a small molecule which, having a specific hydrogen-bonding surface, would combine specifically with the nucleic acid template. This combination would also supply the energy necessary for polymerisation. In its simplest form there would be 20 different kinds of adaptor molecule, one for each amino acid, and 20 different enzymes to join the amino acid to their adaptors. Sydney Brenner, with whom I have discussed this idea, calls this the "adaptor hypothesis", since each amino acid is fitted with an adaptor to go on to the template.

The usual argument presented against this latter scheme is that no such small molecules have been found, but this objection cannot stand. For suppose, as is probable, that the small adaptor molecules are in short supply. Then consider the experiment in which all amino acids except one, (say leucine) is supplied to an organism, so that protein synthesis stops. Why do not the intermediaries—the (amino acid + adaptor) molecules—accumulate? Simply because there is very little of them, and no more amino acids can combine with these adaptors until the amino acids, to which they are at that moment attached, have been made into proteins, thus releasing the adaptor molecule. Thus under these conditions free amino acids accumulate, not amino acids-plus-adaptor molecules. . . .

In any case it seems unlikely that totally free amino acids actually go on to the template, because a free energy supply is necessary, especially when one bears in mind the entropy contribution [that is, as he explains in a moment, the cost, in energy drawn from elsewhere, of preventing random errors] needed to assemble the amino acids in the correct order. Free energy must be supplied to prevent mistakes in sequence being made too frequently.

The adaptor hypothesis implies that the actual set of twenty amino acids found in proteins is due either to a historical accident or to biological selection at an extremely primitive stage. This is not impossible, since once the twenty had been fixed it would be very difficult to make a change without altering every protein in the organism, a change which would almost certainly be lethal. It is perhaps surprising that an occasional virus has not done this, but even there a number of steps would be required. . . .

It is also conceivable that there is more than one adaptor for one amino acid, and the number 20 may be simply an accident (in any case we need a code for "end chain", so perhaps 21 would be more reasonable).

An adaptor molecule had to be highly specific in finding its correct amino acid and in promoting the next chemical step; it thus had something of the character of an enzyme—or, more precisely, of a coenzyme, one of the helper molecules, small and not themselves polypeptides, with which some enzymes must be supplied. It had much more the character of a snip of nucleic acid, for it had to match a location on the sequence of nucleotides, and the obvious way to do that—though Crick did not then say this explicitly, and perhaps did not quite see it yet—was to form hydrogen bonds with complementary base pairs as in the structure of DNA. Crick thought, then or soon after, that adaptors would be made of nucleic acid, although for a while he avoided being too definite about that: let the biochemists catch some first. He also thought they would be very small. Crick had found what might be called the theoretician's version of Hershey Heaven: if an idea works to solve one problem, try it on the next—for perhaps the most striking feature of the adaptor hypothesis was that Crick had generalized the principle of complementary base pairing, bringing it forward from DNA to this next step.

Twice more, in the next five years, an analogous intellectual feat was performed—the postulation, from theoretical necessity, of a new biochemical entity. The next was when Jacob broke out of the genetic impasse that called for the existence of the repressor molecule. The last was when the ideas of Jacob and Monod collided with those of Crick and Brenner to summon up the messenger molecule. Yet the adaptor hypothesis has a special standing. It was the first of these; and it was gratuitous where the others were exigent—for Crick conceived it a step in advance of outright compulsion from the evidence, while the other two resolved cumulative, painful paradoxes. To the adaptor, the appropriate response was curiosity and delight—"Aha!" To the repressor and messenger, the response was more nearly relief—"At last!"

That tribute paid, however, a doubt remains. How large was the element of good luck? To put it the other way around, how much is judgement colored by knowing that the adaptor hypothesis was right? Crick preferred it to the notion he mentioned in the paper, but put aside, that nucleic acid might be padded out by small molecules to form the physical template. Yet these bleached skeletons along the trail—Crick's padded nucleic acid, Gamow's diamonds, Watson's active and resting states of DNA, the litter of others that we don't happen to have picked up and dusted off—remind us that attractive ideas, after all, are cheap and much of the stuff of scientific genius is devising tests. Crick's idea was perfectly in accord with chemical principles; none the less, he proposed no test of the adaptor hypothesis.

As its first effect, the adaptor made the coding problem much worse. The spell of Astbury's numerology was broken. Crick began to realize, somewhat hesitantly, that almost any coding scheme was structurally possible. This was why, in his letter to Brenner, he had called his note for the Tie Club "rather negative."

Indeed, for those who read the paper, the meaning of "template" was subverted. Perhaps Crick should have discarded the word. He kept it, but where it had connoted the physical pattern on which amino acids were put in order, henceforth all that was essential to the term was the idea of the nucleotide sequence. Thus, as Seymour Benzer was driving the formalism of the classical gene down towards reality, Crick freed the nucleic-acid template of its most stringent physical limitation, allowing it to be thought about formally. The conceptual gap was being closed from both sides. We see that now: then, the implications of the change in view took many months to work through.

The idea of the adaptor "discourages a purely structural approach," Crick wrote in the next section of the paper, "and throws us back on 'coding', which, it is important to note, still remains a problem." He had spent much of the fall devising codes of the most general and ingenious sort, and now he offered half a dozen classes of these—not as solutions, but to illustrate the difficulties. One type after another, even those specially built to allow any amino acid to lie next to any other—"easy-neighbor" codes—could be proved impossible, surprisingly, with no more than the sequences of insulin and beta-corticotropin. Crick found himself forced toward the conclusion, though he confessed that he was unable to prove it, that no overlapping triplet codes were possible. Even worse, he wrote:

> The adaptor hypothesis allows other general types; for example, depending on a sequence of *four* base pairs. The insulin A chain data makes this unlikely, but it is difficult to disprove rigorously.

But that was not all:

I have tacitly dealt with DNA throughout, but the arguments would carry over to some types of RNA structure. If it turns out that DNA, in the double-helix form, does not act directly as a template for protein synthesis, but that RNA does, many more families of codes are of course possible. (Incidentally the protein sequences we use to test our theories—insulin, for example—are probably RNA-made proteins. Perhaps a special class of DNA-made proteins exists, almost always in small quantities (and thus normally overlooked), except perhaps where there are giant chromosomes.)

The aside is worth an aside: Crick's notion there was a holdover from a claim by Alfred Mirsky and Vincent Allfrey, biochemists at the Rockefeller Institute, that they had detected protein synthesis in the nucleus of the cell; the notion provides an irreverent illustration—like the bit of eggshell on the head of an eagle chick—of how the tatters of an old idea can cling to the new. Crick concluded:

> In particular base pairing may be absent in RNA or take a radically different form.... Without a structure for RNA one can only guess.
> Altogether the position is rather discouraging. Whereas on the one hand the adaptor hypothesis allows one to construct, in theory, codes of bewildering variety, which are very difficult to reject in bulk, the actual sequence data, on the other hand, gives us hardly any hint of regularity, or connectedness, and suggests that all, or almost all sequences may be allowed. In the comparative isolation of Cambridge I must confess that there are times when I have no stomach for decoding.

❖ ❖ ❖

Even as Crick was formulating the adaptor hypothesis, Mahlon Hoagland at Massachusetts General Hospital discerned how the energy necessary to make each peptide bond is supplied. Moving from Fritz Lipmann's lab down to Paul Zamecnik's, Hoagland had brought with him techniques for tracing with radioactive labels the exchanges of molecular fragments that take place, for example, when adenosine triphosphate is used for energy. Zamecnik knew already that ATP was required for protein synthesis. Having proved that synthesis took place at the microsomal particles, he thought it would be simplest if the interaction between amino acid and ATP also occurred there. The alternative was that each amino acid was activated—acquired its charge of energy—by exchange with ATP before it arrived at the microsomal particle. In other words, in the cell-free system for protein synthesis that they had built from components of rat liver, activation of amino acids would take place somehow in that ill-defined solution from the centrifuge tube that they called the soluble enzymes or the 105,000-g supernatant. "So I asked Mahlon if he would undertake to find out whether the amino acid was activated in association with the microsomes or the soluble enzymes," Zamecnik said in conversation. Hoagland put the methods he had learned with Lipmann to work in the cell-free system perfected in Zamecnik's lab. He soon showed that amino acids are activated by exchange with ATP before they reach the microsomal particles. A temporary intermediate is formed in which the amino acid is hooked to a molecule of adenosine monophosphate—nothing else than an ordinary adenine-bearing nucleotide—by an energy-rich bond. "It was a sort of surprise to me, when the results turned out, that it was the soluble enzyme system," Zamecnik said. "Because it seemed that then you'd need a special enzyme for each amino acid."

On 22 January 1955, aboard the Super Chief on the way to California, in pale green ink on the train writing paper, Gamow wrote to Brenner, saying, in part:

> I do not think I like your proposed code, because of the iregularities which you notice yourself. . . . If one is permited to change the coding procedure as one proceedes, than one can decode almost everything! If you want to have only three determining bases, why not to use a simple tringular code (which I call loose tringles) for a single-strangded (RNA-type) sequence of bases?

He drew a diagram of a saw-toothed template.

> As you code, loose tringles code is rather nonrestrictive and this represents the mane difficulty in any attempt to decode a reasonably long protein. But I have a strong feeling that ▲▲▲ code may turn out to be correct. . . .
>
> In coauthorship of Alex Rich and Martynas Yčas, I have just completed a chapter on protein decoding for the comming volume of "Advances in Bio & Medical Physics". Contains lots of stuff, but no solutions. . . .
>
> Both RNATIE tie and pin are in production. You can get your tie by sending $4.00 to Jim Watson, and pin by sending $6.00 to me.

Gamow closed with sketches of the tie, "hand made," and the pin, "inlade gold."

"Lots of stuff but no solutions"—the review of the coding problem that Gamow had written with Rich and Yčas, and which came out early the next year, was brisk and even lighthearted, but a chronicle of bafflement none the less, and a choice item for the connoisseur of logical games and puzzles leading nowhere. It catalogued the coding schemes that had caught Gamow's eye: his diamond code, his new triangular code (compact triangles or loose ones), Leslie Orgel's code of "major and minor determinants" that Gamow had mentioned in a letter to Crick in the spring of 1954, Edward Teller's stepwise code using the previous amino acids as part of the template—all from that same early period. Each of these codes had originated with an idea for a physical template to distinguish among amino acids. Gamow never achieved the abstraction into classes and families of codes, considered for the moment independently of mechanisms, that his initial papers had inspired in Crick. Each of these codes was offered only to be disproved; Gamow's diamonds by now had been killed by five theorems of increasing ingenuity.

The review served to put into general use the approach and terminology that had become standard in the RNA Tie Club. It ended by discussing the statistical reasoning that suggested that overlapping codes were unlikely—and it would have been a better paper, but a shorter and less amusing one, if it had begun there. Brenner was startled, when he read it, to find in full his own treatment of the statistical distribution of pairs of amino-acid neighbors, with only the leanest footnote of acknowledgment that he had suggested it. By then, however, Brenner had accumulated from the latest reports of sequencing nearly enough pairs for an elegant and absolute proof of the impossibility of all overlapping triplet codes.

❖❖❖

Mid-March, Watson wrote Brenner from Caltech. A visitor from Yale, Donald Caspar, had shown him some new results of x-ray crystallography of tobacco-mosaic virus, proving that its RNA was not jumbled into the core like the DNA in phage, but packed in some highly regular manner that, Watson said,

> places all of the phosphate groups at a common 22 Å radius. This was a surprise (I would have predicted a uniform core—I should mention E.M. [electron microscope] photographs definitely locate the RNA as a central core) but a useful one since it tells us that something is funny about the RNA and that in the virus it must have a different structure than the mess we find following extraction. So following a series of phony imperative deductions, I deduced a radically new type of structure. While it is mad, it is also very pretty and on aesthetic grounds should exist.

This structure was a zipper of two parallel chains, linked by extra oxygen atoms at the phosphate groups in the energy-rich configuration that Lipmann had schooled biologists to look for. Watson drew a diagram down the side of the page.

> Obviously unstable, I suggest that it only exists in viruses and microsomes and that following extraction the P-O-P snap and produced the bastard structure which Alex and I saw in the X-ray pictures. . . . Though idea is likely wrong, the structure is so attractive crystallographically, I think it must be used someplace. [Robert] Corey agrees by saying that if this is not RNA, then RNA will not be as pretty. . . . It could produce an interesting code but I dont want to think about this subject until structure is tested. . . . RNA structure must not be trivial or protein synthesis will remain unravelled for quite a few generations. We must get some clever chemical insite as I would guess that crystallographic approach is not enough.

RNA was not as pretty as Watson's newest structure.

Watson also wrote that when he rejoined Crick for the second half of the year, he intended to concentrate on the structure of plant viruses. The notion that these viruses resembled microsomes, suggested in this letter, survived longer. It was based on no more than the coincidence that both were small particles containing RNA, in intimate functional combination with proteins—and perhaps on the half-conscious hope that plant viruses might do for RNA what phage did for DNA. Yet there was little else to go on.

❖❖❖

Many of the American biochemists interested in protein synthesis got together at a symposium on the structures of enzymes and other proteins, held at Oak Ridge National Laboratory, in Tennessee, the first week in April of 1955. Gamow came to the meeting too, though he remained uncharacteristically quiet. Ideas were offered in unordered profusion: the discussion never attained the simplifying clarity that Crick had reached in the note on the adaptor hypothesis. The ruling idea at the symposium was that amino acids were assembled on an RNA template in a manner that was sure to be limited by the physical geometry—by what Henry Borsook, a senior biochemist from Caltech, called "the topography of the space created by the purine and pyrimidine bases." Most of the models proposed were variations of Alexander

Dounce's idea that had caught Crick's eye in 1952. Paul Zamecnik, however, reported the growing evidence that the first step in synthesis was activation of the individual amino acid by its encounter with an enzyme and adenosine triphosphate, to acquire an energy-rich bond. Coding problems emerged but despite Gamow's presence were never pressed. The terminology of the Tie Club was not used, not even "code."

Still, it's salutary to be reminded that the problems were perceived widely, not just by the circle in direct touch with Crick. Dounce spoke of the restrictions on nearest neighbors if amino acids were specified by overlapping "triads" of nucleotides. He proposed two ways that the restrictions might be eased. The first introduced the concept of degeneracy (though, again, not the term), whereby an amino acid might be specified by several different nucleotide triads. The second mechanism used a combination of nucleotides and the preceding amino acid in the chain—an idea that Dounce had conceived with colleagues at the University of Rochester, and that Gamow promptly told him had been thought of by Teller as well. At another point, Borsook put forward an idea for protein synthesis in which, he said, "some mechanism for transporting the activated amino acids to the template is called for. It might be the amino acid attached to the activating enzyme, or some smaller molecular weight carrier, a coenzyme, or nucleotide." But this was not, after all, the adaptor: it lacked the crucial insight that the carrier molecule must pick out the correct, specific sequence of hydrogen-bonding sites on the nucleic acid. Yet again, after Zamecnik had explained that the amino-acid-activating enzymes were found in the supernatant after the microsomes were centrifuged out at a hundred thousand times the force of gravity, Erwin Chargaff rose to ask whether this supernatant contained RNA, and whether the RNA played "a role, together with the microsomal particles, in what you consider to be evidence of protein synthesis." Zamecnik wrote to me in 1977 that Chargaff's had been "a prophetic question." But Chargaff did not mention base pairing.

The conclusion is inescapable: Crick in Cambridge and Brenner in Johannesburg were thinking well ahead of the biochemical pack. But then, about fifteen minutes later in that same discussion, Walter Sampson Vincent, an instructor in anatomy from the State University of New York at Syracuse, got up to report some experiments with the RNA of unfertilized egg cells of starfish. "Both Dr. Borsook and Dr. Zamecnik have suggested that there should be two RNA fractions in the cell, with differing characteristics," Vincent said. He had found the same thing himself, and proceeded to tell how, at length. His biological specimens—starfish eggs—were unfamiliar; his methods were the well-known ones of Torbjörn Caspersson and Jean Brachet (he had spent a year with Brachet as a postdoc); and worse than that, late in such a meeting, when scientist after scientist has risen to talk about *his* experiments, however tenuously related to the chief topic, the audience gets numb and drifts away. Vincent's data suggested, he said in conclusion, that the nucleus contained two classes of RNA, "one a soluble, metabolically very active, fraction, representing only a small portion of the total." His last words were about that fraction: "One exciting implication of the active, or labile, form would be that it is involved in the transfer of nuclear 'information' to the synthetic centers of the cytoplasm." This astonishing suggestion went unnoticed.

❖ ❖ ❖

The following week, a great deal of notice was attracted by a brief report of a new enzyme, the first to be found that linked nucleotides together into nucleic-acid molecules. In particular, when presented with ribose nucleotides this enzyme synthesized a strand of RNA. The report was read to a meeting of the Federation of American Societies for Experimental Biology, in San Francisco, by Marianne Grunberg-Manago, a young French biochemist of gypsy handsomeness with a vibrant deep voice. She had made the discovery at the New York University School of Medicine, working with Severo Ochoa, a Spaniard by birth and a physician by training who had come to the United States before the war and was now head of the school's biochemistry department. Ochoa, like many other biochemists, had been chasing down the interactions by which enzymes put to work the energy in energy-rich phosphate bonds. Grunberg-Manago had begun—she said in a conversation in Paris, twenty years later—with a crude bacterial extract to which she added the best available preparation of ATP, together with plain, inorganic phosphate made with the radioactive isotope of phosphorus, ^{32}P. She traced a pickup of hot phosphates into ATP, which said that some enzyme was acting, to make the exchange. But the level was very low. She set out to raise it. The best ATP had been none too pure; crystalline ATP of high purity was then first coming onto the market. She tried that in her system—and the activity stopped. "When I use *this* ATP, I have no exchange," she said. "So I think, that's fine, I have something interesting!"

She had a prime instance of what Delbrück once called "the principle of limited sloppiness." First she found that the radioactive phosphates had not been going into adenosine triphosphate at all, but into adenosine *di*-phosphate that had been contaminating the original preparation of ATP. This reaction was new, if not obviously significant. "I thought that at least it's something, might be it's a new enzyme, because we didn't know any enzyme which make an exchange incorporating phosphate into *A*DP." She and Ochoa decided she ought to try to isolate the enzyme. It was elusive. She did not even know what it was turning out as an end product. The work took the best part of a year. She considered dropping it. The break came when she found that the enzyme was not, after all, catalyzing the energy-rich bond in a small molecule that they had set out to find, but instead was making a polymer—a long repeating molecule—that had, in fact, the properties of a nucleic acid.

When Grunberg-Manago got the new enzyme reasonably pure, she found that given a supply of the diphosphate of any ribose nucleotide, the enzyme would string it into an artificial RNA that repeated the same base indefinitely. Thus the adenosine diphosphate she had started with was made into an artificial fibre in which adenine appeared over and over, AAAAA . . . ; it was accordingly called polyadenylic acid, promptly nicknamed poly-A. If the base in the nucleotide diphosphate was cytidine, out came a length of polycytidylic acid, or poly-C; if guanine, then polyguanylic acid, poly-G; and if uracil, then polyuridylic acid, poly-U. Chemically these were indeed ribonucleic acids— but entirely monotonous, like no RNAs that had occurred since life on earth began.

With intense excitement, they thought, of course, that they had found the enzyme responsible for assembling RNA in the living cell. The identification was mistaken, though for at least a year Ochoa would not abandon it. "Severo wanted to call it a synthetase; he thought it would be involved with nucleic-acid synthesis," Grunberg-Manago said. "But then I thought that it's *not* involved in nucleic-acid synthesis, because the polymer it produces is a random polymer." It was an obvious and easy step to show that when the enzyme was given mixed diphosphates (with, say, the bases adenine and uracil) it put them both into the artificial RNA (thus, poly-AU). The crucial point was that the bases appeared in no specific order. They agreed to call it "polynucleotide phosphorylase." What the stuff did in the cell was not obvious. "The role *in vivo*, it's *still* not known," Grunberg-Manago said. "But for me it was immediately very clear that one could test a lot of things with polynucleotide phosphorylase. It was a very useful enzyme."

❖ ❖ ❖

In May, Crick visited Paris, where among other things he met Monod and Jacob for the first time, gave a talk at the Pasteur about an idea that proved totally wrong, and had lunch one day with Boris Ephrussi and Harriett Ephrussi-Taylor. The conversation came back to Ephrussi's warning the summer before, about Benzer and Brenner's plan to demonstrate the parallel sequence of gene with protein, that in fact nobody had proved that a mutation in a gene caused a change in an amino acid. "From a very early stage, almost before I went to Cambridge, before we had the structure of DNA, I realized—I can't remember the exact date—that genes and proteins ought to be colinear; all right? Then we would have discussions about how we would *show* that they were colinear," Crick said. "I remember explaining to Jim even before I knew Sydney. . . . But what none of us realized— We *assumed* that it had been shown that a change in a gene altered an amino acid. . . . And it was only when Ephrussi said to me, 'You don't even know that'—implying it *could* be the way it's folded up—" (In the fall of 1976, Ephrussi mentioned the exchange at lunch, too. "It was in a little café called Cosmos, on the Boulevard Montparnasse opposite La Coupole," he said. "And I claimed that there was *no direct proof* that it's the proteins that are coded by genes. It's an amusing thing, every time now that I meet Francis, he asks, 'Ah, Boris, what is there that you don't believe today?'") Crick went on, "There was no evidence, you see; there was *absolutely no evidence* that a gene made a difference in the sense that you could actually determine the amino-acid sequence and show that with a mutation it was changed."

❖ ❖ ❖

At the end of May, Benzer wrote Brenner from Purdue. "I have been crossing mutants until I feel groggy," he said; the first full report on the *r*II region would appear in a few days in *Proceedings of the National Academy of Sciences*. "There are now so many mutants on the map that one can start looking at the thing through protein colored glasses, but no great truth have yet emerged from this. . . . But the mutation rates still make no sense in terms of simple-minded ideas (i.e., without recourse to very long range 'paragenetic resonance' or 'benzerine')."

❖ ❖ ❖

Mid-June, Stent wrote to Brenner. He had three important items of news to report from the virus lab in Berkeley, he said. A pretty technician whom Brenner knew had got married, alas. A colleague had been appointed to the post at Indiana University that Luria had left in 1950. Thirdly:

> Fränkel-Conrat seems to have done the biggest thing with TMV, since Stanley crystallized it. He can add soluble TMV protein to soluble TMV RNA, aggregate the whole mess into rods *of which 0.1% are infective*!!! Naturally, you don't believe it—neither did I nor anyone else, but unless he has made up the whole thing it now seems that it must be true. You can't beat that for laughs, can you Buddy?

Heinz Fraenkel-Conrat had taken particles of tobacco-mosaic virus apart, which had been done before—and then had successfully put them back together again. The virus is made of one long single strand of RNA and a large number of identical protein subunits—2,130 of these, it is now known, each a single polypeptide chain folded into a shape like the first joint of a thumb. As Watson had glimpsed in the summer of 1952, these subunits join up, side to side but slightly askew, to form a helix with all the thumbs pointing out. The overall shape is a rod with a hole down the middle like a short piece of macaroni. The subunits are held together by weak bonds between certain charged amino-acid side chains. The central hole is twenty angstroms across, and is empty. But the protein units, assembled, have a long continuous internal groove winding up beyond the wall of the central hole. In this groove lies the strand of RNA—which explains the discovery, which had intrigued Watson and everyone else, that all its phosphate groups lie at a common radius. By treatment with mild acid, Fraenkel-Conrat neutralized the charges that hold the protein subunits together, and the rod fell apart and stopped being infectious. He then separated and purified the protein. In a separate, parallel step, he treated virus particles with detergent to strip away the protein to allow the RNA to be recovered. Then he mixed the subunits and the RNA strands in solution once more—and got out normal infectious particles. With the electron microscope, Robley Williams confirmed that whole virus particles were there. The virus assembled itself: the architecture of the complete particle seemed to be an inbuilt consequence of the structure of the protein subunits.

The following year, 1956, Fraenkel-Conrat tried mixing protein subunits from one strain of the virus with the RNA from a different strain. He performed such crosses, and the reciprocal crosses, between several strains, and from seven different combinations he got back infectious particles. In each case, the hybrid virus particles produced a progeny that had reverted to a pure strain: in each case the protein of the progeny virus was the kind belonging to the strain that had donated the RNA to the hybrid. Evidently the RNA alone played the genetic role, able to dictate both its own replication and the construction of specific proteins. The same year, Gerhard Schramm, in Tübingen, separated pure RNA from tobacco-mosaic virus gently enough to be able to show that the RNA alone will give tobacco plants the disease and engender from their cells new, complete virus particles of the same strain. With these experiments, Fraenkel-Conrat and Schramm brought the viruses that contain RNA into the now all-inclusive family of organisms whose heredity is determined by their nucleic acids. They completed the grand sequence of experiments on the contrasting roles of nucleic acids and proteins that had begun with Oswald Avery.

❖ ❖ ❖

At the Cavendish at that time was Vernon Ingram, a British biochemist who had worked in protein chemistry, particularly with enzymes, at the Rockefeller Institute before joining Perutz's group. Ingram was trying to develop chemical methods for putting heavy atoms into protein molecules, the essential step in Perutz's method of x-ray diffraction, but found this tedious. Crick had been thinking about Boris Ephrussi's demand for proof that a mutant gene produced a change in the amino-acid sequence of a protein. In a footnote to the letter in January 1955 that said that the paper on the adaptor hypothesis was completed, Crick asked about something Brenner had mentioned of a mutant hemoglobin bearing a changed amino acid at the tip of one chain. "What was the story about a new end group for an abnormal hemoglobin?" In fact, no such thing had been found at that time. Then, that spring, Crick got Ingram interested in Ephrussi's challenge.

To apply Sanger's methods to the complete sequences of two all-but-identical proteins—say, a mutant enzyme and its wild type—would have taken several years if such large polypeptides could be solved at all in the state of the art then. Ingram thought a simplified technique might do. He and Crick first tried lysozyme, which is the enzyme, one recalls, in such fluids as mucus or eggwhite that protects animals by rupturing bacterial membranes; the chemistry of lysozyme was well studied. "Vernon Ingram and I spent some time trying to look at lysozyme, from different fowls, to see if we could find a difference that we could then pin down in the amino-acid sequence," Crick told me. "And we looked at lots of eggs, because lysozyme is easy to get from eggs—about a hundred eggs from ten distinct varieties of hens—and we never found any difference. We were using just a crude screening. We looked at duck lysozyme, pheasant lysozyme, guinea fowl—and we could easily pick up the differences between these. And I used to go into the lab every morning and *weep*—the lab assistant would produce an onion, and I would look at that, and I would *think sad thoughts*, because that helps, and we would take some tears—and it was easy to show that human lysozyme was different from chick lysozyme. But we never found a difference between two hens. And then Max gave Vernon some sickle-cell hemoglobin."

Sickle-cell anemia is a severe, at times painful, and often fatal disease that afflicts some African and American blacks, showing up in childhood. The red blood cells of people with the disease undergo a gross distortion of shape. In the arteries where the oxygen level of blood is high, the red cells of these patients are plump, soft, dimpled disks—as in a normal person they always are. In the veins where the oxygen level is low, the red blood cells of sickle-cell anemics change into long, somewhat stiff wisps of vaguely crescent or holly-leaf form—hence the name. Sickling destroys many of the distorted red cells as they are pushed through the narrow capillaries of the tissues—hence the anemia. In a crisis of the disease, the rigid cells cannot pass through the spleen, so the child's spleen gets clogged and cannot perform one of its vital functions, which is to make antibodies—and the child is then prey to massive infection. Sickle-cell anemia is hereditary. About two American blacks in every thousand have the disease. About eighty in every thousand, though, have red blood cells that will sickle, but only at much lower oxygen levels still. These people are said to have "sickle-cell trait"; rarely are they bothered by it, for in their veins only an insignificant percentage of the red cells sickle.

The nature of the disease was first understood by Linus Pauling, with what was even for him an astounding flash of physical intuition. Early in 1945, Pauling was appointed to a committee that was to recommend what the American government should do about medical research after the war. The committee's report suggested, among other things, setting up what later would become the National Institutes of Health. On the evening of 6 February 1945, Pauling was having dinner at the Century, a club in New York, with the others of the committee. Twenty-five years later, he wrote:

> One of the members of the group, Dr. William B. Castle, described some work that he was doing on the disease sickle-cell anemia. When he mentioned that the red cells of patients with the disease are deformed (sickled) in the venous circulation but resume their original shape in the arterial circulation, the idea occurred to me that sickle-cell anemia was a molecular disease, involving an abnormality of the hemoglobin molecule determined by a mutated gene. I thought at once that the abnormal hemoglobin molecules that I postulated to be present in the red cells of these patients would have two mutually complementary regions on their surfaces, such as to cause them to aggregate into long columns, which would be attracted to one another by van der Waals forces, causing the formation of a needle-like crystal which, as it grew longer and longer, would cause the red cell to be deformed and would thus lead to the manifestations of the disease.

Pauling said to me, in the spring of 1971 at Stanford University when we first met, that several things had prepared his mind for the instantaneous conclusion that sickle-cell anemia "must be a molecular disease—a disease of the hemoglobin molecule." He had been interested in the structures of proteins for years, of course. He knew that red blood cells contain hemoglobin, water, and little else; he had worked on the structural and chemical properties of hemoglobin, and its affinity for oxygen, before the war. He knew that different kinds of hemoglobin had been found in different species of animals, and could be distinguished by the shapes of the crystals they formed and by the fact that they provoked immune reactions in other species; and, he said, "I was sure that these different kinds of hemoglobin molecules are manufactured under genetic control." In the nineteen-forties, he had been trying to understand the physical chemistry of the interaction between antibody molecules and the molecules of substances that they attack and neutralize; for that reason, "I had been thinking about weak intermolecular forces all this time—and in this case between one hemoglobin molecule and another. So the whole picture seemed clear to me immediately."

Pauling asked one of his students, Harvey Itano, a qualified physician who was working for a Ph.D., to try to show a difference between sickle-cell and normal hemoglobins. The problem then was technically difficult. Itano drew blood samples from people with the anemia who had not recently received blood transfusions. He also obtained samples from people with sickle-cell trait, and of course samples of normal blood as well. He prepared concentrated solutions of the various hemoglobins. Eventually he found that a hemoglobin sample could be put in a phosphate-salt solution, saturated with carbon monoxide rather than oxygen, and tested by electrophoresis, using a technique in which the solution was placed in a tube with an electrode at each end and a slight electrical current run through it. When the current was switched off again, some hours later, Itano found that if the hemoglobin came from someone with sickle-cell anemia, it had moved toward the cathode, showing that the individual protein molecules had positive electrical charges. If the sample came from a normal person, it had moved toward the anode, showing that the molecules had negative charges.

Itano found, in passing, that he could detect no difference between the hemoglobins of normal blacks and normal whites. Furthermore, hemoglobin from people with sickle-cell trait (not the anemia) proved to be a mixture of the normal and abnormal molecules, roughly half-and-half. Comparison of how fast the two substances travelled in the electrical current established that each molecule of sickle-cell anemia hemoglobin had three—or anyway, within the limits of experimental error, from two to four—fewer free electrons than normal hemoglobin. Pauling and Itano, with two colleagues, Seymour Jonathan Singer and Ibert Wells, presented these results in a paper that Pauling read at a meeting of the National Academy of Sciences in Washington, D.C. in April 1949, and that was adapted and published in *Science* at the end of November.

In July of that year, James V. Neel, a specialist in human genetics at the University of Michigan, published a crisp, short paper proving from the incidence of sickle-cell disease and the trait in American Negro families that the cause was a mutant gene, inherited in a Mendelian manner. That is, on whichever pair of chromosomes was involved, one derived from each parent, the normal person carried a pair of normal genes, the person with sickle-cell anemia carried paired mutant genes, while the person with sickle-cell trait carried a normal gene opposite a mutant gene. Pauling and Itano and their colleagues cited the genetic evidence with a sniff: "Our results had caused us to draw this inference before Neel's paper was published." The chemical observation fit the genetics: in the person with sickle-cell trait, both genes directed the making of hemoglobin proteins. They had identified, as they well realized, the first "clear case of a change produced in a protein molecule by an allelic change [that is, a change in one of a homologous pair] in a single gene involved in synthesis."

Pauling did not speculate about how, once the mutation had arisen in the first place, a gene that was so disabling when doubled could have spread so widely. The explanation was offered several years later. The mutation seems to have spread centuries ago in a region of equatorial Africa where malaria is endemic. Malaria is caused by parasites—not bacteria but protozoa, larger and more complex one-celled creatures—and each spends part of its life inside a red blood cell. The person with sickle-cell trait—making both kinds of hemoglobin molecules—has greater than normal resistance to malaria: it has been suggested, though not proved, that the parasite, by itself using up oxygen, reduces the level in the hemoglobin around it to the point where that cell will sickle and burst and the parasite die.

In the five years after Pauling's and Itano's discovery, several efforts had been made to find the chemical basis of the difference between sickle-cell and normal human hemoglobin. The efforts had failed. Analysis of the amino-acid compositions of the two hemoglobins—the bulk proportions of valine, leucine, tyrosine, and so on—proved only that if a difference existed it was too small to be detected that way. The amino acids at the tips of the polypeptide chains were identified, and were the same in the two hemoglobins. The origin of the difference as a mutation of a single gene was undisputed.

"We *knew* about sickle-cell, and sickle-cell had a change of mobility. You see. So we *assumed* from that there's been a change of an amino acid," Crick said. "We didn't know that people had looked for it and hadn't been able to find it." But, Crick went on, after the talks with Ephrussi, "When I thought about sickle-cell, I realized you *can* get changes in charge due to folding" when some amino-acid side groups were exposed and others masked.

Perutz himself had seized upon Pauling's discovery in 1949. He had tried to compare the structures of the two hemoglobins by x-ray crystallography, but detected no difference, because he was using crystals of the hemoglobins in their oxygen-carrying forms. He looked into the molecular mechanism of sickling, and discovered the explanation: when sickle-cell hemoglobin was not carrying oxygen, it was insoluble, and for this reason precipitated into long, filamentous aggregates within the red cell. These were not so ordered as the needlelike crystals Pauling had imagined, but effective in deforming the cell and making it more rigid.

Ingram moved to the Massachusetts Institute of Technology in 1958; he and I talked at the Faculty Club there, over a plate of sandwiches and some beer, one afternoon about a week before I went out to the coast to see Pauling. When Perutz had offered a sample of sickle-cell hemoglobin, Ingram told me, three influences had come together in his mind. "I think it was Francis's strong genetic interest, and the interest in the chemistry of hemoglobin as represented by Max, that made this a reasonable project. And Fred Sanger was there at Cambridge as well"—although in the biochemistry department at that time. In 1955, Sanger published the final paper in his series on insulin, completing the details of the amino-acid sequence by showing how the two chains of the molecule were joined by the bridges pairing the sulphur atoms among their six cysteine residues.

Demonstrations in science can take on exceptional persuasive power when they seem clever, simple, and graphic. Some have also gained by a memorable name—for example, the Waring Blender experiment. Ingram's attack on sickle-cell and normal hemoglobin had all these qualities.

His aim was to break up the protein chains into short pieces, and then to see if any of the pieces varied between the two hemoglobins. He began with purified samples. These he heated nearly to boiling for several minutes, because that separates and opens out the folded polypeptide chains of a protein. As the samples cooled, he added to each a small amount of the digestive enzyme trypsin, and then held them at blood heat for a couple of days. (Later he learned that the digestion needed only ninety minutes.) The trypsin molecule breaks peptide bonds—but acts only on those that it finds next to either of the two positively charged and very similar amino acids lysine and arginine. In fact, trypsin will cut the backbone always on one side of the lysine or arginine residue and not the other. Enzymes that cleave with such extreme specificity are tools of obvious value: Sanger had relied on trypsin to chop insulin into a set, always the same, of pieces that could then be sorted out and analyzed separately. Each hemoglobin molecule contains about sixty of the bonds trypsin can break. Perutz had discovered, though, that hemoglobin molecules were made up of two identical halves. Thus, trypsin ought to chop the molecules up into about thirty sorts of pieces, and always the same thirty. The pieces would average about ten amino acids in length. "The point is that a *small* difference between two complete proteins becomes a relatively *large* difference in a short segment," Ingram said. "And that's the crux." He thought that with a combination of electrophoresis and chromatography, carried out successively on a single sheet of filter paper, a change as small as a single amino acid ought to be detectable.

Accordingly, when Ingram came in the second morning after setting his hemoglobins to be digested, he measured a drop of one sample and placed it close to one edge of a large sheet of filter paper, two-fifths of the way along. He marked the spot. He then put the sheet of paper between a pair of plates of glass, put the sandwich into a bath of mixed strong solvents—the pungent, slightly alkaline chemical called pyridine and the even more pungent acetic acid, the mixture on balance faintly acid and very smelly—and applied a light electric current by attaching electrodes to the two corners of the wet sheet of paper to the right and left of the marked spot. He did the same with each sample. He left them for two and a half hours. Then he took out the sheets of paper and dried them.

At this stage, if one of those sheets of paper had been doused by a chemical that stained protein, an irregular smear would have blossomed, confined along the bottom edge, from the marked point almost all the way to the left side. The smear would have been most dense at the place where the drop was first put down. Right of that, the smear would have stopped, except for one separate blob farther to the right. The electric current had done this. Each hemoglobin fragment had an electrical charge that depended on its particular amino-acid composition; in solution, the positively charged fragments were marched by the current to the left, moving the more quickly the stronger the charge, while the fragments of neutral charge stood at ease all together at the origin, and any negatively charged marched to the right. Various platoons of fragments overlapped—as Ingram knew, of course, without actually having to stain them.

To separate the overlaps, the next step was to give each platoon a half turn and move them all out across the parade ground. Ingram hung each sheet of filter paper from a string, with spring clips, so that its lower edge with the smear of protein was barely trailing in a trough of another mixture of solvents, this time alcohol, acetic acid, and water. As the liquids crept up the filter paper, they carried the hemoglobin fragments along, the lightest of them climbing fastest. When the wet reached the top, Ingram unpinned the sheets, laid them flat, and dried them again.

Stained, now, the protein fragments showed up on the paper as a triangular scattering of blobs, large and small, irregularly rounded, like the outlines of flat pebbles thrown down at the seaside. There were, indeed, thirty distinguishable short polypeptides. One side of the triangular scatter lay straight up from the origin, blob after blob like a backbone, and some still overlapping slightly: these were the fragments whose charge had been neither plus nor minus, and which had therefore stayed put in the first step. To the

INGRAM'S CHEMICAL FINGER PRINTS FROM NORMAL AND SICKLE-CELL HEMOGLOBIN

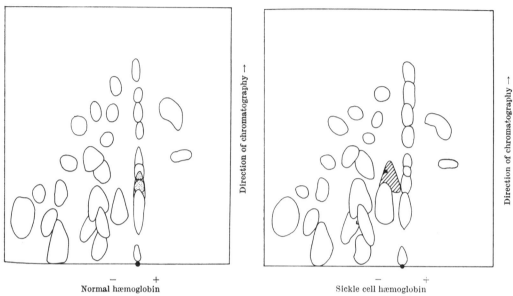

Normal hæmoglobin   Sickle cell hæmoglobin

FIG. 1 "Finger prints" of human normal and sickle-cell haemoglobins. Electrophoresis at pH 6·4, chromatography with *n*-butyl alcohol/acetic acid/water (3 : 1 : 1). The shaded and the stippled spots are those belonging to the peptide showing the difference

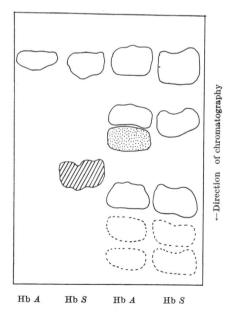

Hb *A*   Hb *S*   Hb *A*   Hb *S*

FIG. 2 Chromatogram of fractions of haemoglobin *A* and *S* digests obtained by large-scale paper electrophoresis. (a) Slowest moving positively charged fractions; (b) neutral fractions

right of that, outside the triangle, were two blobs, the only negatively charged fragments. Ingram found that when he ran the same hemoglobin several times, but with the quantities and timings strictly in parallel, the final patterns of blobs were so nearly the same that the papers could often be superimposed. He called such a chromatogram the "finger print" of the protein. The technique, and the name, are universally used today.

Sure enough, sickle-cell and normal hemoglobins produced different finger prints. On the chromatogram of the digest of sickle-cell hemoglobin, next to the vertical column of neutrally charged fragments but just to the left—thus betraying a slight positive charge—was a large peptide spot that had come out a vivid orange in color. No spot could be seen there in the finger print of normal hemoglobin—but close inspection showed the outline of an orange spot in the backbone of neutral fragments, partly obscured by others.

Chromatography is wonderfully flexible. The difference, once located, could be brought out more clearly. Using larger quantities, Ingram ran the electrophoretic step for each hemoglobin again, and then, with scissors, cut out from each paper a piece carrying the neutral blobs and another with the slowest-moving positively charged ones. Each of the four scraps of filter paper was then separately washed off into solution, and these put on fresh filter papers and put through the chromatographic step. There could be no question: the difference in amino-acid composition was there, and it was confined to one short length of the protein molecules.

"We were *sure* there was a difference, and Vernon found it," Crick said. "In fact, Vernon at one stage said he thought he had found *two* amino acids wrong. We weren't thinking much about overlapping codes by then, and we said, 'Nonsense, Vernon, you go back and look again; you'll find there's one!'" Crick laughed with great pleasure at the memory of his assurance. "So I think we were among the few people in the world, at that time, who realized that you should look for amino-acid differences, in protein sequences, due to mutations, you see. Genetically controlled. All you had to show was that it segregates in a Mendelian way, which had been done for sickle-cell. So it was a perfect case. . . . And you see, Vernon Ingram's discovery made an *enormous* difference. Because people suddenly realized that there was this connection. It now became obvious; and what was clear to us before it was done then became clear to everybody."

The work would now be done in a fortnight, including the hundred hen's eggs. Ingram began in the spring of 1955 and made the first announcement in a paper he read at a meeting in September of 1956. Only a few people were present at the talk—but one was a journalist, and the discovery appeared in *The Times* the next day. The first finger prints were published in *Nature* several weeks later.

❖ ❖ ❖

Mid-July 1955, Brenner got a nine-page, handwritten letter from Crick. "At last I find myself in the happy position of sitting down with your name firmly at the start of a letter. A thousand apologies for not writing earlier." The letter was abrim with news about science, neatly summarized by topics. These were, in part:

First, about coding. As far as I can see things are still as negative as ever (Incidentally the quotation Kai Kā′ūs—was genuine!). Gamow has conceived some enthusiasm for the "combination code", mainly because it gives different frequencies for different amino acids. But its so hard to think of a structural basis, and the supporting evidence is so weak, that I cannot really take it seriously. I have just received, but only read quickly, the final draft of Gamow, Rich, & Yčas. The basic difficulty, as they point out, is that so far there is nothing striking to explain. My own view is that "coding", etc, should be put on the shelf for a bit. I think it may help to *complete* the picture when more experimental facts are known, but I dont think it is likely to go much further on its own.

Re DNA structure. Nothing new, beyond what you have read in the Wilkins & Co letter in Nature recently. . . . I'm told Gamow has a paper in PNAS [*Proceedings of the National Academy of Sciences*] . . . which amounts to paranemic coiling. Jim says that Gamow sent it to Delbrück, who advised against publication. . . .

The Ochoa story, which I expect has reached you. An enzyme preparation which polymerises such things as ADP, to give, with the loss of one phosphate, a polymer i.e. "RNA". Will do it for other bases, and will produce copolymers. Jim says that Alex has got the X-ray powder diagram [that is, a diffraction picture, but not from a crystalline form and therefore diffuse]. . . , and it looks similar to RNA proper!! We have hopes of having Alex here before long (Jim arrived a few days ago. He has a job at Harvard, Biology Dept., but has leave of absence, and hopes to stay here for 6 months) so we may have the story in more detail.

The unit at the Cavendish was extremely crowded; besides Watson and Rich, another visitor was Donald Caspar, the crystallographer from Yale whom Watson had met the previous spring and one of the many who were after the structures of viruses that attacked plants. Crick's friend Leslie Orgel was also coming to Cambridge, to the Department of Theoretical Chemistry. More news:

The Benzer story: now published (just) in P.N.A.S, but I expect you have an advance copy. The cis-trans effect is very interesting. . . .

TMV. Casper [sic] has very reasonable evidence for a peak at a radius of 24 Å, and a hole (10 Å radius) in the middle. Both X-rays and e/m suggest the RNA in the middle. Jim & Casper think therefore that *all* the phosphates of the RNA are at 24 Å, and systematically making contact with the protein shell. They have a draft paper on this. I think the idea is very plausible in outline, but so far the details are vague (too many possible solutions). They are trying to extend the idea to the spherical RNA viruses, but I am doubtful about this. Rosalind Franklin has suggestive evidence that the protein surface of the TMV is not smooth, but rather like a (knobly) screw. . . . Jim is trying to push further work on both cylindrical & spherical plant viruses. . . .

Lysozyme. Attempt by Vernon Ingram & myself to find two hens with different lysozymes so far completely negative. We test *all* eggs on paper electrophoresis. . . . We estimate that we ought to pick up a difference of one charged group. We can *easily* pick up the differences between lysozymes of chick and human tears. . . . It is all rather discouraging. Even if we find a difference we shall still have to show its due to amino acid composition, and also do the genetics (which may mean doing the tears of cocks!) . . .

Visits: went to Paris to see Monod (about adaptive enzymes) and the Pasteur crowd.

That was the visit when Crick met Monod and Jacob and lunched with the Ephrussis.

I think the time is approaching when we should give serious thought to your coming back to England. . . . If only this lysozyme business would get started I would put up a case to the M.R.C. [Medical Research Council] for "molecular genetics".

Almost the same day, Brenner got a typed letter on RNA Tie Club letter-head, beginning, "Dear Val, This is the first official club circular," and listing all the members. There was a handwritten footnote, beginning:

Dear Stanley,
We have recently some progress (with Cys.) on statistical relation between aa's and bases in RNA. . . .

Yours Geo.

Mid-August, Brenner got another long letter from Stent, prattling on about old friends but thinner than usual about new science. There was a coding idea to report, though:

We had another visitation of Gamow (who is spending the summer in San Diego) . . . and he is going stronger than ever on the coding business. Now he has something which, if I understand the results correctly, is not entirely devoid of interest. Having become convinced by what he calls "the Brenner plot" and by Crick's *a priori* considerations that all classes of overlapping schemes are out and that therefore decoding is impossible, he now believes to have shown that a combination of three nucleotides corresponds to a single amino acid. . . . The reasoning involved is rather dubious but, just like his original decoding attempts, it could lead to better things.

The combination code sprang from the observation that the four bases, taken three at a time, combined to make twenty different sets if one supposed that the order of the bases in a triplet didn't matter—so that, for example, AAA specified one amino acid, but the six permutations CAG, CGA, GAC, GCA, ACG, and AGC were all synonymous for another. This was the first serious idea for a non-overlapping code. It yielded exactly twenty amino acids, but Gamow liked it even more because it suggested plausibly why some amino acids were rarer than others: they would be the ones specified by the less richly permutable combinations, like GGG or CCA.

Gamow had thought from his first involvement in the spring of 1953 that the natural occurrence of amino acids should provide the key to the code. Now he returned to that preoccupation. He and Yčas took the amino acids not merely in their rank order but in their approximate frequencies in protein, overall, and compared that with the frequency of the bases—or rather, of the mathematically possible triplets of the bases—as found in bulk RNA. To their delight, the two frequency distributions looked very similar. They published these details of the combination code later that year. Gamow got John von Neumann to help with an appendix to the paper, showing that these frequency distributions were not merely random—a surprisingly abstruse little theorem, and it represents the only significant involvement of any specialist in information theory in the main line of discussion of the biological code.

Crick had noted the possibility of the combination code in his paper for the Tie Club about the adaptor, but had dismissed it because its central idea, that the composition but not the order of the bases set the meaning of the triplet, posed bizarre structural problems. The criticism was correct. Crick had not known of the correlation of RNA and protein. He distrusted it when he did learn of it, and was again correct—though for a reason that he did not foresee and that was not understood until the crucial breakthrough five years later.

Stent mentioned, at the end of his letter to Brenner, that he and others of the phage group were talking of writing a book about molecular biology.

> Max, however, has declared such projects to be sheer nonsense at the present moment, when we are at the stage at which physics was in 1920, i.e. just before the break-through. Unfortunately he has been thinking like that since I've known him, i.e. for eight years, which would already place us at 1928. I thought that Hershey & Chase, Watson & Crick, etc. were not too bad for break throughs but apparently, not yet!

❖❖❖

August 30, Crick scribbled an air letter to Brenner, saying it was time to try to get him a job in Cambridge, so what salary would he like and what was the minimum he would accept? "This last figure in confidence to me. . . . No structure for RNA yet, but the problem smells hot to me." Brenner answered on September 10, enclosing a curriculum vitae and mentioning that he was ready to write up his proof that overlapping codes were impossible. Crick wrote back on October 7, with more energy than hope about the job. "We are overcrowded already to the bursting point" in the unit at the Cavendish; the only university teaching jobs open were junior ones, and the Medical Research Council was likely to prove reluctant to create another post.

> I did not pass on your idea for a salary. . . . My own is about £1500 [$4,200 in those days], plus children's allowances. I agree that one needs about this to live on, but that doesn't of course mean that you would get it! . . . My own view is that the important thing is to get you over here in a reasonable job. People will then get to know you (senior people, that is!) and you should be able to "better yourself" without difficulty after a few years.
>
> I'm sorry if this all sounds rather depressing. Just at the moment I have a rotten cold, as well as being grossly overworked, so dont worry too much. I'll write a more cheerful letter shortly! . . . Am looking forward to your Club screed.

Four days later, the position was overturned:

> My cold is almost gone, and I hope the rather dismal letter I sent you last has not depressed you too much.
>
> For reasons which are too complicated to explain I now feel that it is quite possible that we might be able to arrange for you to come to the unit here, though things would certainly be very cramped. Bragg's support would probably make all the difference. I can do little for a few days because Max Perutz is away with the local 'flu.

On October 20, Crick wrote again:

> I hope you will not think, from the rapid change between my last two letters, that I am constitutionally unstable in my opinion. The two main reasons for the change were
>
> 1. We heard from The Rockefeller that they were going to give us $40,000, spread over 4 years. This changes a position of acute financial stringency into one of comparative plenty, since it supplements our MRC allowance.
>
> 2. A senior member of the unit [it was Hugh Huxley] has decided to leave, thus making it possible for us to put up to the MRC a case for someone in his place.

> I therefore seized the opportunity to put a strong case that you should come here, and be loosely attached to me. This case has much force, because I have not previously asked for anyone to work with me. . . . Max Perutz & John Kendrew are now agreed that if we can arrange it you should join the unit. The major difficulty is now one of space, which still remains acute. . . . It is of course, understood that, whatever the arrangement [for laboratory space], you will be completely free to work on what you please. . . . I think the Rockefeller money will cover any reasonable demands for apparatus.

The following day, he wrote again, hurriedly:

> Your letters arrived the day after mine went off. I was delighted to hear that you were prepared, if necessary, to work in a cupboard! However in case there is only half a cupboard I have been exploring other possibilities. . . . Also let me have a sketch for a research programme—there is no need to stick to this, but I shall need it to make a good case. I was most interested in the rest of your letter, but comments will have to wait for the moment.

On November 4, at more leisure, Crick wrote yet again. Briefly and temperately about the job: recent restrictions on government spending, he warned, were going to make the application to the Medical Research Council "very difficult indeed," but he and Perutz intended to press it. "Don't, therefore, be *too* optimistic, but there is no need to worry about it yet." He then turned expansively to science: "The Unit, in fact, is full of activity (last year was quiet) and everybody is full of confidence." Structures of proteins were moving particularly well. Crick and Rich, after an intensive six-weeks' effort, had devised and written up two general structures for collagen—an enterprise that aroused great resentment in John Randall's laboratory at King's College London, where a group had been working on collagen for more than three years. Kendrew was making "great progress" with the structure of myoglobin; Perutz had been sick, but hemoglobin was now picking up, too. Rich was attempting, once more, to solve the structure of RNA, but making little progress; artificial RNAs made with the new enzyme seemed promising, though the molecules were too short to yield good fibres.

> The facts so far: poly-A gives a good X-ray picture (for one specimen, now lost!) showing very similar spacings to natural RNA, but rather different relative intensities i.e. a similar, but not identical structure. Poly (A + U) looks just like natural RNA. Poly (A + U + G + C) ditto. Present thinking is that, nevertheless, RNA is a two chain structure, crudely like DNA, though possible both chains running in the same direction. I expect we shall have a model-building blitz on RNA in a week or two.

Watson and Crick were considering what general rules might govern virus structures. "Last year My Favorite Experiment was the TMV reconstitution experiment, but I couldn't persuade the people here to try it. (Also to do it protein of one strain + RNA of another)." This idea Fraenkel-Conrat was even then preparing to test. Ingram and Crick were still baffled by lysozyme. "We have seriously considered the abnormal haemoglobins, but been too busy to start."

Brenner, though his side of the correspondence did not survive, had sent a program for research. Crick wrote that it looked fine: "Am extremely pleased to hear your ideas. Its clear that in their general direction they closely coincide with mine." Brenner had also asked about getting the adaptor hypothesis properly published; but, Crick said:

> I am submerged with writing up, and will be for months. I think all the ideas on the adaptor business, except the initial suggestion, are yours, so perhaps you might write it up, and just mention this in the introduction. In any case why dont you do a first draft? I might become a co-author if I contribute something to what you write. Jim is now anxious that he, Leslie and I should write a counter-Gamow note. Dont know how this will develop.

That was all there was to say about the coding problem.

Meanwhile Perutz had been to the London headquarters of the Medical Research Council to ask for Brenner's appointment.

On December 30, Crick wrote, beginning:

> Steady progress. Himsworth [Harold Himsworth, Chairman of the Medical Research Council] has agreed that I may "open informal negotiations" with you for "an appointment of limited tenure" (which means 3 to 5 years) on the Council's Scientific Staff, with salary, say £1150 per annum, starting January 1957.
>
> I need hardly say that I should advise you to accept.

The rest of the letter was taken up with practical information and suggestions about how Brenner should proceed. As the correspondence goes on to show, for the next twelve months Crick kept day-by-day watch over Brenner's move, explaining about finances and contracts, reassuring about the difference between reality and bureaucratic form, nudging the negotiations, securing an allowance for moving costs through Perutz, arranging housing and schools. Crick's bustling cheer never varied; his attentive kindness was unflagging.

❖*b*❖   "You must realize that the code, although it was *the code*, which was an abstract idea, was constantly of course having to be thought of in the actual biochemistry of protein synthesis," Crick said. "Because—well, you can see why, because they were two aspects of the same thing."

In Boston, in the fall of 1955, Zamecnik and Hoagland at Massachusetts General Hospital tried out a new idea with the cell-free mixtures of components of rat liver that they had used to study synthesis of proteins, and that in particular had brought them to discover the step in which a molecule of amino acid was activated by acquiring an energy-rich bond in interaction with a molecule of adenosine triphosphate. Now they thought to turn away from proteins, to see whether the same cell-free system could also be used to study synthesis of nucleic acids. The suspicion was abroad among biologists at the time that protein and nucleic acid synthesis "were somehow closely associated," Zamecnik said. And their cell-free system, after all, was a polyglot mix, consisting of ATP, microsomes, one or more amino acids, and some of the liquid, rich in dissolved enzymes, that was left after the microsomes had been spun out at 105,000 times the force of gravity. So they added ATP labelled with the radioactive isotope of carbon, ^{14}C, then took successive samples and extracted RNA by standard methods. "Sure enough, we got a labelled RNA out." That was exactly the evidence of RNA synthesis they had hoped to see.

"But the labelling happened so fast, we got a little suspicious." Fearing that the result might merely mean that they had failed to wash the RNA thoroughly, Zamecnik devised a control. They would put in ^{14}C built into a different component, which ought to wash away completely from the RNA. So to one flask, instead of the labelled ATP they added a small amount of the amino acid leucine, labelled with ^{14}C, and isolated RNA as before. Years later, Zamecnik wrote about the result: "Strangely enough, the RNA fraction was labelled from the amino acid precursor. In spite of careful washing procedures, the amino acid remained tightly bound to the RNA, and we concluded with some wonderment that it must be covalently bound to the RNA." At that point, with the sudden shift of focus that some scientists find embarrassing, some find exhilarating—but all find familiar—the control became the experiment.

Zamecnik assumed, at first, that the radioactive amino acid was linked to the long, heavy RNA strands of the microsomal particles. However, a colleague, Jesse Scott, reminded them that there was a fraction of the RNA of the cell, ten or fifteen per cent, that did not centrifuge down with the microsomes but remained in the liquid with the soluble enzymes. This RNA had to be in small, light pieces not to be thrown down; it had no known function, was tacitly ignored as garbage, and had never even been named. They began calling it "soluble RNA" and were soon writing it sRNA.

Then Zamecnik started doing the experiments over again—leaving out the microsomes. After centrifuging the crude liver preparation, he kept only the liquid above the pellet. To this he added ATP and radioactive leucine in extremely small amounts, and let the mixture incubate. When he isolated the soluble RNA after ten minutes, he found it just as radioactive as before, and the leucine firmly bound to it. The result was part of a ten-day run of experiments in his notebook in mid-November 1955. From the first positive experiment, Zamecnik later wrote, they regarded soluble RNA "as a possible new and dazzling" intermediate in protein synthesis. But, he went on:

> We were . . . anxious to remove all doubts that (a) the amino acid was covalently bound, (b) that the aminoacyl nucleotide [the two components as bound together] was a direct line intermediate in the pathway from free amino acid to peptide, and not a side path for storage of activated amino acids, and (c) that the entire complex polynucleotide—and not a much smaller triplet or oligomer [short chain] which adventitiously aggregated with this larger type of RNA—was the active intermediate.

Nevertheless, after November Zamecnik put the problem aside for more than six months. When I asked why, he gave two reasons. He was expecting a new associate, Lise Hecht, trained in work with RNA, to arrive that spring, though in the event she came in July; and he himself had just been made chief of the lab and got diverted by administrative matters.

❖ ❖ ❖

On 17 January 1956, Crick wrote "a brief note only" to let Brenner know that a formal job offer was on the way. Crick was planning to attend a series of meetings in the States the next summer, and hoped Brenner could get to them, too. "No further progress on RNA."

❖ ❖ ❖

Early in February of 1956, in Cambridge, Leslie Orgel gave a talk on the coding problem. "He did it very well," Crick reported to Brenner on February 13, and went on:

> This has started us off again, but nothing really exciting. We are looking into the problem: if it is three non-overlapping bases per amino acid, how does one get over the difficulty that there is no comma to show how the bases are grouped in threes? Will let you know how this turns out.

This was the beginning of the comma-free code—an idea of Crick's that was the most elegant biological theory ever to be proposed and proved wrong.

❖ ❖ ❖

On February 18, Crick went up to London for a symposium on nucleic acids, organized by the Biochemical Society. He did not give a paper himself. Among those who did, Maurice Wilkins reviewed at length the evidence, old and new, about the structure of DNA in fibres and sperm heads and with and without associated proteins; he concluded:

> It is only in the last year or two that any convincing demonstration has been made that protein synthesis involves nucleic acids, and that a structure of one of the nucleic acids (and possibly not the whole of it) has been established. . . . It is, therefore, not surprising that little of a constructive nature can be said at present of the implications of nucleic acid structure with respect to protein synthesis. . . . It seems unlikely that theoretical approaches can tell us much about protein synthesis until a great deal more experimental work has been done.

A moment later, Crick got up. His brief comment appears in astringent contrast with the preceding imaginative lethargy. Crick had already decided it was time to get the adaptor hypothesis into print—"I didn't really want to publish the tie-club paper, but it was just a nuisance having nothing that anybody could cite," he told me—and so took this opportunity.

Over the past year, the idea of the adaptor had become more exact. Crick began, as before, with the simple but sharply focussed observation that the double helix of DNA, and any likely structure of RNA, offered no obvious way to form highly specific templates that could distinguish, by charge and shape, between the amino acids. He went on:

> What a particular sequence of bases does provide is a highly specific pattern of sites for *hydrogen bonding*. It is thus reasonable to expect that whatever is absorbed onto the nucleic acids in a specific manner will be capable of forming a definite spatial pattern of hydrogen bonds. This can be envisaged in many ways. One possibility is that each amino acid is combined with its special "adaptor". These might be any sort of small molecule . . . but an obvious class would consist of molecules based on di-, tri-, or tetra-nucleotides. The fact that such molecules have not yet been discovered is not a strong argument against this sort of scheme, since they may be somewhat unstable. . . . The combination between each amino acid and its own specific adaptor could be made by a special enzyme, or enzymes, which would give a better opportunity for preventing "mistakes", such as valine occurring in an isoleucine position.
>
> This idea would thus be in line both with many typical biochemical systems and with the physical-chemical nature of the nucleic acids. The essential point is that each amino acid should have its own *specific* adaptor, capable of forming hydrogen bonds.

Rumor of Crick's proposal at that meeting reached Zamecnik in Boston.

❖❖❖

At the end of March 1956, six weeks after putting the adaptor hypothesis up for more public discussion, Crick was back in London for a three-day symposium on the nature of viruses, organized by the Ciba Foundation. Watson and Don Caspar came along from Cambridge, too. It was not a vast meeting—thirty-four scientists—but the sponsors had brought people from as far as Australia, South Africa, and California. Some there had been working with viruses since well before the war; others, like Crick, were more recently interested. Some worked with bacteriophage, some with plant viruses, some with animal viruses. Some were old acquaintances, like André Lwoff or Rosalind Franklin. Others Crick had not met before—and several of these were skeptical about this new fashion in biological thought, skeptical even yet that nucleic acids and not proteins were the genetic material of viruses, and all but openly hostile to the new men. To that audience, Crick presented, as the first paper on the first day, the general principles of virus structure that he and Watson had worked out in the previous six months.

Their argument was original and presumptuously elementary. Viruses were small. The plant viruses, containing RNA, were smaller than bacteriophage particles. They appeared in the simplest shapes—either rods, like tobacco-mosaic virus, or spheres. Viruses contained a severely restricted amount of nucleic acid. Indeed, the number of nucleotides they possessed, in every case for which this had been estimated, was a small fraction of the number of amino acids in their protein. With a triplet code, not overlapping, "an RNA chain of . . . 6,000 bases . . . can code for a polypeptide chain (or chains) of total length 2,000 amino acids," Crick said. "To form a protein coat, however, we need at least 10 times as much as this, and probably 20 or 30 times as much." In sum:

> The information required to synthesize the virus protein is contained in the RNA. As there is only a limited amount of RNA it can only carry a limited amount of information. Thus the protein molecules of the virus can only be of limited size. Rough numerical estimates show that this amount, used once, is not enough to produce a shell to cover the RNA. Thus the coat must be built up of identical subunits.

One notes the characteristic matter and manner. Six years later, in a letter to Jacques Monod, Crick said, in part:

> I am always slightly surprised that people do not refer more to our ideas on virus structure. The idea of subunits in spherical viruses had been suggested before, but it had made no impression. *All* the modern X-ray work on viruses came about because of our influence, including Rosalind Franklin's excellent work on TMV, which was based on Jim's evidence that the structure was helical. . . . The basic reason for subunits—that the information the RNA could carry is limited—was also quite original. At the time it even seemed daring!

Watson was intent, and Crick to a lesser degree, upon the similarity between spherical viruses and microsomes. The paper suggested in a sentence that microsomes might also be made up of identical protein subunits. But, they said, "In any case, since microsomes contain only a limited amount of RNA, we would predict, by an extension of our first argument, that no very large protein molecule will be found that is not an aggregate." No very large protein molecule of any kind anywhere in any organism—for unstated, behind the perception of the easy similarities between microsome and virus, was the easy certitude that a microsome was a specific intermediate between a particular gene and the protein it dictated. Functionally, the microsomal RNA in the cytoplasm was the executive organ of the DNA in the nucleus. One gene, one enzyme was beginning to be understood as one DNA sequence, one RNA-containing microsome, one protein chain. Though nobody said it, from this view the microsomal particle was very like an adaptor molecule in large. So Watson and Crick imagined that, structurally, the microsome, like the virus, was a shell of protein containing a highly specific packet of RNA. Watson wanted to crystallize microsomes like viruses for x-ray diffraction. Brenner soon wrote to Crick with the same idea. Watson and Crick's theory of virus structure proved correct, crucial, productive for virologists: its great success reinforced—not logically but psychologically—the comparison of virus with microsome structure.

❖ ❖ ❖

On 30 April 1956, Crick wrote to Brenner with a budget of science bulletins. Among them:

> I leave tomorrow for 3 months in the States. . . . I think when I return in August I should start looking into the business of a school for Jonathan [Brenner's stepson]. . . . Re coding. Leslie, John Griffith & I have deduced the magic 20, using a code having 3 bases to 1 amino acid. (Did I tell you about this). A RNA tie-club account of this will reach you shortly. It's clever, but I cant make up my mind as to its correctness. . . . Jim, Alex and I have a structure for Ochoa's poly A. . . . Looks promising. . . . RNA may be several of these twisted together, but the point escapes us. Our latest idea is that microsomes may have spherical symmetry & contain protein subunits. . . . Vernon thinks we have two chicks with different lysozymes. I am doubtful.

The note for the tie club was "Codes Without Commas."

What, I asked Crick, was one to make of the comma-free code? "Aha!" he said. "The nice thing about the comma-free code is that it was a very beautiful idea and it was wrong—and it's always nice to have one or two examples of that, just as a cautionary tale. It was *very* pretty. But you notice that we were always slightly reserved about it."

Crick and Orgel, that spring, had drawn into their discussions John Griffith, the theoretical physicist at the Cavendish whom Crick had briefly enlisted four years earlier to look for physical attractions pairing the bases in DNA. In the code they proposed, three nucleotides specified an amino acid without overlapping and with no restrictions on which amino acids could be neighbors. Their code, they wrote, confronted two difficulties. First, since four nucleotides make 4 x 4 x 4 = 64 different triplets, why were there not sixty-four different kinds of amino acids? Second, when any stretch of consecutive nucleotides was being read off, how were the correct groupings into three selected? This was the punctuation problem. If, for illustration, the four bases of a nucleic acid were indicated by four handy letters of the alphabet, the question was how the cell, synthesizing protein, avoids false readings between, say,

> ... ALA PAL APE ALP ...

and

> ... ..A LAP ALA PEA LP. ... ,

where the overall sequence was identical. Perhaps every gene was read beginning at a marked place. Maybe there was internal punctuation, by special bases or configurations of bases that did not themselves denote any amino acid. Or did the nature of the code, in itself, forestall ambiguity?

The three men had worked out a scheme that was infallible, requiring neither end-stops nor internal commas.

> Using the metaphors of coding, we say that some of the 64 triplets make sense and some make nonsense. We further assume that all possible sequences of the *amino acids* may occur (that is, can be coded) and that at every point in the string of letters one can only read "sense" in the correct way. . . . In other words, any two triplets which make sense can be put side by side, and yet the overlapping triplets so formed must always be nonsense.

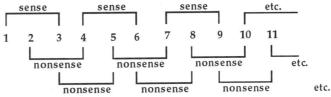

> It is obvious that with these restrictions one will be unable to code 64 different amino acids.

They then showed, with engaging simplicity, that twenty amino acids—but no more—could be specified by such a code. They examined, in two steps, what would happen when an amino acid followed itself in the sequence. To begin with, the triplet BBB, say, was obviously nonsense, because two of them together, BBBBBB, could be read mistakenly, for example as . . B, BBB, BB . , so that the amino acid would not be ordered up twice, after all. They thus rejected the triplets AAA, BBB, CCC, and DDD.

Sixty triplets were left. Once more they looked to rule out ambiguities when the same amino acid, and so the same nucleotide triplet, followed itself. Every triplet of the sixty belonged to a group with two others that were cyclic permutations of itself, like ABC rolling over into BCA and CAB, or DDB with DBD and BDD. Any one from such a group could make sense, but then its two cyclic permutations would have to be nonsense: "For suppose that we let BCA stand for the amino acid β; then ββ is BCABCA, and so CAB and ABC must, by our rules, be nonsense."

But the sixty triplets made twenty such cyclic groups. With only one triplet from each group making sense, twenty amino acids were the largest number that could be coded. And indeed:

> We can write down a construction which obeys all our rules and yet codes 20 amino acids. One possible solution is

where A B ^A_B means ABA and ABB, etc. It is easy to see, by systematic enumeration, that one can place any two triplets of this set next to each other without producing overlapping triplets that belong to the set.

The idea was "precarious" and unsupported by any evidence, they warned, brazening out the weakness. "We put it forward because it gives the magic number—20—in a neat manner and from reasonable physical postulates."

In particular, the comma-free code looked like a perfect marriage with the adaptor hypothesis—at which high point of fancy the paper became solemnly mock-physicomathematical. I know of no passage in the literature of the science that illustrates more vividly the attraction of this way of thinking—the seductive illusion that we can visualize, with perfect clarity, mechanical processes going on in a volume measured in hundred-millionths of an inch.

> To fix ideas, we shall describe a simple model to illustrate the advantages of such a code. Imagine that a single chain of RNA, held in a regular configuration, is the template. Let the intermediates in protein synthesis be 20 distinct molecules, each consisting of a trinucleotide chemically attached to one amino acid. The bases of each trinucleotide are chosen according to the code given above. Let these intermediate molecules combine, by hydrogen bonding between bases, with the RNA template. . . . Now imagine that such an amino acid–trinucleotide were to diffuse into an incorrect place on the template, such that *two* of its bases were hydrogen-bonded, though not the third. We postulate that this incomplete attachment will only retain the intermediate for a very brief time (for example, less than 1 millisecond [thousandth of a second]) before the latter breaks loose and diffuses elsewhere. However, when it eventually diffuses to the correct place, it will be held by hydrogen bonds to all three bases and will thus be retained, on the average, for a much longer time (say, seconds or minutes). Now the code we have described insures that this more lengthy attachment can occur only at the points where the intermediate is needed. If one of the 20 intermediates could stay for a long time on one of the false positions, it would effectively block the two positions it was straddling. . . . Our code makes this impossible.

Then the magic twenty laid to rest at last the ghost of Astbury's numerology:

> While the idea of making three nonoverlapping nucleotides code for one amino acid at first sight entails certain stereochemical difficulties, these are not insuperable if it is assumed that a polypeptide chain, *when polymerized*, does not remain attached to the template. A detailed scheme along these lines has been described to us by Dr. S. Brenner.

The point had first been made by C. E. Dalgliesh in *Nature* nearly two years before. Nobody noticed the subversive consequence: if the polypeptide chain did not stay on the template until it was finished, then in order for it to be complete every time, its assembly would very likely have to begin at a fixed, marked point from which it could next start to separate. Brenner's simple contribution rendered comma-freedom superfluous.

"We were always cautious," Crick said. "There was some lady who wrote a popular book who went absolutely overboard on it. A book called 'The Coil of Life' or something. Speaks of it as if it's an established fact." Hoagland also wrote of the comma-free code as though it were all but established, in an article about protein synthesis that appeared in *Scientific American* at the end of 1959. "But otherwise it *was* rather striking—after all, all you put in were four bases and you got the twenty amino acids coming out"—Crick snapped his fingers loudly—"like that!" The magic number was impressive, I said. "Yes, that's right; but as we said at the time, you need two bits of evidence," Crick said. "Because the magic number is enough to make you publish it, but it isn't enough to *prove* it. If we'd had a *second* bit of evidence which was strikingly suggestive, then I think we *would* have perhaps believed it much more than we did. So I was in the embarrassing position sometimes of finding that people believed it more than *I* did."

❖ ❖ ❖

Delbrück's warning to Stent in the previous summer—"We are at the stage at which physics was in 1920, i.e. just before the break-through"—was as nearly right as such analogies can ever be. Biology was heating up a pressure of discoveries, acute perplexities, ideas true and false, whose like had not been seen in science since the dark night of quantum physics, in Europe in the early twenties.

Crick, in his three months in the United States in the spring of 1956, spent six weeks in Alex Rich's lab at the National Institute of Mental Health, in Bethesda, Maryland, outside Washington. They worked some more on the structure of RNA. Rich had just found that a strand of poly-A would combine with a strand of poly-U and yield good diffraction pictures. The complementary bases paired spontaneously with total monotony, to form a double helix very like DNA, though fatter and possibly with the backbones running in the same direction rather than opposite ways. Crick then spent most of June and July going to conferences. The summer of 1956 was a good one for conferences, and communication was growing so intense—as had happened among the physicists in the twenties, too—that the conference net took on an importance it normally doesn't have. The Cold Spring Harbor Symposium, about genetic mechanisms that year, was merely curtain raiser to a formidable week at the McCollum-Pratt Institute at Johns Hopkins University, in Baltimore, on "the chemical basis of heredity," followed by the Gordon Conference on nucleic acids and then a shorter meeting, chiefly about proteins, at the University of Michigan in Ann Arbor: they merged all into one, a floating poker game with players dropping in or out. Crick joined the tour in Baltimore, where he gave a short, bored talk—little new to say—about the structure of DNA. Watson and Rich then spoke inconclusively about RNA. A month later, on July 17 in Ann Arbor, writing in pale blue ink on onionskin paper, Crick began a letter to Brenner that stretched to sixteen pages. "Let me very briefly deal with your last letter before passing on to the gleanings of the three conferences. I am anxious to get as much as possible on this down on paper today."

Twenty years later, Crick's letter is most striking for the contrast between energy and inconclusiveness: read in tandem with the Cold Spring Harbor volume and the fat book of the meeting in Baltimore, the letter is full of signs of the impasse that Crick himself did not yet recognize. "My own mind is at the moment fixed on protein synthesis," Crick wrote—and then his interesting news was about other matters. Stent had stirred up a lot of attention, mostly skeptical, by proposing a new structure for the nucleic acids that he thought explained, better than Watson and Crick's DNA, how phage reproduced in bacteria. Stent's structure put the two strands running in the same direction and paired the bases like with like. "This was the first pairing Jim ever tried," Crick wrote to Brenner.

> Gunther's idea is that DNA makes RNA . . . which then pairs with another similar strand of RNA. A new DNA chain is then made on this & the RNA chain is destroyed (to explain the distribution of mutants) and the DNA (old & new) then makes more RNA, and so on.
>
> Personally I dont like it, for a number of reasons. But Kornberg's system (see later) might decide.

Heterocatalytic, autocatalytic, the two faces of DNA: while Crick wanted to know how the gene dictated protein, Arthur Kornberg, at Washington University in Saint Louis, had been mapping the metabolic pathways by which the gene duplicated itself. Of all the reports that summer, Kornberg's was "by far the most exciting story," Crick wrote, and went on:

> It really looks as if he may have specific DNA replication in the test-tube. He works with "enzyme systems" as from Coli. He finds
> 1. He needs a primer. This primer is *intact* DNA. Even a *very* light digestion with DNAse destroys its priming ability. Good DNA from several sources works equally well, except for phage (T2?) which works only half as well (this may not be significant)
> 2. He gets *nett* synthesis: usually about 10 to 50% increase in the DNA used as a primer. The exact amount is not yet significant. . . .
> 3. the precursors are the (deoxy) *tri*-phosphates. . . . *All four are required at once.* Omit any one (or two) and the rate of incorporation drops from 50 to 0! (They had a lot of trouble with the original system, as they had varying amounts of the precursors there as contaminants)
> 4. RNAse (the enzyme) added both before and during incubation *makes no difference* (if they really have specific replication this disposes of Gunther)
> They have not yet checked if the base-ratios are copied. . . . Everybody was greatly impressed by the great competence of Kornberg. Its my belief that he really has DNA replication and that he will prove it within a very short time.

Genetics was the one other area moving smartly along, and its lines were by then familiar. Benzer had accumulated "about 1000" mutants in the *r*II region of phage T4. He was beginning to talk about genetic length measurable in nucleotides. He wanted to scrap the word "gene" and replace it with three new terms, "muton" for the smallest spot at which mutation could take place, "recon" for the irreducibly shortest length on the map that could not be split by a genetic recombination even at the fine scale he had reached, and "cistron" for the shortest stretch that comprised a functional genetic unit. (The last was derived from the mating tactic Benzer used to determine which mutations lay near each other on the map, which was technically called the "cis-trans test"—from the Latin prefix *cis*-, meaning nearside, as in Caesar's Cisalpine Gaul.) Crick rolled the new terms on his tongue. Thus, he wrote Brenner:

> Lederberg pointed out that if a "recon" is the position occupied by one base-pair one should not expect more than four different "mutons" there (i.e. corresponding

to the 4 bases) i.e. wild-type + three mutants. So far no more than this have been found.

However the mutation behaviour does not *fit* the W-C simple mechanisms. In general the back-rates [that is, the frequency of reversions from a mutant to the wild form] appear too fast, and the observed rates are all over the place.

Crick tried out a couple of explanations—unconvincing ones that signified only that he had begun to give thought once more to the mechanisms of mutation. (Over the next decade, Benzer's new terms came into a considerable vogue, especially "cistron." But the other two were superfluous once mutations and recombinations could be thought of in terms of base pairs, while the cistron was, in effect, the gene in its principal sense; it is the older usage that has lasted and the newer that has died away.)

Benzer had driven the fine-scale map as far, in principle, as he could by genetical methods alone. He told Crick he was trying to find the protein specified by the *rll* gene, and wanted to come to Cambridge to work with Brenner at last on colinearity, directly matching the map to the amino-acid sequence. Another phage geneticist Brenner knew, George Streisinger, who was working at Cold Spring Harbor, wanted to do the same thing with "Benzer mapping" of mutations in the protein of the tail fibres of phage. "I told both of them to come whenever they felt like it," Crick wrote.

"François Jacob gave a beautiful talk" on the latest work in Paris; "Most of this I'm sure you know, and in any case its in the latest Scientific American.... Its clear that the main problems of bacterial genetics are solved. Lederberg pointed out that he (Josh) only disagreed on a few very minor points."

Microsomal particles were "cropping up everywhere." Crick ticked off instances, beginning with Zamecnik. "(I assume you know his story. *Very* interesting).... I am madly enthusiastic about them, as I suspect they have a simple structure (akin to the plant viruses), and that they really are the key to protein synthesis," he wrote.

> Incidentally my own present picture of protein synthesis is that the "template" (which is inert) is the microsomal particles, and that there is normally a common intermediate, so that protein and RNA are synthesised together. However in certain non-growing cells (e.g. pancreas) this new RNA is rapidly broken down. i. e. *Most* of the RNA (in the particles) hardly turns over at all, but a very small fraction turns over very rapidly.

Not a word about coding. The subject had hardly come up that summer.

> This must be most of the news—if only in outline. There are so many things I should like to do that I can hardly wait for you to arrive. Is there no chance of your arriving a month or two earlier? (i.e. you might fly ahead of your family so that you could find a house for them (ingenious).) In any case we should start thinking about *equipment* now.... Please forgive me if some of this is a little incoherent.

❖ ❖ ❖

Crick did not mention a short, puzzling paper about the turnover of RNA in bacteria that had been presented at Johns Hopkins—squeezed in as an extra just before lunch on the last day—by Elliot Volkin, a biochemist from the Oak Ridge National Laboratory. Volkin and a colleague, Lazarus Astrachan, had been investigating the production of nucleic acids by E. coli during the latent period of phage infection. They knew, to begin with, that the bacteria showed no increase in total RNA during the course of infection, in contrast to the great increase in DNA as the genetic strands for the new phage particles were strung together. To trace what happened from moment to moment, they cultured their bacteria in a broth that contained very little phosphorus. Then, in the basic recipe for one-step growth, they added phage all at once. Three minutes later they added a supply of phosphate made with the radioactive isotope of phosphorus, ^{32}P. The fate of the labelled phosphates could be followed as they were used to build the backbones of RNA or DNA. Five and a half minutes after that, they added phosphate again, two hundred times as much as before—and not radioactive. The effect was to dilute the phosphorus in the culture so greatly that radioactivity would no longer be picked up significantly. The labelling had been done in a five-and-a-half-minute pulse. At intervals from the start of labelling, they took samples of the culture, spun out the bacteria, broke them open by grinding, and separated RNA and DNA.

They noted, first, that within a few minutes a lot of radioactivity appeared in RNA, much more than in DNA, but then tailed off quickly while radioactive DNA steadily increased. This by itself was no more than confirmation of what was known from Zamecnik's published work about RNA in protein synthesis—that, as Crick had said in his letter, "a very small fraction turns over very rapidly."

There was no ready suggestion of the function of Volkin and Astrachan's RNA. When they put the suspension of ground-up bacteria into a centrifuge, the significant proportion of the radioactivity went down with the pellet. Of the nucleic acids remaining in solution in the liquid above the pellet, only a small part were radioactive. This fact soon became extremely confusing in relation to what other labs were doing with the RNA that did remain in solution in that liquid fraction.

Volkin and Astrachan had turned up their most curious discovery, however, when they had analyzed the radioactive RNA to see what bases it was made of. They knew that the DNA of E. coli contained the four bases in almost exactly equal proportions. In contrast, the DNA of phage T2 was in the ratio of 32 thymine and 32 adenine to 18 cytosine and 18 guanine, for a skew of about one and three-quarters to one. The radioactive RNA that Volkin and Astrachan extracted shortly after the labelling pulse showed a skew, in the proportions of uridine and adenine to cytidine and guanine, that closely matched that of the phage DNA.

"We have adopted the working hypothesis that a uniform synthesis of a minor species of RNA occurs," Volkin said. "It is implicit in these findings that the isotopic turnover of the host's total RNA may be a summation of at least two types of RNA turnover: a more active type, similar to phage DNA in mononucleotide composition, and a relatively slower species, similar in composition to whole bacterial RNA." The experiments had met technical difficulties, he emphasized; the results might not yet be clean.

Although Volkin and Astrachan did not mention Jean Brachet's supposition, exactly ten years earlier, that RNA metabolized into DNA, they suspected that they had found in this minor species of RNA the direct precursor of new phage DNA. "Aren't you just saying that the RNA is converted into DNA without breaking up?" asked someone in the audience. "No," Volkin said. "We don't really have any evidence for that. The only thing we can say is that, if our calculated base composition of the 'active' RNA is valid, it is rather close to the composition of the analogous nucleotides in phage DNA."

Crick was not in that audience. He read the results when they were published later that year. The notion of an RNA precursor to DNA did not fit the biochemistry of DNA synthesis that Kornberg had elucidated. The Volkin and Astrachan RNA was an anomaly. It could not quite be dismissed as the inconsequential and self-admittedly sloppy work of people not well known at a lab not highly regarded.

❖ ❖ ❖

Crick did not mention in his letter, either, that towards the end of the Gordon Conference that summer he had given a talk. Zamecnik was at the conference; Hoagland was not. "I put forward two ideas," Crick told me. "One of them I think was quite erroneous, that microsomal particles might be like small viruses and have a lot of similar subunits; I *guess* that's what I said; and the other one I think was a rough thing of the adaptor. But it was very late and at the end of the meeting."

All that winter and spring of 1955–'56, realization had been growing, and not just in Zamecnik's laboratory, that an additional step had to be inserted between the amino acid's activation and its appearance at the microsome ready to be bonded into the polypeptide chain. Biochemists, however, approached this step not dry but wet—not as a problem of language translation and hydrogen bonds in Crick's terms, the physicist's terms, but as a problem of energy, enzymes, and the recombinations of substances in their murky flasks of liver extract.

Robert Holley went in by way of the energy. Holley was a chemist at the New York State Agricultural Experiment Station of Cornell University, in Geneva, New York, but was spending a sabbatical year at Caltech to learn about protein synthesis. He figured that the activated amino acid must react with something or be transferred to something—and that he could follow the reaction by using radioactive phosphates to examine more closely how the energy-rich phosphate bond was utilized. The technique was standard. That spring, using cell-free liver extracts, Holley found indications of a reaction that came immediately after the activation of the amino acid alanine. The new reaction could be stopped completely by the addition of ribonuclease, the enzyme that chops up RNA—though the activation step, just previous, was untouched by ribonuclease. Even before identifying what RNA was involved, or how, Holley sent off a report. It reads badly: the evidence was indirect, the argument inconclusive, the work palpably incomplete. The paper reached *The Journal of the American Chemical Society* on August 3, and was published six months later. First blood, by about a month.

Paul Berg went in by way of the enzymes. Berg was a young biochemist at Washington University in Saint Louis, in Arthur Kornberg's department, where the problem was the replication of DNA; he was already conspicuously surefooted, fast, and precise. As a postdoc, he had been excited by a new and unusual proposal by Fritz Lipmann and a colleague for a preliminary step in the synthesis of fats. But when Berg looked into the reaction Lipmann claimed to have found, he overturned it—and in the most embarrassing way, by proving that the reaction was really due to impurities in Lipmann's enzyme preparations. This got him interested in similar enzymes involved in protein synthesis. He was the first to purify one of the specific enzymes out of the mixture that catalyzed the activation of amino acids. This, as it happened, was an enzyme that activated methionine; he isolated it from yeast supplied by a local brewery. Then in the spring of 1956, Berg noticed that the step that activated methionine took up, molecule for molecule, a corresponding amount of the enzyme. But enzymes are catalysts and don't get used up; they perform over and over again. Since the enzyme was not released, he figured that the reaction could not have been completed by the activation step alone. So he set out to find some naturally occurring acceptor for the activated amino acid, without knowing what the acceptor might be—and without having heard about Crick's hypothetical adaptor. Later that year, Berg found such an acceptor that permitted the enzyme molecule to recycle; in the presence of the acceptor the amino acid became linked to something. When he purified the something, it was RNA—an unusually small piece, not part of the microsome, and apparently specific to the one enzyme and to the amino acid methionine.

Zamecnik and Hoagland, having begun by way of the RNA, were a step ahead. In the summer of 1956, they prepared radioactive leucine, put that into the cell-free system, and extracted and washed the leucine-labelled soluble RNA. Hoagland then added this complex to a fresh flask of the cell-free system. Within ten minutes, most of the radioactive leucine was no longer bound to sRNA, but had been incorporated into protein, at the microsome. They were writing the first papers on these remarkable findings at Christmas-time of 1956, when Watson paid a visit. That was the first time, Hoagland said in a conversation many years later, that he ever heard of the adaptor hypothesis. "In fact, I can remember vividly leaning over a centrifuge in the particular laboratory and talking with Jim about it, and his saying, 'This is the interpretation of your results'," Hoagland said. "And, I can *sense, palpably,* my feeling of resentment, at the time, that Jim would be telling me how to interpret my results—but also the feeling that, God damn it, he's right. You know. It was just—it was just right."

By the end of 1956, then, biochemists had converged from three separate directions on a molecular entity that appeared to function very much like Crick's adaptor. The soluble RNA molecules were an intermediate in protein synthesis, picking up activated amino acids, bringing them to the microsomes, and depositing them there. They were polynucleotides, the sort of molecule that Crick had greatly favored, though nobody had yet shown how they bound to any template. Specific soluble RNAs were being found for each kind of amino acid. Activation and correct matching of amino acid to sRNA was done by specific enzymes.

"Here was one of those rare and exciting moments when theory and experiment snapped into soul-satisfying harmony," Hoagland wrote in *Scientific American* three years later. And yet he and all the others had got their first results with no thought for the coding problem, no idea that theory called for any such discovery. With equal irony, Crick doubted, for several years, that soluble RNAs were the genuine adaptors. He thought they were much too big.

He recognized, of course, that as nucleic acids go they were very small, for otherwise the high-speed centrifuge would have thrown them down with the microsomes. But they were soon characterized more closely, and turned out to be about eighty nucleotides in length and to be comparable in molecular weight—on the order of 25,000 times the weight of a hydrogen atom—to some small enzymes. "We thought it would be a trinucleotide or a bit bigger," Crick told me, meaning himself and Brenner. "We certainly didn't think it would be as big as the present ones." Towards the end of 1957, Crick wrote:

> The RNA with amino acids attached reported recently by Hoagland, Zamecnik & Stephenson [Mary L. Stephenson, a collaborator] would be a half-way step in this process of breaking the RNA down to trinucleotides and joining on the amino acids. . . . The supernatant RNA appears to be too short to code for a complete polypeptide chain, and yet too long to join on to template RNA (in the microsomal particles) by base-pairing, since it would take too great a time for a piece of RNA twenty-five nucleotides long, say, to diffuse to the correct place in the correct particles.

"But that isn't valid at all," Crick told me; "quite an erroneous argument." He had got the mathematics of diffusion wrong.

Crick had ordered up a pair of handcuffs to hold a very small prisoner; when the handcuffs were delivered, they were complete with a large policeman. The adaptor had more to do than Crick and Brenner had originally considered. "When it was discovered, I remember Jim saying to me, 'Protein synthesis is solved!' And then he produced some idea, but I can't even remember what it was," Crick said. "Once we thought about it seriously we realized there was nothing against its being large; and we probably—though I'm really inventing this—we would probably have made the analogy with Kendrew's myoglobin, which after all does look rather large, if all you want to hold is a rather tiny oxygen molecule."

❖ ❖ ❖

Returned to Cambridge in the summer of 1956, Crick wrote again to Brenner, on August 21, all about centrifuges, balances, spectrometers, apparatus for chromatography, glassware, technicians and bottle washers.

> Now about problems: You will be pleased to hear that Vernon started on the abnormal haemoglobins while I was away. He has developed a "finger-print" method. He digests with trypsin; then puts the digest on paper and does an electrophoretic run one way and a chromatographic one the other. Just before I returned he found a spot (i.e. a peptide) in sickle-cell Hb [hemoglobin] which was not in normal Hb. He hopes to fnd the reciprocal, different, spot in normal Hb—he thinks he can see it—and then characterize both peptides [that is, analyze their gross amino-acid contents]. He also hopes to do a preliminary split into the separate chains. . . . If one gene (one cistron) corresponds to one polypeptide chain, one must obviously decide first which of the chains is changed by each mutant. I hope by the time you arrive this work will be well advanced. Incidentally we should like supplies of abnormal haemoglobins, including further supplies of sickle-cell. Are you in a position to arrange this? For exploratory work Vernon needs 5 cc of whole blood as a minimum, but naturally 10 or 20 cc would be better. If you have any sources of supply let me know and we will let you know about transport, etc.
> Assuming the abnormal Hb's go well we should then, I think, press on to

phage. This means obtaining the tail protein pure and I think we should start on this, as a long term programme, fairly soon.

About microsomes: I dont agree that its *essential* to have all microsomes synthesizing the same protein, but I do agree that it would be interesting to try this and also that rabbit reticulocytes [that is, immature red blood cells, which synthesize hemoglobin copiously and almost no other protein at all] are very attractive. We should certainly try them. (Remember, however, that Hb has more than one chain, and it may be that the rule is one microsomal particle[m.p.]→ one polypeptide chain.)

The mails were fast, Brenner's energy formidable. Nine days later, Crick wrote again, saying, in part:

All three Hb's would be very welcome. Cooley's anaemia is supposed to be due to foetal Hb, but I have a hunch it may be caused by an *alteration* in foetal Hb. [Cooley's anemia, today usually called thalassemia, is a hereditary disease in which the sufferer makes too little, or none, of one of the chains of hemoglobin.]

As to transport. The main trouble is that even if people are told not to freeze the cells, they usually do (on the plane) and then the Hb gets mixed with serum if whole blood is used. So would you please send *well-washed* cells. . . .

Looking forward to the news about tail protein. . . .

P.S. Seymour Benzer may come in Sept 1957, which would be fun.

Ingram sent off a report to *Nature* on the finger-print technique and the first detection of a peptide that varied between sickle-cell and normal hemoglobin; it appeared in mid-October. He went on to show that the glutamic acid in the sixth position along each of two chains in normal human hemoglobin is switched to valine in sickle-cell hemoglobin. Valine is neutral in charge; glutamic acid bears a free electron: the change accounts for the difference in charge on the hemoglobin molecules first measured by Pauling and Itano eight years before.

❖ ❖ ❖

Microsomal particles proved recalcitrant. When Watson got to Harvard in the fall of 1956, he began on the particles of *E. coli*, hoping to get their structure by x-ray diffraction. Since the earliest sightings of the particles, they had been reported in all sorts of sizes. Those first seen by Claude in the cells of chicks, with the optical microscope, were larger than those observed by Palade in the cells of rat liver but only with the electron microscope. Other people, working with extracts from *E. coli* and other bacteria, found that the particles sorted by size into sharply distinct classes, as shown by how fast and far they moved in the high-speed centrifuge. Similar results were obtained with cells from the spleens of mice, then with yeasts and pea seedlings. In September 1956, in Atlantic City, at the annual meeting of the American Chemical Society, among the hundreds of papers, a research fellow at Stanford University named Fu-Chuan Chao reported that microsomes from yeast were made of two unequal pieces, which split apart from each other unless a trace of magnesium was present. The ions of the metal, carrying a double positive charge in solution, formed bridges that linked the two pieces. Chao brought order for the first time out of the many reports about the sizes of microsomes, and showed what was necessary to keep them stable—namely, the trace of magnesium. But he also made it less likely that their structure could be simple.

❖❖❖

Late in September of 1956, Brenner mailed out copies of a note for the RNA Tie Club, "On the Impossibility of All Overlapping Triplet Codes." The proof was dazzlingly simple. It sprang from one crucial restriction entailed by every overlapping-triplet code—a restriction that everybody familiar with the coding problem understood perfectly well, but that nobody had thought about from just the angle Brenner proposed.

The trick was to reason back and forth between what was possible in a chain of nucleotides and what was known of particular chains of amino acids. Thus, in any overlapping-triplet code, reading along so that successive triplets shared two nucleotides—ABC, BCD; and so on—any given triplet might have any of four (but no more) different neighbors on one side and four on the other. That is, taking RNA's four bases, adenine, uracil, guanine, and cytidine, any triplet XYZ could be preceded by any of the four AXY, UXY, GXY, and CXY; it could be followed by YZA, YZU, YZG, or YZC. But this was a stringent limitation on coding possibilities. Examination of polypeptide sequences that had been solved so far would show, for each amino acid, how many different neighbors it was already known to have on the side next to the nitrogen atom along the backbone, and on the side away. The number of different neighbors one side or the other, whichever was more, would reveal—divided by four—the minimum number of different triplet synonyms that any overlapping code would require in order to handle that single amino acid.

In 1956, new amino-acid sequences were being published almost every week, and by that fall more than two dozen different protein chains had been analyzed at least in part. By listing all of them, with care to avoid ambiguities where glutamine might not have been distinguished from glutamic acid or asparagine from aspartic acid, Brenner demonstrated that cysteine, for example, had kept company with fourteen different neighbors on the nitrogen side and fifteen on the other, so that at least four different triplet synonyms were required to encode it. Lysine was the most promiscuous yet found, taking sixteen neighbors on one side—but seventeen on the other, so that at least five different triplets were required to specify it. Tryptophan, rare, had so far been detected with three different partners on one side, three on the other, and was the only amino acid that might still, possibly, be coded by just one triplet in any overlapping code. When Brenner listed all twenty amino acids and the minimum number of different triplet synonyms required to encode each, and added them up, the synonyms totalled seventy. "Hence all overlapping triplet codes are impossible."

On October 1, Crick wrote to Brenner, "just a brief note rather than a proper letter," saying, in part:

> The ampules arrived some time ago—I hope Vernon acknowledged them, etc. Your RNA Tie Club note arrived on Saturday. Leslie and I think your argument very neat. We are *slightly* doubtful about some of the sequences you have used, although we agree that the material is not unreasonable—but this is a very minor point. I have no doubt in my mind that the conclusion is correct. Leslie & I are writing up the "comma-less" code for publication, as many people seem to like it.

October 22, Crick wrote "just a brief note" about the money for equipment, and added that Benzer was set to arrive the following September. "Vernon Ingram is hard at work on the sickle-cell peptide. It may differ in *two* amino acids from the wild type." Three days later he wrote yet again, a long, typed letter, "all about phage tails"—the problems of obtaining large quantities of the tail fibres of bacteriophage so that the wild-type and mutant proteins could be sequenced and compared with a fine-scale genetic map. Sanger, Crick said, was trying to develop a way to analyze amino-acid sequences by using extremely small quantities of protein that had been grown with selected amino acids radioactive.

> I stressed to Fred how extremely favourable the phage system might be for this method, since the specific activity of the label will hardly be diluted at all. He seems very interested. . . . The main snag would be that the protein may turn out to be a large one. Incidentally this is all in a very early stage so you should treat it as confidential.
>   My queries to you are,
>   (1) Is there any *direct* evidence that Streisinger's locus [that is, section of the genetic map] (or any other locus . . .) controls the protein of the tail *fibres*? i.e. as opposed to the central plug. . . .
>   (2) Do you think that tail fibres can be got pure, and in what yield? . . . I know you are working on this, but I'm not clear how far you have got. Incidentally how much phage can one prepare in one batch, working in reasonable amounts (i.e. beakers but not vats)? How much of this do you estimate to be tail fibres?
>   (3) What do you think of Sanger's idea?
>   I'm sorry if these questions are a bit naive!

George Gamow had moved that fall to the University of Colorado, in Boulder. On November 24, he wrote to Brenner—a pale blue air letter addressed in turquoise ink, with his own new address in scarlet and the text pencilled half an inch high:

> Dear Val,
>   I have heard from His [Histidine, otherwise known as Melvin Calvin, a biophysicist in Berkeley] that you have written an astounding RNATIE letter, but my copy must have been lost in forwarding. Will you please send another copy to the above (permenent) adress.
>
> Yours as ever
Ala.
>
>
> P.S. Boulder is certainly a wanderfull spot!

The two papers for the RNA Tie Club, on the comma-free code and on the impossibility of all overlapping triplet codes, prefigured the ending of the coding problem as a separate and theoretical enterprise. At first, to be sure, the comma-free code was almost irresistible. It was the one theoretical coding idea that really looked as though it should work. It inspired not only popular but professional enthusiasm, to a degree that is acknowledged but grudgingly today. Crick was pressed to issue the paper more formally. Early in February 1957, Gamow, as a member of the National Academy of Sciences, sent the paper in to the *Proceedings* for the authors. "I was very reluctant to publish it," Crick said. "I only published it because I was asked by three or four people if they could quote the RNA Tie Club thing on it." He quoted it often himself. The comma-free code was the darling of biology—a beautiful creature without a penny of proof to her name—for a couple of seasons.

Brenner's paper had the consequence that without overlapping there were few restrictions, perhaps none, on amino-acid sequences. "This gave the theoretician of those days no real problem on which to work," Crick wrote ten years later. Gamow wrote to Brenner in April 1957, saying, in part, "I think your paper does prove the point, but it is a long time since I have completely lost the belief into the overlaping codes anyway. Of course, without overlap the thing becomes much less interesting from theoretical point of view." Brenner expanded the tie-club note slightly for publication; on June 4, Gamow wrote again:

> This is just to let you know that I have mailed your article for publication in P. N. Ac. Sc. Of course, everybody concerned is completely persuaded no that over-laping code is impossible, but it is nice to have a fullproof general proof of it. . . . I didn't get any new bright ideas on coding recently; apparently the problem is stuck just as the theory of elementary particles.

❖ ❖ ❖

Brenner arrived in Cambridge with his family early in January 1957, and rented part of the Cricks' house for several weeks. Crick himself, that month, travelled briefly to New York and Boston. Watson told him about the discovery of soluble RNAs, so he visited Zamecnik's lab and after excited discussions invited Hoagland, too, to come to England the next fall. The program was to match Benzer's techniques to Ingram's and Sanger's, using Streisinger's mutants, which affected the tail fibres of phage T2 at the tips, so that the protein could be lined up with its fine-scale genetic map, change for change. They would all join forces at the Cavendish. Meanwhile, the first step was to figure out how to separate and purify tail protein by the gram, again and again. Brenner threw himself into that, though he took the time to attend a course of lectures Perutz was giving on protein crystallography. Streisinger wrote, late in the winter, that only about one to three per cent of the total protein of the phage might turn out to be tail-tip. "We will have to grow lots of phage!" Not beakers but vats, after all—thirty-litre batches. On April 2, Brenner wrote to Hoover, Ltd., the manufacturer of household appliances, to ask whether a washing machine could be modified slightly for growing bacterial and viral cultures. "The machine we are interested in is the model with a heater and we do not require the wringer." Two days later, the company gave the lab a free machine with a stainless-steel tub.

Brenner sent an advance copy of a paper of his own on tail-protein genetics to Delbrück. He got back a magisterial discussion of its merits and overbold claims. "What are your plans now?" Delbrück asked. "To get at the *h* and *c* proteins [identifying the proteins by their labels on the genetic map] in some of the mutants, and to break the code? You should have quite some moral support for this next year in Cambridge, what with Benzer coming soon and Streisinger hoping to get there too!"

Other laboratories, by then, had developed similar ambitions. At a Gordon Conference on proteins, the first week in July, Benzer learned that at least four different groups were starting on phage proteins and fine-scale genetics to address colinearity and the coding problem. "Your remark about the battle being joined was more truth than poetry," Benzer wrote to Brenner. Another who had been at the same conference wrote that "phage is involved in somewhat of a furious feud between two protein chemists. . . . These two groups are rather bitter with each other because of priorities, etc, etc. . . . It is clear that the situation is not quite as relaxed as it was a few years ago."

❖C❖   In September of 1957, Crick addressed the Symposium of the Society for Experimental Biology. This was the society's twelfth symposium since the meeting in Cambridge, soon after the war, that had brought together everybody in Europe who was working on nucleic acids. The subject this year was the biological replication of macromolecules. Though someone did present a paper about the synthesis of sugar and starches ("There has been intensive activity and dramatic progress in this field during the past decade or more"), the symposium's substance was genes and enzymes, nucleic acids and proteins.

Crick spoke about the synthesis of proteins. His talk commanded the meeting exactly as William Astbury's had done twelve years earlier—stripping the complex down to the simple by extraordinary force and ease of language. "I have written for the biologist rather than the biochemist, the general reader rather than the specialist," Crick said in the printed version, published the following year. More than fifteen hundred papers on the chemistry of amino acids and proteins had been published in 1956; perhaps two thousand appeared in 1957. The one paper from the period that an amateur of science, or a professional for that matter, can read today for instruction and pleasure is Crick "On Protein Synthesis." Even the errors make the kind of sense they do make because of the principles that articulate the whole. The paper permanently altered the logic of biology.

"In the protein molecule Nature had devised a unique instrument in which an underlying simplicity is used to express great subtlety and versatility; it is impossible to see molecular biology in proper perspective until this peculiar combination of virtues has been clearly grasped," Crick said.

> It is at first sight paradoxical that it is probably easier for an organism to produce a new protein than to produce a new small molecule, since to produce a new small molecule one or more new proteins will be required in any case to catalyse the reactions.
>
> I shall . . . argue that the main function of the genetic material is to control (not necessarily directly) the synthesis of proteins. There is a little direct evidence to support this, but to my mind the psychological drive behind this hypothesis is at the moment independent of such evidence. Once the central and unique role of proteins is admitted there seems little point in genes doing anything else.

Crick, of course, had followed the biochemical development with closest attention, and we have followed him. He now ticked off the important things: twenty amino acids, their activation by specific enzymes, Zamecnik's first clues to the role of soluble RNAs, the imposing accumulation of evidence that microsomal particles were the jigs on which protein molecules were made, the plausible elegance of the comma-free code. This was the occasion when Crick asserted that soluble RNA molecules were so large they couldn't move fast enough and so must be precursors to trinucleotide adaptors yet to be found. This was also the time when he speculated more freely than ever about the similarity of microsomal particles and viruses.

What can we guess about the structure of the microsomal particle? On our assumptions the protein component of the particles can have no significant role in determining the amino acid sequence of the proteins which the particles are producing. We therefore assume that their main function is a structural one, though the possibility of some enzyme activity is not excluded. The simplest model then becomes one in which each particle is made of the same protein, or proteins, as every other one in the cell, and has the same basic *arrangement* of the RNA, but that different particles have, in general, different base-sequences in their RNA, and therefore produce different proteins. This is exactly the type of structure found in tobacco mosaic virus, where the interaction between RNA and protein does not depend upon the sequence of bases of the RNA . . . In addition Watson and I have suggested . . . by analogy with the spherical viruses, that the protein of microsomal particles is probably made of many identical sub-units arranged with cubic symmetry. . . .

It would at least be of some help if the approximate location of the RNA in the microsomal particles could be discovered. Is it on the outside or the inside of the particles, for example, or even both? Is the microsomal particle a rather open structure, like a sponge, and if it is what size of molecule can diffuse in and out of it? Some of these points are now ripe for a direct experimental attack.

Ingram's work with sickle-cell hemoglobin was fundamental to the understanding of protein synthesis, Crick asserted: it showed conclusively "that the gene does in fact alter the amino acid sequence. . . . It may surprise the reader that the alteration of one amino acid out of a total of about 300 can produce a molecule which (when homozygous [that is, when the mutant gene is doubled, causing the anemia]) is usually lethal before adult life but, for my part, Ingram's result is just what I expected."

The manner was deceptive. Like a gymnast who never seems to take a deep breath, Crick made the most agile leaps with nary a grunt to disturb the conversational tone. A twist brought Sanger and the fine-scale specificity of protein sequences to bear, across tens of millions of years, on the fundamental problem of Darwinism:

It is instructive to compare your own hemoglobin with that of a horse. Both molecules are indistinguishable in size. Both have similar amino acid compositions; similar but not identical. They differ a little electrophoretically, form different crystals, and have slightly different ends to their polypeptide chains. All these facts are compatible with their polypeptide chains having similar amino acid sequences, but with just a few changes here and there.

This "family likeness" between the "same" protein molecules *from different species* is the rule rather than the exception. It has been found in almost every case in which it has been looked for. One of the best-studied examples is that of insulin, by Sanger and his coworkers [Crick cited recent papers], who have worked out the complete amino acid sequences for five different species, only two of which (pig and whale) are the same. Interestingly enough the differences are all located in one small segment of the two chains.

Biologists should realize that before long we shall have a subject which might be called "protein taxonomy"—the study of the amino acid sequences of the proteins of an organism and the comparison of them between species. It can be argued that these sequences are the most delicate expression possible of the phenotype of an organism [that is, of the individual's observable characteristics as determined by its genes] and that vast amounts of evolutionary information may be hidden away within them.

The daily concern of evolutionary biologists has been to assess the degree of relatedness among various creatures; the eventual question is to judge the rate at which their ancestors diverged. Crick perceived that relatedness and divergence should soon be measurable by direct comparison of molecular sequences. In the twenty years since, the endeavor to put evolutionary theory on a molecular footing has become a subject of great subtlety and controversy.

But such matters served as the setting for two theoretical assertions. One was new and radical, the other familiar and radical. They appeared in five indispensable paragraphs:

> My own thinking (and that of many of my colleagues) is based on two general principles, which I shall call the Sequence Hypothesis and the Central Dogma. The direct evidence for both of them is negligible, but I have found them to be of great help in getting to grips with these very complex problems. I present them here in the hope that others can make similar use of them. Their speculative nature is emphasized by their names. It is an instructive exercise to attempt to build a useful theory without using them. One generally ends in the wilderness.
>
> *The Sequence Hypothesis.* This has already been referred to a number of times. In its simplest form it assumes that the specificity of a piece of nucleic acid is expressed solely by the sequence of its bases, and that this sequence is a (simple) code for the amino acid sequence of a particular protein.
>
> This hypothesis appears to be rather widely held. Its virtue is that it unites several remarkable pairs of generalizations: the central biochemical importance of proteins and the dominating biological role of genes, and in particular of their nucleic acid; the linearity of protein molecules (considered covalently) and the genetic linearity within the functional gene, as shown by the work of Benzer . . . and Pontecorvo [Crick cited their papers]; the simplicity of the composition of protein molecules and the simplicity of the nucleic acids. Work is actively proceeding in several laboratories, including our own, in an attempt to provide more direct evidence for this hypothesis.
>
> *The Central Dogma.* This states that once "information" has passed into protein *it cannot get out again.* In more detail, the transfer of information from nucleic acid to nucleic acid, or from nucleic acid to protein may be possible, but transfer from protein to protein, or from protein to nucleic acid is impossible. Information means here the *precise* determination of sequence, either of bases in the nucleic acid or of amino acid residues in the protein.
>
> This is by no means universally held . . . but many workers now think along these lines. As far as I know it has not been *explicitly* stated before.

Once information has passed from nucleic acids into protein *it cannot get out again.* If ideas were objects, with properties that suited their import, those words ought to sit on the page like a lump of moon rock on a dinner plate: plain, hard-edged, self-contained, perfectly ordinary, altogether extraordinary. Some scientific sentences do have that kind of existence: e = mc². Crick's original formulation of what he called the Central Dogma will strike the biologist who comes upon it now as too familiar to warrant gesticulation, while to the nonspecialist it must seem too curt to be remarkable—except for that outlandish title. The central dogma is no more identical with the whole of molecular biology than e = mc² comprises the theory of relativity. But if there is one statement from the new science that deserves the general currency of that equation of Einstein's, it is this assertion of Crick's. Most immediately and narrowly, the central dogma defined the difference between the functions— between the two kinds of specificity—of nucleic acids and proteins. In making this distinction, the statement was radical and absolute. Most widely, the central dogma was the restatement—radical, absolute—of the reason why characteristics acquired by an organism in its life but not from its genes cannot be inherited by its offspring. "Once information has passed into protein *it cannot get out again.*"

The central dogma has had a history of some contention. The sequence hypothesis has seemed, by contrast, familiar. The two were entirely dependent on each other.

What was one to make of the central dogma? "Well, that goes back—it's a funny period, and you must realize that I don't think I said anything more than what a lot of people were just assuming without thinking," Crick said. We were sitting in his office in Cambridge, again, on an afternoon late in November 1975; the weather was gray, the room seemed gray, and the trace of Brenner's cigarette smoke was bitter; Crick was in shirt sleeves in the midst of piles of paper and gray folders. We had been talking for nearly an hour.

How early, I asked, had Crick perceived the shape of idea that he came to call the central dogma when he thought of the term? "Well, I can't recollect that," he said. "It must have been in brooding over the coding problem, Gamow and everything. But you realize that what one was called upon to do for that symposium was to write a review article. To write a review article you have to put your ideas down on paper. You then express ideas which you hold but didn't know you *held.*" Who else had held the idea? "Sydney held it, you know; everybody in the club held it, you might say. *Nobody* tried to go from protein sequence back to nucleic acid, because that wasn't on. You see. But I don't think it was ever *discussed.*"

He paused for a moment, sat forward, rested his arms on the desk. "Well, what happened was this. Jim, you might say, had it first. DNA makes RNA makes protein. That became then the general idea. Then we didn't discuss ideas outside that. Well, then, a proper theoretician says, you're concerned with information flow: what are all the possible information flows? We've only been thinking about one of them; what other ones are there? What is often very sensible in making a theory is that when you've got the ideas, you then try and systematize them—not too early; you'll miss some—and then you see what it is you're really thinking. And you notice, we did exactly the same thing with Gamow's code. We took that, which to other people was—well, Gamow's code, with its details; but to us, it was a triplet, degenerate, overlapping *code.* Because we tried to *describe* classes of codes. And then you have a hypothesis which eliminates a whole class."

Why had he called it the central dogma? "Ah! That's a very, very interesting thing! It was because, I think, of my curious religious upbringing." He moved in his chair. "Because *Jacques* has since told me that a dogma is something which a true believer *cannot doubt!*" Crick laughed. "And indeed, a friend said to me the same thing at dinner last night, when I used the word 'dogma,' talked about the central dogma. But that *wasn't* what was in *my* mind. My mind was, that a dogma was an idea for which there was *no reasonable evidence.* You see?!" And Crick gave a roar of delight. "I just didn't *know* what dogma *meant.* And I could just as well have called it the 'Central Hypothesis,' or—you know. Which is what I meant to say. Dogma was just a catch phrase.

"And of course one has paid for this terribly, because people have resented the use of the word 'dogma,' you see, and if it had been 'central hypothesis' nobody would have turned a hair." And yet, I said, the catch phrase had certainly raised a claim of unusual status for the idea. "That's right. And probably rightly, as it turned out," Crick said. "It *is* sort of a super-hypothesis. And it's a *negative* hypothesis, so it's very very difficult to prove. It says certain transfers *can't* take place. That information never goes from protein to protein, protein to RNA, protein to DNA. It's not the same as the sequence hypothesis, which is much more explicit, and says that a certain transfer of information, the overall transfer from nucleic acid to protein, takes place in a certain way. The central dogma is much more powerful, and therefore in principle you might have to say it could never be proved. But its *utility*—there was no doubt about *that*. Because if you *didn't* believe that, you could invent theories, *unlimited* theories, whereas if you just put in that one assumption, that once the sequence information had got into the protein it couldn't get out again, well then, essentially you were on the right track, you see."

Crick said, briefly, that there had been challenges to the idea, which he had mostly ignored. "People would write articles and get thoroughly muddled. But in fact it's interesting, that. The central dogma isn't very clearly explained in that original paper, perhaps, yet it *is* sufficiently clear—to somebody who wants to see what it's about." He mentioned a piece he had published in *Nature* in 1970, clarifying the central dogma. There he had explained, among other things:

> The transfer protein → RNA (and the analogous protein → DNA) would have required (back) translation, that is, the transfer from one alphabet to a structurally quite different one. It was realized that forward translation involved very complex machinery. Moreover, it seemed unlikely on general grounds that this machinery could easily work backwards. The only reasonable alternative was that the cell had evolved an entirely separate set of complicated machinery for back translation, and of this there was no trace, and no reason to believe that it might be needed. . . .
>
> In looking back I am struck not only by the brashness which allowed us to venture powerful statements of a very general nature, but also by the rather delicate discrimination used in selecting what statements to make. Time has shown that not everybody appreciated our restraint.

"It just shows that even when you give people a good idea they don't always recognize it," Crick said.

Neither of us spoke for a minute. When Crick began again, it was in a more inward tone. "You realize that to get that idea, you had to—abstract a lot of things which, later, people then simply cluttered it up with again, things like control mechanisms, or what happened at the origin of life, or secondary, tertiary structures [that is, the folding up of protein chains], and things of that kind, you see.

"I think perhaps you should realize what the major contribution was, and I can't remember who made it or when: the idea that the folding up of the protein *followed from the sequence.* It was essential to have that idea. Anyhow, I think that there were a number of major simplifications which were necessary before you could see the problem; and the first major simplification was to realize that the important thing you had to do was to get the sequence right and after that the folding would follow. And you'll find that people like

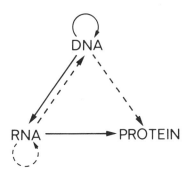

Crick's diagram of the central dogma of molecular biology, 1970. Solid arrows show transfers of information that occur in all cells; dotted arrows show transfers that can occur in special cases.

Haldane, and other people, didn't have that idea. I don't think Beadle had it either. Or anybody. I'm not clear when *we* actually had it, but once you had got the DNA structure, it was clear that you had a sequence of polynucleotides—and what could they code for but the primary sequence? They couldn't code for the folding very easily. It would be bizarre if they did. I mean, you *could* have made a theo—but it wasn't necessary.

"Mind you, I knew quite a lot about protein structure, much more than I knew about DNA: I was *working* on proteins, you see, and I had a very deep insight into proteins. And the *other* thing to realize was, as we discussed yesterday, that some of the amino acids were freaks and some were members of the club. Once you got those two ideas, then coding became a very sensible thing. If you didn't have that, you were just lost, you see. And the central dogma really set all that out. And it put the *additional* point about the *flow* of the information.

"Moreover, at the time, you realize, not so many people were thinking along those lines. I mean, much of the audience when that lecture was given. I ran overtime, and didn't get it over very well. And the ideas were pretty new. They weren't new, though, in the small group where people had been discussing it. So it was propaganda, really, you realize."

If Crick were to list the members of that small group? "Oh, that's very difficult, you know, because there were some that were in and some that were on the periphery and so on and so *forth*. Well, there was Jim; and I've forgotten when Wally Gilbert joined Jim [Walter Gilbert joined Watson in 1960]. And it would be fair to say that until I gave the lectures at the Rockefeller [a series at the Rockefeller Institute in January 1959], these ideas hadn't penetrated to Lipmann, and after the lectures, they had. And there was Alex Rich, who had been at Caltech and by that time moved to Bethesda; and the others who had visited here or were about to; and of course there were what you might call professional coders, like Martynas Yčas and Gamow and so on, mostly people who actually were members of the tie club—and not all of *these* were real, some were just friends of Gamow's who were seeing it all from a distance; and then there were people like Matt Meselson, for example. All the more perceptive people in the phage group, that would come to. Were thinking along those lines to a lesser degree. So you would certainly find that the ideas were quite familiar to Hershey, to Luria, to Delbrück. And the junior members, and so on. Because the phage group was a very imaginative lot of people and they *liked* ideas. People who were *less* receptive to them, probably, in the early days were the biochemists who were, say, working on protein synthesis. Because they felt they were doing real experiments and not all this theory, you see."

❖ ❖ ❖

Soluble RNAs were becoming a vigorous subspecialty within molecular biology. Summary can only hint at the ingenuity of some of the experiments, and the surprises are not ended yet. Berg went on to purify many more of the enzymes and RNA molecules, and led the way to the enumeration of separate activating enzymes and separate species of RNA acceptors for all twenty amino acids. Lipmann, Holley, and others were just behind him. Zamecnik's lab discovered that every sRNA molecule carried, at one end, the three-nucleotide sequence CCA—cytosine, cytosine, and lastly adenine—and that this was where the amino acid hooked on. Brenner's initial surmise that the adaptor had a tail proved perceptive. This bond, too, the last step before the making of the peptide bond, was energy-rich.

Crick himself went to the laboratory bench, the year that Hoagland was in Cambridge. "Francis just *absolutely hated* it. *Hated* the work," Hoagland said. "And I made him learn how to cut the heads off rats, in order to do experiments with rat liver—and he was chasing these rats all over the floor, and he was having a terrible time!" They, also, were trying to find where the amino acid attached, and whether there was more than one, on the soluble RNA molecule. "But that we missed," Crick said. "We did the experiment and the spot was run off the paper or something silly like that. So that—my contribution to experimental work in protein synthesis had been pretty well nil. In the upshot. . . . But the person who got the real proof that the concept was right was Seymour Benzer with Fritz Lipmann." Benzer, Lipmann, and several colleagues, including François Chapeville, eventually devised a direct demonstration that only the soluble RNA molecule, and not its captive amino acid, found the place on the template calling for that amino acid. The experiment was pretty and witty. First they used the cell-free system to make a small sample of the specific soluble RNA charged with cysteine—bearing its sulphur atom at the tip of the side group. Then they used a simple, textbook chemical reaction to knock off the sulphur atom, converting the cysteine to alanine—all the while held fast to the specific cysteine RNA. This molecular hybrid carried the alanine into the positions in the polypeptide chain where cysteine was called for. But that was 1962, when the code itself had been cracked if not fully broken.

Meanwhile, the molecules were changing their name. Soluble RNAs had been christened in general reference to the fraction of the cell-free system where they were found. By the early sixties, they were coming to be called "transfer RNAs," a name more evocative of their shuttling function. It's usually written tRNA.

Holley found a way to separate different species of transfer RNA. He purified a tiny amount of the alanine tRNA of yeast, then adapted Sanger's techniques for sequencing proteins. The single strand of this RNA molecule was seventy-seven nucleotides long. It took Holley seven years and platoons of graduate students to identify them all in order. A striking fact that made the sequencing possible was that nine of the bases were unusual ones—not the canonical A G C U, but variants like inosine, pseudouridine, or dimethyl guanosine that have different arrangements of the atoms in the fringes around the purine and pyrimidine rings. When Holley had the sequence—the first for any nucleic acid—he had the first clue to the structure of tRNA as well, for he found several pairs of short sequences where the bases matched up complementarily, pairing in the Watson and Crick manner to tack the strand together into a cloverleaf configuration. The amino acid attached at the stem of the clover leaf; highly exposed at the turn of the center lobe, farthest from the amino acid, were three nucleotides that paired with the coding triplet on the template. Other workers found the other tRNAs similar. Holley won a share in the Nobel prize in physiology or medicine in 1968.

```
                                        A OH
                                        C
                                        C
                                        A
                                  p G — C
                                    G — C
                                    G   U
                                    C — G
                                    G — C
                                    U   U
                                    G — C
                                        U      A G G C C  U U
           G A  U  G                           | | | | |       A
          U^h         G  C G C G  G_m           C U C C G G      G
           C          | | | |                   U^h        T ψ  C
           G          G C G C     G_m           G
            G  U^h A              —              A
                              C — G
                              U — A
                              C — G
                              C — G
                              C — G
                              U         ψ
                              U        I^m
                                 I  C
                                   G
```

Once distinct transfer RNAs could be separated and purified, people began to try to crystallize them for x-ray-diffraction studies. This proved hard: crystallography has always been at the mercy of balky substances. At last, suitable crystals were made from the phenylalanine tRNA of yeast. By the summer of 1975, Aaron Klug, at Perutz's lab in Cambridge, had an accurate map and a good model of this tRNA. Alex Rich, at MIT, was in hot and sometimes heated pursuit. Crick struck off the aphorism, "Transfer RNA is a nucleic acid molecule trying to act like an enzyme." He has speculated, most recently in a paper published with Klug, Brenner, and George Pieczenik, in the spring of 1977, that transfer RNAs had a crucial part in the origin of life and of the biological code. "That will be a very important paper if it turns out to be right," Crick said that June. "I'm keeping my fingers crossed about it."

The similarities between microsomal particles and the small spherical RNA viruses were melting under scrutiny. In Cambridge a few weeks before the discovery of the structure of DNA, Watson, at Manny Delbrück's suggestion, had introduced himself to Alfred Tissières, a Swiss biochemist trained at Cambridge who had just spent a year and a half at Caltech. Tissières, even then, was interested in microsomes from bacteria. In the spring of 1956, while Watson was again in Cambridge, working on spherical RNA viruses, he invited Tissières to come to Harvard to investigate microsomal particles in *E. coli*. Tissières arrived at Harvard in February 1957, to find Watson setting up his laboratory from the bare walls of the space allotted to him in the Biological Laboratories. When they got started, they found that the microsomes in *E. coli*, even though smaller than the particles in higher organisms, also split into fragments of characteristic and unequal sizes unless stabilized by a trace of magnesium. The larger component weighed about two and a half times the smaller. The particles were two-thirds RNA, the rest protein, and the separate components the same. That was a somewhat higher proportion of RNA than in particles from yeast, rat liver, pea seedlings, and so on. They calculated that there were about ninety thousand particles in each bacterium.

Early in 1958, a three-day symposium on microsomal particles in protein synthesis met at MIT, sponsored by the Biophysical Society, organized by Richard Roberts, who was head of a group at the Carnegie Institution of Washington that was working on the particles as actively as anybody. The most important development at the meeting was semantic: Roberts suggested that for clear and handy distinction between the particles and the amorphous cellular fraction of protein and fat in which they were found, the particles themselves should be called "ribosomes"—short for "ribonucleoprotein particles of the microsome fraction." The new term quickly spread into general use.

Ribosomes, from whatever source, were soon seen not to be truly spherical. The smaller piece, though almost beyond the reach of electron microscopes then, was definitely asymmetrical, so that, Tissières and Watson said in a later report, "A simple model for the structure of [the complete particle] would thus be a capped sphere or acorn shape."

Despite the efforts of several laboratories, by the late fifties nobody had succeeded in building a cell-free system from bacterial extracts. No direct evidence had been found that the ribosomes of bacteria were in fact the site of protein synthesis. "It was a very strange business!" Tissières said in a recent conversation. "It was a question of *mood*. The rumor was, in the States, that everybody who had tried to do incorporation experiments with the ribosomes, from *coli*, had failed."

❖ ❖ ❖

Colinearity proved intractable. Workers at five laboratories at least, in fall 1957—at MIT, at the State University of New York in Albany, at the National Institutes of Health, at the Max-Planck-Institut in Tübingen, and at the Cavendish—were trying to match the fine structure of a phage gene to changes in the protein, insert some experimental support beneath the sequence hypothesis, and go on to break the code. Other biologists were poised to enter the race. Benzer, Streisinger, Brenner, Ingram, and Crick in Cambridge had a slight lead.

"What we did, not having the protein corresponding to the *r*II gene, was to look at other proteins in the phage," Benzer said in conversation in 1971. "We started taking the phage particle apart—heads, tails, tail sheaths, back to heads—and analyzing the various parts. This was shortly after Ingram had invented his finger-printing techniques. Everybody was doing that. We used finger-printing for the proteins in the phage, and every week we thought this one or that one would work. Crick's office in the Cavendish also contained Brenner, me and four other people. It was actually a very lovely time. But frustrating sometimes. I remember putting something down on the lab bench, and coming back the next day, and it was gone. And I asked Brenner, couldn't I have just two linear feet of lab bench that I could call my own, and he said, 'I'm afraid that's impossible.'" In a year's work, they were never able to pin a change in protein finger print to a single, specific phage mutation and so begin to unroll the map. They got nowhere. Neither did any of their rivals. By now, biologists looking on from the middle distance were saying that the results would be interesting only if the map of a gene and the sequence of its protein did *not* march in step.

In what began as a diversion, Benzer and Brenner—with the help of Leslie Barnett, whom Crick had recruited as a part-time technician but who became so accomplished a geneticist that she was promoted to the scientific staff; her name was put on the papers, too—looked at mutations in the *r*II region induced by proflavine, a coal-tar dye Brenner had been curious about even as a student in South Africa. They found that proflavine caused mutations at an entirely different set of sites on the map from those—whether spontaneous or induced by other chemicals—that Benzer had previously mapped. They eventually published a paper about that in *Nature* and then dropped the subject.

That winter, Brenner and Benzer gave some informal special tutoring in genetics, at Crick's house. "You'd hardly believe it; it seems incredible," Crick said. "The people who worked on sequencing proteins, before Vernon Ingram's work, didn't realize that the subject had got anything to do with genetics. And when Fred Sanger realized this, Sydney and Seymour gave Fred and his group, in my sitting room, little lectures on elementary genetics because they had to learn genetics. Up to that, protein sequencing was one subject, genetics was another subject, and there was no reason to believe that there was any connection between the two!" Sanger recalled, though, that the discovery of transfer RNAs, more than Ingram's work, showed him he had to attend to the relationship between sequences in proteins and nucleic acids.

That winter, Rosalind Franklin entered the hospital, with cancer. Early in the spring, to convalesce, she went to Cambridge to stay with the Cricks for several weeks. She went back to work briefly. She died in April, aged 37.

❖❖❖

The fundamental assumption beneath all coding ideas, the relation of DNA to RNA and protein, suddenly came into question. On 12 July 1958, *Nature* published a note from two biochemists in Moscow, Andrei Belozerskii and

Alexander Spirin. They had analyzed the DNA and RNA of nineteen different kinds of bacteria, using paper chromatography to determine the overall composition of bases in each species. They had published most of their experimental results the year before, in the Russian journal *Biokhimia*; in fact, a French group had reported similar results for DNA alone, in 1956. Nobody had paid much attention.

Belozerskii and Spirin's note in *Nature* was surprising in every way. They showed that the base compositions of DNAs varied very widely among the different bacterial species—while the RNAs hardly varied at all. At the top of their table of DNAs, the bacterium *Clostridium perfringens*, which causes gas gangrene, turned up with 15.8 molecules of guanine to 34.1 of adenine, 15.1 of cytosine, and 35.0 of thymine. The Chargaff unities, guanine to cytosine and adenine to thymine, were observed. But in the crosswise calculation—and this was the interesting figure—the ratio of guanine-plus-cytosine to adenine-plus-thymine was 45 per cent. At the other end of the table, the micro-organism *Streptomyces griseus*, which lives in the soil and yields the antibiotic, had 36.1 molecules of guanine to 13.4 of adenine, 37.1 of cytosine and 13.4 of thymine. Here the ratio of G + C compared with A + U was 273 per cent. The seventeen other species that Belozerskii and Spirin listed fell all along the range.

At the same time, the RNAs of these creatures were almost constant. The ratio of RNA nucleotides, G + C against A + U, ranged from 1.06 at the top of the table to 1.29 at the bottom. "The greater part of the ribonucleic acid of the cell appeared to be independent of the deoxyribonucleic acid," the Russians said. The news was alarming. The authors pointed out that across their tables there was, after all, a small tendency for bacteria with DNA much richer in adenine and thymine to have RNA slightly richer in adenine and uracil. That did not help. The paper subverted everyone's ideas about coding. It threatened the sequence hypothesis.

❖ ❖ ❖

The coding problem itself was stuck, just as Gamow had warned. Delbrück, for one, was fascinated by the comma-free code. Together with several mathematically inclined colleagues at Caltech, he investigated the theoretical variations of the idea with relentless sophistication. He reached two interesting observations—and both looked like grave disadvantages. First, in the comma-free code "every misprint of necessity alters the message." This was in contrast to degenerate codes, in which an amino acid was represented by several different triplets: the comma-free code would therefore make the organism more vulnerable to the pressure of mutations.

The second observation arose from the dyadic structure of DNA. Since the base sequences were complementary and ran in opposite directions, there were two different messages, though it was wildly unlikely that these could both dictate protein. A code without punctuation would need some means of making sure that the complementary sequence, opposite the correct message, made nonsense just as surely as did the wrong groupings along the correct sequence. "It turns out that it is indeed possible to construct dictionaries such that the complement to any message composed of words of the dictionary contains nowhere, neither as complements of words nor as complements of the overlaps, a word of the dictionary," Delbrück wrote in 1958. Such codes he called "transposable." "Comma-freedom would . . . seem to be a worthless virtue unless it is coupled with transposability." But the new restriction meant that the bases taken in triplets could specify no more than ten different amino acids; quadruplets were needed. Crick and Brenner, about the same time, reached the same conclusions. The comma-free code was slowly choking on its theoretical consequences—except that Belozerskii and Spirin killed it first, for a code that allotted but one triplet for each amino acid had no tolerance for wide variations in DNA composition.

Another Soviet biologist offered another version of a transposable code, which made complementary and mirror-image triplets signify the same amino acid. Several people, Yčas among them, suggested that the information for the amino-acid sequence is carried not in the nucleic acid alone but in a combination of nucleic acid and protein; nucleic acids and proteins are closely linked in chromosomes of higher organisms. These schemes contradicted the central dogma. Robert Sinsheimer, at Caltech, tried to get around the problem of the huge variation of DNA composition. He suggested that the nucleic-acid message was in a two-symbol code in which adenine was equivalent to cytosine and guanine to thymine; not triplets but base quintuplets would then be required to specify twenty amino acids. "This was rather a desperate measure," Crick wrote later.

Perutz told me, one day at lunch, that in the late fifties the Medical Research Council unit at the Cavendish got letters in steadily increasing numbers from mathematicians and specialists in information theory who wanted to come to England and break the code. "We had to write them that the problem was now not theoretical but biochemical and genetical."

❖ ❖ ❖

A fragment of encouraging news came early in 1959 from Richard Roberts's group working on ribosomes at the Carnegie Institution of Washington. They proved that radioactive amino acids added to a broth of *E. coli* were first incorporated into proteins on ribosomes before they could be traced or chased to other places in the cell. In an experiment exactly parallel to Zamecnik's with rat liver, they showed that in bacteria, after all, ribosomes were the site of protein synthesis performing just the function that they did in higher organisms. By irresistible unconscious logic, assumptions about the function itself were now secure. An anomaly had been removed, a ruling idea reinforced.

❖ ❖ ❖

In January of 1959, Crick lectured at the Rockefeller Institute, then went to Harvard as visiting professor for the spring semester. In March he was elected a fellow of the Royal Society. On the first of June, just after lunch, at a symposium, to an audience of three hundred and thirty-one biochemists and molecular biologists wearing large round name tags and sitting on brown folding chairs in a lecture room at Brookhaven National Laboratory, on Long Island, New York, Crick reported on "The Present Position of the Coding Problem." He was badly sunburned, uncharacteristically brief and gloomy.

Belozerskii and Spirin, he said, had replaced optimism by confusion: their evidence "showed that our ideas were in some important respects too simple." He went on to confront the impasse:

> This large variation of DNA composition is very unexpected. The abundance of the various amino acids does not, as far as we know, vary much from organism to organism; leucine is always common, methionine usually rather rare. The small variation of RNA composition is exactly what might be expected; the large variation reported for DNA needs some special explanation. . . . Listed below are some possible explanations of this phenomenon, though in my view they all, at the moment, appear unattractive. . . .
>
> *1. Only part of the DNA codes protein.* . . . The difficulty of this idea is that the nonsense must make up a rather large fraction of the DNA. If, for example, it is assumed that the base composition of the sense is reflected in that of the total RNA of the organism, then organisms showing extreme base ratios must have a minimum of 35% nonsense in their DNA. . . . A possible reason for the fine dispersion of nonsense might be the provision of "commas." . . .
>
> *2. The DNA-to-RNA translation mechanism varies.* This would allow the RNA-to-protein code to be uniform throughout nature, while permitting the DNA-to-RNA code to vary. This is a formal possibility, but it does not appear at all likely. . . .
>
> *3. The code is degenerate.* . . .
>
> *4. The code is not universal.* It used to be argued that the code would be uniform throughout nature (except possibly for certain virulent viruses) because any attempt to change it would necessarily alter many proteins at once and would thus almost certainly be lethal. However, a counterargument has been given [it had been suggested to Crick by Cyrus Levinthal, from MIT, who was in the audience at Brookhaven] . . . that perhaps under certain conditions (for example, when the organism was in a rich environment and thus did not require too many enzymes) it might he possible to make a change. . . . It does not seem that this point can be usefully argued. . . .
>
> *5. The nucleic acid code has less than four letters.* . . .
>
> *6. The amino acid composition of the protein varies.* Unfortunately a small variation will not do. The organisms with extreme base ratios in their DNA are required to have proteins for which, say, leucine is rare and methionine common. This . . . does not seem very likely. . . .
>
> Finally . . . a few important experimental facts should be mentioned which any theory will have to explain: first, the evidence that the RNA of tobacco mosaic virus controls, at least in part, the amino acid sequence of the protein of the virus; second, the genetic effects of transforming factor, which appears to be pure DNA; and third, the genetic control of at least parts of the amino acid sequence of human hemoglobin.

"The whole business of the code was a complete mess," Crick said. "That review for Brookhaven—you can see there, we were completely lost, you see. Didn't know where to turn. Nothing fitted."

7

"The gene was something in the minds of people as inaccessible as the material of the galaxies."

In the second week of April 1960, a modest symposium on microbial genetics was held in London, at the Royal Institution, and when it was over several of the participants from abroad—notably François Jacob, Ole Maaløe, and, from the Massachusetts Institute of Technology, Alan Garen—took the train to Cambridge to spend the Easter weekend. On Good Friday, April 15, in the afternoon, these three met for a discussion with Francis Crick, Sydney Brenner, and one or two others, in the set of rooms Brenner was using in the Gibbs Building of King's College, Cambridge. The Gibbs Building is a handsomely proportioned gray stone eighteenth-century block that faces the entrance gate of the college across a wide lawn. Brenner's rooms were at the top of the southernmost staircase, two flights up. The ceilings were high, the walls two feet thick, the windows tall, many-paned, and set in deep embrasures; they overlooked the gate and the street beyond to the east, and from the small study looked west out over the back lawn, bounded by a straight reach of the Cam, and across that an arched footbridge and a meadow where cows grazed. The day was cool, sunny, seasoned with rain. Willows by each end of the bridge were pastel green. The discussion was informal. It began by Jacob presenting and Crick sharply questioning the latest results of a line of work that Jacob and Jacques Monod had been pursuing at the Institut Pasteur. During the next couple of hours, what the French had found out and conjectured was put together with what the British and the Americans already knew. Illumination exploded. The fixed idea about the actual mechanism of polypeptide synthesis—namely, that the ribosome, the microsomal particle, in itself specified the sequence of amino acids for the protein chain growing there—which had appeared solid and inescapable for five years or more, was seen to be a trick of perspective that had brought the several pieces of evidence into a plausible but illusory configuration.

As the dazzle subsided and the last wisps of illusion cleared, a new understanding stood in its place. It was the punch line of a great joke. That evening, in exhilarated conversations in the corners of a party that Crick was giving at his house, the experiments to test the new idea were planned. They were performed later that spring in Matthew Meselson's laboratory at the California Institute of Technology, and in a rival series that summer in James Watson's laboratory at Harvard.

"You must certainly have the account of the meeting in Sydney's rooms in the Gibbs building at King's," Crick said, in the course of our conversations in his office in November 1975. "That I regard as one of the key moments." A fortnight later, in Paris, François Jacob said, "I suppose if you were to ask— This is just a psychological phenomenon, but if you were asking the same guys to tell you the same story—I mean, the different people who were involved in one particular story—to tell you about it, you would get different versions, wouldn't you?" The difference is certainly there. It lies not in witness-box facts—who first mentioned what, just how illumination descended—for the experience was as near an instance as one will find of collective intellectual creation. More subtly, what differed was the sense that the various participants had of what the new understanding was really about. When one builds up to 15 April 1960, along the French line of work and then from the English and American side, the facts are identical, yet their prominence shifts lurchingly, figure interchanging with ground. For the English and American molecular biologists, the discovery was first of all about the RNA in that slogan DNA makes RNA makes protein. For the French, RNA was all but incidental and this same discovery was the dènouement of a long train of experiments on a phenomenon called induction—the bringing into physiological expression of characters borne by the genes.

❖❖❖

As elsewhere, even more in France, what came to be molecular biology began small, ill-regarded, pinched for funds and space. As elsewhere, in compensation the beginnings were attended by a zest that was at once playful and rebellious. When Jacob realized, in 1949, that he had to get into basic biological research—for his, like Crick's, was a late vocation—he could discover only two laboratories in France that were pursuing what he thought were the right problems. Both labs were in Paris. One was Boris Ephrussi's, at the Rothschild Foundation's pleonastic Institut de Biologie Physico-Chimique, over near the Pantheon and the Sorbonne. Ephrussi, in the early thirties, with the help of grants from the Rockefeller Foundation, had brought classical *Drosophila* genetics from Caltech to France, and in the process had determined the course of Monod's career. Ephrussi was also, by all accounts including his own, a martinet. The other lab was André Lwoff's, where Monod had been since 1945, across town at the Institut Pasteur. Jacob applied to Lwoff.

André Lwoff. Jacques Monod. François Jacob. Others, of course: most important, in the early fifties at the Pasteur, were the French bacteriologist Élie Wollman and the American immunologist Melvin Cohn. The work took place in an institutional frame that was itself unique. The Institut Pasteur was an anarchic autocracy. Louis Pasteur had set it up in 1888, by public subscription, on the wave of enthusiasm that followed his development and successful demonstration on a small boy of a vaccine against rabies. Pasteur made it a private foundation, independent of the government and the university system, which in France is a great rarity; yet by special law it was recognized as being of public utility; it maintains the central microbiological reference laboratory of France and is a major teaching center. Much of its income came from manufacture of serums and vaccines, some from such unexpected enterprises as the production of the live bacterial cultures for making yogurt, which are delivered daily to anyone in Paris who orders them. The institute opened branches elsewhere in France, in Brussels, and in French colonies in Africa and Indochina. The place developed a particular personality; it attracted loyalties. "The Pasteur Institute is *une maison,* as we say, which is extremely rare in France; here the universities are completely impersonal," Élie Wollman once told me. "The Pasteur has a certain continuity in ideas or subjects, and a kind of spirit."

The institute is heaped up haphazardly on two large square blocks on a slight hill in a nondescript quarter of the Left Bank to the unfashionable west of Montparnasse. There are a blood-transfusion center, a hospital, a modern tower for molecular biology, a low prewar building originally for yellow-fever research, and, says the sign on one building, the *Service de la Rage*—the unit for rabies. The first sight of the place is high iron railings on either side of the street, and set back behind each fence a square brick building in the grand, lugubrious style of Paris at the end of the last century.

The building on the right as one arrives was opened in 1900. Its ambience is of tiled floors in white and terra-cotta, glass-fronted cabinets in varnished oak, tall windows whose top parts can sometimes be opened by a sharp jerk on a lanyard. The laboratories on the ground floor were of an airy vastness that might have housed a couple of 1917 war planes on leave from the front, until Monod had the space subdivided in 1954; on the next floor, the suite of the institute's director would almost have held Lwoff's entire establishment.

Lwoff's fabled attic was three flights up, at a back corner, and consisted of a corridor about twelve yards long, high-ceilinged but windowless and crowded with equipment, with five small laboratories opening off it and tucked beneath the slope of the mansard roof. Lwoff's lab was at one end, Monod's diagonally across at the opposite end. There were also a tiny secretarial office, a kitchen space around the corner by the stairs, and a cold room no bigger than a broom closet. The elevator was small, slow, grumbling, often out of order. "If I had to tell the story, I would begin with that corridor, the work going on at each end—the phage business of Lwoff and Monod's enzymes," Jacob said. "The two things which merged." Jacob was the agent of their merger.

One continuing subject at the Pasteur was bacteriophage, which Felix d'Hérelle had christened there in 1917, which Wollman's parents had worked on there in the nineteen-thirties, which Lwoff picked up only after the war. Lwoff had come to the Pasteur in 1921, at the age of nineteen, as a young medical student. *His* problem in the thirties was the elementary nutritional requirements of micro-organisms. Starting with certain parasitic protozoa— rudimentary one-celled animals—and then moving to bacteria, still simpler, he discovered factors that were essential to growth, and that turned out, when isolated and identified, to be particular vitamins that had then only recently been found to be indispensable to the growth of animals. The discovery startled the world of biochemistry. At the same time, in London, a bacteriologist named Paul Fildes and his colleagues were making parallel discoveries about other bacteria and other vitamins—such simple observations as that *Staphylococcus aureus*, the germ that causes boils, can't grow without vitamin B_1.

Before this, bacteria had been studied as agents of disease, agents of putrefaction, agents of fermentation: Lwoff asked questions of an open, general biological kind about bacteria and got back the answer that they were biochemical systems generally comparable to cells of higher organisms. The point, at that time, was so far from obvious that it flabbergasted many of his scientific elders. He exploited the point. The crude signs of beriberi, pellagra, scurvy, or another vitamin deficiency hardly hint at the vitamin's metabolic role; using microbes as experimental creatures, Lwoff began to analyze exactly at which steps vitamin deficiencies interrupt metabolic processes.

What soon unravelled was an intimate relation between vitamins and another new-found class of chemicals called co-enzymes: a co-enzyme is a small helper molecule, not itself protein, that a particular enzyme, far larger, requires to perform its catalytic function—the chemical equivalent of the diamond bit in the milling machine—and vitamins are either co-enzymes in their own right or precursors. An immediate corollary was that an organism that does not demand a supply of a given vitamin must be able to make its own.

Much beyond these particulars, Lwoff's discoveries of the thirties asserted the biochemical unity—obviously not the identity—of living things. The assertion was crucial to the confluence of ideas that followed. Recall the time: in 1938, Fritz Lipmann, starting with a bacterium that clots milk, began to work out the transport of energy in the cell's manufacturing processes. In 1940, George Beadle and Edward Tatum, using a mold that grows on bread, first realized that what a gene does is specify an enzyme. Salvador Luria and Max Delbrück, in 1943 when they proved that bacteria have genes, closed the circle.

In 1938, a new unit was set up at the Pasteur to investigate something termed microbial physiology, and Lwoff was made its chief and given his attic. For years before then and after, his main collaborator was his wife, Marguerite Lwoff. At the war's end, scientists and students began to come to him. He took on few. Like Max Perutz's group, financed by the Medical Research Council, at the Cavendish Laboratory at the same time, Lwoff's unit had no obligation to teach. Like Crick and the others around Perutz, Lwoff and Monod had their most important scientific intercourse not at home but abroad, and chiefly with the United States. Lwoff's cramped quarters, even earlier than Perutz's, came to be jammed with visitors from America, sometimes from England, there for a seminar, a summer, an academic year or two. The attractions were many: this was Paris shortly after the war, and the Americans brought research grants with them.

Lwoff, Monod, Jacob: the interactions of the two men and then the three make considerations of intellect indistinguishable from those of character at the moment when the metal flows. The immediate clear sense one had of Lwoff—the sense I formed when I met him in 1975; the impression others recorded who had first encountered him forty years before—was of gentle and deep-running courtesy, quick, sharp, ironical humor. He was slim, erect, with a receding chin, a watchful gaze, a glint of mockery at a corner of his mouth. Pictures from the thirties show him with a wild mop of blond curls; when he was seventy-three his hair was short, sober, and gray, but nearly as thick. In writing, courtesy is clarity: Lwoff commanded in English as in French a style of sweetest clarity, and his are the only scientific papers I have read that are sometimes intentionally funny. The art of research, Lwoff once told Jacob, starts with finding oneself *un bon patron*—a good boss. Lwoff was a scientist of great technical finesse, and a perception never bound to a rigid logical line. He was a man of reserve, of severe rectitude, and—the most indelible impression—of balance. In many of these qualities, Jacob was like him in a more nervous key, or came to be so. Monod in many ways was their extreme opposite.

"Those four or five years of daily discussion and argument were really among the most interesting in my life, and I would say the happiest," Jacob said, quietly, of the time of the discoveries he made with Monod. In the work they were equipoised: indeed, if Jacob was the more gifted experimentalist, it is not at all clear that Monod was always the more subtle and original theorist. After the period of their collaboration, Jacob disciplined himself to master a new scientific field. Outside of science, Monod was spectacularly a public man. He spurned the rule of public reticence beyond the circle that most scientists observe, that all enforce upon each other, that none violates without hazarding his reputation. Monod had will, assurance, vanity, an ideological edge, an appetite for polemic, an actor's need for public attention. "There were two personalites in Jacques; or anyway at the simplest level there were two, one of which was repressed for a long time," Jacob said—repressed as the reading out of a gene is repressed in the theory they made together.

Monod was the summation of an extraordinary series of oppositions, paradox raised to a principle. His style was as quintessentially French as Linus Pauling's was American. He was a multiple outsider. The Monods were one of the noted clans of the professional class of France, far-flung, close-knit, sober, devout, ambitious—and Protestants, exiles for a century. Coming from that family, his father was a painter. His mother was American. Monod was a product of the French academic system; he was dissatisfied with it from the time he first came to the Sorbonne in 1928 to the days of riot and near revolution in Paris in May of 1968, when he publicly crossed the barricades to be on the students' side. Monod knew by his teens that he would be a biologist. His great work began only when he was nearly forty. He did that work at the Institut Pasteur, itself founded in deliberate opposition to the universities; at the Pasteur, he was a member of the smallest, liveliest, least typical unit; he rose to be director of the institute—yet by many there, because of his late start as much as his plans for reform of the place, including the sale of the real estate and a move to the suburbs, which was bitterly resisted, he was never accepted as a true Pasteurien. By conviction a democrat, detesting faiths that enslave the mind, by temperament he was autocratic. He was a warm friend to his friends; his will, his style, devoured weaker colleagues; in scientific debate and correspondence he was sometimes cruelly scornful. He was the rare true intellectual among the great biologists of his time, and a man of generous civilization; in his research he followed a line whose great explanatory power derived from its singular narrowness and concentration. Monod was perhaps most perfectly and naïvely French in his conviction that he was a man of implacable, logical clarity, devoid of intuition. "I have to stick to a linear, logical thread," he said to me. "Otherwise I am lost.

"Oh, no, no, I had no doubt that I wanted to be a biologist; and that, I think, is one of the few good intuitions I have ever had—namely, that there is a basic epistemological problem posed by the existence of living beings, that either living beings could be explained in terms which did not contradict or supersede physical laws, or else the interpretation of the whole universe had to be something different. I think I was aware of this—quite early, when I was sixteen or seventeen. And I decided to become a biologist for that specific reason."

Monod and I had first met six years earlier, in September 1969, and we had talked about his work and this book then and several times since. In the first days of December 1975 we had two long conversations, all one Monday evening and then on Thursday at lunch, at his apartment in Paris. We sat, the first evening, in a room with books on two walls, a handsome old refectory table by the window serving as a desk, a battered sofa backed up against the marble fireplace, a television receiver, a gramophone, a couple of big sagging chairs with dark orange covers—a room whose nineteenth-century airs had been rubbed down into comfort. The floor was polished wood, and creaked like a ship at every step. The window looked down from four stories onto the shrubberies and lawns of the acres of gardens at the base of the Eiffel Tower; the tower rose startlingly close, but dematerialized by an icy gray mist, into which its upper reaches vanished, lights and all. Monod had recently spent several weeks sailing around Corsica and Sardinia, and looked tanned and fit; his features, though, were sharpened and his face more deeply lined between the corners of nose and mouth than I remembered, and his hair for the first time was showing a lot of gray. We drank coffee, and he had a glass of madeira. He was relaxed and reminiscent, and smoked cigarettes incessantly. A brown-and-black airedale lay at his feet.

Jacques Lucien Monod was born in Paris on 9 February 1910, his parents' fourth child, third son, but younger by ten years than Philippe, the middle brother. The Monods are of the Huguenot *grande bourgeoisie* of France: generations of clergymen and doctors of distinction ("mainly intellectual pursuits," he said), teachers, lawyers, historians, civil servants; he thought there were now some nine hundred Monods in France. Monod's ancestors had been driven out when Protestants began to be persecuted after the revocation of the Edict of Nantes. His great-great grandfather was a Calvinist clergyman, born in Geneva, who went to serve as a pastor in Copenhagen; there he married a young woman who was herself descended from Huguenots who had fled to Denmark. After the French Revolution, he was called in 1808 to a church in Paris. "And there was a law voted by the Convention that any direct descendants of Huguenot emigrants could become French again by—*de plein droit*, as of right. Simply by proving that this was the case. But also, since the law in Geneva is that anybody who has the so-called *bourgeoisie de Genève*, in direct descent, still retains it—I'm still a *bourgeois Genève!* Therefore a Swiss citizen!" He was delighted now by the accident of ancestry. The iron had entered early. "I am a scientific Puritan," he said, another time.

Monod's father, Lucien Monod, was a painter, who made his living as a portraitist; doubly unusual in that family, he married an American, Charlotte Todd, who had been born in Milwaukee, of New England and Scottish stock, and brought up in Iowa and the Arizona Territory. I asked what sorts of pictures Lucien Monod had painted. Monod pointed to a large watercolor on the wall, a house among trees, in pale colors reduced partway to flat planes—a quiet and conventional picture. "That's our house in Cannes, which I still have," Monod said. Lucien Monod had been badly crippled by infantile paralysis; Jacques, when he was two, caught the same disease, which left him lame in the left leg. The family left Paris in 1914, and during the war lived partly in Switzerland, partly in the South of France. In 1919 they bought a large house in Cannes, overlooking the harbor. Monod spent all his youth in the south, against the backdrop of mountains and sea, sun and olive trees; he always thought himself less a Parisian than a man of Provence, and he became, after the second world war, a skilled rock climber and yachtsman.

No Monod had been a scientist. As a youth, Monod pursued geology, zoology, mathematics, hunted fossils, dissected cats; but the great influences, he wrote at the time he won the Nobel prize, had been two. One was his Latin and Greek master, Dor de la Souchère, who much later, after the second world war, created the Picasso museum at Antibes. The other was his father, who had read Darwin. I asked about that. "Well, he was reading not only Darwin but Stuart Mill, Spencer, Auguste Comte, and all those people," Monod said. "He was a nineteenth-century positivist, and there's no doubt that this had great influence in attracting my attention to science. Which is amusing, because he was an artist and in fact didn't understand science at all." Monod laughed gently. "Didn't know anything about science." Early in his teens, Monod began to study the cello. He continued to play even into the nineteen-seventies. In a corner of the entrance hall of his apartment, a cello stood. Near it, on the wall, hung a pencil portrait, drawn in 1928, of a young man of startling good looks and intensity, playing a cello.

In 1928, Monod came to Paris, to the Sorbonne. His brother Philippe was a lawyer, working in Paris for the Wall Street firm Sullivan and Cromwell, which was then headed by John Foster Dulles; the brothers Monod shared an apartment. "In fact, teaching at the Sorbonne at the time was on the average anywhere between fifteen and thirty years behind the times. I had the extremely naïve ideas of a boy of seventeen; I thought that the way to attack the problems of biology was to start by studying behavior," he said. ("I think that many people who became biologists started with the very naïve idea that one had to begin by resolving the mind-body problem. And I was one of them," he had told me another time.) "But I was profoundly dissatisfied, and very soon realized that the kind of teaching I was having was not going to lead me anywhere. Luckily, I became good friends with a few of my elders, people who were about ten years older and were already in research."

In 1929, preparing for a qualification in zoology, Monod went for the first time to the marine biological station at Roscoff, in Britanny, where many French biologists spent working summers. There, in the next several summers, he met the four scientists who most influenced him. From Georges Teissier, head of the station at Roscoff and professor at the Sorbonne, whose specialty was insects and marine zoology, Monod acquired the taste for quantitative description. (At Roscoff also, Monod met Teissier's wife's sister, Odette Bruhl. She was an archaeologist and a student of the art of Tibet and Nepal.) From Louis Rapkine, a biochemist, he got the belief that biology must describe living processes in chemical, molecular detail. At Roscoff, also, he first met

Lwoff, who was eight years older. "But the most important, surely, was Ephrussi," Monod said. "He confirmed me in my conviction that the key to everything was genetics. Which was a subject not taught in France at that time. And he was decisive in my career: he was responsible for getting me a fellowship to learn genetics at Caltech. Boris is a great man, you know. You ought to talk to him sometime. He's seventy-five, and still working hard."

Boris Ephrussi had been born in a suburb of Moscow. His family was ruined by the Bolshevik Revolution; he himself, though already a university student in genetics, had in the ferment of those years been mixed up with the religious followers of Leo Tolstoy, and in February of 1919 left Russia for Bessarabia, then part of Rumania, where he tried to follow the Tolstoyan way as a farmer. After enduring a year and a half of the Middle Ages, he renounced agriculture and religion and moved to France to get back into science. He began in tissue culture and embryology but grew progressively more interested in the connection between that and genetics. Accordingly, he spent the year 1934 in Pasadena, on a Rockefeller fellowship, learning about fruit flies in the group under Thomas Hunt Morgan. The Rockefeller Foundation then offered to support Ephrussi in developing genetics in France. As a first step, he asked to return to the California Institute of Technology for another year, and to take an assistant to be trained. He found the assistant in the summer of 1935, at Roscoff, where he was working for three months with George Beadle, from Caltech, transplanting eye tissue of mutant colors from the larva of one fly to that of another. The work took two microscopes and four hands attending to the same fly; it ought to have won Ephrussi the Nobel prize, Monod said.

At Roscoff, among other young scientists, Ephrussi met Monod; he offered to take him to California. The opportunity was unique, and a trip like that seemed an expedition. Monod, however, had signed on to spend the summer of 1936 on a voyage to Greenland as the naturalist on board the *Pourquoi-Pas?*, a three-masted sailing and steamship built for Arctic work by the explorer Jean Baptiste Charcot. But Monod had already done a Greenland cruise with the *Pourquoi-Pas?*, in 1934. After some hesitation, he chose to go to California.

"Boris returned to Caltech in '36, carrying me in his bags," Monod said. "I didn't even have a Ph.D. At that time the Rockefeller fellowships were in principle reserved for postdoctoral students. And they made an exception for me, on account of Boris's insistence that there might be some substance in this young guy and that he was really wanting to learn genetics. And while I was at Caltech I didn't do much." We talked for a moment about the legendary group Morgan had built at Caltech. "All the great geneticists of the time, were there—except Hermann Muller. He was in Russia and beginning to have all his troubles; it was the beginning of the purges." Monod worked briefly under Calvin Bridges—"I couldn't get along with him"—and felt the influence of Alfred Henry Sturtevant, both geneticists who had been colleagues of Morgan's for two decades. He met Beadle, and Barbara McClintock, a geneticist who worked with maize and was visiting from the East Coast; he was impressed by Jack Schultz.

"Boris was very unhappy with me because I didn't want to do just the kind of thing *he* wanted me to do! I was very undisciplined, I must say." In Paris in the early thirties, Monod had formed a small group of singers and musicians, who met, to begin with, at the apartment of an elderly uncle of his, to perform choral music by Bach, then rarely heard in France. Monod conducted. The group grew, acquired a name—La Cantate—and when Monod decided that public performances would make them work more seriously, began to give several a season. Within months of arriving at Pasadena, Monod organized a Bach Society and was conducting concerts. Monod's picture began appearing in the local newspapers; when his fellowship came to an end, he was offered a job conducting a choral group and teaching music appreciation to undergraduates at Caltech—for five hundred dollars a year.

(Monod was right: Ephrussi was exasperated. "Monod has incredibly wide culture; he is very gifted in many fields," Ephrussi said in a conversation. "But when I knew him first he was a real spoiled child. The most unstable character you can imagine, intellectually. Gifted, no end. I took him to Pasadena. And this was a complete flop. He was supposed to work—I brought him to California to study genetics. He really made my life miserable, because he turned up in the high society of American millionaires, who tried in fact to hire him as conductor of their local orchestra. I put him to work with Bridges, just to learn the basics of genetics. And he worked a little bit with me and Beadle on transplantation. But most of the time he didn't work. He was periodically going through a crisis. I would come in one day and find him crying, sitting over a piece of paper with names of girls on the paper, and crying, crying—and I said, What's the matter? 'Well, I feel terrible, guilt feelings—you brought me here to learn, I've done nothing.' But the next day he would give a concert, and give an interview to the local paper—and you see my position in respect to Morgan, there was Monod's portrait in tails and white tie, conducting an orchestra, and then big headlines, Monod doesn't give a damn about flies. Stuff like this. And I must say that he didn't learn much, accomplish much, in Pasadena.")

"In Pasadena, as far as I was concerned, as a young graduate student, Morgan was the big boss, very far away," Monod said. "I think Boris saw a little bit more of him; but he wasn't very available to us. While Sturt, and Bridges, and all these people were right there, all the time; you had tea every day with them." The most important thing Monod learned from Morgan's group was the American style: the free, intensely critical discussions, the easy and open relations between colleagues of different ages and ranks. The contrast with the teutonic pattern of French science could not have been greater. "It was then that I discovered what science was about."

On 16 September 1936, the *Pourquoi-Pas?* was wrecked in a hurricane off the coast of Iceland. There was one survivor.

("Then we came back," Ephrussi said. "I came back before him. Then he appeared in Paris, and said he'd been offered a job there [in Pasadena] of conductor, at a salary of five hundred dollars a month." The salary had been inflated in repeated tellings. "And he'd decided that he would rather become a conductor than biologist. So I told him, 'Look, I love science, but I don't feel that science is superior to music if you're really gifted; but my advice is, if you want to become a musician, first thing to do is not to take a job in Pasadena, but to try to get to the Paris Conservatory and learn music. You're a complete amateur.' Well, he went down South to see his family. And the family was of course most alarmed, and his brother Philippe came to see me, and consult what to do with this boy. I remember, it was a very characteristic talk; we talked for about two hours in my office, what the options for Jacques were—and at the end he thanked me and said, 'Well, it's a big problem after all. The problem as I see it now is, is Jacques going to be a new Pasteur, or a new Beethoven?' That's how they regarded this child. Well, he was unusually gifted.")

"When I came back from Caltech, Boris wanted me to work with him following up on this work of Beadle and himself; and we didn't get along," Monod said. "We just *couldn't* work together. He's an extreme authoritarian, and disciplinarian, and as a student if not as a professor I was"—he laughed—"exactly the reverse! Also, I was very interested in doing music, conducting and so on, and wanted to work whenever I felt like it but not whenever he was pressing me to do anything; so we had to break up." Monod moved to Teissier's lab at the Sorbonne. "Already a few years before, I had had a sort of intuitive attraction for very small living things—things that could be manipulated like chemicals. And where one could easily do good measurements, quantitative work. Earlier, before Pasadena, I had been doing work with ciliates"—a class of one-celled animals, fringed with fine fibres with which they propel themselves—"trying to measure growth rates and so on, and I realized that ciliates were difficult materials, so when I had to prepare a Ph.D., I turned to bacteria. And I knew that whenever it became possible I would try to use what I had learnt of genetics, in bacteria." Lwoff suggested that Monod work with something that could be grown in a synthetic, controlled medium—for example, *Escherichia coli.* Monod asked, "Is it pathogenic?" In 1937, the research appeared in no way genetical, for bacteria had not yet been taken to be a province of Mendel's realm. Monod started by measuring the rates of growth of bacteria provided with different types and concentrations of sugars. Following Lwoff, Monod tried out the effects of different vitamins on growth, too.

In the late thirties, La Cantate gave several seasons of Bach, to warm reviews. Monod had developed exceptional skill at directing voices. Odette Bruhl joined the group. They got married in 1938. In 1939, a month before war began, they had twin sons. Monod, because one leg was shorter than the other, had been exempted from the French universal military training. Early in 1940, he argued his way into the army. He told his wife to be ready to flee with the children. When the Germans broke through, she arranged to sail from Brittany to England aboard a fishing boat; at the last moment, one of the boys fell sick and they missed the boat. With the fall of France, Monod was demobilized in August, almost without having served. He was living in the rue Monsieur le Prince, not far from the Sorbonne, and had a laboratory in the zoology department, a tiny room opening off a gallery full of skeletons and stuffed animals. In the autumn of 1940, under the German occupation, Monod went back to bacterial growth.

Bacteria break down sugars with enzymes, to obtain energy from the chemical bonds as well as the carbon to build into other molecules. Some bacteria will only grow in particular sugars. Normal—"wild-type"—*E. coli* are versatile, able to grow in a medium containing any one of several sugars, including glucose or galactose, both of which have hexagonal rings and six carbon atoms; or maltose, a double ring formed of two molecules of glucose; or lactose, formed of a glucose linked to a galactose. The ability to grow on lactose, with its double ring, is one of the defining characteristics of *E. coli.* The bacteria possess the specific enzyme required to break the chemical link between the two rings.

Monod had been trying various strains in the presence of measured concentrations of each sugar in turn, charting the rate and the amount of their growth. Next he tried growing each strain in a medium containing two sugars at a time, in various combinations. In most trials, the bacteria multiplied continuously, their growth levelling off as the sugars were used up. Some of Monod's cultures, though, where lactose was one of the sugars, grew that way but then paused, and after a distinct plateau went through a second burst of growth. The bacteria in these had used only the one sugar in the first burst, then, after the pause, used up the second sugar—which was the lactose. This result, though it seems so unremarkable when described, was unheard of. In December 1940, Monod took the two-step curves to Lwoff, at the Institut Pasteur, and demanded to know what they might mean. Lwoff told him they resembled a case of enzymatic adaptation. Monod responded, "L'adaptation enzymatique? Connais pas!"

Adaptation of micro-organisms to handle a sugar they do not regularly use had first been noticed in yeast in 1900, Lwoff said. Monod objected that the plateau on his curves was an inhibition rather than an adaptation. Lwoff said that the adaptation was thought to lie in the bacteria's developing the capacity to synthesize an enzyme; he told Monod the little that was known, and gave him a couple of books on bacterial enzymes. The story came to be told often: Monod's profession of ignorance began thirty years of work on these elementary metabolic phenomena—work that, he said, looking back, comprised a single straight line. First results were a brief paper and a section in Monod's doctoral dissertation, in which he named his finding "diauxie," from Greek roots meaning double growth.

Shortly after the conversation with Lwoff, Monod joined a resistance group at the university. They circulated the earliest clandestine news sheets. Lwoff was part of a different group, gathering intelligence and passing it on. Such networks were just getting started. Their objectives varied, as did their skill, their discipline, and their security.

Monod got his doctorate late in 1940. After the ceremony, the director of the zoological laboratory where he worked told Lwoff, "What Jacques Monod is doing is of no interest whatever to the Sorbonne."

In the German-occupied part of France, the underground group that gained the reputation as the best organized and the most effective in direct, armed action against the Germans—that is, after Hitler double-crossed Stalin and attacked the Soviet Union in June of 1941—was the Francs-Tireurs et Partisans, run by the Communist Party. Sometime in the spring of 1943, Monod joined the Francs-Tireurs. "That was the *armed* resistance movement," he said. Organization of resistance was intricate and splintered both geographically and politically. The deepest division was between the Communists and those following Charles de Gaulle. That May, in the first attempt at coordination, the Conseil National de la Résistance was formed. Its military branch was the Comité d'Action Militaire, in Paris, a three-man group of whom two were Communists. Monod quickly discovered that the Communists would allow nobody the slightest influence on planning or decisions within the Francs-Tireurs who was not a Party member. "So, after a sleepless night, I joined the Communist Party." Odette was Jewish; Monod sent her and the children to live in the country.

Meanwhile, Monod's brother Philippe was with their parents in Cannes. There, in the unoccupied part of France, *he* joined a group called Combat, which was not Communist and grew to be the strongest resistance group in the South. Because of Philippe Monod's connections with the Americans through Sullivan and Cromwell, he was sent to Geneva to act as liaison with the control point Allen Dulles had set up in Berne (which was, of course, the taproot of Dulles' later directorship of the Central Intelligence Agency).

In Paris, Jacques Monod became executive officer to the chief of staff of the Francs-Tireurs, Marcel Prenant, who was Professor of Anatomy and a lifetime Communist. Another resistance worker, under the alias Pierremain, who is now a senior French civil servant, had known Monod from before the war, through La Cantate; in June 1943, when Pierremain lost contact with his own resistance group because some of them were arrested, he got in touch with Monod and worked with him until the liberation of Paris and then to the end of the war. Under Prenant, Pierremain said in 1977, Monod was responsible both for organization and for information. Monod's alias then was Marchal.

The underground network Lwoff had joined was also broken by the Gestapo, and many were arrested. Lwoff got back into intelligence work through Monod. From time to time, also, he sheltered American airmen in his apartment who were being passed on to the unoccupied zone of France.

At the Sorbonne, Monod was still investigating the growth of bacteria in mixtures of two sugars. By the end of 1943, he decided that his two-step growth curves were a special case of a biochemical model of enzyme adaptation that had been put forward five years earlier by John Yudkin, then at Cambridge. Yudkin had thought along the classical chemical lines of the day. He had proposed that all enzymes were formed from precursors with which they maintained "dynamic equilibrium" in the cell. Most enzymes were not adaptive, but in contrast were produced in normal quantities by growing cells whatever their environment; these were called "constitutive," meaning that they were part of the cell's usual constitution. Yudkin declared that for such enzymes the equilibrium was balanced so that precursor converted to enzyme more easily than the reverse reaction. For adaptive enzymes, though, the equilibrium would be balanced so that only a trace of the enzyme itself was present. But when the adaptive enzyme's substrate was made available—that is, the specific substance it acted upon, lactose in Monod's case—the substrate and the trace of enzyme would find each other, and this would stimulate conversion of precursor to enzyme in quantity. Modifying Yudkin, Monod supposed that the two enzymes required for his two-step growth curves, though each was specific for its sugar, had a common precursor—a "surface," he called it, and said it was perhaps a particular molecule—for which the sugars competed. When glucose was missing or used up, then lactose could act on its target—namely, the trace of its enzyme always present according to the equilibrium theory—and this would swing the precursor over to making the enzyme in quantity. Monod sent in a paper about that.

Yudkin's scheme is hard to follow today for different but overlapping reasons. It lacked the sharp focus of physical-chemical mechanism that distinguishes molecular biology. It lacked biological specificity in the modern sense. Obviously, neither lack was evident then. Indeed, Yudkin's chemical precursors triggered by something simple were an instance of a kind of explanation then widespread. Bacteriophage multiplication was said by some people, outside the American phage group, to be a conversion of precursors; resistance of bacteria to phage or to antibiotics was thought by many bacteriologists of that time to arise by an adaptation similar to the appearance of adaptive enzymes. As I read Monod's paper and tried to fathom what he had imagined when he wrote it, I recalled Delbrück's telling me that Avery's identification of the transforming principle in pneumococcus had raised the possibility that "all the specificity is already there but just needs a stupid kind of molecule to switch it from one kind of production to another."

Yudkin's dynamic equilibrium was an instance of another general doctrine, whereby the large molecules of the cell, the proteins in particular, were said to be in a state of individual flux and instability that was balanced only in the mass. In the thirties, Rudolf Schoenheimer, at Columbia University, maintained that proteins were incessantly exchanging pieces—amino acids or short lengths of polypeptide—with other proteins and the rest of the cell. In the next twelve years, Monod's discoveries overturned Yudkin and severely limited Schoenheimer, but to move away from these conceptions, he said, had been painful.

In the autumn of 1943, a meeting was called in Geneva of representatives of all the armed groups of the French resistance, to coordinate their military actions. Just before the meeting, Philippe Monod heard from his brother that he was the delegate of the Francs-Tireurs from Paris. In November, the Gestapo arrested a minor agent of one of the main resistance networks in France, the Réseau Vélites ("Network of Foot-soldiers": vélites were Roman light infantry), centered on the École Normale Supérieur. Marchal's identity and activities were known to the agent. Monod had to go underground completely, leaving his apartment, never sleeping more than a night or two at one address, staying away from the Sorbonne. On 14 February 1944, the Gestapo caught Raymond Croland, chief of the Réseau Vélites, who knew Monod.

On the run from his own laboratory, Monod was given bench space by Lwoff. "I don't think I was ever searched for, actually," he said. "But the possibility existed because at least one—in fact, several men had been picked up who knew what I was doing and who knew my name and where I worked. But it was known that I lived near the Sorbonne and worked at the Sorbonne, so the Gestapo would have had no reason to hunt for me at the Pasteur Institute." In Lwoff's laboratory, in collaboration with Alice Audureau, a graduate student, Monod that winter began a new set of experiments. He had sharpened his focus still further, to the utilization by *E. coli* of the single-ringed sugar glucose and the double-ringed lactose. "There were some bugs that had been described in the literature, the so-called coliform bacteria, that have every possible character of *E. coli* except that they don't use lactose," Monod said. "And among these strains, many of them can be persuaded to use lactose by culturing them in the presence of lactose for some time. And there were two distinct schools of thought: some people who believed that the appearance of the new enzyme was an adaptation promoted, exclusively, by the presence of lactose, and others—though fewer, as a matter of fact—who believed that the change was due to a sudden mutation, in rare individual bacteria. And selection." Both explanations claimed indisputable experimental support. Acquisition of the ability to use lactose had first been shown to be a

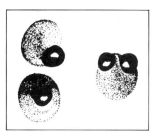

Monod's demonstration, in 1943, of mutation in bacteria: small colonies are mutants that have begun to utilize lactose. (Drawn from photograph, too poorly printed to reproduce, in "Mutation et adaptation enzymatique chez *Escherichia coli-mutabile*," by J. Monod and A. Audureau, *Annales de l'Institut Pasteur* 72 (1946): 871.)

mutation in 1907. "And Alice Audureau had isolated a strain of bugs who had these characteristics. They are not rare—and as a matter of fact, she isolated this strain from the guts of André Lwoff himself! Which is the reason for the name I gave to the strain; it's well known in the literature as *Escherichia coli* ML. And it means *mutabile* in *Lwoffii*." He laughed with a sound like a rusty pump. Audureau taught Monod to isolate bacterial mutants.

"What Alice and I did was to show that in fact the acquisition by these bugs of the capacity to metabolize lactose involves the *two* phenomena. That is to say, first, mutation, which can be observed on the agar plates, plus an adaptation which is then provoked by lactose." When they spread a sample of strain ML, unable to use lactose, evenly across glass dishes of gelatin containing lactose as the only source of carbon and energy, they found that after twenty-four hours' incubation at blood heat, a number of vigorous colonies were visible to the naked eye. The colonies were of widely different sizes. Since the doubling time of the bacteria was constant, the colonies had started at different moments. When they repeated the experiment but using gelatin that contained both sugars, ML colonies digesting the glucose started to grow immediately—and then sometimes, *inside* such colonies at one or two points but not uniformly throughout, new colonies appeared of bacteria that could be picked off and shown to be utilizing lactose. The point so far was precisely the same as that of the fluctuation test, using resistance to phage as the mutant characteristic, published a couple of months earlier, in the United States, by Salvador Luria and Max Delbrück. What Monod's demonstration lacked in statistical sophistication, it partly made up by the graphic simplicity of the bacterial colonies of various sizes, which he could show by photographs.

The mutant bacteria retained, with perfect stability through any number of generations, their ability to use lactose. But none the less, in the next part of the experiment Monod and Audureau put the mutant bacteria into a culture medium containing both sugars. Now they got the original phenomenon of diauxie, two bursts of growth, with a lag of an hour or more between the time the glucose was all used up and the moment when the bacteria, though proved to be able to metabolize lactose, actually began to do so. "*This* was the phenomenon of induction which I later decided to study," Monod said. "And I remember very well doing some of the last experiments *after* I had had to plunge into what was called the *illégalité* and those were done in André's lab, with Alice Audureau."

❖ ❖ ❖

Geneviève Noufflard was a young student at the Conservatory who first met Monod in 1942 in the organ loft of the church of St.-Germain-des-Prés during a recital. She was still living in Paris in 1976, where she taught music and played the flute—middle-aged and slightly stocky, busy, gentle, with fluent English and a quick, gay smile. La Cantate broke up when the war started, but at the Conservatory her teacher of the history of music organized a choir of students, in which she sang and which she also served as secretary, and which Monod was brought in to conduct.

One evening late in 1943, they left a rehearsal together. "We went down the street, and he had a briefcase, one of those soft leather things—I can still see it! —and he said, 'I have something to ask you.' And he said, 'I am going; I am leaving tonight to something very dangerous; and I would like you to keep this for me, and if I haven't asked you for it within'—I don't know, five days, ten days, I don't remember what was the delay—'will you please do such-and-such things so that my wife is told.' And of course I was terrified during all those days; but what he told me, shortly later, was that he had been to Switzerland, which was of course absolutely smuggling through. And I only heard all the detail from Philippe Monod, last week."

By Christmas of 1943, Noufflard's parents and her grandmother—who was Jewish and from Alsace, where she was well known for her French patriotism—were in comparative safety under false identities in Normandy; in January 1944, one evening after another choir rehearsal, she asked Monod to get her into the Francs-Tireurs. "I never was a Communist; he was, then. But the Communists were doing the best work and had the same object—and besides, they worked with much more *rigor*, in the way of secrets and all that, so I trusted them." Monod told her in detail of the probabilities of arrest, torture, slave labor, and death. "I mean he did everything he could to dissuade me. And when he saw that I was decided, he said, 'Okay, all right, come tomorrow.'" She became one of Monod's couriers, through streets patrolled by German soldiers and collaborating French police. One of the hourly anxieties of the resistance was to find hiding places for papers. Monod, until he had to drop out of sight, sometimes hid papers at the Sorbonne—inside the hollow leg bones of a skeleton of a giraffe near the door of his lab. "The long bones of the giraffe" became a standing joke between them.

On 1 February 1944, an organization was created called the Forces Françaises de l'Intérieur to take overall military command of the armed resistance. Monod became chief of the Troisième Bureau of the FFI for the Paris region, meaning that he was the staff officer in charge of operations, the planning and action side—choosing ends, finding means. Monod's character, his intellect, had a unique kind of strength for that work, his comrade Pierremain said. "He gave the most meticulous attention to organization—but it was always for the purpose of immediate action."

Monod asked Noufflard to become his secretary. "That was of course much more interesting, because I did all the typing for him, which I did like this"—pecking gestures with two forefingers— "and all this occurred in this building, only on the floor below, where I lived all my life." She searched for her spectacles, and brought out a bulky folder to show me situation summaries, operations reports, spare photos hardly bigger than postage stamps for false papers (Was this she? "It used to be," she said), plans and orders, ration tickets, and several small maps, the only ones she had kept, that had been drawn on tough tracing paper and passed on to Monod by nameless agents;

one of these showed the layout inside a German munitions depot underground at Bourret, outside Paris, on 15 May 1944.

In April 1944, Monod became chief of the Troisième Bureau—chief of staff for operations—at the national headquarters of the Forces Françaises de l'Intérieur; now the reports and plans covered all of France. Within the French resistance the struggle between Communists and Gaullists grew more intense. Working through Switzerland, Monod began to arrange parachute drops of arms and supplies to prepare for the Allied landings. He changed his war name to Malivert. After the war, Lwoff pointed out to Monod that the hero of that name in an early novel by Stendhal was sexually impotent. "He did not know it, and looked rather disturbed for having selected this name," Lwoff wrote to me. Later another friend asked Monod about that, and Monod explained that he had chosen the name out of his rational conviction that his actions could have no real effect.

On June 6, the Allies landed in Normandy. Monod's weekly situation reports, dictated to Noufflard, captured almost unbearable pressures: the extreme clarity, even grandeur, of the strategic understanding, the extreme energy of the language, moving as if will alone could triumph against the extreme difficulty of knowing what was possible, of finding the materials, of coordinating the forces. "One of the main things was the blowing up," Noufflard said. "Very many blowing up trains and blowing up railroads. And all sorts of things, trying to get hold of mail, and sabotaging factories that made armaments, and liberating people in prisons. And in the end we couldn't meet at this house anymore, because it had been used a little too much to be safe. So I found a painter's studio"—it was in Montparnasse, in a building where a number of artists worked—"and he used to come there, every afternoon and often again in the evening. And of course we had to have an excuse—I mean a reason if somebody suspected. So I acted painter, and I made a *huge* lot—I had made *this* at that point." She turned over a small sketch among the papers in the folder. "It is a little like him. And I had made big charcoal studies of him all over the room, so"—she gestured with a sweep of her arm—"and I used to think that if somebody would come and say, 'What is he doing here?' I could say, well, he's coming for—that."

At the studio they kept a large map of France, on which the railway and navigation networks were marked where bridges and junctions and locks had been blown. Monod also began to build up a nationwide command radio network, with scarce and obsolete materials and despite alert German monitoring. On Bastille Day, 14 July 1944, his report on the Germans' slow retreat northward and eastward ended:

> One may think that our forces, which are numerous and well led, could transform that retreat into a rout, and that with the pressure of the entire population, Paris could be restored as a French city even before the last German has left and before the first allied soldier has entered. This magnificent prospect is not a simple dream; it could be a reality; for it to be so requires that the armed forces of the FFI bring to the solution of the problem the passion, the ardor, and the confidence that such an unparalleled task merits.
>
> To conclude, we note that of the 80 to 100 tons of arms promised to the Paris region for at least six weeks, and which were to arrive in the past week, there have been dropped exactly three.

On August 1, the Second Armored Division of the Free French, under General Philippe Leclerc, came ashore at Omaha Beach in Normandy; in the landing was François Jacob, a medical student who had escaped from St.-Jean-de-Luz, south of Bordeaux, with a boatload of Polish troops at the fall of France in 1940, and who had fought with the Free French through Africa; a week later, Jacob fell gravely wounded, hit by more than a hundred pieces of a fragmentation bomb.

August 10, the railwaymen of Paris went on strike. On the 15th, the Paris Métro workers, police, and gendarmerie struck. On the 18th, the post and telegraph workers went out; the strike became general. The Communist Party called the populace to arms. "Of course you know about the liberation of Paris and how it occurred," Noufflard said. "But one time he came, and he was very excited, and he dictated to me an order—and this is something I treasure." She showed me the sheets of white paper, pinned together, on which she had taken down the order in pencil, and the single pink flimsy page that was a carbon copy, the bottom one of many by the look of it, of what she had typed from her notes. Dated 20 August 1944, when the Allied forces were still several days away, this was an order to the citizens of Paris not to fall into the error of trying to capture buildings or to hold strong points, but to create many moving armed diversions—and to build barricades in the streets. The order read, in part:

> National Headquarters, Third Bureau, to the Paris region. . . . In the city perhaps even more than in the countryside, the F.F.I. can have only one tactic: that of the mobile guerilla. . . . Therefore:
> 1. Multiply armed patrols in vehicles throughout Paris and the suburbs.
> 2. Wherever possible, and to begin with on the main arteries frequented by enemy patrols, put up barricades strong enough to stop cars, trucks and armored cars. These barricades should be provided with gaps to allow passage to friendly patrols.
> 3. They should be defended by armed groups, who have the mission to interdict free movement to enemy vehicles.
> 4. The patriotic militia without weapons, and the population, should be invited by wall posters and by loudspeakers mounted on vehicles to join in the construction of the barricades. . . .
>
> For the Chief of the National Headquarters
Chief of the 3rd Bureau
MALIVERT

"And after he dictated the order, he said to me, 'They won't do it—it's *too bad!* Because, you know—it could work!' And the next day, it was all full of barricades everywhere," she said. In the next forty-eight hours, with fighting still raging, the Parisians were able "to free little parts of streets, between two barricades, where French people could consider themselves free, because only tanks could come over barricades, not light things. And spreading like this, whole districts had been freed," she said.

"And then, our staff had arranged with some of the army officers who had worked with Vichy—some were trying to save their skins; some were just good people who had become involved—had organized that we should move in, in to the Ministère de la Guerre, on the rue St.-Dominique, as soon as—of course, not before!—as soon as Leclerc and the armies of liberation would arrive in Paris. That night, when—you've heard of it, surely; I have a recording of it—when our secret radio gave the news that Leclerc was arriving, they said on the radio that every church bell must ring now; the curate should do it, or just anybody can ring bells. I was in the house of another comrade of ours, and we opened the windows and heard the bells starting all over Paris. And so I came here very quickly, because I knew he would come to go to the Ministère de la Guerre.

"And this is a horrible memory, because I am not a courageous person." She laughed. "Bullets going like this"—her fingers flew through the air—"I hated. Four of us went in the dark to the Ministère de la Guerre, walking like this"—her head ducked; they were hugging the wall—"and it was fighting all over, and we had VZZZT! the bullets; and I was holding his, the end of his jacket. And we finally arrived there. As we crept into the building, all we could hear were our own footsteps and the sound of rifles outside. We tiptoed up a marble stairway and into a big room full of crystal. And that was really, it was the most— I can't start telling you, but it was a *fantastic* evening with all of those officers welcoming us, all dirty, having been on the barricades all day, and with those FFI brassards; and one of *them* sending a sentry to get some champagne he had recuperated on German officers who were living in his place; and we, with those—I was the only woman there—toasting to the victory. It was a strange end."

After the liberation of Paris, even while fighting was fierce in northern France, the political conflicts grew bitter and complex among the Communists, the various non-Communist groups of the resistance, and the Gaullists who had arrived to establish themselves as the legitimate government. The regulars of de Gaulle's Forces Françaises Libres took over direction of military action, and were in continual struggle with the Forces Françaises de l'Intérieur, the armed resistance and its leadership, which attempted for months to maintain its independent authority nationally and in certain regions where it had been strong. A separate command staff for the forces of the interior was set up in the Ministry of War. Monod continued in his previous job, with the rank of commandant, but saw his effectiveness, like that of his colleagues, progressively weakened. The uneasy coexistence came to a stop when de Gaulle dissolved the separate command at the end of 1944.

In December, Monod left for the headquarters of General Jean de Lattre de Tassigny, of the French First Army. De Lattre perceived his quality; trusted by the Communists as well, Monod took on the touchy task, in the midst of war, of integrating the armed irregular units and their officers into the First Army—particularly the Francs-Tireurs, which were simultaneously the most effective fighting units and the most independent.

That winter, in an American army mobile library, Monod began catching up with some of the scientific publications that had been unavailable during the Occupation. He came across the issue of *Genetics* that carried Luria and Delbrück's report of the fluctuation test. His own demonstration that bacteria mutate was confirmed and generalized. "Bacterial genetics was founded." Soon after that, he read Oswald Avery's paper identifying DNA as the transforming principle in pneumococci. Geneviève Noufflard joined a French unit attached to the American Third Army, to act as interpreter. In her folder of papers was a snapshot of Monod in uniform—lean, fine, taut with ardor. At the bottom of the folder, also, was a slender white stick, a baton. "This was his. For conducting. The choir—but it's broken."

❖ ❖ ❖

The airedale came over and sniffed. "You know, being in the Communist Party—in *a* Communist party, under the circumstances of working underground—teaches you a lot about how—about the style of the Party, and made me, as time went on, have more and more doubts about a party organized this way, and working with that style," Monod said. "Especially with their *absolute contempt* and *hatred* of anything that was not the Party. They were allied with all the other resistance movements, but they had *absolute contempt* for them. And it was quite clear, although nobody said so, that if they could, immediately after the war they would crush them. I became aware of this gradually. Because nothing of this was ever said; but it was a permanent and obvious implication from their entire style of work. I decided that I would never leave the Party until the war was finished, in any case, but that when it was finished, I would think about it.

"So when I was demobilized from the army, my one idea was to go back to the lab and think about nothing else. I drew a curtain over the memory of the wartime. And I refrained from doing what any good Communist was supposed to do, namely, even if he was already a secret member, he had to become an official member. I refrained from going to a cell and getting my card. As a matter of fact, nobody insisted, actually, because they liked what they called the *'sous-marins.'* The U-boats. But I decided I would see how the Party, and the Russians, were going to act, at the end of the war. And let's put it that after six months of working in the lab and watching politics, I decided to quit quietly. So in fact I left in '45. I behaved as an honest Party member very precisely up to the day I was demobilized. Then I broke off." He had become a Communist at the time when the Party was weak and joining was most dangerous; he dropped out just when many in France were rushing to join.

In the autumn of 1945, Monod joined Lwoff at the Institut Pasteur and went back to feeding milk sugar to colon bacilli. The following summer, Lwoff and Monod flew to New York for the Cold Spring Harbor Symposium. The subject was heredity and variation in micro-organisms: this was the occasion, Monod wrote years later, when "the new discipline of molecular genetics acquired both body and soul." The symposium, the first at Cold Spring Harbor since the war's interruption, was a synod of "the phage church . . . with its three bishops: Max Delbrück, Al Hershey, and Salvador Luria. Max was pope. *Primus inter pares,* it was he who defined the dogma."

Mutation was the dominant theme. The triumphant principle was George Beadle and Edward Tatum's "one gene–one enzyme"—although all that had so far been shown, Delbrück pointed out tartly, was that one mutant gene meant at least one enzyme deficiency. Delbrück and Hershey presented papers on spontaneous and induced mutations in bacteriophage. Their results suggested that if bacteria were infected by two different phages simultaneously, sometimes a burst of daughter phage showed characteristics derived from both parents, as though phage genes had recombined inside the host bacterium. Luria applied the fluctuation test to spontaneous mutations by which bacteria become resistant to antibiotics. Lwoff's paper, complementary to Luria's, was about spontaneous biochemical mutations—those by which bacteria lost, or gained, the ability to make enzymes essential to growth on limited diets, of which Monod's double growth curves for *E. coli* were just one example.

The greatest excitement of the meeting was the announcement that bacteria sometimes mated. The idea that mutations in bacteria were the same as gene mutations in higher organisms was so powerfully attractive—seven years before the structure of DNA—that Luria, at the end of his paper, had been forced to an awkward warning that "the similarity of the genetic systems involved can only be considered, for the time being, as a useful working hypothesis." Proof was lacking. Then, in the discussion after Hershey's paper, Joshua Lederberg, a graduate student, twenty-one years old, said that he and Tatum, his research supervisor at Yale, had demonstrated that bacteria sometimes exchange genes—the mixing that is the evolutionary function of sex in higher organisms.

Lederberg had designed the experiment. The design was clever, crucial to his immediate result—and the model for scores of later experiments in the main line of molecular genetics. Genetic recombinations in bacteria had been sought before but with no success, which proved nothing, for if rare they might be hard to spot. Lederberg first proposed to take two mutant strains, each with a different biochemical defect—such as the inability to synthesize some amino acid or vitamin—and to put them together in a batch of a broth fortified by glucose, mixed amino acids, and yeast extract. After letting the cells multiply, he would wash them free of broth, then sow them thickly on plates of gelatin that contained only glucose, none of the health foods. There, only a bacterium that had put together a complete set of wild-type genes could multiply. However rare, such a colony would be the only thing to show up. Even so, if genetic recombinations were indeed sufficiently rare events, once in ten million, they would be impossible to distinguish from true mutations that reversed a defective gene back to the wild form, which went on at those low frequencies in any case. Lederberg therefore planned to cross strains each having several defects. The likelihood of multiple back-mutations would be vanishingly small.

At this point a lucky chance intervened. To isolate bacteria with multiple biochemical deficiencies, with the methods available before the discovery that Lederberg was about to make, was a long, tedious job. But Tatum, for his own experiments relating genes to enzymes, had already isolated a number of multiple biochemical mutants in *E. coli*—not, however, in *E. coli* strain B, the one used by Delbrück and the phage group, but in another strain, designated K12, that he had found handy because it was regularly used in bacteriology courses at Stanford University, where Tatum and Beadle had been working until 1945. Lederberg suggested to Tatum that they look for genetic recombination in crosses of multiple mutants of K12. They found them almost at once. They grew two mutants—the first unable to make for itself the amino acid methionine and the B-complex vitamin biotin, the second unable to synthesize proline and threonine—together in a broth that supplied all the deficiencies. The culture was centrifuged; the bacteria were washed and sown onto agar that did not make good any deficient nutrients. A few colonies none the less appeared—about one for every ten million bacteria put onto the plates. Bacteria mated.

This experiment, and some simple variations of it, Lederberg and Tatum reported at Cold Spring Harbor that summer of 1946. By the time, a year later, that Lederberg handed in his doctoral dissertation, he had carried out the bacterial matings to show that six of the genes of *E. coli* K12, including the gene for the capacity to digest lactose, were linked in a fixed linear spacing exactly like genes on the chromosomes of fruit flies or people. He was convinced that genetic recombination in bacteria was a true sexual union, and indeed a fusion of the two cells, pooling their genes, just as when sperm fuses with ovum. He named the process "conjugation." But nobody had seen it under the microscope. Lederberg also found that other widely used strains of *E. coli*, including strain B, would not yield recombinants. They are now known to be sexually sterile. If Lederberg and Tatum had used almost any strain other than K12, they would have failed to discover bacterial conjugation.

Monod was greatly tempted to become a proselyte to phage. But enzymatic adaptation was beginning to get the attention of embryologists as well as geneticists; in 1947, he was asked to review the subject for a symposium on growth. "Putting that report together was decisive for me. Looking through all the literature again—including my own—I understood clearly that this remarkable phenomenon remained almost entirely mysterious," Monod said. The problem of enzymatic adaptation lay at the intersection of genetics and biochemistry. "Its significance appeared so profound that there could no longer be any question of my not pursuing it."

In 1947, the reasonable explanation of adaptation of *E. coli* to digest lactose still was that the sugar converted an enzymatic precursor from one form to another. But Monod lacked tools. The first task was to get out a sample of this or any other adaptive enzyme. In the spring of 1948, he extracted a lactase—the broad term for an enzyme that breaks lactose into glucose and galactose—from *E. coli* ML, in concentrated though not purified form. Lederberg, meanwhile, had found the ability to metabolize lactose, or its absence, to be particularly useful in analyzing the genetics of *E. coli*, and had laboriously hunted down scores of mutants that were impaired in this trait. Lederberg also announced a simple, sensitive chemical test for the presence and strength of the enzyme in colonies of bacteria or in extracts.

Early in September 1948, the Lysenko affair blew up—the most grotesque scandal in the history of science, and all the more bizarre for entailing, among some intellectuals in the West, a self-willed intellectual delusion. Trofim Denisovich Lysenko was a Russian botanist, poorly educated, who turned the faking of scientific results and a debased skill at Marxist polemic into a claim on Josef Stalin's patronage that made him the all-but-absolute ruler of Soviet biology. Lysenko had begun in the early thirties by advertising astounding yields of wheat from seed improved by non-Mendelian methods. He had soon gone on to assert that the genetics of Mendel and Morgan was a bourgeois science: Marxists, he said in Stalin's name, were to view science like any other cultural activity as a consequence of a society's organization of the means of production, so that true communist science would be locked in the class struggle with capitalist science. In particular, Lysenko and his collaborators dogmatized that dialectical materialism proved that an individual living creature interacts with its environment in such a way that characteristics it acquires in its lifetime can be inherited by its offspring.

Stalin made Lysenko boss of Soviet agricultural research. Opposing scientists were put down, argued down, shouted down, at meetings in 1936 and 1939. Lysenko placed his followers in controlling positions in laboratories. His opponents were fired, jailed, sent to labor camps. Hermann Muller had been working in Moscow at the Institute of Genetics; in 1937, warned to leave, he joined a medical unit in the Spanish Civil War, then after the fall of Madrid accepted an appointment in Edinburgh, and finally returned to the United States in 1940. Nicolai Ivanovich Vavilov, the greatest Russian geneticist, was harangued by Lysenko, arrested, tried as a spy and traitor, condemned to death, reprieved, but died in prison at the end of 1942.

After the war, Lysenko and his school advanced increasingly fanciful ideas of heredity, the modification of species, and agricultural practice. But his agricultural innovations were proving disastrous. Soviet science was obviously lagging behind the West. Other scientists began to resist Lysenko, and found support even as high as the Communist Party ideologue Andrei Zhdanov. Lysenko appealed directly and secretly to Stalin.

A public debate was organized of the Lenin All-Union Academy of Agricultural Sciences, to open on 31 July 1948. A few days earlier, with Stalin's help, Lysenko packed the academy with thirty-five of his people. He delivered a diatribe against Mendelian and Morganist genetics, calling them abstract and idealist, fascist, racist, incompatible with Soviet science and Marxist materialism. Lysenko's opponents were forced to speak, too. They were vilified by his supporters. At the end of the meeting, Lysenko announced that his speech had been read, personally corrected, and approved by Stalin. Its full text was published the next day by all the main Soviet newspapers. The Soviet Academy of Sciences met and announced support for Lysenko. Russian biologists abjured publicly their belief in genes and chromosomes. Laboratories were shut down. Books were banned. Firings were nationwide. Classical genetics was suppressed.

Western Communist intellectuals faced the gravest test of their faith since the Hitler-Stalin pact of 1939. In England, J. B. S. Haldane, for many years a member of the Communist Party, wrote, "Genetics is my profession. If it is attacked, I will defend it." He broke with the Party over Lysenko. Others, including Desmond Bernal, equally long a Communist, kept mum. In the United States, Muller resigned from the Soviet Academy of Sciences in protest. In France, where the Communist Party was larger and more intellectually vigorous, the controversy stayed on the front pages for weeks. *Combat*, the radical but non-Communist Paris daily that Albert Camus had started as a wartime clandestine newspaper, ran a series of articles under the banner "Mendel or Lysenko." In one of these, on September 14, the biologist Marcel Prenant—who had been Monod's superior for a while in the Francs-Tireurs et Partisans—claimed that Lysenko had been distorted in the Western press, that he "respected the basis of classical genetics," but that he had succeeded in making acquired characteristics heritable, to the immense benefit of the Soviet people.

The next day, the last article in *Combat's* series appeared, with the headline "'The victory of Lysenko has no scientific character whatever,' says Dr.

Jacques Monod." With merciless efficiency, Monod quoted Western Communists who had denounced the scientific absurdity of Lysenko only the year before, and quoted *Pravda* to demonstrate the vicious unreason of the debate in Moscow. He recounted Lysenko's fraudulent rise to Stalin's favor, chided Prenant sarcastically for embarrassing contradictions, explained that Lysenko's triumph was essentially ideological and dogmatic. If the highest authorities of the Soviet regime backed Lysenko and humiliated his opponents, that was because "the doctrinal fanaticism, the sterile casuistry of a Lysenko corresponded with their modes of reasoning, satisfied their way of thinking." Monod's condemnation was root and branch: good sense, objective truth, rational thought, and the very future of world socialism had been mortally corrupted by the Soviet rulers. The article resounded through France. It was all the more devastating because it was Monod's first public attack on the Communist Party. The enmity was to be bitter and permanent. The article also brought him the friendship of Camus—an "intellectual encounter," Monod called it, that he valued most highly.

"In fact, this whole business was very important to me," Monod said. "Because the phenomenon was so extraordinary. There were absolutely *no roots* to it. I mean, no material roots, no experiments, absolutely nothing. Nothing but ideology. There wasn't even the beginning of some sort of conspiration"—he pronounced the French word as though it were English—"or even of *entente* between some dissidents, against which this would have been used. The scientists who were condemned were just as good Communists as the others, there's no question about that. It was a purely theological affair. And that started me thinking a great deal about the origin of this ideology. Trying to understand why such an *insane* phenomenon could occur. Explaining it all on the basis of Stalin is not enough. Because it had started before Stalin. The attack on the genetic approach, based on a sociological ideology, you see, was already present in 1925 and '26. In fact—well, this whole trend of Western socialism goes back very largely to Rousseau. To the idea that man is good and society is bad. And therefore if you introduce the idea that what defines man as a species, and different men as individuals, is very largely biological rather than social, that goes against the creed."

Our conversation carried well past midnight. I apologized for that. "Not at all: it's amusing to me," Monod said. He got a leash for the dog. We went out, and walked around the block to the Champ de Mars, stopping near one pier of the Eiffel Tower. Monod snapped off the leash and the dog leapt away. The night had cleared somewhat. The grit of the path ground underfoot. Monod lit a cigarette and turned up the collar of his trench coat. We arranged to meet in three days for lunch. I remember thinking that he made a characteristically French gesture out of smoking. As I left to walk across to the friends I was staying with, Monod hunched his shoulders with his hands in his pockets and called to the dog.

❖*b*❖   Early in 1949, Melvin Cohn came to work with Monod. That spring, Lwoff abruptly changed his scientific problem. In September, François Jacob came to Lwoff to ask to work for him.

The subject Lwoff now took up for the first time was bacteriophage, but a strange kind of phage infection called "lysogeny," which bacteriologists at the Pasteur had been puzzling over since the early twenties. They had noticed that a phage that would multiply virulently in one strain of bacteria could sometimes be set to infect a second, closely related strain—and would vanish. It would not commandeer the bacterial biochemistry to make more of itself, would not lyse the hosts and burst forth. Yet something was in there, none the less. The bacteria themselves would multiply unimpeded. Most peculiarly, they were immune to any further infusions of the same phage into their culture. But soon, free virus would be found in the culture, detectable in the usual way: drops of the broth were put through a superfine filter and then into a dish containing the original susceptible strain of bacteria, where plaques would appear. Most of the work on lysogeny at the Pasteur in the thirties had been done by Eugène Wollman, collaborating part of the time with his wife, Elisabeth Wollman. A few bacteriologists elsewhere, including Macfarlane Burnet in Australia, had produced parallel results. But the disappearance and reappearance of these phages was unpredictable. D'Hérelle himself, from the start, denied the phenomenon; it clashed with his vision of phage as a bactericidal panacea. Delbrück flatly refused to believe in lysogeny; a bacterial culture with a few persisting phage particles contradicted his fundamental proof that a bacterium infected by phage bursts, minutes later, and releases about a hundred new particles. He counted the work in France and Australia as worthless; he said that the experiments had been contaminated. Lysogeny had certainly never been seen in *E. coli* strain B and the T phages.

Lwoff had known Eugène Wollman in the thirties as a scrupulous and dogged experimentalist. Eugène and Elisabeth Wollman, Jews, had been seized by the Gestapo at the end of 1943—he late in the afternoon of December 10 in the hospital of the Institut Pasteur; she a week earlier. They were sent to Auschwitz and never again heard from. Their son, Élie Leo Wollman, had fought with the maquis in the south of France, and had joined Lwoff's unit at the Pasteur in 1945 a few weeks before Monod.

When Lwoff began looking into the phenomenon, those who gave it credence argued that lysogeny was a mutation of bacteria so that they sponaneously generated virus particles, or that lysogenic bacteria gradually leaked virus without bursting. Lwoff used a lysogenic strain of *Bacillus megaterium,* a benign soil bacterium, and, for testing, a second strain susceptible to the phage. His techniques were of remarkable delicacy: he cultured bacteria one at a time. He once explained that this was because he was no good at statistical reasoning. But Lwoff is dry and droll. He also explained that although *B. megaterium* is a large creature, as bacteria go, he had to grow each individual in a microdrop, observed through a microscope; each time the bacterium doubled, he would remove one with a micropipette aided by a micromanipulator. He soon showed that virus particles were released only when, rarely and without apparent cause, a lysogenic bacterium would burst after all and free a litter of phage. Next he found—as Eugène Wollman had before him—that when lysogenic bacteria were artificially broken open with the enzyme lysozyme, to digest the bacterial walls without affecting the phage, no phage particles could be found within.

Lwoff then surmised that in lysogeny the set of genes of the original infecting phage merged into the chromosome of the bacterium, where it behaved indetectably like the neighboring genes along the string. He called the phage genes, fully integrated and dormant in the host chromosome, the "prophage." He supposed that occasionally some stimulus in the environment upset the merger and induced the production of bacteriophage. He committed himself to the surmise in a paper of 1949. Then he had to prove it. He told the story in 1966, in the collection of essays dedicated to Delbrück on his sixtieth birthday:

> Our aim was to persuade the totality of the bacterial population to produce bacteriophage. All our attempts—a large number of attempts it was—were without result. . . . Yet I had decided that extrinsic factors must induce the formation of bacteriophage. . . . Our experiments consisted in inoculating exponentially growing bacteria into a given medium and following bacterial growth by measuring optical density [that is, the turbidity of the bacterial mist, even though the particles were not visible individually]. Samples were taken every fifteen minutes, and the technicians reported the results. They (the technicians, that is) were so involved that they had identified themselves with the bacteria, or with the growth curves, and they used to say, for example: "I am exponential," or "I am slightly flattened." Technicians and bacteria were consubstantial.
>
> So negative experiments piled up, until after months and months of despair, it was decided to irradiate the bacteria with ultra-violet light. This was not rational at all, for ultra-violet radiations kill bacteria and bacteriophages, and on a strictly logical basis the idea still looks illogical in retrospect. Anyhow, a suspension of lysogenic bacilli was put under the UV lamp for a few seconds.
>
> The Service de Physiologie Microbienne is located in an attic, just under the roof of the Pasteur Institute, with no proper insulation. The thermometer sometimes rises in a manner that leaves no conclusion other than that the temperature is high. After irradiation, I collapsed in an armchair, in sweat, despair, and hope. Fifteen minutes later, Evelyne Ritz, my technician, entered the room and said: "Sir, I am growing normally." After another quarter of an hour, she came again and reported simply that she was normal. After fifteen more minutes, she was still growing. It was very hot and more desperate than ever. Now sixty minutes had elasped since irradiation; Evelyne entered the room again and said very quietly, in her soft voice: "Sir, I am entirely lysed." So she was: the bacteria had disappeared! As far as I can remember, this was the greatest thrill—molecular thrill—of my scientific career.

A hundred or so phage had burst forth from each bacterium. A new definition of virus had also been turned loose—an unsettling definition that said that viruses are sometimes far more intimately and permanently associated with the genetic material of their host cells than had ever been imagined.

Induction of phage was immediately shown to occur with some other combinations of lysogenic bacteria and phage, and to be produced by other stimuli than ultraviolet light, including some chemicals, and among them chemicals known to cause cancers. Lysogeny began to supply unexpected answers to long-standing problems. For example, work elsewhere showed that the bacillus for diphtheria is lysogenic for certain bacteriophage—and that among the genes of this phage is the one that causes the bacteria to make the toxin (first isolated in 1888 at the Pasteur) that causes the disease's lethal effects.

The first question was obvious: What about analogous phenomena in cells of higher organisms? Were there, for example, human diseases caused by the emergence of viruses that had been lying dormant—not as viruses, just as the proviruses, the genetic portion—and therefore indetectable? Were cancers caused by viral genes that had been hidden away on the cell's own chromosomes perhaps for many years—even for many generations—until they were induced, maybe by a chemical in the environment, and came forth to transform the cellular biochemistry? "You know, this played an important role in the thinking of cancer people," Lwoff once said to me. "Because at that time the so-called viral theory of cancer was not widely accepted; there was no model for it. And lysogeny brought a model for the interrelation between a virus and a cell. And also a model for the possible mode of action of carcinogenic agents, which could disturb something in this balance." Lwoff therefore proposed, in 1953, that "inducible lysogenic bacteria might become a good test for carcinogenic, and perhaps anti-carcinogenic, activity." But answers were not obvious. By the mid-seventies it was proved, for instance, that ordinary domestic cats carry, integrated in their DNA, a virus for leukemia that has been with them for thousands of generations and that an ancestral cat originally caught from a baboon. The argument about human cancers raged fierce and undecided.

Stranger questions arose. When viral genes integrated themselves into the host's chromosome, and then re-emerged, did they ever bring out with them bits of the host's own genes? Shortly after Lwoff's first paper about induction of lysogenic bacteria appeared, Joshua Lederberg was experimenting with two strains of *Salmonella typhimurium*, an unpleasant intestinal germ that gives mice the equivalent of typhoid fever. Lederberg by then had a laboratory at the University of Wisconsin and graduate students of his own. He was interested more deeply than ever in bacterial conjugation. With a student, Norton Zinder, he was looking for conjugation in a different species from *E. coli*.

At first they thought they had it. They started with separate cultures of two Salmonella mutants that had multiple nutritional defects: one was unable to synthesize three amino acids—phenylalanine, tryptophan, and tyrosine—so that growth stopped unless these were supplied in the soup; while the other mutant couldn't make its own methionine or histidine. When Lederberg and Zinder mixed about a billion of each strain and put them on a plate of jelly that contained no amino acids at all, a few colonies none the less grew. The bacteria in these had made good the mutual genetic deficiencies of their ancestors as though the genes had been pooled, just as in the original demonstration of conjugation in *E. coli*. Lederberg and Zinder tried combining other Salmonella mutants—including, for example, ones that were resistant to the antibiotic streptomycin. With these, too, genetic recombinations appeared and thrived.

To check that conjugation had really happened, Lederberg and Zinder used a trick for growing the two parent strains in a single batch of broth while keeping them physically apart. They employed a glass tube, bent into a U shape and with a superfine glass filter at the crook—like putting a wire-mesh fence down the middle of a monkey cage. Given the device, the procedure suggests itself: the U tube was filled with broth that contained no amino acids, the mutant strains of Salmonella were put one in each arm, and then the broth was pumped back and forth through the filter. The U tube had just been invented by Bernard Davis, who had proved with it that direct contact was necessary between conjugating *E. coli* in Lederberg's original cross.

Now, though, with no contact possible between the two strains of Salmonella, recombinant bacteria none the less appeared and multiplied. Lederberg and Zinder had found the opposite of what they expected. Mosquitoes were getting through the fence. The filterable agent responsible for the transfer of genes was characterized. For example, it was not a transforming principle like Avery's, not naked nucleic acid, because it was impervious to DNase and ribonuclease. Lederberg and Zinder proved step by step that it had to be a bacteriophage. The Salmonella they had started with turned out to have been lysogenic, and the phage particles that were occasionally induced came forth carrying a few assorted host genes in them, across and into other bacteria that they infected lysogenically.

Lederberg and Zinder called the phenomenon "transduction." The discovery was to prove immensely fruitful. It offered, for example, the first hint of the reason why bacteria become resistant to drugs as quickly as they do: they need not wait for a mutation to arise if they can pick up a gene for resistance carried over from another strain. This is why a hospital is a dangerous place to be sick: your germs may fall among bad companions. Transduction offered the more distant possibility of a means to insert known genes deliberately into a cell where they will function as the host's do. The phrase for that, genetic engineering, was years off when Lederberg and Zinder announced transduction at Cold Spring Harbor in 1951.

In 1951 also, Esther Lederberg, Joshua's wife, found a lysogenic strain of *E. coli*—not strain B, to be sure, but strain K12, which was even closer to home because it was the same one in which her husband and Tatum had originally discovered bacterial conjugation. She isolated the phage involved and called it by the Greek letter *lambda*. She also proved that lysogeny for phage *lambda* could be transmitted in bacterial crosses like any other genes. Jean Weigle, at Caltech, a Swiss physicist turned biologist, promptly tried the obvious experiment and announced that phage *lambda* could be induced.

Lwoff's discovery of induction of lysogenic bacteria raised yet another and still subtler set of questions. How did the genetic information of the phage merge with that of its host, what prevented it from being expressed, and by what mechanism did it re-emerge?

❖ ❖ ❖

While Lwoff was defining, if not yet explaining, the nature of lysogeny, Monod had been redefining the nature of adaptive enzymes. But he did not do so in any single, decisive experiment. Monod's was far from the Jim Watson style of science. Rather, there was a long series of experiments, each precisely calculated to test one point before he went further. He was mountain-climbing, setting each piton carefully before moving to set the next; he was mountain-climbing in a mist that would only occasionally swirl and break open to let him see more than a step or two ahead. Monod, I was told by everyone I met who had worked with him, had a gift for the design of experiments that would isolate exactly the point he wanted to test: this was the sense in which he was correct when he said that he was a man of rigorous logic, not intuition.

The early experiments built the experimental system: this kept evolving, but it was always based on the enzymes with which E. coli metabolized sugars, in particular lactose. "You see, one of the great advantages of the *lac* system, and that's one thing I realized very early, is that it's an unessential system; therefore you can do anything you like to it, without killing the cell," Monod said. "That's a very basic virtue of the system. We had to have a system which we could handle any way, genetically and otherwise, and have perfectly—living bugs. This is because bacteria can live on lots of other sugars. Lactose is just one possibility among many." The other great advantage was that a lot had been done with the genetics of lactose utilization, especially by Lederberg. Monod's enzyme system was as crucial for all his work as, say, E. coli and the T-set of bacteriophages were for Delbrück and Luria, or the cell-free liver-extract system for Paul Zamecnik, Mahlon Hoagland, and Fritz Lipmann. The logic of Monod's system, once developed, made the march of the papers inexorable; the technical detail through which that logic must be grasped—if it is to be grasped and not merely admired from a distance—is somewhat complex, but those complexities of detail soon fall away. Then the logic of the system—more, the logic of the demonstrations Monod and his colleagues produced with it—is a rare pleasure to follow.

Monod worked with a long succession of other scientists—some French, many visitors. Of the French in the lab, many were women: Alice Audureau, Madeleine Jolit, Anne-Marie Torriani, Germaine Cohen-Bazire in the forties and early fifties, others later. Lwoff's group, marked by its visitors, got the reputation at the Pasteur of being thoroughly Americanized, even to the hours they worked and the fact that nobody went out to lunch. Americans were welcome because they were already doing—particularly those who came out of the phage group—a kind of biology that others in France had not yet recognized.

Melvin Cohn came to Monod's lab at the beginning of 1949, more interested in Paris than in the work they might do. He stayed six years. Cohn was an immunologist who had just taken his doctorate at New York University; his supervisor there, Alvin Pappenheimer, had met Lwoff and Monod at Cold Spring Harbor in 1946 and had later suggested that the exquisite sensitivity of immunological methods would help get at what went on in enzyme synthesis. When Cohn arrived, Monod had recently extracted from E. coli the enzyme that split lactose into its two halves, galactose and glucose. It was not yet obvious, Cohn told me over breakfast in a New York City coffee shop in the spring of 1971, that lactose metabolism in E. coli comprised an ideal system with which to investigate enzyme synthesis. "There were a lot of potential systems in biology—but none of them that could withstand primitive equipment, the heat of Paris summer, and so on," Cohn said.

In the chemist's diagrams, galactose and glucose looked like this:

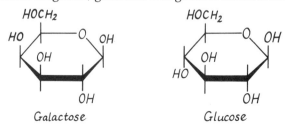

Galactose   Glucose

These sugars differed only by the manner in which an oxygen atom was attached at one of the carbon atoms around the ring. Yet that slight difference was enough to determine the way that the two attached and lay together to form lactose. Compare lactose with maltose. The drawings are highly schematic—structures in the chemist's sense, not the crystallographer's. The bonds joining the two rings at the oxygen atom are covalent, but are not double bonds, and so are not rigid.

Maltose

Lactose

The orientation of the two halves of the lactose molecule presented to the enzyme a unique conformation to distinguish and cleave. This was the bond at what by convention is labeled the *beta* carbon of the galactose ring, so that the point where the enzyme attached was called the β-galactoside linkage. Thus, the complete reaction took place in the sequence shown schematically on the next page.

The same linkage shows up in a range of compounds other than milk sugar, where a molecule of galactose is linked at that carbon to something other than glucose—and it was the availability of these that led Cohn and Monod to settle finally on the lactose enzyme system. The compounds are collectively termed β-galactosides. Monod and Cohn purified the lactase from *E. coli* and found that it would catalyze the breakup not only of lactose but of several other β-galactosides. So they renamed the enzyme "β-galactosidase"—and that is another indispensable term of the art. They knew from Lederberg's mating of *E. coli* that β-galactosidase was genetically controlled. It was, of course, a protein molecule and enormously larger than the lactose whose split it catalyzed. Later it was shown to be made up of four identical polypeptide chains, each 1,173 amino acids long. The three-dimensional structure of the enzyme, and how it finds and snips the β-galactoside linkage, are unknown. Cohn injected some of the enzyme into rabbits to provoke their immune response, and got a specific antiserum; when mixed into a solution suspected of containing some of the enzyme, the antiserum would combine with and precipitate even the faintest traces. This was an essential tool.

The individual names of the β-galactosides are necessary only to the specialist, but as a class the compounds leapt into importance when Monod and Cohn realized that some of them could mimic the effect of lactose, switching *E. coli* to producing the enzyme. So Cohn went to Cambridge and then to Bonn, to synthesize a whole series of these compounds, some simple, some

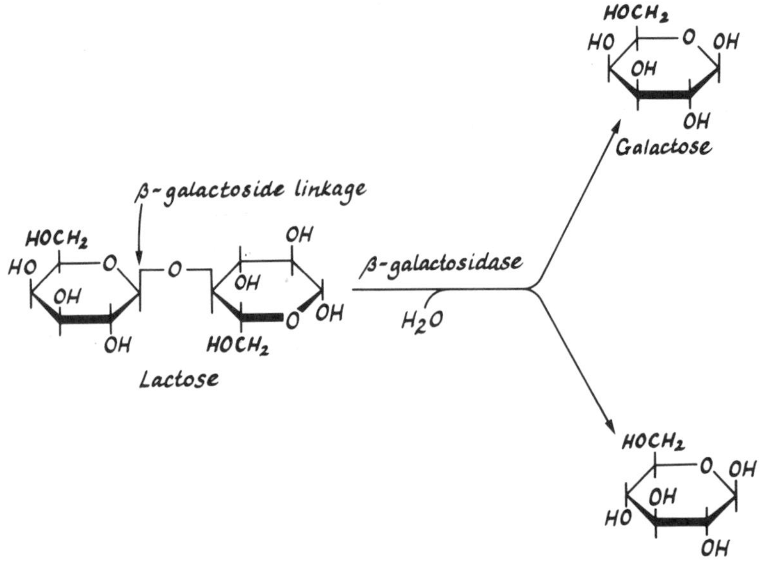

Galactose

β-galactoside linkage

β-galactosidase

H_2O

Lactose

more complicated. The most essential, as it turned out, were among the ones here, though the fact that there were many other systematic permutations is obvious from a glance at their diagrams.

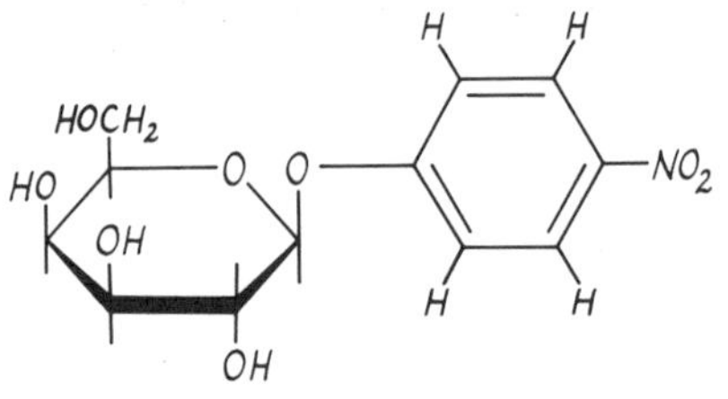

Para-nitrophenyl galactoside

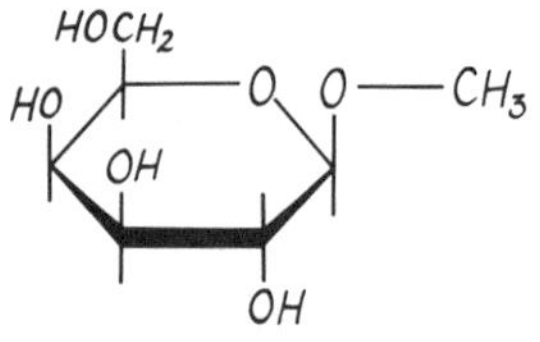

Methyl galactoside

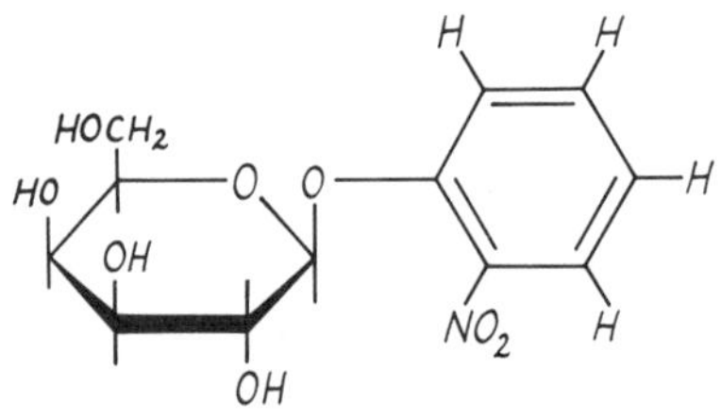

Ortho-nitrophenyl galactoside

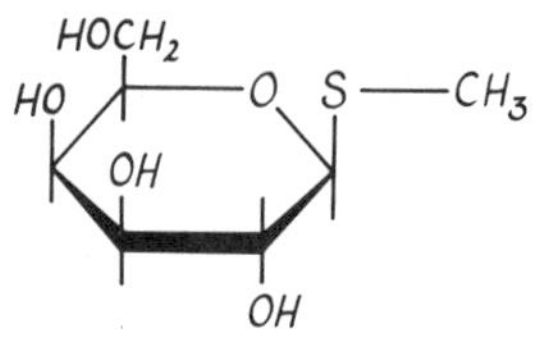

Thiomethyl galactoside

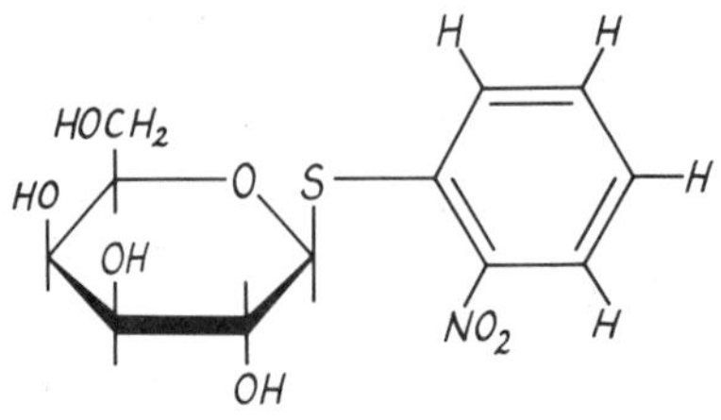

Thio (ortho-nitrophenyl) galactoside

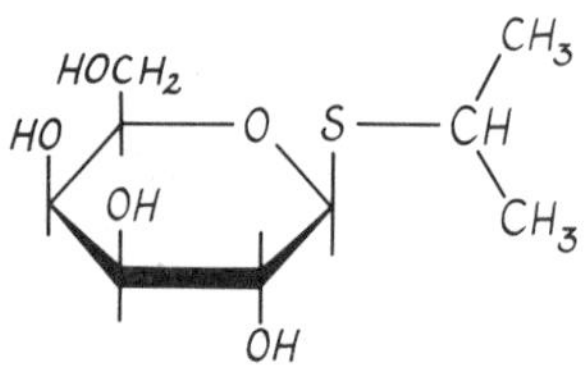

Isopropyl thio-galactoside

Such decoy substances, fed to E. coli, produced a startling variety of responses. Some, including the top two above, were split by β-galactosidase much more readily than lactose itself, and yet were incapable of causing the bacterium to start making the enzyme. Thus, where a single molecule of the enzyme broke ninety-five lactose molecules per second—not, in fact, a particularly fast performance in the narrow room of the bacterial cell—it would break three hundred twenty molecules of para-nitrophenyl galactoside in the same time. But the fastest turnover was found with the next substance in that list, ortho-nitrophenyl galactoside: twenty-six hundred molecules of this were split every second by each molecule of enzyme. Furthermore, this compound is colorless, but its nonsugar half, a hexagonal ring called orthonitrophenol, when split off by itself is a brilliant yellow. Lederberg had already observed that the compound thus provided a direct and vividly sensitive way to assay the activity of β-galactosidase enzyme in any sample of bacterial juice.

In complete contrast, the last two substances listed above, thiomethyl and isopropyl galactoside, in which the crucial linkage was made to an atom of sulphur rather than oxygen, were impervious to the enzyme—and yet, very surprisingly, were much more powerful than lactose itself in provoking the bacterium to start making the enzyme.

The variations went one further still. The third compound listed, thio (ortho-nitrophenyl) galactoside—chemists at this point begin to use initials to head off bewilderment, but we won't need the term again—was exactly like the sensitive indicator molecule above it in the list except that sulphur once more replaced oxygen at the linkage. As one would suspect, it also cannot be metabolized by the enzyme—but the surprise was that *this* compound acted as an antagonist, turning off the production of β-galactosidase even when lactose and no other sugar was present.

They had a complete and unrivalled tool kit. One substance would indicate how much enzyme was present. That could be checked independently with the specific antiserum. Some substances could be metabolized by β-galactosidase but would not start up its production. Other substances, the most interesting, would start up production of the enzyme but were not themselves affected by it. These they called "gratuitous."

Gratuitous stimulants to enzyme production were what Cohn had set out to find, he said—for a theoretical reason, but first of all to solve a practical problem in the design of the experiments. When E. coli began to produce β-galactosidase and metabolize lactose in the way that Monod had been studying since 1940, the cells always started more or less starved. "But those cells that are not as starved as the others get a head start," Cohn said. "So you're studying a response that is heterogeneous in the population in the time you're looking at it, and not what you want to see, which is the true kinetics, the true time-course of the response in the single bacterial cell."

With the bacteria out of step, the pattern of the biochemical response appeared more blurred than it actually was. "Now, when we discovered that you could give *coli* substances which it could not metabolize, which would start up the enzyme— Well, then you allow it to grow in a neutral medium using a sugar that does not induce β-galactosidase, or using any carbon and energy source that is neutral for this enzyme, and *then* add this enzyme-inducing substance that the bacterium can't metabolize; then for the first time you are able to study more or less the true course of the response in the individual bacterium." He thought about that. "The population ages semi-synchronously." Again and again, the initial problem in doing biochemistry or genetics with micro-organisms has been to get them to march in step. Lwoff's cultures of single bacteria were a radical avoidance of the same problem.

In short, with the array of artificial β-galactosides, Cohn and Monod could decouple the production of enzyme at will from its natural stimulus and from the natural substrate, lactose. The decoupling was one of the two fundamentals of the experiments that the lactose enzyme system made possible. The other fundamental was that, thanks to Lederberg, the system could be manipulated genetically. To take advantage of genetic recombination, they switched from strain ML to K12. With the system, Monod and his collaborators could dissect in minute detail the relation of genes to the moment-to-moment, live, responsive functioning of enzymes (a different question from the mapping of the relation of gene mutations and amino-acid sequences). This they proceeded to do.

Even as they started, the fact that the decoupling was possible at all changed profoundly their understanding of the nature of the phenomenon they had been calling enzymatic adaptation. Simply, since β-galactosidase synthesis could be started up gratuitously, by a substance that the enzyme could not metabolize, it was no longer plausible that in the normal case the starting up was triggered by a trace of enzyme combining with newly added lactose to convert an enzyme precursor. This was the theoretical point that Cohn had set out to test.

By that time, Cohn said, "The term 'adaptation' was causing a lot of confusion, because people would say, You're giving a bacterium a substance it can't metabolize; you're giving it the signal to make an enzyme it can't use; it's a very unhappy situation for the bacterium: how can you call that an adaptation? So in order to get on with the subject, we decided to change the name of what we were studying. We called it 'induction'—'induced enzyme synthesis'—so that we could keep away from any evolutionary or teleological arguments about whether it's adaptation or not. But historically it's the same phenomenon that was called adaptation."

❖ ❖ ❖

In September 1949, François Jacob came to Lwoff and asked to work in his group. Jacob was born in Nancy in 1920. When he was three years old, his family moved to Paris. The family was not professional or academic; his father was in the real-estate business. One grandfather, though, was a general in the French army and an accomplished mathematician. "And I had an uncle who was an M.D., whom I liked very much. And so it was more or less decided by the family that I should study mathematics; but at the time when I was supposed to move to the university, I was bored by mathematics, so I decided to do medicine. To be a surgeon."

Jacob is tall, thin, with dark hair cropped close and combed back, an air of reserve, a smile that suffuses his face with gentleness. His life, his style, his science have been more private than Monod's. We met a couple of times at the Institut Pasteur, and then at his apartment overlooking the Luxembourg Gardens, in brilliant spring weather. After a few minutes of our first conversation, I saw that he was always moving slightly, shifting in his chair, quick to get up to the blackboard, pacing, sitting again, crossing his legs, twisting one leg around the other. After he was nearly killed in 1944 in Normandy, the surgeons removed twenty pieces of bomb fragment but left another eighty, which he still carries. After the war, he finished his last four years of medical training in two. "Which means that I'm a very bad doctor! But then for many years I didn't know, very well, what to do; because I couldn't do surgery, because you can't do surgery with a bad arm." He worked for a spell as a journalist; he started research in antibiotics for a munitions firm that was trying to change to pharmaceuticals. "I was very rapidly disgusted by industry. I knew nothing much in medicine, but in science I knew absolutely nothing. But I had read two or three things"—exceptionally, he did not read Erwin Schrödinger's *What Is Life?* until years later—"and had got the feeling, the *right* feeling, that heredity, nucleic acid, and bacteria were going to be interesting. So, then I tried to figure out what were the interesting labs in France for that. And by inquiring among some friends I finally came to the conclusion that there were two labs, and one especially pleasant and very good, which was Lwoff's."

Jacob enrolled for the *Grand Cours* of the Institut Pasteur, a course in microbiology to which all principal scientists there contributed lectures, Lwoff and Monod among them. It was, and still is, the only thing of its kind in France. Afterwards, Jacob came to Monod and asked for a place in his lab. Monod told him (he told me years later), "I have no space. And anyway, I'm not the *patron.* Go see Lwoff."

Jacob once described the next encounter. He found Lwoff in his laboratory, eating his lunch with his secretary and his lab technician. "I told him of my wishes, my ignorance, my eagerness. He fixed me for a long time with his large blue eyes, tossed his head, and said to me, 'Impossible; I haven't got the least space.'" Jacob kept coming back—three or four times by Lwoff's memory, seven or eight in Jacob's account. Each time he got the same blue-eyed look, the same shrug, the same refusal. He tried for a last time in June 1950. The eyes were bluer than usual. "Without giving me time to explain anew my wishes, my ignorance, my eagerness, he announced, 'You know, we have discovered the induction of the prophage!' I said, 'Oh!' putting into it all the admiration I could and thinking to myself, 'What the devil is a prophage?' . . . Then he asked, 'Would it interest you to work on phage?' I stammered out that that was exactly what I had hoped. 'Good; come along on the first of September.'" Jacob went down the stairs, out to the street, into the first bookstore to find a dictionary.

Jacob began work in the attic of the Institut Pasteur at a moment when biology was effervescent with change. "When I arrived, the lab was divided in two parts," he said. "There was a corridor between them; at one end, Lwoff was pouring UV light on lysogenic bacteria, including phage, and at the other end of the corridor Monod was pouring galactoside derivatives on *coli*—derived from the intestines of André Lwoff—to get enzyme induction. And the great joke was about induction, because each was 'inducing' according to his fashion, convinced that the two phenomena had nothing in common except the name." Even to describe the research in such parallel terms suggests a similarity of mechanism that went unglimpsed by anyone for another three years at least.

Jacob started a thesis on lysogeny in still another kind of bacteria. Then, also, Élie Wollman returned to Paris from two years at Caltech. Wollman had been working in Delbrück's lab, with Gunther Stent, to determine the exact timing of the first steps by which phage infected their hosts. One day while going through the files in Pasadena, he came across a copy of a paper by his parents about lysogeny, across which was written, "Nonsense!" In Paris, Wollman joined Lwoff and Jacob to pursue lysogeny. After Esther Lederberg, in 1951, discovered lysogeny and bacteriophage *lambda* in strain K12 of *E. coli*, Wollman and Jacob switched to those organisms. Because genetic recombination took place in that strain, lysogeny could be approached by genetic methods. In the fall of 1951, Seymour Benzer and Gunther Stent arrived at the Pasteur.

The next April, Lwoff, Stent, and Benzer went to Oxford for that meeting of the Society for General Microbiology where Luria was on the program to talk, but was prevented by the U.S. State Department's refusal to give him a passport. Instead they heard Watson read out Hershey's letter about the Waring Blender experiment, in which phage shells had been separated from freshly infected bacteria, showing unexpectedly that phage protein remained outside while it was DNA that penetrated. The occasion was one of those crossroads that people reach at the same moment but from very different directions. For Watson, the Hershey-Chase experiment was first of all about the importance of DNA. For the delegation from the Pasteur, including Benzer, Hershey's letter was about bacterial and phage genetics.

William Hayes, from Hammersmith Hospital in London, was also at the Oxford meeting. Lwoff and Stent met Hayes there, and first learned—months before Watson heard of it—that certain crosses of mutants of *E. coli* strain K12 suggested that easily traced characteristics, like resistance to streptomycin, were passed from one bacterium to another in one direction but not in the reciprocal way. This was the news of sexual differentiation in bacteria—the distinction between genetic donors and recipients that Hayes had found working alone in London, and that was independently discovered by the Lederbergs with Luca Cavalli-Sforza in the United States and Italy. Word of Hayes's discovery was carried back to Paris. In May, Wollman went to London to do some experiments with Hayes. At the end of July it was he and Stent who spread word of Hayes's work at the phage meeting at Royaumont. And this, in turn, was the stimulus that excited Watson to spend the fall of 1952 writing the paper with Hayes that claimed, incorrectly, that Joshua Lederberg's published data showed *E. coli* to have three chromosomes.

Only Hayes, at that early point, interpreted the discovery aright. When bacteria conjugated, he said, they did not pool their genes, as Lederberg supposed, but rather one, the donor, or male, passed a copy of its genes, often only part of them, to the recipient, or female. Oddly, the ability to be a donor was something a male bacterium could pass to a female: bacterial masculinity was itself a genetic element that could be transmitted by conjugation. Hayes discovered not just bacterial sex but bacterial sex change. He called the genetic element the sex factor, often written F for short. The bacterium that had got it was F^+.

❖ ❖ ❖

"Let me mention one thing, which is now *completely* forgotten and unknown," Monod said in the course of our long conversation that evening early in December 1975. "There is one concept in biology that everyone accepts as quite obvious today, which wasn't obvious at all until 1953 or '54. Namely, that macromolecules in the cell are stable molecules. The Schoenheimer myth that proteins in particular are things that are in a so-called dynamic state—exchanging pieces, say amino-acid residues and so on, with the rest of the cell—was *absolutely overwhelming*. When I started studying, systematically, enzyme induction, which stems from the work that I did underground with Alice Audureau in 1943, '44, it wasn't at all clear that enzyme induction corresponded to the synthesis of a new molecule. It could be activation of a precursor, which was one of the assumptions we studied. It could be conversion of one protein into another. In fact, that's one of the assumptions I made myself. And designed an experiment to test. And it's only after some *very* painful work with Mel Cohn that we came to the conclusion that enzyme induction consisted in the actual synthesis, from its elementary units—namely, amino acids—of the whole molecule of a protein which once formed was stable.

"Until you knew *that*, you couldn't even begin to build a theory of what was happening. In other words, for many years I worked on this system without knowing for sure that I was working on protein synthesis. Although that's what I was hoping. Not that I had any serious reason at the time, any experimental reason, to believe it. Just because it would be much more interesting if it turned out to be that way."

One could imagine, in 1951, that the β-galactosidase as formed in response to different inducers might show slightly different specific properties—might be, in other words, not one but a family of enzymes with slightly different molecular structures. Cohn and Monod put immunology to work to test the possibility. The experiments were numerous, thorough, tedious. The results were negative.

Monod then made his first direct attempt to distinguish between enzyme produced by conversion of a precursor and enzyme built up from amino acids entirely anew. Eleven years after he had taken the double growth curves to Lwoff in puzzlement, Monod penetrated to the events that lay behind the curves. He did the experiments with Germaine Cohen-Bazire and Alvin Pappenheimer, Cohn's former chief, now in Paris on sabbatical. The work had two parts, related in technique yet so different in style that they consort oddly in the same paper, a thornbush with one bright flower.

In the thorny part, they isolated a series of mutants of *E. coli*, each unable to make a different one of the amino acids. They didn't get mutants for all twenty, but accounted for eleven. Then they grew each mutant in a medium that offered a limiting quantity of the amino acid it particularly required, and a great excess of all other nutriment. Growth stopped when the one amino acid was exhausted. Then they put in inducer, and waited twenty minutes. In two cases they got insignificant traces of β-galactosidase; in the other nine no β-galactosidase was synthesized at all. Then they put in a small quantity of the deficient amino acid. In every case, synthesis of β-galactosidase then became copious until the amino acid was exhausted once more. The results were without ambiguity, and the paper said so: "Each of these amino acids behaves as though entering into the composition of new protein. . . . One must say that the formation of the molecule of β-galactosidase only occurs in conditions permitting synthesis *de novo* of a complete protein and not by just adding a few amino acids to a pre-existing protein."

Then in the spring of 1952, they looked at the rate of induced enzyme synthesis, in normal bacteria, and produced a single, striking demonstration. To a culture of *E. coli* that was growing normally on a sugar other than lactose, they added a gratuitous inducer. From moment to moment they took samples. The bacteria began synthesizing β-galactosidase almost immediately, and continued at a fixed rate exactly proportional to their overall growth. The unfluctuating steadiness of the rate of synthesis—a single straight line on a graph—in response to an inducer that the enzyme could not use as a substrate, said immediately and even more strongly that β-glactosidase could not be converted from some precursor but must be synthesized from scratch. The promptness with which the bacteria responded to inducer was also important, though the problem to which this was part of the answer—namely, the functioning of the ribosomal particles in protein synthesis—was not even in sight yet. Synthesis of β-galactosidase began about three minutes after the inducer was added.

❖ ❖ ❖

In September of 1952, Benzer, Stent, and Pappenheimer left Paris and another contingent of Americans arrived. The genetic techniques and bacterial strains that Benzer had learned to use in Paris—in particular, *E. coli* K12 lysogenic for phage *lambda*—were the ones he put together a year later to begin mapping the fine structure of the rII region. Benzer's Paris apartment was taken over by David Hogness, a postdoctoral fellow also from Caltech, who had trained as a physical chemist but had done his thesis in genetics. Hogness had come to work with Monod. In Lwoff's attic, the two ends of the corridor were getting closer together. Now lysogeny as well as galactosidase was being investigated in *E. coli* K12. Terminology had merged; thinking was on a converging course. By the early fifties, Lwoff later wrote, "Jacques Monod used to say that the induction of enzyme synthesis and of phage development are the expression of one and the same phenomenon. The statement looked paradoxical, but was, paradoxically, a remarkable intuition."

❖ ❖ ❖

Adaptive, inducible enzyme systems were rare. Most enzymes were of the sort called constitutive which the cell synthesized at normal rate without the stimulus of a substance added from outside. Constitutive enzymes included, for example, those that catalyzed the steps in making the various small building blocks of the cell, the essential metabolites such as amino acids. From the beginning of Monod's research it had been an obvious question whether there was a single, universal control mechanism that would account for constitutive enzymes, too. Joshua Lederberg had collected mutants of *E. coli* in which the cells' metabolism of lactose was altered. Among these were strains in which the synthesis of β-galactosidase had become consitutive, meaning that synthesis of the enzyme went on at top speed all the time even with no lactose, no other inducer or substrate added to the system. These mutants lent force, if not clarity, to the conviction that the two types of enzyme system were closely related.

Monod's idea of the relationship was prejudiced by his concentration on an inducible system. By the winter of 1952–'53, he once said, "It did not appear unreasonable to suppose that the synthesis of constitutive enzymes was governed by their substrate as produced internally by the cell, which would be implicated if the mechanism of induction was, in reality, universal." So he chose a typical constitutive enzyme, tryptophan synthetase, the one responsible for the final step in a tortuous path that leads to the amino acid tryptophan, and tried on it one of the methods he had developed with Cohn. The substrate of that enzyme was the small molecule one step before tryptophan in that pathway; by the not unreasonable supposition, this substrate would also be the enzyme's inducer. Monod conjectured that a chemical analogue of the substrate, supplied in quantity, would block induction and turn off synthesis of the enzyme. The handiest analogue of that substrate was tryptophan itself. Sure enough, when he and Cohen-Bazire put an abundance of tryptophan into the system, synthesis of the enzyme was powerfully inhibited.

They did not follow out this discovery. The first to pick it up was Leo Szilard, at the University of Chicago. Almost since Szilard had joined the phage group in 1947, he had been fascinated by Monod's original demonstration that *E. coli* fed a mixture of glucose and lactose will deplete the one sugar before using the other; in 1950 he had devised a way to show that if the glucose level in the mixture was maintained at an extremely low level, the bacteria would use both sugars at once. In the summer of 1953, Szilard and his regular collaborator, Aaron Novick, were working with a mutant of *E. coli* that was unable to make its own tryptophan. The mutant, when starved of the amino acid, made a lot of a substance that was an earlier step in the tryptophan pathway. The instant tryptophan was supplied again, production of that precursor turned off. From Szilard's work and then others', it emerged that the cell's synthesis of tryptophan synthetase and many another constitutive enzyme was not stimulated by the presence of some inducer; rather, it was turned on by the absence of the end product of the particular biosynthetic pathway—by the absence, in this case, of tryptophan. When tryptophan was in good supply in the environment, synthesis of the enzymes needed to make it was parsimoniously turned off. As Szilard emphasized, this was a classic case of feedback control.

In a letter to *Nature* sent off on 8 October 1953, and published two months later, Monod, Cohn, and three of their visitors from the United States and England formally proposed that the world's biologists abandon the term "enzyme adaptation" in favor of "enzyme induction."

In the summer of 1956, Henry Vogel from Yale University suggested that, by analogy, the turning off of constitutive enzymes by their end products be called "enzyme repression." Monod in the spring of 1953 had found the first example of an enzyme whose synthesis was repressible, but had taken it as verification of a generalized hypothesis of induction that was false.

❖ ❖ ❖

Their minds at play with the symmetries of induction and inhibition, and feeling the itch for order, the recurring compulsion to bring both mechanisms under a single rule, Monod and Cohn in the early fifties sometimes argued the logic whereby two negatives make a positive, whereby induction might require not provocation of synthesis but the inhibition of an inhibitor of the synthesis: thinking of the analysis of the subtleties of the game of poker by Edgar

Allan Poe—for Edgar Poe is one of the odder resonances between French and American culture—they called this logic the theory of the double bluff.

❖ ❖ ❖

The culmination of the painful work to show that enzyme molecules were newly made from amino acids, and once made were stable, came in an experiment of characteristic conceptual simplicity. Monod, Cohn, and Hogness carried it out in a few days towards the end of 1953, though the design, the techniques, the controls—and the writing up—stretched over a year and a half. In the first phase of the experiment, they grew *E. coli* cells for several generations at blood heat in flasks of broth that contained as the only sulphur source the radioactive isotope ^{35}S. This got incorporated into the two sulphur-bearing amino acids, cysteine and methionine, and so built into proteins. No galactose inducer was present. Succinic acid—not a sugar but chemically close—supplied energy and carbon. Everything besides sulphur that the cells needed was provided in excess. As the hot sulphur was used up, growth slowed and stopped. The bacteria were left starved of sulphur for an hour, so that all of the radioactivity was built into proteins.

To start the second phase, the experimenters diluted the starved culture with fresh broth; conditions were reversed, so that the sulphur was not radioactive but inducer was present, a generous amount of the gratuitous inducer thiomethyl galactoside. Immediately, the bacteria began to grow and to synthesize β-galactosidase. Growth stopped again, as did the synthesis, after several generations when the supply of cold sulphur was used up. Then the cells were centrifuged, washed, ground up, all the protein extracted, and finally, in several steps of purification, the β-galactosidase isolated. The essential last step in isolation was immunological, using Cohn's anti-β-galactosidase serum from rabbits, which specifically precipitated that protein and nothing else.

In one control, Monod, Cohn, and Hogness tried isolating β-galactosidase from cells fed radioactive sulphur but never induced. They found a slight trace of enzyme, and it was, of course, radioactive. In a second control, they put in radioactive sulphur and inducer together, and grew β-galactosidase that was saturated with radioactivity. The two controls marked the possible range. But the main experiment put the question from the negative side, and that was its power: when inducer was present, the free sulphur was not radioactive. The result was definitive: β-galactosidase was synthesized in full quantity, but showed only a faint trace of radioactivity, no more than in the first control. They reported:

> The results of the precursor experiment indicate that β-galactosidase is synthesized exclusively from material that is assimilated after the addition of the inducer and hence proteins existing in the non-induced bacteria play no significant role as precursors.
>
> A second conclusion can be drawn from the results of the precursor experiment, namely, that non-β-galactosidase proteins are stable, not being degraded to amino acids by any mechanism. For if the state of the proteins within the cell consists of a continual synthesis from and breakdown to their constituent amino acids (*i.e.* state of "dynamic equilibrium"), then one would expect the β-galactosidase synthesized in phase II of the experiment to be labelled with ^{35}S as a result of the breakdown of the radioactive proteins.

Hogness once described that work to me. "It was part of the dogma of how things got regulated and how things got adjusted, at the time, that there was a lot of protein turnover, breaking down to amino acids. The data was for mammalian cells. But there was a lot of confusion in that data. So when we showed that there was no protein turnover in exponentially growing [that is, doubling at full rate] bacterial cells—it later turned out that there is some turnover when they go into lag phase, that is, when they've achieved full growth—it caused, among the more classical biochemists, disbelief. But the result was very clear."

"Our conclusion was against the whole *Zeitgeist* of the time," Monod said. "And I remember it was early in 1954, I lectured in the States, giving seminars in various places, including in Berkeley—this is where I met Gamow—on the subject that the interpretation of the so-called 'dynamic state' of the proteins was wrong. That proteins inside a cell can be perfectly stable. Which we had demonstrated with Mel. You've got to realize that it raised an absolute *furor!* There was this Hegelian idea, you know, that this dynamic state was a sort of secret of life. . . . At that time, the only people who were fully aware that this business of the dynamic state of protein molecules couldn't be right, were the crystallographers. Because it couldn't be right if they got good crystals. To which, of course, the cell physiologists or biochemists said, 'But you are looking at *dead molecules!*' You see? Let me try to find a quotation that I know I have in one of my old papers."

He went to a bookshelf, took down one from a set of leather-bound volumes. The floor squeaked. He leafed through the pages to the paper he had written with Hogness and Cohn. "I remember this sentence raised a furor with many of my colleagues." He skimmed for a moment, then began to read. "'This suggests'—this is the interpretation of our results—'very strongly indeed that turnover rates measured under these conditions'—that is, the Schoenheimer type of conditions—'express the dynamic state of the tissue, rather than the state of the protein molecules within the cells. In any case, there is, to our knowledge, no experimental evidence that the proteins within the cells of mammals are any more dynamic than those of *Escherichia coli*'—which we had just demonstrated were not dynamic, but stable molecules. And, mind you"—he put the book away—"if it were not true that macromolecules, proteins, are stable, molecular biology would not be what it is."

Matthew Meselson heard Monod's lecture at Caltech that spring, and began to wonder whether it would be possible to distinguish newly synthesized enzyme molecules from older ones in the ultracentrifuge by labelling one stage of growth with heavy hydrogen in the broth.

Had Gamow mentioned coding? "Yes," Monod said. "I do remember seeing Gamow, in California, very soon after he had written this paper [about the code] and he was telling me about it. And because he was insisting so much on his so-called diamond structure, and I felt that this could not be right, therefore—as often happens in science—I argued with him about the diamond structure, which was really of no importance, and missed discussing the really basic point, is there colinearity or not?"

At the end of that summer, Cohn left Paris for Arthur Kornberg's Department of Microbiology at Washington University in Saint Louis. He and

Monod remained close friends, writing frequently; Monod's letters to Cohn are an anthology of experiment, speculation, and gossip.

❖ ❖ ❖

At the Institut Pasteur, early in 1955, François Jacob and Élie Wollman uncovered a new phenomenon in bacterial mating that gave physical reality to the differentiation into male and female, genetic donors and genetic recipients, that had been argued out for three years. With that, they began to resolve very rapidly the paradoxes that had swarmed over bacterial genetics. The explanation emerged from an idea of Wollman's for an experiment of spectacular simplicity.

Genetic recombination in bacteria had always been a rare event. In June of 1953, at the symposium at Cold Spring Harbor where Watson presented the structure of DNA, Jacob gave a paper in which he and Wollman summarized what had been learned about lysogeny so far. Hayes read a paper on sexual differentiation in *E. coli*, and mentioned that he had isolated a new substrain of K12 that produced recombinants a thousand to ten thousand times more frequently than normal. The new bacterium got tagged *Hfr* Hayes, the name to signify high-frequency recombination and to distinguish it from a similar strain found by the Lederbergs' colleague Cavalli-Sforza in Italy. Jacob brought the news of high-frequency recombination back from Cold Spring Harbor to Paris. He also brought back a Waring Blender as a present to his wife.

"The discovery of *Hfr* strains was a great breakthrough to the subject," Wollman said in a conversation in the fall of 1976. We were in his office at the Institut Pasteur, a corner room, correct, carpeted and curtained, with bookcases behind him; he had been vice-director of the Pasteur for ten years. He was short, with wavy hair, a long straight nose and a long stubborn jaw, a wide mouth firmly set; his manner was an attractive blend of stiffness and thoughtful openness. Had the solution of the structure of DNA, in the same months as the discovery of *Hfr* strains, made much difference to the work in Paris? Wollman hesitated, and then said, "As a matter of fact, it didn't. At first. Well, it was received by us as a very great advance, of course, but it had no direct implications at the time." *E. coli* K12 *Hfr* Hayes was a male, or donor, bacterium. Within a few years it had spread, a benign plague transmitted by post, to genetics laboratories all over the world. Wollman and Jacob wrote to Hayes asking for a sample; he mailed them some. From this they grew up their own stocks and began to look for mutants affecting one characteristic or another. With recombination now frequent, evidence mounted up that the genes for some characteristics passed from male to female regularly, others not. One characteristic regularly transmitted was the genes for phage *lambda*.

"Go a little earlier now," Jacob told me. "The conclusion of Lwoff's work was that in lysogenic bacteria, the genetic material of the phage is present in a hidden form, which he called the prophage. And induction of phage development was activation of this prophage. In some way this thing was sleeping, and when you put ultraviolet on it, he got awoken.

"Now the question of course was, what is exactly the structure, and where is it located, what's the relationship inside the cell, with other components? Okay. And it very soon appeared that you could solve that only by genetic means, and obviously by conjugation in bacteria. But the problem was that this conjugation that had been found by Lederberg, originally, occurred at very low frequencies. You mixed cells and you had about one in a million recombinants. Until Hayes and Cavalli found these *Hfr* strains—that is, males which mate at high frequency. These strains gave ten per cent recombinants. Okay. And the first experiment— We wanted to find out what was the behavior of the prophage, which meant doing crosses between lysogenic and non-lysogenic bacteria, to see what happened among the progeny. And the very first thing we found was that if you were crossing a male, lysogenic—is that all right? you follow?—with a female that was nonlysogenic, a large proportion, half or more, of the females lysed and produced phage." When they tried the cross the other way, males that were *not* lysogenic to females that were, none of the bacteria produced phage.

"It was a very strange phenomenon. It meant that you had a cell which carried this dormant potential phage and which was unable to produce phage, but when you crossed it with a female which knew *nothing about* phage, the prophage was going to the female and then was activated. And this is what turned out to be exactly the same thing that Monod was doing at the other side of the corridor." Jacob laughed. "It was a very good tool, because it was just like a light turning on to show that mating was occurring and that the prophage was going into the cell." On reflection, it showed one thing more: whatever control it was that kept the prophage from being expressed was not donated to the female along with the genes for the phage itself.

Late in 1954 or early in 1955, these many elements came together in Wollman's imagination, and he saw a way to investigate the exact timing of the events of bacterial mating and the early moments of genetic recombination. He borrowed the idea of the Waring Blender from Hershey, put it to use with Hayes's strain of *E. coli.* As usual, the method was to cross mutants that had multiple differences that could be traced out in the next generations.

He and Jacob prepared two cultures. The first was of *Hfr* males that had healthy appetites but certain infirmities: that is, they could synthesize the amino acids threonine and leucine and could live on either of the sugars lactose and galactose, but they were sensitive to phage T1, to the poison sodium azide, and to the antibiotic streptomycin, which last would stop them from multiplying themselves, but not from mating. The second culture was complementary: females that were finicky feeders but otherwise robust; they could not synthesize the two amino acids nor digest either of the two sugars, but were resistant to phage T1, to sodium azide, and to streptomycin.

Wollman and Jacob mixed some of the two cultures, and spread diluted samples from the mixture on glass plates of jelly that contained streptomycin. The dishes also contained either lactose or galactose and various combinations of amino acids, chosen to select and grow up colonies whose female ancestor had acquired the ability to digest lactose or galactose, to synthesize threonine, and so on. With a supply of mutants suffering from assorted nutritional oddities, these experiments could be done quickly and in many similar combinations.

The first thing that Wollman and Jacob showed was entirely predictable: bacteria took time to find their mates. At intervals from the moment of mixing the cultures, they drew samples, immediately diluted them to stop any new matings, but then plated them so that bacteria that had started conjugation could finish. In fact, the frequency of recombinations increased steadily for an hour before levelling off. So then they tried the matings and platings all over again, but now every time they removed and diluted a sample, they put it into the blender—which had found no place in a French kitchen: "My wife was completely disgusted with this instrument"—at high speed for two minutes, to separate males from females. In this series, with mating broken off after various set intervals, they would see how long it took for the genes to be transferred. Wollman, and soon everybody else, called the experiment *coitus interruptus.*

Thus they discovered that the characteristics were not transferred at the same time. Rather, each characteristic had its own moment—some early, some late—before which it would not show up in recombination at all. Jacob went to the blackboard, drew a horizontal line for the bacterial chromosome, marked it off at intervals. "We knew a certain number of markers. On the bacterial chromosome. Let's call them A, B, C, D, et cetera, on the male. And on the female, A minus, B minus, C minus, et cetera—so that in the progeny you could recognize the recombinants. All right?" He drew the second line.

"So we thought, let's try to separate; maybe we find a moment in time, and before that there is not yet any mating, but after that something has happened.

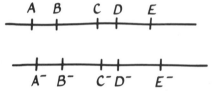

And when we did this kinetics, we just realized that if we interrupted at ten minutes, the character A from the male had gone to the female, but not the others. That at fifteen minutes, A plus B had gone into the female; and at twenty minutes, A plus B plus C." The result was in flat contradiction of the hypothesis Lederberg was then advocating vigorously, that conjugation was a fusion in which all genes were pooled; those that recombined rarely, Lederberg had said, were eliminated after the fusion.

Lwoff's people had tea together every afternoon in the corridor near the box where the ultraviolet lamps were set up. One of those afternoons, Wollman and Jacob discussed the result with Monod, and Jacob suggested that the male transferred a copy of his chromosome into the female like a long piece of spaghetti, which was broken at whatever point had been reached when the two cells separated. They called it the "spaghetti hypothesis." In other words, Jacob said, "When you put the male and the female together, something happened and the chromosome of the male was going through to the female according to a very precise time schedule. And this was a tremendous tool." He made another quick sketch of bacteria conjugating, and I copied it down.

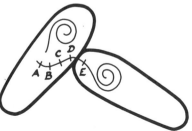

The experiment had begun, Wollman insisted, without a glimpse of where it would lead. "Purely, at the time, the idea was only to make a kinetic"—that is, timed—"study of the conjugation process," Wollman said. "I had *no idea* that it would be a unidirectional transfer of chromosome. The experiment was done without preconceived idea. Absolutely!" His voice had been low; now he was vigorous, definite, and animated. "We did it just to know whether, when you separated the bugs at different times during mating, the same phenomena that you had seen before, which was the unequal transfer of genetic markers"—that is, chosen characters different in two strains—"to recombinants would still happen or not." They had thought they might find one of several explanations, Wollman said: a complete heterogeneity of mating behavior in males, perhaps, or even Lederberg's fusion followed by gene elimination. "Or *anything*. The idea was just to see how different markers behave when you make interrupted—ah, short pulses of matings. Because the fact itself could *not* be anticipated. I mean, nobody had ever thought before of a chromosome which would be transferred from one end, and be injected during two hours, in bugs which normally divided *every twenty minutes!* This was—this was unforeseeable!"

Wollman and Jacob had stumbled upon a way to measure off the genes on Hayes's bacterial chromosome as directly and physically as a child squeezes toothpaste onto a brush or a carpenter unrolls a coiled steel tape measure. As they saw instantly, and reported in a note in mid-June 1955 in the weekly *Comptes rendus* of the Académie des Sciences, they had the means to make a genetic map of biochemical characteristics expressed in units of time. Perhaps the most surprising part of the discovery was how long that time could be: even when they started they found some characters that were not transferred from donor to recipient until fifty minutes from the onset of mating.

❖❖❖

Georges Cohen had known Monod since 1944. "When I first met him he had the uniform of a French army commander. We were both members of the Communist Party during the war. I was in the Party because there was nothing else to do during the German occupation. It was two or three days after the liberation, in Paris, at the Collège de France. I had received notice that the so-called intellectuals of the Communist Party, in the university, should meet, to discuss matters, and then this man came in, and this was Jacques Monod." Cohen, when we talked one Saturday afternoon late in 1976 at his laboratory, was short, plump, balding, cheerfully helpful, with a deep, gravelly, bubbly Parisian voice, an American accent to his English, and a huge wet cigar. Cohen was ten years younger than Monod; he had trained as a biochemist, and was already working at the Pasteur when Monod got there in 1945, though not in Lwoff's unit. Cohen left the Communist Party at the same time Monod broke publicly, and for the same reason—the Lysenko affair. In 1954, Monod was made chief of a newly organized unit for cellular biochemistry at the Pasteur. The next year he moved to the ground floor of the same building, to a large set of laboratories refurbished to his specifications and the pleasant corner study where I had first met him. Cohen joined Monod's new unit shortly before the move.

"I was working on the effect of amino acids on growth, things like that, and he proposed me to work on enzyme induction, for a change," Cohen said.

"And as you know, if you add small effectors, small compounds like thiogalactosides"—the gratuitous inducers—"instead of lactose, to a growing culture of *coli*, the bacterium answers by making big amounts of galactosidase. Which is the phenomenon of enzyme induction Monod was working on. So he says, why don't you make the compound, the inducer, radioactive, add it to the culture, and see where it goes during the course of induction. Whether we can find it bound to a given fraction of the cell, or a given macromolecule, after breaking the cell. So I went to Germany"—he went to the University of Bonn in November 1954—"learned how to synthesize, in an organic chemistry laboratory, the necessary compound, and came back and did the experiment. And as a result, we didn't find at all the mechanism of induction. That would have to wait five more years to be found. This was a very naïve approach to the phenomenon, as we come to see now in retrospect. But instead, as a result, I discovered the permeases—the protein in the cell membrane that is specifically responsible for the entry, into the cell, of a family of compounds."

When Cohen added his radioactive galactoside to cultures of *E. coli*, it was quickly taken up and strongly concentrated within the cells if they had already been induced, but was accumulated only in negligible amounts by cells that had not previously been given an inducer. In ten days of intense work and by genetic analysis whose intricacy we need not pursue, Cohen and Monod showed that synthesis of β-galactosidase was not by itself sufficient to enable intact *E. coli* cells to use lactose. A different, preliminary factor was necessary, for in its absence the cell did not accumulate lactose or other galactosides in the first place. They showed that this new factor could only be a specific protein, another enzyme. Its place was in the cell membrane. Its function was to pull lactose through the membrane—an active transport even from a soup where the sugar molecules were far between into a cytoplasm where they were already concentrated.

Somewhat later, they named the new enzyme galactoside permease. Synthesis of the two enzymes was turned on by the same inducers and at the same time. Yet the permease was governed by a gene that was distinct from the gene for β-galactosidase, for it proved possible to breed bacteria that could make one enzyme but not the other, either way around. The gene for β-galactosidase was already tagged Z in genetic maps, so Monod and Cohen labelled the gene for the permease Y. The evidence for the existence of the permease was entirely indirect. They could not isolate the protein itself; it was isolated only ten years later and by another laboratory. They quickly found that other permeases are necessary and specific to transport other sugars into the cell; many more have since been identified.

❖ ❖ ❖

On 4 April 1955, Francis Crick wrote from Cambridge, saying, in part:

Dear Professor Monod,

My wife and I were thinking of making a trip to Paris some time early in May, and I was hoping that I might have the opportunity of making your acquaintance. For some time now I have held the view that induction was due to a change in the folding up of the protein *during the synthesis,* basing this on general arguments about what we know about protein structure (which isn't much) and protein synthesis (which is even less). As far as I can tell all the experimental results can be fitted into a picture of this sort if considered carefully enough, and I notice that your own ideas must be along similar lines, to judge from a reference to antibody synthesis in your last paper.

This was the trip when Crick also lunched with the Ephrussis. Pauling had argued since the mid-forties that the astonishing specificity—and hence multiplicity—of antibodies had to be due to the protein chain's adjusting its configuration according to need. This was yet another variation on the theme of adaptation, and even Crick was susceptible. He gave an informal seminar at the Pasteur on May 18, and met the people there for the first time. "I remember very well that we all knew Jim, which was unmistakable," Jacob said. "Once you have seen Jim once you couldn't miss him, and we just had no idea what Crick was, so Crick was just an appendix to Watson for us—until we saw Crick, and then it was clear that Crick was not an appendix to Watson."

❖ ❖ ❖

Discovery of the permease led Monod to the next essential realization—a succinct example of his particular style in science, for the realization was obvious, provided one came at it exactly aright. Scientists reach an extreme, sustained identification of the patterns of their thought with the patterns that they perceive in, and project into, the phenomena they are trying to elucidate. Lawrence Bragg was a lifelong student of optics: the power and focus of his spatial imagination were peerless. Linus Pauling, more than any other, brought quantum mechanics into the study of molecular structures: he was believed by his contemporaries—Bernal found him uncanny—to apprehend problems directly in their quantum relations. The phenomena that Monod dealt with, though one can't say they were more difficult, were different, for they were not in the first place spatial but instead temporal—the unfolding of an interlocking, often highly reflexive, sequential pattern. In each of the discoveries he made in the ensuing ten years there was a moment of total absorption as he resolved the bacterial cell like a partly cut diamond slowly in the light, then a gleam of perception so quick—and so quickly resolved into its place in the sequential pattern—that to Monod himself, on his testimony at least, there had been nothing much to call an intuitive leap, merely an extension, subject to test, of the inevitable logic of the system itself. The discoveries were about control. They had to be indirect.

"The real interest to me of the discovery of the permeases— But this is also one of these logical things that are terribly difficult to put forward," Monod said. He started again. "It is the discovery of the permease which allowed me to demonstrate that the mutation from inducible to constitutive is due to a genetic element that is distinct from the elements which define the proteins."

The logic is there, entire but compressed and unitary, just as it was the first moment he perceived it. Break the logic down: Gene Y for permease was distinct from gene Z for β-galactosidase, because bacteria could be bred that made one protein but not the other. When both genes were present, induction of the enzymes was simultaneous. In a constitutive mutant, making enzyme without the presence of inducer, if both genes were present, both enzymes were made at full rate—and the arresting fact was that by no mating was it possible to get one enzyme inducible while the other was constitutive. "We showed that the inducible-to-constitutive mutation affected the galactosidase and the permease simultaneously." So a third gene—at least, a third genetic element—controlled inducibility itself. Yet it was in the chromosome—part of the bacterial DNA. It could be mapped. They labelled it *I*. "And then I soon found that you could show that all these genetic elements were very closely linked. By fairly straightforward genetics. With the instruments devised by François and Élie it was easy to show that they were closely linked." Interrupted-mating experiments placed I, Y, and Z so close that they were probably contiguous on the chromosome. It was hard to tell whether I was between the two genes for the structures of proteins or to one side.

❖❖❖

In 1956, Wollman and Jacob published a first, rudimentary timed map for Hayes's strain K12 of *E. coli;* a line of seven characters, spaced out, with the ones affecting synthesis of threonine and leucine close together at eight minutes from the start of conjugation, and the gene for the β-galactosidase that they had been scoring, at twenty-five minutes. Immediately after the lactose genes on the map, at twenty-six minutes, appeared the genes for phage *lambda*.

Again, but more vividly than before the invention of interrupted matings, in a cross between a male carrying the *lambda* prophage and a female that was not lysogenic, the instant the prophage was transmitted it was induced. "Just like a light turning on." But this was a dramatic single case of a more general fact, for when the male's chromosome entered, the female bacterium became for a period of several hours a cell endowed—almost as though it were a fruit fly or a maize plant—with two sets of genes. After that, cell divisions brought the daughter cells down to single chromosomes again. But in that interval, geneticists could investigate in bacteria questions that previously could be asked only about higher organisms, such as the possible dominant character of one version of a gene over its opposite number on the other chromosome, or the exchange of genes by crossing-over of segments between the chromosomes. If bacterial genetics had been born at least twice before, this time the science reached sexual maturity.

Jacob went to the States again in the summer of 1956, and Wollman went too, for Cold Spring Harbor and then the McCollum-Pratt Symposium at Johns Hopkins University, in Baltimore. The tour overlapped with Francis Crick's. In the long letter that Crick wrote to Brenner in July, after the meetings, he said, among all the other bulletins:

> *François Jacob* gave a beautiful talk on the coitus interruptus. Most of this I'm sure you know, and in any case its in the latest Scientific American. He also has a lot of detailed work on the loci of various lysogenic phages [that is, mapping the places where the prophage was integrated into the bacterial chromosome]. . . . Its clear that the main problems of bacterial genetics are solved. Lederberg pointed out that he (Josh) only disagreed on a few very minor points.

That same year, Thomas Anderson visited Paris and took a superb electron micrograph, now famous, of a long, slender, rod-shaped donor cell lying with a short, rotund recipient. Male was connected to female by a short, narrow tube.

By 1957, Jacob and Wollman had enriched the map to about twenty genes. Using different strains derived from Hayes's original, they confirmed that the sequence and relative timing of the genes was all but invariable—but also found that the starting point for transfer of the sequence was displaced in some strains, and that other males transferred the sequence backwards. Jacob then suggested that *E. coli* had a circular chromosome that broke open, at a place characteristic of each strain, before mating. It was a revolutionary idea for its day, greeted with skepticism. Lederberg, for one, disagreed strongly. The recent genetic map of *E. coli* is drawn as a dial marked off from zero to ninety minutes and carrying locations for more than two hundred and fifty genes, some so close together that they must be shown on expanded segments that piece out a larger ring around the first one.

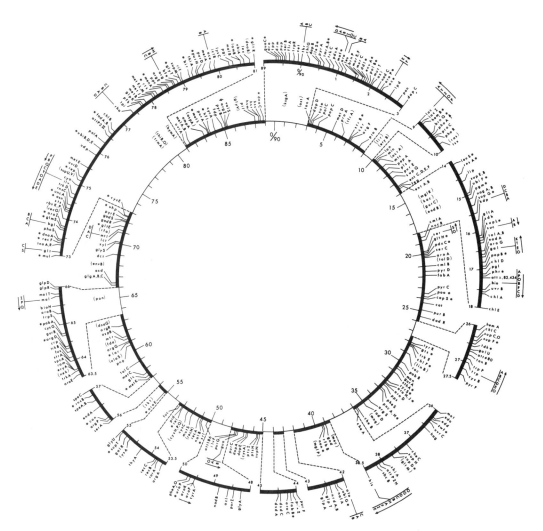

The genetic map of the *E. coli* chromosome as it was known at the end of 1972. The inner circle represents the entire chromosome, to the scale of the ninety-minute clock shown within; crowded patches are enlarged on the outer rings. Since this map was prepared, many of the genes shown have been placed more correctly, and hundreds more have been located; the circular presentation has grown impractical. (Drawing reproduced from "Linkage Map of *Escherichia coli* Strain K12," by Austin L. Taylor and Carol Dunham Trotter, *Bacteriological Reviews* 36 [December 1972]: 504–524, where, also, the gene symbols are explained. Later map: "Recalibrated Linkage Map of *Escherichia coli* K12," by Barbara J. Bachmann, K. Brooks Low, and Austin L. Taylor, *Bacteriological Reviews* 40 [1976]: 116–167.)

❖C❖   Once more I have been reconstructing—trying to evoke in vicarious experience—a state of understanding shared by a small group of scientists on the
eve of a revolution of their ideas. The experiment that came next quickly acquired a memorable if jokey name: the PaJaMo experiment. In technique, it
was no more than another permutation of genetics of the lactose enzyme system. In consequences, the PaJaMo experiment ranks—where? To say one can
place it not far behind Oswald Avery's evidence that the transforming principle must be DNA, ahead of Alfred Hershey and Martha Chase's demonstration that DNA, not protein, was the genetic material of bacteriophage, about
on a level with Paul Zamecnik's proof that the ribosome was the site of protein
synthesis, would be accurate but inadequate. A demonstration, a model, a
theory, has its first importance in the peculiar quality of its relation to work
around it. In this relation the PaJaMo experiment was unexpectedly powerful.
It forced the solution of two problems. At home, it turned upside down the
logic of regulation of enzyme synthesis that Monod had been investigating,
and led to a general theory of the repressor and of groups of genes controlled
together, to be termed the "operon." Afield, when accepted, the PaJaMo experiment broke the impasse in Crick and Brenner's comprehension of how information in the sequence of bases in DNA came to be expressed as a sequence of amino acids in protein, and thus led to the theory of the messenger
and the solution of the coding problem.

The PaJaMo experiment began the collaboration between Monod and
Jacob. They had met in 1949 and had eaten lunch together most working days
since, but had only once done an experiment together—a brief attempt in 1951
to compare the damage ultraviolet light did to the synthesis of β-galactosidase
and to the multiplication of the virulent phage T2 in *E. coli*. Monod's most
valued collaborator had been Cohn; Jacob was still working closely with
Wollman. What brought Monod and Jacob together was, in the first place, the
fact that the systems and tricks that each had worked out now seemed useful
in the other's problems. What developed was an intellectual bond of highest
intensity that lasted seven years and that struck onlookers with awe. In that
time they published twenty-two papers together.

"It was an extraordinarily intimate relationship," Monod said. "I think we
for— From 1958 up to 1964, we used to talk at least two hours, two or three
hours a day." He paused. "And it would be very hard for me, and I think also
for François, to say who had the first idea of the— Well, I know that the first
glimpse of the operon, as a concept, was François's. And it immediately
clicked because I had found the genetic linkage between permease and galactosidase." He gave other examples. Jacob had for the most part conducted
the genetical experiments; Monod had done all the biochemistry. In their discussions, roles were fluid. "I would say—but that's a question of temperament—that François is much more intuitive than I am; and I'm more of a
strict logician than he is. So that in our discussion very often it would be the
way, that François would come up with some not-very-well-formed idea of
some sort, and we would discuss for two hours in my office just what was the
content of that idea, and how it would be translated in terms of experiments,
proofs and so on." It had been suggested, I said, that Jacob was the experimentalist, Monod the theoretician. "Well, I have often been considered as a
theorist who never touched an experiment; but it isn't true, in fact. It's more a

question of—I like to say that I have no intuition, only logic." But that couldn't be true, either. "Not completely; although I don't know what intuition is—nobody does—while I think I know what logic is." I floated my pet definition of intuition. He paid it no attention. "Then there are a very many odd sorts of differences—for instance, I'm not visual," he said. "I am not visual at all. François is extremely visual. I remember when we were looking at petri dishes, hunting for the kind of mutant we wanted, he *always* spotted it before I did. Invariably." I began to say that there were few who had worked together so closely in science. Monod interrupted. "1 think Watson and Crick and François and myself are something—" But he did not finish his sentence. A friend had told me, I said, that it was a kind of love. "Yes," Monod said. That the interplay had sometimes been acutely difficult he did not mention.

By late 1957, when Jacob and Monod's collaboration began, certain provocative symmetries were evident between the genetics of bacteriophage and of inducible bacterial enzymes. The *lambda* prophage was a set of genes lying together on the chromosome of *E. coli* strain K12; they were not normally expressed but could be induced, as by ultraviolet light. The genes for the lactose enzyme system were a set lying together on the chromosome and were not expressed unless induced by lactose or an analogue. Mutants existed in which the lactose enzymes were constitutive, their synthesis proceeding at full rate even with no inducer added. Of course it was a trivial turn of phrase to point out that *lambda*, too, was constitutive whenever phage particles infected a cell of any strain of *E. coli* that was sensitive rather than lysogenic. After all, for years Lwoff, Monod, Jacob, and all those around them had been discussing induction in the two systems as though they ought to be comparable but in fact were not. "The great joke was about induction," Jacob said. "Because in both cases you did elicit either phage synthesis or enzyme synthesis, and everybody was convinced it had nothing to do—that they were completely different phenomena." Yet it was undeniably intriguing when the *lambda* genes became constitutive in another way: when a lysogenic (in other words, inducible) male piped the chromosome carrying the prophage into a non-lysogenic (constitutive) female, the genes began to be expressed the moment they entered.

Not until June of 1958 would the total correspondence between the two systems be recognized. Symmetries now obvious were not compelling before then. A chief reason was Monod's persistent sense of the logic of induction as a positive process: he was convinced that the inducer turned something on. The only universal mechanism for control of enzyme synthesis that had been suggested so far was Monod's own extension of positive induction—his idea that production of a constitutive enzyme was kept turned on by an inducer made internally by the cell itself. That logic was consistent with every experiment that Monod had done since 1940. It was consistent with the great changes of view—the overturning of the received idea of the biochemistry of the cell as a state of dynamic equilibrium, and all that followed—in which Monod's work had been instrumental. The mental set continued even when enzymes normally constitutive were shown to be repressible systems, turned off by an abundance of their own end product. Induction and repression then looked like opposites. It still seemed unavoidable, by that logic, that the mutation responsible for the change from inducible to constitutive, in the lactose system, led the cell to make its own inducer internally that kept synthesis of the enzymes turned on without added galactosides.

In the spring of 1957, Monod and Jacob designed an experiment to test that. The experiment was also, in its formal relation to previous work, a test of the symmetry between lactose and *lambda* systems. Monod and Jacob proposed a cross with the enzyme exactly parallel to what Jacob and Wollman had done with the phage when they mated lysogenic males with nonlysogenic females and found the phage genes expressed as soon as transferred. That is, they planned to cross males that made β-galactosidase, but only when induced, with females that lacked the β-galactosidase gene but were none the less constitutive. That they could breed—and recognize—such odd females was known from Monod's work with permease. At the blackboard again, Jacob drew a diagram and put in the standard symbols.

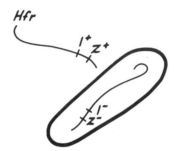

The male, with the gene Z for β-galactosidase, was the wild type and was written Z^+. The female, unable to synthesize the enzyme, was Z^-. They made the male inducible, and that again was the wild type and symbolized I^+. The plus-or-minus convention followed strictly the presence of the character as observable in the wild type, or its absence, and did not relate to the presence or absence of some supposed gene product: at this time, Monod still thought it likely that the constitutive mutants, written I^-, had the talent to make something, perhaps their own inducer, that the wild type lacked. The question was, Jacob said, his chalk squeaking—

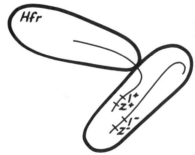

"When you transfer the genes, will the cell as soon as it makes enzyme also be inducible, or not?"

In the fall of 1957, Arthur Pardee, from the Virus Laboratory of the University of California at Berkeley, came to work for a year with Monod. Arthur Beck Pardee was an enzymologist of eccentric, impatient originality. A scout not a homesteader, with a flair more for the particular than the general, his mark—"Pardee was hereabouts"—will be found on close examination of several discoveries later staked out by others. It had been Pardee, with Howard Schachman and Roger Stanier, who had shown in 1952 that *E. coli* cells contain microsomal particles. "Pardee was a phenomenologist of great talent," Schachman said a quarter-century later. Pardee met Monod briefly in 1952 when he gave a lecture at the Institut Pasteur on changes in bacterial enzymes after infection by phage. They did not stay in touch, but Pardee often found his own work running parallel to papers out of Paris. In 1954, curiosity piqued by Monod's first demonstration that synthesis of β-galactosidase could not take place in the absence of essential amino acids, and therefore appeared to be not the conversion of a ready precursor but the building of new protein, Pardee had noticed in the paper's fine print that Monod had also found *E. coli* unable to make the enzyme in the absence of the base uracil. Uracil meant RNA. Pardee therefore tried experiments that showed that induced synthesis of the enzyme stopped within minutes when synthesis of RNA was interrupted. Some sort of connection was obligatory. The idea beckoned that RNA was a byproduct of protein synthesis, the nucleotides perhaps hooking up simultaneously with amino acids on DNA and then peeling off to be disposed of by the ribosomes. Stent, Pardee's neighbor, thought the idea was great; in print Pardee suggested only that "continuous formation of RNA is essential to protein formation." At about that same time, Pardee's lab, independently of Monod's work with Georges Cohen, came across inducible bacterial permeases.

Then, late in 1954, Pardee found a striking new example of feedback inhibition of an enzyme by its eventual end product. This enzyme—later rechristened aspartate transcarbamylase—catalyzed the fifth step back from the end of the metabolic pathway by which the bacteria elaborated pyrimidine nucleotides. When the pyrimidine cytosine was plentiful, the enzyme was blocked, turning off the rest of the pathway. By the fall of 1955, Pardee proved that the block occurred at this enzyme and not later or earlier in the pathway; he showed that the inhibition was of the enzyme's activity—though he did not stress the distinction between such a mode of control and one that interfered with an enzyme's synthesis. That fall, also, Edwin Umbarger at Harvard Medical School noticed another instance of feedback inhibition of an enzyme by an end product, this time in the pathway that synthesized the amino acid isoleucine. Umbarger sent off a preliminary note to *Science* a month before Pardee's full paper was ready for *The Journal of Biological Chemistry.*

Before Pardee came to Paris, he was developing with Gunther Stent a way to disrupt the DNA in bacteria by feeding them radioactive phosphorus, freezing the bacteria in liquid nitrogen for several weeks to give the radiophosphorus time to decay, and thawing them again, to see whether, after the DNA was killed, enzyme synthesis continued. It did not. The technique was coarse, but it raised questions, once more, about the relation of DNA to RNA and especially to ribosomes. "Pardee's a strange guy," Stent said two decades later. "He's the Douanier Rousseau. He's extremely primitive. But nevertheless, extremely gifted. I mean, his ideas seem utterly and ridiculously naïve; but he always comes out with good things." In correspondence in the spring of 1957, Monod suggested that Pardee combine radiophosphorus decay with bacterial conjugation to gain precision in the DNA to be disrupted. Writing on May 16, Monod said, in part:

It also seems to me that the problem could be repeated using the cross:

Hfr Lac$^+$ x F$^-$ Lac$^-$

in which only the Hfr have been marked with ^{32}P. What do you think?

That is, a strain of males that possessed the genes to utilize lactose would be crossed with a strain of females that did not have them—and only genes donated by the males would be attacked by radioactive DNA.

Pardee arrived at the Pasteur late in September of 1957 expecting to do that experiment. But then Monod proposed that he first try the straightforward mating without the radiophosphorus, to get acquainted with the mating system and to be sure that once the females acquired the gene they made the enzyme and that its appearance could be timed and measured. "This direct measurement of enzyme activity was a dramatic departure from the previous technique for studying mating, which depended upon the appearance of colonies on selective plates a day or so after the mating event," Pardee wrote in 1978, in a brief essay for a volume that Lwoff edited as a memorial to Monod. Pardee's first problem was a technical one that had bothered Monod, to distinguish between β-galactosidase made by mated females and by males. Within a few days he showed that he could do so by a standard tactic, using females that were resistant to streptomycin; after interrupting the mating with a blender, he added streptomycin to the culture to stop the males from synthesizing anything. That established, Pardee soon showed—in an experiment never published—that within minutes of the transfer of the Z$^+$ gene from donor to recipient, the recipient began synthesizing the enzyme at full speed. "This result in itself was of great interest, since it demonstrated that the gene becomes active vitally as soon as it enters an appropriate environment," Pardee wrote.

Monod then suggested that Pardee try the mating that would test the genetics of constitutivity. Pardee had not considered that phenomenon before. "I had in fact built the genetic strains that were needed for that experiment, before Arthur came," Monod said. "So when he wanted a good subject, I said, Well, let's do that, I have the strains, they're already here; I was hesitating to start the experiments because they were difficult to actually perform; and I was sure Arthur would do them better than I would, which he did."

It was Monod's practice, before any experiment, to trace out rigorously the results that were possible and what each would mean. Chess problems perhaps provide the closest thing in a nonscientist's experience to this all-encompassing imaginative pursuit of checks and balances: moving the white rook to that square to block this pressure from the black queen will open that diagonal for the black bishop, and so on. I failed to ask Monod whether conducting music—conducting contrapuntal, choral Bach—gave anything like the same stretch to the mind; no doubt the question would have meant little. As soon as the gene Z$^+$ reached the constitutive female who lacked it, she ought to start making enzyme, and without added inducer—unless the gene I$^+$, piped into the female at the same time, changed her from constitutive to inducible, in which case she ought not to synthesize enzyme without inducer, yet if inducer was then added, three minutes later she ought to start up.

They were playing an opponent more imaginative than they. What happened was not predictable. Pardee grew the two strains to high concentration in a medium that contained no inducer. Then, into flasks of fresh broth, still with no inducer, he mixed a quantity of each strain, fewer males than females so that virtually every male would find a mate. Neither parent could synthesize β-galactosidase, the males lacking inducer and the females lacking gene Z^+. "We knew exactly the time at which, in these crosses, the gene was coming in," Jacob said. "And we looked at how long does it take, when once the gene has come in, before the enzyme is made. And at three minutes you get the steady, linear synthesis." Then the surprise. "It turned out that it first makes enzyme, without inducer, for thirty minutes"—he misremembered the time—"and then it stops and then you require inducer."

Pardee found that three minutes or so after the genes for the lactose system were transmitted, β-galactosidase synthesis began, steady and copious. The cells were functioning constitutively. Ninety minutes later, synthesis stopped almost as abruptly as it had begun. The cells were no longer constitutive. Simultaneously, for a control, he had done the parallel experiment where he put in streptomycin to prevent the male parents from synthesizing anything and then put in a gratuitous inducer. Induced, the recipient cells continued synthesis indefinitely.

Pardee did the first experiment of that kind on 3 December 1957. "The experiment was intended to get at the idea of what constitutivity is," he said in a conversation at his house in a rich suburb of Boston, one evening early in 1978. "The kinetics"—the rate measurements— "of that first experiment were very crude, just a few points. See, we were only asking, the first time, the broad question, Do you get enzyme without any inducer, or don't you. In this I-minus-type cross. And then the natural control is to have the inducer there to see how much enzyme one might expect"—in other words, to provide a comparison. On December 12, Monod wrote a long letter to Mel Cohn, full of new work and the comings and goings of scientists; he described Pardee's rapid development of the technique and its great promise. "Already Pardee has in his hands a system which is without doubt the best, at the moment, for studying the expression of a gene immediately after its introduction into a foreign cytoplasm." From the first crude trial to refine the methods and the measurements took two and a half months; the definitive, publishable run of the experiment was done in the middle of February.

That genes Z and I were separate was dramatically confirmed. That they functioned very differently was now apparent. The first remarkable result of the experiment was that gene Z, responsible for the structure of β-galactosidase, expressed itself with extreme rapidity, and almost at once at its maximum rate. This observation was the pivot on which turned the proposal, three years later, of the messenger. The second remarkable result was that when both versions of gene I were present at once, inducibility prevailed—but slowly. In the terminology of peas and fruit flies, the inducible character was dominant; the dominance was expressed after a delay.

That winter, some time after Pardee did the first experiment to test the nature of constitutivity, Leo Szilard visited Paris, gave a talk, and in discussions with Monod and Jacob forced the logic further. By the mid-fifties, Szilard was working intensely with the Pugwash conferences and with his other attempt to bring scientific opinion to bear on politics, the Council for a Livable World. Yet his interest in theoretical problems in biology was still strong, and he travelled a lot, visiting laboratories wherever he went. "He was talking to everybody, and in a very special way," Jacob said. "He took you apart, put you on the chair, took his notebook and began to ask you to talk and he was writing everything you were saying, and immediately when you began talking he forced you to talk in *his* language. When he finished talking with you he said, *'Sign!'* Three years later he was coming to you to say, 'You told me that day that such-and-such occurs, is that still true? Or what?'" But had Szilard's interventions been productive? "Yes. He was a very remarkable man," Jacob said. "And he was very much interested in this business of activation versus repression."

The enzyme system that Szilard knew from his own work with Novick in 1954, the pathway by which *E. coli* synthesizes tryptophan, was Monod's classic instance of repression by end product: availability of tryptophan acted to inhibit an enzyme involved in making tryptophan. This inhibition affected the synthesis of the enzyme and was thus formally the mirror image of the induction of β-galactosidase by the availability of lactose. Szilard began to think intently about that sort of comparison in the course of a conversation with Werner Maas, in Chicago in mid-April of 1957. Maas, an enzymologist at the New York University College of Medicine, was in town for the annual meeting of the Federation of American Societies for Experimental Biology, and was staying at the Quadrangle Club, the faculty club of the University of Chicago. The two men had breakfast together one day. Szilard knew already that Maas had discovered, with Luigi Gorini, still another instance of feedback inhibition in *coli*; indeed, the previous fall, Szilard, Novick, Maas, and Gorini had conceived an experiment about that, and now, over the coffee cups, Maas spoke of the results and the speculation they had provoked him to.

The feedback effect, as it happened, was the control by the amino acid arginine of the pathway that builds it. The last three substances in that pathway are the small chemical building blocks ornithine, citrulline, and arginine itself; each is made from its precursor by the help of a specific enzyme. Maas had come to Chicago to read a paper reporting that arginine inhibited the synthesis of the enzyme, ornithine transcarbamylase, that converted ornithine to citrulline. But the experiment thought up with Szilard tested the converse biological point that it was lack of arginine in the cell that stimulated synthesis of ornithine transcarbamylase. For this, Gorini and Maas had taken a mutant that was blocked one step further back, unable to make its own ornithine—and so required arginine in its culture medium. When they grew the mutant in conditions where it was amply fed except for arginine, they found that synthesis of the enzyme—for converting ornithine to citrulline, and even though the cells had no ornithine—began at once and continued at maximum rate until a concentration was reached twenty-five times normal. Maas had written the experiment up in a short note the week before coming to Chicago.

Going further than the note, Maas told Szilard that the effect was exactly like an induction—except for the essential fact that the substrate was not there. "From that I reasoned that arginine brings about a repression," he said to me in New York in 1978. "And that if you limit the arginine, you set off enzyme synthesis. And so, on the basis of a unitary hypothesis of regulation, I had to say that you didn't need the substrate for induction. And that led me to the idea that induction was removal of repression."

Szilard got excited by the idea. He told Maas at once that he thought the French data on induction of β-galactosidase could be explained that way. He suggested they write a paper together. Lacking more direct evidence, Maas declined; but in September he reported the experiment at a meeting of the American Physiological Society, and there, answering a question from the audience, he said:

> I think that most people feel the distinction between adaptive and constitutive enzymes is an artificial one. It has been suggested that in the case of constitutive enzymes the inducer is always present, whereas in the case of adaptive enzymes it is ordinarily not present. . . . Alternatively, it is possible that adaptive enzyme formation is due to the inhibition of a feed-back inhibitor normally present, and that inducers act by preventing the action of such feed-back inhibitors.

On October 7, Szilard read a paper about induction and repression at a meeting of the German Chemical Society, in Berlin. The paper was the basis of his talk that winter at the Pasteur, and he presented it at other laboratories during the next few months. In his talk, and in hours of discussion with Jacob and Monod, Szilard urged upon them the logic of the similarities between the two sorts of control of enzyme synthesis. "You have to put this into the picture," Jacob said. "The argument about repression, that is inhibition, versus induction, which means activation, and whether or not you could put both in one common mechanism, which is inhibition and activation by release of an inhibition. And that was Szilard's main argument, which was a very good one." The discussion went on in Monod's office. Others besides Jacob remembered that it grew hot.

What real difference Szilard's intervention made is open to question. Like Jacob, the chief thing Monod remembered was Szilard's insistent question: Why not suppose that induction could be effected by an anti-repressor rather than by repression of an anti-inducer? But the question was not new to Monod: he was reminded of his conversations years before with Mel Cohn. Pardee had already got the first, fundamental results before Szilard came through. He did not remember Szilard's seminar, nor did he take much part in the discussions. Szilard had no direct influence on his experiments or on his thinking. "I did an experiment at his suggestion," Pardee said. "It was a total failure. It was technically impossible, it would have taken nine hours to—and everything had gone to hell by then. But—I did it, anyway, to make him happy. I can say unequivocally that he didn't suggest this answer"—the repressor—"to me personally. My recollection is that it came up as a possibility in lab discussions. But you know, whether Monod had been primed by Szilard in previous discussions, I can't tell you. . . . On the other hand, a lot of us were involved in this. My lab had made a discovery of repression prior to that, as had Monod's lab. You know, actual experimental evidence. So repression was very much in all of our thinking at that time."

Part of the argument was about the nature of the repressor and how it acted. What kind of molecule? Nucleic acid? Protein? An enzyme? Was it the primary product of its gene? In Szilard's paper in Berlin, he had put forward a molecular mechanism; though the paper went unpublished, two years later he described the mechanism again. It was an awkward model. He proposed that a polypeptide chain, once assembled and folded up, remained attached to the site where synthesized; he imagined that repressors were molecules of a novel kind, hybrids of two parts, one a small metabolite appropriate to its pathway,

❖

say arginine or a galactoside, the other perhaps a polynucleotide. Such a repressor, he thought, would connect to the enzyme by the metabolite, and to the site of synthesis by base pairing, and prevent release of the enzyme. But no detail of Szilard's model survived at the Pasteur—except, perhaps, the note that the repressor might be nucleic acid rather than protein. "I didn't have the feeling that he had any better idea what was going on than the rest of us, as far as mechanism went," Pardee said. The interesting questions seem to have been more general: Once the repressor was produced what did it act upon? Something else in the cytoplasm or reflexively on the DNA? "We had to consider that," Jacob said. "But I don't think it went very far."

On 28 February 1958, Monod wrote to Cohn. He had been using mutants to analyze the fine structure of the genetic region controlling the lactose enzymes, he reported, but the work was a struggle with hundreds of bacterial strains, with Benzer's new nomenclature, with contaminations, and so on. All the same, he had found that the constitutive mutation lay extremely close to the gene for galactosidase, somewhat farther from the permease—and, it seemed, in between the two. Then there was the biochemistry, where the news was all confusion—except that Pardee's experiments stood out.

> The fundamental result is that when one injects the genes $z^+ i^+$ of an Hfr [male] into a female $z^- i^-$, one gets practically immediate synthesis of enzyme. Immediate means immediately after the injection of the gene, the kinetics [that is, the timing] of which one can follow simultaneously with a blender. There is not more than two or three minutes between the injection of the gene and the beginning of synthesis of enzyme.
>
> If one now follows the synthesis of the enzyme in these crosses, one sees at first with stupefaction that after a while instead of increasing it diminishes and practically stops at about the 120th minute.
>
> If one now makes the reverse injection, gene $z^- i^-$ into a female $z^+ i^+$, one gets no synthesis of enzyme at any time!! (even after three hours of incubation)

From the asymmetry of these two results, Monod wrote, three important points followed. He listed them, and the letter is a strip of stop-motion photographs that freeze the leaping argument mid-air between experiment and published paper. First, one could now be sure that in bacterial conjugation there was no mixing of cytoplasms, because if there were, the two crosses would have been exactly symmetrical and would have produced the same result. Jacob and Wollman had been asserting for years that only genes were transferred, and still the confirmation was welcome.

> Second important point: these experiments demonstrate that constitutivity is not—as I have recently inclined to believe—a character exclusively to do with the gene. One must admit that it is expressed cytoplasmically, for without that, once again, the two reciprocal crosses must have given the same results.

The gene, once transferred, had to have a product that appeared in the cytoplasm. At the instant of dictating that paragraph, Monod was still by habit considering constitutivity —i^-—to be the positive character. In the next frame one sees the balance shift.

> Finally, the fact that the result was obtained in the particular direction observed seems explicable in only one way: one must admit that in the pair of alleles [that is,

different genes for the same character] i^+ i^-, it is the gene i^+ that is active and the gene i^- (constitutive) that is inactive. Two possible interpretations.

a)—Inducibles synthesize an enzyme that synthesizes an inhibitor (repressor) of the synthesis of the enzyme.

The circular statement, I think, was unintended: Monod saw the repressor not as the primary product of the gene i^+ but as made by one enzyme to turn off synthesis of another, namely β-galactosidase. He did not pause to consider where the turning-off took place.

> Induction will then be due to a displacement of the repressor by the inducer. The arrest of synthesis in the crosses will be due to the progressive accumulation of the repressor. (I forgot to tell you that at the 120th minute when constitutive synthesis is stopped . . . [the cells] become inducible).
>
> b)—The other interpretation: the inducibles synthesize an enzyme that destroys an internal inducer. The constitutives have lost this enzyme, and so accumulate the inducer.
>
> Formally, the two hypotheses are equivalent, but the first seems much superior.

The first was the hypothesis most easily generalized; it fit the fact that almost all enzyme systems were constitutive and it suggested why, in inducible systems, constitutive mutants almost always made the enzyme at a higher rate than the induced wild type.

> In sum, if all this is confirmed, that will be the proof of what in the old days we called the theory of the "double bluff" that we discussed so often without ever taking it too seriously.

No mention of Szilard.

That winter, Mahlon Hoagland, from Paul Zamecnik's group in Boston, spending the year with Crick in Cambridge, came over to Paris for a fortnight at the Institut Pasteur. "It was dead winter," Hoagland said in conversation two decades later, but the first results of the PaJaMo experiment were already known in the lab. "There was a lot of kidding and fooling around going on, in which they said, 'Mahlon, you know, there's something just screwy, about this whole business. You just can't make ribosomes fast enough to account for the rate of protein synthesis after the initial entry of the β-galactosidase gene.' And we argued about it; but they said, basically, 'Of course we know that ribosomes are the important sites of protein synthesis, but we're awfully tempted to say that it isn't the ribosomes, it's the DNA itself, and that proteins are made on DNA.' And I said, 'Well, you know that this would be stepping way backwards, from the advances that we've made,' and they said, 'Well we really know that, but, gee, the whole thing is absolutely bewildering.'" The exchange defined a dilemma: the French data demanded an alternative to ribosomes, the American evidence asserted there was no alternative. Each side knew of the other's results, of course; yet at the Pasteur the sharp focus remained fixed on the action of the gene itself in the living bacterial cell, while at the Cavendish and at Massachusetts General the central reality was the function of the ribosome as shown in the cell-free systems. For each side, the other's results had an aura of the provisional. This difference in focus persisted for two years.

The first report appeared in May 1958, three pages stripped to essentials in the weekly *Comptes rendu* of the French Academy of Sciences. Monod looked at the initials of the three authors—PJM—and had the passing thought that if they had formed a Hebrew word, in which vowels are not written, it could be pronounced only one way. The idea that constitutive mutants made an internal inducer was dead, the three reported.

> The results described suggest exactly the opposite hypothesis. The facts are explained if one supposes that the gene I determines . . . the synthesis, not of an inducer, but of a "repressor" that *blocks* the synthesis of β-galactosidase, and which the exogenous [external] inducers displace, restoring the synthesis.

In several respects they were theorizing ahead of the leading edge of what they knew for sure. Monod was deep in one series of experiments to find mutants that could be mapped in gene Z and that modified the β-galactosidase protein, in order to prove that this gene really did determine the enzyme's structure. Pardee, Jacob, and Monod were of course already trying more complicated variations of the basic matings, forecast from the first design of the experiment the previous fall, to see, for example, whether the second gene that apparently determined an enzyme structure in the lactose system, gene Y for the permease, behaved as it ought in the same way.

❖❖❖

The second week in June 1958, a symposium was held in Brussels on the relationship between nucleus and cytoplasm in the cell. Jacob and Pardee both presented papers. Between them they ranged over the whole matter of transfer of genes between bacteria and the regulation of their expression; each at the end came back to the puzzling promptness with which enzyme synthesis started up after the gene was introduced. The time lag was, Jacob said in his paper, "extremely short, not exceeding a few minutes. Such a lag is rather small for what could be expected a priori for the synthesis of stable macromolecular intermediates." Oddly, he nowhere mentioned ribosomes by name: yet, "Such a rapid synthesis would suggest that few, if any, high-molecular [-weight] intermediate structures have to be built in order for the gene to become expressed." Pardee made the other half of the same point. Many had supposed, he said, that DNA made "an intermediate carrier of information (perhaps RNA)." In that case, though, why did the cell's rate of enzyme synthesis remain constant instead of increasing steadily as more intermediates were produced? He did not mention ribosomes, either: yet, "The templates might be unstable and rapidly reach a steady-state concentration as a balance of formation and inactivation." That, however, he judged "not too plausible."

The alternatives were less plausible still. "We discussed it many times; this was soon after we had done the PaJaMo experiment," Monod said. "And we had come to the conclusion that the intermediate whatever it was had to be a short-lived intermediate. Had to be unstable. The logical choice, as I gathered at the time, was that the gene was acting itself, which seemed absurd, or there was an unstable intermediate. But I stopped at that; I couldn't go any further."

Ideas were coalescing even as people were moving. Pardee was going back to Berkeley; Wollman was going there too, to work for a year with Stent again. Jacob had accepted an invitation to go to New York in September 1958 to give a lecture to the Harvey Society about the genetics of lysogenic viruses. He had two other papers to give that summer, at congresses in Stockholm and Montreal, so he wrote the Harvey Lecture towards the end of June, after returning from Brussels. In the pressure of preparing it, just a few days before setting off from Paris, Jacob was forced to the great clarification that united *lambda* and galactosidase. "One Sunday afternoon I was working at home on that lecture, and I suddenly realized the whole parallelism."

He was writing, in particular, about the immunity of lysogenic bacteria to additional infection by the same phage. Immunity had always been the innermost puzzle of lysogeny. Immunity was itself a genetic trait. It had been mapped to one locus within the sequence that made up the prophage. When male *E. coli* K12 that were not lysogenic for phage *lambda* conjugated with females that were, no phage synthesis was induced: immunity thus behaved like a dominant gene. When the females in those crosses were superinfected by putting more *lambda* particles in the broth, although the phage DNA penetrated, nothing ensued: immunity thus was expressed, as had been understood for years, as a gene product in the cytoplasm.

Now Jacob suddenly recognized, in the facts long known, that immunity as a trait corresponded exactly to inducibility in the lactose system. "You could make a completely parallel story to the two," Jacob said. "You could predict from system A to system B and *vice versa*." For the enzyme system, he wrote for the Harvey Lecture:

> The most likely explanation is that, in the inducible strains, the synthesis of β-galactosidase is inhibited by a cytoplasmic repressor whose production is genetically controlled. The induced synthesis would result from the release of the repression by a specific inducer. The analogy between this phenomenon and immunity of lysogenic cells is such that we can hardly escape the assumption that immunity also corresponds to the presence of a repressor in the cytoplasm of lysogenic cells.

He saw a half step further still. A complete phage particle contained a number of different proteins as well as the phage DNA. Thus the suspected repressor of prophage was a gene product that blocked the expression not just of a single gene but of the considerable set that determined the phage, whether they lay in the bacterial chromosome or were freshly injected. Evidently a string of genes could be switched on or off simultaneously by something that acted at one locus—at one spot on the genes themselves. Monod had glimpsed in the lactose system that several genes could be controlled as a unit. That the control acted at the DNA could be seen clearly, then, only for *lambda*.

The central PaJaMo experiment, a great deal of supporting evidence, and the sustained argument, with acknowledgement to Szilard, for a generalized model of regulation of the synthesis of enzyme proteins by repressors were all in hand by the summer of 1958. Even as the paper was being written, the pursuit of the consequences of the PaJaMo experiment was taking two directions. Pardee went home with the design in his pocket of an experiment to test the alacrity with which genes responded to regulatory signals. When Jacob got back to Paris, he and Monod began debating what the repressor acted upon.

"From the PaJaMo experiment came two things," Jacob said. "One was the messenger, and one was the regulatory system. And these are two different aspects of the same thing." That created difficulties in explaining it, I said. "Yes. But I think we can discuss that in two parts," he said.

"Now, the messenger was the following." He went to the blackboard again, then turned back and gestured with the chalk for emphasis. "The idea at that time—and it was summarized in Crick's lecture, I think already in 1957, to the Society for Experimental Biology, the central dogma business—was that DNA makes RNA makes protein. But it was also known that proteins are made on ribosomes. And that ribosomes contain a large proportion of RNA. And therefore, at that time, RNA was taken as equal to ribosomal RNA. Now, the PaJaMo experiment, as well as the phage experiment"—on the board he chalked a right angle, the vertical and horizontal frame of a graph—"showed that the gene comes into the cell, let's say, at time zero." He placed the tip of the chalk at the corner of the graph. "And two minutes later, enzyme synthesis starts." He moved the chalk two minutes to the right, then as he talked drew the upward-sloping straight line.

"Now, two minutes was very short. That's one thing. And the second thing was that if you had this synthesis and then if you removed the inducer, here"—he marked a point about an hour along the line of enzyme synthesis—"within two minutes synthesis *stopped*. Now, the onset could work, just conceivably, with the RNA being ribosomes, but it was not absolutely obvious that it could turn *off* as fast as that, with very stable structures. So then the next experiment, and actually it was done by one of Pardee's students, Monica Riley, was to see what happens if we just destroy the gene itself. If the ribosomes, which we know are stable, have received from the gene all the information to make the enzyme, you should be able to destroy the DNA—and since the ribosomes are still there and are stable, they should go on making the enzyme." He came back to his chair.

"So the experiment was to destroy the gene. Which is not so simple in bacteria. This could be done by building a lot of radioactive phosphorus into the DNA of the males, then transferring just this one piece of DNA into the female bacteria, then letting them start making enzyme, then freezing the cells to stop the metabolism while the ^{32}P decay occurs and kills the gene, then thawing the bacteria—and see whether or not there is more enzyme made." Whose idea was *that* sequence? Jacob raised his eyebrows. "That was the obvious experiment to do," he said. It put interrupted conjugation together with the technique for killing DNA, in the combination that Pardee and Monod had worked out a year earlier. "But then it turned out it was technically difficult to do." Riley and Pardee, in Berkeley, needed the best part of a year to make it work.

"The other apsect is the regulatory part: then we have to come back to the preparation of this Harvey Lecture," Jacob said, and laughed. He did not quite put it to his audience in New York that the repression had to have a target and that the target was likely to be DNA. The reasoning was subtle. But it began, he said, "that Sunday afternoon, when I suddenly realized that you could make exactly the same model for both systems, and that in addition the phage had one very specific property which led to a particular model. Because the phage—you have to make phage particles, with many proteins, *and* their DNA replication, so making phage is an enormously complicated thing. Now at that time—that was as I said in the summer of 1958—it was clear that the two experiments that were done by transfer of genetic material, both of them showed that you had genes that were silent. In one case the gene of the enzyme was silent, and in the other case all the genes of the phage were silent. Now, at that time, as we said, the idea was that the gene product was ribosomal RNA." Ribosomes specific for each protein, I said. "Yes. Now, this meant that for phage you had to make a large number of specific phage ribosomes, which had to remain silent." To be ready to jump, I said. "Yes. And therefore the most likely thing was that regulation was operating on the gene itself, and not on the product. And from this I began to make some kind of a model."

But if control operated directly on the gene, the mechanism would seem to be either on or off: the model would also have to explain how the rate of synthesis of enzymes varied. Writing the lecture, Jacob happened to observe one of his sons playing nearby with an electric train. The train was a simple set with a switch but no rheostat for controlling the strength of the current to change the speed. "And my son wanted to slow down the train, and I saw he was doing *that*"—Jacob pinched an imaginary switch between thumb and forefinger and made a quick back-and-forth gesture, on and off. "Just an oscillation. And depending on the speed of the oscillation he could regulate the rate." Thus Jacob got the first glimmer of how expression of a gene might be controlled.

All this, Jacob said, "was one of the very last days I was in Paris. I left, and I went on cooking this model in the planes. And when I came back between Stockholm and Montreal, Monod was in his boat somewhere."

The Harvey Lecture took place on September 18. In New York, Jacob saw Wollman off on the train to California, gave the lecture, and flew home right away. He had not seen Monod since late the previous spring "I remember very well that I got off the plane in the morning and at two o'clock I was in Jacques' office to tell him about the idea. And then he began to laugh. He had two or three objections." Monod's laughter was harsh, his objections scornful. "And he was strongly against the idea of the control operating at the DNA. And I was so sleepy that I said Okay, we'll talk tomorrow about the details. And then I came back tomorrow and we began to argue, strongly." That argument started their close collaboration.

Two days earlier, I was talking with Monod about the same set of discoveries. He had not thought, he said, that the phenomenon of induction "was hooked up as closely to DNA as it was to turn out. That came as a complete surprise to me. Oh, yes. Nobody expected it, you know. Absolutely no one. Everybody recognized that the regulation occurred, of course. We had proved it, and others had proved it. And that the regulation occurred on the basis of some sort of genetic determination. But everybody was assuming that the regulation was occurring somewhere lower down." Lower down? "Further along in the string of information. That it would turn out to operate right at the level of the gene itself was a— In fact it was hard for François and myself to really come to that conclusion and state it in black and white. It seemed so incredible. You know, the gene was something in the minds of people—especially of my generation—which was as inaccessible, by definition, as the material of the galaxies. That experiments we were doing would involve an actual physical interaction between a compound in the cell and actually the *gene itself*, was something extremely difficult to come to."

"Yes, but we were much more prepared than he was, because of the prophage," Jacob said. "The prophage *was* a shock, because that was the first time you could put something back and forth in the chromosome." Jacob had argued that point at the defense of his dissertation in 1954, with Ephrussi, who was on the jury. "I said that prophage forced one to admit that something can go back and forth in the chromosome. And this Ephrussi disliked deeply. . . . So I think for the phage people it was much less shocking, the idea to have regulation at the gene level.

"There were two things Monod objected which I remember," Jacob said. "First, there were the two enzymes, galactosidase and the permease, and we knew that the two genes were linked, and in my model I assumed that in the linkage there was only one regulatory system to turn on and off the whole thing. That you had one switch. In fact, I think the main idea I had had that Sunday afternoon was the idea of a *switch*." He snapped his fingers on the word. "The prediction of my model was that the two enzymes had to move absolutely together: under a variety of conditions, with certain mutants, or adding a certain inducer, or varying the concentration of inducer, and so on, the prediction was that the amount of the two proteins produced had to be strictly parallel. And he had things, substances, which did induce permease and not the galactosidase. And I told him that I didn't believe in this nonparallelism. And it turned out, a year later, to be another permease which had nothing to do with this system.

"And the second argument—the second argument is very funny. Because Jacques was very much impressed by the differences in *rate* of synthesis of enzyme you can obtain depending on the concentration of inducer. That is, if you add inducer at time zero, and you plot the amount of enzyme made against the total amount of protein. Now, you always have a straight line, but when you are below the saturation level of inducer that gives the maximal rate, you get different straight lines, different rates. And he thought that couldn't be something acting at the gene—a simple yes-or-no. He thought this had to be some mechanism at the protein-synthesizing machine itself.

"But this was one of the points I had been cooking—the switch idea. He said that with a yes-or-no switch you don't get different rates. But this is exactly what I had realized you *can* do, one of those days in June"—the occasion when he had noticed his son regulating the speed of a toy train by oscillating the on-off switch. "And this was my argument which in September I exposed to Jacques. For hours! After I had slept."

The idea of the switch began a two-year period when experiments suggested themselves faster than they could be tried—a fantastic time, in the memory of those who were there, when everybody was swept into the exhilaration and "when every experiment worked."

Perhaps not every one. In a conversation about these matters that stretched over the last weekend in May of 1977, Jacob warned that inevitably an account of the history of a complex, fast-moving science will oversimplify. "Out of the sound and the fury you get a police novel"—a *roman policier*, a detective story, moving towards a neat solution. Science, as it is happening, is not much like that. We talked in the study of his flat on the fourth floor of a modern Parisian building. Sheets of plate glass surveyed the Luxembourg Gardens. A bookshelf displayed a chunk of an alabaster head, classical Egyptian. Elsewhere stood a group of three highly stylized small statues, two little women with their arms hugged under their breasts and flat, Modigliani-shaped faces, with another face like that, a fragment—Cycladic pieces from 2,500 BC, Jacob said. Doors were folded open to a sitting room, marble-floored, where in the far corner a tall gray stone statue, elegantly thin, looked out. A Khmer goddess of the twelfth century, Jacob said; her back was to the room because her front was battered. He was concerned about balance: that I give enough weight to Lwoff's work and powerful example, and that I not overplay the seductive glamor of Monod's life and personality. He was restless, thin, somewhat hunched over—hawk's head ducked between his shoulders. Repeatedly he leaned across and stabbed the tape recorder off. Each time, I took notes and after a few minutes turned the machine on again. He talked in some detail, but disjointedly, about Monod.

Most of what Jacob said about Monod he wanted off the record; but I needed to understand so that what I did write would not be incorrect in detail and emphasis. Two years later, however, in a volume of essays commemorating Monod, Jacob wrote of these things, saying in part:

> He had an unusual feeling for the interplay between theory and experiment, and was a virtuoso of the hypothetico-deductive method. Not only did he very rapidly perceive the experiment necessary to check a particular point; he also squeezed the results to the very limit of their significance. . . . At the same time, his attitude toward theories always amazed me. I have a certain taste for changing fixed ideas, . . . even if I have contributed to setting them up. Jacques, in contrast, did not like to get rid of his theories. He had a strong tendency to stick to his model, sometimes slightly beyond the point of reason.
>
> As is frequent with so rich and strong a personality, several different and somewhat contradictory individuals coexisted within Jacques Monod; two at least, if one considers only the scientist. Each took over in turn, depending on his mood and on the circumstances. The first of these individuals—let us call him Jacques—was a very warm and generous man of great charm; a man interested in people as well as in ideas, constantly available to his friends, ready to discuss their problems and find a solution; a man of great rigor and insight, always to the point, asking cogent questions, and sharply self-critical. The second individual—let us call him Monod—was incredibly dogmatic, self-confident, and domineering; a person unceasingly in quest of admiration and publicity, demanding to be the focus of attention; a person making definitive black-and-white value judgments on everything and everybody, fond of teaching fellow scientists the *real* meaning of their own work but sweeping away as nonsense any objection they might timidly offer. Jacques was able to bring all his personal activities to a halt and go out of his way to help a friend in a difficult situation. Monod could quite easily turn a friend into an enemy with a few words. In private, one dealt almost always with Jacques. In larger gatherings, one sometimes had to deal with Monod. Working with the former was an exceptional pleasure. Arguing with the latter could be a difficult experience. . . .
>
> The afternoon that I returned from New York, I was received by Monod. The next morning, however, when I again entered his office after an eighteen-hour sleep I found Jacques. I started once more to tell my story, and the discussion between us went on until the end of the afternoon.

Pardee wrote to Monod on December 8 that Monica Riley had begun the experiment to destroy the gene with radioactive ^{32}P while the bacterial culture was frozen. "Already she is having trouble—but not for long I imagine." The work was technically exacting and took ten months. Another graduate student in his lab was timing the induction of β-galactosidase, and its inhibition by those chemical analogues that acted as antagonists. Pardee's comment opens a glimpse into the uncertainties that continued to haunt them:

> We still find a three-minute lag before enzyme formation commences. However, it does not seem true that induction, removal of the inducer for 5 min or more, and readdition of inducer abolishes the lag on the second round of induction. In other words, we don't seem to have any evidence for formation of a stable template as I had thought before.

At the beginning of December, Monod was asked if he would allow his name to be put forward for election as the next professor of biochemistry at the Sorbonne. The honor was ironical. It would mean "practically renouncing all personal work" in science, he wrote Cohn. "But in compensation, in two or three years I would be Dictator of French Biochemistry. . . . My first impulse is 'No'." He changed his mind on condition that he could keep his laboratory at the Pasteur, and in mid-January wrote Cohn that he detected "a vast plot intended not to name someone else but more exactly to name no matter who else rather than me. If I am not chosen, definitely, I'll breathe a sigh of relief." He was elected to the chair in February—still with mixed feelings.

By intercontinental mail, Monod, Pardee, and Jacob wrote and rewrote the long and definitive version of the PaJaMo paper: they had extended the basic experiment to include the gene for the permease, too, and were still trying new variations of matings and controls. It was March before they sent it off to *The Journal of Molecular Biology*; the journal had just been founded in Cambridge by John Kendrew, and was a sure sign, like Monod's chair at the Sorbonne, that the rebellious young discipline was settling into professionalism. This second paper got far wider attention than the three pages in French a year earlier.

Also that winter, Georges Cohen did an experiment with Jacob that brought regulation of the class of enzyme systems turned off by their own end products firmly into the model. They showed that the entire sequence of enzymes responsible for the several steps in synthesis of tryptophan was turned on and off as a unit by a separate gene, tagged R for regulation. The genes that dictated the amino-acid sequences of the enzymes lay together on the chromosome; the regulatory gene showed up a distance away on the map. Mutations in R produced bacteria whose synthesis of the entire sequence of enzymes—and so of tryptophan—was no longer inhibited by an abundance of tryptophan. More, in bacterial matings, while two chromosomes were present, gene R was dominant over the mutant version. Thus the repressor produced by this gene, like other repressors, was a substance that diffused through the cytoplasm to its target. The tryptophan and lactose systems differed only in that the repressor of the one was activated by a small molecule, the end product, while in the other the repressor was deactivated by a small molecule, the substrate.

Monod began to try to extract the repressor substance itself, but with no appearance of energy and with no success. He suggested an experiment to

Pardee that would determine whether the repressor required an enzyme to make it or was itself the primary product of its gene.

Sometime in the fall of 1958, Monod visited Cambridge and went to lunch at Crick's house in Portugal Place. "That's when they first told us about the PaJaMo experiment," Crick said to me. "And we were very *puzzled* with the experiment; it didn't *fit in* with the ideas we had. Because we were messed up by the ribosomes. And we tried various ways to get out of it." At the Pasteur, by 1959, a younger scientist said, "The saying in the lab was, 'Ribosomes are crap!'" How was that in French? "Oh, we didn't speak French! Or not necessarily. 'Les ribosomes sont une poubelle,' we used to say. Okay? The garbage can. You have to put something in it for it to work."

Jacob having conceived of the operon, Monod having found the first evidence for it in the genetic linkage between permease and galactosidase, the next step was a purely logical one. "We had the proof that the repressor was a diffusible substance, because of the dominance relationships proved by the PaJaMo experiment," Monod said. "Therefore, the regulation—that was the simplest hypothesis—had to be at the genetic level. Therefore there had to be something which received, which was the acceptor, for the repressor that we had to postulate. Anything that does something specific—that was the repressor—had to be recognized by, or be able to recognize, some specific spot. And *that* was the 'operator.'"

At that point, Jacob and Monod had a conversation that Georges Cohen also took part in; all three remembered it vividly. The discussions took place in Monod's office; Jacob's lab was still in the attic. "I spent my days in the staircase because the elevator was always blocked," Jacob said. "I almost had an office in the stair."

"Jacob was an extremely good experimentalist," Cohen said. "In order to build a theory, you have to bring *data* to somebody. And Jacob brought a great amount of data. Because his power was— When he discussed with Monod and they were building up a theory together, Monod would be telling Jacob, 'If you're right, if the theory is right, it must have the following consequences, which can be tested experimentally.' And Jacob would *immediately* find an ingenious way to test the thing experimentally. That Monod would never have found by himself." Did Cohen have an example? "Yes. I know a particular example. A constitutive strain of bacteria in galactosidase is one that doesn't make repressor. Okay? So since you have no internal repressor you make galactosidase, all the time. Now, Jacob had the idea that if there is a repressor, this repressor must have a target. So if the target is something that is DNA in nature—then it can mutate. So Monod immediately told Jacob, he ought to be able to find another type of constitutive mutant where it is the *target* that is affected by the mutation, not the repressor. You'd have normal repressor, but it would have nothing on the chromosome to bind *to*.

"Now, the usual constitutives are recessive, not dominant, because making a repressor is dominant over not making it." But in a bacterial mating where the target—the operator—on one chromosome was mutated, "then the repressor from the other chromosome won't be able to act on it, because it's mutated, and that's going to be dominant—you're going to make galactosidase even in the presence of repressor. So Monod told Jacob, 'You must find this new class of mutants which is constitutive dominant, instead of the class we know.' And immediately Jacob found the way to sort those out. It was the proof of the target. Which was the operator."

"I remember, I was discussing with François some of his results with phage," Monod said. Microbiological mutations are infinitely varied: Jacob had discovered a mutant *lambda* that was virulent—would multiply and burst forth—when set to infect cells that carried the *lambda* prophage and so were normally immune. The explanation could have been that the DNA of the virulent mutant lacked the target for the prophage repressor that normally conferred immunity. "Most of our conversations for two or three years were comparing the phage system and the *lac* system, and basing the whole thing on the idea that we had to find in what ways they were the same. So I tried to think, if we had mutants of the phage-virulent type in the *lac* system, what would they be? And concluded they ought to be operator-constitutive mutants. Which was— The discovery of the operator-constitutive mutant was a very decisive step in establishing the whole theory of the operon and the regulation. It was an obligatory conclusion of our model: it had to be that way."

"When we made this model saying that there was a gene making a repressor that made another gene silent, we had a very clear prediction," Jacob said. "We could consider this as a transmitter and a receiver of signal. "And this regulator gene"—gene I in the lactose system; in the *lambda* system, the gene for immunity—"was the transmitter of the signal. So the prediction was that there should be a receiver of the signal. Therefore we should expect a mutant of this receiver. Now, if you look at that"—once more he was up and at the blackboard—"there, this is a chromosome. You have two chromosomes after the mating; you have several genes." He drew the diagram.

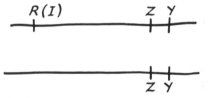

He marked gene Z. "That's the structural gene, that dictates a certain protein." He marked the gene Y as well. Then he marked still another and labelled it I. "Now you have another gene here, which is a regulator gene. The idea was, this was making the product which acts as the signal somewhere. Now suppose you have two of these things"—he marked the regulator gene on the second chromosome—"and suppose you break this one"—he crossed out the first regulator—"of course the signal is cytoplasmic, and so you still have the second regulator gene, which will work on *both* chromosomes, like this and like that.

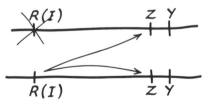

"Therefore that mutant with the defective gene for the repressor is recessive. But let's say now that *this* is the receiver, the operator." He marked its place on both chromosomes.

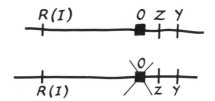

"And suppose you break the receiver." He crossed out one operator. "Then whatever else, whether or not the transmitter is making the signal, the signal will not recognize the receiver, will not act on the operator. And therefore the protein will be made. Therefore *this* mutation is dominant."

Then further crosses, which placed functional genes for galactosidase and the permease, Z^+ and Y^+, on opposite chromosomes, tested the further prediction that mutation of the operator allowed the expression only of genes lying on the same chromosome—technically known as the "cis" position—and had no effect on genes on the other chromosome. The operator itself made no product that diffused through the cytoplasm. The mutation in it was something about the DNA itself, adjacent to the structural genes of the lactose enzyme system, that changed so that the repressor no longer acted on it.

"The analogy I used to represent these things in the old time was, let's assume you have two planes which are flying, full of bombs. One is loaded with A bombs, the other with H bombs. They are flying around, and as long as they receive the signal *don't drop don't drop don't drop,* they don't drop. You have two transmitters. If one transmitter is broken, they still receive the other, *don't drop.* But if one of the receivers is broken, then that one will drop. Because it doesn't get the signal. And you will know which one because it will be either H bomb or A bomb and you can do *that,* too, by putting the right genetic markers on the system."

Mid-September 1959, Jacob talked about all these ideas at a conference in Copenhagen, called to discuss the intractable problem of colinearity between base sequences in genes and amino-acid sequences in proteins. The meeting was small and informal, but of quality. Ole Maaløe was chief organizer. Watson, Benzer, Meselson, and eleven others came from the United States; Monod, Jacob, Georges Cohen, and a younger colleague, François Gros, came from the Pasteur; Crick, Brenner, and Kendrew from the Cavendish. Niels Bohr, now nearing seventy-four, gave a lecture. Watson sat through most talks reading a newspaper at full stretch, noisily turning the pages; after several days, when he himself got up to talk, every scientist in the room pulled out a newspaper and opened it.

Jacob returned yet again—but for many in the audience the idea was fresh—to the remarkable speed with which, once the structural gene was introduced, enzyme synthesis started up and reached its steady top rate. The speed could be explained if proteins were synthesized directly from the DNA of the gene. But no examples of that had ever been found. Higher organisms, with their DNA segregated in the cell nucleus and protein synthesis in the cytoplasm, presented grave objections. The speed with which synthesis began raised equally grave objections to the doctrine that ribosomes were the intermediary bringing the information from DNA to make a specific protein. A couple of minutes was hardly time enough to build such complex structures.

Worse, since the repressor itself functioned at a particular site on the DNA that was part of a cluster of genes, if ribosomes were the intermediates, how were they turned off as promptly as they were turned on? The brief time lag could be explained only, Jacob said, by the possibility that had been emerging for eighteen months in Paris: an intermediate large enough, of course, to carry the information but small enough to be quickly made—and unstable, so that it broke down just as quickly after use. He called the unstable intermediate "the tape" or just "X." He did not say what it could be made of.

"Everybody was there," Jacob said. "I remember I gave a talk, and discussed this paradox, and at that time nobody reacted to that. Just nobody." Crick in particular still resisted the idea.

When Wollman came back to Paris from Berkeley in the fall of 1959, he found Jacob's lab and Monod's tightly and inextricably interwoven. Wollman no longer fit—or perhaps refused to fit. This, though he did not say so to me, was a matter of abiding bitterness.

In October, Jacob and Monod published a theoretical note to establish the sharp distinction between the familiar genes that determined protein structures and the new class of genes that regulated. It even looked to them, then, as though the product of the regulatory gene were not a protein but RNA. But the fact to be underlined, they said, was that in every known case, when several structural genes had their expression controlled by the same regulatory gene—"that is to say, in all probability by a unique repressor"—the structural genes were grouped tightly together. "The hypothesis therefore emerges that the action of the repressor depends on a structure common to the group of genes whose expression it governs." They could imagine several ways that that might work, but the best fit to the evidence was that the group of genes had among them a single element: the operator, target of the repressor.

On October 27, Monod send Cohn a copy of the note, observing that it was the result of nearly a year's discussions with Jacob. He went on:

> We have found a mutant that is constitutive dominant and seems to act only in the 'cis' position [that is, it affected only the chromosome on which it lay]. But we hesitate as yet to conclude that it is really a mutant that conforms to our theory.
>
> We are thinking up all sorts of experimental ideas to verify the proposition that repressors are soluble RNA. So far nothing positive.

In Monod's lab, François Gros had been looking at the effects on enzyme induction of feeding bacteria base analogues. That is, he was using nucleotides in which one or another atom of the base had been changed, so that if the cell built them into nucleic acids its metabolism would thereupon be altered. That winter, Gros took a mutant of E. coli that could not synthesize the base uracil. He fed it β-fluorouracil, which was nothing more than uracil with an atom of fluorine added at the fifth position around the ring. The fake uracil blocked synthesis of galactosidase at once. This was still another item of evidence that interference with nucleic-acid turnover affected protein synthesis very quickly; because uracil was used in RNA where DNA used the related base thymine, Gros's finding once more implicated RNA as the intermediate.

By mid-January they were sure they had the mutants they were looking for, and Monod reported them to Cohn in detail and in triumph. He then said:

> If this extraordinary story is true and if one can find other confirmation, this will be the proof that induction is genetic, that is to say that the effect of the inducer is to unblock a set of repressed genes.

Thus on 29 February 1960, Jacob and Monod and two colleagues published another note, to announce that they had found mutants in the lactose system that proved that the operator they postulated could be assigned a specific locus on the chromosome. These were the operator-constitutive mutants that caused the structural genes for β-galactosidase and permease on the same chromosome (but not on a separate chromosome) to be expressed even in the presence of repressor. The known genes in the lactose region, they said, appeared in the order Y-Z-O-I. But the burden of the note was theoretical:

> The hypothesis of the operator implies that between the classic gene, an independent unit of biochemical function, and the entire chromosome, there exists an intermediate genetic organisation. This comprises *coordinated units of expression (operons)* made up of an operator and the group of structural genes coordinated by it. Each operon, by the intermediary of its operator, will be submitted to the action of a repressor whose synthesis will be determined by a regulatory gene (not necessarily adjacent to the group). . . . This hypothesis explains the correlation seen very widely in bacteria between functional association and close genetic position for systems of sequential enzymes. It brings with it other verifiable consequences, notably that the enzymes of a sequence governed by the same operator cannot be induced separately.

The strength and economy that create grace in the papers Monod and Jacob wrote together are qualities of the style of thought that the French like to call Cartesian, since Descartes was the noted master of it. It's as rare in French scientific writing as in any other.

❖ ❖ ❖

"Well, we *were* unsympathetic, in the sense that we recognized they'd got an interesting experiment, but we weren't clear there wasn't some snag, you see," Crick said. "It wasn't until we were converted on the road to Damascus that we saw it all, you see!"

Despite Crick's cool response in Copenhagen to the problematic implications of Jacob's results and to the proposal of an unstable, information-carrying tape, he was acutely aware that the central questions of molecular biology were at an impasse. He had said so with gloomy vigor at the symposium in Brookhaven, Long Island, six months before. He had spoken then about the baffling confusion of the coding problem; he had dwelt on the upsetting report from the two Russian biochemists, Andrei Belozerskii and Alexander Spirin, that among a score of different species of bacteria the base compositions of their DNAs varied very widely while the RNAs varied hardly at all; he had listed the six solutions he could imagine to account for this phenomenon and had found them equally repugnant. The relationships between DNA, RNA, and protein had been put so much in doubt that he had wondered even whether the DNA-to-RNA translation mechanism, or the genetic code itself, would turn out not to be universal.

The one assumption, far less fundamental than those, that Crick did not re-examine at Brookhaven or in Copenhagen was that ribosomes, in particular the ribosomal RNA, carried the specific information from the gene for the or-dering of the amino acids in each kind of protein. A variety of new experi-ments in several labs had only confirmed that in bacteria as in higher organisms the ribosomes were the site where proteins were assembled.

Crick's picture of the ribosome then was captured in a note that he and Brenner mailed out to the RNA Tie Club in December 1959. Titled "Some Footnotes on Protein Synthesis," the note collected their most recent musings. Five years earlier, to the same restricted audience, Crick had called for small, specific adaptor molecules to sort out the twenty different amino acids and bring them to the right places in the growing chain. The adaptor had been found in the short and specific lengths of soluble RNA. Crick had been bothered by the fact that the soluble RNAs contained eighty or ninety bases where he had expected no more than about ten; now he and Brenner tried out a couple of ideas about the more complex functions that might account for the excess. Then they turned to ribosomes. Here too, Crick's expectations had been upset. He and Watson had thought ribosomes would be built like spheri-cal viruses containing RNA that infect plants. At Harvard, though, Watson and Alfred Tissières had just found that ribosomes were made of two unequal pieces, each containing protein and RNA. This was "a surprise," Crick wrote.

> Nobody seems to have produced a very good reason for the two parts . . . except the idea that the [larger] is a box and the [smaller] is its lid, and that the two come apart momentarily, from time to time, to let finished protein molecules escape.

But still they wondered why the two parts were different.

> Rather surprisingly, a simple answer is possible. It is only necessary to assume that the particle cannot easily be produced from protein subunits alone, but needs RNA to stabilise the arrangement. This is in any case true for the rods of TMV [tobacco-mosaic virus]. Suppose for the purposes of explanation, that it were true for the small *spherical* RNA viruses. . . . On this view, then, each of the two portions of a ribosome consists of *part* of a spherical shell; the size of each would be deter-mined by the size of the RNA, so that the protein/RNA ratio would be the same for both parts, as is found to be the case. . . . At first sight the structure might be

> but this makes it difficult to see why the two parts should not be equal. Other alternatives are:

> ... This sort of idea would explain very well why the molecular weight of the RNA has, apparently, two fixed sizes (leaving aside the difficult question of the RNA breaking down into further sub-units) and not, as one might expect, a continuous range of sizes.

This is a man talking to himself: the problem, not quite stated, was that the end products, polypeptide chains, come in a continuous range of sizes.

> It [the invariance of size] raises considerable difficulties when one comes to consider the RNA as a specific template. One is naturally inclined to the idea that only *part* of the ribosomal RNA acts as a template, but further than that we shall not venture at the moment.

In the summer of 1959, Matthew Meselson had come from Caltech to Cambridge to visit Brenner for a few days. Meselson's specialty had been the use of heavy isotopes and a high-speed centrifuge to separate different generations of large biological molecules according to their densities—which was how he and Franklin Stahl, in the first use of the technique two years before, had shown that DNA replicates semi-conservatively. Meselson was trying to get away from centrifuges. For the past year, he had been investigating a particular chemical—5-bromouracil, a modified version of the base thymine—that was known to cause mutations, trying to show what physical difference it made when phage incorporated it into their DNA. The work had been fruitless. With Brenner, Meselson tried a variation of the experiment, which didn't succeed either. Meselson invited Brenner to come to Pasadena the following summer.

By the spring of 1960, Brenner and Crick should have known—though neither could remember, seventeen years later—of experiments completed but not then published by which Meselson and a graduate student, Cedric Inglis Davern, had shown that ribosomal RNA was almost perfectly stable. Applying density-gradient centrifugation to RNA, Meselson and Davern grew *E. coli* for many generations in broth containing heavy isotopes of carbon and nitrogen, then switched the bacteria suddenly to the broth containing the normal, light isotopes; allowed them to multiply and harvested successive samples; extracted the ribosomes from each and disrupted them gently to free the RNA; put that into a heavy solution of a cesium salt, and then used the centrifuge to separate the RNA molecules into bands of different density. "Fully labeled molecules persisted for at least three generations while unlabeled molecules were being synthesized," they wrote. "No partially labeled RNA could be detected." Ribosomal RNA, once made, did not turn over. Davern toured Europe in the summer of 1959. In Cambridge, he met Brenner, though he was not sure, by 1978, that he had mentioned the experiment back then; in Paris, though, Davern gave a seminar about it, and Jacob was in the room. Meselson and Davern sent their paper to *The Journal of Molecular Biology;* the first draft reached Cambridge at the end of January 1960, and a revision arrived on Good Friday. From Paris, early in February, Monod wrote admiringly of these results in a letter to Pardee.

No one remarked that the French call for an unstable tape to carry information was, in form, like Crick's adaptor hypothesis: the recognition of a physiological function that cried out for an anatomical base. For one thing, where Crick had been explicit about the kinds of molecules the adaptors might be, Jacob and Monod offered no such details about the tape. For molecular biologists, as Monod himself recognized, the proposal lacked substance.

❖ ❖ ❖

Early in 1953, in one of a series of papers that looked into the way phage-infected bacteria manufactured DNA, as traced by the pickup of radioactive phosphorus into nucleic-acid backbones, Alfred Hershey had mentioned briefly that he had found a small fraction of the RNA in the cell that was synthesized actively and extremely rapidly just after the infection, reaching its peak within five minutes. "Enzymatic tests showed that the active material was authentic RNA," he wrote. "Its metabolic activity clearly distinguishes this material from the bulk of RNA that was formed in the cell before infection."

In 1956, Elliot Volkin and Lazarus Astrachan, at Oak Ridge National Laboratory, in Tennessee, had published preliminary results obtained much like Hershey's with phage-infected bacteria and radioactive phosphorus, suggesting that the cell possessed a minor species of RNA that turned over rapidly—and that resembled not the bacterial but the phage DNA in the proportions of its four bases. This was not microsomal (ribosomal) RNA; it was not soluble (transfer) RNA. Their first report, completed in March, and the paper that Volkin had read on the matter at the McCollum-Pratt Symposium in Baltimore in June 1956, had been apologetic to an extreme about the possibility of contaminated experiments, uncertain about conclusions. Besides that, they thought that their radioactive RNA, resembling phage DNA in base composition, was converted into DNA for new phage particles.

In the next two years they had buttressed their experimental results with some well-conceived controls and had published again. Their conclusions, though, got mushier. "The new RNA" they had found after phage infection "may be directly related to the synthesis of the new protein." Or, since radioactive phosphorus was also most rapidly incorporated into proteins and fats shortly after phage infection, "the activity of RNA at that time may not be directly related to synthesis of a new protein but may reflect a general reorganisation of cellular material."

None the less, by the late fifties, other biologists recognized Volkin and Astrachan's high-turnover RNA as an experimental anomaly that could not be ignored, a minor species of intellectual irritant. Crick and Brenner knew of it. I asked Jacob whether he and Monod had been aware of it. "The question is, what does that mean, to be aware?" he said. "To be aware is to have it on the front of consciousness, and clearly we were not, at this level, even if we knew."

❖ ❖ ❖

In mid-March of 1960, Brenner visited Paris and was told about the operator, the mutants that demonstrated what happened when the operator was impaired, and the theory of the operon. "I can remember Jacques distinctly saying that maybe this very fast expression in the PaJaMo effect was due to DNA making protein," Brenner said. About that time, Monod gave a lecture in the great amphitheatre at the Institut Pasteur; François Gros attended. "For some time Monod was very dubious of his own ideas, because there was no substratum for his hypothesis," Gros said in a conversation in 1976. "He raised the problem as a kind of paradox. And he said, either there must exist an RNA which is actively turning over, but we know that such an RNA does not exist, or perhaps proteins are made by some unknown fashion on DNA."

Back in Cambridge, Brenner spotted a preliminary notice—one paragraph out of thousands in a collection of abstracts—of some work just finished by Sol Spiegelman, at the University of Illinois. Brenner knew Spiegelman well; the work was a puzzling ambiguous extension of Volkin and Astrachan's. Spiegelman, with Masayasu Nomura and Benjamin Hall, had observed like Volkin and Astrachan that *E. coli* infected with phage T2 made an RNA that in several respects matched the phage DNA. But now Spiegelman asserted that when he disrupted the cells in a medium that contained a high concentration of magnesium, this newly synthesized, phage-specific RNA "was found to occur mainly in the ribosome fraction."

❖ ❖ ❖

"You see, there was a meeting in London, I guess, of the Microbiological Society, and all these people came up to Cambridge, and we decided we'd get together and have a discussion on, you know, just on this business," Brenner said. We were in his laboratory one Saturday late in February 1976. He had been setting up an experiment in a square foot of clear space on a crowded bench, transferring various phage stocks, two by two, from large, labelled bottles into small test tubes in a rack, referring to a written list on the bench, using a fresh glass pipette for each transfer, sucking up a quantity into the pipette with his mouth, stopping the mouth end with his thumb, easing out the excess to reach the measured amount, then adding to each test tube, with larger pipettes, a portion of a turbid broth of *E. coli*: precise work in the quick, casual rhythm of the chef in his kitchen. "Genetic engineering," he had said, rolling his eyes. He put the rack of tubes into a warm-water bath and was free to talk. "We went to my rooms in the afternoon to discuss it; it was the quietest place. It was just a small group. Was it Good Friday?"

Six or seven people climbed the two tall flights of stairs to Brenner's rooms in King's: Alan Garen and his wife, Susan Garen, were there; Leslie Orgel; and Ole Maaløe—though Maaløe once said to me that he did not remember being there.

"There was a symposium in London. Then I went to Cambridge; I gave a seminar there," Jacob said in his office in December 1975. "I think it was in Brenner's private apartment in King's; that's right. Then Francis and Sydney immediately jumped over that." Had the idea changed any since the meeting in Copenhagen the previous fall? "No."

"Yes, I was present all right because I was in fact quizzing—I was leading the discussion," Crick said in his office in November 1975. "Good Friday, that I remember. Probably look in my diary, because there was a party that evening. And it started by my quizzing François about the so-called PaJaMo experiment. We were quite familiar with it. And what he essentially said was that they'd tightened it up, and many of the possible criticisms had been overcome."

"Well look, if I can say how we saw the problem," Brenner said. "We were worried about the ribosomes. Okay? The idea then was that ribosomes were the factories for protein synthesis; they contained some kind of transcript of the DNA." The first problem, he said, was that ribosomal RNA varied so little in its base composition compared with DNA. The second problem was that ribosomal RNA, carefully extracted from the ribosomal protein, came in just two sizes or lengths, which appeared to belong—the point was all but settled— to the larger and smaller pieces of the complete ribosome. "Whereas if this was a template for protein synthesis, one would expect all kinds of varying lengths. So we were worried about that; and we were worried about it from the coding point of view. And the other thing is that it had been known, you know, that ribosomal RNA didn't turn over very rapidly. The Meselson-Davern paper was something that fitted into this, which was that the molecules once made were completely stable.

"Now that was *that* background, but there was another background, which I think you have to know about," Brenner said. "The phage background. Because it had been shown that in fact there was no, or very little, synthesis of ribosomes during phage production. There was a paper by Spiegelman and Nomura, of which I had seen an abstract in the *Federation Proceedings*, okay? Who again showed that there was a little bit of synthesis of RNA after infection by phage." By chasing this RNA into the ribosomal fraction of the cell extracts, however, Spiegelman raised in acute form the problem of whether "the cells were making a small number of very special ribosomes," Brenner said.

"All right, now, the Paris people had done the PaJaMo experiment. And there were the problems that had arisen out of that. Namely, that the β-galactosidase is expressed very, very rapidly indeed. But the problem was that one could not exclude that there weren't a *few* ribosomes made, made very rapidly, that were then capable of *prodigious* protein synthesis. So that there was one consistent set of interpretations—namely, that you made a small number of very effective ribosomes, both for induced enzyme biosynthesis and also for the phage-infected cell. Right?"

Later in our conversation, Brenner came back to that. "You see, the Paris people were interested in regulation. We essentially were interested in the code. So we had a slightly different approach," he said. "There was this problem that only a small number of the ribosomes are involved. You must remember, the PaJaMo experiment was a matter of *how much* you had to synthesize. All they showed was that it happened very fast, that it took off linearly from the beginning, with no exponential rise." That is, the graph slanted up in a straight line, no upturning curve showing an increase in the *rate* of the individual cell's enzyme synthesis as more ribosomes were formed. "Well-ll, no one could really believe, you know, that those beautiful Paris curves were really straight lines. That you couldn't draw something through the points that had a bent slope, you see. And if there were very *few* ribosomes involved—in the limit, *one*—you couldn't tell the difference. Okay? I mean it now seems ludicrous that one entertained all those ideas, but there was no choice. We have a very similar problem, with all this excess DNA in higher organisms. I've just given a lecture, comparing our perplexities with those here; it will turn out in five years' time that people will be *amazed* that one thought of all these elaborate schemes to explain it. *Whatever* elaborate schemes you're thinking of. I mean, somehow you've gotta see the clue."

"We probably didn't see the Meselson paper until after that," Crick said. "The discussion opened simply with François saying that they'd done the PaJaMo experiment with the timing more careful and so on. You know they were all very good experimentalists; the experiment had been improved to the point where was not reasonable to doubt it anymore."

Jacob talked about the switch, the operator and the operon, which had appeared only in French. He told them about mutations in the gene Z that altered the galactosidase protein—work that also had appeared only briefly and in France; Crick said he'd prefer to see a chromatographic finger print showing changes in amino acids directly. The perfect experiment, Jacob then said, would put a gene that controlled the synthesis of a protein into a cell that lacked it, and later take it out again. He reported two lines of work, his with Monod in Paris and Pardee's with Monica Riley in Berkeley, that added up to that ideal. Both were now completed, though the paper reporting them all was still being discussed by transatlantic mail.

In Paris, they had repeated the PaJaMo experiment—but leaving out of consideration the regulatory gene, I, and following only the structural gene for β-galactosidase, Z. The aim now was not to exhibit the repressor but to deal directly with the timing of expression of the enzyme. Only three elementary models were possible. In the first, transfer of structural specificity from gene to protein required no stable intermediates and the gene acted directly as template. In the second, again no stable intermediates were formed but the gene acted by means of an unstable intermediate. In the third, the gene did form stable intermediates, such as specific ribosomes. The results in Paris showed with great precision that the gene Z^+, put into a cell that lacked it, functioned without significant delay at its maximum rate. Crick now accepted this as proved. That eliminated the idea of a stable intermediate—unless the gene made a limited number very quickly and then stopped. If such an attempt to save the stable intermediates seemed unlikely, so did the notion that the gene acted directly. But the refined PaJaMo experiment could not distinguish further among the three models.

The experiment Pardee and Riley had done in Berkeley was new, technically amusing, and persuasive. It amounted to removal of the gene from the cell after it had begun to function. They had grown male bacteria, carrying gene Z^+ but sensitive to streptomycin, in a broth where the available phosphorus was the radioactive isotope ^{32}P. The bacteria, with their DNA heavily labelled, were then centrifuged out, resuspended in a nonradioactive broth, and mixed with females that were resistant to streptomycin and that carried genes Z^- and I^-—that is, they could not make β-galactosidase but were constitutive. The strain of males transferred the gene Z^+ ten minutes after conjugation began.

They interrupted mating thirty-five minutes after mixing the two strains, added streptomycin to stop growth of the males, and allowed the females to develop until fifty minutes after penetration of the first Z^+ gene. They diluted the preparation with the sort of minimal broth in which bacteria were stored alive without growing. They added glycerol, a syrupy emollient. Then they took one sample to test for enzyme activity. They put other samples into small glass ampules, sealed the ampules by fusing the glass at the neck, and lowered them into a vacuum-insulated flask of liquid nitrogen. The bacteria were frozen almost instantly at 196 degrees below zero centigrade.

Protected from bursting by the glycerol, the bacteria were not killed, but their vital processes were arrested while the radiophosphorus in the DNA from the male bacteria continued to decay. An atom of ^{32}P decays by emitting a beta particle, which is a high-speed electron, whereupon it is transformed into an atom of sulphur. The transformation, and the recoil of the atom as the electron leaves, breaks the bonds of the backbone of the DNA at that point. It happens at random. They figured that fewer than five hundred atoms of ^{32}P had been transferred per bacterium. Half of those decayed in fourteen days. The Z$^+$ genes were being killed.

From day to day, Riley raised ampules of the frozen bacterial suspension from the liquid nitrogen and thawed them. She separated the bacteria from the protective storage medium and resuspended them in fresh broth containing streptomycin. Then she added a potent gratuitous inducer of β-galactosidase synthesis. For comparison, she and Pardee ran the complete experiment allowing different times after mating before freezing. Also, they ran the whole thing in parallel without the radioactivity.

The nonradioactive bacteria sampled before freezing were synthesizing enzyme copiously. So were the radioactive ones before freezing. Whatever theoretical model the lactose system in *E. coli* conformed to, it was functioning in all its parts. Thawed after ten days, samples of nonradioactive bacteria synthesized β-galactosidase just as vigorously as those never frozen. But the bacteria whose Z$^+$ genes had suffered ten days of radioactive decay made the enzyme at less than half the rate they had before. Inactivation of the gene, Jacob said in Cambridge, abolished protein synthesis without delay. Stable intermediates between the gene and its protein—in other words, ribosomes whose RNA carried the information to specify the sequence of amino acids—were ruled out. Continual action of the gene was necessary, either directly or by way of an intermediate that was unstable and so had to be steadily renewed.

"That's when the penny dropped and we realized what it was all about," Crick said.

"Sydney and Francis put the thing together with something which was known which has to do with the phage DNA synthesis," Jacob said. "That there was some kind of very bizarre RNA made after phage infection by T2." What Hershey had first found, I said. "Yes. Hershey and two others, I forget their names—"

"The one thing that had been kind of stuck was the Volkin and Astrachan RNA," Brenner said. "They had shown that you made an RNA whose apparent base composition, okay, was the mimic of the phage DNA." He lit a cigarette. "And of course Volkin and Astrachan themselves interpreted it that this was a precursor to DNA. . . . From my point of view what happened at that day was that I clicked together the Volkin and Astrachan RNA with these possibilities."

"Well, then what was clear suddenly was that the Volkin-Astrachan RNA was the message," Crick said. "Which they hadn't realized in Paris. In other words, that people had already *discovered* the messenger RNA and hadn't *realized* it. You see. And the Paris people—whatever they put in their review later—had not at that stage realized that that was evidence for their ideas. All right? And once you got— The fact that they'd got a jolly good experiment in Paris, on the one hand, and that people some time ago had found the very thing that they were postulating, then of course we accepted the idea like *that!* And then of course it immediately followed"—Crick's words were tumbling with recollected excitement—"that the ribosome was an inert—was a *reading head*, and had got nothing to do with the message. And with that it just went on. You see. The false idea was just removed in one blow."

The interplay of ideas speeded up. The six or seven people in the room were looking at the same biological model, brand new, vividly imagined, almost palpable in the air before them, yet still from different sides. For Jacob, the Volkin and Astrachan RNA supplied what the experiments with Pardee had demanded. For Crick and Brenner, the combination at a stroke redefined the function of the ribosome: all ribosomes were equivalent instruments for converting the message on the strand of unstable RNA that came from the gene. For Crick especially, the coding problem leapt from its sickbed with a glad shout: whatever the RNA of the ribosomes was doing, if it was not specific for proteins its base composition need not, should not, match the DNA of the genes, and so the infuriating Russian results lost most of their importance. Little problems got well too. The fixed sizes of ribosomal RNA no longer mattered; the silk glands of insects, which produce just the one protein, largely made of glycine and alanine, but whose ribosomal RNA matches that of the insects' other organs, ceased from troubling.

For Brenner, PaJaMo times V. & A.—one needs a fast notation—rebounded on the Spiegelman, Hall, Nomura experiment with phage DNA. Brenner thereupon vaulted from enzyme to phage system just as Jacob had, twenty months earlier. Spiegelman had found his RNA that resembled phage DNA closely associated with ribosomes: correctly understood, the experiment gave the first glimpse of how to catch the unstable RNA in the act of carrying information from gene to ribosome. "They all came together," Brenner said. They would use phage genes to take the French results with enzymes the crucial step further, by applying Meselson's techniques. For the moment, as Brenner began to explain it all, Jacob the master of cross-system reasoning was left behind.

"What I do recall is, suddenly I was talking to Francis, and no one else"— Brenner laughed—"was following me. *He* picked it up straight away. I mean, that's the way it suddenly hit me between the eyes, that of all things, yes! there had to be an RNA, which was *added* to the ribosomes. And I was trying to explain to them how the Volkin-Astrachan thing fitted in to explain the Spiegelman-Nomura experiments. . . . I think that based on the PaJaMo experiment, where it was one of the possibilities that one had something added, it suddenly dawned on me that there was a complete interpretation of the *phage business*"—and not just enzyme induction—"which I don't think the Paris people had ever considered. The most important outcome of that was that *here* was a *way* of *actually proving it.* I just followed straight on to design the experiment we later did."

Later Brenner said, "Messenger had been lying around for a long time. Because—everybody knew that in some sense the genes had to send a message to the cytoplasm. But its exact physical form is what we were after."

"We were just lost until the messenger was discovered," Crick said. (He told the English historian of science Robert Olby, in 1968, that the failure to find the messenger much earlier had been the one great howler in molecular biology: "The only thing one is thankful for is that it wasn't all done by some-one, as it were, outside the magic circle, because we would all have looked so silly. As it was, nobody realized just how silly we were.")

"We planned the experiment that day," Jacob said. "That's when we decided, Sydney and myself, to go to Caltech. I had been invited by Del-brück to come and spend a month there, and Sydney had been invited by Meselson."

Jacob was spending the night at Brenner's house. They went there, still talking. Brenner doubled back to the Cavendish to look up how one extracted ribosomes. He sketched the experiment they could do at Caltech. Crick also went home. Probably that afternoon, at any rate within the next few days, he dashed off a paper that he proposed that he and Brenner publish. The draft survived: written at a sitting, impetuous and colloquial, putting in order the arguments and evidence of the previous hours. He ended:

> We now boldly combine the conclusions of the two experiments [Pardee, Jacob, and Monod; Volkin and Astrachan] and generalise them to produce the following hypothesis.
> (1) genetic RNA has the same over-all base-ratios as genetic DNA.
> (2) it passes into the ribosomes, but it is only a minor component (10–20%?) The major part of ribosomal RNA is *not* genetic RNA
> (3) genetic RNA is (at least in some circumstances) "unstable", that is, it may have only a limited life.
> We thus arrive at a picture of the action of ribosomes which differs in important respects from the one we used to hold. On the old picture a ribosome made a single type of protein and continued to do so while it remained intact (though possibly at varying rates). On the new picture a ribosome may be making one protein at one moment, and a quite different protein a few minutes later. . . . Thus the population of microsomes may be looked upon as the machinery for protein synthesis through which flow the instructions from the genes. . . . This obviously allows the control mechanism to operate at the gene level, but we shall not pursue this aspect of the problem here.
> It is regrettable that one cannot formulate a more explicit hypothesis about the instability of the postulated genetic RNA. Three possibilities suggest themselves.
> (1) genetic RNA is broken down each time a protein is made ie. it is only used once. This seems to us unlikely.
> (2) genetic RNA is broken down after, say, 10 copies of the protein have been made. (One can imagine a ribosome being filled up with new protein molecules until it disaggregates, thus releasing the new protein and allowing its genetic RNA to be destroyed.)

The ribosome was still a pot with a lid to make proteins in. Then fantasy took over:

> (3) genetic RNA may be inherently stable but there is a special mechanism which breaks it down in certain circumstances. . . . Alternatively some genetic RNA's may be stable or others unstable. Or they might be a special breakdown or in-activation mechanism for each particular type of genetic RNA.
> It is clear that although these ideas are attractive they run into formidable dif-ficulties. . . . It is surprising that if genetic RNA exists and often turns over that it

has not been picked up before. We suspect that some of these difficulties may have arisen because it is unstable, and others because it is perhaps apt to leak out of ribosomes unless special precautions are taken [that is, during experimental manipulation]....

Although the *combination* of ideas presented here is new to us we cannot fairly call the ideas themselves new.... The ideas arose during discussions at Cambridge with Dr. François Jacob ... and also with Dr. O. Maaløe.... It is because it has so radically altered our own thinking that we have thought it worth while to put this particular formulation onto paper.

"There was this party in our house," Crick said. "And some people didn't know anything of what was going on, but were just enjoying the party. And other people were in little groups, just standing there, 'Could we do density shifts?' and this, that, and the other. So it was a very exciting evening and Ole Maaløe realized that a lot of *his* results on the abundance of ribosomes now made sense." But Brenner and Jacob got to the party late. "François's leg was troubling him," Brenner said. "I just have a vague memory of Francis explaining it all to Maaløe again." Maaløe, when asked, did not remember the party clearly. Nor did Jacob. "You think there was a party? At Francis's? It must have been filled with pretty girls, and I was busy—" Jacob laughed. "Then we did the experiment," Brenner said.

Back in Paris by the following Tuesday, Jacob wrote to Meselson, and that was the first communication of the idea to the rest of the world. He said, in part:

I met Sydney in Cambridge last week and we have planned a few experiments to do together with you if you agree, in June. This experiment comes from the fact that the operator theory of repression, that is the repressor acting directly at the genetic level, should not only prevent the synthesis of the proteins, but also excludes the idea of stable template for protein synthesis. A possible model would be that no genetic information is contained in the RNA particles, but that the gene sends to these particles RNA unstable molecules which act as message to the particles, brings the genetic information to them for the synthesis of a particular protein and are destroyed at the time of the protein synthesis. This hypothesis can be probably checked easily in bacteria infected with phage T2. I think that Sydney will write you in more detail about this.

May 7, Brenner did so, saying, in part:

This letter is to tell you about exciting developments here and also to discuss an experiment that we should do in Pasadena. I don't know whether you know of the recent Monod-Jacob work and so I will just mention the salient features.

First of all it seems very likely that control mechanisms e.g. repression may operate at the genetic level. Jacob has found constitutive mutants of the β-galactosidase which are different from the i mutations and act only in the cis position [that is, only on genes in the same chromosome].... It seems that the best way to explain this finding is to assume that the templates have to be continually renewed and are destroyed either after one or a small number of synthetic steps. Into this picture we can fit the finding [Pardee and Riley's] that the continued expression of the β-galactosidase requires the presence of the gene. Our picture is as follows: Proteins are synthesised in the ribosomes. Most of the RNA is structural and carries no genetic information. Thus RNA seems to have a base composition which is constant throughout nature. Into the particles are then fed the RNA templates which have a limited life, and which are direct copies of the DNA genes.

The obvious candidate of such template RNA is the Volkin RNA found after T2 infection. This has an A/G [adenine to guanine] ratio which is the same as the phage DNA and is unstable showing turnover. In order to prove this theory directly we have to investigate the distribution of the Volkin RNA amongst the ribosomes of the cell. Speigelman [sic] has shown that most of it does in fact occur in the ribosomes but we do not know whether there are a few ribosomes with all the RNA of this type or whether all the ribosomes have a small fraction of RNA. We can also look at the problem of a synthesis of the head protein in a phage infected cell. After infection, cells continue to synthesise protein at the same rate as before and by the twelfth minute one half of this synthesis can be accounted for by the head protein. If there were a few particles which synthesised phage proteins, the absolute rate of synthesis would have to be tremendously high; on the other hand, in the new theory we might expect perhaps half of the particle[s] to be synthesising head protein at a reasonable rate.

The experiment he and Jacob wanted to do was, he wrote,

to study by density gradient centrifugation the distribution of the new RNA amongst the particles . . . It would help us if you had available all the density labels that we might need, heavy bases, heavy amino-acids, etc. You know exactly what we will have to use.

Early in May, François Gros went from the Pasteur to Harvard to work for three months with Watson on the effects of certain antibiotics on the growth of ribosomes.

That month, also, Crick rewrote his memorandum from the Easter weekend as a note for the RNA Tie Club, though he never sent it out. "What are the properties of Genetic RNA?" asked the title; then, "Is he in heaven? Is he in hell? That damned elusive Pimpernel," sang the epigraph. This version ticked off more evidence in fewer words. In particular, the full paper by Nomura, Hall, and Spiegelman had reached Cambridge. A single bacterial cell contained, by their estimate, about thirteen thousand ribosomes. Upon infection, it made enough of the new, phage-specific RNA to amount to no more than a couple of hundred new ribosomes. None the less, Spiegelman was convinced that he had accounted for Volkin and Astrachan's RNA as new ribosomes that were, on the conventional model, specific to phage proteins. To reverse Spiegelman's interpretation of his own results was, no doubt, a delicious confirmation of the new idea. "We knew he was wrong," Brenner said.

Support was also trickling in from the outlying shires. Martynas Yčas, Gamow's coding ally, had moved to the State University of New York at Syracuse, and from there, that May, he sent Crick an advance copy of a report of a transient kind of RNA that he and Walter Vincent had found in yeast cells; the base ratios of the newly synthesized RNA, they said, matched the yeast's DNA rather than its ribosomes' RNA. It had been Vincent who found, early in 1955, a rapidly turning-over RNA in the nuclei of starfish eggs that he thought might be "involved in the transfer of nuclear 'information'" to the sites of protein synthesis in the cytoplasm.

Jacob went to Caltech by way of Berkeley and Stanford, where he gave seminars. When he arrived he gave another. They were full of experiments, full of theory, the year's work in Paris, the dénouement in Cambridge. The unstable, information-bearing, intermediate RNA was received with doubt by almost everyone except Pardee and Riley in Berkeley. Brenner got to Pasadena

on June 6. Delbrück told them, "I don't believe it." Meselson had been working since the beginning of the year with Jean Weigle, a Swiss colleague, trying
to show that genetic recombination—the events whose frequency biologists
since Morgan and Sturtevant had used to map genes and latterly mutations
within genes—resulted from actual physical breakage and joining of DNA
molecules. Weigle was away in Geneva; Meselson's scientific attention was
preoccupied by finishing those experiments. Brenner and Jacob were put in
Weigle's room. "My intellectual contribution to their experiment was not very
great," Meselson said. "It was really an act of great intellectual and physical
exertion on the part of both of them, not of me. I enlisted Sydney to help me
by counting plates, late at night, and I helped them during the day." The first
preliminary trial began on June 7. Brenner and Jacob intended to leave by the
end of the month.

The plan was to distinguish whether, after phage infection, new RNA went
to new ribosomes, or whether there were no new ribosomes, just the preexist-
ing ones "for hire"—Brenner's phrase at the time—to the new message when
it came along. So that old ribosomes could be labelled, the bacteria would be
grown with heavy carbon and nitrogen, the bacteria switched to a broth containing normal, lighter isotopes and simultaneously infected with phage, and
new RNA labelled with radiophosphorus. Then ribosomes would be separated from bacteria, put into a cesium-chloride solution, and spun at thirty-
seven thousand revolutions per minute for thirty-six hours or so.

The essence of Meselson's technique, once more, was that in this enormous
centrifugal force—on the order of a hundred thousand times the force of
gravity, enough to flatten an elephant into an oil slick—the cesium chloride in
the solution became distributed in a gradient that was denser towards the bottom of the tube; anything of like density in the tube would sink or float to the
level that exactly corresponded to it. Thus, ribosomal particles grown heavy
before infection would form a band farther down the centrifuge tube than any
made after infection when the isotopic labels were light. Radioactivity could
then be checked in each band of ribosomes.

That, anyway, was the scheme Brenner and Jacob had outlined in Cambridge. For twenty days in Pasadena, they could make nothing work at all.
The realities of the experiment confronted them with all sorts of cumbersome
details and two fundamental, cumulative difficulties. The first was to find the
heavy ribosomes. The heavy isotopes, ^{13}C and ^{15}N, were scarce; Linus Pauling
had obtained the heavy carbon for Meselson from the Soviet Union, of whose
Academy of Sciences he was a member. Meselson had a couple of tiny flasks
in his office. Brenner and Jacob had to grow heavy bacteria in teaspoonfuls of
broth; they got so few bacteria that they couldn't easily separate the
ribosomes. To overcome that, and to provide bands for comparison in the
density gradient, they grew much larger quantities of *E. coli* in exactly the
same conditions but with normal, not heavy, isotopes, and mixed a fifty-fold
excess of this culture—they called it a "carrier"—with the heavy culture just
before grinding up the bacteria and taking out the ribosomes to centrifuge
them.

In Meselson's classic procedure, any band of nucleic acids in a cesium-
chloride gradient was spotted by its strong absorption of ultraviolet light at
the telltale wavelength. But in order to measure variations in radioactivity of

the solution, Meselson had devised a method whereby the bottom of the centrifuge tube—not glass but plastic—was pierced and the contents distributed drop by drop into a line of test tubes for separate counting. But then, when the ultraviolet absorption of each drop was measured, the light carrier ribosomes, being plentiful, would show up—but there were nowhere near enough heavy ribosomes to be detected that way. So Brenner and Jacob found they had to accept a high count of radiophosphorus (meaning new RNA) in combination with a low absorption of ultraviolet (meaning no light ribosomes) as circumstantial evidence that a particular drop had a concentration of heavy ribosomes.

"You could *never see* the actual experiment," Brenner said. He made the running notes and records of their work on pale green squared-off paper in a notebook, or on loose-leaf sheets ruled vertically in red and provided by Caltech; Meselson wrote on special centrifugation forms he had had duplicated in purple ink; here and there appeared long columns, measurements drop by drop, in Jacob's characteristically French numerals. "During these experiments, in the heat and humidity, François was in intense pain from his legs," Meselson said. Several times they tried plotting the ultraviolet absorption and radioactivity, drop by drop, on a graph, hoping to see by shape what the numbers didn't make evident, peaks that might be bands, that hinted a concentration of what had two days earlier been invisibly small components of living cells. Brenner later put all these raw data and first hesitant interpretations, fifty sheets or so, into a green cardboard binder. What I could see in them was fog—thick fog, wet, cold, silent. The nonscientist does not often break through the sharp-edged view of the world that science presents to sense the swirling gray uncertainty in which that confidence must take shape.

The ribosomes would not grow properly labelled, and in the cesium-chloride solution they fell apart. "We put them into the Spinco"—a standard make of ultracentrifuge of the time—"and we just set them running and we got bands after bands after bands," Brenner said. Brenner tried fixing ribosomes with formaldehyde, an embalming fluid. For three days, they tried to label ribosomes with heavy water, but the last lines of the last sheet of *that* experiment read "Inoculated with bacteria" and then "No growth." Then Brenner wondered whether the information-carrying RNA could be spotted in a completely different way, by the effects of x rays on protein synthesis. "I mean it was wrong. But the idea was this, that maybe we could damage a template so it would jam the ribosome." So for a while, each time they switched on the centrifuge they went down two flights of stairs to a windowless basement where Renato Dulbecco—he had been Watson's benchmate as a graduate student of Luria's, ten years before—had a lab with an x-ray machine.

Then on June 26, by Meselson's notes, they stopped the centrifuge after spinning three tubes for thirty-six hours. Jacob measured the ultraviolet absorption and radioactivity. In the gradient for the first tube, the ultraviolet absorption rose sharply once and less sharply again. Radioactivity was smeared almost evenly across the entire gradient; four drops—79, 80, 81, 82—counted slightly higher. Drop 81 was over a third above the level of the smear. When they put all that on a graph, the four hot drops rose up looking more like heavy ribosomes than anything they had yet seen. Brenner marked drop 81 on the graph with the Greek letter *phi* for phage. The third centrifuge tube produced still more interesting results. Here the absorption of ultraviolet showed the light carrier ribosomes well aggregated into two peaks of equal height close together, while the radioactive RNA, though still appearing messily everywhere, peaked distinctly once. Meselson warned them the experiment might take six months to debug, and left for New England and a Gordon Conference.

Jacob and Brenner had three days. They talked unceasingly. Brenner looked back at the data for the two encouraging gradients. Meselson's forms showed that they had made up the first preparation with slightly more magnesium than usual—and the other, tube 3, with two and a half times the concentration of the first. Meselson had written down the raw quantities of everything put into the centrifuge tubes. Brenner converted those to actual concentrations of magnesium, which he noted at the top of each sheet. But magnesium ions held the parts of a ribosome together. This explained why the third tube showed two bands for the light ribosomes: one, the less dense, which had emerged clearly in this tube only, was whole ribosomes, while the other, in the same place in the gradient as the peak in tube 1, was a band of pieces—and this peak was smaller in tube 3 than in tube 1. Furthermore, when Spiegelman had used a high concentration of magnesium, he had tracked the RNA produced by phage infection to the ribosomes. "So I said, 'Let's put in a lot!'" In Jacob's remembrance, this conversation took place at a beach, and Brenner jumped up and down on the sand in excitement.

They had enough of the heavy isotopes for a set of three centrifuge tubes, at three concentrations of magnesium. For comparison, they would run three similar tubes using ribosomes from bacteria grown entirely in normal isotopes, to check that the radioactivity then showed up at a different density in the gradient. So, once more Brenner and Jacob grew bacteria in a small flask of broth where sixty per cent of the carbon was ^{13}C and ninety-nine per cent of the nitrogen ^{15}N. They transferred the bacteria to a test tube of a medium that kept them alive but starved them, put the tube into a water bath at blood heat, added phage T4 in abundance, and started up growth by adding glucose and a fresh broth made with ordinary carbon and nitrogen—but containing no phosphorus. Two minutes later, they added phosphate compounded from ^{32}P. Jacob missed the tube and spilled the first dose of hot phosphate into the water bath. Seven minutes after infection, they poured in the fifty-fold excess of the carrier bacteria and immediately harvested the mixture of cells. To get out the ribosomes took several steps of washing; then spinning out the bacteria at low speed; grinding them in a mortar with alumina, a grit; adding DNase to get rid of DNA; spinning out the debris at low speed; then centrifuging the remaining liquld at high speed for two and a half hours to throw down the ribosomes.

All these things were done in a cold room, at a few degrees above freezing, because at higher temperatures the ribosomes were more likely to disintegrate. Even the mortar was cooled. Then the three tubes of cesium-chloride solution were made up, a milligram of purified ribosomes—thirty-five millionths of an ounce—in each. Magnesium was added to each tube, in the form of the soluble compound magnesium acetate—the first tube to a three per cent concentration, the second to six per cent, the third to ten per cent. They made up the three control tubes to the same concentrations. They put three of the six into Meselson's usual centrifuge, the other three into a second machine downstairs in Dulbecco's lab. They started them up. Then they found that Weigle's water bath was contaminated by the spilled ^{32}P, so they rinsed it out and hid it behind the Coca-Cola machine in the basement to cool off.

In Brenner's remembrance, the next day was the day they went to the beach; a young woman at the lab took them in a Volkswagen.

The day after that, their last in Pasadena, they stopped the centrifuges. Now they had to separate the bands for assay. Brenner set up a rack of dozens of small, numbered test tubes. He took a plastic tube from the centrifuge and gently pierced the bottom with a needle. As the cesium-chloride solution dripped out, beginning of course at the dense end of the gradient, he caught the drops two at a time in separate tubes. He got eighty-one drops. That was the control gradient—no heavy isotopes, three per cent magnesium. Next, the tube containing heavy ribosomes and three per cent magnesium. That yielded seventy-three drops, collected two by two; and so on for all six tubes.

Two sets of measurements now. First, Brenner measured ultraviolet absorption, sample by sample, while Jacob wrote down the values. The numbers showed immediately that in each of the first two gradients the light ribosomes had formed two compact bands. Plotted on graphs, a series of tiny open circles connected up by a pencil curve, the absorption looked a child's cartoon of a two-humped camel's back. They completed all six gradients. Second, they counted the radioactivity, to find the new RNA—the crux of the experiment. Brenner transferred each two drops to a tiny disc of aluminum, called a "planchette," with a raised rim like a miniature petri dish, and dried them in a small electric oven. These were taken to a machine, again in the basement, that registered how long it took for a given number of atoms of ^{32}P to decay on each planchette. The machine was automatic, slow, and printed the results chip by chip on a paper tape. Brenner and Jacob sat in two armchairs. They started with a blank planchette to count the background radiation.

The first gradient, the control with no heavy isotopes, showed a sharp rise in radioactive RNA that corresponded well with the second hump on the camel, which they reckoned was the band of whole ribosomes. It's easy to lose grip on what the abstract counts were supposed to signify: with no dense ribosomes, the new RNA formed after phage infection had gone where it should, clean enough. The chief technical problem of the experiment appeared to have been overcome. Then they loaded the counter with planchettes for the next gradient, the first with dense ribosomes. This time, if the radioactive RNA banded with the whole light ribosomes as before, that would all but conclusively disprove the messenger idea. This time, they wanted to see a peak of radioactive RNA shifted in the denser direction, clear away from the light ribosomes, to betray an invisible band of dense ribosomes.

As the paper tape emerged from the counter, the first few counts from the dense end of the gradient were moderately high, but after that the activity fell. They decided later that some unattached RNA had sedimented at the very bottom of the tube. The light ribosome bands were more than halfway up the gradient, the first at about drops 42–43 and then the big peak of whole ribosomes at drops 50–53.

That summer, John Kennedy was campaigning for the presidential nomination of the Democratic Party. Though it was a quiet week, Brenner remembered that he or Jacob occasionally wandered down the corridor to a room with a television set, then hurried back.

The counts began to rise—or rather, the time required for a planchette to score four hundred ^{32}P decays began to shorten. From each interval they calculated the counts per minute. At drops 32–33, the time dropped below two minutes: 202 counts per minute, after taking away 15 for the background radiation. This was not much, but the word "pic" sounds the same in French or English. At drops 38–39, the count reached 322 radioactive events per minute. The next planchette raised the count to 440. The next, carrying drops 42–43, shortened six seconds further, raising the count to 500 exactly. Brenner began to worry that the counts should begin to slide down the other side of the peak. "We waited for the next one, because we knew that now it had to turn over; otherwise it would be just a big mess." They were only four planchettes away from the band of whole light ribosomes. The next planchette dropped fractionally to 456 counts per minute. "So we had to wait for the *next* one, and that plunged down, and it was fantastic." Dulbecco heard the cheer go up.

Brenner and Jacob counted the rest of the gradients and calculated all the corrections, but that gradient was the one they plotted on the graph right away, and the one they published. They telephoned Meselson at the Gordon Conference. Meselson told Watson, who said tensely that *his* lab at Harvard was doing a related experiment. The next morning, Brenner and Jacob gave a seminar. Delbrück listened silently. Then Jacob flew back to France. "I was so exhausted that I went home," he said. "In fact, it's mainly Sydney who did the experiment." Brenner flew to San Francisco.

Up the peninsula at Stanford, Arthur Kornberg—the year before, he, Melvin Cohn, Paul Berg, David Hogness, and others of the microbiology department at Washington University in Saint Louis had moved to Stanford like a baseball team bought by a new owner—took Brenner aside to warn him that the idea of an unstable intermediate RNA carrying information would not stand up. "Remember, no control experiments had been done," Brenner said. "Nothing. We just did this experiment. You know, we believed the result. But no one else did. And it wasn't at all publishable in that shape—I mean, I came back to Cambridge and spent four months doing the controls."

❖ ❖ ❖

"I thought that in your account I saw one reason, perhaps a basic reason, why science is almost always done best by *young* men," Jacob said. Science was a game. It demanded absorption to the point of innocence. "We were like children playing."

❖ ❖ ❖

François Gros had arrived in Watson's laboratory in May of 1960, planning to spend that summer pursuing one problem, and almost immediately found himself pursuing another. A control took control. At the Pasteur, he had been working with an antibiotic that stopped bacterial ribosomes from maturing. The work had caught Watson's eye. One of the first things Gros tried at Harvard was to see how the ribosomal RNA was built up and broken down when the bacteria were treated with the drug. To track the RNA, he labelled it by feeding the bacteria radioactive uracil—the base peculiar to RNA—for twenty seconds, and then took successive samples in the next few minutes to follow the pulse of radioactivity through the cell's various components. For the control, he did the same thing with cells not blocked by antibiotic.

In *that* experiment, the radioactivity betrayed the presence of an unstable RNA that matched neither ribosomal nor soluble RNAs. It might be no more than a precursor to ribosomal RNA. Or it might be the bacteria's native equivalent to the rapidly metabolized, phage-related RNA of Volkin and Astrachan. And of course Gros knew, and Watson may have heard from Alan Garen at MIT, of the idea of an unstable information-bearing RNA that had been elaborated in England a few weeks earlier. So Watson and Gros began to try to show that they had a candidate for the unstable information-carrier in normal bacteria, not infected by phage. This was the work Watson told Meselson about at the Gordon Conference.

Watson indeed grew heatedly competitive about the discovery. The work at Harvard took months, and brought in several other scientists, and was at last minimally successful: that is, they showed that uninfected bacteria possessed a transient RNA that seemed to behave like the new RNA formed after phage infection. By putting in enough magnesium, they too got the unstable RNA to associate with whole ribosomes. They did not analyze its base composition nor prove that it was not a precursor of new ribosomes. Meanwhile, without fuss, Yčas and Vincent published in the *Proceedings of the National Academy of Sciences* their finding, in yeast, of an unstable RNA whose base composition matched that of the yeast DNA. Corollaries to Brenner, Jacob, and Meselson's demonstration mounted up.

By far the most important part of Watson's work on the messenger, though, was his recruitment of Walter Gilbert to molecular biology. Gilbert was a theoretical physicist at Harvard when, in the summer of 1960, Watson stopped by the physics department to visit him. "He said, 'There's something very nice going on in the lab; whyn't you come look at it?', so I came around and looked at it and I joined the experiment," Gilbert said. "Jim and I and François Gros did all the experiments together—just ran them continuously day and night; it was a very exciting period." Nearly eleven years later, on a Saturday afternoon in February, I had walked through rain and ice to the Biological Laboratories at Harvard, past the bronze rhinoceroses, and in. Gilbert's office was on the third floor, at one corner, a narrow room. He came in, set out an umbrella to dry. I noticed a baggy sweater and leather sandals. He was short, smiley, round-bellied, red-cheeked, red-lipped, balding, with an energetic black beard—the very picture of Santa Claus as a young man. The conversation that followed took a spiral form, moving around a certain set of related topics and then returning to them, each time in more detail. We talked for several hours, the gray outside his window getting darker and colder.

"The experiments we did at that time were, conceptually, terribly trivial," Gilbert said after a few minutes. "To take a radioactive compound that's going to be a precursor of RNA—uracil, radioactive phosphate. Feed it to bacteria and look for an RNA species which is made quickly and broken down again." Yet the work at Harvard had proved difficult—and not just technically within a clear conceptual frame, as at Caltech, but in some more fundamental way. Gilbert came back to the difficulty later.

"The major problem really was that when you're doing experiments in a domain that you do not understand at all, you have no guidance to what the experiment should even look like. Experiments come in a number of categories. There are experiments which you can describe entirely, formulate completely so that an answer must emerge; the experiment will show you A or B; both forms of the result will be meaningful; and you understand the world well enough so there are only those two outcomes. Now, there's a large other class of experiments that do not have that property. In which you do not understand the world well enough to be able to restrict the answers to the experiment. So you do the experiment, and you stare at it and you say, Now, does it mean anything, or can it suggest something which I might be able to amplify in further experiment? What is the world really going to look like?

"The messenger experiments had that property. We did not know what it should look like. And therefore as you do the experiments, you do not know what in the result is your artifact, and what is the phenomenon. There can be a long period in which there is experimentation that circles the topic. But finally you learn how to do experiments in a certain way: you discover ways of doing them that are reproducible, or at least—" He hesitated. "That's actually a bad way of saying it—bad criterion, if it's just reproducibility, because you can reproduce artifacts very very well. There's a larger domain of experiments where the phenomena have to be reproducible and have to be interconnected. Over a large range of variation of parameters, so you believe you understand something. And this was what was happening with the messenger experiment: one learned how to get something out of the cell, how to isolate something, how to avoid— In order to isolate that something, for example, you had to prevent its being broken down by enzymes. And even to formulate that, you have to see it by accident and then discover what you have to do in order to see it reproducibly. So there's a whole domain of creating the phenomenon."

❖ ❖ ❖

Brenner, back in England, devised controls. In the central experiment at Caltech, the old, dense ribosomes had not themselves been seen, only the single band of new radioactive RNA at the place where, by their density, they ought to be. So Brenner had first to prove that heavy ribosomes went where they ought to in the density gradient. He grew two batches of cells: a large culture, not labelled at all, to supply light ribosomes in bulk, and a small culture labelled not only with heavy isotopes but with radiophosphorus. He used no phage, for the radioactivity this time was intended to locate stable ribosomes rather than unstable intermediate. When the cultures were mixed and the ribosomes extracted and centrifuged, the heavy ribosomes obligingly produced two bands forming a camelback on the graph that was the twin of the one made by the light ribosomes in the central experiment, but displaced towards the dense end of the gradient. The hump that corresponded to the heavy whole ribosomes was exactly where it belonged.

Then there was the possibility—an asymmetrical, inelegant, preposterously odd possibility, seen from the present day—that phage protein was made directly on the phage DNA and that the only task of the new, phage-related RNA was to stop the existing ribosomes from making any more of their specific bacterial proteins. "You can see this was in our thinking, if you will look in that messenger paper—which now seems quite crazy—look at the theories which are being eliminated," Brenner once told me. One of these alternatives was "that the phage DNA does it, you see—the phage gene made a poisonous RNA that killed ribosomes, and *that's* why you found the Volkin and Astrachan RNA in ribosomes, and then the phage substituted an entirely new mechanism." The idea was the last flicker of the two years' suspicion at the Pasteur that the PaJaMo results might demote the ribosome. "I can remember Jacques distinctly saying that maybe this very fast expression in the PaJaMo effect was due to DNA making protein." The possibility was not formally ruled out by the experiment at Caltech. So Brenner had to show that phage protein actually was made on the bacterial ribosomes. He did so by infecting bacteria with phage and supplying radioactive sulfur to be built into amino acids. On October 14, he wrote to Meselson about that, "I don't think there is any doubt at all that all the phage proteins are made on ribosomes. The experiments also show very nicely no *new* ribosomes are made." These and other experiments of mounting ingenuity filled scores of sheets that summer and early fall. Meanwhile, François Gros went back to Paris, and he at the Pasteur and Gilbert and Watson at Harvard devised *their* controls. As fall wore on at the Cavendish, Perutz, for one, began to be alarmed at the time that was being lost.

❖ ❖ ❖

That fall, Jacob and Monod christened the unstable, information-bearing intermediate between gene and protein "messenger RNA." Out of the amorphous biochemistry of RNA, three distinct anatomies had now been differentiated: rRNA, tRNA, mRNA. Two had begun their scientific careers as purely intellectual creations, function postulating structure. Jacob and Monod wrote the idea of the messenger, and the term, into a theoretical summation they were preparing. Meanwhile, Brenner had written up the experimental work; in December, Jacob came to Cambridge. "But we couldn't finish it so I went to Paris for a few days to finish it," Brenner said. Jacob and Monod cited both the Pasadena and the Harvard experiments—though they finished their long paper first.

Into the paper Jacob and Monod poured the entire theory and experimental support of regulation of protein synthesis at the level of the genes themselves. To summarize the conclusions they proposed a model. Though details have been corrected since then—notably the substance of the repressor—the principal features of the model have never been put more neatly:

> The molecular structure of proteins is determined by specific elements, the *structural genes*. These act by forming a cytoplasmic "transcript" of themselves, the structural messenger, which in turn synthesizes the protein. The synthesis of the messenger by the structural gene is a sequential replicative process, which can be initiated only at certain points on the DNA strand, and the cytoplasmic transcription of several, linked, structural genes may depend upon a single initiating point or *operator*. The genes whose activity is thus co-ordinated form an *operon*.
>
> The operator tends to combine (by virtue of possessing a particular base sequence) specifically and reversibly with a certain (RNA) fraction possessing the

proper (complementary) sequence. This combination blocks the initiation of cytoplasmic transcription and therefore the formation of the messenger by the structural genes in the whole operon. The specific "repressor" (RNA?), acting with a given operator, is synthesized by a *regulator gene*.

The repressor in certain systems (inducible enzyme systems) tends to combine specifically with certain specific small molecules. The combined repressor has no affinity for the operator, and the combination therefore results in *activation of the operon*.

In other systems (repressible enzyme systems) the repressor by itself is inactive (i.e. it has no affinity for the operator) and is activated only by combining with certain specific small molecules. The combination therefore leads to *inhibition of the operon*.

The structural messenger is an unstable molecule, which is destroyed in the process of information transfer. The rate of messenger synthesis, therefore, in turn controls the rate of protein synthesis.

Then they itemized once more the evidence, assumptions, and speculations built into the model.

At the end they came to the function of the ribosome.

The property attributed to the structural messenger of being an unstable intermediate is one of the most specific and novel implications of this scheme. . . . This leads to a new concept of the mechanism of information transfer, where the protein synthesizing centers (ribosomes) play the role of non-specific constituents which can synthesize different proteins, according to specific instructions which they receive from the genes through M-RNA.

There, in a sentence, was the first published statement of the pivotal realization that had come to Crick, Brenner, and Jacob in their conversation on Good Friday, eight months earlier.

Crick has wondered ever since whether he and Brenner should have published a note about that, independently of the French, after all. "I think we *did* have a new idea, and I think in actual fact, we had a *radically* new idea, and I think looking back on it one would have been wise to have done something," Crick said. "But that was—at the time, it was that I made a decision" not to publish. "So—actually the reverse happened, and they simply incorporated it into their review!" He laughed. The new idea? "That the ribosome was an inert reading head and didn't carry—and that the ribosomal RNA wasn't the messenger RNA. That was the absolutely crucial idea." Now Crick's tone was reflective, even wondering. "Once you got the idea that the ribosome was a reading head, then the whole world changed. You see."

Half a century earlier, Thomas Hunt Morgan had found embryology intractable and had turned to genetics. When genetics brought him a Nobel prize, he spoke in his prize lecture, the summer of 1934, of the central problem of physiology, a problem "new and strange": the field between the genes that theory postulated and the traits observed in the organism, the unknown territory "where the properties implicit in the genes become explicit in . . . the cells." Morgan knew, though, that a synthesis of genetics and embryology was still remote. He told Boris Ephrussi so that same year. Twenty-six years after that, from the pinnacle of the grand structure that Monod—Ephrussi's student—and Jacob founded on bacterial genetics, they could see, at last, a little way even into that unknown territory. They concluded their review:

The fundamental problem of chemical physiology and of embryology is to understand why tissue cells do not all express, all the time, all the potentialities inherent in their genome [that is, in the totality of their genes]. The survival of the organism requires that many, and, in some cases most, of these potentialities be unexpressed, that is to say *repressed*. Malignancy is adequately described as a breakdown of one or several growth controlling systems, and the genetic origin of this breakdown can hardly be doubted.

According to the strictly structural concept, the genome is considered as a mosaic of independent molecular blue-prints for the building of individual cellular constituents. In the execution of these plans, however, co-ordination is evidently of absolute survival value. The discovery of regulator and operator genes, and of repressive regulation of the activity of structural genes, reveals that the genome contains not only a series of blue-prints, but a coordinated program of protein synthesis and the means of controlling its execution.

Jacob and Monod titled their paper "Genetic Regulatory Mechanisms in the Synthesis of Proteins." It reached *The Journal of Molecular Biology* on 28 December 1960.

8   *"He wasn't a member of the club."*

❖*a*❖   The distraction for Sydney Brenner, that fall of 1960, had been that he and Francis Crick had tempted each other back into the problem of mutation and the problem of the code. Nothing in biology has been more fundamental and instructive than mutation—random change within continuity—and the effort to understand it. Mutation is the basis of Darwinism. As a phenomenon, mutation generates the heritable variations among closely related organisms on which natural selection acts; for theory, the rate of mutation is a leading term in the equation as yet unsolved that would predict the rate of biological evolution. Mutation has been essential to Mendelism at every step, not only as the source of heritable variations—round peas or wrinkled, vermillion eyes or the wild type, rough plaque or smooth—but as the tool for understanding. From Alfred Henry Sturtevant to Seymour Benzer, Mendelians have mapped not the genes themselves—*that* had to await the discovery, by Frederick Sanger and then by Walter Gilbert in the mid-seventies, of ways to read off directly the sequence of bases on a strand of DNA—but the relative positions of changes in the genes. The Mendelian map was a map of defects and differences. Mutation fairly launched molecular biology, when George Beadle and Edward Tatum proposed, on mutational evidence, that one gene makes one enzyme, when Salvador Luria and Max Delbrück in the United States, Jacques Monod in wartime France, devised proofs that micro-organisms mutate and therefore have genetics, and when Linus Pauling realized that sickle-cell anemia is an inherited flaw in the hemoglobin molecule. Then in 1953, James Watson and Francis Crick offered, as an immediate consequence of their structure of deoxyribonucleic acid, the first molecular explanation of mutation.

Watson and Crick put the genetical implications of their structure, one recalls, in their second paper in *Nature*, appearing at the end of May of 1953. Here, among other things, they asserted that mutations were a change in the sequence of bases; they speculated that spontaneous mutations were caused by the rare shift of one or another base to an unusual tautomeric form at the instant when replication of the double helix involved it. By this means, they thought, the momentary displacement of a hydrogen atom allowed, say, ade-

433

Adenine (imino form)   Cytosine

Guanine   Thymine (enol form)

nine to form hydrogen bonds with cytosine instead of its correct complement, thymine. Thus the wrong nucleotide would be built into the growing
replica—and in future cycles descendant from that strand the wrong pair
would persist. Such an error in copying could be reversed in some later generation, though as rarely, by a comparable tautomeric shift at the same site, in
the instant of copying, that accidentally brought a nucleotide of the original
sort in to pair.

Many mutations, of course, are not spontaneous but induced—by x rays, as
Hermann Muller demonstrated in 1927; by ultraviolet light, as Lewis Stadler
discovered in 1928; by various classes of chemicals, as began to be recognized
in the mid-forties. An explanation of the mutagenic action of certain chemicals
was developed, pursuing Watson and Crick's surmise, by several people in
the late fifties, in a complicated little dance of collaboration and competition.

The dance began in 1955, when Arthur Pardee and Rose Litman, at
Berkeley, announced that 5-bromouracil, if present in the culture medium of
bacteria infected with phage T2, produced a strikingly high proportion of

Thymine   5-bromouracil (BU)

mutants among the daughter phage. Now, 5-bromouracil is an analogue of the base thymine, which is to say it is like thymine except that it contains an atom of the heavy element bromine at the fifth position around the ring. And it was already known that some micro-organisms will pick up bromouracil supplied in their food and build it into DNA in place of true thymine. Analogues to other bases soon were found that caused mutations, too.

These new agents of phage mutation inevitably provoked Benzer's curiosity. In 1957, Benzer and Ernst Freese, a biophysicist who had just come to Purdue from the California Institute of Technology, mapped mutations induced by bromouracil in the *r*II region of bacteriophage T4. The observation that most surprised them, they said, was that the mutations induced by bromouracil almost always appeared, on the linear map of the fine structure of the gene, at points different from those where spontaneous mutations occurred. They also noted that they had begun experiments to see whether bromouracil would induce their phage mutants—both sorts, spontaneous and bromouracil—to mutate in reverse, back to the wild type.

Just before that, Matthew Meselson and Franklin Stahl, starting their collaboration, got interested in 5-bromouracil because it offered a means to grow phage with DNA much denser than normal. Stahl wrote to Brenner and got a strain of phage that could not synthesize its own thymine. But phage proved too fragile in the centrifuge, so they switched to bacterial DNA freighted with heavy nitrogen, which worked fine. They noticed, though, that the bromine atom was placed where it would influence the distribution of charge around the ring in such a way that the false thymine might shift into its unusual tautomeric form more readily, and thus more readily pair as though it were cytosine. This was the explanation that Meselson was testing of the mutagenic effect of the stuff—and it was correct, though all his tests were fruitless—when he visited Brenner briefly in the summer of 1959.

Meanwhile, however, Brenner had been looking into the effects of another potent mutagen, proflavine—not a base analogue but a dye, bright yellow, one of a family of dyes in the yellows and oranges derived from the colorless coaltar chemical acridine. So while Benzer was spending his year at the Cavendish in 1957–'58, vainly hoping to demonstrate the colinearity of gene and protein, he and Brenner, with the help of Leslie Barnett, mapped the mutations induced by proflavine within the *r*II region in relation to each other and to those induced by bromouracil. Their most curious observation was that the two mutagens were mutually exclusive: not once did they hit the same site on the map. But how the acridine dyes acted to produce mutations was unknown, except that, unlike the base analogues, they were not built into DNA.

The first public attempt to fit such findings into a coherent theoretical frame—though many had discussed them—was made by Freese, who had moved on to Watson's lab at Harvard. The data and the possible explanations Freese confronted were a jungle. Fine-structure mapping of mutants by the hundreds was obsessional, time-consuming work. Results—the spots on the plates—were not always unequivocal. Examples suggest the flavor of the ambiguities. Some mutants were not totally crippled but could grow feebly on the indicator strain of *E. coli*. Benzer called these "leaky." When a stock of mutant phage showed reverse mutations, one could not always tell from the size and form of the plaques whether the wild type was restored. Some reversions were obviously partial—for instance, forming tiny plaques; some surely looked like true back-mutations; others were "pseudo wild." Another potential snare was double mutants. Further, certain sites on the map, for no reason known, were tens or hundreds of times more susceptible to mutation than others. Benzer had termed these "hot spots." They invited speculation—were they due, for instance, to some structural discontinuity that exposed particular nucleotides?—while they complicated interpretation of the statistics.

Finally, in principle a change in the sequence of nucleotides in the DNA of a gene could arise in several very different ways. One base might be substituted for another. A base might be added, or one deleted; several bases or even a long stretch might be added or deleted. More subtly, an apparent reversion to the wild type might be accounted for by a second mutation at a different site. That is, if a nucleotide substitution switched an amino acid at one point in the polypeptide, to make the chain bend or cross-link differently and thus distort the enzyme's structure, there might be a number of points where a further switch would make a compensating structural change sufficient to counteract the physiological effects of the first. Such mutations were called "suppressors"—though nobody was sure they occurred.

Freese slashed a way through the tangles by following the line of experiments he had begun with Benzer to see which phage mutations could be induced to revert to the wild type by which mutagens. He published two papers about the molecular basis of mutations in the spring of 1959. He gave a third at the symposium that June at Brookhaven National Laboratory—appearing the morning after Crick made his confession of bafflement at the gloomy position of the coding problem. Freese analyzed reversions to strengthen the idea that mutations fell into two exclusive classes depending on the mutagen. One class comprised mutations caused by base analogues, such as bromouracil, and by nitrous acid, another chemical known to be mutagenic—for these mutations, all along the map, reverted readily to the wild type when a stock of a mutant was dosed by a base analogue, no matter which. In contrast, of the spontaneous mutations only a small proportion, about ten per cent, reverted when treated with base analogues. And of forty genetically different proflavine mutants that Freese tested with base analogues, only one reverted—in what at that rate might itself have been a spontaneous mutation. The proflavine mutants and the large spontaneous set comprised the second class.

To explain the mechanism, Freese's first argument was that "suppressor mutations do not occur"—that the wild type was not restored by some other, compensating mutation at a distance on the chromosome. He proposed two distinct classes of base substitutions in DNA. The first was the simple copying error whereby one purine was switched to another (A for G or the reverse) or one pyrimidine for another (T for C or the reverse). Such errors he called "transitions." Watson and Crick's tautomeric shifts would produce transitions. Base analogues produced transitions. Nitrous acid was understood to work by knocking amino groups—that is, NH_2 groups—off the fringe of the DNA bases that had them, namely cytosine, adenine, and guanine. "Some of the new bases have altered pairing properties and cause the change of a nucleotide pair in subsequent DNA replication," he said at Brookhaven. Thus the mutations that nitrous acid caused would also be transitions—neatly explaining why they were reversible by base analogues.

But since most spontaneous mutations could not be induced to revert by base analogues, it seemed likely that only the few that could, the ten per cent, arose as Watson and Crick had imagined. However, Freese conceived an alternative form of base substitution in which a purine was replaced by one of the two pyrimidines or a pyrimidine by one of the two purines, leading in the next cycle to a complementary change on the replica strand. He wanted to call such substitutions "transversions." It seemed obvious that these could not be reversed by base analogues. Most spontaneous mutations, he thought, were transversions. So were proflavine mutations. The mode of action of proflavine was still unknown; the molecular mechanism Freese could sketch for transversions was much vaguer than for transitions. But he thought that the scheme offered tantalizing glimpses into the coding problem: perhaps classes of amino-acid changes could be discerned that correlated with the classes of base substitutions.

Freese's talk called forth vigorous discussion—Meselson, Benzer, Crick, others—and the tribute of substantiating detail. The significant new evidence came from Brenner. "I think I should communicate a result that has been found by Mrs Alice Orgel," Brenner said. She was Leslie Orgel's wife, and was working at Cambridge. "All bromouracil mutants so far tested are *not* induced to revert by proflavine." This was a complementary set of experiments that Freese had not got around to, and it sharpened further the differentiation between the two classes of mutagenic action. "One point should be emphasized," Brenner then warned. "In the work done with Benzer on the mutagenic spectrum"—the map—"of proflavine-induced mutations, not one was found to coincide with a bromouracil site. I think this is highly significant."

Whereupon the problem of mutation slowed almost to a standstill for eighteen months, except that several laboratories worked away at trying new mutagens and testing the mutants in back-mutation. Sometime in the winter of 1959–'60, Leonard Lerman, a biochemist from the University of Colorado working at the Cavendish, came up at last with a way that acridines might affect DNA. A dye molecule, Lerman thought, binds to DNA by sliding between adjacent base pairs—like a plastic poker chip inserted into a stack of coins—forcing the consecutive bases 6.8 angstroms apart rather than 3.4 angstroms.

"But the rII stuff all got very doldrumy, around about that time," Brenner said.

Then messenger RNA turned the coding problem upside down. Crick and Brenner understood that at once. "Oh, well, I mean the *moment*, look, the *moment* I came back from Pasadena, and told Francis this was true," Brenner said. "And we'd discussed this before, straight after the April—that that must mean that the messages are direct copies of the DNA. And it therefore must be true that the code was degenerate." The interlocking tumblers of the puzzle clicked into new positions, each freeing the next.

In the first move, messenger RNA resolved the dilemma posed by the fact that, when different creatures were compared, the base compositions of their DNA varied greatly, while the bulk RNAs of these species varied much less and the amino-acid compositions of their proteins differed but little. For if ribosomes were stable and unspecific and only the transient messenger RNA carried the information from DNA, the narrow range of RNA compositions from species to species was both reasonable and trivial. All that mattered was the base composition of the fraction that comprised messenger RNAs.

Next, since messenger RNAs now had to vary exactly as the DNA, the only plausible way to get from DNA to RNA to protein was to suppose that the code had synonyms. But see what that move opened up. If the code was degenerate, and was indeed carried in base triplets, then many more than twenty of the possible sixty-four triplets made sense—and the code could not be comma-free. If not comma-free, the message would have to be read from a marked, fixed starting point. The messenger hypothesis as yet unpublished, it and its consequences diffused by letter and by word of mouth.

Crick spent part of the fall of 1960 in the United States, and got back to the Cavendish on November 14, a couple of weeks after Brenner finished the messenger controls. In the course of the preceding months, Alice Orgel had run through the set of spontaneous mutations in the rII region that were not revertible by base analogues, and found that these indeed reverted with proflavine, as Freese's scheme demanded. She and Brenner also showed that several other acridine compounds were mutagens. It remained true that mutants produced by one class of chemicals, either acridines or base analogues, could not be made to revert by the other. It also remained true, as Brenner had said to Freese from the audience at Brookhaven, that the two classes of mutagen acted at different sites on the genetic map. The peculiarity of the rII gene, though, was that the protein produced had still not been isolated, most probably because it was some enzyme that did not form part of the finished phage. When Brenner and Leslie Barnett tried genes known to control proteins that *were* part of the finished phage—like the tail-fibre gene with which for years they had hoped to prove colinearity and solve the code—they found with some surprise that although base analogues made mutations readily in these, acridines did so rarely or not at all.

Sprung from the prison of comma freedom, Crick and Brenner came to a possibility that contradicted Freese. The starting point of the idea was the nagging, total dissimilarity between the maps of mutations caused by the two classes of agents. "For two years that sat and gnawed at us," Brenner said. Over lunch at the Eagle one day soon after Crick's return—the last such lunch to figure in the history of biology—Brenner surmised, and Crick immediately took it up, that acridines act as mutagens by causing the insertion or deletion of a base pair. The mechanism was easy to visualize: an acridine dye, they thought, taking Lerman's idea a half-step further, sometimes slipped between adjacent bases along one chain of the DNA but not the other, and as replication proceeded either the gap was filled in or the base opposite was dropped.

The surmise was theoretical and the evidence as yet fragmentary, but this time Crick and Brenner got into print without hesitation. By mid-December, they completed a short letter, with Leslie Barnett and Alice Orgel, for *The Journal of Molecular Biology,* and Kendrew put it in at the end of the first issue of the new year. They gave the letter a strong Crick title: "The Theory of Mutagenesis." Their speculation, they said, led to a prediction and a problem.

The prediction: adding or deleting a base pair was "likely to cause not the substitution of just one amino acid for another, but a much more substantial alteration, such as . . . a considerable alteration of the amino acid sequence, or the production of no protein at all." This last could be why, with acridines, no phage with mutant tail-fibre protein showed up—if the mutations were not merely crippling but lethal. "This may have serious consequences for the naive theory of mutagenesis."

The problem: reversion to the wild type caused by nearby suppressor mutations was no longer excluded. With Freese's two kinds of base substitution, if suppressor mutations had occurred—that is, if the forward mutation could take place at one base pair and the reverse one at a different base pair—the classes of mutagenic chemicals would not have been mutually exclusive. But under the new hypothesis, acridines added or subtracted base pairs, while chemicals of the base-analogue class produced substitutions. Neither class could easily reverse the other and so base substitutions that produced suppressors were not ruled out.

> Indeed from what we know (or guess) of the structure of proteins and the dependence of structure on amino acid sequence, we should be surprised if this did not occur. . . . Thus our new hypothesis reopens in acute form the question: which back-mutations to wild type are truly to the original wild type, and which only appear to be so? And on the answers to this question depend our interpretation of all experiments on back-mutation.

❖ ❖ ❖

At the turn of the year, Perutz got a letter from Vladimir Alexandrovich Engelhardt, who was director of the Biological Section of the Soviet Academy of Sciences. The International Congress of Biochemistry, the fifth of these triennial jamborees since the war, was to meet in Moscow early in the following August. Engelhardt was in charge of the largest of the many divisions of the congress, the one given over to "biological structure and function at the molecular level." He asked Perutz's help in planning the daily panels of invited papers, and enclosed his idea of a program. Perutz, reading the draft program, saw that Russian conceptions of molecular biology were five years out of date. He replied that he would need complete freedom to invite the principal speakers and topics.

The exciting meeting of the summer, however, promised to be the symposium at Cold Spring Harbor, early in June. The topic was "Cellular Regulatory Mechanisms."

❖❖❖

The shuttles of corroboration were weaving the messenger hypothesis into the fabric. In February, Benjamin Hall and Sol Spiegelman published a paper that gave support of a completely new kind. When a solution of DNA was heated nearly to boiling and then cooled suddenly, the hydrogen bonds between the purine-pyrimidine pairs that held the two strands of the double helices together were ruptured and the strands unwound and floated loose. The change could be demonstrated in several ways: for example, the viscosity of the solution decreased drastically as the structure collapsed, and the bases were now exposed to attack by chemicals to which the double helix was virtually impervious.

A similar unfolding and loss of structure had long been known to afflict proteins when they were heated in solution, and was called "denaturation." But denaturing of DNA was first noted by Julius Marmur and Paul Doty, of Harvard, in 1959. Polypeptides, if not actually cooked but denatured gently, could fold themselves up again as the solution was cooled, and regain their original structure. Marmur then made the interesting discovery that when he cooled a solution of denatured DNA slowly, the single strands spontaneously reconstituted double helices. At least, some of them did. When the DNA came from bacteriophage, which have very little DNA, reconstitution of double-helical molecules was quick and nearly total. When the DNA came from bacteria, which each have thirty times as much as a phage, renaturation was slower and restored about half the double helices. When it came from calf thymus cells, which have five thousand times more nucleotides than bacteria, renaturation took place hardly at all. Marmur reasoned that for successful reunion, two single strands in the solution had to find each other which were perfectly complementary, or very nearly so, in their base sequences. Thus, Marmur also discovered that reconstitution would take place only between DNA strands that came from the same or closely related organisms. Renaturation has since become the indispensable technique for testing or detecting complementary sequences—the way to find and fish out particular genes. Marmur and Doty published preliminary reports of the phenomenon in the spring of 1960.

At once, Spiegelman saw a clever way to detect whether DNA really made RNA. He and Hall had already been experimenting with E. coli and phage T2. That had been the work that had caught Brenner's eye, pushing further Volkin and Astrachan's demonstration that the RNA produced by the cell after infection had a base composition, overall, that corresponded to the phage rather than the bacterial DNA. They had learned to separate the new-made, phage-related RNA from the rest of the RNA in E. coli. So now they purified some of this RNA; mixed it with denatured, and so single-stranded, DNA from phage T2 itself; cooled the solution slowly—and got out hybrid molecules of which one strand was DNA and the other RNA. The experiments were ingenious and unusually intricate, requiring an arsenal of purification procedures, two kinds of radioactive labelling, and five-day centrifuge runs with cesium-chloride gradients. In the most important control, Hall and Spiegelman showed that the phage-related RNA would not form hybrids with DNA from other sources—not even the DNA from phage T5, which was very similar in overall base composition. They proved that the messenger carried a specific message.

❖ ❖ ❖

Late in February, Crick became a baptized proselyte to the phage church: he came into the lab and immersed himself in experiments in phage genetics. During the winter, he and Leslie Orgel, with Brenner as well, had been speculating that messenger RNA carried the genetic information in a fashion different from any seriously considered before. They began from some model-building done months before the messenger hypothesis by Jacques Fresco and Bruce Alberts, biochemists at Harvard, who showed that strands of RNA in solution might be able to form double helices despite a good deal of ir-regularity of base pairs, if a base or a short run of bases not properly paired could make bends or loops to swing out of the helix. Crick had been interested in the idea at the time. He and Orgel now imagined a couple of such structures that might allow complex codes in which bases signified one thing for the protein when paired, something else when single and looped out. These notions got as far as a draft paper, dryly titled "On Loopy Codes," but in fact they were never developed in much detail and the paper was never published.

The loopy episode was important for one consequence only: thinking about a single-stranded messenger taking up such an irregularly paired structure. Crick imagined it doubled back like a hairpin, or rather like the cord on a tele-phone handset that has become coiled around itself. That picture led to the idea that suppressor mutations might occur to pairs of bases opposite each other—and therefore spaced symmetrically in relation to where the strand doubled back. That idea, in turn, drove Crick into the laboratory. He was im-patient, anyway, to know whether mutants generated by acridine compounds ever reverted to the wild or pseudo wild by suppressor mutations. Now, in particular, if suppressors could be found, their mapping would quickly show whether they came in symmetrical pairs. If suppressors showed up but not in pairs, their genetic behavior would test the molecular mechanism proposed for mutation by acridines—although in February of 1961 it was by no means obvious what, in detail, the genetic behavior of mutants made by adding or dropping bases ought to be.

So Crick learned to do experiments in molecular genetics himself. "Which he was not very good at; but he did a lot," Leslie Barnett once told me. "He's just not very practical; he would drive a microscope eyepiece right through the slide—very unhandy. And yet he brought a fresh approach to it." Crick's first impulse, Barnett said, was to question every accepted detail of tempera-tures, concentrations, timings. "You know, I would say. 'Now we're irradiat-ing these for six minutes.' '*Why* are you irradiating them for six minutes?' And I'd say, 'Well, as a matter of fact, Sydney said that was a good time, and I just did it.' And he'd say, 'Why don't we see what really is the optimum? Because Sydney is inclined to have all this in his head.' He wouldn't accept anything like that. He'd say, 'We're going to take the magic out of it.'" Among those working at the Medical Research Council unit that spring were Gunther Stent, from Berkeley, and Bruce Ames, from the National Institutes of Health. A phage lab had been improvised in the ground-floor corridor of an annex to the university's zoological museum, standing across a courtyard from the Cavendish but since pulled down. Finding possible suppressors, proving them, and mapping them was laborious, and was practical at all only because Benzer had already driven the rII system so far. "Francis really didn't like

❖

doing lab work," Barnett said. "He was the one with the ideas, and then I was just really his hands, in a way." Crick recruited other hands—to examine petri dishes and count plaques. He was in the lab himself through the week and often on Saturdays and Sundays from February until late in June. He performed thousands of crosses and filled a hundred-odd notebook pages—driving a blue ball-point rapidly, attentively—with procedures, counts, and comments. "It's true that I don't much like doing experiments, but I revelled in the phage work," Crick said. "Odile remarked that she'd never seen me so fit and so cheerful!"

By the end of April, they had found several mutants that could be made to revert by suppressors that were nearby on the map. The first week in May, Crick opened a new line of experiments, and on a Monday morning he set it out:

> Decided to see if one r [one rII mutant] had many *different* suppressors.

"There you are. Eighth of May," Crick said, pointing to the entry. "That was the crucial step."

The notes that follow are laconic, yet the laboratory comes to life in them. First, the symbols in brief: "r" is any rII mutant; "P13" is one particular proflavine mutant; "B" is the strain of *E. coli* on which rII mutants produce their characteristic large, clear plaques; "K" is the strain, K12 lysogenic for phage *lambda*, on which the wild type grows but the r mutants do not, allowing revertants to be identified easily.

> Chose P13, as the distance from the suppressor found was larger than the other two cases studied.
> Plated P13 on B. (On Friday or Saturday?). Picked 10 different (r) plaques. Inoculated 10 tubes with these, and grew. . . . Plated 0.1 of each of these on a separate K plate. Incubated.
> From each plate picked *three* different plaques. The first one picked was the nearest to the centre of the plate. The other two were picked from plaques which looked a little different from the first one picked. These were *streaked* on B: grown 5 1/2 to 6 hours: picked from a single plaque: These were numbered. . . .

In the next fortnight, nearly a dozen new mutants of the mutant P13 were spotted that had the effect of restoring the wild type fully or partially. The original P13 had been generated by proflavine, three years earlier when Benzer was in Cambridge; the new mutants were spontaneous, which is to say, helped along by six minutes of ultraviolet light. Crick renumbered *P13* as *FC0*, and numbered the mutants of the mutant *FC1*, *FC2*, and so on. The next step was to map them all.

❖ ❖ ❖

Brenner, Jacob, and Meselson on the messenger experiment, the paper repeatedly delayed, at last came out in *Nature* on 13 May 1961—a year and four weeks from its first conception. On the following pages appeared the paper by the Harvard group; Watson had got his way. "Ours just sat in the bloody office there for four and a half months"—about three months—"while we were waiting for them," Brenner said. "And you can interpret it as you like, whether having read our paper they did other things in the meantime—it doesn't matter." Jacob and Monod on genetic regulatory mechanisms, the great theoretical review, appeared a fortnight later.

❖❖❖

Brenner stopped in Washington at the end of May and gave a talk at the National Institutes of Health about messenger RNA. The 1961 symposium at Cold Spring Harbor began on June 4. It stood at the confluence of twenty years of genetics and biochemistry. Watson led a large delegation from Harvard; Monod and Jacob, an even larger one from the Institut Pasteur. Brenner and François Gros were down to give papers on messenger RNA; Jacob was to present the regulation of the activity of genes; Szilard would preside over a session on feedback control of enzyme action. More than forty papers would be read on closely related matters. Monod had been one of the organizers and was to give, with Jacob, the concluding summation and prospect.

Brenner took the matter of RNA in protein synthesis in the most general sense. "The dominant idea in molecular biology is that DNA carries information encoded in the form of specific sequences of nucleotides which determines the amino acid sequences of proteins," he began—to an audience that hardly needed to be reminded that the dominant voice in molecular biology was Crick's. "Exactly how this specification is accomplished is the main problem of present-day molecular biology and its solution, the breaking of the genetic code, is the main ambition of many workers in the field." And he went on:

> There are many approaches to the coding problem. One approach involves ignoring the transcription machinery completely, considering it as a black box with the information from DNA going in at one end and the polypeptide chain coming out of the other. Attempts to deduce the code from the cryptic messages of amino acid sequences have ended in failure. . . . There seems to be more hope in attempting to correlate the genetic fine structure with the amino acid changes produced in proteins by mutations. . . . Together with specific mutagenesis, it might well be possible to deduce the code from an exhaustive set of such changes. . . . The other, and much more difficult approach, is to investigate the nature of the transcription machinery. Except in broad outline, very little is known about the mechanism of information transfer, and as it has turned out, most of what we thought we knew was incorrect.

Brenner reviewed "the classical model of information transfer," itself barely six years old, and its grave difficulties, and presented the alternative, the messenger. At the end, he warned, "To prove that the messenger does carry information for specific proteins would require an in vitro reconstruction experiment"—that is, to put an RNA into a cell-free system and get out the polypeptide it specified—"which no one has yet succeeded in doing."

Several people in his audience, including Alfred Tissières and François Jacob, stirred at that, for they were trying it or planning it. Gordon Tomkins, from the National Institutes of Health, took notice for another reason, but said nothing. Brenner then acknowledged the precariousness of generalizing the messenger from phage-infected bacteria to uninfected ones—let alone to higher organisms. But, he said, "We strongly believe that mechanisms as basic as the transcription process are bound to be unitary in Nature." He extemporized fervidly on the unity of Nature and the beautiful ubiquity of the messenger, in his rising inspiration mentioning even Hermes, the messenger of the gods.

At which moment, Erwin Chargaff stood up at the back of the room and shouted out, "Don't forget, Hermes was also the god of thieves!"

❖❖❖

"There were large elements in the community who did not believe in the hypothesis at all—that is, that there was an intermediate that was *not* the ribosomal RNA," Gilbert said. "The original experiments have an element of interpretation in them. They didn't actually prove the hypothesis as the hypothesis was stated. One couldn't do what one can do now, take a known piece of RNA and make a known protein with it. One couldn't do that then. So the argument was much more indirect." Who were the doubters? "Chargaff had strong views on the other side," Gilbert said. "He was an anti-molecular biologist in many ways, in the last decade. That's partially due, probably, to a feeling of bitterness over DNA. . . . His writing is quite amusing. He writes anti-molecular-biology polemics." Among the strongest dissenters had been "the Carnegie group"—Richard Roberts, who had given the microsomal particles their new name, and his colleagues at the Carnegie Institution of Washington with whom he had proved that ribosomes in bacteria were the site of protein synthesis just as in higher organisms. "They were wedded to their ribosomes and wedded to a different view as to how the cells behaved. For years, there was a running fight about what did the pulse-labelling experiments see, what did the phage-RNA experiments mean. And that conflict actually went on probably until 1967, '68, in animal cells. The animal-cell people kept on worrying about where the messenger was, was it in the ribosomal RNA or somewhere else, until finally people began to learn to take animal-cell ribosomes apart and find messengers there."

Gilbert came back to the question several minutes later. "People become convinced of things at very differing degrees. Large chunks of the community will be three to five years behind other chunks of the community in what their beliefs about the world are."

Jacob and Monod's review of genetic regulation appeared just at the time of the Cold Spring Harbor meeting. Its impact was immense and was hardly lessened by the fact that several of its elements—notably the repressor and the messenger—were already published and well discussed. The paper is one of the most famous in the science. Gilbert read it at once, of course, and was at the symposium. "Most of the crucial discoveries in science are of such a simplifying nature that they are very hard even to conceive without actually having gone through the experience involved in the discovery—or at least experiencing the state of knowledge of the field at the time," Gilbert said. "Jacob's and Monod's suggestion made things that were utterly dark, very simple. But as their suggestion was based on genetic evidence, and involved what many people felt was a recondite or abstract form of argument—" I asked what he meant.

"Well, first of all, it was French. Which meant that it was made as a theoretical argument bolstered up in part by somewhat true, and somewhat shaky, experimentation," Gilbert said. "There are several elements in the French approach. One is their tremendous, in many ways, over-theorizing." He put his feet up on a chair. "So if you wanted to be cynical about it, you might easily dismiss the whole thing. Secondly, a large strength of molecular biology has been its ability to use the genetic approach to attack a molecular problem. . . . In fact, all of the interplay between the phage people and the biochemists really comes out to the exploitation of the genetic approach. To the phages and thus to the DNA molecules. On the other hand, there are a number of biochemists also doing molecular biology who shy away from the genetic arguments, feel that there is something not quite proper about them. And so, even after Jacob and Monod suggested there were control genes and that these genes might make repressors, there was a strong 'Show me!' aspect to that— What were the repressors?

"I had just gotten into the field at that time. It was a very beautiful paper. It was clear that one wanted to know what these things were. As time went on, and no one could put their finger on any of the repressing substances, it became more and more of a mystical question— Were these things really there? Was there any reality behind that suggestion that a molecule actually went and sat on the DNA?"

One classical geneticist had been attempting to illuminate the mechanism of repression since the mid-forties. Barbara McClintock published her first paper on the genetics of maize in 1926, and ever since had dedicated herself with an ascetic joy—as a woman in American science in the nineteen-thirties, she sometimes worked without pay; all her life she has resisted acquiring possessions—to that single specialty. "It was *fun*," she told me once. "I couldn't wait to get up in the morning." She was first at Cornell, later at Cold Spring Harbor, where the Carnegie Institution had its genetics department and gave her a job. "I've known Monod since he was a little boy in science," she said, and brought out a photograph of him with Boris Ephrussi, on the roof of a building at Caltech, that she had taken while working there in the winter of 1936. McClintock's corn plants had begun to give her hints of patterns of gene expression and of genetic control systems by 1944 or '45. She had put everything else aside to pursue them. She introduced the problem at the Cold Spring Harbor Symposium in 1951, but her ideas were not understood. "You must remember, at that time even suitable *terms* were hard to find. I was trying to say that there was something else besides genes; later I began to call them controlling elements." She published a paper about the phenomena in *Genetics* in 1953, which was ignored. At the Cold Spring Harbor Symposium in 1956, where Jacob and Wollman talked about interrupted bacterial matings, McClintock gave a talk titled "Controlling Elements and the Gene." She had found in maize such things as inhibitors, activators, modifiers, and inducers; but the paper got little attention.

On reading Jacob and Monod's first short paper on the operon, which appeared in French late in 1960, McClintock realized that the control systems they put forward for *E. coli* resembled those she had been studying in maize. Greatly excited, she called a staff seminar at Cold Spring Harbor at the end of November, in which she analyzed the parallels. Then she wrote a paper about them for *The American Naturalist*, and sent a draft copy to Monod in the spring of 1961. In their great synthesis for *The Journal of Molecular Biology*, which appeared a few weeks later, Jacob and Monod had not cited McClintock's work—an unhappy oversight, Monod told me. ("They did not understand the technical aspects of maize genetics," McClintock said.) They were glad of her prompt support. That summer, in Monod's concluding summary for the Cold Spring Harbor meeting, which he and Jacob wrote together, they said:

> Long before regulator genes and operator were recognized in bacteria, the extensive and penetrating work of McClintock . . . had revealed the existence, in

maize, of two classes of genetic "controlling elements" whose specific mutual relationships are closely comparable with those of regulator and operator. . . . Although, because of the absence of enzymological data in the maize systems, the comparison cannot be brought down to the biochemical level, the parallel is so striking that it may justify the conclusion that the rate of structural gene expression is controlled, in higher organisms as well as in bacteria and bacterial viruses, by closely similar mechanisms.

"The bacterial work was done in the late fifties," Gilbert said. "Now, twelve years later, the whole picture of control is much more complicated than any they imagined at the time. It's clear that there isn't a single, unique answer."

❖ ❖ ❖

By the time Brenner got back to Cambridge, the *FC* family of mutants numbered more than fifty, and Crick had picked up suppressors of suppressors and was working for suppressors of suppressors of suppressors. He had mapped his original suppressor mutations of the *r*II mutant *FC0*, and found as he segregated them that on their own they were themselves typical acridine *r*II mutants. So for these he repeated the analysis—to find and map, for example, a subfamily of still newer mutants that would act on *FC1* to make it revert to wild or pseudo wild; a similar subfamily that would revert *FC6*; and so on. Then he pushed the trick one rank further. *FC42* and *FC47* were each suppressors of the first-rank suppressor *FC7*; each when isolated by suitable crosses proved to be a typical acridine *r*II mutant—and for each he had begun to generate mutants that reverted it in turn.

The next step would be to see whether a suppressor derived from one mutant would also suppress other mutants of that same rank—whether *FC42* would suppress not only *FC7* but *FC6* and *FC1* and so on. Crick had been invited to the annual meeting of the French Society for Physical Chemistry, the last week in June, and from there was going to Tangier for a six-week vacation, before the biochemical congress in Moscow. "When I went away, I left Sydney with the following job. I said, the mutants can be divided into plus and minus, all right?" These were the alternating ranks. *FC0*, the original mutant, was labelled *plus*, though Crick could not know whether in fact it represented the addition or removal of a base pair. *FC1* and the others of that rank, therefore, were *minus*. *FC42*, 47, and the others of *that* rank were *plus*, and *their* suppressors *minus* again. "Some of the pluses and minuses hadn't been crossed. Putting them together was called 'uncles and aunts'—when you think about it, because you put together not a plus, and if it had got a minus—" He drew the simplest imaginable diagram.

"These two had *been put* together 'cause that's how you *made* it, but *these* two hadn't been put together"—

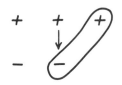

—"so you can see that was called uncles and aunts! It was the relation one up and sideways! And I left Sydney with the job of doing the uncles-and-aunts experiment."

The French physical chemists met near Chamonix, in a large, isolated hotel in alpine pastures six thousand feet up, almost at the tree line, glacier rising above. The subject of the meeting was DNA; the papers were routine; it was the brilliance of the setting that brought visitors from England and America. Ten of those present had come from Cold Spring Harbor. Many were going on to Moscow in August.

Crick read no paper. When one speaker mentioned mutations produced by proflavine, Crick got up to report briefly the state of the work at the Cavendish on the problems of mutation and the code.

> There is little doubt that the mutations produced by acridines are in some way different from those produced by *all* the other mutagens studied so far. They tend to map at different places and they usually do not produce a *mildly* altered protein, as far as we can judge. In addition, recent work in our laboratory on the *r*II gene of phage T4 shows that the back-mutants are usually not due to true reverse mutation, but to close suppressors. . . . The suppressor mutants are, by themselves, good *r*II mutants and map fairly close to the acridine mutant they suppress. They, too, can "revert," either spontaneously, or with acridines, by the action of close suppressors. . . .
>
> All these results are compatible with our suggestion that the action of acridines is to add or delete bases. An attractive additional hypothesis is that the code is read in short groups, starting from one end of the gene. The exact starting point is supposed to determine which group is read. The deletion of a base would then alter the active reading from this point onward. The double mutants produced by the reversion of acridine mutants would then, on this hypothesis be altered not just in two, separated, amino acids, but in a short stretch of amino acids in sequence. If this were so it would be very important for decoding.

❖*b*❖ The Moscow River takes a sweeping meander south and southwest before gathering itself and turning left again into the long reach that will carry it northeast through the center of the city. On the far side of the river, just at the elbow of the turn, steep bluffs command a superb prospect. From these, Napoleon first looked down across the river to the golden domes and the fortress seven kilometers away. Here now rise the towers of the University of Moscow. They were erected in the early nineteen-fifties and supply a routine example of Stalin's taste for neo-Victorian gigantism; but they make up the largest complex of meeting rooms and auditoriums in the city. When the scientists at the fourth International Congress of Biochemistry, in Vienna in 1958, took up the Russian invitation to hold the fifth in Moscow in August of 1961, it was obvious that the university was the only place that could house it.

The congress was the first big international scientific meeting to come to Moscow. Stalin had died in March 1953, five days after the discovery of the structure of DNA. Sputnik had gone up in October 1957. Western scientists had discovered they knew nothing much about Soviet science, and had become curious. Five or six thousand foreign biochemists were to attend the congress; more than twenty-two hundred papers were to be read. The president of the congress was Alexander Ivanovich Oparin, director of the Biochemical Institute of the Academy of Sciences of the Soviet Union; Oparin was famous abroad for certain speculations in the nineteen-twenties and early thirties about the origin of life in the very different chemical environment of the earth two billion years ago. Soviet scientists were encouraged to offer papers, and over a thousand of them did. Younger Russian scientists who had some English or French were urged to volunteer to be interpreters; those who came forward were given six months of free, intensive language lessons. One of these interpreters was Zhores Alexandrovich Medvedev, who was then a senior biochemist at the agricultural college in Moscow, studying how plants assemble proteins.

Of the many divisions of the congress, the one on molecular biology was nominally organized by Vladimir Engelhardt, but after the exchange of letters at the beginning of the year, Perutz had arranged for the papers by Westerners in that division. The announcement of revolutionary discoveries can hardly be planned ahead: Perutz got together a program that, although it had the right people and showed off some hard-won progress, was in truth to be seen as a roll call of problems—familiar problems, problems of demonstrated intractability. Crick had agreed to be chairman of one of the sessions. Watson was to preside over another. Strong delegations of molecular biologists were coming to Moscow from the Institut Pasteur, from the Cavendish Laboratory, from the Max-Planck-Institut in Munich, from Harvard, MIT, Caltech, the New York University Medical School, the Rockefeller University, the University of Wisconsin, the University of California at Berkeley.

The second week in August, the visitors began to arrive, many with their wives. There were eleven hundred American biochemists. "The Americans dribbled in through Leningrad and other places first," Matthew Meselson told me ten years afterwards. "We did that, and Moscow was generally overcast, and rather chilly, though it was August. There were some smells, sour smells—I don't know where they were coming from but I remember them. Some Americans were very exhilarated by being in Moscow. I was. Because it was new, and I liked some of the new things. I liked the food. And I liked the people—I met some people I was very interested in meeting, who had been making heavy isotopes. Other Americans were very unhappy, and were complaining about their hotel rooms, and I was somewhat ashamed of those people. It was all great fun. We were staying at a giant hotel in the center of the city. They'd take us in big busses from the hotel to the meeting place. It was incredibly big. I thought there were perhaps ten thousand people."

The second week in August, 1961, was in almost every way imaginable an unsettled time for scientists to meet in Moscow. World politics had suddenly become more fevered than they had been for five years. Soviet biology was tormented by a new climax of the ideological struggle that had paralyzed it since the mid-thirties. Three days before the congress began, Russia's second cosmonaut, Major Gherman Titov, had come back to earth in Vostok II. Four days after it began, Khrushchev and Walter Ulbricht of East Germany started building the Berlin Wall. At the beginning of August, Trofim Denisovich Lysenko had been named by Khrushchev to be the head of the Lenin All-Union Academy of Agricultural Sciences.

Lysenko's career had now come the second time around an eccentric spiral. He had held this job before, under Stalin. After Stalin's death, some Soviet biologists grouped to fight back, helped by physicists like Andrei Sakharov but led by those who eventually called themselves molecular biologists—Engelhardt, Perutz's correspondent, one of the chief among them. The struggle grew intense. At the end of 1955, Lysenko was dismissed as president of the All-Union Academy of Agricultural Sciences. Oparin—he of the origin of life—had become one of Lysenko's chief supporters, and had been made secretary of the biological section of the Soviet Academy of Sciences, but was now replaced by Engelhardt. Lysenko counterattacked, politics against science. In 1958, he established himself with Khrushchev much as he had with Stalin, promising a quick technological cure for the systemic sickness of Soviet agriculture and a firm restoration of ideological authority over Soviet biologists. Lysenko became director of the Institute of Genetics; Oparin became director of the Institute of Biochemistry; Engelhardt was replaced but given an institute of his own. Genetics continued unteachable. "The assertion that there are in an organism some minute particles, genes, responsible for the transmission of hereditary traits is pure fantasy without any basis in science," wrote Isaak Izrailevich Prezent, Lysenko's cleverest theoretician, in 1961. Lysenko's reappointment early in August 1961 as president of the Academy of Agricultural Sciences put him back in control of the nationwide network of research institutes concerned with agricultural biology—and also made him, at two removes, Zhores Medvedev's superior. His domination of Russian biology continued until Khrushchev's fall in 1964.

But Lysenko did not appear at the Congress of Biochemistry. Crick, Meselson, and others visited the Soviet Atomic Energy Research Laboratory and learned of experiments in molecular biology going on there under protection of the physicists. "I thought when I was in Moscow that Lysenko had been defeated for some time," Perutz once said. "We were completely unaware that all this was still going on. I did meet many Russian scientists that week who were certainly not Lysenkoists. It was only eight or ten years later, when I read Medvedev's book about Lysenko, that I realized how perfectly frightful the conditions for research were for them then. What did emerge from the congress was how impressed, and I think stimulated, the Russian biologists were by the revelation of the enormous advance the West had made in the preceding decade—working from the molecular basis of classical genetics."

Medvedev began writing the history of the Lysenko era in Russian biology at about the time of the congress. He became one of the earliest and most outspoken of the dissidents among Soviet scientists. That book was published in the Soviet Union in 1967. Yet when Medvedev allowed it to be published in the United States two years later, he was fired from his job, and a year after that became one of the first dissident intellectuals to be put into an insane asylum—and then one of the first to be freed, after a campaign of protest in Russia, promptly and brilliantly organized by his twin brother, Roy, which also reached the West. In January 1973, Medvedev was permitted to go to England, to take up a year's visiting fellowship offered by the Medical Research Council; six months after his arrival in London, the Soviet government cancelled his citizenship. The council therefore put him permanently on its staff.

Early in 1976, I went to see him at the laboratory where he worked in the northern suburbs of London. That turned out to be a large, bleak building atop a long ridge; we talked in his study, a cubicle on an upper floor, and talked some more over tea in the senior common room. Medvedev's face was gray in the English February, and rather puffy; his eyes were pale, his chin was strong, and his mouth, in repose, had a recalcitrant set. His chief scientific interest had long been the biochemistry of aging, he said—originally with plants, now with animals. "By 1969 I made a very good laboratory," he said. "When I got the place it was not yet built, but that was a great advantage, because I was able to supervise everything and to order the equipment. So I was very unhappy when I was dismissed after publication of Lysenko book, abroad.

"When Lysenko was dismissed, after Krushchev's fall, it was permitted quite seriously to criticize him from many different points, and a lot of articles appeared," Medvedev said. "And I submitted my book for publication and it was in fact approved. A special commission considered it from many points of view and made a nearly unanimous decision about this book. So my violation was only that I published it abroad when the state publishing house failed to do so. But this still was, in '69, a quite unusual step for a scientist. I expected that when the book appeared in the West I would certainly have a lot of troubles, with officials; I expected that I could be demoted from the position as head of the laboratory; but I thought that I could continue as a senior scientist at the same place. But part of my problem was that I was also the author of another book—it was about international cooperation among scientists and about the position of scientists in the Soviet Union"—he took a paperback with a brilliant red cover from a low shelf behind him—"and this was already circulating in *samizdat*, along with something else of mine about censorship of mail and other problems. And this book in *samizdat* at the same time as the publication in the West of the one about Lysenko, made officials take the action against me, trying to put me in a madhouse because they said I showed signs of 'incipient schizophrenia'—which finally my brother and I described in still *one more* book." Medvedev laughed. His photograph on the front of the paperback on the desk showed a face as still and closed and direct as a cocked fist. Any bureaucrat, East or West, would recognize that look instantly.

When had Russian biologists—when had Medvedev—first begun to consider the Watson and Crick structure of DNA? "The problem was that the Watson and Crick model, at first it was not connected with protein synthesis," Medvedev said. "So at the beginning their model didn't make very strong impression in the Soviet Union. The discovery was very genetically oriented—and genetics was suppressed, still. It hadn't started to recover." The structure of the genetic material seemed irrelevant to research that was permissible. "People knew about the model, but they were not able to realize its consequences. Yes. *My* interest in this model started from the first works by Hoagland and others with transfer RNA—when the special RNA was discovered which transferred amino acids for incorporation. From this point I started to understand how a possible mechanism can function. This made great impression on my whole conception. I realized that there is a biochemical direct connection between nucleic acids and proteins."

The civil war that split Soviet biology for a third of a century is unfathomable, almost unbelievable, to those who did not experience it. In the final cycle of the controversy, Medvedev said, the coming of the International Congress of Biochemistry to Moscow in August 1961 proved catalytic. Did Lysenko have any role at the congress? "No. But the problem was that Oparin, who was president of the entire congress because he was president of the Soviet Biochemical Society"—the hosts—"as well as director of the Biochemical Institute, and Oparin's deputy, Norair Sisakyan, Academician, both were Lysenkoists. And a number of other people in the Biochemical Society who were active in the congress, they were mostly Lysenkoists. They fought against the idea that DNA had a role in heredity! To admit the role of DNA would force them to discard absolutely the inheritance of acquired characteristics." Western biologists I had talked to, I said, who had gone to Moscow for the congress in 1961, all had supposed that Lysenkoism was a dead controversy, a scandal of the Stalinist past. "No, no, no," Medvedev said. "Lysenko was a very hot issue, at this time, among our biochemists. And because of DNA and all of its consequences that were becoming clear then, and that the congress made us realize, molecular biologists in the Soviet Union started to be in the first group to do battle against Lysenko. At this time. And they used the opportunity of the Biochemical Congress to make quite clear their disagreement with Lysenko and with his whole biology."

A year to the day after the conversation with Medvedev, I was in London again, to listen in at a two-day discussion of the problem of chromosome structure organized at the Royal Society by Aaron Klug. The society inhabits an eighteenth-century residence of noble proportions, restored and suitably rebuilt, in Carlton House Terrace. Scientific meetings there are held on the splendid principle that all who are interested are welcome. This one was crowded. At noon, I met a scientist I wanted to interview, and so was late getting to the lunchroom in the basement; about the only place free was at a corner of a long table by the far wall. On sitting down, I heard that the two men to my left were speaking Russian to each other and occasionally to an older man directly across from me. He, hair white and thick, jowls heavy, napkin tucked up, was concentrating with pleasure and a vigorous tattoo of knife and fork on the cold roast beef and salad on his plate. The badge pinned to his lapel said Vladimir Engelhardt.

Perutz had told me once that Engelhardt's English was excellent—and so it proved, except that now he was hard of hearing. One of his younger colleagues, who had given a paper that morning, translated a few questions. The answers came back in English high and loud. After lunch, Engelhardt's companions left us, and we went upstairs to find a quieter place. Engelhardt had once been a man of immense frame; he was still tall, his stooped shoulders still broad, his head large and strong-featured, his eyes direct and of the palest blue. The Fellows' Room on the main floor was almost deserted. We sat by the french windows. The furniture had evidently been bought from the lounge of some seaside hotel, but the view is one of the most civilized in the world, out across the Mall and St. James's Park to the domes and cupolas of Whitehall, and in the distance the Palace of Westminster and the tower of Big Ben.

The section of the Congress of Biochemistry that Engelhardt organized with Perutz in 1961 could not even be titled "molecular biology," he said. The term was too newfangled. "We had to exist under camouflage." He had the same trouble for years with the name of his research institute, and with the terminology of molecular biology in scientific papers. The organizing committee of the congress, the boards of editors of some of the biological journals, "were still occupied more or less by people—of rather conservative, I would

say, opinions." His tone was precise. "And I should say that even the most—
Now they are even regarded as progressive, and they are quite good scientists,
but even the most progressive, still they were very reluctant to this new name,
to these new views. That study on the molecular level can discover any important things, was regarded as a kind of fancy—or a premeditated exaggeration,
of a few people who were themselves working" in that field. He laughed.
Oparin was opposed to molecular biology; even Belozerskii was opposed,
Engelhardt said. And Lysenko? "Ha?" I repeated the question, louder. "He
was, yes, yes, yes, yes; he was."

On the wall near us hung a portrait of Baron Gottfried Wilhelm von Leibniz; nearby were Tycho Brahe and Edmund Halley; across the room hung a
portrait of Sir Isaac Newton, President of the Royal Society from 1703 to 1727,
his face pale yet highly flushed, his eyes large and watery. Leibniz stared at
Newton, Newton gazed grumpily out the window, their bitter dispute about
which of them had discovered the calculus unresolved.

"Lysenko was still in his position as the leader of the biological science at
that time. But practically, during the biochemical congress, no influences were
felt direct," Engelhardt said. "In chemistry, he was not very strong. Once he
visited our institute. Four of us were there, the vice-president of the Academy
of Science, the president of the Medical Academy, Lysenko, and myself. We
spoke something about DNA. He says, Lysenko says, 'All this DNA, DNA!
Everybody speaks about it, nobody has seen it!' *I* say, 'But dear Trofim
Denisovich, I could show you a preparation of DNA; it is well known by
chemists.' 'Show me, please.' I asked my secretary to get a small quantity to
show him. They brought a small test tube with a little of the preparation.
'Look here, that's the DNA,' I say. Lysenko looks. 'Ha! You are speaking nonsense! DNA is an acid. Acid is a liquid. And *that's* a powder. That can't be a
DNA!'"

Engelhardt laughed gustily. "That shows his knowledge of chemistry." We
stood and for a moment looked out the window. The grass in the park was beginning to show green. "All my life I have been honest biochemist. At the
decline of my age, I changed my profession, and became molecular biologist."
We talked for a minute or two about research projects Engelhardt had led in
the thirteen years after the fall of Lysenko. I asked how old he was. He said,
with delight, "Eighty-two!" As we left the room, I took his arm for a moment,
and through the heavy stuff of his sleeve felt the sparrow bone beneath the
down-soft old man's flesh.

❖ ❖ ❖

The fifth International Congress of Biochemistry began on Thursday, 10
August 1961, a fair and pleasantly warm day; Meselson's chill clouds came
later. The molecular biologists had been assigned the largest auditorium at the
university for their main program, but that first morning the hall was
preempted by a press conference for Major Titov. John Kendrew was down to
give the opening talk for that panel, a report of the slogging progress in Cambridge in the x-ray crystallography of the structures of proteins, chiefly
Perutz's work with hemoglobin and Kendrew's own, well advanced, with the
related but simpler respiratory protein myoglobin. In the event, Kendrew did
not come to Moscow and Perutz read the paper instead. He unveiled a shiny

new model of the hemoglobin molecule, in contoured layers of white and red plastic, which brought the resolution down to five and a half angstrom units, or slightly better than twenty-two billionths of an inch. That was still fairly coarse, as molecular dimensions go, not nearly sufficient to betray the positions of individual amino acids. It was good enough to show the molecule's four polypeptide chains and how they fitted together, and the positions of the four heme groups—each carrying an iron atom, the sites where oxygen loaded on and off the molecule. The chains were unaccountably irregular, the heme groups unexpectedly far apart, the logic of the structure inscrutable. None the less, "We are undoubtedly on the eve of a major advance in our understanding of one of the central problems of biology—the relation between the structure of the molecules making up living cells and their biological functioning." The understanding Perutz had in mind, for the hemoglobin molecule, was nine years yet to come.

Perutz had arranged thirty-odd principal reports to be delivered and discussed in five and a half days. The congress, all told, offered twenty-two different sections of biochemistry and seven other comparable symposia of invited speakers, from evolutionary biochemists to technologists of the food industry. Once it was under way, some two hundred papers were got through each morning and another two hundred each afternoon, in Russian and English, French, German, Polish, Spanish; in large halls and small classrooms; before audiences of hundreds, or dozens, or only a few other scientists. There were thirty or more papers to choose among in any half hour and, in molecular biology alone, scores every day outside the central schedule Perutz had organized.

Meselson had been asked to give a paper on a later day of the congress; otherwise he skipped most of the meetings to talk to scientists, visit laboratories, or go sight-seeing. On the second or third day, though, he stopped in a classroom to listen to a short report read by an American scientist whom he knew only slightly, Marshall Nirenberg. The room was large but almost empty. Two others there were Walter Gilbert and Alfred Tissières. Nirenberg was a junior scientist on the staff of the National Institutes of Health. He was a tall man, hesitant of speech and incipiently clumsy; his hair was dark, his face pale, soft, and sullen. The listing that had attracted Meselson's attention suggested that Nirenberg and a colleague, Heinrich Matthaei, had been looking at the role of RNA in synthesis of proteins. The title of the paper was awkwardly uninformative: "The Dependence of Cell-free Protein Synthesis in *E. coli* Upon Naturally Occurring or Synthetic Template RNA." The title gave no hint that Nirenberg and Matthaei had solved the coding problem.

❖ ❖ ❖

Marshall Warren Nirenberg took his doctorate in biochemistry at the University of Michigan in 1957, and came to the National Institute of Arthritis and Metabolic Diseases to do postdoctoral research. After three years of that, he was invited to stay on. His aim in science, Nirenberg once told me, had been naïvely simple. "The key thing—my long-range objective was to study how information flowed, how DNA could in some way direct the synthesis of proteins." The National Institutes of Health are the Pentagon of American biology, an uneasy yoking of competing services set down on a green hillside in a northern suburb of Washington, D.C., where they have grown and spread to no apparent plan. I sought out Nirenberg there for the first time in 1971. "People have always underestimated Marshall," a friend had warned. I perceived why. Nirenberg's speech was breathily shy, his manner earnest, wide-eyed, enthusiastic, confiding, and devoid of apparent subtlety. When he started on his own in the fall of 1959, messenger RNA was of course unheard of. Nirenberg had known, he said, Volkin and Astrachan's results. "Template RNA was just a thought. Though it seemed very likely from the work that had gone on before, there was no molecular evidence for it, no clean evidence," he said. "I felt that it was one of those things bound to be right; and I guess everybody else felt that way. But no evidence. The way to get evidence was to set up a cell-free system, and then to use nucleic acid as a template to direct protein synthesis. Obviously, this could be done. I had no experience in protein synthesis. I felt that I was competent to do this kind of work. And I tried setting up systems that people had published."

Johann Heinrich Matthaei trained as a plant physiologist at the University of Bonn, got his degree in 1956, worked there nearly four years more before coming to the United States on a NATO postdoctoral research fellowship. His research had been with plant cells. "But the other thing was that I had reviewed papers on cell-free protein synthesis for a journal in Germany, so I was aware of the problems," Matthaei once said. "When I went to the States my main idea was to synthesize a specific protein *in vitro*, but of course in the back of my mind I knew very well that it was important to try to prove the existence of a genetic code." Some years later, Matthaei settled at the Max-Planck-Institut für Experimentelle Medizin, in Göttingen; I talked with him there in 1976. Göttingen and her university are a sedately prosperous old German couple, she with flashes of remembering that once she was charmingly pretty, he still proud of brilliance past, both content to grow dowdy and conservative. The Institute for Experimental Medicine, near the center of the town—but one could be describing almost any of the Max Planck Institutes that have been planted all over West Germany since the war—is a squat, square tower with balconies around the outside and generous spaces within.

Matthaei was tall and gaunt. He walked with a limp from a skiing accident. His cheekbones were high, his jaw square, his hair receding, his face tanned. In the United States in 1960, Matthaei had started out at Cornell in the laboratory of Frederick C. Steward, a plant physiologist famous for having learned to grow carrots from single cells of other carrots rather than from seed. "I suggested to him that I should try to synthesize a specific carrot protein!" Matthaei said. He also began looking for a lab where he could do more interesting work. He went to see Fritz Lipmann at the Rockefeller Institute and several people at the National Institutes of Health.

Matthaei met Nirenberg in August of 1960, and moved to his laboratory on November 1. Nirenberg was thirty-three, Matthaei thirty-one. They shared a laboratory cubicle, a few feet of bench—Matthaei the visitor on a training fellowship, Nirenberg nominally his boss. Head of their section was Gordon Tomkins, an enzymologist two years older than Nirenberg; Tomkins died in 1975. One floor up was the chief of the laboratory, Leon Heppel.

As it happened, Heppel was skillful, to a degree unique in the United States just then, at making artificial RNAs of different defined compositions. He had spent the winter of 1953–'54 with Roy Markham at the Molteno Institute in Cambridge—the laboratory to which Watson's fellowship and his work with tobacco-mosaic virus had been notionally attached for a while. Markham and his colleague John Smith had been refining, to exquisite sensitivity, the techniques of electrophoresis and paper chromatography needed for analyzing the base compositions of short chains of nucleic acids, in particular RNAs. Heppel learned these methods and himself devised new ways to use enzymes to build the chains up or cut them down again. Early in 1956, when Marianne Grunberg-Manago and Severo Ochoa found polynucleotide phosphorylase, which for a while they mistook for the enzyme that strung together the cell's natural RNA, Ochoa wrote to Heppel for help in analyzing the polynucleotides they were getting out. The two laboratories collaborated by mail for a year. In the summer of that year, Heppel took on a young biochemist, Maxine Singer, who was just finishing her thesis at Yale.

Enzymes for nucleic-acid work were still scarce, and some of the other ingredients expensive. Polynucleotide phosphorylase acted only on nucleotide diphosphates, which, if bought commercially, then cost around a thousand dollars a gram, twenty-eight thousand dollars an ounce. Over the next four years, as a byproduct of experiments in nucleic-acid synthesis, Heppel and Singer accumulated a library of artificial RNAs. They had homopolymers: polycytidylic acid—poly-C—where all the bases in the string were cytosine; polyadenylic acid—poly-A—all adenine; polyuridylic acid—poly-U. These became widely available by 1960. (Polyguanylic acid—poly-G—was not available: because of the way its bases stack up it was hard to make and hard to handle.) More rare, Heppel and Singer had co-polymers made up, for example, of U and C in random order—poly-UC—and so on. They had co-polymers that were enriched—say, with more U than A; they had polymers in which one end of the string had a different composition from the rest. They kept them all in a freezer.

When Nirenberg first tried to set up a cell-free system that would build amino acids into proteins, nothing had yet been published in English that could show him how to do so with bacterial extracts. Spiegelman had announced a bacterial cell-free system in 1958, and there had been other attempts, but none of them would perform without the presence in the mixture of an ill-defined mess of fragments of bacterial membrane: everyone else in the trade dismissed those systems as contaminated with intact bacteria that were actually doing the protein synthesis reported.

But in the late fifties, Paul Zamecnik, at Massachusetts General Hospital, with a postdoctoral colleague named Marvin Lamborg, developed a cell-free system from *E. coli* almost exactly like the rat-liver cell-free system his lab had first produced five years earlier. Both systems needed ribosomes; both needed the solution of nucleic acids and enzymes of the fraction of the cell extract that remained in the supernatant after two hours in the centrifuge at a hundred thousand times the force of gravity; both needed ATP for energy plus the enzymes to build more; and both needed a supply of amino acids. Optimal concentrations of these ingredients were similar for the two systems. Performance was similar. "And Lamborg really got the bugs out of the system," Zamecnik said. "We had less than a hundred thousand live *E. coli* in our system; and we were willing to rely on it, because we could *add* that many and not influence its performance."

In the summer of 1959, Alfred Tissières at Harvard got interested in extending his work with bacterial ribosomes into a cell-free system. "To start with, we thought that it was really logical to go and talk to Paul Zamecnik," Tissières said in 1977. "He told us that they were actually doing some experiments, and they didn't know whether they were working, but maybe they were working—the thing is that Paul Zamecnik is a very careful man, and he would never say anything until he's absolutely dead sure." By the fall, Zamecnik was sure; his and Lamborg's paper was ready at the end of the year, and they gave a copy of the manuscript to Tissières. It was published in the summer of 1960. Tissières, with a graduate student named David Schlessinger and joined in the spring by Françoise Gros (wife to François), tested and improved the system somewhat; fine tuning lay, once more, with the trace of magnesium needed to hold the ribosomes together. Tissières talked about the *E. coli* cell-free system at the Gordon Conference on nucleic acids that summer. "It created a tremendous amount of stimulation for people to work with *coli* systems," Tissières said. Neither Nirenberg nor Matthaei, however, was at the Gordon Conference. Tissières's paper appeared in *Proceedings of the National Academy of Sciences* in mid-November.

Nirenberg kept a journal during that time—not the formal record of lab work but a personal daybook of scientific reading, plans, problems, and ideas. He wrote it in pale blue ink, sometimes in pencil; his hand was simultaneously unformed and overcomplicated. The journal consisted for the most part of lists—lists jumbling together every sort of thing that was on his mind to get done, the banal, unconsciously poignant lists that a man makes thinking he must drive himself harder. In that journal, scattered among hundreds of schemes for experiments, that fall, are glancing references to the idea of using synthetic RNAs as "templates" or "messengers" for protein synthesis. The idea was hovering at the fringe of Nirenberg's attention; it was not worked out.

The cell-free system had been used brilliantly to analyze protein synthesis, distinguishing the machinery, the biochemical requisites, and the sequence of steps. It had so far been mute about the gene. The next goal was obvious: to put genetic information into a cell-free system and get out the specific protein. That was hotly sought in several laboratories more experienced than Nirenberg's. Perhaps, for example, it would be possible to induce synthesis of an enzyme, cell-free. Nirenberg's first plan had been to make the enzyme penicillinase, which resistant bacteria use to chop up the antibiotic, in a cell-free system. He chose this one because it was inducible, yet was much smaller than, say, β-galactosidase, and was closely studied in several labs. "I prepared DNA and RNA from different strains of bacteria, and then tried to add this to a cell-free system," Nirenberg said. "And I tried two things. One, to measure increased enzymatic activity, hoping in a naïve fashion that maybe I'd be able to synthesize an entire enzyme; and also tried radioactive tracers, just to check to see that the experiments were getting amino-acid incorporation into protein."

The attempts achieved little until Matthaei arrived. Matthaei was technically dextrous and accurate—and, just as important, they redefined the problem slightly. Backing off from the attempt to make any particular protein, they decided that the first need was to set up the cell-free system so that it would respond when they added any RNA that carried genetic information. They thought of such RNA as template, although in the loose, general sense that hard usage had worn the word down to. Matthaei said, in our conversation fifteen years later, "I always have to state, many people don't expect this: we knew beforehand *we are going to make an assay for messenger* RNA." They had no particular idea where to find template RNA in the cell, except that it would be associated with ribosomes, nor how to isolate it.

Neither Nirenberg nor Matthaei was in touch with any of the people at Harvard, or MIT, or Caltech, or Cambridge, or the Pasteur who knew of the work completed but not yet published about messenger RNA. They had access only to what had been published—Volkin and Astrachan's papers and the PaJaMo experiment. Yet at the same time, they were not intently focussed on the ribosome as the vehicle, itself, of genetic specificity. Isolation and inexperience allowed their theoretical imaginings to retain a benign fuzziness. "I always thought that it's likely, the message is likely to be associated with the ribosome, but not the ribosomal RNA," Nirenberg said. But their immediate technical aim had not been vague, both men insisted.

"I think we should not be really honored for the basic construction of our cell-free system," Matthaei said. "We were very much following actually other people's proposals in the literature." The basis for the work was necessarily Zamecnik's paper with Lamborg, the previous summer, Matthaei said. But the version from Tissières's lab was just out, and that was the recipe that Nirenberg followed and later most often cited. Tissières was far more generous with technical tips, and had taken one experimental step beyond Zamecnik. He had tried the effect of the enzyme deoxyribonuclease and found that it too, like RNase, turned off the system.

Nirenberg and Matthaei needed a couple of months to get the system working. They made a minor improvement, to stabilize the components so that batches could be made up and stored, frozen, without great loss of potency. They put the components together—still, fundamentally, ribosomes, the solution of nucleic acids and enzymes left above the centrifuge pellet, and the energy supply, plus the pinches of this and that needed to adjust and stabilize. Then they added nineteen cold amino acids and the twentieth, which happened to be valine, made with radiocarbon. They promptly found the radioactivity showing up in the fraction they believed to be protein. The system would begin to run down after thirty minutes, but would continue building some protein for an hour and a half or more. They reached this stage by Christmas, Matthaei said, and he went to visit his family, not yet moved south from Cornell.

Not in anybody's version of a cell-free system so far had the protein produced ever been specifically identified. The next step was to run the system and add ribonuclease. Protein production thereupon stopped dead. When they left the RNA alone but put in deoxyribonuclease, protein synthesis was not inhibited instantly; it fell sharply after about thirty minutes. The system was sensitive to extremely small amounts of enzyme. Matthaei would put thirty-five hundred-millionths of an ounce of crystalline DNase into a drop of cell-free system amounting to a fifth of a teaspoonful. The implication was strong: DNA present in the system had been making RNA, so that when the DNA was destroyed, the RNA, or some essential part of it, soon ran out and protein synthesis slowed way down. This was the condition Nirenberg required for an assay for template: to make the system respond noticeably to information-carrying RNA introduced from outside, they had first to stop the production of RNA within.

Then they tried to start the stalled system up again. They made no attempt at finesse, but used an RNA shotgun. "What we found is, when we prepared crude RNA, unfractionated RNA from ribosomes, and then added it to a cell-free system, it stimulated amino-acid incorporation," Nirenberg said. "But the stimulation we got was very, very small. We were measuring fifty counts, above a background of two hundred and fifty counts, or a hundred counts above the background. Not very impressive." Hearing that Szilard was in Washington, Nirenberg went to show him the work. "Szilard was really very kind, I must say. He spent an entire afternoon with me going over the data." Back at the bench, Nirenberg and Matthaei sharpened the sensitivity of the system by incubating it each time to reduce still further the protein synthesis that went on before the crude RNA was added. At that stage, they knew no way to isolate information-bearing RNA from the bulk of the ribosomal RNA in the cell; in their own understanding the differentiations were not so absolute as they soon became. "But we felt, of course, that there should be a difference," Matthaei said. "We felt that there should be some *non*informational RNA, to put it this way." As controls, in the hope of showing that the system was responding to something more specific about the added RNA than merely its overall chemical or physical presence, they tried decoys—a polysaccharide that was slightly acid and that came in comparable strands, DNA from salmon sperm, and some polyadenylic acid, poly-A, that Heppel gave them. None of these things stimulated the cell-free system to build radioactive valine into protein.

Towards the end of March, they sent a brief provisional report of their modified cell-free system and its curious requirement for extra RNA to a journal recently founded to disseminate news flashes, *Biochemical and Biophysical Research Communications.* In all this time, apparently, not a word about the discovery of messenger RNA reached either Nirenberg or Matthaei. Yet the term "messenger" was creeping into use like a new piece of slang; at the end of their bulletin Nirenberg and Matthaei speculated, "It is possible that part or all of the ribosomal RNA used in our study corresponds to template or messenger RNA." The suggestion was little noticed.

"I drew up a list of possibly two hundred experiments that had to be done," Nirenberg said. "And among them was to try the template specificity of many different kinds of RNA. Natural RNAs, from viruses, as well as synthetic polynucleotides. And systematically started to study them." In a later conversation, he told me, "Then I got hold of some tobacco-mosaic virus RNA, some viral RNA. Which I thought would be a pure template. Pure message. And we added it to the extracts, here, and it was just *superb*." This was early in May. "We hadn't seen anything even approaching the activity. With respect to amino-acid incorporation. The template activity. As the viral RNA." Nirenberg was shaking his head with wonder. "I mean it was just *beautiful*. Thousands and thousands of counts, ah—showing up in protein. It was superbly active. As template. For protein synthesis. So, I called Fraenkel-Conrat, on the phone."

Heinz Fraenkel-Conrat, in Berkeley, with a host of collaborators including Wendell Stanley, had recently worked out the complete sequence, 158 amino acids long, of the protein subunits of tobacco-mosaic virus; the paper announcing the sequence could not have escaped Nirenberg's notice, because it began on the page of the November 1960 *Proceedings of the National Academy of Sciences* where Tissières's report of the bacterial cell-free system ended. "I figured that perhaps we were making protein associated with tobacco-mosaic virus—coat protein. And that maybe Fraenkel-Conrat had a stock of the RNA, and had mutants of various kinds to show it," Nirenberg said. Fraenkel-Conrat was in fact energetically collecting mutations of the virus and identifying the exact amino-acid change in each. By using mutagens known to cause base transitions, he was learning which amino acids could be changed to which others by a transition of one, or of two, bases—and planned thereby to group the amino acids in interrelated families and solve the genetic code. Pooling resources, Nirenberg and Fraenkel-Conrat could put viral RNA into the cell-free system repeatedly, while making first one and then another amino acid radioactive, and finger print the proteins to see whether the radioactivity appeared in the spots on the filter paper called for by the sequence. So Matthaei prepared bottles of components and samples of the radioactive protein they had already got with viral RNA, and packed them in a bag with dry ice. Nirenberg flew from Washington on Sunday, May 14, to spend four weeks in Berkeley. In the rush, apparently neither man saw the messenger experiments in that week's *Nature*. "I actually left Heinrich with a whole series of protocols to do. While I was gone."

Matthaei kept laboratory notes in books bound like ledgers, tall, with pale gray, thick, stiff covers and heavy pages. He wrote in ink, economically and clearly. "I would say I was the hard, accurate worker, neat worker." A ledger was open on a low white table between us. "Fairly neat." We were turning its pages. "Of course, my, because we were in a hurry, my notes don't all look very neat—" Matthaei had a self-depreciating laugh as another man might have had a polite stammer. The pages were very neat. "We were living—we were not living for a museum!" He had invented a complicated system whereby plans for experiments were grouped and subgrouped, lettered and numbered in logical progression in one volume, the results and calculations entered in another volume cross-indexed to the first. "But if you ask me who had the ideas, and who was, ah—now, running the team spiritually, I can only tell you it was both of us," Matthaei said. "When we joined, we were both convinced that the main thing, if we look at cell-free protein synthesis, the main thing which has to be shown is that one can make a specific product. . . . Then it was really the constant struggle for the *better* idea, you know. From when I joined him until the end of our good friendship."

By Monday afternoon, 22 May 1961, Matthaei had worked his way down to an experiment to see if an artificial RNA could stimulate the system to build amino acids into protein. He had poly-U, poly-A, and poly-AU. They had chosen these because they were relatively easy to synthesize, marginally easier to handle. A myth has taken root among many molecular biologists who can recall that year—the myth that Nirenberg and Matthaei tried polyuridylic acid in the system as a control, expecting a negative result, because Heppel happened to have some of the stuff in his freezer and because he or Gordon Tomkins suggested it. In the myth, the discovery that followed was blind luck. But Matthaei made this trial months after Nirenberg scribbled down in his journal the first glimpses of the possibility of using synthetic RNA to stimulate protein synthesis. They had indeed used some of Heppel's artificial RNAs as negative controls, but that was back in March when they were trying to show that their assay did not respond except to informational RNA.

On May 22, Matthaei's experiment was designed the other way around, to detect response even to the most elementary functional message. His preparations now contained sixteen of the amino acids made with the radioactive isotope of carbon, ^{14}C. Both men insisted vehemently that they had expected synthetic RNAs to switch this system on. "Of course. Of course!" Nirenberg said. "I mean—we had an assay for messenger RNA; it worked with viral RNA, the viral RNA was very up; it worked with unfractionated RNA from *E. coli*; and we were looking for simple templates." Matthaei was still more explicit. "Before I did the major experiment, . . . and that's documented in the protocols, I knew *exactly*. . . . If it has just one base, it should code for only one amino acid. And that precisely was proven, and of course we were excited, because we knew *exactly* what we had. And we knew what we'd wanted to get."

Once set up, the work went simply. With simultaneous blank runs, Matthaei established the base levels of radioactivity and incorporation. He had two tubes into which he put poly-U, two for poly-AU, two for poly-A. He began that afternoon at half past three, incubated the tubes for an hour, precipitated the protein out and washed it, and counted the radioactivity. Background subtracted, "Then with poly-U we got almost twelvefold stimulation; the two numbers agree. And then poly-AU gave little stimulation, but it gave some, . . . and poly-A alone was—doubtful, you know. And so *this*" —pointing to the two numbers for poly-U—"was *the* effect."

Matthaei worked day and night for the rest of the week. He had to phone around to obtain labelled amino acids he hadn't already got, and run the experiment with all twenty in at the start. Then he had to find which of the twenty accounted for the radioactivity in the final product. Heppel had given them a milligram—less than four hundred-thousandths of an ounce—of poly-U, Matthaei remembered. I said that was a lot. "Today if you know how little you need it's a lot, but in those days not!" Matthaei could not afford to run twenty parallel tubes with nineteen cold and a different hot amino acid in each, but tested them in groups. By late Thursday night he was down to three candidates, and was testing phenylalanine and tyrosine in one tube, lysine in another. "And here it says, 'Determine whether phe or tyrosine.' You know, this was the final experiment; I had grouped them all; the final group was these two," Matthaei said. "And now I tested them individually."

He began experiment 27Q at three o'clock in the morning of Saturday, May 27. Put in the nineteen unlabelled amino acids, put in the radioactive phenylalanine, put in precisely ten micrograms—35 hundred-millionths of an ounce—of poly-U. Set up a control with everything in save the poly-U to establish the base level of radioactivity. Set up another control with everything in, poly-U in, add ribonuclease to establish that destroying the RNA prevents incorporation. Put the three tubes—a fifth of a teaspoonful of liquid in each—in a warm-water bath, just under blood heat. Set an alarm clock for sixty minutes. Note all the steps in the ledger, neatly. Wait. Precipitate the protein with trichloracetic acid; wash it; put the three samples onto planchettes; put the planchettes into the radiation counter. Enter the numbers in the spaces provided. "'Heinrich, it's going to have been one of the most exciting times in both of our lives,' he said some time after," Matthaei said. "When we already had sometimes difficult times.

"And when Gordon Tomkins came in, well I guess maybe nine, eight or nine in the morning, I already told him I now know it's only *this* one which is coded. I got a twenty-six-fold stimulation for phenylalanine, and practically nothing for tyrosine." Polyuridylic acid, made entirely, monotonously, from nucleotides with the base uracil, the first artificial RNA that Grunberg-Manago had made, six years earlier, prompted the cell-free system to put together a unique proteinlike substance, but an idiot protein, resembling some artificial fibre—a polypeptide chain composed entirely, monotonously, of one kind of amino acid, phenylalanine. Matthaei had identified the first word of the genetic code—that poly-U translates to poly-phe.

The myth in a more invidious version perceives the blind luck of the discovery not in Nirenberg and Matthaei's intentions but in their experimental conditions. "A breathtaking and likewise fortunate experimental finding in retrospect," Paul Zamecnik called it at a meeting on protein synthesis in 1969. By then it was known that if the trace of magnesium in Matthaei's test tube had been fractionally less, the experiment would not have worked at all. In the lower concentrations of magnesium typical in living cells, ribosomes are able to attach to strands of messenger RNA and begin knitting polypeptide only at one of the triplet sequences of bases AUG or GUG, or less effectively UUG or CUG—of which poly-U obviously has none. But the two men had systematically varied the conditions when they were trying to make the cell-free system stable and responsive to RNA from outside. In the event, the magnesium concentration for the crucial experiments, though nearly half again as high as in Zamecnik's bacterial cell-free system, was precisely what Tissières prescribed. Ah, say others, but poly-U for many reasons was more tractable and more efficient as a messenger than other artificial RNAs, and besides that, poly-phenylalanine was most peculiar stuff, so inert and insoluble a chemical that it precipitated almost of itself and was impossible to miss. But these facts are without force once it is established that poly-U was tried as one in a planned set. They amount to carping that Nirenberg and Matthaei made their discovery the easy way.

Everyone in that lab understood, of course. Beyond, the news spread with surprising slowness. The Cold Spring Harbor Symposium was a week off. That spring, Nirenberg had applied to go to the symposium, but had been turned down: unknown, unpublished. During the week before the symposium, Matthaei went to the talk that Brenner gave at the National Institutes of Health. Szilard was in the audience. Nirenberg was still in California. That was the first time Matthaei heard of the messenger experiments. "And Sydney Brenner made a provocative statement: one couldn't possibly hope to test messenger RNA in a cell-free system! I said, 'Well, how can you know?' Well, he asked back, whether I could tell him anything, any more information, but I didn't do it at that time. So he was mad with me, later." Tomkins went to Cold Spring Harbor, but sat mum. Monod, in his concluding summary of the symposium, ticked off all the evidence about messenger RNA that was publicly available, and then warned, "Formal proof of the structure-determining function of 'mRNA' will be obtained only when the synthesis of a specific protein, known to be controlled by an identified structural gene, is shown to take place in a reconstructed system containing messenger-RNA from genetically competent cells, while all other fractions were prepared from cells known to lack this particular structural gene."

Nirenberg learned of Matthaei's success when he got back from Berkeley. He never did learn of the messenger papers. He never cited them in his own publications. In July 1977, I asked him what reaction he had had to the results Brenner and Jacob had gotten in Meselson's lab. "I'm quite totally unaware of this," Nirenberg said. "Which experiment?" I described it. "That didn't influence me at all." I said that it had been published just when his and Matthaei's crucial work was beginning. "I don't even remember it."

Nirenberg's own trials with tobacco-mosaic virus looked good but were inconclusive. "We *thought* it was the coat protein." Collaboration with Fraenkel-Conrat continued by mail. The evidence remained partial and conspicuously inconsistent, but it could be argued that the viral RNA bore information for much else besides the coat protein. By the spring of 1962, Fraenkel-Conrat and Nirenberg had persuaded themselves to publish a paper, with Matthaei and a postdoctoral colleague of Fraenkel-Conrat's named Akira Tsugita. They came down, "It appears justified to conclude that TMV-RNA directs the synthesis of a protein similar to TMV-protein in a cell-free *E. coli* system." But the characterization had been incomplete, the conclusion was wrong, and several years later the paper had to be formally retracted. Tobacco-mosaic viral RNA is not, in fact, an efficient template for directing the synthesis of viral coat protein. In the infected tobacco cell, the ribosomes cannot read it, because eukaryotic ribosomes, in contrast to those of bacteria, are unable to initiate synthesis at sites within the RNA strand. The tobacco cell synthesizes another, shorter, RNA—a messenger for coat protein.

Months before the crucial discovery, Matthaei had enrolled for a two-week course on bacterial genetics at Cold Spring Harbor that started immediately after the symposium. He prepared a good quantity of the product of the poly-U reaction in order to start proving that it was indeed made up of phenylalanine put together as a polypeptide. Now he had to hand the work over to Nirenberg. "It turned out polyphenylalanine has *so unusual* properties that it must be terribly easy to characterize it. And I told Marshall, I want to do that, and then he took it away and—to do it himself." The phrase was flat. It had echoed many times in his mind. "He might still have given me the chance to do it; well, I wouldn't have minded him to do that; the only point is, I knew he was sloppy, and—and you know, if you like your material and if you like to— You know I often got *very clear-cut* results." They wanted to publish the work as a bloc, and that took two months' intensive labor with controls, checks, refinements. A pair of papers—one on the technical requirements of their system, the other telling what happened when they put RNAs into it—was sent to *Proceedings of the National Academy of Sciences* on August 3.

❖❖❖

"Marshall was not on the grapevines; it wouldn't have been his personal inclination to talk a lot anyway; he's not that way," Maxine Singer said. "There was no reason why anybody should have known about Marshall. He was one of this young, new NIH crew. At that time NIH didn't have the reputation as a scientific institution it has now. The impression university people had was that it must be just another government one-shot sort of thing. So Marshall was very much not part of all that. He did not come out of the phage group—just had no connections with those people." As biologists got ready for Moscow, only the faintest and most coincidental whispers were in the air. Tomkins was in Boston June 30, and talked about Nirenberg's discovery to Alex Rich; but Rich had four papers of his own to complete before leaving in the second week of July for Europe, where he drove with friends, sleeping bags, and tents from Copenhagen through Scandinavia to Moscow. "I recall very well that we didn't know anything about Nirenberg," Tissières said. "People had told me, 'Nirenberg is doing experiments, but you can't believe what he is doing, anyway.' Anyway, we go to Moscow, first person I meet in Moscow, I think I meet Jim Watson and Wally Gilbert. And they tell me there is something tremendously important. Okay. So, we go I think the following day to this paper. They had talked to Nirenberg. We go to this paper of Nirenberg, there was *nobody*." Tissières, Gilbert, and Meselson were almost alone.

Nirenberg had about fifteen minutes. Nobody listening to him that day in Moscow was likely to have been reminded of Oswald Avery's careful proofs, nearly twenty years before, that the transforming principle of pneumococci could be nothing else than DNA. Scientists mostly consider an interest in the history of their subject as a symptom of declining powers. Besides, the flannel-tongued compression of Nirenberg's report, its reliance on scientific grunt and gesture (the tables and charts he showed, which take up more inches in the printed version than the words of the text), and the six-months' scramble of experiments it presented did not recall the strength and patience of the earlier work. Yet in almost every way the one demonstration was the lineal descendant of the other. The microbiological methods, the enzymatic tools, the types and multiplicity of controls—taken all together, the form of proof—that Nirenberg and Matthaei deployed were Avery's, extended and made routine in the years since. Their aim was Avery's too: to understand the specificity of the nucleic acids, by putting in something known to see what came out. Avery's proof that the transforming principle was DNA set the agenda of biology henceforth; Watson and Crick's elucidation of the structure of DNA confirmed rather than concluded that agenda. But the questions that Avery had put, Nirenberg and Matthaei began at last to answer. The comprehensibility of DNA as the transforming principle, the content of the aphorism that DNA makes RNA makes protein, became plain in their demonstration of the first word to be identified in the genetic code: "One or more uridylic acid residues appear to be the code for phenylalanine."

I said to Meselson that Crick did not recall Nirenberg's presentation of the paper in the Moscow classroom. "Crick wasn't there. I heard the talk," Meselson said. "I think that no one whom I knew well heard it. I specifically remember that Francis was not there. I heard the talk. And I was bowled over by it.

You know, there's a terrible snobbery—but it's self-protective. But there is a terrible snobbery that either a person who's speaking is someone who's in the club and you know him, or else his results are unlikely to be correct. And here was some guy named Marshall Nirenberg; his results were unlikely to be correct, because he wasn't in the club. And nobody bothered to be there to hear him. Anyway, I was bowled over by the results, and I went and chased down Francis, and told him that he must have a private talk with this man."

Meselson went to Crick. Crick had not heard of Nirenberg or Matthaei. "I was in Tangiers for six weeks before the Moscow meeting," Crick said, a short while before I talked to Meselson about it. "What I remember very clearly is running into Nirenberg, and having heard that he had *done* it—it could well be Meselson who told me—and talking about it with him and arranging for him to give it again to a wider audience. I was the chairman of a session there, and I arranged with the organizers that Marshall Nirenberg would be added to the program."

Crick went to Perutz and Engelhardt. "Francis, I think, had the last session," Perutz said. "I added Nirenberg to the program—as I remember, he came at the very end; at least certainly it was a tremendous climax." Matthaei had not come to Moscow. "While Marshall was reporting in Moscow, he called me on the phone," Matthaei said; Nirenberg did not specifically remember the phone call. "I suggested to Marshall that we should add a note in proof, which states that we now found another coding polynucleotide." Matthaei had determined that polycytidylic acid—the one whose every base is cytosine—directed the building of another monotonous polypeptide, in which every amino acid is proline. The second word in the dictionary was that poly-C makes poly-pro. They added some fine print to say so at the end of the second paper in *Proceedings of the National Academy of Sciences*. Nirenberg's second reading in Moscow took place in the great hall, to hundreds of scientists. His delivery, this time, "electrified the audience," Crick later wrote. (Whereupon, in the interest of historical accuracy, Seymour Benzer, who had been in Moscow too, mailed Crick a photograph taken of that audience, in which several people appeared to be asleep.)

"I don't remember anything else important happening there besides Nirenberg's announcement," Meselson said. "I went to visit Lysenko's lab; I wanted to see it. He wasn't there; there were cobwebs all over the place. . . . And I remember, I ran up to Nirenberg—yes, it must have been after the big presentation, not the little one—I ran up to Nirenberg and I embraced him. And congratulated him. Something which I normally—it's not my nature. It was partly also a feeling that he had been unjustly treated and I felt a little bad about that. Not very unjustly treated—but. It was all very dramatic. And it gave some people who were in this field the immediate itch *to get out of Moscow*, to get back to the lab."

❖ ❖ ❖

Jacob and Gros, who read papers that same day in Moscow, heard Nirenberg's talk with chagrin. "We tried, François Gros and myself, to extract a specific messenger which was a galactosidase messenger, and have him translated *in vitro*," Jacob said. All through the previous winter, he and Gros had attempted to isolate messenger RNA from induced *E. coli*, put it into a cell-free system, and make enzyme. "Which we never got," he said. "But we used to talk as a *joke* of putting poly-A or poly-U—" He laughed. "Oh, yes; but it was a joke; I mean we were absolutely convinced that nothing would have come from that." Tissières felt a congruent chagrin. "Of course, the idiotic thing as far as *we* were concerned, we had Paul Doty in the lab next door," he said. "We were on very good terms; we were talking about science constantly, with Paul Doty. He *had* poly-U. But we never put it in." Brenner had not gone to Moscow. "But I had chaps doing this *here*, you see," he said. "Doing it with turnip-yellow-mosaic virus. You see, we said if the messenger was true, it should be possible to add a message to ribosomes." But had they not used artificial RNAs? "No, we didn't use— But I mean *that* thing, that was done as a control," Brenner said. "Because they were doing it with tobacco-mosaic virus. And poly-U was suggested by Gordon Tomkins as a control. Okay? It didn't occur to us to use synthetic polymers." Severo Ochoa had not gone to Moscow. He had intended to use artificial RNAs as messengers, and was even then refining a cell-free system like Nirenberg's. "When we heard the news from Moscow we *immediately* tried it. And other polymers, copolymers we had in the icebox. We got immediate results with four or five."

❖❖❖

When Crick arrived home, he found that Brenner had not got around to the experiments combining acridine mutants in the *r*II region, the mutants that suppressed them, the suppressors of suppressors—different levels in the hierarchy of the *FC* family—to test the theory that the dyes acted by adding or deleting bases in the DNA of the phage. "When I came back from Moscow, Sydney hadn't *done* uncles and aunts. So I started doing it, putting the pluses and the minuses together"—he pointed to the diagram again—"to show that two pluses didn't work, and that pluses and minuses often worked." To state the case more fully, the experiments soon found that any two mutants, each of which could separately make a third revert to the wild type, could not suppress each other; that a mutant that could suppress another could suppress most mutants at the same level of the hierarchy as that other; and so on. All the evidence fit the simple idea that an acridine mutation, by adding or deleting a base, shifted the reading of the ensuing sequence so that different amino acids were called for. Whether the original mutant was in fact an addition or a deletion, its suppressor was the opposite and so put the reading back into phase, while *its* suppressor worked in the same direction as the original mutant, and so forth. They began to talk about the "reading frame"; then or later they called all these "frameshift mutations."

In the course of breeding scores of double pluses and double minuses and checking the glass dishes to see, as theory predicted, that none of these recombinations had produced the wild type, Brenner one day noticed a small, fuzzy plaque among all the plates of large, clear, *r* plaques—a strain of phage unmistakably of the wild type. The only saving explanation was that one of the doubles had reverted spontaneously. "And then we realized that *of course* you could revert *this* way," Crick said—that if the code was a triplet code and the messenger read from a fixed start, then a third added base, or a third deletion, could combine with the first two to restore the reading to the correct frame. "Sydney noticed that the double plus had reverted, and he tried to unscramble

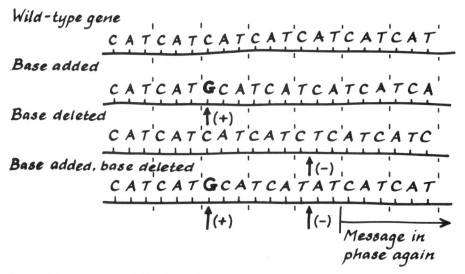

it. And it was very difficult. It had reverted extremely closely." In order to prove that the wild type had been restored by three separate mutations that shifted the frame in the same direction, the new, spontaneous, third one had to be split from the others and isolated by back-crosses; but the new one was so near on the DNA to the others that this was hard to do.

"So we decided that rather than do that, we'd have to make some triples." Crick reached for another sheet of paper. "I can just do it on the back of this. And what you've got to construct, you see, is a plus-plus-plus." The diagram was yet simpler than the last.

"But you see, there's no way of doing that." Crossing two double pluses together would produce a quadruple, still a mutant, not a triple. "So what you do, by hard work, you construct that, and then you construct *that*—two separate ones."

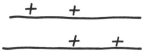

Two different double pluses had to be found that shared one mutational site. There was no way to breed them intentionally, no trick to select them preferentially. Thousands of plaques were examined, thousands of mapping crosses made. "It's very tedious," Crick said. "But all you have to do, once you've gone through all the labor of constructing that, and constructing *that*, these two"—pointing to the shared mutational site—"being the same, you just have to do *one cross*."

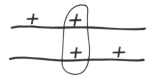

Brenner went to Paris again that fall, to do an experiment with Jacob: they wanted to settle the lingering question whether induction of β-galactosidase really caused the bacterium to make new, extra messenger RNA. "I mean it sounds odd," Brenner once said. "That was in fact an experiment that we thought terribly important. It was the closing-up of the PaJaMo loop." The experiment worked fine.

Before he left Cambridge, Brenner wrote down for Crick three sets of possible crosses of double-plus mutants. "Then he went off to Paris," Crick said. "Well, we hadn't worked out then the difference between shifts of frame to the right and shifts of frame to the left. Although I knew that there was a difference, I'd never thought about it. So then I checked whether these three crosses would work, and I predicted correctly that two of them *wouldn't* work, because they'd have produced a nonsense shift. So we just made the other one." The phage lab was still set up in the corridor of the annex to the zoological museum, across the courtyard from the Cavendish. Leslie Barnett did the experiment. "And all we had to do was look at one plate. And see if it had any plaques on it. So we came in late at night, ten o'clock at night, or something," Crick said. "And there were the plaques on the plate! So I said to Leslie, 'Let me check; we may have got the plates mixed up,' and she checked it, and then I told her, 'We're the only two know it's a triplet code!'"

They got a paper into the last week's issue of *Nature* for that year: "The General Nature of the Genetic Code for Proteins." The fundamental issues that Crick, Gamow, Watson, and Brenner had embarked upon in the summer of 1954 were now resolved. "The genetic code is of the following general type," the paper declared, and went on:

(a) A group of three bases (or, less likely, a multiple of three bases) codes one amino-acid.

(b) The code is not of the overlapping type....

(c) The sequence of the bases is read from a fixed starting point. This determines how the long sequences of bases are to be correctly read off as triplets. There are no special 'commas' to show how to select the right triplets. If the starting point is displaced by one base, then the reading into triplets is displaced, and thus becomes incorrect.

(d) The code is probably 'degenerate'; that is, in general, one particular amino-acid can be coded by one of several triplets of bases.

After the uncles and aunts, decorously renamed, the conclusion:

At the recent Biochemical Congress at Moscow, the audience of Symposium I was startled by the announcement of Nirenberg that he and Matthaei had produced polyphenylalanine . . . by adding polyuridylic acid . . . to a cell-free system which can synthesize protein. This implies that a sequence of uracils codes for phenylalanine, and our work suggests that it is probably a triplet of uracils.

It is possible by various devices, either chemical or enzymatic, to synthesize polyribonucleotides with defined or partly defined sequences. If these, too, will produce specific polypeptides, the coding problem is wide open for experimental attack, and in fact many laboratories, including our own, are already working on the problem. If the coding ratio is indeed 3 . . . and if the code is the same throughout Nature, then the genetic code may well be solved within a year.

In another conversation, Crick reflected upon that paper and the work that went into it. The paper and the work are admired throughout the science as a classic of intellectual clarity, precision, and force. "Essentially, we had the idea that acridines might add or delete bases; we didn't actually go and do work then to test the hypothesis. I *was driven* to try and test loopy codes. By the time I had got there, I had time on my hands and could learn to do it; I had learned about the *r*II system—and *that* led us to realize that we could study the acridiné mutants by this thing, because we'd got suppressors from acridine mutants. So that was the origin of it. It's the usual business—it seems very straightforward, and when you actually look what happened, you did it for a lot of *silly* reasons"—he laughed—"that led you to the right thing.

"I don't think all that was terribly important, you know," Crick then said. "I mean, people say it showed it was a triplet code, but it was pretty obvious it was *likely* to be a triplet code. I think it *was* an important contribution to mutagenesis. But the fact is, if we'd shown that the code was a *quadruplet* code, *that* would have been a discovery. But since everybody who knew assumed that it couldn't be doublet, there were too many amino acids, and therefore it was probably triplet, and since anyhow the triplet nature would have come out in the chemical work before long, although it did hot up the subject and make it feel it was going somewhere, I think you could have deleted the whole work and the issue of the genetic code would not have been very different. It would not have affected Nirenberg's discovery and most of the other work. This I think is the test, you know, that historians should apply. If you delete a bit of work, would it make a difference?"

Perhaps not, for the code. Yet in the line of experiments and theoretical observations that stretched back to that first suggestion about tautomeric shifts and base mispairing in the second paper with Watson on the structure of DNA, Crick set the theory of mutagenesis on its foundations. And the theory of mutagenesis has produced, among other things, the most important practical benefit from molecular biology so far. Bruce Ames, three years after his sabbatical visit to the Cavendish, began collecting and classifying mutations and mutagens. He found that some known carcinogens were potent bacterial mutagens—though others were not. To develop a system, using *Salmonella* rather than *E. coli*, to pick up mutations induced by substances suspected of causing cancer, he made several crucial improvements. He took up an earlier suggestion that cancers may be caused not simply by a chemical from outside but by the intermediates and derivatives that mammalian cells may metabolize the original chemical into. Therefore, to supply that metabolic machinery, he added to his plates of agar a purée of fresh raw rat liver. Also, he found a bacterial strain that was thin-skinned—that is, it lacked the polysaccharide capsule, outside the membrane, that protects normal *Salmonella*—so that substances penetrated more easily. With that, and fine tuning, Ames had what has proved to be a fast, cheap, simple, and sensitive way to screen chemicals to see whether they may cause cancer. By the late seventies, the Ames test was in use in industrial and governmental laboratories throughout the Western world.

Not one year but five were needed to solve the rest of the genetic code. At New York University, Ochoa threw his laboratory—he had the enzymes and the skill—into synthesizing artificial RNAs of many sorts of compositions, and as fast as he got them he tested them in the cell-free system. Nirenberg came back from Moscow and did with Matthaei and another colleague a small, neat experiment that showed that when poly-U directed the synthesis of poly-phe, the transfer RNA—Crick's adaptor—specific to phenylalanine was indeed an intermediate. That fall, even as that work was being finished, Nirenberg went to a meeting at the New York Academy of Sciences where Ochoa gave a talk. "It floored me, that Ochoa had made such advances," Nirenberg said. "Things developed much faster than I would ever have dreamed that they'd develop." In Bethesda, Nirenberg turned to Heppel and Singer for more artificial RNAs and for instruction in their biochemistry, and they began making new ones. The competition between Nirenberg and Ochoa became as intense as any in the history of science. Matthaei, discontented with his share of the direction, the work, and the credit, went back to Germany in April of 1962 to try decoding independently. "I felt I should start working on my own; it was too unpleasant." Brenner coined the term "codon" for a triplet of bases that specifies an amino acid; introduced partly in satirical reference to Benzer's "cistron," "recon," and "muton," Brenner's "codon" is the one that survives in universal biological use. Crick never went back to experiments but urged on all contestants by letters and visits.

Chemical difficulties aside—but some of these, like the recalcitrance of poly-G, were fierce—Nirenberg, Ochoa, and others who joined the race had two problems. First, the RNAs of mixed bases that they were using were random, known only in overall proportions; these could not be used to distinguish specific triplets—for example, to pick out the correct codon among UUC, UCU, and CUU, when the cell-free system responded to poly-UC by incorporating the amino acid serine. But second, the race itself generated misdirection, sloppy work, and hasty publication. At the end of 1962, Crick wrote a fifty-four-page review—with the title, more astringent and dandified than ever, "The Recent Excitement in the Coding Problem"—in which at one point he said, in palpable contempt of work ill conceived, "There are so many criticisms to be brought against this type of experiment that one hardly knows where to begin." Well begun, he ticked them off by entire categories, worst of which were "the necessity for the rigorous experimental proof of each codon" and the failure to allow for synonyms, the degeneracy of the code.

In December of 1962, Crick, Watson, and Maurice Wilkins went to Stockholm to receive the Nobel prize in physiology or medicine, Perutz and Kendrew to receive the Nobel prize in chemistry, John Steinbeck to receive the prize for literature. Crick's prize address was "On the Genetic Code," but it was not an occasion for castigating his colleagues.

The sixth International Congress of Biochemistry met in New York City in the summer of 1964, and there Nirenberg announced a second discovery, once more a technique that broke the impasse. With Philip Leder, a postdoc in his lab, Nirenberg found that lengths of artificial RNA as short as three bases—trinucleotides—when put into the cell-free system instead of a longer message, were enough to make the ribosomes bind with the kind of transfer-RNA molecule complementary to that one codon and carrying the specific amino acid called for by that one-word message. The essential technical trick was to label one species of amino acid at a time and to put the incubated mixture through a Millipore filter, trade name for a filter made of cellulose nitrate with pores in this case less than half a micron in size, meaning less than eighteen millionths of an inch. The filter, they found, held back the ribosomes with the triplets and any transfer RNA bound to them, but let unbound transfer RNAs pass through. The radioactivity on the filter was then counted. The control was to repeat the experiment without the trinucleotide. Meanwhile, Har Gobind Khorana, at the University of Wisconsin, had perfected the precise and intricate biochemistry needed to make long strands of RNA with known, simple, repeating sequences. His first was the messenger UCUCUC . . ., which produced the polypeptide serine-leucine-serine-leucine . . . .

Between Nirenberg's technique and Khorana's, the code was almost entirely elucidated by the time of the Cold Spring Harbor Symposium of 1966. All twenty amino acids were accounted for, leucine and arginine each having six different synonymous codons, the others ranging down to methionine and tryptophan each with a single codon. Crick proposed what has become the standard form in which to present the genetic code, a table in which the first, second, and third bases of the sixty-four triplets are arranged along the left, top, and right margins, marking off columns and rows that intersect in the body of the table at the amino acid for each codon. The importance of the table for biology has been compared to that of the periodic table of the elements for chemistry.

Three codons, the triplets UAA, UAG, and UGA, had no amino acids assigned to them. One by one, in experiments in phage genetics by Brenner and independently by Alan Garen at Yale, and then last by Brenner and Crick in 1967, these three triplets were proved to be nonsense codons, whose function was to signal the end of the polypeptide chain. These experiments were the exuberant last exfoliation of molecular genetics at its rococo extreme. "Finding out about the machinery without touching the biochemistry," Brenner described it. "There was a culture—well, a *cult*, almost—that became typical of molecular biology. What became prized was ingenuity. You know? And I think that an important thing about the best of molecular biology would be those papers which, although they are experimentally intricate, nevertheless do things in such a way that the kind of deduction about what is inside the animal, the organism, is unique. Okay? And of course, the simpler the methods used, the more highly prized. In other words, the *r*II genetics, the acridine mutants, the nonsense mutants, will always be a classic: why? Because it was just done on bits of paper. . . . I mean, it doesn't need all the bloody tubes and counters and so on. And I think that the cult got founded around these ideas of how to solve the code without ever opening the black box. Okay? I mean, Gamow had the idea you could do it without doing *anything*. But I mean, we realized you had to do *some* experiments." And he added, "What has finally happened is that *all* of that has been demolished by the fact that in the last five years we are learning to do molecular genetics *directly* by sequencing the DNA. Sanger, and these people, who show how you actually sequence the DNA and find the base change. Whereas the attempts to use genetics could be interpreted as just cheap ways of trying to sequence the DNA."

In 1968, Nirenberg, Khorana, and Robert Holley won the Nobel prize in physiology or medicine. Several years later, I asked Meselson if Nirenberg was now a member of the club. He paused for a long time, and then said, "I don't know." Then he said, "Anyway—sure; in a sense, he has a club of his own."

2nd → 1st ↓	U	C	A	G	3rd ↓
U	Phe Phe Leu Leu	Ser Ser Ser Ser	Tyr Tyr Non2(ochre) Non1(amber)	Cys Cys Non3 Trp	U C A G
C	Leu Leu Leu Leu	Pro Pro Pro Pro	His His Gln Gln	Arg Arg Arg Arg	U C A G
A	Ile Ile Ile Met	Thr Thr Thr Thr	Asn Asn Lys Lys	Ser Ser Arg Arg	U C A G
G	Val Val Val Val	Ala Ala Ala Ala	Asp Asp Glu Glu	Gly Gly Gly Gly	U C A G

The compact table in which the code is always displayed was proposed by Francis Crick. The four standard bases of RNA—uracil, cytosine, adenine, and guanine—are shown as U, C, A, and G. The initial base of any triplet is indicated on the left, the second base along the top, the third down the right within the box defined by the first two. The triplets UAA, UAG, and UGA signal the termination of the polypeptide chain, and were called nonsense codons when it was realized, from the growth patterns of certain phage mutants, that they do not signify any amino acid at all; the names amber and ochre, for two of them, perpetuate accidents of terminology in the laboratories where the mutants were first found.

III

PROTEIN

Structure and Function:
The solution of how protein
molecules work

9   *"As always, I was driven on*
by wild expectations."

❖α❖   A scientist, as he grows older, may go one of several ways. He may, and many
do, withdraw from the laboratory and turn to something else. Possibly he will
remove no farther than to supervise the work of younger colleagues. He may
assume the direction of a large laboratory, perhaps one of the grand institutions. He may start to write books. He may be taken up to deliberate at the
committee tables, to murmur down the telephones of the politics of big
science. The scientist will offer various reasons for the change, but an underlying fact is that active research demands so much uninterrupted time, such
stamina, such intense concentration. "I'm getting perhaps too lazy to do the
really hard work you have to do in the lab to make wet experiments go," a
friend said. "Although grad students and postdocs in my lab do those experiments, I probably am not going to do many more like that. Because to do good
molecular biological experiments you have to be in the lab day and night; you
can't just do it on an eight-hour day. It's just not possible."

Not all great scientists take that way, to be sure. Of the famous pairs of
molecular biologists, James Watson abandoned research of his own even as
early as the time he got his Nobel prize, in 1962, found it more profitable to
write *The Double Helix* and the multiple editions of his textbook and then more
gratifying to be boss of Cold Spring Harbor. Francis Crick continued to do
science much as he had always done it, although later he transferred his atten-
tions to neurobiology. Jacques Monod, soon after he got his Nobel prize, in
1965, became preoccupied by his vision of what the science signified, which
led him to write *Chance and Necessity*; then in 1971 he took over direction of the
Institut Pasteur to rescue the place from administrative disorder and financial
crisis. François Jacob elected that most difficult intellectual climacteric, to learn
a second newly developing science, and so began to teach himself to grow
mammalian cancer cells in glass dishes as though they were newly fertilized
embryos—hardest of all wet experiments then and still. John Kendrew, the
first man to determine the three-dimensional structure of a protein down to
the locations of its individual atoms, soon after his Nobel prize became an adviser to governments; when his advice led to creation of a European Molecular

475

Biology Laboratory in Heidelberg, he became its first director. Kendrew's colleague of a quarter century, Max Perutz, saw his tiny unit within the Cavendish Laboratory grow to an enormous, independent institution; yet Perutz told me, more than once, "I'm a casual administrator. I try to do it all in passing in the corridors or at lunch in the canteen. I'm still determined to remain an active research scientist."

He who stays in the laboratory may yet make a stealthier and more unhappy withdrawal. Power and fame or, more mildly, a sense of new responsibilities, may have attracted those who give up doing science themselves; the demands of experimentation may have become unacceptable; but deeper than any other reason for stepping back is the inexorable loss of faculty, loss of luck, loss of will. At lunch one day in Paris, early in December of 1975, I asked Monod whether he missed doing research directly. "Oh, I miss it," he said; then what began as a shrug became instantaneously more thoughtful. "I do more than miss it. It's too short a question." He paused, began again. "No, I don't know that it is actually working at the bench that I miss—miss so very much, although I do, at times; but it is in fact not being in this permanent contact with what's going on in science, *in the doing,* which I do miss." I was reminded of a parallel conversation in which Watson had tried to claim the opposite, that he could stay close to what was happening in science. But if one was not actively working, Monod said, "Then you don't have that. And also if you're overburdened with general responsibilities, it becomes not so much a question of time but your subjective preoccupations. There's a displacement—the internal conversation that you keep running in your head concerns all sorts of subjects, things that have got to be done, rather than just thinking about situations [in research]. That's what bothers me most."

When his term as director was up? "No, it's too late to go back to research." Why? Monod paused once more, and then said, "Well, you know, I always had a sort of amused and—amused, pitiful sympathy for the wonderful old guys who were still doing something at the bench when it was quite clear that whatever they did, it would be less than one hundredth of what they had been able to do before." We spoke of examples—of scientists whose work became gradually less interesting as they aged, of others who lost their critical judgement and fooled themselves into believing they had solved problems that were beyond them.

There had been counter-examples, I suggested. "Very few," Monod said. "Can you mention one?" We talked for a moment about Oswald Avery. Louis Pasteur was an obvious instance. Then I mentioned Perutz. "*Max* is—yes," Monod said, conceding with evident pleasure. A minute later, Monod said, "But of course, if one has done a piece of work that really has marked the trend of scientific history in his—time, it is very difficult to do that *twice.* For instance, ever since Fritz Lipmann started working in protein synthesis he has been doing some very good work. But it is still of very little historical importance as compared with the fact that he really understood the energetics of the chemistry of life. The fact that he understood what ATP is about is really the thing that marked his whole career."

A couple of months before that, I had lunch with Roger Kornberg, a biochemist a full generation younger, for whom the problems and the men of the fifties were of another era—except that he grew up with them, as he is the eldest son of Arthur Kornberg, whose Nobel prize was for the first discovery of enzymes by which DNA replicates. At the age of twelve, Roger was in his father's lab at Stanford doing biochemical experiments, thinking it play. At twenty-eight in 1975, he was just going to Harvard after four years' work with Crick and Aaron Klug on the problem of the structure of chromosomes in higher organisms, where DNA is intricately knotted with protein. He was tall; moved lightly; had a long face, a long thin nose, a wide and expressive mouth, a graceful, subtle, welcoming intelligence.

We had begun, that day, with the difference that molecular biology had made in the sort of explanation that is acceptable. "What matters is, what constitutes a solution to a problem in various people's minds?" Kornberg said. "And I think there the distinction doesn't lie between molecular biology and biochemistry—because it is true that there was a school of molecular biology, that of Delbrück, Luria, and so on, that was mainly concerned with the genetic side and put forward a formalism in the same sense that the biochemists of the thirties were solving *their* problems in terms of a formalism. But there *is* a clear distinction—it being that for people who either think in terms of genetics, as opposed to structure, or who think in terms of biochemistry as a sequence of chemical reactions, as opposed to structure, it is true that the problem is solved when they have defined all the components. Or, in the case of genetics, when they have mapped all the components, and defined their relation to one another.

"And it really is true, that with Pauling and the advent of structural chemistry in molecular biology or biochemistry, there has been a change in what some people are willing to accept as a sufficient explanation, from the formalistic side to a kind of mechanical one. In other words, from being able to define all the components and feel that you've solved the problem because you've encompassed it, and are not missing any of the ingredients, to the other side where you say, But the transformation is a mechanically almost inconceivable one. How do you do it with nuts and bolts; how do you do it with squares and blocks and the sorts of things that we know molecules are made of?"

Out of the biochemical shadows emerged the mechanics, I said: Erwin Chargaff damned molecular biology as oversimplification, called it a sign of the fall of civilization, because, indeed, it marks the end of a certain long and civilized tradition in biology. "It's the end of a certain form of abstraction," Kornberg said. "It removes another level of mystery from life, and Chargaff is a great lover of mystery." Could a scientist, could Chargaff, be that, except as a rhetorical posture? "Oh, but—that's one of the most remarkable characteristics of most scientists," Kornberg then said. "At some level, they fondle their problems."

The idea startled me. "Oh, but it's true, they do!" Kornberg said. "Almost every scientist winds up working on a problem he can't bear to solve. And that's where his life in science ends. It's probably being very cruel to the older scientists, but I really believe it's true. Or sometimes it's a gradual loss of energy, of the ability to focus the energy on the problem. Or perhaps it's a loss of edge—of the *hunger*. Some younger scientists—a few—have that quality that Francis has exemplified; he was *ruthless* in solving problems, I mean he would just carve them up and solve them in the most brutal way, much to the dismay of people like Chargaff who enjoyed the mystery of those problems and didn't want to see it disappear, to them the mystery was the beauty of it. . . . It proba-. bly does happen to all aging scientists."

Yet the opposition between types of explanation—chemical formalism and molecular dynamics—should not be pressed too far, Kornberg said. "It's interesting—you can see the movement both ways." He named biochemists gone molecular. "And in the other direction, you look at someone like Max Perutz, whose previous interest was purely structural, in knowing where all the nuts and bolts were. And now he's thinking more in terms of the binding of oxygen in hemoglobin, and he's doing spectroscopy and taking other experimental approaches to hemoglobin that were previously entirely in the province of the chemical people rather than the structural people. So, though I think the distinction is important, and is real, one has to be careful not—as people often do—to confine a person to an approach." Perutz's pursuit of the relation of structure to function in the hemoglobin molecule, I said, had been growing in power and interest for the best part of forty years. "Fantastic. Max is a most impressive man," Kornberg said. "Someone like Francis sprang fully armed from the head of DNA. And Max—as you say!—is a developer. He started out without any intellectual or experimental equipment for the work in hand and just seems to acquire it, year by year."

One day in Cambridge at the end of a long discussion with Kendrew—in 1971, before his new lab at Heidelberg had been created—I mentioned Perutz's determination to stay active in research. Kendrew nodded, with a slight smile. "This is right about Max, you know, and I think he wants to go on doing the same thing," he said. He was leaning back, smoking a pipe, feet up; the cut of his tweed, the trim of his white hair, the clip of his speech were unobtrusively established. "I happened to come back from a meeting in London last Saturday, and came in here in the evening to pick up the mail, and there was Max in a white coat mounting a crystal; the thought occurred to me that if I had come into the lab thirty years ago, on a Saturday evening, Max would have been in a white coat mounting a crystal—*just* the same. And he loves it, of course, and nothing would persuade him to do differently." Kendrew's voice became wistful. "Other people want to do different things, and they may be other things in research or they may be outside research."

Perutz and I first met early in May of 1968—for him a time of mingled triumph and frustration, though neither was evident in his quiet clarity. Over the telephone, he had been reluctant to appoint a time to talk; just back from the States, very busy. But that was the shy spring in English biology after the publication of Watson's memoir. Crick, I had already been told, was touring Greece for the next five weeks. Perutz's defense was to ask for a fifty guinea donation to the laboratory's recreation fund. His voice was deep, somewhat muffled, and he spoke with a slight Viennese accent. I went up to Cambridge on a fine morning several days later. When Perutz came out of his office to meet me, he was wearing a light cardigan that he later changed for a white lab coat, and had on open leather sandals over wool socks. He was short, thin, sallow, with a fringe of dark hair around a notable cranium; he had whippy wrists and large, broad-fingered, working hands; the eyes behind his spectacles were small, deep-set, and brown. A worn leather book satchel was open by his desk. The floor it stood on was brown cork tile.

Perhaps the most spectacular advances in molecular biology in the last year or so, Perutz said, had been in determining the structures of enzymes by protein crystallography—and x-ray crystallography of proteins was his own consuming interest. "One hopes that crystallographers might be able to do for enzymology in the second half of this century what they did for mineralogy in the first, bringing a logical system based on atomic structure into this science. But enzymes being a thousand times more complex than minerals, it will be more difficult and will take more time."

How do protein structures compare in difficulty with DNA? "Probably again a factor of a thousand. DNA was comparatively simple and could be solved by the method of trial. There was only a little information from the x-ray-diffraction pattern. What Watson and Crick had was three independent values—the width of the double helix, the distance between the stacked parallel bases, and the height of one turn of the helix. On the basis of these three, they managed to solve the structure by model-building. Solving the structure of a crystalline protein by this method is impossible, because it does not repeat. The essential structural feature of DNA is its repetitiveness. The same pattern repeats periodically around the fibre axis, and the only nonperiodic part is the base sequence—which in fact we cannot solve by x-ray analysis. Enzymes do not have a periodic structure; the amino acids are laid down by the information in the genetic code in a nonperiodic manner and the structures fold up in a completely irregular way. I shall show you some protein structures presently that will give you an idea of the degree of complexity. So that to determine one of their structures involves determining several thousand independent values. Say they contain between a thousand and ten thousand atoms. And each atom is located in space by three coordinates. And there's no way around this—you've got to slog it out."

To say that molecular biology was developing from easier problems to harder ones was trivial in itself; but the development in the case of this science had a pleasing logical orderliness as well. DNA had conceptual priority as powerful as it was obvious. Perutz agreed that this was so. "Genes have only one function: to encode the structure of proteins. But turning this around, enzymes are but the expression of the genetic code. Having solved the genetic code, you want to know what is being said. The gene, in the code, lays down the sequence in which the twenty different kinds of amino acids are connected on the protein chain. The protein chain may contain several hundred amino acids. That is all the gene does. But the protein folds up into a very specific and extremely complex three-dimensional lacework. This folding up appears to be inherent in the sequence of amino acids. The chain can't help itself."

When the way had been found to get out the structure of one enzyme, did that illuminate the structures of others? Did structure explain function? "You may have a method for solving protein structures, but it is still a difficult and laborious job to solve any enzyme in particular," Perutz said. "In 1957, when Kendrew got the first glimpse of the structure of the first protein to be solved, myoglobin, a friend said to me, 'Surely when the structure of one protein was solved, then we'll know what they're like.' Recent results have shown how ridiculous this was." Myoglobin is the substance that stores oxygen in muscle tissue. The next protein structure to be determined was hemoglobin, two years later by Perutz himself, a molecule closely related to myoglobin but four times as large. These first structures were not finely detailed. The looping paths of polypeptide chains could be seen unmistakably, but individual amino-acid residues and the alignment of their side chains could not be distinguished. The next protein and first enzyme to be solved was lysozyme, in 1965, by David Phillips, a Welshman working under Sir Lawrence Bragg's supervision and with Kendrew's help at the Royal Institution, in London. Phillips got his structure to atomic scale, 2.0 angstroms' resolution; the same had by then been achieved for myoglobin. Protein crystallography had been almost exclusively a British specialty. But in the past twelve months, Perutz said, four more enzyme structures had been published. One was ribonuclease, solved by Frederick Richards, at Yale. The other three were enzymes that digested proteins by cleaving peptide bonds between particular amino acids: carboxypeptidase, solved by William Lipscomb and collaborators at Harvard; chymotrypsin, solved by David Blow and Paul Sigler in Perutz's lab; and papain—first found in the juice of unripe papayas, whence the name, and widely used as a meat tenderizer—by Jan Drenth in Groningen, Holland.

"What structural analyses do is to teach us how biological catalysts work. Knowing how *one* works doesn't really tell us anything, yet, about how any of the others do," Perutz said. "Yet they are probably not all of different types. It seems likely that there are certain fundamental patterns which are repeated in related enzymes, and this is what one would like to find out. There is already some evidence of similarities. Myoglobin has a chain that is folded in a way very similar to the folding of hemoglobin—which my colleagues and I have now solved, here, to atomic resolution. Then the group at Munich recently solved the structure of insect hemoglobin, which is genetically *enormously* far removed from these mammalian hemoglobins"—in evolutionary terms about a billion years separated them from their common origin—"and it turns out that there the chain is again folded in exactly the same way as it is in the mammalian myoglobin and hemoglobin. Another which cropped up recently is an enzyme contained in milk, called alphalactalbumin, concerned with the synthesis of milk sugar; this turns out to have a structure similar to lysozyme—which is, of course, a quite different enzyme, but which is also concerned with sugars, with the breakdown of polysaccharides," as in the cell walls of invading bacteria. "Then there are indications that among the enzymes that break down proteins, there is a whole group very similar in structure."

Perutz talked for several minutes about the techniques of x-ray crystallography of proteins. Then he went on, "You end up with a series of maps, like microscope sections through a tissue, only on a thousand times smaller scale. Each map is a section through a protein molecule, and on the section *density* is plotted. From these densities you can see the coiling of the protein chain, the way the side chains are arranged, the kinds of chemical interactions that occur between them, and often you can actually see the business end of the enzyme—the so-called active site, where the catalysis takes place. This you can do because chemists manage, as it were, to deceive the enzyme, by feeding it a chemical which is very like the substance which it normally attacks and decomposes, but which has some slight chemical modification so that the enzyme cannot act on it. It combines with the enzyme without being affected by it. It locks the enzyme." In Bragg's metaphor, the bulldog was thrown a tarry bone. "You can put this in the crystal, and it will tell you where the site is. Thus you can actually discover how the enzyme works. This is interesting, because these turn out to be living mechanisms. The digestive enzyme just solved by Lipscomb and his group at Harvard contains an atom of zinc at the active site. The enzyme grips the substance it is going to split like a vise—one wing of the enzyme, as it were, closing in on it: it actually moves. Then presumably moves again when the substance is split and released.

"The most spectacular is my own protein, hemoglobin, which undergoes a drastic change of structure in the course of its function. Hemoglobin is the protein of the red cells that carries oxygen from the lungs to the tissues, and also facilitates the transport, on the return journey, of carbon dioxide. It now turns out that the whole molecule undergoes a drastic rearrangement every time it takes up oxygen, and returns to the original state every time it gives up oxygen. The molecule is a highly complex chemical mechanism, which changes structure in response to a specific chemical stimulus, which is oxygen." Perutz's explanations were entirely spontaneous; yet he spoke not merely in connected sentences but in connected paragraphs. "The change of structure enables the hemoglobin molecule actually to increase its affinity for oxygen as it takes oxygen on—which is not what one naïvely would expect. And the reverse structural change as the oxygen goes off enables the molecule to take up carbon dioxide directly, but more importantly, to take up hydrogen ions. Taking up hydrogen ions"—hydrogen atoms from solution, each missing its one electron and therefore carrying a positive charge—makes the blood more alkaline. Making the blood more alkaline enables the serum to take up bicarbonate"—which is made from the carbon dioxide and water produced as food is burnt in the tissues—"and transport this bicarbonate back to the lungs, where it turns again to water and carbon dioxide, and is exhaled. Jacques Monod has called hemoglobin an honorary enzyme, because it acts like one.

"It looks as though this property of hemoglobin, of changing its structure in response to specific chemical stimulus, is not unique—and as though enzymes that are involved in metabolic regulation act in this way. For instance, an amino acid is built up from simpler substances by a whole series of enzymes, each one catalyzing one small chemical step which is required in the synthesis of the amino acid. But a bacterium wants to make an amino acid only when it is needed, when it's run out. When there is enough, it wants to stop that and use the food for something else. So there is a feedback control, whereby the action of the first enzyme in this biosynthetic series is inhibited by the end product of the last. It is as though you had a factory making motorcars which stopped the manufacture of steel the moment the cars cannot all be sold: the presence of completed cars at the end of the production line inhibits the steel mill. It seems that this is done by an enzyme at the start of the series that has two sites: one responding to the end product, called the regulating site; the other, the catalytic site, which makes the starting product of the series. The only way one can interpret this feedback control is that the enzyme must be able to exist in two distinct states, one active and the other inactive, so that the end product operates the switch that turns it from active to inactive.

"The greatest discovery in molecular biology in 1967, and the one that's received the least publicity, is the genetic repressor—isolated by Walter Gilbert and Benno Müller-Hill, at Harvard." Gilbert and Müller-Hill had isolated the molecule capable of switching on or off the row of genes that control synthesis of the enzyme that splits milk sugar—galactosidase. The mechanism of repression was a key and a lock, Perutz said. "The milk sugar provides the key, and the lock is a protein that is put into the circuit of the gene, and normally locks it. But when the key goes in, the lock drops off the DNA, and the synthesis of messenger RNA begins. Thus, the repressor is a protein which—once more—appears capable of changing from an active state, where it combines with one specific region of the DNA, to an inactive state where it drops off that region—in response to one specific chemical stimulus. All these transformations are interesting because they are the most elementary manifestations of the property of a living system that can turn chemical energy into movement."

Perutz suggested that we look at a model of hemoglobin. He led the way into a large room across the hall. The room, though not full, was cluttered. A shelf ran down the length of the window wall, strewn with sheets of stiff, clear plastic that were marked in black with irregular swirling curves which resembled geologists' contours—density maps of sections through the molecule. There were stools to be shoved aside, and large tables. A pair of photographs hung on a wall. One showed a sculpture of a seated figure, all spindly cursive limbs, stylized and elegantly secret; the other was a writhing open knot of what looked like intestine, labelled "sperm-whale myoglobin 1957." The resemblance was amusing.

On the tables were models. They were surprisingly large, and at first difficult to take in; they were responsible for the room's impression of clutter. We went over to one. It was built, like the others, within a cubical frame of steel straps that defined the edges, about four feet long. The floor of the cube was thick, unpainted plywood. From that, within the cube, a jungle sprang up of thin metal rods. Looping through these like creepers were several lengths of flexible, plastic-coated cable, red or white. The rods only provided support. The trick was to ignore them totally, though they obscured almost everything else, and concentrate on where the colored cables led.

This model before us, Perutz said, was a single molecule of hemoglobin of horse in its oxygen-carrying state. The living hemoglobin molecule, human or horse, was made of 574 amino acids, in four chains, ten thousand atoms strong, and totalling 64,500 times the weight of a single hydrogen atom, or about 4,000 times the weight of a single atom of oxygen. All of this was required to transport just four molecules of oxygen each made of two atoms. At those ratios, the breath of life requires about 280 million hemoglobin molecules in each red blood cell, and there are about 160 trillion red blood cells in a pint of blood.

Perutz picked up a pointer to trace the twisting, swooping paths of the four chains in the model—two identical chains termed "alpha," another identical pair called "beta." These were constructed out of thousands of short, straight bits of brass wire socketed together at angles, each bit precisely in scale with the length of a chemical bond, and the sockets where the wires intersected showing the positions of principal atoms. The tracery was clamped at hundreds of places to the supporting rods. The red cords and white ones, Perutz said, were not the real structure but only aids to the eye, Ariadne threads tracing beta and alpha chains. I saw what he had meant by lacework, except that this was more a brier patch, leafless and with the thorns provided by hundreds of those short wires, along the chains, sticking out free at one end, their bare tips representing the positions of the fringe of hydrogen atoms sprouting off the amino acids. "Not the usual balls and spokes; spokes alone," Perutz said, and one could see why that was necessary, except that the modelling technique reduced visual contrasts to a difficult minimum.

On a closer look, there were some leaves in the brier patch. Individual amino acids, clusters of wires connected to other clusters by the characteristic skew twist of the peptide bond, had little round key tags dangling from them. These bore messages like ARG 40 or PRO 119. The sequences of the amino acids in the alpha and beta chains of adult human hemoglobin—their primary structure, meaning no more than the chemist's one-dimensional string—had been published in 1961 independently in two laboratories, Perutz said, one in Munich and the other in New York. The alpha chains had 141 amino-acid residues in unvarying order, the beta chains 146; the sequences of the two sorts of chains were similar. "The sequences can be brought into register by leaving appropriate gaps in one or the other," Perutz said; the biggest gap in either list was only five residues long. The amino-acid sequences of the two sorts of chains of horse hemoglobin, when they were got out, proved to be identical in length to their human counterparts, and similar in sequence: for example, the alpha chains of humans and horses differ in only seventeen amino acids. The single chain of Kendrew's sperm-whale myoglobin, 153 amino-acid residues long, was recognizably, though more distantly, related. So much for primary structure.

Perutz's pointer followed a straight run along one, then another chain of the model. These were lengths of alpha helix, interrupted by sudden corners. When Linus Pauling had first published the alpha helix, Perutz had gone straight to his x-ray patterns and found at once the evidence that confirmed that hemoglobin contained stretches of alpha helix. Then when Kendrew got the first clear structure of myoglobin, even at the low resolution of the visceral skein in the photo on the wall, lengths of alpha helix could be seen—eight such lengths all told. Alpha helix, pleated sheet, and so on were termed, by convention, the secondary structure of protein molecules.

The tertiary structure was the folding up of each chain, stabilized by cross links and with the side groups of the amino acids taking specific positions. Finally, for hemoglobin, there was the way the four chains fitted together into two identical alpha-beta halves and then the whole. The model before us showed all of that, down to the positions of all but a few individual atoms in the molecule in its oxygenated state. This was the biggest molecule for which so detailed a structure had been achieved—though Perutz said at once that because the molecule had two identical halves, the problem had not been of larger scale than that of the digestive enzyme Lipscomb at Harvard had recently solved.

As Perutz confronted his model, I began to perceive that knowing all he did know about the structure of hemoglobin he was baffled that the structure failed to explain how the molecule functioned. The structure was intricate, the bafflement elementary. Why was the hemoglobin molecule so large, so complex, so irregularly but specifically folded up, when all that it had to do was load and unload four oxygen molecules each made of two atoms? By what mechanism did the structure change as oxygen came on or off? And to what effect?

There in the model, caught in the brier patch, were four red balls for the four iron atoms of the hemoglobin molecule. Perutz reached for a row of toggle switches at a corner of the iron frame, and began snapping on other aids to the eye. A string of colored lights ran along part of a chain. Another color traced another chain. He turned those off again. Four red lights came on: the iron atoms were bulbs. Two were near the top of the model, two were well beneath, all were widely spaced apart. Around each iron atom, a ring of four dim blue lights appeared. These represented nitrogen atoms. The disc with the iron atom in the middle, those four nitrogens around it, and carbon atoms ringing those, was called a "heme group."

The heme group makes blood red. The heme group is not built of amino acids, is not itself protein. Hemoglobin has four peptide chains and four heme groups, myoglobin one and one. In the model, each heme group was attached to its chain about three-fifths of the way along, at a residue of the amino acid histidine—in the eighty-seventh position in each alpha chain, in the ninety-second of each beta chain. The folding up of the chains put each heme group in a pocket at the surface of the molecule: a neatly twirled bundle of spaghetti entrapping four flat bits of tomato sauce. The heme groups are the active sites where the oxygen molecules attached, one directly to each iron atom.

Iron rusts, I said, remembering the obvious. But oxidized iron forms a strong chemical bond with oxygen, Perutz said, while the combination of oxygen with hemoglobin is easily reversible: the respiratory function of the molecule depends on the fact that these four iron atoms combine with oxygen lightly, without being oxidized—in plain terms, without rusting—so that when blood reaches the tissues the oxygen unloads. The size and complexity of the hemoglobin molecule, its specific structure and its changes of structure, he said, appeared to be necessary so that the reaction with oxygen would be reversible with acute sensitivity. The combining had to respond to the exact amount of oxygen—technically, to the partial pressure of the gas independent of the other gases present—at the membranes of the lung; the decombining had to respond to the lower partial pressure of oxygen in the tissues. But the mechanism of hemoglobin was not only sensitive, Perutz said, it approached perfect efficiency in loading oxygen on and off. Here was the real mystery that the structure ought to illuminate.

"This incredible apparatus!" Perutz was staring at his molecule, and evidently at the one in his mind. The model itself looked dusty and tired. I was surprised to be told that I was one of the first outsiders he had discussed it with. He and his collaborators had completed calculation of the structure by the previous September. They had delayed publication until the measured positions of atoms were checked against the limitations that physical chemistry imposed on which ones could approach how closely to which: the checking that Crick and Watson had done for DNA with plumb line and ruler in a few days had required for hemoglobin a computer program and several months. Then they built the model—and created an equivalent model in the memory of the computer so that at command the computer could draw on its screen a picture of any segment of the molecule desired, adjust the viewing angle as instructed, and then produce the adjacent view necessary to create three-dimensional slides.

Three weeks hence, in a lecture before the Royal Society, Perutz was going to display the solution and his model for the first time. I asked if I might attend. He said he would be in London the day before, a Wednesday, to set up. We agreed to meet at the Royal Society and perhaps go on to lunch.

Carlton House Terrace, five minutes on foot from the smoking whirlpool of Trafalgar Square, is a serene Georgian row that combines beauty and reticence. The line of its portico disguises how very large are the mansions within, how varied the things now done there. The Royal Society's near neighbor just then was Crockford's, a red-velvet-lined gambling house; another neighbor housed a clandestine operation of the Foreign Office. Perutz had driven up from Cambridge in his lab's panel truck. The hemoglobin model stood at the front of the society's great meeting room when I got there, and he was checking the throw of a stereo-slide projector. He set a box of cardboard spectacles on the table by the door, and we went out, through the series of grand halls and into the glistening day. Over grilled salmon and a plain tomato salad, we returned to the problem.

Suppose there were two molecules of hemoglobin, Perutz said, a rich one with three molecules of oxygen bound on, a poor one with none: a free oxygen molecule between them would stick to the rich hemoglobin with a likelihood of seventy to one. The reverse was true as oxygen was coming off, as in the tissues: the poor hemoglobin with just one molecule of oxygen was seventy times as likely to lose it as the rich hemoglobin was to drop any of its four. "It's like having a power shovel that will always be fully loaded, and then fully unloaded, in every cycle. You can see that the effect is extremely important physiologically, for if the discharge of each oxygen didn't weaken the attachment of the others remaining, most of the oxygen would be carried back to the lungs unused and a man would asphyxiate even if he breathed normally." Evidently, hemoglobin's drastic change of structure was intimately related to this functional efficiency. "Is hemoglobin a kind of oxygen tank? No, it turns out that it is a kind of molecular lung." The difference was movement. "The hemoglobin molecule is an organ in miniature."

After lunch, we took a taxi across town to put Perutz on the train for Cambridge. We talked for a moment about James Watson and his book. Perutz soon came back to his obsession. He had been studying the structure of hemoglobin by x-ray crystallography for thirty-one years, he said; he spoke about how he had got interested in the field while he was still a student in Vienna. "All the work so far has not yet told us how or why hemoglobin works as it does. We have not got the glimmer of an idea. We haven't even got a working hypothesis."

❖ ❖ ❖

On a winter's afternoon nearly three years later, part of a weekend at Perutz's house in Cambridge punctuated by visits to the laboratory, he told me that he had gone into x-ray crystallography and hemoglobin by "almost pure accident." Max Ferdinand Perutz was born in Vienna in 1914, the son of a textile manufacturer. He trained as an organic chemist. "The institute there was headed by one of these great classical organic chemists of the German school, Ernst Späth; he was an Austrian, but he worked in the manner of the German school. He elucidated the constitution of alkaloids with some success, with huge teams of research students, and was the prototype of the laboratory tyrant, who had his name on every paper published and who called all his favorite students together after his lectures and gave them talks inculcating the right moral attitude towards science. No, I can't remember what he said in those talks."

Perutz found alkaloids boring and biochemistry in Vienna backward. Enzymes had long since been discovered; James Sumner had crystallized urease and established that enzymes are proteins in 1926; their catalytic properties were being studied elsewhere; but little awareness of the existence of enzymes seemed to have penetrated to Vienna. However, in 1934 Perutz went to a course of lectures by a young Viennese scientist, "on organic compounds of biological interest." From these, Perutz first learned that vitamins were being investigated at Gowland Hopkins's biochemical laboratory in Cambridge. "I'm not sure that we heard anything about enzymes in those lectures." Here, he thought, might be found a subject for his doctoral dissertation.

"In the summer of 1935, I learned that Hermann Mark, the professor of physical chemistry in Vienna, was going to attend a Faraday Society meeting at Cambridge. So I asked him if he could find me a place as a research student at the biochemistry lab here. Now, when he returned, I found to my chagrin that he had forgotten all about me; that surprised me, because he was one of those people who never forgets anything. But he said he had visited a man called Bernal, who was head of the crystallographic laboratory, and *he* had said that he would like to have a research student, so why didn't I join Bernal? I remonstrated that I knew nothing of x-ray crystallography, and I think in essence he replied, 'You'll learn it, my boy.'

"He must have told me that Bernal was trying to apply x-ray crystallography to biologically interesting problems, because I know I arrived in Cambridge in 1936 with the idea in mind of working on one of those. However, when I arrived, Bernal didn't have any crystals of biological interest, and instead he put me to work on some horrible minerals." These were an iron-silicon compound called rhodonite, from blast-furnace slag. Perutz was set to find their space group and unit cell, the crystallographic bounds within which the individual components of the molecular structure must then be placed. He spent several months on that, then tried some radioactive minerals as well. The year was Perutz's introduction to the techniques of crystallography, at the height of the work of Sir William Bragg, then at the Royal Institution; Lawrence Bragg, then at the University of Manchester; Linus Pauling, at the California Institute of Technology; and Desmond Bernal in Cambridge. This was the classical crystallography of minerals and of "the method of trial": solutions intuitively found, then rigorously tested—but a crystallography that had hardly begun to take on biological substances except for William Astbury's attempts with fibres at the University of Leeds and Bernal's own discovery with Dorothy Crowfoot, at the Cavendish Laboratory, in Cambridge two years earlier, that protein crystals preserved in the liquid from which they precipitated could be made to give rich, sharply defined x-ray-diffraction patterns.

Perutz wrote, thirty-five years later, that "Bernal headed . . . a sub-department housed in a few ill-lit and dirty rooms on the ground floor of a stark, dilapidated grey brick building. These dingy quarters were turned into a fairy castle by Bernal's brilliance and his boundless optimism about the powers of the x-ray method. He would occasionally tell Lord Rutherford, the Cavendish Professor of Physics, of his first crystallographic excursions into the fields of biology, but no echoes of these encounters reached us students. We were but a side show among the glittering spectacle of atomic physics." Perutz told me, "Within a few weeks of arriving, I realized that Cambridge was where I wanted to spend my life."

Today, graduate students in science and those who have just got their degrees expect to spend a couple of years in laboratories in other universities, often in other countries. These journeymen, the American postdocs especially, have become an essential pathway of communication in the guild of science as they carry word of good work and how to do it from lab to lab. Medieval in spirit, familiar in small scale from the itinerant careers of scientists at German universities in the nineteenth century, the practice grew great only with the grants and fellowships after the second world war, in time for Watson's generation. Monod's fellowship to Caltech in 1936 was a rare expedition. Perutz was the one student he knew of in Vienna "who had such adventurous plans." Only his father's money made their realization possible. He left Austria for Cambridge a year and a half before the *Anschluss* with Nazi Germany. "I suppose I thought, like so many others, that what we feared could never happen."

In love with Cambridge if not with minerals, Perutz went back to Vienna for the long vacation in the summer of 1937; only then did he start again to look for a substance of biological relevance to work on. A cousin was married to a young professor of physiological chemistry in Prague, Felix Haurowitz, a specialist in proteins. That September, Perutz went to Prague to ask Haurowitz's advice. "It was in a conversation with him that the idea of hemoglobin first arose," Perutz said. His own notion, as well as he could remember, was that he might try to determine the molecular structure of hemin, which is the chloride—the salt—of heme itself, separated from the protein part of the hemoglobin molecule, and in the oxidized state; hemin forms reddish-brown crystals. Haurowitz had begun studying the chemistry of hemoglobin in the early twenties. He convinced Perutz that hemin was uninteresting—after all, the molecule was small and its chemical formula known—and that hemoglobin itself and entire was the structure that must be solved. He had no crystals to offer, but told Perutz of a physiologist at Cambridge, Gilbert Adair, who was a leading authority on hemoglobin and who might be able to give him some crystals. "When I returned to Cambridge, I approached Adair," Perutz said. "He was very nice and helpful, and grew some crystals of horse hemoglobin for me, and came along one day, very apologetically, saying, 'These are probably much too small'—but in fact they were magnificent, just the thing for x-ray analysis."

A supply of crystals of the digestive enzyme chymotrypsin arrived in Cambridge that fall, sent to Bernal by John Howard Northrop at the Rockefeller Institute. "I started on both these together," chymotrypsin and hemoglobin. "Determining their unit cells and space groups. And had no particular preference to begin with." Crystalline chymotrypsin, however, turned out to be twinned, which meant that in the orderly packing of molecules that made up each crystal there were two components, laid down back to front in relation to each other. Because of that, in the x-ray photograph, diffraction patterns as if from two crystals, one turned from the other, were superimposed, with no way of separating them. So Perutz's choice of hemoglobin to go on with emerged from the suitability of the crystals Adair had given him: garnet-red, flat diamond-shaped plates, about half a millimeter, or a fiftieth of an inch, in the long dimension, and almost uniquely favorable for x-ray crystallography—for good reasons that Perutz discovered in a few weeks that fall, for other reasons that he did not find until sixteen years later.

Thus, at the end of 1937 in Perutz's hands, two grand lines of scientific investigation joined: the physiology of hemoglobin and the physics of x-ray crystallography. Yet Perutz came to the problem of hemoglobin—to the founding of his kind of structural molecular biology—as much an outsider to the century-long traditions of physiological chemistry as Max Delbrück and Salvador Luria were outsiders in that same era to the comparable and related traditions of biochemistry. "I knew probably next to nothing. Because I had never so much as been through a course in biochemistry. I was a pure chemist. And had spent my first year in Cambridge learning crystallography," Perutz said. "You know, I could rationalize now that I found hemoglobin to be interesting, and that much more was known about the function of hemoglobin than of chymotrypsin at that time. But, you see, at that time the great unsolved problem was the structure of *proteins*. It didn't matter too much which protein you chose—or so it seemed. You wanted to solve the structure of a protein."

Was there any idea then of relating structural change to function in hemoglobin? "Not at all. Nothing like that had been discovered. I think I had by then learned something about enzymes, and about their magic catalytic properties. And I think, probably inspired by Bernal's conversation and by his lectures, I got the idea, which I am sure must have been his idea, that you could discover the mechanism of their catalytic function if you could succeed in determining their structure. This idea of Bernal's was daring, and much more imaginative than anything the enzymologists were thinking of—then or even twenty years later, when they were still extremely skeptical that the crystallographers' hopes for solution of structure would actually tell them something about enzymes' function. But the other thing you must realize is that in the late thirties, certainly in the Cambridge circle, we knew nothing about the possible role of the nucleic acids, and really the Secret of Life—in capital letters—consisted of the function of enzymes. They were the real stuff of the living cell, and if you could solve the structure of enzymes, you could solve the workings of living cells in simple physical and chemical terms."

Hemoglobin—the pigment of blood, its respiratory function and its crystals—runs a brilliant red thread through the history of physiological chemistry. For early physiologists as for any small child, the immediate fact of blood was its intense color. The awareness was ancient, first set down by Galen, that the color of blood changes from the scarlet that pumps from a torn artery to a darker, purpler tinge that bleeds steadily from a cut vein. Red blood cells were first seen by Antony van Leeuwenhoek, the Dutch micro-scope builder who began his renowned series of observations in 1673 and whose exuberant curiosity also led him to discover sperm cells, protozoa, bac-teria, and scores of other previously invisible phenomena. Even as the modern chemistry of gases and of combustion was first developing, in the eighteenth century, it was realized that when blood flows through the lungs, oxygen passes into it and carbon dioxide out. In 1799, the chemist Humphry Davy, at a laboratory near Bristol, heated blood near to boiling and tried to measure the oxygen and carbon dioxide given off. In 1802, the chemist William Henry, in Manchester, formulated the simple law that perpetuates his name, for the solubility of gases in pure liquids: the amount of a gas dissolved in the liquid, when the system reaches equilibrium, is directly proportional to the partial pressure of that gas in the mixture of gases above the liquid. But no such straightforward rule obtained for oxygen and carbon dioxide in blood. In 1837, Heinrich Gustav Magnus, a diligent chemist at the University of Berlin, surveyed a vast confusion of reports, tried some experiments with chemicals and vacuum pumps, and settled at least that blood whether from artery or from vein contains large amounts of both oxygen and carbon dioxide, that the carbon dioxide released in the lungs is not formed locally by oxidation but has been carried there by the blood, and that more oxygen and less carbon dioxide is contained in arterial than in venous blood. Yet precise measurements still were technically formidable, while hemoglobin and its power to bind oxygen had not been discovered.

The earliest observations of crystals of the pigment of blood were reported in 1840 by a physiologist named T. L. Hünefeld, at the university in the Prus-sian town of Greifswald, who found light red, flat, sharp-edged crystals in a sample, pressed between glass plates, of dried menstrual blood. Hünefeld also reported that he had seen crystals in blood from swine. In the summer of 1847, Karl Bogislaus Reichert, a young professor of anatomy at Dorpat, a small city in eastern Estonia, on a visit to Germany dissected a pregnant guinea pig that had suddenly died, noticed some odd clumps of what looked like drying blood in the midst of the moist fetal membranes, put them under the micro-scope, and beheld, to his surprise, red tetrahedral crystals. Reichert wrote that these seemed to be albuminous—protein-like—and went some way toward demonstrating this. Hemoglobin was thus the first protein to be crystallized, ninety years or so before the enzyme urease.

More reports of crystals of the coloring matter of blood quickly followed Reichert's; the substance acquired a variety of names; its study began. In 1851 or shortly before, Otto Funke, of Leipzig, devised the first techniques for pre-paring blood crystals deliberately. With blood of an old but healthy horse, it was enough to spread a drop on a microscope slide, allow it to begin to dry, and then to add a drop of water, whereupon, under the microscope, blood cells could be seen to rupture and red needle crystals to appear and grow until the entire field of view was covered with a network of them. By adding ether or alcohol instead of water, Funke got bigger crystals and from the blood of other species.

In 1853, Ludwik Teichmann, a medical student in Göttingen, trying to repeat Funke's experiments, got from the blood of dog, man, rabbit, ox, swine, dove, and frog reddish-brown crystals of a substance that was not protein and that contained iron: he had found hemin, and gave it its name. Terminologies proliferated. Chemists grew numerous and more sophisticated. In 1864, Felix Hoppe-Seyler in Tübingen—the biochemist who, six years later, himself repeated the experiments of the young Friedrich Miescher that first found nucleic acids before publishing the discovery in his journal—named the protein pigment crystallized from blood "haematoglobulin or haemoglobin." He used the shorter form henceforth, and it soon prevailed. The word came apart like the molecule: "haem," or "heme," originally from the Greek for blood, was now confined to the dark red, iron-bearing, nonproteinaceous part of hemoglobin, and "globin" was the colorless protein left behind.

The color of blood was now looked at more precisely, with epochal consequences. Two years earlier, Hoppe had made the first report of the absorption spectrum of blood in the visible region of light. The methods of spectroscopy were much what Newton had done to discover the composite nature of sunlight. In a darkened room, Hoppe shone a concentrated beam of sunlight through a narrow slit, then through a finely ground prism, then onto a white sheet of paper. He got, of course, the rainbow—with numerous sharp, narrow, dark lines, in fixed positions, that had first been seen in the spectrum of sunlight in 1814 by the optician Joseph von Fraunhofer, and that had come to be used as reference points where today physicists would speak of particular wavelengths. Hoppe interposed, in the sunbeam, a solution of blood thinned way out with water and held in a glass vessel with flat walls a centimeter apart. Immediately, two distinctly marked dark stripes sprang into the spectrum, near its center—one in the yellow and the other in the green. He located the stripes between Fraunhofer lines D and E; the wavelengths are around 5600 and 5350 angstroms. Hoppe showed that no other constituent of blood besides hemoglobin produces this absorption pattern; he speculated about the discovery's forensic uses, and proved with it that hemoglobin is the only pigment in red blood cells.

Hoppe's spectroscopy provoked the experiments that at last made the connection between the change in color of blood and the change in its content of oxygen, and so at last established hemoglobin as the substance on which respiration depends. Hoppe's report was pointed out to George Gabriel Stokes, in Cambridge, a mathematical physicist of eminence—Lucasian Professor of Mathematics and Secretary of the Royal Society—and the widest interests.

"I had no sooner looked at the spectrum, than the extreme sharpness and beauty of the absorption-bands of blood excited a lively interest in my mind, and I proceeded to try the effect of various reagents," Stokes wrote in 1864. The directness of his methods and the clarity of his reasoning were exactly matched in his writing, so that the process of discovering is freely expressed—but more than that, its pleasure. "The observation is perfectly simple, since nothing more is required than to place the solution to be tried, which may be contained in a test-tube, behind a slit, and view it through a prism applied to the eye," Stokes said. "It seemed to me a point of special interest to inquire whether we could imitate the change of colour of arterial into that of venous blood, on the supposition that it arises from reduction"—that is, in the chemist's term, from removal of oxygen. Hoppe had demonstrated that the pigment was attacked by acids, so Stokes devised a reducing solution—of ferrous sulphate, which is an astringent iron salt, balanced with tartaric acid—which he rendered somewhat alkaline by ammonia or carbonate of soda. The iron in this solution would capture oxygen. If a little of this "be added to a solution of blood, the colour is almost instantly changed to a much more purple red as seen in small thicknesses, and a much darker red than before as seen in greater thickness," Stokes wrote; he went on:

> The change of colour, which recalls the difference between arterial and venous blood, is striking enough, but the change in the absorption spectrum is far more decisive. The two highly characteristic dark bands seen before are now replaced by a *single* band, somewhat broader and less sharply defined at its edges than either of the former. . . .
>
> If the purple solution be exposed to the air in a shallow vessel, it quickly returns to its original condition, showing the two characteristic bands the same as before; and this change takes place immediately, provided a small quantity only of the reducing agent were employed, when the solution is shaken up with air. If an additional quantity of the reagent be now added, the same effect is produced as at first, and the solution may thus be made to go through its changes any number of times.

He tried other comparable reducing solutions with the same effects.

> We may infer from the facts above mentioned that the *colouring matter of blood*, like indigo [the vegetable pigment, whose chemistry was familiar], *is capable of existing in two states of oxidation, distinguishable by a difference of colour and a fundamental difference in the action on the spectrum. It may be made to pass from the more to the less oxidized state by the action of suitable reducing agents, and recovers its oxygen by absorption from the air.*

Stokes had not yet seen Hoppe-Seyler's latest paper, offering the name "haemoglobin," and so proposed to call the pigment "*cruorine . . .* and in its two states of oxidation . . . *scarlet cruorine* and *purple cruorine.*" Experiments presented, Stokes turned to what he dryly termed "physiological speculations," and wrote, in part:

> It has been a disputed point whether the oxygen introduced into the blood in its passage through the lungs is simply dissolved or is chemically combined with some constituent of the blood. . . . Now it has been shown in this paper that we have in cruorine a substance capable of undergoing reduction and oxidation, more especially oxidation, so that . . . we have all that is necessary to account for the absorption and chemical combination of the inspired oxygen.

The repeated reduction and oxidation of cruorine in solution exactly illustrated its physiological function, Stokes said in conclusion.

> As the purple cruorine in the solution was oxidized almost instantaneously on being presented with free oxygen by shaking with air, . . . so the purple cruorine of the veins is oxidized during the time, brief though it may be, during which it is exposed to air in the lungs, while the substances derived from the food may have little disposition to combine with free oxygen. As the scarlet cruorine is gradually reduced [by the reagent], so part of the scarlet cruorine is gradually reduced in the course of the circulation, oxidizing a portion of the substances derived from the food or of the tissues. The purplish colour now assumed by the solution illustrates the tinge of venous blood, and a fresh shake represents a fresh passage through the lungs.

Stokes wrote of oxidizing: yet even as he noted that hemoglobin could be made to go through its changes any number of times, he remarked on the looseness of the combination. Hoppe-Seyler and his German contemporaries saw that, too. They knew as well that hemoglobin in solution outside the living organism, treated with certain chemicals or exposed for a while to air, will combine with oxygen in a manner much more difficult to reverse, forming a brown substance that was named "methaemoglobin." (In chemistry, "met" as prefix indicates a closely similar compound of the same molecular weight.) Furthermore, by the end of the century, the Scottish physiologist John Scott Haldane had demonstrated that carbon monoxide even in small proportions competes with oxygen to bind with hemoglobin in a similar loose, reversible manner. By then, too, many analyses had indicated that hemoglobin bound oxygen at the simple ratio of one atom of heme iron to one molecule of oxygen. The nature of the bond was not understood.

By 1871, hemoglobin crystals had been obtained from upwards of fifty different species from fish and snakes through mammals and man. In 1909 appeared an extraordinary volume, *The Crystallography of Hemoglobins*, by Edward Tyson Reichert, a physiologist at the University of Pennsylvania, and Amos Peaslee Brown, a mineralogist there. Reichert had conceived the ambition to plot the evolutionary relationships among species by the divergencies among their protein molecules. His essential idea was merely seventy years ahead of the technology: only with the advent of Frederick Sanger's methods for sequencing amino acids could students of evolution begin to measure the similarities among proteins, and only with Sanger's means of sequencing nucleotides in DNA, beginning in 1976, could such measurements of genetic similarity begin to be accurate. But Reichert understood the enormous scope for diversity if proteins were large, specific molecules; he settled on crystal forms—and recruited his colleague Brown—as the means to get at degrees of difference, and on hemoglobin as the easily crystallized protein universal among animals. Their book surveyed the nineteenth-century literature of hemoglobin; catalogued crystals of the stuff from a hundred and nine different vertebrate species—Philadelphia had a good zoo—complete with drawings and measurements of the crystal forms; and ended with six hundred large, clear, well-printed photomicrographs of hemoglobin crystals.

Iron and other metals presented subtle problems to any theory of chemical bonding, for reasons—fully worked out only recently—that have to do with the variable dispositions of the outer electrons of their atoms. The metals' capacity to gain, lose, and rearrange their electrons is the basis not only of their chemical reactivity but of their electrical conductivity, their magnetic properties, their hardness, even their curious propensity to form complicated compounds that are brilliantly colored—as with cobalt in vitamin B_{12} (pink) and, once more, with iron in hemoglobin.

The iron atom, in particular, can exist in compounds in two states of valency depending on how many electrons it has lost or is sharing with the chemical partners of the moment. An iron atom chemically compounded that is minus two electrons thus carries a double positive charge, symbolized Fe^{++}; it is said to be in the ferrous state. An iron atom minus three electrons carries a triple positive charge, Fe^{+++}, and is said to be in the ferric state. True oxidation of iron, as in rust, shifts an electron over to the oxygen, changes the iron atom into the ferric state, and forms a bond that takes great energy to break; this is why oxidation is not easily reversible.

Investigators could distinguish between ferrous and ferric iron compounds by, to cite one method, putting the substance into solution, running a light electrical current through the liquid, and seeing how quickly the iron accumulated at the negative terminal. By this means, in 1923, James Bryant Conant, who later became president of Harvard but was then a young chemist on the premises, showed that in deoxyhemoglobin the iron is ferrous, which itself had long been known—and that when oxygen combines with hemoglobin the iron remains in that ferrous state. Thus, contrary to Stokes's conclusion, oxyhemoglobin is the product of no oxidation. That was a surprise to most chemists.

Conant urged that the process be called "oxygenation" and "deoxygenation" instead. Carbon monoxide combines with hemoglobin similarly, leaving the iron ferrous. But methemoglobin, Conant showed, is the true oxidation, for there the iron loses an electron, becomes triply charged, takes the ferric state. Conant's experiment explained the reversibility of the binding of oxygen to hemoglobin, yet only shifted the mystery: how hemoglobin managed the iron's binding to oxygen while remaining ferrous was not explained.

Perutz's familiarity with the physical chemistry and even with the crystallography of hemoglobin was haphazard at first. "As one does, you know, one gradually picks up a subject when one works on it." He knew of Reichert and Brown's book by the fall of 1937, but primarily for its "marvellous atlas of haemoglobin crystals," which offered welcome confirmation of his own classification of the crystal structure of hemoglobin of horse. He could not recall giving the book's historical compendium more than a passing glance. In the data of the crystallographic catalogue, although smothered in detail, was the fact that oxy- and deoxyhemoglobin from the same species, the same creature, crystallize in different forms. Perutz could not remember whether he spotted this then.

That winter, Perutz attended a course of lectures on the respiratory proteins. Besides hemoglobin, these comprise myoglobin, several enzymes, and a related set of substances called the cytochromes. All carry heme in conjunction with protein. All cooperate within the cells of animals in the stepwise oxidation of food molecules to provide energy, though the details of these steps were worked out only later. Indeed, so few of the others had by then been isolated that hemoglobin was as yet the only one closely studied. The lectures were given by David Keilin, who was Quick Professor of Biology in Cambridge as well as director of the Molteno Institute. Keilin had discovered one of the respiratory enzymes, cytochrome C. "Keilin was a true biologist. Who had taught himself biochemistry," Perutz said. He learned from Keilin the spectroscopy of the heme-bearing proteins, and beheld the demonstration that hemoglobin changes spectra as it takes oxygen on and off. He learned from Keilin or from Adair of the difference between the ferrous and ferric forms, oxyhemoglobin or methemoglobin. The crystals Adair had given him were of horse methemoglobin; as it turned out, they had the same crystallographic form as crystals of oxyhemoglobin, and the substance was somewhat more stable.

Adair also gave Perutz a set of his papers to read. In the three decades before the first world war, two of the most fundamental facts of hemoglobin had been in question: the size of the molecule and the rate at which it took on or released oxygen. After the war, as a graduate student, Adair had begun a series of meticulously conducted, rigorously argued researches that settled both questions. Between these was a relationship of great subtlety, expressing the central mystery of hemoglobin.

The earliest way to estimate the size of large molecules was by total chemical analysis of the substance—and the first clear results for any protein had been obtained for hemoglobin. In 1885, O. Zinoffsky, in Basel, performed a scrupulous analysis of horse hemoglobin down to the elements that compose it and produced the formula $C_{712} H_{1130} N_{214} O_{245} S_2$ Fe, meaning that for each iron atom the molecule contains 712 carbon atoms, 1130 hydrogen atoms, and so on. Multiplying up the atoms' weights and adding them together, he put the minimum molecular weight of hemoglobin at 16,730, where the unit, of course, is the weight of a single hydrogen atom. That was enormous by the conceptions of the day, yet Zinoffsky looked for evidence that the true molecular weight might be a multiple of this value.

At the turn of the century a new way to determine the size of large molecules came in: by measurement of osmotic pressures. A thin membrane of some substance that would act as a selective filter was chosen so that water, salts, and other small molecules passed through the membrane but large molecules were stopped. Into a closed column fashioned of this membrane, a solution of hemoglobin, say, was poured. A thin glass tube was inserted and the column sealed at the neck around the tube. The column was then immersed in a jar of distilled water. Water molecules moved both ways through the membrane—but because inside the column the water molecules were interspersed with others that were turned back at the membrane, water entered the column more rapidly than it left. Pressure grew until it reached the equilibrium where water was pushed out as fast as it came in. That pressure could be read by the height of the liquid rising in the glass tube. At a given concentration of the substance studied, many molecules of moderately large size would lead to higher pressures than just a few molecules of rather greater size: thus, the lower the osmotic pressure at equilibrium, the fewer but larger the molecules in solution. A simple proportional formula calculated the size.

In 1905, E. Waymouth Reid, in England, got by these means a molecular weight for dog hemoglobin near 48,000. But two years later Gustav Hüfner and a colleague, in Tübingen, got high osmotic pressures for hemoglobin that neatly bracketed the chemists' value for the minimum possible molecular weight with only one iron atom per molecule—16,700. The coincidence was persuasive. That was the basic figure physiologists accepted. Some thought, indeed, that the molecules clumped together to form a range of different sizes.

Adair overturned all this. He detected a score of possible sources of error overlooked by earlier workers. Experimentally, errors ranged from decomposition of protein molecules during the several days needed to reach equilibrium, to the use of membranes that did not truly hold back all the large molecules or that did not freely pass small salts. Lapses of theory included failure to recognize the effect that any net electric charge carried by the large molecules would have on the behavior of the salts in the solution. Hüfner, he noted, had used membranes of parchment that were relatively impermeable. Adair made his own membrane columns out of collodion, a cellulose. He worked at low temperature to stabilize the protein. He added salt to equalize electric potential inside and outside the membrane. He set stiff criteria for the reliability of measurements. He began to get results in 1921 and submitted the decisive paper to the *Proceedings of the Royal Society* in April 1924. It was not published until nearly a year later. The chemist turned historian John Edsall recently wrote, "We may suspect that the referees were disturbed· by Adair's then highly unorthodox conclusions concerning the size of the hemoglobin molecule, and therefore delayed the publication of the paper."

From the results of many trials with hemoglobin from several species variously prepared—at least one sample was from his own blood—Adair concluded that the osmotic pressure in dilute solutions was only about a quarter of what Hüfner had reported. The molecule had to be four times larger than the accepted value. The molecular weight of hemoglobin was 67,000. Each molecule had four heme groups and four iron atoms.

Within a year, Adair's conclusions got dramatic verification from an entirely separate line of experiments. In the north of Sweden, at the University of Uppsala, beginning shortly after the war, the physical chemist Theodor Svedberg had been developing a new instrument of his own design that he christened the ultracentrifuge. The machine was a simple idea that required superb engineering. Rotor and bearings allowed great rotational speeds to be built up, monitored, and maintained for hours and days. Cooling systems kept the experiment at constant low temperature. The individual cells for solutions were made of glass or quartz, and a high speed camera was set up so that one cell was photographed repeatedly as it passed by. Svedberg's aim was to study the sizes of large biological molecules or particles by determining their distribution in a high centrifugal field.

The first substance Svedberg reported was hemoglobin, which he chose for its "eminent physiological importance"—and for its color, which made it easy to observe. He examined a variety of preparations of methemoglobin and of carbonmonoxyhemoglobin, and ran the centrifuge at several different speeds, usually for thirty-six or forty hours. By that time, he wrote, "A state of equilibrium is finally reached when sedimentation and diffusion balance each other." The speeds of this early machine were far short of those soon to be reached; nothing so sophisticated as a density gradient of the solution itself was achieved; but the photograph through the glass of the cell, at equilibrium, allowed the concentration of hemoglobin at several distances from the center of rotation to be measured. "The determinations indicate that hemoglobin solutions are built up of molecules containing four groups of molecular weight 16,700; that is, that the molecular weight of hemoglobin in aqueous solution is probably 66,800." Furthermore, hemoglobin molecules in solution behaved not as a collection of aggregates of different sizes but as though they all had the same mass. Svedberg finished his first paper on hemoglobin in the summer of 1925, in ignorance of Adair's results. It appeared early in 1926 in the *Journal of the American Chemical Society*.

Eighteen months later in the same journal, Svedberg reported the building of a radically improved ultracentrifuge "capable of running at a speed of 42,000 r.p.m. and giving an effect 104,000 times that of gravity." Once more, he tried hemoglobin. Under such force, the molecules were actually spun out of solution in a few hours, and the speed with which the protein boundary moved through the cell could be determined from pictures taken every thirty minutes. Svedberg's previous results, and Adair's, were well confirmed. The impact of the work was immediate, of the device profound and long-lasting.

A singular molecule of definite size, molecular weight 67,000, four heme groups, four iron atoms, binding four molecules of oxygen. These facts were basic to determining hemoglobin's structure. Adair and Svedberg having fixed them, Perutz could ignore the history of competing concepts that had gone before. But even as Adair was establishing the physical facts of the molecule, he turned to their physiological implications—the problem, even more vexing, of the relation of the four heme groups to each other and to the rate at which the molecule loaded oxygen on and off.

The essential measurements of the rate of dissociation of oxygen from oxyhemoglobin had been made at the beginning of the century, in Copenhagen, by Christian Bohr—and they were strange. Christian Bohr was Niels Bohr's father. He had trained as a physician and had apprenticed as an experimental physiologist in a celebrated laboratory, that of Carl Ludwig in Leipzig: for biologists in the grand tradition, one traces these lineages of training—of style—just as one observes, say, that Franz Liszt studied the keyboard with Karl Czerny, who learned from Beethoven. Christian Bohr took up the physiology of blood and respiration in 1883. Others at that time, of whom Hüfner in Tübingen was the chief, supposed that hemoglobin behaved simply, each heme group independent of the others so that the blood, when saturated with oxygen as in the lungs, unloaded it at first very slowly as the oxygen pressure began to drop, then more rapidly, at last very rapidly. Hüfner published an equation for this relationship in 1890, with data that agreed tolerably well. In life, it turns out, myoglobin—molecular weight 17,816, with a single heme group binding a single molecule of oxygen—behaves almost exactly as Hüfner thought hemoglobin did: the curve on the graph fits Hüfner's equation for hemoglobin but, in fact, represents the performance of myoglobin in solution at normal bodily conditions of temperature and acidity. The graph is best understood as an unloading curve; one should read it from upper right to lower left in order to follow the decrease in saturation of oxygen, left-hand scale, with the dropping of oxygen pressure, bottom scale.

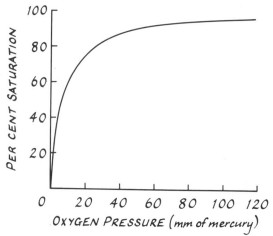

Christian Bohr's reverence for experimental fact, learned in Ludwig's lab, became famous. It led him to map out what really happened to the proportion of hemoglobin carrying oxygen when the partial pressures of oxygen and carbon dioxide were systematically varied through many values. The work was laborious and slow. He carried it out with two younger colleagues, Karl Hasselbalch and August Krogh. They used whole blood, from dogs, thinking to get closer to physiological conditions than they could with hemoglobin solutions. They treated the blood to prevent clotting and rotting, and put it into an apparatus, designed by Krogh, of three interconnected glass chambers with tubes and cocks by which pressures of oxygen and carbon dioxide could be fixed and measured; they sealed the whole thing and set it rocking, by a motor, to mix the blood and gases. In 1904 the three men published the classic paper on the matter.

Their values for the percentage saturation of hemoglobin, as oxygen pressure was lowered, produced a curve of a new form. Where Hüfner's was hyperbolic, what Bohr found was sigmoid, S-shaped. Even when the two curves are drawn together, a moment's reflection may be needed before the significance of the difference between them begins to emerge. The significance is cumulative. It is, literally, vital. Once more, the slope reads from upper right to lower left. At the pressure of oxygen in the lungs, hemoglobin demonstrated fully as lively an affinity for it as Hüfner had said, as great an affinity as myoglobin is now known to demonstrate. Blood left the lungs saturated with oxygen. But Bohr and his colleagues discovered that as the partial pressure dropped into the middle range found in capillaries and tissues, the dissociation of oxygen from hemoglobin accelerated precipitously.

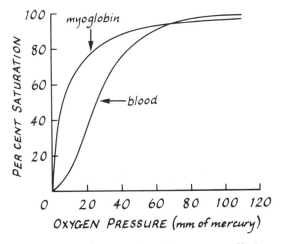

The biological importance of the drop in oxygen affinity cannot be overstated. As blood passes through the capillaries, the percentage of the oxygen it carries that it can deliver into the tissues is from four to five times greater than it would be if hemoglobin were replaced by myoglobin. The rate of diffusion of oxygen into the tissues is correspondingly greater. Comparing Bohr's curve with Hüfner's, John Scott Haldane wrote, in 1935, "A man would die of asphyxia on the spot if the oxygen dissociation curve of his blood were suddenly altered so as to assume the form which Hüfner supposed it to have in the living body."

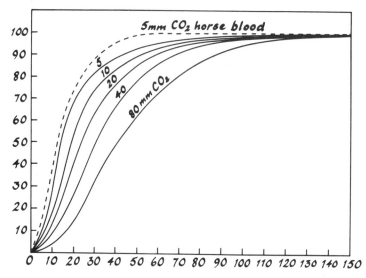

The family of oxygen-dissociation curves of hemoglobin, redrawn from the paper by Bohr and colleagues in 1904.

Beyond that, Bohr's sigmoid curve for the dissociation of oxygen from hemoglobin had no obvious chemical basis. In particular, the drop in oxygen affinity signified by that S shape was much faster than could be the case if hemoglobin were a molecule with a single heme group independently combining with oxygen. As the discharge of oxygen speeded up, to get rid of further molecules of oxygen became easier—in defiance of the logic of simple chemical models. Then, back in the lungs, as oxygenation began to rise, combination with further molecules of oxygen got easier. The implication was that the heme groups interacted. Their binding of oxygen was cooperative. How they interacted was a profound mystery.

Bohr and his colleagues took the measurements further. To determine the oxygen affinities of blood at various partial pressures, they had held the pressure of carbon dioxide steady. But then they did the experiments again, for each of a set of different steady partial pressures of carbon dioxide. They displayed these results as a family of oxygen-dissociation curves. The form of each member of the family was sigmoid—but as carbon-dioxide pressure stepped up, the affinity of hemoglobin for oxygen fell off ever more sharply in the middle range of oxygen pressures. It followed, the three wrote, that as blood reached the tissues and its oxygen content began to drop, the carbon dioxide it was picking up, at the same time, itself drove the affinity of blood for oxygen down further.

This unexpected chemical coupling released oxygen much more quickly and completely from the red cells to the plasma and thence into the tissues than would otherwise have happened. The very strong downward influence of carbon dioxide on oxygen binding had been overlooked before, they said, because previous investigators had almost always worked at high partial pressures of oxygen. When these were high, hemoglobin became saturated in any case, even in the presence of greater partial pressures of carbon dioxide than ever encountered in atmosphere or organism. And *that* fact, though it seemed negligible by comparison with the discovery of the sigmoid form of the dissociation curve and of the profound effect of carbon dioxide on it, led physical chemists towards one more conclusion: carbon dioxide, unlike carbon monoxide, did not eject oxygen from the hemoglobin molecule by direct competition for a place bound to the heme group.

The influence of carbon dioxide on the binding of oxygen by the blood was the burden of the paper. Ever since, that influence has been called the Bohr effect—though it was realized six years later that the influence is more indirect and chemically subtler than Bohr knew. The Bohr effect may have been misnamed: many years later, Krogh, himself grown to world eminence as a physiologist, claimed in a conversation reported at second hand by Edsall that it had been he himself, and not Bohr, who had demonstrated the effect of carbon dioxide on the dissociation of oxygen from hemoglobin.

An entire chapter of physiological experiment and chemical speculation, ardent and inventive, followed Bohr. Much of it was British. In 1907, Joseph Barcroft, at Cambridge, began research into hemoglobin which he continued until his death forty years later. In 1910, he showed that the effect of carbon dioxide in lowering the oxygen affinity was only in small part due to a direct action of carbon dioxide on hemoglobin, as Bohr thought; rather, for the most part the Bohr effect is caused by protons—hydrogen atoms, each stripped of its electron—that are set free within the water of the solution as carbon dioxide combines with it. Also in 1910 at Cambridge, Archibald Vivian Hill attempted a mathematical statement of hemoglobin's oxygen equilibrium. He faced the paradox: the accepted molecular weight was 16,700, and each such molecule had one iron atom and one heme group—yet Bohr's curves required the oxygen-binding centers to interact with each other. Hill proposed that in solution hemoglobin molecules formed an assortment of molecular aggregates—a mixture of multiples of the accepted molecular weight. He was then able to express hemoglobin's binding of oxygen as a fairly simple equation with two constant factors. Hill's equation, Barcroft wrote, fitted his own data. Then the Haldanes joined in, John Scott Haldane and his son John Burdon Sanderson Haldane at Oxford.

Speculation was interrupted by the war and put to rest by Adair in 1925. Establishing the correct size of the hemoglobin molecule, though it did not solve the mystery of heme-heme interaction, allowed the phenomenon to be stated clearly for the first time. This Adair did next. A single molecule with four heme groups, four iron atoms, taking up four molecules of oxygen: Adair perceived the obvious, that such a molecule could form compounds with oxygen in a series of steps, where all four of the heme groups carried oxygen molecules, or only three, or two, or one, or none. He put that observation as a biochemical equation, with four constants, one for each step in the binding of oxygen by the molecule; he adjusted the constants so that the equation did indeed generate the sigmoid curve for oxygen dissociation. Adair's equation was the formal assertion of the unexplained fact of hemoglobin's cooperativity. Adair was also the first to determine the oxygen equilibrium curve accurately by experiment.

"Gradually I learnt more from Adair, who gave me a collection of his papers to read," Perutz said. But when had he begun to get closely concerned with Bohr's work? "Oh—really, not until I had a structure. You see, without a structure— I mean, I read about these physiological properties, heme-heme interaction and the Bohr effect, and they were—they were interesting, but I was without a clue about their possible significance. Really, only— Let's see, it was not until 1965 that I built a tentative atomic model of hemoglobin. And *then* I first started speculating about function. And my first speculations were wildly off the mark."

Also in the winter of 1937–'38, in Keilin's lectures, Perutz learned of the pretty physical geometry of heme itself. The iron in the heme group sits in the middle of a square of four nitrogen atoms. These are linked together by carbon atoms in covalent bonds in four small rings; then those in turn are linked to form a larger, flat, stable ring that looks when diagrammed like nothing so much as a crochet-work antimacassar and is called a "porphyrin"—the word itself harking back to the Greek for "purple." The porphyrin ring, with variations in its outer carbon and hydrogen atoms, appears very widely in biological molecules. Hemoglobin's only rival in biological fascination has been the

magnesium-centered porphyrin chlorophyll. In the iron porphyrin that is the heme group, the distance from the midpoint of the ring to any of the four nitrogens is about two angstroms. Set in the ring like a gemstone in a claw mounting, the iron atom bonds to the nitrogens. It forms a fifth bond, perpendicularly to the plane of the four, to attach to the globin, the protein; and on the far side the sixth bonding position is available for oxygen, say, or carbon monoxide.

That winter, also, Perutz encountered for the first time the work of Linus Pauling. Pauling by then had already elaborated the principles of valency and resonance by which, with such tremendous effect, he brought quantum mechanics to bear on the nature of the chemical bond. By the early thirties, indeed, theorists had reached the point where they could relate the chemistry of the metals to their magnetic behavior. The theory can be sketched at almost any level of abstruseness, but the fundamental discovery, here, was that electrons spin. Every electron is a dynamo in miniature, generating by its spin a minute magnetic field.

Electrons, the developed theory said, are most likely to be found in certain relationships to the nucleus of the atom and to each other. These are not exactly electron orbits circling the nucleus in the too-literal planetary model that Niels Bohr had imagined in 1912. They are better to be understood as standing waves—like the natural vibrating of a fixed string Erwin Schrödinger proposed in the opening sentences of the celebrated paper in which he introduced the wave mechanics early in 1926. Schrödinger's mathematics predicted the likely positions of the electrons accounting for these standing waves; each position has a certain energy, shape, and symmetry and is called an orbital. The possible orbitals could be enumerated and a hierarchy of them could be distinguished, grouped in sets and then in shells around the nucleus; the farther out the electron, the less tightly it is bound. The chemical behavior of each element is determined by the orbitals occupied by its electrons, particularly the outermost ones. The specific wavelengths at which an element or compound absorbs radiation, or emits it when heated—in other words, its color and spectrographic signature—are equally the direct result of the orbitals occupied by its electrons. In fact, the sets of orbitals in each shell of the hierarchy were given letter symbols based on the appearance of the spectra relating to them: *s* for "sharp," *p* for "principal," *d* for "diffuse," and so on. The abrupt change in spectra when hemoglobin took oxygen on or off was now seen to signal a substantial rearrangement of electrons in their orbitals.

The physicist Wolfgang Pauli had perceived in 1925 that no two electrons in the same atom can be in the same quantum state, an assertion since known as the Pauli exclusion principle. By that principle, a given orbital might be occupied by one electron or by two, never by more—and if occupied by two electrons they spin in opposite directions. Thus paired, spinning electrons cancel out their magnetic effects. A substance with all its electrons paired that way is said to be low spin, and also diamagnetic—*dia* from the Greek for "twofold." A diamagnetic substance is pushed out by a magnetic field. But when an orbital contains only one electron, and especially when several orbitals each contains only one electron, spins are not paired and the magnetic effects do not cancel out. Such a substance is said to be high spin and paramagnetic; a paramagnetic substance is pulled into a magnetic field.

Pauling began to think about the structure of hemoglobin and the physical chemistry of its binding of oxygen in 1935. He got into it, he said to me one time, "because the nature of the bond wasn't known, and this of course interested me because of my interest in the chemical bond." In particular, he was curious about the interaction between the heme groups. "Then I decided that it ought to be possible to get information about the bonding of oxygen to hemoglobin by studying magnetic properties."

That fall, Pauling set a graduate student, Charles Coryell, to measure the magnetic susceptibility of various heme-containing molecules. Forty years after that, Pauling told Perutz that his aim when he planned the experiment had been first of all simply to remove the last doubt that hemoglobin combines with oxygen by a true chemical reaction that takes place directly with the iron. The alternative was that the combination was an unspecific adsorption of molecules of oxygen to the surface of hemoglobin. That idea had been advocated by some earlier chemists; yet except for Pauling's remembrance of its lingering, it appears to have been abandoned by the late twenties. In any case, what Pauling and Coryell proved about the combination of oxygen with iron in hemoglobin was more detailed, more puzzling, and, as it eventually turned out, more important.

The experiments were simple. A glass tube divided at the midpoint was filled at one end with water and at the other with a solution of the substance to be measured. The tube was hung vertically from one arm of a sensitive balance scale, set up so that the lower end of the tube, but not the upper, dangled between the poles of a powerful electromagnet. The magnet was turned on and the tube weighed. The tube was then inverted so that the other tip was between the poles, and weighed again. A diamagnetic substance—and water is diamagnetic—in the lower chamber of the tube would be pushed out of the field of the magnet, while a paramagnetic substance would be drawn into it, so that the extent of paramagnetism of the substance compared with plain water was revealed by the difference in weight of the tube one way up and then the other. From that difference, calculation gave the number of unpaired electrons per molecule.

Coryell got magnetic readings for oxyhemoglobin, carbonmonoxyhemoglobin, and deoxyhemoglobin early in 1936. Carbonmonoxyhemoglobin was pushed out by the magnet and therefore had no unpaired electrons. This meant that the six bonds of the iron atom—to the four nitrogen atoms, to the globin molecule, and to the carbon atom of the carbon monoxide—are all covalent, shared-electron bonds. Next, the molecule of oxyhemoglobin, like that of carbonmonoxy, contains no unpaired electrons. "Each iron atom is accordingly attached to the four porphyrin nitrogen atoms, the globin molecule, and the oxygen molecule by covalent bonds," Pauling and Coryell reported.

But it was known that the normal oxygen molecule, free in the air, is itself paramagnetic, with two unpaired electrons. The first surprise was that these disappeared in the formation of oxyhemoglobin. "It might well have been expected, in view of the ease with which oxygen is attached to and detached from hemoglobin, that the oxygen molecule in oxyhemoglobin would retain these unpaired electrons," they said, with the bond to hemoglobin being formed by a pair of other electrons the oxygen had available.

> However, this is shown not to be so by the magnetic data, there being no unpaired electrons in oxyhemoglobin. *The oxygen molecule undergoes a profound change of structure on combination with hemoglobin.*

Deoxyhemoglobin, Coryell found, tipped the balance decisively when the electromagnet was turned on. The stuff is paramagnetic to a pronounced degree. "There are present in each heme four unpaired electrons," they said. This meant that the iron atom is not attached to the four nitrogen atoms and to the globin molecule by covalent bonds. The iron atom is instead bonded to its neighbors more weakly, essentially by the bonds called ionic or electrostatic.

> It is interesting and surprising that the hemoglobin molecule undergoes such an extreme structural change on the addition of oxygen or carbon monoxide; in the ferrohemoglobin [that is, deoxyhemoglobin] molecule there are sixteen unpaired electrons [four to each iron atom] and the bonds to iron are ionic, while in oxyhemoglobin and carbonmonoxyhemoglobin there are no unpaired electrons and the bonds are covalent. . . . Such a difference in bond type in very closely related substances has been observed so far only in hemoglobin derivatives.
>
> It is not yet possible to discuss the significance of these structural differences in detail, but they are without doubt closely related to and in a sense responsible for the characteristic properties of hemoglobin.

"The significance of Pauling's experiment was not apparent to him, but became apparent only thirty years later," Perutz said in a conversation in the fall of 1977. "In old-fashioned terms one would have said, in deoxyhemoglobin the bonds are ionic, and weaker, and in oxyhemoglobin they are covalent and stronger. That's what he does say. Though nowadays people don't usually put it that way. Now, what Pauling *didn't* say, although it was known to him, was that this is accompanied by a change in the bond *distances* from the iron to the nitrogens—weaker bonds being longer. And when I asked him, two years ago, why he hadn't pointed that out, he said, 'I didn't think it was significant.'"

Pauling's work on hemoglobin had one other consequence: it first excited his interest in the structure of proteins.

❖*b*❖   To grow crystals of proteins for x-ray diffraction is still today an art, or at least a knack: Perutz has a glass thumb, and has often grown and mounted crystals for graduate students less adept than he. One Saturday morning early in January 1971, we met at the room, downstairs in his laboratory in Cambridge, where crystals were prepared. Hardly larger than a walk-in pantry, the room had workbenches down the two long walls, shelves above, no windows, and an odor of beeswax. To the right, on the wall, was a cupboard, and on it "Haemoglobin crystals growing. Open and close very gently please." Inside, gently, were racks filled with small, stoppered bottles. "The question is, how much salt?" Perutz said, as he took down one rack. "If you put in too little, you get nothing; if you put in too much, you get a shower of little crystals, too small to use." Some proteins refuse to crystallize at all. Hemoglobin is usually docile.

A useful way in to the relationship between structure and function has been the investigation of unusual hemoglobins. From 1947, when Pauling first proposed that sickle-cell anemia was a "molecular disease" caused by a mutant molecule, Perutz and his succession of colleagues made the study of variant hemoglobins an ancillary specialty. A biochemist colleague, John Kilmartin, modified certain hemoglobins by chopping off one, or two, or three amino acids from the ends of chains by suitable chemical treatment. Some hundred and fifty different abnormal human hemoglobins had been detected since the early nineteen-fifties, Perutz said. He received blood samples for possible structural determination flown in on ice from all over the world, sometimes two or three a month.

That morning, Perutz had two chores. He began with "a new and interesting abnormal human hemoglobin—it's called 'Milwaukee'." Abnormal hemoglobins are usually named after the places where they are first found. "The two beta chains are unreactive to oxygen, because the valine residue in position 67, which is very near the iron atom of that chain, is replaced by a glutamic acid, which blocks the iron atoms. Yet the molecule does show some degree of interaction between the heme groups." To find out why, Perutz wanted to prepare crystals of the deoxygenated form.

As he talked, he went over to a large, steel-framed, glass-walled box that stood on the bench. The thing was a caricature bust of a mechanical man: the double barrel of a binocular microscope and its knobs, sticking up through the box's lid, donated an attenuated nose, forehead, and eyes to the odd image, while out of the front of the box hung two long, flaccid, quivering rubber arms, with hands of different colors, one pilgrim gray, the other by chance repair a diseased yellow-pink. "Nitrogen box," Perutz said. The pink arm, partly inflated, rose with the majesty of a gloved policeman stopping us to allow children to cross. "It's filled with nitrogen, under slight pressure in case of leaks and pumped constantly across a platinum catalyst to remove even traces of oxygen. This you need because crystals of deoxyhemoglobin would otherwise avidly take up oxygen and change their structure." He checked the microscope first, to be sure nobody had changed the focus, then poured talcum powder generously over his hands, pushed one of the rubber gloves right side out, poured in talc, and began working his hand into it, into the nitrogen box, up to the elbow. The first hand gone, the second glove was awkward.

The sample of hemoglobin Milwaukee was already in the box, Perutz said. He had received ten cubic centimeters, about a tablespoonful, of a solution prepared by a hematologist in Buffalo. "The blood comes from still another place in America, Saint Louis I think, where there's a man who has it." With his beta chains unreactive to oxygen, I asked, how did the man survive? Also in the box, according to an inventory posted on the wall above it, were crystals, quartz capillaries, small tubes for storing crystals, ground-glass-stoppered test tubes, microscope slides, pasteur pipettes, pasteur pipettes cut off, pasteur pipettes drawn out, syringes (disposable and glass with long needle), tweezers, teats (one with hole), scissors, cotton threads, filter paper, filter-paper wedges, razor blades, dissecting needles, rack for specimen tubes, Kleenex, and beeswax. "Oh, he's a heterozygote," Perutz said. "He has the normal gene as well and is not seriously ill; but for a small fee he will donate blood from time to time."

"First I have to dissolve the components of the reducer." Perutz's gloved hands were moving, inside the box, with a deliberate economy that avoided all the fragile catastrophes. He stirred two powders into a small flask of water. "Ferrous sulphate plus sodium citrate makes ferrous citrate. If you add this to hemoglobin, it takes the oxygen off. But this oxidizes in air, itself, so it's a good idea to mix it, too, in a nitrogen atmosphere."

One rack in the box held six small test tubes. "Now I add a concentrated salt solution to precipitate the hemoglobin. Each tube here contains the same amount of hemoglobin. The idea is that the first tube may produce a lot of little crystals, and in the last tube you probably get none at all because the salts were too dilute, and in the middle somewhere you get, you hope, crystals of a good size and without faults." He used a hypodermic needle and a graduated syringe to measure a different amount of salt solution for each tube, checking quantities against a chart. "Next I'm putting in the iron-citrate solution, the reducer I mixed, a tenth of a cc. to each tube." He sealed each tube and set them all in a large jar with a ground-glass stopper sealed in turn with what he said was a silicone grease. "Now they go in a cupboard for a month to see if they will grow. Best not to disturb them: don't come in every day to see if they are growing." Precipitation was not precipitate.

Next Perutz was to mount for x-ray diffraction some crystals of horse hemoglobin from which the terminal arginines of the alpha chains had been chopped off. The loss, he said, affected the salt bridges between chains that held the molecule in shape; the result was a hybrid structure between the oxy and deoxy forms. "Years and years I've spent with horse hemoglobin. It's a plague." This set of test tubes had been put to grow crystals nine weeks earlier. "Here's the tube with the most concentrated salt solution: the crystals are all too fine." The tube contained what looked like ruby port, with slush in the bottom. "Here, this one, only slightly less concentrated, no crystals at all. Somewhere in between—yes, here.

"First I take out the clear liquid on top," he said, doing so with a long pipette, rubber fingers squeezing rubber bulb, squeezing again to drain the liquid into a fresh tube. "This comes in handy later on." Used tools he put to one side, to take up clean ones. "It's a lot easier to do this than to manipulate one of those boxes for radioactive materials where the hands are mechanical and you work with controls outside." He sucked up crystals from the tube and dumped them with a little of their liquor onto a microscope slide, which he then maneuvered under the objective. In the light shining up from beneath, the crystals glowed like garnets. He went after a little one with a dissecting needle, teased it out from the clump of trash, then another.

Through the microscope, the featureless dark wall of the tunnel opened out and the crystals floated silently above a white nimbus, filling one's entire capacity for attention—translucent, flat, faceted, brilliant russet-red. A few days' work would usually determine the elementary classification of a crystal's symmetry and repetitive structure. To form a crystal at all, the molecules had to be identical, so that they packed, one by another, in the same alignment. Crystallographers, knowing this, were less surprised than biologists at Svedberg's determinations and later Sanger's.

Under the microscope, a well-formed crystal allowed the angles between its faces to be measured. When polarized light was shone upwards and the crystal turned, it would go dark as the layers of its hemes sliced across the direction of polarization, then glow translucent once more as they were moved to run parallel to the polarization and allowed the light through. "Most of the bigger ones here are not single crystals and so won't give good x-ray patterns," Perutz said. This light was polarized; contrasting patches of light and dark within the larger crystals betrayed their flaws.

Perutz shoved two or three crystals to one side, opened another tube. "More saline—more crystals, but smaller and possibly worse. Yes, this is fiddly work. But very soothing. If you're nervous or worried, I can recommend it as a tranquillizer. It's entirely manual, but demands absolute concentration." He fell silent, examining a particular crystal, pressing it with a needle. It clicked. "There. I split that double one—cracked the one part but left the other intact." Taking up the intact portion in a drop of liquid in a pipette, he transferred it to a capillary tube made of glass thinner than Christmas-tree balls, thin as a flake of ash, so thin his fingers threatened to crush it. The walls were a thousandth of a millimeter thick. Just one man in the world, a glassblower in Berlin, had the knack.

Now, slowly, the crystal slid down inside the capillary tube, within the pale liquid; got stuck on an air bubble; then moved again. X-ray crystallography was already a quarter-century old when Perutz learned it; but the first x-ray-diffraction picture of a protein crystal had been taken only three years earlier, in Cambridge, by Bernal and a young colleague, Dorothy Crowfoot. Before that, Astbury had obtained some tantalizing blurs and rings in x-ray pictures from fibrous proteins; Astbury and others had tried to get diffraction patterns from crystalline proteins, but had failed. "There were even papers published saying that protein crystals gave no x-ray-diffraction patterns,"

Perutz said. "Bernal had discovered why they failed, and that the way to get good pictures was to keep the crystals wet. So Bernal and Crowfoot had taken x-ray photographs of pepsin, by sucking the crystals up into a thin glass capillary with their mother liquor; and they got beautifully sharp diffraction patterns, extending to the spacings of the order of interatomic distances."

More than thirty years after that, in a volume of essays prepared for the pleasure and honor of Linus Pauling, Dorothy Crowfoot Hodgkin wrote of the "ancient history of protein X-ray analysis." The crystals of pepsin had been grown by a biochemist named John Philpot, when he was working in Svedberg's laboratory in Sweden. Philpot had left a bottle in the refrigerator while on a skiing holiday, and when he got back was astonished to find crystals more than two millimeters long, hexagonal double pyramids of pepsin. Hodgkin went on:

> He showed them to Glen Millikan, a visiting physiologist from California and Cambridge, who said, "I know a man in Cambridge who would give his eyes for those crystals." Philpot naturally offered him some crystals to take back in his coat pocket and so Millikan took them to J. D. Bernal.
>
> It was very lucky for protein crystallography that Millikan took the crystals in the tube in which they were growing in their mother liquor. This enabled Bernal to make his first critical observation: that the crystal lost birefringence [that is, stopped refracting polarized light in the way of true crystals] when removed from their liquid of crystallization. He observed only a vague blackening of the film when X-rays were passed through the dry crystal. . . . The wet crystals gave individual X-ray reflections, which were rather blurred owing to the large size of the crystals. . . . but which extended all over the firms to spacings of about 2Å. That night, Bernal, full of excitement, wandered about the streets of Cambridge, thinking of the future and of how much it might be possible to know about the structure of proteins if the photographs he had just taken could be interpreted in every detail.

"Protein crystallography began with the paper by Bernal and Crowfoot in *Nature* in 1934," Perutz said. The paper, in fact, could give no detail at all about pepsin's structure, though Svedberg's value for its molecular weight, from the ultracentrifuge, was found compatible with the size of the crystal's unit cell—that is, its minimum repeating block of molecular pattern. "It took twenty-three years before the first protein molecule was solved," Perutz said. "But Bernal was one of the most imaginative scientists I have ever met. He could see the possibilities in his first step."

With a decisive flick, Perutz snapped off the end of the capillary. "The next step is to introduce a thread of cotton which will suck out the liquid around the crystal." He did that, slowly and craftily. Then he aligned the crystal by inserting a dissecting needle. A moment's divergence of angle and the tube would have splintered. "What I've done is to mount the crystal between two bits of cotton—actually pipe-cleaner fluff—that will hold it in position. Then I moisten it with a drop of the solution saved earlier that contains the reducer." He sealed the end of the tube with a drop of beeswax, which he had melted by pressing a treadle that passed a current through a bare wire within the nitrogen box. The crystal was ready to be set before an x-ray tube.

There, a crystal so mounted could be turned, slowly and precisely, in the thin x-ray beam as one might roll an uncut diamond between thumb and finger, up to the light, to glimpse its fire. In 1895, at the University of Würzburg,

Wilhelm Conrad Röntgen was experimenting with the discharge of electricity through gases held in a glass tube at very low pressure. One day he had nearby a screen coated with a fluorescent compound used to detect ultraviolet light; while the tube was discharging, he noticed the screen brightly fluorescing. Röntgen put his hand between the tube and the screen: he saw the shadows of his bones. He placed photographic plates, well wrapped, near the screen; they became fogged. A key placed on a box of plates left its shadow. He traced the effect to a spot on the wall of the tube where the glass glowed greenish, directly opposite the cathode, or negative terminal, of the electrical rig. He named the new kind of radiation X-Strahlen, or x rays, for their mysterious properties. In 1901, Röntgen was awarded the first Nobel prize in physics.

In 1897, at the end of April, in a Friday Evening Discourse at the Royal Institution in London, Joseph John Thomson, Cavendish Professor of Physics, announced the existence of negatively charged corpuscles a thousand times lighter than the hydrogen atom. Thomson's corpuscles were also called cathode rays and, later, electrons. That was a time when physicists' ideas of the nature of light and all kinds of radiation, of energy, matter, space itself, were severally and radically in question just at the dawn of the quantum theory and relativity. In 1898, George Gabriel Stokes in Cambridge suggested that x rays were pulses of radiation emitted when electrons, ejected from the cathode, hit an anticathode, or target.

Thomson, in his lectures in Cambridge during the next decade, talked about the generation of x rays. One of his students was William Lawrence Bragg; half a century later, Bragg wrote of Thomson, "He pictured an electron moving with its associated lines of force, and the 'whip-crack' which would run along these lines if the motion of the electron is suddenly arrested at the anticathode of the X-ray tube." Bragg also heard lectures in the physics of optics; one of these was devoted to the nature of white light and the geometry by which, when it passed through the close-ruled lines of a diffraction grating, it was spread into the spectrum of visible wavelengths. Whether x rays behaved as waves like light, or as particles was not settled then. Bragg's father, William Henry Bragg, at the University of Leeds, had excellent experimental reason to argue that x rays were particles.

In 1912, it occurred to a young physicist in Munich, Max von Laue, that x rays, considered as waves, had wavelengths far shorter than visible light—about as short, in fact, as the spaces between atoms in a crystal. In that case, he realized, the repetitively spaced layers of the crystal would affect the x rays just as a diffraction grating affected light, scattering the waves in patterns. Laue suggested to two students, Walter Friedrich and Paul Knipping, that they try to find such patterns on a photographic plate. Laue had not yet worked out even where to set the plate in relation to the beam and crystal. The tubes then available forced long exposures and frequent stops to cool.

They began with a crystal of copper sulphate, very low in crystallographic symmetry but possibly high in secondary fluorescent radiation, which they thought might be important but which they soon realized was irrelevant. After developing several plates slightly and vaguely blackened, Knipping lined up a plate on the far side of the crystal from the beam and got the first x-ray-diffraction photograph—a smeary spiral of spots. They had proved decisively the wave nature of x rays. When they changed to a crystal of highly symmetrical form, a cubic crystal of the simple mineral zincblende—zinc sulphide: zinc and sulphur one to one—and set it so that the x rays fell perpendicularly on a cube face, they got a handsome crisp pattern, four ways symmetrical, obviously relating to the symmetrical arrangement of the atoms in the crystal.

Sir Lawrence Bragg told the history of the discovery and development of x-ray diffraction many times, always with fresh, fascinated pleasure in the interplay of luck and intelligence at each step. He told the story synoptically in conversation the time I went to see him in London, at Perutz's suggestion, early in 1971, and most completely in the book he was writing then and just finishing when he died that summer. Laue's paper appeared early in 1912 in the *Proceedings of the Royal Bavarian Academy of Sciences*. The elder Bragg saw the paper and showed it to his son during the long vacation. They were both reluctant to abandon "the parental hypothesis" that x rays were particles. Towards the end of the vacation, Bragg set up an experiment in his father's lab to test whether the spots could be due not to waves but to x-ray particles being shot down avenues between the atoms in the crystal structure. The experiment was fruitless.

"When I returned to Cambridge and pored over Laue's paper I could not but be convinced that he was right in ascribing the effect to diffraction," Bragg wrote. "At the same time I had my great idea, which explained Laue's results in a much more simple way than he had in his original paper." Laue had assumed that the two kinds of atoms that composed the crystal of zincblende alternated at the corners of a repeating cube; then, in order to calculate the positions of the spots on the photograph as the result of diffraction, he was obliged to suppose that the x rays were of five specific and otherwise arbitrary wavelengths. Bragg's great idea—he also used to call it his "brain wave"—was a unitary geometrical vision of the exact relationship between the arrangement of atoms in the crystal and the x rays that passed through and onward to the photographic plate. The relationship, as he said, was extremely simple. It was also spatial—a construction that was, of itself, essentially speechless, which must be one reason why Bragg, seeing it whole in his mind that autumn day in Cambridge in 1912, had felt the silent delight that he always afterward remembered.

"My idea was that the formless x-ray pulses could be regarded as reflected by sheets of lattice points in the crystal." Lattice points? Think of a vulgar patterned wallpaper; choose any one feature of the design—tip of a rosebud, eye of a bird; find every place where that same feature repeats: those points make a lattice. Choose any other feature of the same design, find every place where *this* repeats: *these* points make, in that design, an equivalent lattice. Now think of the same thing in three dimensions. A lattice point in a crystal is an atom that is found repeatedly in the same patterned setting of other atoms. Sheets of lattice points? The points of any particular lattice can be connected up by imaginary lines, and in several different ways, from the more or less rectangular or hexagonal to the increasingly oblique. In three dimensions, these imaginary connections form not lines but planes.

Unit cell? Think of the wallpaper, of the minimum portion of the pattern which the printer could stamp out repetitively to build up an entire roll, an

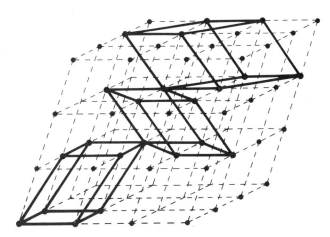

entire Victorian wallful. In three dimensions, the repeating block is the unit cell. For x-ray crystallography, it is the unit cell that is classified by its symmetry or lack of it as triclinic, monoclinic, orthorhombic, hexagonal, and so on to cubic; and the fundamental dimensions in the crystal are the lengths of the axes of the unit cell. None the less, whereas the lattice points are assigned to real atoms, the unit cell, like the lattice sheets, is an abstraction, a geometric convenience for analysis.

More courteous than most writers on crystallography, Bragg was sparing in the use of these abstractions.

> The object is to get as accurate a map as possible of the positions of the atoms in a structure. A crystal structure lends itself to the analysis of the diffraction phenomena, because the units of pattern are regularly arranged and diffract in an identical way, but in general the crystal is only a means to an end and "X-ray analysis of atomic arrangement" is a more appropriate title than "X-ray analysis of crystals". For inorganic substances such as the minerals the crystalline structure is an essential feature, because the pattern is a continuous one with no molecular boundaries. But in the vast field of the complex organic compounds, the crystalline structure is of quite secondary importance. The arrangement of atoms in the organic molecule is the object of the research.
>
> When the X-rays fall on the crystal structure, they are scattered by the atoms, or to be more precise by the distribution of electrons around each atom. Heavy atoms with many electrons scatter powerfully, light atoms much less.

Bragg's understanding was of a relationship of motions, a happening in three-dimensional space. "I was led to think whether Laue's results might be due, not to a few definite wavelengths in the incident x rays as he supposed, but to the action of the crystal grating on the formless pulses of radiation which Stokes had supposed to constitute x rays. I realized that such pulses would give diffracted spots, because they must be reflected by the planes of the crystal lattice in definite directions."

Diagrams are static and flat, and several must therefore be superimposed in the mind to convey what happened. As a wave fell on a sheet, the wavelets scattered by the points built up a reflected wave—an established first principle of the wave theory of light.

"When I applied this test to Laue's pictures it was at once clear that all the spots were indeed in positions explained by reflection in the crystal planes."

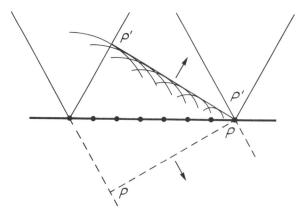

To check that x rays behaved that way, Bragg shone a narrow beam of x rays onto a sheet of mica.

> Mica cleaves so readily that one would expect the sheets of atoms parallel to the cleavage to be well marked. I found that for any angle of incidence of the rays, the mica always gave a reflected beam as if it were a mirror reflecting the radiation. I vividly remember taking my plate, still wet from the fixing dish, down to J.J. Thomson's room and showing it to him. It was very gratifying to see my professor's great excitement.

But the parallel planes of points in a crystal lattice were stacked up; the stack created the effect equivalent to a diffraction grating. When a beam of x rays penetrated the crystal at an angle to these planes, its waves met the atoms and were scattered in all directions. The rays that penetrated deeper obviously had farther to go. That extra path length was crucial. The rays coming off the successive planes were now, in most directions, out of step with each other—out of phase, their troughs and crests no longer marching synchronously. Their interference—trough of one with crest of another, and so on—cancelled each other out. Only at certain specific angles did waves reflecting from successively deeper planes drop behind precisely enough to fall back into step and build a strong reflected beam.

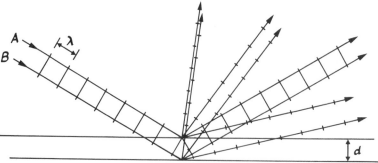

Bragg saw that the angle at which the glancing x-ray beam was reflected from any particular set of parallel sheets of points, the spacing of those sheets, and the wavelength of the entering-and-departing train of waves, bore an unvarying relationship to each other. The relationship was inherent in the geometry.

The relationship was stated in a formula, which expressed the angle at which the waves from each plane have fallen exactly one step behind those from the plane above. The scattering also produced further orders of reflection—that is, other strong beams—at those specific higher angles where the waves were displaced by exactly two steps, exactly three steps, and so on.

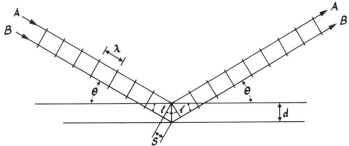

For those with a geometer's eye, the formula reads directly off the diagram.* But the crucial fact is simply that if the wavelength and the angle are measured, the spacing of the lattice sheets emerges. "This optical principle has come to be called 'Bragg's Law', though its only novelty was its application to X-rays and crystals!" Bragg wrote, long afterwards. But he also saw that when the pattern repeated in three directions, it built up discrete diffracted beams—and so produced discrete spots on the photographic plate. He drew the diagram below himself, for his first paper on his discovery, read to the Cambridge Philosophical Society on 11 November 1912.

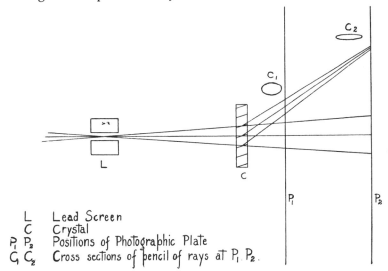

L   Lead Screen
C   Crystal
P_1 P_2 Positions of Photographic Plate
C_1 C_2 Cross sections of pencil of rays at P_1 P_2.

The spacings and intensities of the spots in Laue's x-ray picture of zincblende made perfect sense, Bragg found, if the crystal lattice was taken to be face-centered cubic—that is, with the zinc atoms at the corners and at the center of each cube face, and sulphur atoms forming a similar lattice displaced

---

*The wavelength of the reflected wave train is conventionally denoted by the Greek letter *lambda*, while *theta* is the glancing angle at which the radiation strikes and leaves the crystal plane in phase, and *d* is the spacing of the planes: $\lambda = 2d \sin \theta$. The further orders of reflections come off at angles whose sines are whole-number multiples of that of the first reflection. The formula that describes any reflection, where *n* is its particular order, thus becomes $n\lambda = 2d \sin \theta$—Bragg's Law.

from the zinc cube halfway along one of its diagonals. The first substances whose atomic arrangements Bragg solved from x-ray photographs that he took himself were the salts sodium chloride, calcium chloride, and calcium bromide. All three were simple cubic lattices with atoms of the two sorts alternating at the corners. The arrangement of atoms in sodium chloride, common salt, established, he reasoned back from the density of the crystal and the molecular weight of the substance to determine that adjacent atoms were 2.8 angstroms—a little more than a hundred millionths of an inch—apart. Bragg thus, in passing, first established a scale for measurement of all crystal spacings and all x-ray wavelengths.

Bragg's law is the rigorous statement of the intuitively evident fact that the scattered wavelets from each lattice point will reinforce each other in certain directions, cancel out in others. Think of the ocean beating through spaces in a barrier reef and spreading into the still lagoon within—in overlapping partial circles rising to peaks of spray where crest meets crest. Just so, but in three dimensions, as spreading partial spheres, the trains of diffracted x rays now overlapped, reinforcing or cancelling out. When intercepted by a flat photographic surface, the diffracted beams produced an array of spots varying, sometimes markedly, from pale to dark. The variations indicate, cryptically, the distribution of scattering efficiency—that is, electron density—across the lattice sheets.

A set of such photographs, taken with the crystal realigned each time so that one of its axes was parallel to the beam, presented to the crystallographer all the repetitive features of the molecules in the crystal—but transformed, and requiring to be read back into the object that produced it.

The glistening dark red jewel of modified deoxyhemoglobin of horse aligned and sealed in its capillary, Perutz set the capillary gently in a larger tube, with a ground-glass stopper, standing in a rack. Perutz straightened up and sighed with annoyance. "Sorry, but I'm beginning to sweat." He began stripping out of the rubber gloves. "Before I mount the next ones I've got to take my cardigan off." I left him to his tranquillizer.

❖ ❖ ❖

Rutherford died on 19 October 1937, after surgery for hernia.

Perutz that winter got x-ray patterns for chymotrypsin and hemoglobin that allowed him to measure their unit cells. Methemoglobin of horse proved to have a rarely favorable crystal form. The wet crystals showed sharp spots far out towards the edge of the pattern, which meant that they were reflected by Bragg planes, within the crystal, as close together as two angstroms; this proved that the molecules were identical, one to the next, down to atomic dimensions. The unit cell could be chosen in such a way that each one comprised two molecules of hemoglobin, which was small and handy. Figured from the size of the unit cell and the density of the crystal, the molecular weight of hemoglobin came to 69,000, in reassuring agreement with the 67,000 of Adair and Svedberg. Further, there was but one molecule per lattice point, which meant that all the molecules were lined up in the same direction—a fact that was to make analysis much easier.

When the crystal was set up so that one of its axes, that one conventionally labelled b, was vertical and perpendicular to the x-ray beam, the spots on the top half of the film exactly mirrored the positions and intensities of those on the bottom half. Such symmetry did not show up along the other axes. Down the middle of the unit cell in the direction of the *b* axis, therefore, ran an axis of twofold symmetry: that is, around that axis the unit cell divided into two halves that were identical—and again, to atomic dimensions. This was the decisive simplification. Perutz's diffraction photographs proved, before the chemists could show it, that the hemoglobin molecule had two identical halves.

Bernal and another of his colleagues, Isidore Fankuchen, an American crystallographer, wrote with Perutz a short letter to *Nature* about the first results with hemoglobin and chymotrypsin. The letter appeared on 19 March 1938. Eight days earlier, Hitler had invaded Austria. Perutz's parents left Vienna immediately for Prague. From visiting student, Perutz was transformed in a few hours into a refugee. "How can anyone imagine what an appalling shock this is?" Perutz said. "It affects your relations with everyone around you. Before, you were a guest. And now—what are you? Do they really want you to stay? Well, some people did, and others very clearly did not." Perutz also knew that his money would soon run out.

How to take hemoglobin any further was not evident. Protein molecules were a thousand times the size of anything whose structure had been solved by x-ray analysis before. With simpler structures, "X-ray analysis is essentially a cut and try method," Bragg said in a lecture in 1952; and he explained:

> The normal procedure is to make a guess at the arrangement of the atoms in the crystal and check whether this postulated arrangement would diffract the X-rays in the manner actually observed. The guesses are inspired guesses based on previous experience of the relative arrangements of atoms, on what the chemists can tell us about the molecule in the case of an organic compound, and on any other scrap of information which narrows the range of likely structures, but it remains a guess.

The largest molecule then analyzed by those methods was penicillin, by Dorothy Crowfoot Hodgkin and Charles Bunn; their finding took four years of work and was published in 1949. Penicillin had a molecular weight just over 300.

Pauling's response to that problem in 1938 had been to turn to the exact analysis of the structures of amino acids and the peptide bond, in order to use these as the blocks for model-building. In a discussion just before the war, Bernal told Pauling that that approach was too indirect and too slow. Bernal favored straightforward crystallography. He had found an obscure, nineteenth century paper that recorded the swelling and shrinking of protein crystals when moistened or dried or when penetrated by dye molecules. Hemoglobin and chymotrypsin crystals shrank when dried, but no longer produced good patterns. There were other examples. Crowfoot, working in Oxford but in close touch with Bernal, was able to measure the progressive shrinkage of crystals of a protein from milk, lactoglobulin, as they dried and to correlate that with the change in the x-ray pattern; she described this in a letter to *Nature* accompanying the one by Bernal, Fankuchen, and Perutz. Bernal decided, and said in their letter, that drying protein crystals showed changes in the spacing and intensities of their x-ray spots that ought to "make possible the direct . . . analysis of the molecular structure once complete sets of reflections are available in different states of hydration." Perutz settled down to try.

In Prague, in the first week of April 1938, Felix Haurowitz took from a cupboard a sealed bottle of purified hemoglobin of horse that he had set to grow crystals. He used a pipette to move a few drops of slushy liquid to a microscope slide, then popped a glass cover slip over them and put the slide under the microscope. He had expected to see light-colored needle crystals of oxyhemoglobin. Instead, the slide carried bluish-red, flat, large hexagonal tablets of deoxyhemoglobin. (Later, he decided the bottle had been contaminated by bacteria that used up the oxygen.) As he looked at the crystals, a wave of change moved across the field of view. Beginning at one side, progressing across the middle, the hexagonal tablets disappeared, dissolving back into the liquid. Then beginning again at the same side, new crystals formed and quickly grew—the lighter-red needles of oxyhemoglobin. Air had penetrated from one edge of the cover slip. In every other way, the conditions from which the two forms of crystals had emerged were, of course, literally identical.

Haurowitz then examined individual hexagons of deoxyhemoglobin in polarized light, letting air at them. Even as the color of a crystal began to lighten in a ring around the outside and a tangle of tiny needle forms began to show there, the ring became translucent to the polarized light—while the center of the crystal remained dark longer. Inescapably, the binding of oxygen by hemoglobin changed the crystal lattice totally.

Haurowitz reported these observations in a paper published in Germany that summer, with a picture, taken through the microscope, of the crystals transforming. He noted that a closer look at the change of structure would require x-ray studies of deoxyhemoglobin in addition to the start on methemoglobin that had been reported from Cambridge. Perutz found that Haurowitz's description and photograph brought the central problem of the structure of hemoglobin so vigorously alive that thirty years later he told me that Haurowitz had shown him the transformation through the microscope in September 1937, six months before the discovery in fact was made. "I did repeat Haurowitz's experiment, and grew the crystals of horse deoxyhemoglobin, and took x-ray pictures of them, but they had an impossible unit cell and space group," Perutz said. "The crystals were hexagonal, and molecules were arranged along a hexagonal screw axis so that they were all lined up in a row. And the total length of the unit cell was the length, therefore, of six molecules of hemoglobin—that is to say, three hundred and fifty angstroms. There was no way of measuring x-ray pictures of such crystals in detail at that time."

Perutz spent the summer of 1938 in Switzerland. Small and not conspicuously robust, he was a leather-tough, passionate mountaineer and skier from his student days in Austria. Rationalizing his passion and his crystallography—and in order to get a travel grant for the vacation—he took up the study of glaciers. The summer's recreation was published as a crystallographic study of glacier structure and flow, in *Proceedings of the Royal Society* for August 1939. He published three more papers on glaciers after the war. By the summer of 1938, Czechoslovakia was under pressure from Germany. Perutz's parents fled to Switzerland, where they found temporary asylum. On September 30, at Munich, the British and French governments yielded Hitler the Sudetenland, and the partition of Czechoslovakia began. Haurowitz got away to Turkey, where he worked at the University of Istanbul throughout the war; he emigrated to the United States in 1948 and went to Indiana University, where he taught the course on proteins and nucleic acids in which he said that genes were necessarily made of protein—and in which James Watson got an A.

At the beginning of October 1938, Bragg arrived in Cambridge to succeed Rutherford. Bernal moved to Birkbeck College, London. Bernal was a scientist of outstanding intelligence and vision, and of encyclopedic learning—friends called him Sage—animal vitality, and charm. There was no possibility of his being elected to any Cambridge chair, or even to a fellowship at one of the colleges—in part because of the Cambridge University politics of the day, in part because of politics on a larger frame, his Communism. Bernal took Fankuchen—also a Communist—and all of biological crystallography with him except Perutz. With Bragg arrived several colleagues, including a new head for the crystallographic laboratory.

Several weeks passed before Perutz gathered his courage and his hemoglobin patterns and went to see Bragg. "Max was the man who had got the best photographs of a protein at that time, x-ray pictures," Bragg said in our conversation. "And our fancy was fired with these. He wanted to go on, and I was all for his going on. Tremendous worker. As he is still. Incredibly patient and—diligent, extraordinarily diligent. Our ideas, mainly his ideas at the time, about what the molecule was like, were absolutely wide of the mark. Silly sorts of models." Bragg at once got in touch with the Rockefeller Foundation, through Wilbur E. Tisdale, at the foundation's International Education Board in Paris, to ask for a grant to support Perutz's research. "I wish to take a part in this research myself," Bragg wrote to Tisdale on November 28; and he went on, in part:

> My research team at Manchester was largely responsible for making the first extension of X-ray analysis from very simple to complex crystalline patterns, and I should now like to have a share in extending it still further to the very complex patterns of proteins.
>
> I wish to employ Perutz as my assistant, and to give him a salary of £275 [$1,320] a year. It is easier for me to ask him to stay on and do this research if I can promise him a post for three years ahead.

Bragg enclosed prints of some of Perutz's x-ray photos and a one-page "programme of research." He secured Perutz's grant, to begin January 1, and £100 for a new and more powerful x-ray tube. Perutz was thus able to guarantee his parents' support and secure them permission to immigrate to England; they arrived in March.

❖ ❖ ❖

The way back from the spots on the film, riding the waves that made them, into the crystal was in theory straightforward, for there existed a mathematical technique that was perfectly matched to the problem. This was Fourier synthesis, named after Jean Baptiste Joseph Fourier, a mathematician of Napoleonic France who first set down its principles. Any physical event that repeats at regular intervals can be represented by a Fourier series, which reduces it to a set of simple wave forms, with equivalently simple mathematical statement. Fourier methods offer mighty intellectual leverage. They are found everywhere in the natural sciences and technology. They are elegant and by no means recondite. None the less, the high palisade that modern education generally erects between the hard and the soft, between those who have a clear sense for mathematics and those who don't, runs on the wrong side of Fourier analysis.

Fortunately—just as one can point out the gross syntactical features of a foreign language without requiring that it be learnt—the principles behind Fourier techniques can be conveyed, for the historical point, without close study of the equations. Notes from musical instruments supply the classic start. A tuning fork that strikes *A* sounds a beating variation in air pressure at 440 cycles per second that will drive the eardrum, or a recording pen, to produce the simplest possible periodic curve, the sine wave. The amplitude—the loudness—of the note may vary, but the simplicity of the shape, and the fundamental frequency, will remain. *A* played on flute or oboe is unmistakably the same pitch, yet has the timbre characteristic of the instrument. Timbre is produced by the overtones, the harmonics, and each of these has a frequency that is a whole-number multiple of the fundamental tone. The complex periodic form of the note that the instrument plays can thus be analyzed into a series of forms, each a simple wave.

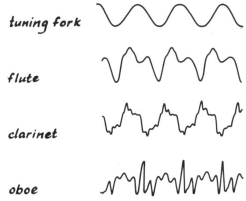

Any periodic phenomenon of whatever complexity, if it can be measured, can be broken down by Fourier analysis in this manner, though often the labor requires a computer.

The essential syntactical feature to be grasped is that each of the simple wave forms composing a Fourier series is defined by a pair of factors, one its amplitude and the other its phase. Amplitude measures the strength of that component; amplitude is represented by the height or depth of the wave at a given point. Phase, more subtly, measures the position of that component wave form, its relative displacement ahead of or behind the others in the set; phase is represented by the angle the wave makes at that given point.

A Fourier analysis breaks the entire complex periodic phenomenon down into the series of those pairs of factors, amplitude and phase, one pair for each of the simple component waves. An equation results. In the musical example, the equation places the overall variation in amplitude—the note one hears—on one side of an equation, and sets equal to it the sum of all the paired terms for amplitude and phase of the fundamental tone and each harmonic, going on until all the harmonics present have been listed.*

Fourier methods presented x-ray crystallography an almost magical promise, but one attended by an apparently intractable problem. The problem lay in the notion of phase and phase angle. The flute, once more, will suggest why. The note of the flute is made up almost entirely of the fundamental and the first harmonic, double the fundamental frequency; the harmonic in the flute's case has an equal amplitude but its phase is displaced by one-sixth of a period. The syntactical point is that amplitude and phase are independent of each other; to solve the equation requires that both terms of each pair be found.

The magical promise lay in the geometry of diffraction, whether of light or of x rays, for there Fourier methods have an unexpectedly direct meaning—a pleasing literalness. When the terms of a Fourier series are summed, the original, complex periodic event is restated. But this can be done physically. Thus, if the pure fundamental tone is sounded, and simultaneously each pure harmonic, at the correct amplitude and in the correct phase relation, then the note of the flute, the oboe, or whatever is perfectly synthesized. A diffraction grating bears to the spectra of diffracted light that it produces the same relationship as the device that dissects the note of the flute does to the fundamental and overtones. However, when light of a single wavelength shines through a diffraction grating, each one of the discrete spectra that results is not a harmonic of the fundamental wavelength. It is itself one of the successive terms of a Fourier analysis. But the analysis is not of the original light, for that was at a single wavelength and as simple as the beat of a tuning fork. Rather—here's the unexpected part—the analysis is of the diffraction grating. At this dénouement, lecturers on optics produce delightful demonstrations in which beams of light are recombined like musical tones to produce an image of whatever diffracted them.

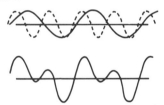

An optical grating is a line of slits, and mathematically one-dimensional. But as Bragg had seen at the beginning, a crystal is a three-dimensional grating. Fourier analysis in three dimensions quickly becomes cumbersome, but the principle is the same. In 1915, the elder Bragg noted the mathematical consequence of the fact that a crystal structure is a physical phenomenon that repeats periodically. In a lecture to the Royal Society, he said:

> If we know the nature of the periodic variation of the medium we can analyse it by Fourier's method into a series of harmonic terms. . . . The series of spectra which we obtain for any given set of crystal planes may be considered as indicating the existence of separate harmonic terms. We may even conceive the possibility of dis-

---

*A closer look at the principles behind the mathematics of Fourier analysis will equip the non-mathematical to comprehend the years of difficulties that Perutz finally broke through: a discussion, short and highly schematic, appears as a note at the end of this chapter.

covering from their relative intensities the actual distribution of the scattering centres, electrons and nucleus, in the atom.

The direct physical correspondence between the observed diffracted beams—the spots in the picture—and the terms of a Fourier series was not fully realized until 1925, when several investigators, the younger Bragg among them, came to the understanding independently.

With stunning simplicity, where the film intersected the x rays the amplitudes of the diffracted rays were themselves the amplitudes of the Fourier terms, while the relative phases of the diffracted rays were themselves the phase factors. The amplitudes for all the terms—in other words, for all the harmonics, corresponding to all the sheets of scattering points in the crystal—could be known from the brightness of the x-ray spots. The amplitude at each spot was the square root of the spot's intensity; it thus might have either the positive or the negative value, be above or below the base line, and which sign was right was part of what could be known only from the phase angle. "If one also knows the phases, the crystal is solved," Sir Lawrence Bragg once wrote; then, his explanation felicitously visual, "All that is necessary is to sum the Fourier series, and the positions of the atoms will appear as condensations of density of appropriate amount and extension." Since the electrons (and not the nuclei) did the scattering, the summation produced a portrait of the crystal as a cloud of electrons that varied in density, repetitively.

Yet there, in the central principle of the Fourier method, stood the difficulty that barred the way: the spots on the x-ray photograph yielded the amplitude of each Fourier term, but not its phase. Since the phases might have any values, the spots by themselves could be explained by an infinite number of distributions of electrons. The phases must be found. Fourier methods resolved the problem of x-ray analysis, Bragg wrote, "into one of *phase determination*, and the advances which have made it possible to apply analysis to more and more complicated structures centre around new ingenious ways of finding the phases."

From the late twenties, Fourier methods began to be applied to crystals of somewhat larger, organic molecules. Each structure solved, however, still began with the informed, inspired guess of a likely arrangement of atoms. The guess, if only approximately right, told enough about the values of the phases so that a trial Fourier series could be calculated. "Fourier series have a surprisingly obliging way of trying to tell the investigator something with the most sketchy basis of information," Bragg wrote another time. Even that tentative series would improve the positions of some of the atoms, and from that refined structure more phases could be fixed and a new series calculated that would in turn allow still further refinement, a progressive sharpening of focus. That procedure worked well for structures whose atoms were numbered by the dozen. For proteins, Bragg's cut-and-try was out of the question. Missing the essential clue to the phases, the information in the diffraction patterns from hemoglobin was otherwise not too meagre but far too copious. Even a small protein may require many tens of thousands of measurements; the difficulties go up very rapidly with the number of atoms.

At the end of January of 1939, Bernal gave a Friday Evening Discourse at the Royal Institution. These occasions are a pleasant eccentricity—for they still go on—founded by Michael Faraday in 1825. In good Prince Albert's golden day, when aristocrats and the new manufacturing rich patronized the sciences like the arts, once a month in season, after dinner, an audience in evening dress assembled to hear an eminent scientist explain and to watch him show off his latest discoveries. Faraday himself astonished the top people with demonstrations of the relationship between electricity and magnetism. J. J. Thomson announced the electron at a Friday Evening Discourse. By the nineteen-thirties the showmanship and the science had declined towards bird watching with slides. Bernal discoursed about the structure of proteins. Little was known, much was hoped, and the hopes lived on the many schemes that glimpsed consistent uniformities of structure and size in the proteins. Svedberg, for example, thought by 1939 that he had discovered that the molecular weights of proteins seemed to fall into definite classes that were multiples of one another. "This suggests very strongly that all proteins are built from some common unit," Bernal said. "What that unit is is more difficult to determine."

It was not clear, he said, whether the amino acids were linked together by peptide bonds and in chains, or joined by other links and in rings or cages. "The evidence that the X-ray study of the crystalline proteins . . . provides for the elucidation of their structure is abundant, but it is extremely difficult to interpret. Photographs of crystalline proteins show hundreds of spots and marked differences of intensities," Bernal said. "Unfortunately, however, direct analysis of these photographs is rendered impossible by the fact that we can never know the phases of the reflections corresponding to the different spots. The ambiguity introduced in this way can only be removed by some physical artifice."

He suggested two artifices. One was the observation of changes in intensity of the spots as the crystal was dried. He explained that. The other was "the introduction of a heavy atom." This meant, though he did not explain, that if an atom of some extremely heavy element could be introduced in regular, repeated positions into the protein molecule, its extremely large number of electrons might influence the scattering of the beam, and so the intensities of the spots, enough to betray the phases.

❖ ❖ ❖

When German armies overran Holland and Belgium in the spring of 1940, the British rounded up all enemy aliens living in coastal areas. Cambridge is some fifty miles inland from the North Sea coast. On Whitsunday, the end of May, Perutz was arrested. "It was a very nice, very sunny day—a nasty day to be arrested." Along with others from Cambridge, he was taken to a temporary camp at Bury St. Edmunds, then to another in a suburb of Liverpool, then moved to the Isle of Man, in the Irish Sea—and then, in his westward progress, transported to Canada, early in July, where he was interned in the citadel high above Quebec City. The views were superb—of the Northern Lights, of the American frontier to the south. "Freedom was always in view, at least."

By fall, the number in the camp grew near a thousand. The internees were diverse—ordinary people, often refugees in the first place, many students, and a remarkable concentration of intellectuals and scientists. In the citadel with Perutz were, among others, the physicist Herman Bondi and the cosmologist Thomas Gold. Perutz organized a camp university, arranging courses of lectures in which such men taught their specialties to fellow internees. Perutz taught x-ray analysis. "Our star performer being Klaus Fuch"—Perutz gave a small and wintry smile—"whose lectures on theoretical physics were better than any I have heard before or since." The absurdity of imprisoning these people, many with skills they were eager to put to war, was patent even in the emotional state of the time; after a quiet campaign in England on behalf of the internees, most of them were brought back that winter. Perutz returned to the Cavendish and the swelling and shrinking of hemoglobin early in January of 1941.

Over the next year, Perutz learned to dry crystals of methemoglobin of horse slowly, and to stop the drying at intermediate stages. He took x-ray pictures from all angles at each stage. He measured the intensities of thousands of spots. He performed intricate calculations. He drew diagrams. The results were sparse. He got hold of some crystals of oxyhemoglobin that appeared favorable for x-ray analysis, and tried to compare them with methemoglobin. Then he was given some crystals of methemoglobin of horse that had been grown from a solution without salt. The new crystals in their salt-free liquor had exactly the same form as the first, but yielded some reflections of different intensities. The differences allowed Perutz to judge whether certain of the Fourier phases were positive or negative. Those were enough to suggest a curve representing the electron density on a line through the molecule—a curve with four peaks, equally spaced, about nine angstroms apart. "This suggests a haemoglobin molecule consisting of four equal and parallel layers of scattering matter," Perutz wrote in the spring of 1943 in a note for *Nature*. "There might be four layers of polypeptide chains with the main chains folded in the plane of the layers and the side chains protruding at right angles to them." That might allow those side groups that carried a charge to face outwards, bottom and top of the stack, since they were attracted to water, and the uncharged side groups, being hydrophobic, to face inwards. There were suggestions, also, that the molecule might be shaped like a dumbbell.

Perutz asked repeatedly for a job in the war. "And this finally came in the form of a mysterious call to work on the mechanical properties of ice," he said. He had been chosen because of his work on glaciers. "I was told that I should try and find ways of making ice more resistant to cracking—making it less brittle. But nobody disclosed what this was for. So I worked away on this until I got a letter from Hermann Mark, who was now in Brooklyn, whom they'd enlisted in this same project. And Mark, being a polymer chemist, knew that polymers lose their brittleness if you embed fibres in them. So he tried to embed wood pulp, the sort of stuff you make newspapers from, in ice, in order to make a watery mush of, I don't know, five per cent wood pulp; and froze that—and it worked. It abolished all the brittleness and made ice, weight for weight, as good as concrete. Or so we thought. And then, gradually, it transpired that this was needed to make an aircraft carrier of ice."

The cover name was Project Habbakuk. The idea was to make an aerodrome of ice, equip it with refrigeration machinery, and tow it to the middle of the Atlantic to serve as staging post for aircraft flying from the United States to Britain. Habbakuk's originator was a journalist named Geoffrey Pike, who had the ear of Bernal, who had the ear of Lord Louis Mountbatten, chief of combined operations of the Royal Navy, who had the ear of Winston

Churchill, who gave the project high priority and took it and Bernal to the Anglo-American summit meeting in mid-August 1943 in Quebec, where they had the ear of Franklin Roosevelt and his military advisers.

Perutz went to London in top secrecy, and was given a Cambridge engineering student and some commandos for assistants and a laboratory that was the bottom level, five stories underground, of an immense icebox—a meat-storage plant at Smithfield Market. England was not cold enough; at Quebec it was decided to build the ice craft in Newfoundland and to transfer the planning to Washington. "So in September '43, we were all supposed to be whisked to Washington. And then it was remembered that I was an enemy alien—and the Americans wouldn't let me in." A detective came to Perutz with some perfunctory questions. Two days later, he was handed a British passport. "And then, because everything was so fantastically urgent, I was sent across the Atlantic with about a dozen other VIPs by air. First we flew in a Sunderland flying boat to Shannon—neutral territory, so all the military got into plain clothes—and then in a Yankee Clipper from Shannon to Gander, and from Gander to New York. This was my first transatlantic flight; a great experience. We all signed a dollar bill; I put it away carefully but have never been able to find it. And we landed in New York Harbor. Where the immigration officer became very suspicious about a man who carried a passport saying 'British subject by naturalization September 3, 1943'—and I arrived there on the fifth. He thought this was some put-up job by the British. Questioned me no end of time." He spent the weekend with Hermann Mark, then took the train to Washington. He thought the British experts there would be working sixteen hours a day. "But to my surprise the entire team came to meet me at the station, and at four o'clock in the afternoon, all terribly homesick—and none of them had anything to do. The project was being examined by the U. S. naval engineers. And they just hung around there waiting. So I did the same for several months."

The problem was creep. "While it was true that weight for weight the Pikrete, as we called it, was as good as concrete in *fast* tests of tensile strength or of shear strength, it was subject to slow creep, which concrete is not. Like a glacier. A glacier actually flows, very slowly but still it flows, under a shear stress of only one kilogram per square centimeter—about the same as one atmosphere of pressure, and that is a very small stress. So the ice craft would have needed strengthening, with steel—or, lowering the temperature. To reduce creep. And of course to lower the temperature you had to put on board much more refrigeration machinery. And finally the Navy Department came to the conclusion that the amount of steel needed for reinforcement and to refrigerate all the ice needed to build one of these aircraft carriers, even in the winter of Newfoundland, would be greater than the amount needed to build the entire thing of steel."

❖❖❖

Perutz came back to the Cavendish early in 1944 to a pile of data. His grant from the Rockefeller Foundation had grown to allow him to hire two assistants. They had been measuring spots. They judged the intensity of each spot by comparing its blackness with a reference film-strip carrying a score of spots graduated from light to dark and numbered to define a scale of 1 to 100.

Each crystal was seen in three patterns, as viewed down the axes. The unit cell offered 62,700 separate lattice points even for an attempt to resolve the structure no more sharply than to a spacing of 2.8 angstroms; the favorable symmetry of the methemoglobin molecule, however, made many of these points equivalent and reduced the number of different reflections to be measured to 7,840.

Without phases, the only apparent hope to get any structural information from the data was by the method of Patterson synthesis. In 1934, Arthur Lindo Patterson, then at MIT but originally a student of the elder Bragg, had discerned that the mathematics—at least formally—did permit a Fourier summation based on a diffraction pattern to be made without a phase angle in each term, and with the observed intensities of the spots used directly instead of their square roots, the amplitudes. No phases; not even the problem of knowing for every square root whether it had the positive or the negative value—Patterson's Fouriers made use only of data that read directly from the diffraction picture. Plowing a very deep furrow, Patterson comprehended that such a Fourier synthesis was itself something more than a formal possibility: it had a physical meaning.

Even crystallographers found the meaning difficult to visualize. Where a straight Fourier analysis, given the phases, produced in sum the positions of the scattering points in the unit cell, Patterson's method produced values for the relation—in space and in scattering power—between each possible *pair* of scattering points in the unit cell. That quickly became an unmanageably large number. The hemoglobin molecule, hydrogen ignored, contains about five thousand atoms. Symmetry reduced this to twenty-five hundred positions in space, each, of course, defined by three coordinates. Thus, where straight Fourier analysis would aim to produce these sets of coordinates, seventy-five hundred independent values, Patterson's method produced, at full stretch, a set of values for each pair of atoms—and twenty-five hundred atoms make 3,123,750 pairs.

A cocktail party for a hundred strangers requires only a hundred invitations; for each person at the party to get to know every other requires five thousand introductions, ten thousand attempts to remember a new name. The guests at a Patterson party had their shoes nailed to the floor: the values his method produced were for the length and direction of reach of every handshake and its strength of grip. Such values—vectors, to the mathematician—could be diagrammed. In short, Patterson bought freedom from the phase problem at enormous conceptual and computational cost.

Patterson methods have an important place in x-ray crystallography. Thus, the first structure of an amino acid, glycine, was solved in 1939 by Robert Corey, in Pauling's lab, using a Patterson synthesis. But for large biological molecules the method failed—and twice crucially. Rosalind Franklin, in 1952 and '53, was beaten to the solution of the structure of DNA in part because of her obstinate dedication to the pure crystallography of Patterson methods, her refusal to consider model-building until too late. Perutz spent 1938 to 1952 attempting Patterson methods on hemoglobin, out of dedication to the molecule rather than the method. The most valuable thing he gained was Patterson's devoted friendship. Perutz's hope and Bragg's was that by taking Patterson sections and projections through the molecule on various planes, to reduce drastically the number of pairs being considered at once, they might stumble onto regularities of structure that would let them get further. "I had that great and childish faith that there might be some underlying simplifying features. Which the three-dimensional Patterson would reveal," Perutz said. "Because otherwise, of course, it wouldn't tell you anything." The illusion of order led them on.

John Kendrew came to the Cavendish early in 1946 and asked to work with Perutz. John Cowdery Kendrew was born in Oxford in 1917, and raised there; his father was a climatologist at the university. He trained as a physical chemist in Cambridge and took his degree in the spring of 1939. He had wondered even then about switching to biology, with no clear idea of what kind. Soon after the war began, he joined the Air Ministry, where he worked at first on the introduction of airborne radar. In 1940, he turned to operations research as a civilian scientist for the Royal Air Force, first in England, then in the Middle East, then in Southeast Asia.

Bernal was also in the Far East by then, as a scientific adviser to Mountbatten, who had been named supreme allied commander for Southeast Asia. Kendrew spent time with Bernal, and Bernal sang to him of the allure of proteins and x-ray spots. On a military trip to California, Kendrew met Pauling and learned of *his* interest in the structures of amino acids and proteins. "Being a chemist, one always thought of Pauling as one of the great figures of the century, in chemistry," Kendrew once said in conversation. "So to find that he had also moved over to biology, obviously this had its influence. But I think a much bigger influence in my case was Bernal, because I spent much more time talking to him."

At war's end, Kendrew might have gone to London to work with Bernal, except that he had some scholarship money unused in Cambridge and so approached Perutz instead. Perutz taught him the elements of crystallography, arranged that he take a doctorate as a research student of Bragg's, then had to find him a research topic. Hemoglobin was the same slow hard problem; the few other proteins whose crystals were known were either already being done or were unsuitable. One day near the Physiology Laboratory, a couple of blocks from the Cavendish, Perutz met Sir Joseph Barcroft. Barcroft reminded him that the hemoglobin of fetuses behaves differently, in biochemical functioning, from that of adults of the same species, and suggested that Kendrew might try to find a structural basis for the difference by x-ray crystallography. He offered a supply of blood from fetal sheep. Kendrew's thesis subject was settled.

Early in 1947, Perutz with his two assistants finished a fifty-page paper on the results with hemoglobin so far. Great labor and dogged ingenuity had produced meagre conclusions. "The methaemoglobin molecule appears to be a cylindrical disk of average height of 34 A with a slightly convex circular base of 57 A diameter," they wrote, and included a drawing of the cylinder. Beneath it was drawn an aspirin tablet stood on edge, a heme group in the orientation at which they thought it sat in the molecule. "This is merely a simplified, diagrammatic picture, giving as it were the fuzzy outline of the molecule, whose surface could not possibly be as smooth as this drawing suggests." They found four marked peaks of density in certain sections through the molecule, and decided that these meant that it was built in four layers like a cake.

Later that year, Bragg turned to the Medical Research Council for support for the x-ray studies of hemoglobin. The council agreed to set Perutz and Kendrew up as a two-man unit. In the English way, it became clear gradually, as the unit grew, that Perutz was its chief.

By the end of the year, the two had completed a paper comparing fetal with adult hemoglobin of sheep. The x-ray studies proved all but conclusively that the two were different proteins, but could say nothing about the molecular structures that underlay the differences.

In the spring of 1948, Perutz finished the three-dimensional Patterson synthesis of methemoglobin of horse. He was writing it up when, in May, Pauling visited the lab. A few weeks earlier, in bed with flu in Oxford, Pauling had worked out his folded-paper model of what he later called the alpha helix; this visit to the Cavendish was the time when Perutz showed him the data that suggested rodlike structures running through the molecule, and Pauling privately thought, but did not say, that these were probably the helical form he had found for the polypeptide chain.

Barcroft died in 1947; in June of 1948, a symposium on hemoglobin was held in Cambridge in his memory. Perutz and Kendrew reported their studies, but also prepared a paper together about the application of x-ray crystallography to large molecules of biological interest—an adept and stylish review of their specialty for an audience of scientists who here were laymen. They mentioned the phase problem with patent anguish: "the fundamental difficulty of all X-ray analysis, . . . encountered in its most acute form in the analysis of very large molecules." In simple cases it was possible sometimes to "deduce or guess the phases of some or all of the reflexions." For instance, the symmetry of the crystal might be such that along the appropriate axis the diffracted rays were all exactly in phase, or exactly out of phase, with the incident beam; this reduced the problem to one of finding merely the signs—plus or minus, peak or trough—of the amplitudes of each reflection.

> In such cases it may be possible to apply various tricks to find out the signs of some reflexions. For example, if one atom of the unit cell has a very much larger atomic number [in other words, is much heavier] than any of the others, its contribution to each reflexion will also be very large by virtue of its high electron density, and will effectively determine the sign of all but the weakest reflexions; once the position of the heavy atom is known direct calculation will often give the signs of enough strong reflexions to enable a first approximation to the crystal structure to be computed. This is the *heavy atom* method; another is the method of *isomorphous replacement*.

Isomorphous replacement, they explained, required a series of closely related compounds, which differed only by a substitution of a single atom and which crystallized in the identical unit cell (hence, "isomorphous," from the Greek for "same form"). In such a series, the pattern of spots was laid out identically but with progressive changes in the intensities, from one crystal to the next, that could be attributed to the different scattering power at the atom that had been switched. The classic use of isomorphous replacement had been the solution, in 1927 by James M. Cork, an American working with the elder

Bragg at the Royal Institution, of the structure of the alums, a series of simple minerals typically joining an atom of metal—which can be one of several elements—with aluminum and sulphur.

The general principle of substituting atoms or introducing heavy ones was current in Bernal's lab at the time that he and Crowfoot took the first diffraction pictures of pepsin. When Crowfoot moved to Oxford and began to work on the structure of insulin, she took her first x-ray photos of the zinc salt of insulin back to show Bernal, who then sent her a note that suggested she ought to attach an atom of cadmium to the molecule in place of zinc and grow crystals of that. Her attempts with cadmium and later, in 1941, with iodine were halfhearted, but produced discouraging indications that a heavy atom introduced into such a large molecule made no detectable difference to the intensities of the strong reflections. Though these experiments were never published, Perutz knew of them.

Hemoglobin was several times larger than insulin. So Kendrew and Perutz warned, in 1948 in their primer on crystallography for physiologists:

> Where a macromolecule (such as a protein) forms crystals . . . none of the methods devised for guessing phases in single crystals can be applied. For example even the heaviest of heavy atoms would make a negligible contribution to the reflexions from so large a unit cell, so it would exert no control over the phases.

That was the received opinion among crystallographers. Perutz and Kendrew themselves, therefore, had taken the only alternative.

> The physical meaning of this so-called Patterson synthesis is one of the most difficult conceptions in crystallography. It would hardly seem justified . . . to dwell on a method as abstract as Patterson's, were it not for the supreme importance which this method has now assumed in the analysis of macromolecular structures.

Supreme importance: the week before the Barcroft symposium, Perutz turned in the report of the complete three-dimensional Patterson synthesis of methemoglobin of horse. He read it before a meeting of the Royal Society at the end of the year. "The photographing, indexing, measuring, correcting and correlating of some 7000 reflexions was a task whose length and tediousness it will be better not to describe," he said. To get from those data to a set of contour maps then required that a Fourier series of seven thousand-odd terms be summed for each and every point at intervals one angstrom apart throughout the volume of the unit cell—which meant, even helped by the symmetry of the molecule, 58,621 points and therefore a set of computations with more than four hundred million separate terms. Punched-card computing machines were then coming into use in science, and one of these made the calculation possible. Hemoglobin still had four layers; Perutz now thought that each layer was packed with parallel rods of polypeptide chain—hatbox had become cigarette case—or with a longer chain repeatedly folded back on itself. The implication, unstated, was that each layer contained a heme group. And he said:

> If the globin molecule consisted of a complex interlocking system of coiled polypeptide chains where interatomic vectors occur with equal frequency in all possible directions, the Patterson synthesis would be unlikely to provide a clue to the structure. On the other hand, if the polypeptide chains were arranged in layers or parallel bundles, interatomic vectors within the layer plane or in the chain direction should occur particularly frequently and should give rise to a vector structure

showing a corresponding system of layers or chains, which could then be interpreted without difficulty. All the more plausible hypotheses of globular protein structure put forward in recent years have been based on systems of the latter kind.

He cited suggestions by Astbury and Pauling.

Hence it was not unreasonable to hope that the Patterson synthesis might lead to interpretable results which would justify the great effort involved in its preparation.

Kendrew had meanwhile been looking for a protein of his own. Of the half-dozen reasonably small ones, most had by then been claimed by others. "One picked the next best," Kendrew said. He picked myoglobin. "There was another element to it, that I was collaborating with Max, and he was already on the hemoglobin thing." Myoglobin carried one heme group and one iron atom and was a quarter the weight of hemoglobin; had a similar spectrum; formed oxy-, carbonmonoxy-, and metmyoglobin. It combined with oxygen even more avidly than hemoglobin; as we have seen, its oxygen dissociation curve was hyperbolic and not sigmoid, as expected of a molecule in which cooperativity between hemes could not occur. The stuff was found in muscle; its function was to take oxygen from hemoglobin and store it until needed. "The proteins were obviously in some way related." Kendrew began with myoglobin of horse. Though the crystals were impossibly small, by mid-1948 he thought he could say that myoglobin's basic structure was like that of hemoglobin—a flat layer of rods. "Max had his famous hatbox model, which was a cylinder with four layers in it, each layer being a chain, you see. And I came out with a tentative model, simply a penny, one chain. They were both quite wrong."

"For a long time the idea that the molecule contained some kind of regular structure of protein chains, which would give a strongly defined character to a Patterson synthesis was a guiding star which encouraged the investigations. As events turned out, it was a false star," Bragg said in a lecture in 1963. "If this had been realized at the time the problem would have seemed so hopeless that the quest might well have been discouraged, but fortunately this was not the case."

❖❖❖

"My main function, I think, was to keep them at it," Bragg said in 1971. Perutz in the summer of 1948 had led an expedition that camped for two months on the ice of the Jungfraujoch, attempting to drill a hole a hundred and fifty meters deep in order to measure the relative rates of flow at top and bottom of the ice. He had picked up again his interest in the physics of glaciers. He found that glaciers flow just as some metals do when they are rolled. Below a critical stress such a metal is hard, above that soft; in the same way, ice seems rigid near the surface, where stresses are least, but becomes plastic where they are greatest, near the glacier bed. "He was very keen," Bragg said. "You see, I rather think that Perutz might have gone off into other lines, if I hadn't had such tremendous enthusiasm about this."

Bragg began to get deeply interested in Perutz's deductions about the dimensions and folding of the polypeptide chain. Though the finality of the Patterson impasse was not faced, Bragg turned Perutz and Kendrew to the general question of the form of the chain. The three started to catalogue helical models—the sprawling and indecisive catalogue that was vitiated by an error of physical chemistry, by an assumption that was too restrictive, and by a misleading datum. The datum was the spot in Astbury's prewar diffraction patterns from alpha keratin that seemed to demand a turn of the screw every 5.1 angstroms. The assumption was that the helix had to have an integral number of residues per turn. But the shaming error was their ignorance of the fact that the peptide bond, between amino-acid residues, had the character of a partial double bond, so that little rotation could take place there, the atoms to each side all having to lie in one plane. Bragg, Kendrew, and Perutz, on polypeptide configurations, was finished at the end of March 1950, and appeared in *Proceedings of the Royal Society* in October. As that work was under way, in the fall of 1949, Francis Crick joined the unit at the Cavendish and began to teach himself crystallography.

That fall, also, Perutz's imagination was fired by Pauling's discovery that sickle-cell and normal hemoglobins moved in an electric field as different molecules. "It of course interested me greatly that there should be a chemical difference, and as I thought perhaps a structural difference, between two genetically different hemoglobins," Perutz said. "The paper was marvellously exciting, and inspired me to try and determine what the difference in structure was. Which, in a way, was absurd—because I had no idea how to determine the structure of one hemoglobin, let alone two." That fall, he got a small sample of red blood cells from a sickle-cell patient from a friend at the Rockefeller Institute, Chandler Alton Stetson, who said he had noticed, under the microscope, that the transformation of a cell in sickling looked as though a crystallization were spreading from a single origin. Perutz also obtained some sickle-cell hemoglobin in solution. He found at once that sickled cells behaved in polarized light as though they contained crystals, while normal cells did not. Some simple chemistry demonstrated that the two kinds of hemoglobin were soluble to exactly the same degree when oxygenated, but that sickle-cell hemoglobin, deoxygenated, was only a tenth as soluble as normal deoxyhemoglobin, and thus was many times more likely to precipitate in crystals as the cells lost oxygen to the tissues. The great surprise was that sickle-cell hemoglobin in the oxy form could be made to grow crystals that were identical with those from normal human blood so far as x-ray analysis could detect: the same unit cell, the same x-ray patterns in every detail and intensity.

This was Perutz's first examination of an abnormal hemoglobin. With several colleagues, he published two letters in *Nature* about it over a year and a half. They were fundamental to an understanding of the disorder, to an understanding of the whole class of disorders of hemoglobin. They greatly annoyed Pauling, who thought their conclusions had been implicit in his own. They produced no clue to the structural problem.

Crick's first essay in protein crystallography was to demolish the illusions of his colleagues that the five years of work with Patterson projections and the three-dimensional Patterson synthesis had found fundamental regularities, the layers of parallel rods, in the structure of hemoglobin. Crick sampled variations of Perutz's simple models and showed that any of them would produce much more pronounced Patterson vectors—much higher peaks on the contour map—than Perutz in fact had found. Perutz had proved only that the molecule had to be disorderly. Perutz's data, correctly understood, Crick said, showed in particular that the model could not contain long parallel stretches of polypeptide chain; straight lengths were not likely to be more than fifteen to twenty angstroms, broken up by corners. The reasoning was simple, dead central, and devastating. Crick argued the case first in a seminar within the lab, several months after his colleagues had published the paper on the possible helical chains. "Here was this new fellow, whom you couldn't stop talking—you know how Crick argues, he *hammers*; and he was telling us that what we were so proud of, was wrong," Perutz said. This was that seminar that Crick titled, at Kendrew's suggestion, "What Mad Pursuit."

Pauling published the alpha helix, the less stable gamma helix (never found in life), the pleated sheet, and other structures in April and May of 1951. James Watson arrived in Cambridge early that fall. Perutz's group was at a peak of activity that Watson hardly mentioned in *The Double Helix*—but he was never much interested in the structure or function of proteins and had yet to learn crystallography. Kendrew had just finished a study of the use of the new digital computers for calculating Fourier and Patterson syntheses. Crick was writing up "What Mad Pursuit" for publication, now basing his calculations on alpha helices packed in Perutz's model. Bragg and Perutz were finishing a series of three more papers on hemoglobin. In the first they quietly conceded that the molecule could not, after all, be a highly regular structure of parallel polypeptide chains. In the other two, Bragg used Perutz's x-ray data to establish the external form and true dimensions of the molecule: they now found it to be ellipsoidal, and with several possible sets of dimensions, on the order of 55 angstroms by 55 by 65. (The actual dimensions were later found to be 50 angstroms by 55 by 64.) And that "rather modest result," Bragg wrote years later, "is noteworthy as being the first definite quantitative piece of knowledge to be won."

❖C❖   The diversion into sickle-cell anemia brought Perutz a letter, early in 1953, from a young biochemist at Harvard, Austen Riggs. They had not been in touch before, but Riggs had read Perutz on sickle-cell hemoglobin and enclosed a copy of a paper of his own. Riggs had tried combining human hemoglobin with a reagent, an organic chemical called para-chloromercuribenzoate that contains an atom of mercury in each molecule. Several people before Riggs had shown that similar chemicals attach themselves to hemoglobin specifically at the sulphur atoms—more exactly, at the sulphydryl groups (symbolized –SH)—at the tips of the side chains of the amino acid cysteine. Riggs was first to try para-chloromercuribenzoate. Mercury has a high affinity for sulphur. His aim was to block the sulphydryl groups to see if that affected the oxygen dissociation curve and thus, by implication, the interaction among heme groups.

He found that two molecules of the reagent combined with each molecule of human hemoglobin. In fact, human hemoglobin contains six cysteine residues; two of these were evidently more reactive than the four others, probably because they were at the surface of the molecule. Riggs showed that blocking these flattened the oxygen-dissociation curve considerably. In passing, he tried the reagent on sickle-cell hemoglobin, and got the same result—and thought to bring his work to Perutz's attention. For Riggs, the important conclusion was that "sulphydryl groups of globin play a large part in the mechanism of heme-heme interaction." In Perutz's context, the significance of the experiment was almost exactly the reverse: even with the mercuribenzoate in place, the oxygen-dissociation curve was still strongly sigmoid. "I jumped when I saw that, because it was clear to me that if it left the biological properties intact, then it would also leave the structure intact," Perutz said. "And there was a hope then that one might crystallize hemoglobin with the mercury attached and the crystals might prove isomorphous with the crystals of the native hemoglobin."

The promise of Riggs's finding was understood at the Cavendish as nowhere else. In the course of 1951, preparing for the papers written with Bragg that fall and winter, Perutz had taken the measurements of hemoglobin's x-ray diffraction one tedious but important step farther. "I tried to determine what is known technically as the absolute intensity from a hemoglobin crystal. That means the fraction of the incident x-ray beam which is actually diffracted." In other words, where he had previously measured the relative degree of blackening of the x-ray film, now using a geiger counter he measured the fraction of the incident x radiation that reached each spot. These figures helped Bragg and Perutz get the shape of the molecule in the fall of '51, and then, for a paper published in 1952, to determine directly the electron density along the line of one particular slice through the molecule. "We snatched at such small successes to keep ourselves in heart to carry on," Bragg later wrote.

Absolute measurements of intensity once made, Perutz also used them to calculate the overall effectiveness of the protein molecule as a diffractor of x rays. "Perhaps I should say, first of all, that the method of isomorphous replacement was nothing new," Perutz said. "You might say that essentially it was used by Bragg when he determined the structure of common salt. He compared the intensities from crystals of potassium chloride with sodium chloride. They have the same structure, but in the one the scattering from the potassium ions just about matched that from the chloride ions, whereas in the other the chloride ions scattered much more strongly than the sodium ones did." Heavy atoms had been used to get several striking crystallographic solutions since. Most recently, and with a drama that had aroused the enthusiasm of every crystallographer, the structure of the complex, asymmetric, almost featureless cluster of six crazily tilted carbon rings that makes the poison strychnine had been solved by Johannes Martin Bijvoet in Utrecht in a virtual dead heat with a team using the methods of classical organic chemistry under Robert Robinson in Oxford. "The reason we didn't try it in proteins before was that we didn't realize that one heavy atom could produce a noticeable effect in the diffraction pattern, amidst the enormous mass of thousands and thousands of light atoms," Perutz said.

His calculation of the absolute intensity was a surprise. "If you do this sort of experiment for, let us say, the crystal of common salt, you will find that for the very strongest reflections, quite an appreciable fraction of the incident beam, more than a tenth, is deflected into the diffracted beam," Perutz said. Bragg had established that at the beginning. "But I found that in a protein crystal, the absolute diffraction was very small indeed—suggesting that even in the most powerful reflections, the scattering contributions of something like ninety-nine per cent of the atoms obliterated each other by interference and only that minute fraction, say from a hundredth of them, were reinforcing each other and giving rise to a reflection. And *that*, of course, was why we had to use such long exposures to get diffraction pictures from protein crystals. Now, I think it must have been at that point that it became clear to me that a heavy atom introduced into the protein structure would make a measurable difference in intensities of the diffracted rays."

Four years earlier, Perutz and Kendrew had written that "even the heaviest of heavy atoms would make a negligible contribution to the reflexions from so large a unit cell." That doctrine now was called into question. "This is because the electrons, say in a mercury atom—eighty electrons in a mercury atom—are all concentrated at one point," Perutz said. "So they'd all be reinforcing each other. While the electrons of the light atoms are spread out over a large volume, so that, as I said, most of their scattering effects obliterated each other by interference. I probably did some rough calculations to see what kind of effect the heavy atom would have—and I realized that it might enable one to determine the phases."

What about the iron in hemoglobin? "The iron atoms are not very heavy, for a start. Iron has an atomic number of 28—it contains twenty-eight electrons—while mercury has eighty. But the other trouble was that you couldn't replace them, you know; they were just fixed points. There are two methods in x-ray crystallography by which you can determine structure with heavy atoms. One is known as *the* heavy-atom method, and its principle is that if you have, say, an organic chemical—a structure of carbons, hydrogen, nitrogen—you attach to it an atom which is so heavy that it outweighs the scattering contributions of all the light atoms and dominates the phases of the diffracted rays. There is no hope of doing any such thing in a protein, because there would be no atom heavy enough to dominate all the phases. But you see there is this other method of introducing a heavy atom and comparing the intensities in its presence and its absence. That does not require that the heavy atom dominate the phases, merely that it should produce measurable changes in the intensities. And I realized that *that* method might work." But he published none of these calculations. "I had no idea how I could introduce a heavy atom into hemoglobin."

Bragg first heard of Riggs's paper in a conversation with Francis J. W. Roughton, who was Professor of Colloid Science and another in Cambridge's line of authorities in the physiology of hemoglobin. Bragg saw the point at once, and went to tell Perutz—"in great excitement," he said later, "only to have Perutz tell me very coldly that *he* had given this information to Professor Roughton!"

Perutz knew nothing about para-mercuribenzoate. But a few months earlier, Vernon Ingram had joined the unit. I once asked Ingram about his work with Perutz. We talked, over sandwiches, at the faculty club at the Massachusetts Institute of Technology, where he had been teaching since 1958. He was spare and English (though born in Breslau; his parents, refugees, brought him to England) and had a succinct and humorous intelligence.

"The actual suggestion for the experiment came from Max," Ingram said. Riggs had used human hemoglobin, which Perutz, of course, knew offered no crystal form—no unit cell—favorable to x-ray analysis. It was essential to try methemoglobin of horse. "Max said that hemoglobin had this –SH group, one per half molecule, and that therefore it should be easy to attach the mercury atom to it, because it combines very readily with –SH groups; and therefore let's try and crystallize a mixture of hemoglobin and the mercury derivative. He made the suggestion. I did the chemistry. And then I can't remember who did the actual crystals—either he or David Green." Green was a graduate student of Perutz's.

"And that was the first crystalline protein with a heavy atom on it—and enabled Max to produce the first Fourier projection of hemoglobin. The very first real projection of a protein. It wasn't a three-dimensional model. But it was the starting point." It was the start also for Ingram of a lifetime's interest in hemoglobin—with his next step, of course, the demonstration, at Crick's urging, that sickle-cell anemia was due to a single substitution of an amino acid. We talked about that and about the wonderful persistence of hemoglobin as a central object of study. "Because hemoglobin was a human protein, there was a wealth of medical information concerning its diseases," Ingram said. "In other words, your pool for finding mutants was enormous. And the inherited hemoglobin diseases forced themselves on peoples' consciousness, because they were very important in clinical terms."

"Certainly Pauling wanted to solve the structure of globular proteins as much as any of us did," Perutz said. "He certainly realized its importance and had great ambitions to do so." We were sitting, after lunch, at a corner of the table in the dining room of the house in Cambridge that Perutz and his wife bought in 1960; the house was small, but in such almost indefinable qualities as the brown of the woods in the furniture and the colors and thicknesses of the books on the shelves it had the same hint as their speech of the Middle European in the midst of altogether fluent Englishness. "There was Harker in Brooklyn, working on ribonuclease, and Carlisle in London working on the same protein. There was Dorothy Hodgkin's work on insulin, and parallel work on insulin being carried on by a former pupil of hers, Barbara Low, at Columbia. And that's about all. The subject really had a bad name, and the people working on it were widely regarded as cranks by other crystallographers as well as biochemists. Because it was hopeless. You could never solve it.

"One mercury atom on each half-molecule. When I took the first x-ray pictures of these crystals, I saw that they were indeed isomorphous with normal hemoglobin of horse: that is to say, they had exactly the same cell lattice and gave a very similar-looking diffraction pattern. But when I examined the picture closely, I saw that there were marked changes in the relative intensities of the diffraction spots. This perhaps was the most exciting moment in all my research career"—Perutz's voice took on a buttery glow of satisfaction—"because I realized at that moment that now, in principle, the protein problem was solved." The moment was early in the summer of 1953. Five months before, Watson had been exalted by a similar rush of conviction on seeing the x-ray picture—a far simpler picture seen by a simpler eye—that told him the structure of DNA could be solved. Perutz paused, then said, "I of course thought that within a year I would know what the structure is. That was another matter."

Pauling's workshop in the structure of proteins, expanded to admit DNA, took place in Pasadena that September. To an audience of the world's top crystallographers, Perutz first announced his discovery. The meeting was a triumph for Bragg's group. It marked Bragg's departure from the Cavendish: after fifteen years in Cambridge, he was moving to London, to direct the Royal Institution. His reasons were two. He was sixty-three, and wanted something new while still active. His father had directed the Royal Institution for nineteen years until his death in 1942, and now the place was in scientific decline and financial trouble, so that Bragg had a strong sense of obligation. He asked Perutz and Kendrew to come with him; each hesitated, then refused, though both became advisers to the place, and Kendrew worked hard to help Bragg set up a new protein-crystallography group there.

"The way Bragg put it was that I had found the Rosetta Stone for the deciphering of diffraction patterns. That, I think, is a very good simile," Perutz said. "Because I had the key; but I couldn't immediately unlock all the secrets. First of all, we ought to go into a physical problem of x-ray analysis which comes in at this stage. I ought to make this very clear, because it was really of vital importance. Certain restrictions on values of the phases are introduced by crystal symmetry. Suppose we take a quite symmetrical structure—not a protein but, say, urea. In this structure, the molecules in the unit cell are related by what is known as a center of symmetry. That is to say, there is a certain point in the unit cell where for each atom on *one* side of this point, there is a diametrically *opposite* atom in an exactly equivalent position. What that means for the phases of the diffracted rays is that they, too, *must* be so placed that their wave crests are related to the center of symmetry—so that for each crest on one side there must be an equivalent crest on the other side."

At the center of symmetry, all the waves were in step, having either a crest or a trough there. Building out from there, "you can forget about exact phases, and can give your Fourier terms simply signs: you can say for each reflection that it is either plus or minus. And of course that means that your task of determining structure is enormously simplified. Instead of having to find a phase, which can have any value from nought to three hundred and sixty degrees, all you have to do is to find the sign for each Fourier pair. Up to 1935 or so, very few structures were solved that did not have a center of symmetry.

"Now, proteins, of course, consist of amino acids—which are inherently unsymmetrical, so that a protein crystal cannot possibly contain a center of symmetry. However, it may contain a symmetry *axis*. Suppose you take a protein molecule like hemoglobin which consists of two identical halves related by a symmetry axis. Then when you look down that axis—in other words, project the structure along that direction onto a plane—again there will be, for each point on one side of the axis, an equivalent point on the other side. That means that the x-ray reflections from lattice planes which are parallel to the symmetry axis—but only from those—will, again, be restricted by the symmetry: they again must all have a crest or a trough at the symmetry axis. All right? And, yes, you can align the crystal so that you take only the reflections from the planes parallel to the symmetry axis.

"Now. When you apply the method of isomorphous replacement, and use a single heavy atom, what it allows you to do in the first place is not to determine the *phases* of all the diffraction patterns, but only to try and determine the signs of those reflections which are from planes parallel to the symmetry axis. The argument runs like this: you have, say, two mercury atoms, as we had in hemoglobin, in equivalent positions one on each side of the symmetry axis. To apply the method, first we must determine the positions of the heavy atoms." Here, Patterson methods were uniquely valuable: one or a small number of heavy atoms made large and easily interpreted differences to the diagram, locating themselves relative to the symmetry axis. "The contribution from the pair of heavy atoms alone to each of the different reflections will then be known directly, and will be either plus or minus."

Once those were calculated, and then the x-ray patterns from the crystal with and without the heavy atom compared, "if the intensity of the reflection *rises* from having the heavy atom, then you know that they are reinforcing each other and so that the sign of the protein reflection by itself"—the object of the search—"is the same as the sign you've got for the heavy-atom reflection. And if it *falls*, you know they're opposite. So, at this stage, what the method gives you is the signs of all the protein reflections from lattice planes parallel to the symmetry axis.

"And this is all that I got from that first stage in 1953." Pages of tables of amplitudes with and without heavy atoms generated an odd, spidery diagram like a bright child's drawing of a choppy sea: a nine-layer stack of irregular wavy lines, sprinkled with more than two hundred plus and minus signs. These were the abstract essentials of a paper that Perutz completed with Ingram and Green early in 1954. It was the crucial paper of four, all about the structure of hemoglobin, that Perutz wrote with various colleagues and that Bragg communicated in March, as a block, to the *Proceedings of the Royal Society*—a formidable technical *tour de force*, a fundamental contribution to the theory of diffraction, and for all that an anticlimax.

"What I was able to calculate, then, was not the three-dimensional structure but an image of the hemoglobin molecule projected down the symmetry axis," Perutz said. The molecule viewed in that direction was 63 angstroms thick, corresponding to some forty atomic diameters; what Perutz got was the structure squashed flat, with all its features superimposed in inextricable confusion—the tantalizing trace left on the page of an old book by the pressing of an unknown lacy flower.

The Fourier projection appeared as the final one of the four papers. Perutz published this one with Bragg. They presented "an electron density map of a single row of molecules in projection." The view through the row resembled a radio astronomer's map of the band of sky towards the center of the Milky Way—and with reason, for the optical principles and the techniques of Fourier summation in the two sciences are similar, except for the spectacular irrelevance of scale, and had been Bragg's peculiar gift to the Cavendish. The contours in the molecular map were drawn at intervals at which density increased by one electron per square angstrom. They thought they could distinguish two furrows, either of which could be where the molecule split in half, and they marked these *X-X* and *Y-Y* on the map. They warned:

> This map shows molecules of irregular outline and complex internal structure. Because of lack of resolution and the great depth of material projected neither polypeptide chains nor haem groups can as yet be recognized. . . . The picture of the haemoglobin molecule which now emerges from the Fourier projections cannot yet be interpreted.

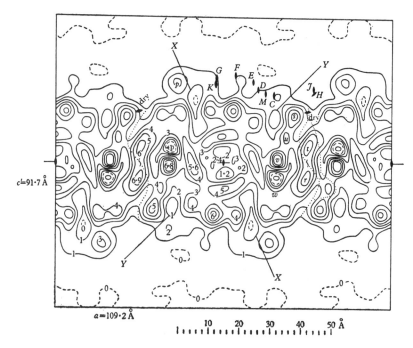

Fourier section of hemoglobin in projection, 1954. (Drawing from "The Structure of Haemoglobin: Fourier Projections on the 010 Plane," by Sir Lawrence Bragg and M. F. Perutz, *Proceedings of the Royal Society* A 225 (1954): 320.)

❖ ❖ ❖

"Perutz and I worked together, you see, entirely on this. Producing wild ideas," Bragg said to me. "And then he got his heavy-atom idea, and the last paper I wrote with him we did a partial solution, a very, very partial solution, using his heavy atom. And then it stuck. . . . Stuck with a projection. Which didn't tell us a thing. But it was thrilling, because it was working; you could see it was self-consistent. That's the key word there. Everything fitted."

"As always, I was driven on by wild expectations," Perutz said. "And thought that the structure of the protein would be sufficiently separate—distinct—so that one could make it out from the projected image. And this hope was completely disappointed: the projected image was entirely enigmatic."

Then, at the Cavendish one day in the spring of 1954, "while I was trying to interpret these pictures, with and without mercury, Dorothy Hodgkin dropped in, and said to me casually, 'If you could get *two* heavy-atom derivatives, you could determine the structure in three dimensions.'" Hodgkin told Perutz to look at a passage toward the end of Bijvoet's paper of three years earlier about the structure of strychnine. There, Bijvoet had suggested that phase angles, and not just the signs, could be obtained directly if the method of isomorphous replacement were extended so that one compared the diffraction patterns from three crystals—one of the native substance under study and two isomorphous derivatives having the heavy atoms attached in different places. "Dorothy Hodgkin's remark sent me off on the next phase," Perutz said.

"There were all kinds of dreadful complications. In retrospect, six years doesn't seem such a terribly long time, but when you're actually living six years and trying, not very successfully, to apply this method which you *knew* could solve the structure of proteins, it seemed an eternity." The problems began the moment Perutz tried to crystallize hemoglobin a second time with para-mercuribenzoate aboard: the crystals no longer had the same form as from the native protein, for now the *b* axis was twice as long. "Which meant that successive unit cells were no longer equivalent and there must be some difference between the structures of the two." More generally, hemoglobin seemed to offer no more hooks, other than the pair of reactive cysteine residues, to attach other heavy atoms.

Then in the autumn of 1954, Perutz got very sick. He had suffered since school days from a recurrent digestive disorder, but had blamed it on nerves and had been too shy to see a doctor. The illness became serious just before the war, and got much worse on his trip to the United States in 1943, from which he returned in bad shape. No cause could be found. One doctor prescribed a mixture of drugs that subdued the problem; but by 1954 the remedy no longer helped. Perutz grew hardly able to work. At the end of the summer of 1955, he thought he would have to resign his job. Investigations found nothing. Medicines made him worse. Luckily, a colleague then suggested that the illness might be coeliac disease, which is an allergic response of the digestive tract to the gluten in wheat flour—a gravely debilitating reaction to an almost unavoidable component of foodstuffs. By cutting out all bread and flour, he began to get better. It turned out that he had become sensitive to many other foods; gradually he learned what these were, and that he could stay well on a diet limited with ascetic purity to meat, fish, mild cheese, a very few kinds of vegetables, and an occasional cup of tea. His wife, when they were invited to dinner, had to telephone ahead to explain the rules.

Kendrew, meanwhile, had tried for several years and had failed to grow crystals of myoglobin of horse that were any larger than the fine needles he had started with. He turned to other species. He had glanced at whale myoglobin even in 1948. Diving birds and mammals—penguins, whales, dolphins—because they must store large amounts of oxygen, have muscles rich in myoglobin. Five per cent of the wet weight of the meat of the sperm whale is myoglobin: its muscles are colored nearly black by the stuff. Through a chance acquaintance of Perutz's, a chunk of sperm whale was sent to Kendrew from Peru, where the meat was considered a delicacy. This yielded superb crystals, of good size and very favorable crystallographic structure. Kendrew switched to sperm whale. None the less, myoglobin at first seemed a poor subject for heavy-atom methods, for it offered no free sulphydryl groups to anchor a heavy-atom compound.

Perutz and Kendrew engaged Ingram to help search for rational biochemical methods to get heavy atoms into their molecules—for example, by some compound that could be installed at the iron in the heme. None of these worked. The chemical problem was first solved by an American biochemist, Howard Dintzis, who came to the Cavendish from Harvard in 1954 on a postdoctoral fellowship. Dintzis knew more about the chemistry of heavy-atom complexes than either Perutz or Kendrew. He set up racks of bottles of crystals of myoglobin or hemoglobin, trying one after another a great variety of compounds that might in principle stick to protein side chains and carry in an atom of gold or silver, platinum, iodine, mercury, or what not. After letting the crystals soak for a few weeks, he took x-ray photos. By such energetic empiricism, Dintzis built several compounds of myoglobin with heavy atoms at different single sites on the molecule and with crystals all of the same form. Another postdoc then made still others. By mid-1955, x-ray patterns of the several crystals showed that they had managed to get heavy atoms onto the surface of myoglobin at five separate sites. Meanwhile, Bragg had set up for protein crystallography at the Royal Institution, with Kendrew's help, and had recruited a young Welshman working in Canada, David Phillips. Kendrew now used both labs, the Cavendish and the Royal Institution, to x-ray the heavy-atom myoglobins. He first computed a projection of the molecule, and found it as baffling as Perutz had found the projection of hemoglobin. Only a three-dimensional Fourier synthesis would do.

The next obstacle to fall was the mathematical one of locating the heavy atoms in the third dimension. Determining two coordinates, x and z, of a heavy atom on the plane of a projection had been straightforward; how to find the y coordinate, the one that placed the heavy atom in the depth of the molecule, was not obvious. Perutz took the problem to Crick, and to the crystallographer William Cochran, who was at the Cavendish though not in the Medical Research Council unit; each man made a suggestion, and Perutz put these together into a method he thought would work. "However, I didn't publish anything, because I got ill," Perutz said. "This method was just waiting in my notebook." Harker, at Brooklyn Polytechnic, then pubished a different method, still not satisfactory. Then Bragg, casting back to the years before Fourier methods were widely used in crystallography, produced a neat and graphic technique. Putting these together, Kendrew was able to locate his heavy atoms at least relative to each other, which was enough.

Kendrew could see from that point forward that the solution of myoglobin was inevitable. "In the end, the first low-resolution picture of the molecule—in *theory*, one could have done the whole data-collection and calculation for that one in about two weeks," he said. They were working first to a resolution of six angstrom units—a focus sharp enough, they thought, to resolve neighboring polypeptide chains, though not to pick out amino-acid side groups or individual atoms. "It was only three hundred reflections, you see. Three hundred spots. From each of three or four different heavy-atom derivatives. That didn't take so long." Looking years ahead to solutions at higher resolutions, Kendrew urged some friends at the Rockefeller Institute to find the amino-acid sequence of myglobin, using the methods by which Sanger had at last completed the sequencing of insulin. Kendrew was also pursuing new methods for measuring intensities. The first step was to replace the matching of the blackness of spots by eye, against a chart, with a semi-automatic instrument, developed at King's College London, that any lab technician could use. Then he turned to the Mathematical Laboratory of the University of Cambridge, which was developing one of the first big high-speed digital computers, for help in calculating the three-dimensional Fourier synthesis—the condensations of electron density, the diffraction grating, the structure at last. All told, what in theory could have been done in two weeks in fact took two years.

Hemoglobin was still stuck on the vexatious doubling of the *b* axis of the crystals with mercury attached. "This seemed most baffling and its cause quite obscure," Perutz said. "We thought it might be different horses, for example; it might be some impurities; all *kinds* of hypotheses. I got quite desperate about it. There was Kendrew going ahead splendidly, he was probably solving his structure, and I—after having produced the method—was getting absolutely nowhere!" Then, in the spring of 1957 and once more able to work full time, Perutz found the answer. It was embarrassingly simple. "A chance observation gave me the idea that the effect might have something to do with the acidity of the solution. Once I had the idea, I of course solved it in an afternoon.

"It turned out that there is a reversible transition between two different crystal lattices in hemoglobin, with the transition point at just about the degree of acidity where I usually grew the crystals." Proteins fairly commonly take alternative crystal forms; their spheroidal molecules can pack together in different ways that are energetically almost equivalent, the form of the moment depending on trivial variations in acidity or salt. "I discovered I could simply reverse the effect by adding acid or alkali to the solution; and on one side, the effect disappeared. Also, it turned out that in heavy-atom derivatives this effect was more marked—it had never troubled me before in making crystals, you see."

The more recalcitrant problem was to get other heavy atoms into the molecule. "None of these derivatives which were so successful in spermwhale myoglobin attached themselves to hemoglobin at all," Perutz said. "It didn't seem to have these niches that had served so well for myoglobin." Dintzis went back to the United States in 1956. "After he left, Ann Cullis, who was my research assistant at that time, found he had made large numbers of preparations which he had never looked at. So we went systematically through all the little bottles of crystals he had prepared, and one of them showed intensity differences of a kind we had not seen before. So in fact Dintzis had solved the problem but had never looked at the crystals in which he had solved it! There were several hundred preparations."

The one that worked was an odd chemical quirk. The two reactive sulphydryl groups on hemoglobin had such an avidity for alien compounds that they attached themselves to any heavy atom introduced; to get heavy atoms to attach elsewhere, Dintzis had figured, he would have to block the sulphydryl groups first, which he regularly did with a chemical, acetamide, that did not contain a heavy atom. Then to one bottle primed that way, he happened to put in another mercury compound, a very simple one called mercuric acetate. These mercury atoms attached themselves at another place.

Later, they learned that hemoglobin of horse contained a second pair of cysteine residues, buried in the molecule and not normally reactive; blocking the first pair with the acetamide somehow loosened the structure and allowed the second pair to react to the mercuric acetate. Dintzis had also tried a compound that carried not one but a pair of mercury atoms; when that bottle was opened, each sulphydryl group carried the pair. Perutz got a fourth heavy-atom hemoglobin from a colleague in California. He made a fifth himself by combining two of Dintzis's tricks. "So Ann Cullis and I took the x-ray photographs of the five different derivatives and we had an army of young ladies who measured the intensities!" They had more than twenty thousand reflections to measure.

❖ ❖ ❖

Kendrew got out the structure of sperm-whale myoglobin to six angstroms' resolution in the summer of 1957: that was the first protein to be solved. His report, with five collaborators including Dintzis and Phillips, appeared in *Nature* for 8 March 1958. The Fourier synthesis produced, as always, a set of contour maps. Each was a section through the unit cell, sixteen all told, spaced nearly two angstroms apart. These were drawn on clear plastic sheets and stacked up, held in alignment by bolts and spacers, the whole block about a foot long and eight inches thick, so that one could look through the cloud of electron density from any angle. From that, in turn, Kendrew constructed a model out of plasticine—the writhing, visceral knot whose photograph I had noticed the first time I met Perutz. The stack of plastic sheets was lying on its side on the window ledge in Kendrew's office during our conversations. Peering into it, one dimly made out wandering, ropey clusters. From one perspective, a stubby dark bar coalesced: the heme group, a nebula seen edge on. The model itself stood on top of a filing cabinet. It was a foot and a half long, on a varnished wooden base, a convoluted dust-darkened sausage speared on a supporting stick as by a giant's toothpick, and entrapping a flat round muffin.

"To go from that to that"—Kendrew waved at the Fourier maps, then at the model—"was for me quite a major undertaking." Myoglobin's unit cell had contained two molecules. "It was very hard to see the boundaries between them. There was even some guesswork, that could have been wrong—though as it turned out it was right. But never having seen the blessed thing before, we had no idea what to expect. And on the whole expected something much more regular than this."

The report in *Nature* was marked by a comical tone of surprise, of eyebrows raised at eccentricity. The myoglobin molecule, the collaborators said, was known to have only one terminal amino group—only one right-hand end to a protein sequence. They went on:

> It is simplest to suppose that it consists of a single polypeptide chain. This chain is folded to form a flat disk of dimensions about 43 A. x 35 A. x 23 A. Within the disk chains pursue a complicated course, turning through large angles and generally behaving so irregularly that it is difficult to describe the arrangement in simple terms; but we note the strong tendency for neighboring chains to lie 8–10 A. apart in spite of the irregularity. . . . If we attempt to trace a single continuous chain throughout the model, we soon run into difficulties and ambiguities, because we must follow it around corners, and it is precisely at corners that the chain must lose the tightly packed configuration which alone makes it visible at this resolution. . . . Also, there are several apparent bridges between neighboring chains, perhaps due to the apposition of bulky side-chains. The model is certainly compatible with a single continuous chain, but there are at least two alternative ways of tracing it through the molecule, and it will not be possible to ascertain which (if either) is correct until the resolution has been improved. . . . The haem group is held in the structure by links to at least four neighboring chains; nevertheless, one side of it is readily accessible from the environment to oxygen. . . .
>
> The arrangement seems to be almost totally lacking in the kind of regularities which one instinctively anticipates, and it is more complicated than has been predicted by any theory of protein structure.

"The other thing—and what turned out to be quite untypical—was that everything fitted the idea that all the more-or-less straight lengths of the chain were Pauling's alpha helix," Kendrew said. "One didn't *know* that they were alpha helix, but they fitted that. And there was independent physical evidence from which one thought that it contained quite a lot of helix. It turned out that it contained seventy per cent. Of course Pauling was delighted about this; it vindicated him as having discovered the key to protein. And then in 1959"—when Kendrew got the structure at higher resolution—"one was able to prove that it absolutely was an alpha helix and not any other kind of helix. So Pauling was even more delighted.

"And the only thing is that every protein that's been done since, with the exception of hemoglobin, has had far less! I mean, I think it's simply not correct to say that the alpha helix is the fundamental structure of proteins. Because most proteins really have remarkably little. . . . In a way, I was terribly lucky that mine did have a lot of alpha helix, because if you have a protein like lysozyme, with practically no helix in it, it's almost impossible to tell anything at six angstroms' resolution—because you can hardly see the chain at that resolution." Side groups they could not discern, let alone identify. "So if the first one to come out had been something like lysozyme, probably not a very good Fourier anyway in those days, we would have been in some despair and wondered whether we'd got it right at all. But this contained recognizable features that made one feel that it was correct. One was the heme group itself; there had to be a heme group, and there it was. And the second thing was that without even building any models there was this evidence of chains, you see." He gestured again towards the stack of plastic sheets.

"The big surprise was that it was so irregular," Kendrew said. Crick's strictures proved justified. The glimpses of neatly piled rods that had lured them on had been false. More than that, Kendrew had quietly completed what

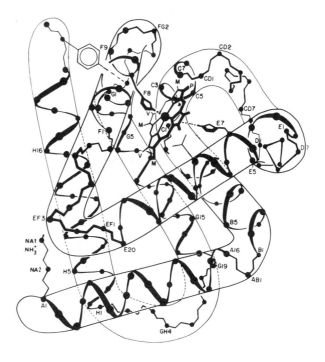

Sanger had begun: no simple and general geometry governed protein structures, any more than a general rule of chemistry governed amino-acid sequences. The most striking feature of the structure of myoglobin was its total lack of symmetry.

❖ ❖ ❖

Perutz got out the structure of hemoglobin of horse to low resolution—5.5 angstrom units—in 1959. Kendrew completed at the same time the analysis of myoglobin at 2.0 angstroms. They were pressing the limits of technology. The higher harmonics of the x-ray diffraction, which supplied the finer details of the crystal structure, lay farther out in the pattern of reflections. That pattern, in full three dimensions, was a sphere of which any particular photograph was the two-dimensional sample intersected by the flat film. For low resolution, Kendrew had needed only the three or four hundred reflections that appeared within a small sphere at the center of the pattern. To double the resolution of the image of electron densities, the radius of the sphere had to be doubled: it now enclosed eight times as many reflections. To drive the detail in the maps of myoglobin from 6.0 to 2.0 angstroms, a factor of three, increased the number of reflections to be worked with for each crystal form by the cube of that factor, twenty-seven. The outermost reflections were fainter and required longer exposures; crystals began to break down from radiation damage.

In all, for myoglobin at this stage, about a quarter of a million spots had to be recorded, measured, and corrected, from different heavy-atom derivatives and from exposures of different durations. Kendrew and his colleagues achieved an average error in the determination of amplitudes of less than four per cent. The calculations had far outrun the capacity of the Cambridge computer. A new model came on line in 1959—one of the fastest computers, with one of the largest memories, that had yet been built, but no more than sufficient.

Phillips, meanwhile, working at the Royal Institution with Uli Arndt, a physicist but also an instrument maker touched with genius, had built a device for crystallography that dispensed with film altogether: an automatic diffractometer that measured directly, with a proportional radiation counter, the intensities of successive x-ray reflections and recorded them on punched tape ready for the computer. Kendrew used the device for the last measurements for myoglobin at 2.0 angstroms.

In hard focus, myoglobin confirmed the observations made two years before. The spacings and angles of successive side chains were now distinguishable, and proved the surmise that the straight runs were alpha helix. The resolution was still not sharp enough to split neighboring covalently bonded atoms, for these lie one to one-and-a-half angstroms apart. None the less—and unexpectedly—by careful study of the shapes of the lumps of density projecting from the polypeptide chain, Kendrew was able to identify many of them unambiguously as belonging to one amino acid or another.

At the Rockefeller Institute, the sequencing of myoglobin was well under way: crystallography now came to the aid of chemistry, confirming or clearing up many assignments in the sequence. The heme group could now be seen very distinctly. Contrary to Kendrew and Perutz's expectation, it was not perfectly flat: the iron atom appeared to lie more than a quarter of an angstrom, all of two and a half billionths of an inch, out of the plane of the group. They could also see that the iron atom was attached to one of the helical segments by the side chain of an amino acid which they identified as histidine—a link that had first been suggested by Pauling on chemical grounds thirty years earlier.

Hemoglobin to low resolution—at last, twenty-two years after the conversation with Haurowitz in Prague that started Perutz up the mountain, and complemented by the solution of myoglobin—offered two chief surprises. To build a model, Perutz rolled out sheets of a heat-setting plastic corresponding to the thickness of each of the Fourier sections through the molecule, at a scale of two angstroms to the centimeter. From each sheet, he cut—like cookies from a thick dough—the shape of each island on the contour map where the electron density exceeded a set value, 0.54 electrons per cubic angstrom. The shapes were then stacked up at the correct heights and in the right map positions. Discs for the heme groups were attached at the right orientations. The model was then baked so that it set permanently. For comparison, Kendrew made the same kind of model of myoglobin.

The first surprise was the similarity of hemoglobin's four subunits to the chain of myoglobin. One protein was from horse, the other from whale. One was four times the size of the other. Considerable differences in amino-acid composition were known; variations in sequence were later detailed. Yet, in overall folding, the two beta chains of hemoglobin were essentially identical to the single chain of myoglobin, and the alpha chains were very similar. Differences in structure occurred at the tips of the chains, at a bend at the top and another at the side, and at the faces where the four units of hemoglobin came together. These were details. The important result, Perutz reported, was the resemblance. The folding that Kendrew had first discovered had since been found in myoglobin from seals as well.

Its appearance in hemoglobin of horse was portentous. It suggested a universal form for these proteins in vertebrates, Perutz wrote.

> How does this arise? It is scarcely conceivable that a three-dimensional template forces the chain to take up this fold. More probably the chain, once it is synthesized and provided with a haem group around which it can coil, takes up this configuration spontaneously, as the only one which satisfies the stereochemical requirements of its amino-acid sequence. This suggests the occurrence of similar sequences throughout this group of proteins, despite their marked differences in amino-acid content. This seems all the more likely, since their structural similarity suggests that they have developed from a common genetic precursor.

In the complete molecule, the two alpha chains confronted each other square on like two men, identical twins, broad-shouldered and broad-bellied, seated upright on straight chairs, toe to toe. The two beta chains faced each other similarly but inverted, heads down. In brief, the arrangement was not square but tetrahedral.

The heme groups were set in four separate pockets at the surface of the molecule—and this was the other and greatest surprise, for they were about as far apart as they could be. The iron atoms lay at distances of 33.4 and 36.0 angstroms between symmetrically related pairs. On the half-molecule made of an alpha and its closest beta chain, the iron atoms were 25.2 angstrom units apart. The two reactive cysteine residues, where the mercury attached, were on the beta chains, in the same stretch of helix as the beta heme groups; these mercury atoms were 30 angstroms apart.

> Little can be said as yet about the relation between structure and function. The haem groups are much too far apart for the combination with oxygen of any one of them to affect the oxygen affinity of its neighbors directly. Whatever interaction between the haem groups exists must be of a subtle and indirect kind that we cannot yet guess. . . . The structure of reduced haemoglobin is still unknown, but it would not be surprising if loss of oxygen caused the four sub-units to rearrange themselves relative to each other, rather than to change their individual structure to a marked degree.

The problem of the physiology of hemoglobin had become acute.

Hemoglobin of horse in the deoxy form, whose crystal structure Haurowitz had caught in the moment of transformation at the addition of oxygen, had been barred from x-ray analysis by its impossible unit cell, 350 angstroms long. To distinguish individual spots would require a specially built vacuum camera with a long distance between crystal and film. In the mid-fifties, Perutz had collaborated briefly with a biochemistry student at Oxford, Margaret Jope, who was doing her thesis on the crystallization of human hemoglobin. She gave him crystals of human deoxyhemoglobin. These turned out to take three different crystal forms; two were useless for x-ray analysis, but the third was excellent. Perutz put that one aside, mentally, for future use. Its turn came when the structure of hemoglobin of horse in the oxy form was solved; in the fall of 1959, Perutz set a research student of his own, Hilary Muirhead, the task of determining the structure of human deoxyhemoglobin.

At the end of December of 1959, Bragg set out to get Nobel prizes for the crystallography of proteins and of DNA. In Stockholm, committees of Swedish scientists, one for each of the science prizes—physics, chemistry, and physiology or medicine—begin at the close of each year to consider the awards to be announced the following autumn. Winners from past years are sent engraved forms in Swedish and French inviting them to nominate candidates. Bragg, with the prize for physics for 1915, was the eldest statesman of Nobel politics. In 1932, for instance, he had apparently been the first to suggest that a prize be awarded to three scientists jointly: he wrote to the Nobel Committee for Physics urging that precedent be broken by giving the prize next year to Erwin Schrödinger, Werner Heisenberg, and Paul Dirac, together, for their overturning and rebuilding of quantum theory. (In the event, Heisenberg alone got the prize in physics for 1932, Schrödinger and Dirac shared it the following year, and the first three-way split was the prize for physiology or medicine in 1934.) Early in January 1960, Bragg wrote to Erik Hulthen, Chairman of the Nobel Committee for Physics, proposing Perutz, Kendrew, and Dorothy Hodgkin for that prize, for x-ray crystallography of proteins. He wrote to Arne Westgren of the Nobel Committee for Chemistry, suggesting that since Perutz, Kendrew, and Hodgkin had been working with the structures of biologically important chemicals, but by physical methods, the two committees get together to decide the jurisdictional question. The prize for chemistry, he argued, should go to Watson, Crick, and Maurice Wilkins for the discovery of the structure of DNA. (Rosalind Franklin was two years dead.) Bragg asked Perutz to assemble a complete set of his and Kendrew's current papers, which he forwarded to Stockholm. Bragg also canvassed the support of other scientists for his candidates—and in particular with a long letter to Pauling, who replied grumpily that he thought a prize to Watson and Crick was premature while Wilkins did not deserve one at all. Perutz's structure of hemoglobin at low resolution and Kendrew's of myoglobin at high were published back to back in *Nature* of 13 February 1960.

A NOTE ABOUT FOURIER ANALYSIS

Fourier analysis takes things apart so they can be put back together again. A Fourier analysis asserts a relationship between a complex periodic phenomenon and the simple harmonics that make it up, as in the following equation—here stripped to a skeleton in order to bring out the underlying principle. To the left of the equal sign, the overall variation in amplitude is represented by A. This is the entire complex periodic phenomenon in question—be it air-raid siren, picket fence, jazz-drummer's beat, or the arrangement of atoms in a crystal. Sir Lawrence Bragg, to illustrate a point in a crystallographic paper, once published a Fourier analysis of the time table for the trains leaving for Cambridge on Sundays from Liverpool Street Station, London. On the right of the equation, the phenomenon A is broken down into the series of paired terms for the amplitude and phase of the fundamental and each harmonic:

$$A = A_0 + A_1\cos(2P_1) + A_2\cos(4P_2) + A_3\cos(6P_3) + \ldots + A_n\cos(2nP_n) + \ldots .$$

In that formula, A_0 is the mean amplitude—in the case of the musical note, the mean pressure of the air. It represents the base line, the value around which the periodic fluctuations occur. The next term, to continue with music, is the fundamental note, with amplitude A_1 and phase P_1 and with the fundamental frequency; the next term is the first overtone, with amplitude A_2, phase P_2, and a frequency twice the fundamental—and so on for as many terms as necessary to exhaust the harmonics present.

That equation, to be sure, is no more than a picture postcard of a high mountain range, exasperatingly deficient of road directions for how to get there. Even for the scientist who knows the road, the equation written that way must seem perilously oversimplified.

A closer examination of the factor for phase, alone, in any one Fourier term will suggest why the problem of getting at the phases from the x-ray pattern of a crystal was so baffling. Once more, what is to be grasped is not the full poetic power of this foreign language but a bare syntactical relationship. The factor containing the phase in the typical term—the nth term—which was compressed in the equation as written above as $\cos(2nP_n)$, is more usually written in an expanded form:

$$\cos\left(\frac{2\pi nx}{a}\right) - \alpha_n$$

There, a_n is the phase angle that states the position of that nth harmonic curve in relation to the others, a is the fundamental interval—in the musical example, the wave at 440 cycles per second—and x is a fraction of that interval, representing the instant or point along the interval where amplitude and phase are measured for that term and every term in the series.

Thus, a fuller and more usual statement of the Fourier series is:

$$A(x) = A_0 + A_1\cos\left(\frac{2\pi x}{a}\right) - \alpha_1 + \ldots + A_n\cos\left(\frac{2\pi nx}{a}\right) - \alpha_n \ldots$$

—a sum that must be worked for each value of x in turn. Typically, the fundamental wave might be divided into 100 parts, so that 100 separate sums of n terms would have to be done.

10 *"I have discovered the second secret of life."*

❖𝑎❖ The first direct evidence of the relationship between the molecular structure of hemoglobin and its physiological function came with an enigmatic discovery in Cambridge in the fall of 1961, and in a conversation in Paris just afterwards. In August of that year, at the Congress of Biochemistry in Moscow, Max Perutz had briefly reviewed the models of myoglobin and oxyhemoglobin. When he got back to Cambridge, Hilary Muirhead was completing the first step in the solution of the structure of human deoxyhemoglobin. She had hung the first mercury compound onto the reactive cysteine residues; she had calculated the projection of the molecule down the symmetry axis and was finding the positions of the two mercury atoms. She got these in October. Where the distance between them in oxyhemoglobin had been thirty angstroms, in deoxyhemoglobin it turned out to be thirty-seven. The beta chains had moved. If one ignored the gap in the argument between horse and human hemoglobin, then upon discharging oxygen the molecule shifted so that two stretches of helix, each carrying a mercury atom and a heme group, were farther apart. On taking up oxygen, the hemes moved closer again. Thought of one way, the change was absurdly small, some seventy billionths of an inch: yet that was more than ten per cent of the overall length of the molecule. Haurowitz's sighting of the change in crystal form, twenty-three years earlier, Muirhead now drove down, for the first time, to the molecular level.

Soon after Muirhead's discovery, Perutz went to Paris at François Jacob and Jacques Monod's invitation to give a talk at the Institut Pasteur. "When I told Monod about this observation, he was absolutely delighted," Perutz said. "Though at this stage we had no idea exactly what it meant. This was the moment when at least half a dozen people knew that hemoglobin must undergo a change in structure—otherwise the mercury atoms couldn't have changed."

❖ ❖ ❖

That summer and fall of 1961, after Monod got back from Cold Spring Harbor, the interior conversation running in his head had been increasingly preoccupied by a new line of speculation about how enzymes were regulated. The demonstration of hemoglobin's change in structure as it functioned gave Monod an essential corroboration. Hemoglobin, not an enzyme, took on a curious status: "an honorary enzyme," Monod later called it, and an independent model for a very general hypothesis.

Two problems had emerged at the Pasteur from the success of the theory of the operator and repressor—from the scheme of regulation of enzyme activity in bacteria, formulated during the previous two years, whereby a repressor molecule, itself the product of a regulator gene, intervened at the genetic switch, the operator, to switch off the synthesis of an enzyme or functionally linked set of enzymes by blocking the expression of the structural gene or linked genes. The first problem was to characterize the repressor more thoroughly, to find how it worked, to isolate the actual repressor molecule if possible, and in any event to justify more rigorously the radical assertion that the repressor acted upon the very genes, by sitting on the DNA or dropping off it. For this was the pivotal realization in the theory of the repressor, and enforced a total reorganization of the understanding of interactions within cells. Thus, the first shock, Monod told me, had been that the effect he had been studying for twenty years—the induction in *Escherichia coli* of the enzyme β-galactosidase by the presence of its substrate, milk sugar—took place at the level of the gene. "That was a fantastic realization," he said. "I never was very interested in the messenger! Our most intense interest—François and myself— at the time of writing the '61 paper and immediately afterwards, lasting one or two years, was to further enquire and justify the drastic assumption that regulation occurred at the operator level. At the genetic level."

In arriving at the theory of the repressor, the experimental weight had rested on the bacterial genetics of which Jacob was master: now, though some difficult genetical questions were to arise, the weight appeared to shift towards the biochemistry and enzymology that were Monod's domain. Monod's was the responsibility, in particular, for isolating the repressor. The previous June, presenting the theory at Cold Spring Harbor, Jacob had said, "No positive evidence has been obtained as yet concerning the chemical identity of repressors, but since they are presumably primary products of the regulators, the assumption that they are polyribonucleotides [that is, short strings of RNA] appears as the most reasonable guess." In later years, to some American biologists—notably to Walter Gilbert and to Mark Ptashne—the reasonable guess of the French came to seem like hedging their bets. Be that as it might, by the end of 1961 Monod and Jacob were coming to see clearly that repressors were proteins. Experimental evidence, though at first indirect, was pointing that way. "The RNA repressor was based on completely stupid reasons," Jacob said when I asked about that. From new genetic evidence, "it was clear in the winter of '61–'62, that the phage repressor was not RNA. That it was protein or at least contained a strong protein component. And then immediately I isolated the same thing in the *lac* system." They also knew clearly that the repressor molecule was going to prove elusive. Calculations, though preliminary, suggested that any given repressor was present in the cell in a very few copies.

More generally, looking back, Monod said, "The repressor couldn't be anything but protein. Because of its capacity to recognize structures. Which is a privilege of proteins in any case." The repressor had to relate physically to two different substances: to the specific sequence of DNA, the operator, where it was supposed to block the reading-out of the structural gene, and to the specific chemical, the small molecule to which it reacted by attaching to the operator or by letting go. This duality, inherent in the nature of the repressor, was the other and profounder conceptual revolution. "For almost twenty years, I had been working with what I would call 'mechanistic' models of enzyme induction," Monod said. "That is to say, on the idea that there was some sort of *necessity* that the inducer of an enzyme"—of its synthesis—"was either its substrate or something sterically"—structurally in three dimensions—"closely analogous to the substrate. Now, discovering the operator, and the operator-repressor interaction, and the fact therefore that the *inducing* interaction has *nothing to do* with the structure of the enzyme came as a complete shock to me."

Monod's assumption, the natural and universal assumption, had always been that induction of an enzyme's synthesis required a direct effect of the inducer—milk sugar in the case of his lactose system—on the enzyme. Discovery in the early fifties of close chemical relatives of milk sugar, the galactosides, that in some cases are powerful inducers and in others are inhibitors, only intensified, then, the picture of a structural interaction between inducer and enzyme. The theory of the repressor abolished that. "I realized that for twenty years I had been working on a very basic problem, but still not the problem I thought I had been working on," Monod said. At Cold Spring Harbor, Jacob and Monod had pointed down this road.

> An important line of argument, which we will not develop here, is the following: since the regulator gene does not control the structure of the protein [that is, the enzyme protein], and since the effectors [inducers] appear to react with the product of the regulator, rather than with the products of the structural gene, there need exist no steric relationship between the protein itself and the effector.

Despite the great explanatory power of the theory of the repressor, by no means did it apply to all enzyme systems. This was the second general problem at the Pasteur in the fall of 1961. With *E. coli*, the availability of lactose as a food or of certain artificial analogues as decoys induces synthesis of the enzymes that digest lactose: the sugar molecule at the start of the pathway deactivates the repressor. In the same organism, as Monod had found with Germaine Cohen-Bazire in the winter of 1952–'53, the presence of the amino acid tryptophan inhibits synthesis of the sequence of enzymes that build tryptophan: the molecule at the end of that pathway *activates* the repressor. The action was opposite, yet the scheme the same. Inhibition of a metabolic pathway by its own end product was, of course, a classic instance of a feedback loop, as had first been clearly identified by Aaron Novick and Leo Szilard in 1954, with that tryptophan pathway.

The next cases of end-product feedback were spotted by Arthur Pardee, in 1954, at the University of California in Berkeley, and by Edwin Umbarger, in 1955, at Harvard Medical School, and sent in for publication almost simultaneously in the fall of 1955. Pardee's is an early enzyme in the long pathway by which *E. coli* makes pyrimidine nucleotides. The enzyme interchanges groups from two molecules, aspartic acid and carbamyl phosphate, to put together the next substance required in the assembly line. The enzyme is now called aspartate transcarbamylase. Pardee established that the activity of this enzyme—and not the enzymes ahead or behind in the pathway—is turned off by the presence of the pyrimidines uracil or cytosine.

Umbarger's enzyme controlled the pathway that synthesized the amino acid isoleucine. This was a long path, too—five steps, mediated by five enzymes, starting with the amino acid threonine and subtracting or adding specific fragments until isoleucine is constructed. The first step is to knock off the amino group, NH_2, at one end of the threonine molecule. The enzyme that accomplishes that first step is called, obviously, threonine deaminase. Even before these details were entirely clear, Umbarger realized that when isoleucine as well as threonine is present in the bacterium, no more isoleucine is made. And, Umbarger thought, the inhibition takes place specifically at the first enzyme in the pathway—for no more threonine is deaminated. Umbarger's first brief note antedated Pardee's two long papers by a month.

Others found more examples. Closest to home, Monod's colleague Georges Cohen, at the time working in the enzymological laboratory at the Centre National de la Recherche Scientifique in a southern suburb of Paris, found yet another similar instance of feedback inhibition, one that was deliciously intricate because the pathway branched in two places to lead to three different amino acids—and possessed two different enzymes that catalyzed the same reaction at the earliest branch.

It was becoming apparent for bacteria, at least, that the final small molecule synthesized in any given metabolic pathway is a strong and specific inhibitor of its own synthesis. Some of these feedback inhibitions, as with tryptophan, could be shown to take place by repression at the gene. Others were different—like Pardee's pathway for pyrimidines or Umbarger's for isoleucine. Piqued in curiosity and pride, in the fall of 1959 Monod suggested to a graduate student he had taken on, Jean-Pierre Changeux, that for his dissertation he look into the feedback control of threonine deaminase by isoleucine. "I didn't know exactly where it would lead, but it was a good subject for a budding biochemist, because it meant doing some good enzyme chemistry."

Almost from the start, Changeux had got some peculiar results. The initial question was obvious: since isoleucine, five steps along the path, is not, in fact, much like threonine structurally, how could it compete at the enzyme's active site? When Changeux tested the variation of the enzyme's activity at different concentrations of substrate and of inhibitor—good enzyme chemistry—he found the interactions abnormal. The irregular sinuosities on the graph were struggling to be sigmoid. The enzyme molecule appeared to be built of subunits. Their binding of the small molecules appeared to be cooperative. But at the same time, Changeux found that the enzyme, when heated, kept its power to snip the amino group off threonine—but quickly lost all response to isoleucine.

Changeux's first note of this desensitization in his lab book appeared on 29 November 1959. Three years later, a dispute about priority arose, and Monod wrote to Pardee: "As a matter of fact, Changeux's first observation (which was quite accidental) . . . came as a complete surprise and even as a shock to both of us." Still more, the desensitized enzyme, its regulatory competence gone, now interacted with its substrate normally, with no trace of cooperativity. These data, and the dissimilarity of structure between threonine and isoleucine, forced Monod and Changeux to the idea that substrate and inhibitor bound at different sites on the enzyme molecule.

In the fall of 1960, Monod visited Pardee in Berkeley, and learned that he and a young colleague, John Gerhart, had recently performed some closely similar experiments on the enzyme aspartate transcarbamylase—including heating that destroyed the enzyme's sensitivity to its inhibitor even while it kept its catalytic power. "My recollection was that Monod was quite surprised to learn this. Surprised and pleased," Pardee said in a conversation early in 1978. "And he didn't tell me of any results *he* had of that sort, at that time." The next spring, Pardee visited Paris. "I gave a talk at the Pasteur about all of this stuff, and I sat down and talked with Changeux at great length about what we had been doing." Pardee and Gerhart, for one thing, had their enzyme pure, which allowed more precise and more advanced experiments.

On another line, Changeux demonstrated that threonine deaminase is subject not just to the one but to both modes of control—repression of its synthesis at the gene as well as inhibition of its activity at the enzyme itself. He proved this by finding a mutant that lacked the repressor and by then performing the bacterial matings to show that the regulatory gene is at a distance on the map from the structural gene. Changeux's repressorless mutant produced about fifteen times more enzyme than bacteria of the wild type. Evidence of the ubiquity of regulation by repressor was welcome at the Pasteur; none the less, when Changeux extracted threonine deaminase from the mutant it turned out to be sensitive to inhibition of its activity by isoleucine, exactly like enzyme from the wild type. The two control mechanisms were altogether independent. Feedback inhibition was possibly a fine tuning.

Changeux came to Cold Spring Harbor with the group from the Pasteur in June of 1961, and read a short paper, tongue-tied with technicality, about isoleucine feedback. They found when they got there that others, Umbarger in particular, had independently reached many of the same conclusions about regulatory enzymes. Pardee was not there. Umbarger, his paper immediately before Changeux's, spoke of his own work, Pardee's, and others', but not the work in Paris, and said, "It is thus quite clear that the site[s] at which catalysis and endproduct inhibition occur need not be, and perhaps never are, identical." But he thought that the sites in some cases would overlap, or at least that the inhibitor, when present at its site, would physically interfere with the catalytic act at the other site.

Changeux and Monod had taken the argument further by a small step that had great consequence. The inhibitory and catalytic sites, they said, were entirely distinct. "Following Jean-Pierre's work I realized—or rather, we realized, because he clearly thought of it— Ah, but it was clear from the experiments," Monod said. "That the interactions, say, between the inhibitor of an allosteric enzyme and the substrate could not be a direct interaction. Had to go via the protein." And, asserting proprietorship, Monod had branded the phenomenon "allostery," from the Greek *allo*, for "other," and *stere*, for "solid" and so three-dimensional: thus "other shape."

Sitting in Monod's bedroom, at Cold Spring Harbor, he and Jacob prepared the meeting's closing summary. Monod delivered it the next day. In the revised and polished version later printed, it is a magisterial survey not merely of the week that included, among other things, the theory of the repressor and the experimental evidence for the messenger, but of the full state of molecular biology two months before Marshall Nirenberg's announcement of the cracking of the genetic code. "The most remarkable feature" of feedback inhibition, Monod and Jacob wrote, "is that the inhibitor *is not a steric analogue of the substrate.*" They went on:

> We propose therefore to designate this mechanism as "allosteric inhibition." Since it is well known that competitive behavior toward an enzyme is, as a rule, restricted to steric analogues, it might be argued that an enzyme's concept of steric analogy need not be the same as ours, and that proteins may see analogies where we cannot discern any. That this interpretation is inadequate is proved by many observations which were reported here. Umbarger and others have shown that in general, only *one* enzyme, the first one in the specific pathway concerned, is highly sensitive to inhibition by the endproduct. If steric analogy were involved, the different enzymes of the pathway would then be considered to hold private and dissenting opinions about stereochemistry. . . .
>
> Since the allosteric effect is not *inherently* related to any particular structural feature common to substrate and inhibitor, the enzymes subject to this effect must be considered as pure products of selection for efficient regulatory devices. . . . A particularly interesting possibility is suggested by this discussion. Namely that, since again there is no obligatory correlation between specific substrates and inhibitors of allosteric enzymes, the effect *need not be restricted to "endproduct" inhibition*. . . . It is conceivable that in some situations a cell might find a regulatory advantage in being able to control the rate of reaction along a given pathway through the level of a metabolite synthesized in another pathway. Wherever favorable, such "cross inhibition" might have become established through selection. In other words, *any* physiologically useful regulatory connection, between any two or more pathways, might become established by adequate selective construction of the interacting sites on an allosteric enzyme.

Nothing was said by anyone at Cold Spring Harbor about the actual mechanics of allosteric effects. A month later, though, Pardee and Gerhart turned in to the American publication *The Journal of Biological Chemistry* a paper on their latest work, in which they proposed that their enzyme "has a second site, distinct from the active site, for which the end product has a high affinity. The bound end product perhaps inhibits by deforming the enzyme." But that paper did not appear until March of 1962.

In the fall of 1961—only "after a lot of groping"—Monod was struck by the climactic unification of the ideas he and Changeux had been developing. At some point, he realized an obligatory implication of the concept of allostery. The binding sites, entirely nonoverlapping, would have to interact exclusively by a reversible change provoked in the overall structure of the protein molecule when it bound its specific, regulating small molecule.

The American enzymologist Daniel Koshland, Jr., had proposed in 1959 that the active site of an enzyme might be modified, locally, by its interaction with the substrate, but such changes, though plausible, had not been proven to take place. In the fall of 1961, Monod knew of Changeux's work with threonine deaminase—but this yielded no direct evidence that the regulatory enzyme changed its conformation overall. Through the visits back and forth in the preceding twelve months, Pardee and Gerhart's findings were, in general, well known to Monod. Finally, Perutz's published structure of hemoglobin, showing the wide separation of the heme groups, was of course known to Monod; he saw perfectly well the resemblance of cooperative effects in hemoglobin and in regulatory enzymes. And thus Perutz's news, that fall, of the movement of the mercury atoms as hemoglobin loaded oxygen on and off could not have been more interesting. Monod had the high pleasure of seeing an essential prediction of a new theory borne out.

Any coherent account imposes order and must make the next realization seem more obvious than it was. For two years and more, Monod's chief concern had been the repressor. "We understood that there had to be a triple interaction between the repressor, and the genetic segment that we called an operator, and the small molecule we called the inducer, which starts the whole thing going," Monod said. "And if there was a reaction, there had to be some sort of chemical modification of the inducer itself. And we couldn't find one. And one day, I remember perfectly well"—it was in November 1961—"having lunch in our old place, I was thinking about these experiments that Jean-Pierre Changeux had shown me which strongly indicated the *indirect* regulation of enzyme *activity*—and it suddenly hit me that we did not need to assume any chemical modification of the inducer, that the mechanism of induction [of enzyme synthesis] could be of the very same general nature as the one we were developing for the regulation of enzyme activity."

Formally, in their own logic, the two situations were the same: neither for the repressor acting at the genetic level nor for the regulatory enzyme was there any obligatory structural relationship between the small molecule that controlled the activity of the protein and the molecule at which that activity was directed. At first, Monod took the similarity to mean that the cell must possess an allosteric enzyme that the inducing small molecule—lactose or one of its analogues, say—would activate, the enzyme then in turn controlling the repressor on the gene. Monod, that fall, thought up various ways to detect that enzyme, for example by labelling one of the artificial inducers with a fluorescent tag to see where it went. But Jacob found no genetic evidence for the existence of such an enzyme in the lactose system. Nor was it essential to the formal similarity of the two modes of regulation.

"Let me give you two examples," Monod said. "There's a guy called Monod, working on the induction of the galactosidase enzyme, and the one thing we knew, all along, was that galactosidase was specific *to* galactosides and that it was induced *by* galactosides. Therefore the obvious, naïve idea was to say that there was an intimate obligatory relationship between these two facts. Okay—that turns out to be wrong. On the other side, there are enzyme chemists, working with enzymes, and whenever they turn up with something that inhibits that enzyme specifically, they assume—it's in all the books—that it is due to a direct interaction between the substrate of that enzyme and the inhibitory substance. That was the classical interpretation of any so-called competitive inhibition in enzyme chemistry. Now my student comes up with an experiment that seems to show clearly that this kind of interpretation is out. And therefore the only alternative is that the interaction in reality is not a direct one—but one that goes by a change of shape in the protein." Monod turned the similarity back onto the repressor: he concluded that it, too, functioned by a change in overall conformation of the protein, caused directly by the controlling small molecule.

"I call it—when I talk of this as a phase in my intellectual career—call it the discovery of allosteric systems," Monod said. "Of the principle of allosteric systems. Of the basic principle, which I think I—I really discovered, that in specific regulatory effects, which are all-important, the interaction is always indirect. And the secondary realization that if it were not *that* way, we wouldn't be around to talk about it, because no cell could be constructed otherwise. No system of intracellular information could function unless it was independent of immediate chemical—ah, requirements." He was thinking out the words slowly. "What you need is a system which is built for its utility, to the interacting chemical fluxes that have to be regulated. But which cannot be dependent on, for instance, the capacity of two small molecules to interact together." He called this the concept of *gratuity*—the freedom from any chemical or structural necessity in the relation between the substrate of an enzyme and the other small molecules that prompted or inhibited its activity. "It has to be a system which has the basic property of an electronic system," he said. Allosteric proteins were relays, mediating interactions between compounds which themselves had no chemical affinity, and by that regulating the flux of energy and materials through the major system, while themselves requiring little energy. The gratuity of allosteric reactions all but transcended chemistry, to give molecular evolution a practically limitless field for biological elaboration.

"I have discovered the second secret of life!" Monod told his colleagues; years afterwards they remembered his excitement. Two decades of work coalesced; as the theory of the repressor had done, yet in a manner more fundamental and embracing, allostery brought diverse and apparently contradictory modes of regulation under an overarching singular vision.

From the moment of global understanding—and such moments have their own unity of opposites, in which the excitement is an oceanic calm, a thrilling sense of well-being, for all *is* right with the world—to marshal the evidence and the arguments took more than a year. The first great paper on allosteric proteins, written with Changeux and Jacob, was submitted to *The Journal of Molecular Biology* at the end of 1962. A second followed two years later, with Changeux and Jeffries Wyman, who was a physical chemist and mathematician from Harvard but at that time working in Rome with an Italian group studying hemoglobin. "The first paper I think was more important than the second," Monod said. It was more biological.

"The first paper *was*, really, on the idea of indirect regulation. Which I think is the really important idea. The second paper is a physical-chemical interpretation of this fact in terms of the geometry of the molecule." In the first, the authors proposed "a general model schematizing the functional structures of controlling proteins." The model now was explicit about mechanism:

> These proteins are assumed to possess two, or at least two, stereospecifically different, non-overlapping receptor sites. One of these, the *active site*, binds the substrate and is responsible for the biological activity of the protein. The other, or *allosteric site*, is complementary to the structure of another metabolite, the *allosteric effector*, which it binds specifically and reversibly. The formation of the enzyme-allosteric effector complex does not activate a reaction involving the effector itself: it is assumed only to bring about a discrete reversible alteration of the molecular structure of the protein or *allosteric transition*, which modifies the properties of the active site, changing one or several of the kinetic parameters which characterize the biological activity of the protein.
>
> An absolutely essential, albeit negative, assumption implicit in this description is that an allosteric effector, since it binds at a site altogether distinct from the active

site and since it does not participate at any stage in the reaction activated by the protein, need not bear any particular chemical or metabolic relation of any sort with the substrate itself. The specificity of any allosteric effect and its actual manifestation is therefore considered to result exclusively from the specific construction of the protein molecule itself, allowing it to undergo a particular, discrete, reversible conformational alteration, triggered by the binding of the allosteric effector. The *absence* of any inherent obligatory chemical analogy or reactivity between substrate and allosteric effector appears to be a fact of extreme biological importance, and in a sense it is the main subject of the present paper.

Pardee's regulatory enzyme was published by now, and the cooperative relation of the binding at its several sites was classically sigmoid; Changeux had pressed on to experiments with analogues of his inhibitor; four other candidates for allosteric, regulatory enzymes had recently turned up. But in the case Monod and his colleagues built, hemoglobin had a unique standing. Its cooperative effects, sixty years established, Perutz had shown by the wide spacing of the heme groups to be possible only by indirect action, mediated through the protein. Monod cited what Perutz had told him about the change in distance between the sulphydryl groups in the molecule.

> Thus, in the case of haemoglobin, there is complete evidence that the regulatory effect, i.e. the cooperative binding of oxygen, is related to a reversible, discrete conformational alteration of the protein, i.e. in our nomenclature, to an allosteric transition. Actually, thanks to the considerable work which has been devoted to it, the haemoglobin system provides the most valuable model from which to start in the further analysis and interpretation of allosteric effects in general.

Hormones, as chemical messengers in multi-celled organisms, might perform, they suggested, as allosteric effectors. They argued in detail "that the specific effects of small molecules in activating or inhibiting, at the genetic level, the synthesis of messenger RNA and protein are mediated by an allosteric transition of the repressor." For a moment, they were circumspect: "The most serious objection to the concept of allosteric control is that it *could* be used to 'explain away' almost any mysterious physiological mechanism."

The basis in molecular structure for allosteric effects was uncertain. Their mediation through the protein "would not necessarily involve a conformational alteration *sensu stricto*," the paper said (the Latin tag was Monod's), and went on:

> It might conceivably be due, for instance, to a redistribution of charge within the molecule without *detectable* alteration of its spatial configuration. In a protein molecule, however, any redistribution of charge might be expected to involve or to facilitate a true conformational alteration. Actually, as we have seen, indirect evidence suggests in many cases and direct observations prove in a few instances that allosteric transitions involve the breaking, or formation, or substitution of bonds between subunits in the protein.

The likelihood was—again with hemoglobin a leading instance—that allosteric proteins were not single polypeptide chains but were built up of subunits and that allosteric transitions worked by rearranging these.

But the burden of the paper was biological, indeed evolutionary. It was the final act in the tearing down and rebuilding of classical biochemistry by molecular biology. Monod, bringing Changeux and Jacob with him, wrote of allostery with a fervor astonishing in a scientific report:

While possession of this almost universal key raises serious latent dangers for the experimenter, it is of such value to living beings that natural selection must have used it to the limit.

"Chemical equilibrium means death," Monod once said to me: perhaps the essence of his vision of the processes of life. He worked out the consequences at length in the paper.

A living system is constantly fighting against, rather than relying upon, thermodynamic equilibration. The thermodynamic significance of specific cellular control systems precisely is that they successfully circumvent thermodynamic equilibration (until the organism dies, at least). . . . Still, the arbitrariness, chemically speaking, of certain allosteric effects appears almost shocking at first sight, but it is this very arbitrariness which confers upon them a unique physiological significance, and the biological interpretation of the apparent paradox is obvious. The specific structure of any enzyme-protein is of course a pure product of selection, necessarily limited, however, by the structure and chemical properties of the actual reactants. No selective pressure, however strong, could build an enzyme able to activate a chemically impossible reaction. In the construction of an allosteric protein this limitation is abolished, since the effector does not react or interact directly with the substrates or products of the reaction but only with the protein itself. . . . By using certain proteins not only as catalysts or transporters but as molecular receivers and transducers of chemical signals, freedom is gained from otherwise insuperable chemical constraints, allowing selection to develop and interconnect the immensely complex circuitry of living organisms.

The second paper, two years later, was about protein structures that might plausibly account for allosteric transitions. It was rooted in heme-heme interaction from its opening sentence. Monod acknowledged the debt: several months before turning the paper in, he sent Perutz a draft with a note to say, in part, "Since I rely rather heavily on hemoglobin as a justification of the model, I would appreciate it very much if you could take the time to read the paper and let me have your comments." The paper was filled with equations for the most part founded on the similarity between cooperative binding in allosteric enzymes and the oxygen-dissociation curves of hemoglobin. The central conclusion was that allosteric proteins must of necessity be made up of polypeptide chains in multiple identical subunits, symmetrically joined: the complex took either of two alternative states, *R* for relaxed or *T* for tense: what happened at one binding site created a dislocation that forced a compensating change at the symmetrically opposite, equivalent site, so that transition from *R* to *T* or the other way restored a symmetry. Cooperativity became the necessary physiological expression of symmetrical molecular anatomy. Crystallographers, at least, found the conclusion self-evident.

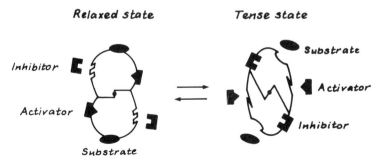

Allosteric proteins, I said to Monod, provided the first example in molecular biology after the discovery of the structure of DNA in which function was so inextricably linked to structure—in which structure itself was a functional concept. "I would tend to say so, although I may be a little bit arrogant in saying so," Monod replied. "But—yes, I think you're right. The only other example I know—let's try to be precise—where a very general and very basic biological effect is interpreted in terms of the general structure of the molecule that effects it—that is true. And what is so nice is that in both cases it's related to symmetry. That pleases me. In epistemological terms I think that the only time when science gets to its ground level is when it can interpret something in terms of symmetry."

Of all Monod's ideas, allostery was the one that Crick most admired. "It's quite ludicrous, for example, for Monod to say he's not a theoretician," Crick said to me in 1971. He laughed with delight. "I would say particularly, Jack Monod's work on allostery is a very powerful theoretical concept. Never mind the mechanism of it, which did have to be worked out. That meant that *you could connect any metabolic circuit with any other metabolic circuit,* you see, because there was no necessary relation between what was going on at the catalytic site and the control molecule that was coming in." Crick's voice rose. "Well, that's an *extremely* powerful idea, never mind the details."

♦ ♦ ♦

Muirhead and Perutz got the first three-dimensional Fourier synthesis of human deoxyhemoglobin in October 1962. The resolution was to 5.5 angstroms, as before; only two heavy-atom derivatives had been measured, so results were as yet imprecise. They looked for two sorts of structural change: in folding of individual chains and in positions of the chains in relation to each other. They detected no changes in folding larger than what might be due merely to experimental error, imprecisions of measurement. The alpha chains had not moved, so far as they could tell. The beta chains had moved dramatically. They had slid apart, twisting slightly as they did so, opening a gap between them, a large cleft in the molecule, losing contact with each other, slightly changing their contacts with the alpha chains—and the distance between their iron atoms had increased from 33 to 40 angstroms. Perutz began to build a plastic model of this form to show off at a meeting in New York the first part of November. "I was interrupted by the announcement of the Nobel prize. I had to apologize for bringing the half-finished model."

Early that week, a telephone call from Stockholm turned out to be a Swedish journalist telling Perutz that he and Kendrew had won the prize for chemistry; but both Pauling and Hodgkin, in previous years, had received similar calls that had proved to be mistaken, so Perutz said nothing at home. Next day, rumors thickened; the BBC had been alerted to stand by. On the first of November, his secretary burst in waving a pair of envelopes containing telegrams, one to Professor Perutz, the other to Professor Kendrew. When opened, these were from the Papal Academy of Sciences, inquiring how many reprints were wanted of papers, now to be published, that the two men had given in Rome many months before.

Hours after that, telegrams arrived from the Secretary of the Royal Swedish Academy of Sciences. The Nobel committees, with a lightness of touch they had not been known to possess, had got together to give prizes for the two discoveries, rooted in x-ray crystallography, that had been made in the Cavendish Laboratory in 1953 in the six months before Bragg's departure. The prize in chemistry for 1962 went to Perutz and Kendrew equally for the method of solving the structures of globular proteins by isomorphous replacement with heavy atoms, and for the solutions of the first two proteins, myoglobin to high resolution and hemoglobin to low; while the prize in physiology or medicine went to Crick, Watson, and Wilkins in common for the solution of the structure of deoxyribonucleic acid. Bragg's campaign had been entirely successful, except for Dorothy Hodgkin, and, in the event, she got the prize for chemistry two years later. The prize in physics in 1962 was awarded to the Russian theoretical physicist Lev Davidovich Landau; but he had suffered grave brain damage in a car crash earlier in the year and could not come to Stockholm. The prize for literature was awarded to John Steinbeck.

Stockholm in the pearly glow of December: a city of iron beauty, snow and granite seen across black water, flags everywhere outside and candles within, the twentieth-century festival of enlightenment at the winter solstice. The prizes were presented late in the afternoon of Monday, December 10. King Gustaf VI Adolf spoke quietly and personally to each man. He gave Perutz a framed diploma, hand-lettered in Swedish and painted with a semiabstract picture in the style of the nineteen-twenties, representing streaming rays, spots, helices, and a single seeing eye above the crumpled ribbon of a polypeptide chain; a gold medal; and a bank draft for some eight thousand pounds. (Perutz paid off the mortgage on his house and bought a car.)

A banquet and a ball were given that night. After dinner the winners were cheered four times, and replied with short speeches. Perutz said, in part, "I was deeply moved by the Royal Swedish Academy's decision to elevate me, a man of modest gifts, to the Olympian heights of a Nobel Laureate. On hearing the news, a friend who knows me only too well sent me this laconic message: 'Blood, toil, sweat and tears always were a good mixture.' . . . Despite the twenty-five years for which I have been at it, the task which I have set myself has only just begun." His goal, he said, had become "to explain the physiology of breathing in terms of the architecture of the haemoglobin molecule. For, as Sir Francis Drake said in a prayer well-known to English schoolboys: 'When thou givest to Thy servants to endeavor any great matter, grant us also to know that it is not the beginning, but the continuing of the same, until it is thoroughly finished, which yieldeth the true glory.'"

Watson replied for Crick and Wilkins as well, but said, "As it is difficult to convey the personal feelings of others, I must speak for myself. This evening is certainly the second most wonderful moment in my life. The first was our discovery of the structure of DNA. At that time we knew that a new world had been opened and that an old world which seemed rather mystical was gone.... The last thing I would like to say is that good science as a way of life is sometimes difficult. . . . We must thus believe strongly in our ideas, often to the point where they may seem tiresome and bothersome and even arrogant to our colleagues. I know many people, at least when I was young, who thought I was quite unbearable. Some also thought Maurice was very strange, and others, including myself, thought Francis was at times difficult." Few heard him, because he did not speak into the microphone. Watson was thirty-four. He published only one more research paper after that day. The pictures of the ball that were seen around the world were those of Crick doing the twist with one of his young daughters, and of Watson in the arms of one of the pretty Swedish princesses in a low-cut gown.

Next day, Perutz and Kendrew, in their Nobel lectures, presented the rudiments of protein crystallography and the details of their models so far. With the structure of deoxyhemoglobin to low resolution two months old, in Stockholm Perutz addressed the problem of the relation between the molecule's structure and the oxygen dissociation curve almost for the first time pulicly. "The haemoglobin molecule may . . . be regarded as an enzyme with two functions and several active sites, which interact in a complex and sophisticated manner. The explanation of this behaviour is one of the main objects of our research." But to find the explanation of function "in terms of the structural changes of which these new results have just given us a first glimpse," he warned, "it may well be necessary to solve the structure of at least one of the two forms in atomic detail."

Monod, Changeux, and Jacob on allosteric proteins reached *The Journal of Molecular Biology* that week. A few days later, Monod sent a copy to Pardee; Pardee replied at once with several minor corrections and a major claim that Gerhart and he had been first to perceive that inhibition of an enzyme by the end product of its pathway required a second active site on the molecule. Monod answered equally promptly, politely but implacably, citing dates and details of Changeux's work and the progress of his own thought. On 10 January 1963, Pardee wrote again, by hand, saying that he was satisfied that the discoveries were independent. "I must admit that I was until now rather unhappy about the simultaneous discovery of the separate site idea, because simultaneous discoveries have happened to me so often in 'important' cases that just once I'd like to come up with something new all alone."

"Pardee did some *very* parallel work, actually," Jacob said to me. "He certainly was extremely close to the same thing." What Pardee lacked, perhaps, was the knack of generalization.

❖*b*❖ The third protein structure and first enzyme to be solved was lysozyme from the egg whites of hens, by David Phillips and colleagues at the Royal Institution. Fired by the ambition to sharpen the focus of the structure of myoglobin to the point where individual atoms could be distinguished, Kendrew and Phillips in 1960 began to collect the data for resolution to 1.5 angstroms, the virtual limit. They used an interim version of the automatic diffractometer to measure the radiation at each spot, without film. The data for the ultimate model of myoglobin were ready by the time of Kendrew's Nobel prize; afterwards, much to Bragg's disgust, Kendrew never calculated the model. Phillips looked for a protein of his own. A visiting American, Roberto Poljak, suggested lysozyme: groups elsewhere were working on it, including Corey and Pauling at Caltech, but they all were reported to be having trouble making heavy-atom derivatives.

Lysozyme was not particularly interesting enzymologically. It had been discovered and named in 1922 by Alexander Fleming, who later found penicillin, and who remarked of lysozyme that it had "the power of rapidly dissolving certain bacteria." But it was not, for example, one of the great and

well-studied family of proteases—pepsin, trypsin, chymotrypsin, and so on—
that cleave polypeptide chains next to specific amino-acid residues. Nor was it
a regulatory enzyme. It catalyzed the breaking of a bond between two slightly
different sugars that formed alternate links of a polysaccharide that was the
main component of the cell walls of the bacteria; in action it was not unlike β-
galactosidase.

How lysozyme worked chemically was not known when Phillips started
its structure; it was attractive crystallographically. It was a small protein,
smaller even than myoglobin, with a molecular weight of about 14,600 and a
single chain of 129 amino-acid residues. The determining question was its
crystals, and these were good. Poljak shook up some bottles of crystals with
various heavy-atom reagents, let them sit, took some pictures that immediately looked promising. They used three isomorphous derivatives, first got out a
structure to low resolution, then by the spring of 1965 had a Fourier synthesis
at high resolution, 2.0 angstroms. "That was a fair amount of work, but relatively little trouble," Phillips said.

The maps showed an egg, the polypeptide chain looping back and forth
like a strand of wool loosely skeined. A deep cleft opened into one side.
Lysozyme had not much alpha helix, a lot of irregularity. The cleft evidently
contained the active site. How a length of the polysaccharide was held there,
which bond exactly it was that the enzyme broke, and what the enzyme did to
provide the energy that activated the break the map did not reveal.

Phillips meanwhile had set a research student to determine the three-
dimensional structure of a simple, close analogue of the enzyme's substrate, a
triplet of one of the sugars without the other, that was known to inhibit the action of the enzyme. Each ring in the triplet, they found, had a bend at one end,
shaping it like a chair. Then they tried soaking the analogue into the crystals
like heavy-atom compounds. An enzyme interacts with its substrate in a fraction of a second; x-ray diffraction takes hours of exposure; but an analogue
that blocked the enzyme, they figured, would stay in place and reveal how it
bound to the active site. The extra computation was negligible, for once the
enzyme itself had been solved, the structure with the analogue in place required only calculation of the difference between the two diffraction patterns.
A trial at low resolution proved that the tactic worked. The triplet bound at
the top of the cleft in the enzyme, leaving an empty pocket below. Phillips
asked for high-resolution studies of the binding of several different inhibiting
analogues.

The Royal Society appointed an international discussion meeting on the
structure and function of lysozyme, to be held at the Royal Institution on 3
February 1966. Perutz organized the program. One he invited was John
Rupley, an enzyme chemist at the University of Arizona; by the end of 1965,
Phillips had heard from Rupley about work still unpublished that showed that
the true substrate of lysozyme must be the hexasaccharide, for the enzyme
split that thirty thousand times as fast as it split any triplet; the bond cleaved
lay between the fourth ring and the fifth. By then, also, Phillips's group had
built, twig by twig, a brier-patch brass-wire model of lysozyme itself. At the
end of January they got the data for bound analogues.

"The consequence was the most rewarding three days that I've ever spent,"
Phillips said. "We had the model for the enzyme. We then had these maps
showing how these inhibitors were bound. We were committed next week to

talking about all of this at the Royal Society. So, I got out the maps, took my jacket off, and spent three days, in the lab, looking at how these inhibitors fit and wondering how one might go on from that to see how the substrate fit. And just for once—if you've ever read Medawar's remarks on the dishonesty of a scientific paper—just for once, the logical argument, which I go through in my paper, about extrapolating from the inhibitor binding to the substrate binding, really is a fairly accurate account of the thought processes."

Phillips put the triplet in place at the top of the cleft. Then he added the bits of brass for a fourth sugar ring. This in its normal shape brought several of its atoms too close to certain amino-acid side groups. He relieved the overcrowding by twisting the chair conformation into a half-chair, or sofa. After that, two more sugar rings strung on without crowding or distortion. All six were placed to form convenient hydrogen bonds with nearby side chains in the cleft, to hold the hexasaccharide in position.

Phillips then looked for the catalytic mechanism. "At this point, I must admit, I knew really rather little about proposed enzyme mechanisms," he said. The connection between successive sugars was made from the first carbon on one ring, through an oxygen atom shared by the two, to the fourth carbon in the next ring. The cleft in the enzyme was for the most part lined with amino-acid side chains that were electrically uncharged. But close to the linkage to be broken between the fourth and fifth sugar rings, the cleft had the amino-acid residues glutamic acid and aspartic acid. One was the thirty-fifth residue along the string, the other the fifty-second, but the loops brought them into position so that their side chains sandwiched the link to be broken. The aspartic acid, in particular, appeared to be part of a local net of hydrogen bonds among acidic residues that held the link in place and held it next to a strongly negative charge. Phillips supposed that when the substrate was in place, the negative charge on the aspartic acid promoted formation of a positive charge across from it, on the bridging carbon atom of the fourth sugar ring.

Such a positively charged carbon atom is called a carbonium ion; it weakened the bond between the carbon and the bridge oxygen atom. That bond would be weakened further, Phillips thought, by the strain of the twisting of the fourth sugar ring out of the normal chair conformation. At the same time, on the other side of the bond the glutamic acid would be just right to donate a proton to the bridge oxygen atom, so that it could switch its bond from the carbon of the fourth sugar to the proton—upon which the fifth and sixth sugars would float free.

At that point, Phillips and the model were visited by Charles Vernon, a biochemist from University College London, who was to give a general paper on reaction mechanisms in the chemistry of sugars, as background to possible mechanisms of lysozyme. "The program having been organized before we knew we were going to get so far, crystallographically," Phillips said. "And he took one look at this mechanism and the model, and said, 'My goodness—that's a carbonium-ion mechanism with a transition-state distortion of the substrate—just the sort of thing I was going to tell *you!*' I'd have to say that was the high moment in my research work so far."

Bragg opened the Royal Society discussion. He was seventy-five. Seven weeks earlier, he had been in Stockholm for the fiftieth anniversary of his own Nobel prize in physics, and had seen the prize in physiology or medicine given to Lwoff, Monod, and Jacob. Of lysozyme, he recalled that, two years earlier, he had predicted that five to ten years would be needed before any other protein was solved to the detail Kendrew had reached with myoglobin. "My own laboratory has proved me wrong, which is perhaps the most pleasant way in which it could have happened."

Perutz closed the discussion. The audience was weary and half asleep. "For the first time we have been able to interpret the catalytic activity of an enzyme in stereochemical terms," he said. He saw his hearers come to attention. He went on to compare lysozyme with myoglobin and hemoglobin. "What these three proteins have in common is their distribution of polar and non-polar"—charged and uncharged—"side chains." The interior of the lysozyme molecule as well as that of the globin chains was made up largely of amino-acid residues that were uncharged and that did not readily allow electrical charge to flow or leak away.

> In each case the active centre lies in a pocket formed by this non-polar medium, and the pocket contains some polar residues which are essential for activity. . . . The non-polar interior of enzymes provide the living cell with the equivalent of the organic solvents used by the chemists. . . . In myoglobin and haemoglobin the non-polar medium prevents a permanent transfer of negative charge from the ferrous iron to the oxygen molecule [that is, the iron's rusting]. . . . The non-polar medium in the cleft of lysozyme increases the strength of the interaction between the car-bohydrate [the polysaccharide] and the carboxylic [charged, acidic] groups of the enzyme. . . . Preliminary X-ray data suggest that other enzymes are constructed on similar principles.

"My generalization has turned out to be true," Perutz wrote to me in 1978. "The strong electrical interactions made possible by the structure of enzymes are probably the most important factor in catalysis."

❖ ❖ ❖

"By the time the repressors were actually isolated, which was late in 1966, they had become a—holy grail?" Walter Gilbert said. "Like isolating the neutrino. That is, here's this mythical thing, you've postulated its being there. Those of us who were involved in the isolation, of course, believed in its existence in a way that other people did not. Problem: we had to make enough right guesses in order to do that isolation. In order to make those right guesses, we had to be convinced of its existence—in a form which almost took a chunk of the interest out of it. We had to know how it worked, well enough to say, That's the way to do it." Gilbert had wanted to isolate the repressor from early 1961, when he was first moving from physics to biology and first read Monod and Jacob's paper on genetic regulation. People in other labs had the same idea, among them Brenner in Cambridge and, of course, Monod and Jacob. Another was Mark Ptashne, who was drawn to research in biology, and to the question of the repressor, by reading Monod and Jacob's paper even as an undergraduate at Reed College, in Oregon.

But repressors are scanty in the cell. When the difficulty of isolating them began to be felt, Monod performed a series of trials, inducing synthesis of enzyme with very small amounts of inducer, to see how few molecules were actually required per cell to inactivate all the repressors of the lactose operon present. Though the work sheets were not dated, that was about 1962. He calculated that an individual *E. coli* cell contained seven or eight molecules of the *lac* repressor, which was less than two thousandths of one per cent of the cell's protein. The repressor produced by the bacteriophage *lambda*, to prevent itself multiplying and bursting forth from a lysogenic strain of *E. coli*, would be no less scarce.

Even as Monod and his colleagues were developing the theory of allosteric proteins, other laboratories were analyzing and announcing other enzyme systems that were regulated at the gene but in more complex ways than Monod and Jacob's repressor scheme. Doubts about the repressor began to rise, variants and alternatives to be formulated. Perhaps repression took place at the level of the messenger RNA rather than of the gene. Perhaps the product of the *lac* regulator or phage repressor gene was a true enzyme or else a co-repressor that influenced protein synthesis in any of several conceivable, more conventional ways. Gunther Stent, then Brenner, published papers elaborating these suspicions. "The *real* question that the people who were in the field were really worried about was—whether the whole thing was right," Ptashne said in conversation. Though the worries were never acute, genetical controversy required biochemical settlement.

After Gilbert's first experiments with messenger RNA in 1960 and '61, he took an occasional glance at the *lac* repressor, but for three years his main concern was the way messenger RNA is read by the ribosome. The experiments showed, among other things, that the ribosome picks up one end of a messenger and works along it, being in contact only at a single locus—two adjacent codons, six nucleotides—at any instant, and knits out the sequence of bases into the sequence of the protein chain. "One's picture in 1960 of how proteins were made on ribosomes, really was that the ribosome held everything in a fixed position," Gilbert said. "The RNA was held there and you sat everything down and you made the protein." He was one of those who discovered that a single messenger could be read by many ribosomes at once, moving along the strand of RNA one after another. This cellular stratagem was called polyribosomes; Alexander Rich, at MIT, found the same thing at about the same time, and there was a priority fight. "That was one of those points when one could see one's own ideas change very abruptly; and it's very hard to recapture, afterwards, exactly how one had thought before," Gilbert said. "And the sudden transition, in '63, depending on very explicit experiments, that took one to the view that the whole thing is a moving process. You had this nice long messenger; and these ribosomes picked up one end and ran along it."

In the summer of 1965, Gilbert turned to isolate the *lac* repressor. Ptashne, by then, was a Junior Fellow of Harvard's Society of Fellows, and had a bench in Meselson's laboratory. He began at the same time to seek the *lambda* repressor. The two men were a corridor apart, and for eighteen months worked in intense and intimate competition. The experiments that succeeded were entirely different, yet years afterwards they described the psychological circumstances in words that were nearly the same.

In the ferment of alternative explanations, they agreed—and Gilbert later wrote—that the repressors must be sought by "means that would not depend on the specific models that might be imagined for the actual mechanism of repression." Gilbert's plan was to make a radioactive version of one of the inducers of galactosidase, put it into the culture, and see what in the cells it bound to. "One had no way of assaying for the repressor's function," Gilbert said. "One didn't know what it was going to look like at all. And therefore you had to devise experiments to play on unknown properties, and guess sufficiently about the unknown properties before you did the experiment, and hope that you could see enough in the experiments to lead you a little further."

Gilbert figured that normal repressor and inducer would interact too weakly for the bound complex to be detected. So he began with an inducer he thought would bind more tightly than normal to the repressor. He spent nine months isolating from the bacteria something that interacted with the supposed inducer. "Finding it first, then losing it, spending three months breaking them open and searching through them for the stuff that I had once found and lost, and learning all sorts of things about the ways in which one can handle proteins. Things that can go wrong. And finally found the material again—and then could prove that it wasn't the repressor."

He was working with Benno Müller-Hill, a postdoctoral fellow from Munich. They turned the problem the other way around. They reverted to a standard inducer, isopropyl thiogalactoside, the most active of the range of artificial inducers of the lactose system that Melvin Cohn had found at the Pasteur a dozen years earlier. Back then, Georges Cohen had actually tried to find what reacted with radioactive isopropyl thiogalactoside, but discovered permease instead. The difference now was that Müller-Hill searched out a mutant bacterium that made a repressor molecule that bound the inducer more tightly than normal.

They had the mutant by the spring of 1966. Then, by bacterial mating, they produced cells with twice the normal number of regulator genes. They ground up the bacteria, made a crude extract of proteins, and put that into a cellophane sac that had pores too small for proteins to pass through, although water and inducer molecules could move in and out freely. Then they immersed the sac in water containing the radioactive inducer, and waited long enough for the concentration of free inducer to equalize within and without. If some inducer had bound to repressor molecules, the level of radioactivity would be greater within the sac by that small amount. "There was a time in which I would go around giving seminars on why I *hadn't* found the repressor," Gilbert said. "A great problem making an interesting seminar out of totally negative experimentation.

"We finally managed to identify something. Then you go through a period of lovely— You barely have command of the experiments, the effect may come and go, and you do sudden-death controls: if it turns up in the wrong tube you're dead. You go through a series of these for a couple of weeks until finally you're convinced that maybe the controls really have worked and once you're convinced of that, then you're in much better shape, because *then* if an experiment doesn't work you shut it off and go back and do it again the next day." The first hint of a result was very faint. "About four per cent more inside the sac than outside. That's just down in the area— Well, you can try to repeat that again and again and again; or you can go ahead to the next stage and say, well, if there's an effect there, you can try to make it better by fractionating the material and purifying it a little bit. Get off the ground. It is generally a better technique not to focus on the experiment itself but to focus on the idea and try to improve the configuration."

Gilbert and Müller-Hill announced the isolation of the product of the regulator gene of the lactose system in a paper turned in to *Proceedings of the National Academy of Sciences* at the end of October 1966. Then they began to try to make enough repressor, relatively pure, to be able to show how it worked. They already knew that the protein had at least two separate interacting sites to bind inducer, and so no doubt at least two subunits, as allosteric theory demanded—for the curve for induction of galactosidase synthesis in the mutant bacteria was sigmoid.

"Certainly the reason I went into molecular biology, probably one of the major reasons I went into science at all, was because it seemed to me that the repressor was the great problem," Mark Ptashne said. We sat talking in his lab office at Harvard, on a hot summer's day; he was stubby, neatly muscular, fit, and was wearing aviator-style spectacles, T-shirt, sawed-off blue-denim shorts, and sandals—more exposed skin than appeared prudent in a laboratory. "And it seemed to me that people who claimed to be trying to isolate the repressor and prove or disprove the theory, weren't really serious. Weren't really willing to take the kind of risks that were necessary."

What kind of risks were those? "*Well*," Ptashne said. "*Psychic* risks." The risk of committing one's full effort to a difficult problem? "Well yeah! Here I was, a nobody, and, you know, Wally was coming in as a— I don't think either one of us could have kept it up without the goad of the other guy competing. In a friendly way." He hoisted his calves onto a chair next to him. "So the point was, here all the hotshot scientists have tried this and tried that and it doesn't work, and here we are, and he's new to the field and I'm a graduate student, just completed, and we worked night and day, and the thing is, had either one of us succeeded completely and the other guy failed—" On the way in, I had noticed a paper pinned to the door from the corridor, a cartoon of a thermometer with very little red creeping up and the caption "Ptashne Stradivarius Fund." He had recently bought a violin on a mortgage. "I mean, the thing that was really tough about it was, as I say, I mean, nothing to show for it and always worrying, even though I was close to Wally, if he had succeeded, completely—and he had, of course, Benno working with him at the time. And *he* must have been worried." Ptashne talked with animation, and often smiled slightly to himself. "And then there was an incredible period when things would come and go. I'd have a result and then I couldn't repeat it, and he'd have a result and couldn't repeat it, and so it was just—absolutely hair-raising."

Ptashne's plan—not his first plan: "I mean, this was about, you know, hysterical scheme number ten"—was to create a combination of disabled bacteria and mutant phage such that a high proportion of the protein synthesized would be the product of the *lambda* gene for the repressor. Then the cells could be fed radioactive amino acids to label the newly synthesized proteins. On that plan, he built a finely balanced tower: the risks, the hard labor, the play of thought are breathtaking, as is the luck. The experiment is entertaining. The paper that reported it—refreshingly, in the first person—snaps with cocky energy.

Firm foundation of Ptashne's precarious tower was the fact that in *E. coli* lysogenic for phage *lambda,* although manufacture of phage protein was generally repressed—the essence of lysogeny, after all—one phage gene that had to be expressed was that for the repressor itself. In form, *lambda* repression exactly matched repression of the lactose operon, which did not turn off the regulator gene. The corollary was obvious, and had just been explicitly proven: if the cell was superinfected by more *lambda,* then although the repressors already present turned off most of the invading genes, the added repressor genes continued to function. Thus, potentially, the production of repressor could be amplified manyfold.

Ptashne first found that irradiating *E. coli*—a highly sensitive mutant— with huge doses of ultraviolet light damaged the cells' own DNA so badly that it could not be copied properly into messenger RNA, although the machinery for making proteins was less severely hurt. This step he proved by infecting irradiated cells with phage to which they were not lysogenically immune; the phage DNA started up the cells' protein synthesis again.

Second, he found a mutant *lambda,* not inducible, whose repressor gene was somewhat resistant to ultraviolet light, to put into the bacteria before irradiating them. Potentially, in bacteria bearing the mutant phage, irradiated and then superinfected, synthesis of bacterial proteins, and of phage proteins other than the product of the repressor gene, would be turned way down, at the same time that synthesis of repressor was amplified. In practice, even with still further refinements, when Ptashne grew the treated cells with radioactive leucine to mark new proteins, then ground up the cells, extracted proteins, and spread them out chromatographically, he found many bands of protein to be radioactive. However much repressor might be there, the tricks for turning off other protein synthesis had evidently been no more than partly effective.

To pick out repressor, if it were there, Ptashne added another level to the experiment, using two different radioactive labels. He irradiated a batch of bacteria and divided them into two cultures. One culture he infected with *lambda* phage that made repressor as usual. The other he infected with *lambda* carrying mutations that blocked synthesis of repressor. The first culture he fed leucine made with tritium, the triple-weight, radioactive isotope of hydrogen, ^{3}H. The second, the one whose cells could not make repressor, he supplied leucine made with radioactive carbon, ^{14}C. Each of these radioactive isotopes decays by emitting from the nucleus a beta particle, which is an energetic electron or its equivalent of opposite charge, a positron; but the beta particle thrown off by tritium has only a tenth the energy of that from radiocarbon. The difference in energy is easy to detect.

After feeding time, Ptashne mixed the two cultures, extracted the proteins, spread them out chromatographically, and looked for the single one that was labelled with tritium but not with radiocarbon. It was there: a single protein made only in the culture that had a functional gene for repressor. To make sure, he ran the parallel experiment with labels reversed, and found a protein in the same place labelled with ^{14}C only. Then he tried the same procedure with yet another phage mutant, which made a modified repressor—one that functioned but only at certain temperatures; the unique label betrayed a modified protein. Ptashne announced the isolation of the *lambda* repressor in a paper completed two months after Gilbert and Müller-Hill's.

The two papers are regarded as classics. Yet by themselves they proved little: only that the product of the regulator gene, or of the phage repressor gene, was a protein, which was believed already on less direct evidence. The experiments did not show that these gene products were themselves the repressors and not merely some enzyme or cofactor; they did not show whether these gene products acted on the DNA, on the RNA, or on some enzymatic pathway.

Gilbert and Müller-Hill, by their isolation technique, could get enough of their protein to try to test its action directly, in a test tube; the attempt failed, lacking other biochemical requirements that became clear several years later. Ptashne had much smaller quantities of his protein, and it was mixed with other proteins, though none of those others was labelled with tritium. He did the only experiment he could, which was to find out whether his protein bound to the phage DNA or not. He took two samples of DNA: one from phage of the wild type, whose DNA thus carried the site where the repressor was supposed to act; the other from a mutant phage that was irrepressible, thus missing the DNA site for repressor to bind. He added his mix of proteins including the radioactive presumptive repressor into each sample. Then he centrifuged each. In each tube, the DNA moved down. The protein molecules were much lighter than the DNA: in the tube prepared from phage of the wild type, the tritium-labelled protein travelled with the DNA, but in the tube with the mutant DNA, lacking the binding site, his protein stayed at the top. "Which showed that the operator is DNA and that the product of the repressor gene was directly seeing those base sequences in the DNA," Ptashne said. "Which was the fundamental thing." Gilbert and Müller-Hill then performed the parallel experiment and proved that the lac operator was DNA.

❖ ❖ ❖

Monod and Jacob failed to isolate the substance whose nature and mode of action was their theory's keystone. The failure disappointed Jacob profoundly. The working relationship between Jacob and Monod had begun to break up from 1964 or '65. That was understood in the world of biology; a glance at the lists of their scientific publications showed as much. The Nobel prize to Lwoff, Monod, and Jacob, at the end of 1965, was the first Nobel to be awarded to any French scientist for thirty years. Perutz once remarked that he had been lucky to be in Cambridge, where so many Nobel prize winners worked that a new one was not made too much of. The three Frenchmen became national figures. More than a year passed before Monod began to publish new research again. Monod had always had a public face: a concern for great issues, as in the Lysenko controversy, but also a zest for combat and a flair for performance. Jacob was a man of public reserve. They were often of similar opinions; upon returning home from Stockholm, for example, they gave a joint interview in which they called for the near-total reorganization and decentralization of French university education and French scientific research; but Jacob soon retired to the laboratory.

Even before the prize, their science had begun to diverge. A partial division of labor had always obtained. To begin with, Jacob was not a biochemist and had no taste for that sort of work. As they were collaborating most closely on the regulation of enzyme synthesis, a subtler difference of scientific emphasis evolved. Jacob was more concerned with the genetics, which led towards the idea of the operator and the operon; Monod was more involved with the repressor, which led to allostery—and yet, of course, operon, operator, repressor, and inducer were engine, ignition, lock, and key, the interacting elements of a unitary phenomenon, and of course both men were deeply engaged all through. According to several who were at the Pasteur in the nineteen-sixties, isolation of the repressor was not pursued vigorously. "The biochemists should have got the repressor—*they never did it seriously,*" said one geneticist. "But who did they have to do it?" an American scientist asked—for the fact was that neither Monod nor any of his colleagues had the temperament or turn of mind for the sorts of approach that eventually succeeded. Anyway, Monod was entranced with his grand design.

Jacob's name was on the first and more biological paper about allostery. He had nothing at all to do with the paper about the molecular mechanisms and symmetries that Monod completed with Changeux and Wyman at the end of 1964 and published in 1965. After the Nobel prize, Monod and Jacob wrote only two more research papers together.

Each man began to write a book. Monod's was *Le Hasard et la Nécessité* (*Chance and Necessity*) in which he tried to explain to the lay reader, in highly schematic outline, the chief findings of molecular biology—the grand design—and then to show that certain consequences for belief and ethical behavior, atheist, anti-Marxist, rationalist, necessarily followed from this biological outlook. He first spoke of these consequences in public after his election to a chair of the Collège de France—an institution unique to that nation, an honorary elite, the academy beyond the Academy of Science, and even more ancient than its approximate counterpart on the side of letters, the Académie Française. In his inaugural lecture—an occasion of pomp—in November 1967, he addressed the rise of modern biology as the working out of the theory of evolution in physical terms; attacked vitalism; quoted liberally from Heidegger, Democritus ("Everything in the universe is the creature of chance and necessity"), Pascal, McGregor, and Nietzsche; spoke of the evolution of ideas and ideologies, of the alienation of modern man from scientific culture, and of the ethic of knowledge:

> The ethic of knowledge is radically different from religious or utilitarian systems, which find in knowledge not the end in itself but merely a means to an end. The only end, the supreme value, the "sovereign good," in the ethic of knowledge is not, let us admit, the happiness of mankind, still less its temporal power or comfort, nor even the Socratic "know thyself"—it is objective knowledge itself.

The call was a romantic one. Some old friends reading it later—Salvador Luria, Gunther Stent—recognized the substance of Saturday-morning arguments at the Institut Pasteur in their youth. The lecture was saved from sententiousness by stylistic grace and Monod's sense of the ridiculous—those quotations from McGregor, for example, turned out to be sayings ("Each of Science's conquests is a victory of the absurd") Monod had coined himself and ascribed to his mother's Scottish grandfather. Great slabs of the lecture were reprinted in the French press, where they inspired a lugubrious correspondence. Monod returned to these ideas, this time in English, in lectures the next year at Pomona College, California, and then began to elaborate his text back into French.

Jacob's book was *La Logique du Vivant (The Logic of Life)*. It was engendered by no public occasion. It was an historical meditation on the resolving power of ruling ideas in science over three centuries, in particular the science of heredity and heredity's other aspect, embryological development. Jacob pursued these from the end of millennia of myths of monstrous miscouplings and prodigious births with the clarification of the idea of species, through the introduction of time and so of evolution into biology, then down through the cell to the gene and the molecule. His book was like one of those grand fireworks the Japanese make that go off in the night sky in a cascade of illuminations, chrysanthemums into willow trees into showers of arrows, each as it fades flowering into the next. It displayed a subtle delight in detection of ambiguity. When the two books appeared, both first in 1970, Jacob's was overshadowed by the great popular success of Monod's in France and Germany, and then in Britain and the United States by the fact that critics had prepared for Monod a noisy ambush.

In the spring of 1968, the students of Paris rose in insurrection—"the events of May." They rose, to begin with, against conditions in the university. Slogans and responses hardened into violence that brought out riot police, troops, clubs, armor, and tear gas, that for a few weeks looked likely to bring in the mass of the workers, and that almost brought down the government of Charles de Gaulle. Monod saw, to begin with, that the students were protesting abuses whose reform he had long urged. He intervened, both publicly and in secret negotiation, to try to temper the vengefulness of the administration towards the students and at the same time to hold back the slide of the student movement towards the infantile far left. On a memorable evening, Monod appeared on the stage of the Théatre de l'Odèon—less than three blocks from where he had lived in 1942 when he first joined the Francs-Tireurs—to debate the radical leaders of the students. He was scorned. "They treated him like shit," said a Frenchwoman who was there. In the lurid light of bonfires and burned-out cars, amid the felled chestnuts of the Boulevard St.-Michel and in the stink of gas, he pleaded for a truce to allow wounded students to be evacuated. His pleas were in vain.

In 1971, Monod became director of the Institut Pasteur. He brought the place into the present day: he rebuilt its chaotic administration, restored its shattered finances, put its activities as a producer of vaccines on a commercial basis, reformed its relationship to the government's financing of research while preserving its independence. To do so took him altogether out of science; his intelligence was no longer a force felt by the young scientists in the laboratories he had created. His style of administration was autocratic, not collegial, and repulsed old friends. A particular decision to move the institute to the suburbs, selling off the historic but very valuable acres in Paris, was bitterly opposed and at last defeated by the senior scientists. Monod's successes at the Pasteur were undoubted but embattled.

His public activities grew manifold. When Zhores Medvedev was imprisoned in a Soviet insane asylum, Monod was one of the Western scientists who organized the protests that were instrumental to his release. Publication of *Chance and Necessity* put Monod on the lecture platform and the television screen. He campaigned for birth control when even to advocate it was still a crime in France. He accepted the presidency of the militant women's-rights organization called Choisir (Choice) a group that campaigned for, among other things, legalization of abortion. He took provocative part in a sensational abortion trial, a test case, arranging to be named as the one who had paid the abortionist.

The flamboyance, the autocracy, the mixture of idealism and vanity, offended many of Monod's old friends; they estranged Jacob, who saw the worst face of his old friend's character now dominant. Several scientists at the Pasteur, including André Lwoff, hated *Chance and Necessity*. The science was unexceptionable and worth popularizing: the problem was Monod's claim that the science forced his philosophical conclusions. "Starting from molecular biology to derive an ethic seemed no different from Teilhard de Chardin," Jacob said—and Teilhard had been a figure of derision for Monod. But most regretted was Monod's departure from science—and in particular, especially for Jacob, the failure to get the repressor. I perceived in Jacob a tic of greatness that I had seen before, in Bragg and in Pauling: he had wanted it all.

❖ ❖ ❖

Bragg, on retiring from the Royal Institution, kept in touch by a yearly visitation of Perutz's laboratory on behalf of the Medical Research Council, to whose secretary he would then write a two-or-three-page letter about what was going on. Thus, in mid-January of 1968, he reported, among other things:

> Perutz is working on his vast haemoglobin model, tidying up the details and getting the best atomic positions. There is enough detail for a complete book on chemistry in one such molecule! . . . The whole question of publication is difficult. How can one convey all the information in a huge molecule. A list of coordinates is of no help unless one builds a model. The 'paper' sent to colleagues should be a model!

That was the model of oxyhemoglobin of horse at high resolution that Perutz first made public before the Royal Society four months later. The Croonian Lecture for 1968 took place on the afternoon of Thursday, May 30. The society's great lecture theatre, where Perutz had set up the model and the stereoscopic projector the day before, now was crowded. Extra folding chairs were brought in; someone said it was the largest audience the Royal Society had ever gathered for a talk of the kind. At the front of the room, Perutz looked small, solemn, slightly hesitant.

His platform manner was not so flowing and vivid as his personal conversation, for in the reverse failing of somebody like Crick, Perutz seemed to fear that it would be a discourtesy to tell his peers too much of what they already ought to know. He enjoyed the showmanship of the colored lights in strings and clusters on the model that picked out the protein chains and the heme groups. The dozen three-dimensional slides, which he explained had been drawn by a computer onto an oscilloscope to be photographed, were sensationally effective in making clear the molecular details, because all the supporting rods and clamps had been stripped away and we saw floating above our heads only the links and joints of the structure itself, at the best angle of view for each detail.

The details were the lecture: the structure seen at last, down to the positions, within one angstrom unit, of almost every one of the molecule's 574 amino-acid residues and, indeed, the angles of their side chains. After touring the molecule loop by loop and corner by corner—his thirty years' hard labor—Perutz focussed more closely on those features that promised, still enigmatically, to explain how hemoglobin worked. "The model rules out certain mechanisms of haem-haem interaction which have been proposed, but it still does not reveal what the actual mechanism is," he said.

Each heme group, as in myoglobin, lay in a pocket formed close to the surface by the near approach of several stretches of polypeptide. The pocket was waxy, which is to say that the side chains of the amino acids that composed it were almost without exception nonpolar, uncharged, and therefore hydrophobic. Water was excluded. Data had been accumulated by 1968 on the amino-acid sequences of hemoglobins and myoglobins of several different species of mammals; though these varied considerably overall, the pocket of the heme group was made almost identically in every one. "The invariance of the residues surrounding the haem group implies that they are nearly all essential for the function of the haemoglobin molecule," Perutz said. As physical chemistry, the reason was clear: the low level of charge and of conductivity in the immediate setting allowed the iron atom to combine with a molecule of oxygen lightly—oxygenation, not oxidation.

"The fold of the polypeptide chain in haemoglobin and myoglobin appears to be one of the fundamental patterns of Nature." Like the heme pockets, the rest of the interior of the molecule excluded polar, charged residues. Simply, the molecule seemed to fold up into its highly specific shape because amino-acid residues that were hydrophobic turned away from water to the inside. Comparison with abnormal human hemoglobins showed that when a polar amino acid was substituted inside the molecule where a nonpolar one normally stood, the effect on the stability of the molecule was disastrous. But in these matters the new model merely confirmed principles already assumed.

The reversible change in shape of the molecule in its reaction with oxygen was now seen more exactly. Within any one of the alpha or beta subunits, at this higher resolution, change still appeared to be very slight. The shift took place between the subunits, and most conspicuously at the contact between the folded-up alpha chain of each half-molecule and the beta chain of the opposite half. On deoxygenation, that beta subunit rotated, relative to that alpha chain, by 13.5 degrees and simultaneously moved by 1.9 angstroms along it; the symmetrically opposite alpha and beta chains did the same; the motions all together were what carried the beta heme groups 7 angstrom units farther apart.

Again, the model confirmed the known. It went slightly further. The faces where the subunits came together could now be studied closely, and the amino-acid side groups enumerated by which the subunits interacted. These contacts involved a large number of weak bonds. In the face between an alpha chain and the beta chain of the *same* half-molecule, thirty-four residues took part, putting about a hundred and ten atoms each as close as four angstroms to at least one other—a large area of contact in which many side groups interlocked. The contact between the alpha chain of one half and the beta chain of the opposite half, however, was made by nineteen residues involving some eighty atoms in close touch. Thus, here where the gross movement took place as oxygen was unloaded, the area of contact was smaller—and smoother, as well, constructed so that the two subunits slid past each other. The geometry of these movements had now been elucidated. Their mechanism remained obscure.

The model presented one curious structural uncertainty, Perutz said: the tips of the four polypeptide chains could not be seen clearly in the electron-density map, and therefore could not be placed unequivocally in building the structure. By convention, the amino acids of a polypeptide are numbered starting from the one whose raw end is an amino group, NH_2, and therefore called the amino terminus or N terminus, and proceeding knuckle by knuckle (valine 1α, leucine 2α, serine 3α, alanine 4α, alanine 5α, and so on) to the other end whose stub is a carboxyl group, CO, called the carboxy or C terminus (arginine 141α). Amino groups are positively charged, carboxyl groups negatively. Four chains have eight ends.

The only terminal residue visible on the map of oxyhemoglobin was the valine at the beginning of each alpha chain. The valine at the beginning of the beta chain cast no shadow of its density on the map; nor at the carboxyl ends did arginine 141 of the alpha chain or histidine 146 of the beta chain. These were therefore not fixed but wobbling loose—and so made a blurred contribution to the diffraction pattern. At least, the tips of the chains were loose in the high salt concentrations at which the crystals had been grown. Perutz thought that in normal, bodily conditions the terminal residues of oxyhemoglobin would very likely be tied down by salt bridges or hydrogen bonds either to each other or to nearby loops of the chains.

For all the copious detail, the structure left the cooperation of heme with heme and the Bohr effect both unexplained. Perutz concluded by saying that movement at the molecular level, in enzymes, genetic repressors, even the molecules of muscle tissue, made the hemoglobin molecule the exemplar of a very general and fundamental problem in biology. He confessed his bewilderment with his own results with hemoglobin so far. The only way forward was to map deoxyhemoglobin to at least the same high resolution as the model of oxyhemoglobin he had just presented. "What we have done is merely the anatomy at the atomic level. It is vital to advance to the physiology."

❖ ❖ ❖

The Bohr effect was first to yield to the illumination provided by the structure of hemoglobin. It yielded grudgingly, a fraction at a time, and in terms that would have been alien to Christian Bohr himself. For Bohr and his colleagues, the effect they discovered was that when blood moved from lungs into tissues and the concentration of carbon dioxide, mildly acidic, began rising, the affinity of hemoglobin for oxygen began dropping, so that the hemoglobin molecule unloaded its oxygen more efficiently. Or, stating Bohr's observations the other way around, as hemoglobin unloaded oxygen it made the blood more alkaline, neutralizing the effect of the carbon dioxide and allowing more to be absorbed, so that the blood transported carbon dioxide from tissues to lungs more efficiently. For Barcroft and the physiological chemists who came after him, hemoglobin's making the blood more alkaline meant that as the molecule threw off oxygen it became able to take up protons—the positively charged nuclei of hydrogen atoms—from the surrounding liquid. An average of two protons had to be taken up by each hemoglobin molecule. The structural question was to know what specific structural changes attracted protons to what specific places in the molecule.

Perutz and his biochemical colleague John Kilmartin unravelled the Bohr effect by pulling at the loose ends of the polypeptide chains. Late in 1968, Kilmartin found a clever, direct treatment to make the amino termini of the chains—alpha, beta, or both, at will—unable to react chemically in the normal range of blood's alkalinity. He had to mask the molecule's reactive sulphydryl groups with one chemical, then add a second that combined with and blocked the amino ends of the chains, then use a third to strip the mask off the sulphydryl groups again. The resulting modified hemoglobin he spread out chromatographically, and got three fractions. Then by separating the alpha from the beta chains of a sample taken from each fraction, he was able to show that one fraction had the amino ends of all chains blocked, another had alpha chains blocked but beta chains open, still another had beta chains blocked but alpha chains open. Perutz then found that crystals of all three types gave diffraction patterns almost identical with those of normal oxyhemoglobin.

Kilmartin's treatment had not, of itself, modified the structure more than minutely. None the less, he found that with the alpha chains thus blocked at the amino ends, the Bohr effect was reduced by about a quarter. In contrast, when the amino ends of the beta chains were blocked but the alpha chains were open, the Bohr effect was untouched. The reversible transformation of the molecule on deoxygenation altered the charge at valine 1α by enough—on average in equilibrium in the normal environment of the blood—to attract an extra proton to about half the deoxyhemoglobin molecules present at any instant.

The precise mechanism of that shift in charge, however, took years to prove. Perutz first suggested that weak, electrostatic bonds of the kind called a salt bridge formed in the deoxy structure to tie the amino end of each alpha chain to the carboxy end of the other one, which lay five angstroms away. The bridge would neutralize positive charge on the amino end, until disrupted by transformation back to the oxy structure—upon which a proton would like as not be thrown off. But the high-resolution map of deoxyhemoglobin, when it came out, showed no such salt bridge. Subtle alternatives were argued. Only in 1976 was it shown that Perutz's first guess was correct and that a chloride ion—an atom of chlorine negatively charged, drawn from the surrounding solution—provided the salt bridge.

So the amino termini of the alpha chains accounted for a quarter of the Bohr effect. Perutz and Kilmartin then looked back into the electron-density maps to try to find the tips of the beta chains. The high-resolution map of oxyhemoglobin showed a weak, diffuse peak for the last amino-acid residue but one, tyrosine 145β, and no sign of histidine 146β. The carboxyl terminals were free to move. The low-resolution map of the deoxy form, though, showed a faint loop of density extending from the carboxyl end of the beta chain back towards the reactive sulphydryl group, on the same chain at position cysteine 93. Once more they turned to modified hemoglobin. They put in an organic reagent that occupied the reactive sulphydryl groups, but without a heavy atom. Hemoglobin thus altered was known to have its Bohr effect reduced by half. Then Perutz grew crystals of the oxy and deoxy forms of the altered hemoglobin. He did not have to solve the complete structures of these, but only to compare intensities of the diffraction spots and map the differences. In the oxy and deoxy forms, even at low resolution, the blocking chemical showed up where it belonged. Otherwise, the oxy form was the same as normal hemoglobin. The deoxy form showed that the tips of the beta chains were not in their normal position but nearly twice as far from the reactive sulphydryl groups.

Perutz thereupon built a model of the relevant parts of the chains. The model suggested that in normal deoxyhemoglobin the side ring—positively charged—of the terminal histidine of each beta chain was held in position by a bond to the residue in the same chain that adjoined the reactive cysteine. Again, the transformation to the oxy form broke the salt bridge and the positive charge at the terminal residue was no longer neutralized. In 1970, the high-resolution structure of deoxyhemoglobin clearly showed this salt bridge.

I next saw Perutz the first Friday in April of 1970, in London once more, where he had been presiding over a session of a symposium at the Royal Institution in honor of Bragg's eightieth birthday. In an optimistic mood, he ticked off progress on the Bohr effect. A quarter of it was due to the amino groups at the beginnings of the two alpha chains. Half of it was due to the terminal histidines of the beta chains. There, Kilmartin had just succeeded, with an enzyme, in snipping off that single amino acid from the tip of each beta chain, to find that the Bohr effect was exactly halved.

The remaining quarter? Well, Perutz said, he had tried still another way to see the tips of the chains. Suspecting that they took fixed positions under normal bodily conditions, he had recently grown some crystals without using salt to precipitate them, and the week before had looked at difference maps. But after all, he said, "They did not show a trace of those terminal residues. But the most prominent feature shown was a hole in the middle of the molecule when I took away the salt." He had quickly formed an idea of what might fit the hole, and how that might account for the rest of the Bohr effect by a further reversible neutralization of local charge groups. "If this last hypothesis is right, we've got it all."

The Bohr effect proved subtler than Perutz had ever expected. Although the kinds of mechanisms that had to be found were known, in principle, to Perutz and his collaborators by 1970, details of shifts in charge on specific locations of the molecule to explain that last fraction of the effect were still in question in 1978. The work on the Bohr effect required innumerable experiments and structure determinations on an ever-expanding variety of hemoglobins, whether mutant or artificially modified. The pursuit was painstaking to the point of tedium. Yet at every step it exemplified the interplay between biochemical problems and techniques and the visualization of structure and structural change at atomic dimensions; specificity in this strongest sense is, of course, what molecular biology is about.

"You know, people were surprised and even incredulous when we first suggested that these mechanisms could depend on such simple, local, slight effects as the shift of charge by the positioning of a side group, or the making and breaking of salt bridges," Perutz said eight years later. "But these turned out to be essential to the functioning of enzymes as well, and have come to be taken for granted." Over the course of that decade, Perutz and Kilmartin's

patient teasing out of the Bohr effect was of quiet but decisive power in changing the way people visualized biochemical problems. That also is what molecular biology is about.

That April day in 1970, though, Perutz was excited about a paper he had
just heard at the symposium: the structure, presented by Robert Huber, of the
Max-Planck-Institut for protein research in Munich, of the hemoglobin of the
larva of a fly. The fork in the evolutionary tree separating insects from vertebrates occurred some seven hundred and fifty million years ago. The amino-
acid sequences of insect and vertebrate hemoglobins were largely dissimilar.
Yet the tertiary structures—the overall folding up—of these hemoglobins were
startlingly alike. The conspicuous difference was that the structure of insect
hemoglobin was not transformed when loading or unloading oxygen.

"The heart of the hemoglobin problem is the trigger that starts off the conformation change," Perutz said. "This is hard to discover, because you have to
compare the electron-density maps of the two molecules, the oxy and deoxy
forms. And each of the two has its own sets of experimental errors, arising
from calculation of the phases for the Fourier synthesis. The errors were
cumulative in the comparison, making fine-scale differences uncertain.
"Huber's paper on insect hemoglobin threw brilliant light on the heart of the
problem. For *this* hemoglobin does not change its conformation. So when you
compare the oxy and deoxy states, you are comparing two maps with the
same errors in phases." We were eating a hasty lunch; Perutz had ordered a
well-done steak and now was meticulously trimming the fat from it. "So what
does Huber see? That *the heme group itself* undergoes a slight distortion! This is
just what we are looking for. And though I could not detect these changes in
our hemoglobin—even knowing they may be there—I may be able to detect
changes that are consistent with those changes." Perutz's enthusiasm was intense. "Kendrew should have been able to find these changes with myoglobin," he said. After a few minutes more, he left to catch his train.

❖ ❖ ❖

For all that the enterprise of science is by nature collective, the experience
of discovery varies greatly among great scientists. Perutz has been at an extreme, not only in the longer rhythms of a lifetime given over to one problem
but in the chiefly solitary, contemplative manner in which he has come upon
discoveries. The totem of his intelligence was neither the tiger nor the shark
but the elephant: will and a colossal attention span. Behind the industry and
an unflagging, paradoxical urgency—"led on by wild expectations"—looms a
brooding weightiness of mind.

Huber's report that when insects respired there was a structural shift at the
heme group itself brought Perutz to search for the same thing in horse and
man. That spring, the electron-density maps for deoxyhemoglobin at high
resolution, 2.8 angstroms, were completed. Perutz pondered them. Over the
last weekend of July 1970, he perceived the trigger.

At that moment he was the first to see, as well, the extraordinary significance of Pauling's discovery, in 1936, of the magnetic difference between
oxy and deoxyhemoglobin. Pauling had found (to recapitulate) that oxyhemoglobin was diamagnetic, with its heme groups carrying no electrons unpaired
in spin and with each iron atom covalently bound in six directions—to the
four nitrogen atoms in its porphyrin ring, to the oxygen molecule, and to the

histidine of the polypeptide chain on the opposite side of the ring from the oxygen. In contrast, deoxyhemoglobin was strongly paramagnetic, each heme bearing four unpaired electrons and grasping its iron atom by ionic, or electronic, bonds, looser and weaker.

A fortnight later, as it chanced, I came to Cambridge. Perutz was in the midst of experiments, he said; he was agitated by a barely suppressed eagerness, able to speak connectedly of nothing but his discovery. The cascade of realizations had begun when he looked at the portions of the new electron density map that represented the deoxy heme groups. They appeared not flat but markedly domed. "When I looked at the bit of electron density which represented the flat pigment group," he said, "the iron atom turned out to be three-quarters of an angstrom out of the plane of the ring. In the *oxy* form, the iron atom must lie in the plane of the four nitrogen atoms to which it is covalently bonded. Whereas in the deoxy state, the iron atom is ionically bound. I *knew* that much: but what I did not realize was how much *larger* the ionically bonded iron atom is than the covalently bonded. I had not read, or had forgotten, the literature.

"So two weekends ago, I read up the literature. The *ionic* iron's radius is larger by about twelve per cent. The other thing I had not realized was that the *covalent* iron's radius was just right to fit in the square of the four nitrogen atoms. While three-quarters of an angstrom's displacement out of the ring, as shown by the maps, agreed closely with the displacement calculated on the basis of the ionic atom's greater diameter. In a way it was good that I had not known this in detail—it might have influenced my measurements!"

The physics by which Perutz described the mechanism was to grow more technical over the next seven years; his first, naïve apprehension was superbly easy to imagine. In the deoxy form, the iron atom was too large to fit the porphyrin ring and was squeezed out slightly. On combining with oxygen, the iron shrank just enough to slip snugly into the ring. The bonds, now covalent, were shorter—the point Pauling knew but didn't mention in 1936. The porphyrin ring acted as a mechanical amplifier, changing a very small increase in diameter into a sharp linear motion.

Kendrew and his colleagues had said that in myoglobin, on loading oxygen on or off, the iron atom did not seem to move into or out of the plane of the porphyrin ring. At Cornell University, Lynn Hoard, a crystallographer trained by Pauling, had asserted that Kendrew must be wrong and that there was a perceptible shift into the plane of the ring in hemoglobin compounds where the bonding became covalent. In the search of the literature that followed the flash of understanding, Perutz came upon Hoard's paper. "Hoard predicted it all," Perutz said. "What Hoard wrote was that the pigment ring was very flexible."

Several months later I asked Kendrew about that. "The question is, does the iron atom shift as between the oxy form and the deoxy? I never worked, myself, with the deoxy form of myoglobin; a colleague of mine has done some work. It's technically very difficult," Kendrew said. "I think it may be that the information was there all the time." I asked what he meant. "I'll tell you the kind of problem we had," Kendrew said. "When we first got the structure of myoglobin"—at low resolution—"we wanted to build a model. And nobody had ever, at that stage of things, determined the structure of a heme group. In isolation. All we could do was to make the best possible guesses as to what the structure ought to be—just as, when we wanted to see where some side chain was, like phenylalanine, what we did was to put into the Fourier a model of a phenyl ring, with the dimensions taken out of the textbook. So what I in fact did for the heme group was to go and consult the best theoretical chemist I could find around the place and said, 'Now if we're going to build a model of this thing, what dimensions do you suggest?' Now, he made it flat. Five years later, somebody obtained the structure of an isolated heme. And found it wasn't flat. More like an umbrella. And some ten or twelve degrees of twist. We'd put it in as flat. Then we started trying to refine the structure—and of course, having made the assumption it was flat, it stayed flat forever after. Wrong assumption. If you look at our published map of myoglobin, the *best* resolution, you'll see the heme group is flat. But I'm damned sure it isn't. It's been artificially pushed that way. And this is what I mean by saying the data probably contain more information than is apparent."

Two other new structural features were revealed by the solution of deoxyhemoglobin at high resolution. First, six salt bridges, three pairs, could now be discerned, holding the halves of the molecule together. These electrostatic links were not present in the oxy form. Such bonds were weak: where a covalent bond between two carbon atoms took some fifty kilocalories of energy to break, the salt bridges Perutz had now found each had a strength of about two kilocalories. The total agreed well with biochemical estimates of the energy that had to be put into hemoglobin when it combined with oxygen. "People have talked for years about the interaction energy in oxygenation, about twelve kilocalories," Perutz said. "Now we know what the energy is used for—two kilocalories to break each salt bridge."

Second, the heme pocket on each beta chain was partially blocked by the side group of a nearby residue, valine 67β, that moved out of the way in the oxy form. "The alpha chains have no difficulty in accommodating the oxygen molecule. In fact, one might have said that it rattles," Perutz said. "But in the beta heme pockets, you can't have even a single atom sitting with the iron atom without prising it apart. This told me the order of the oxygenation. The alpha chains appear to be oxygenated first, and that change somehow makes it easier for the beta hemes to react." But that possibility was incidental to the mechanism—to the trigger. "The essential question is, what effect does the addition of oxygen have?" Perutz said. *"The iron atom is attached to the protein chain by a rigid bond so that when it clicks in and out of place it takes something with it."* Thus the iron atom transmitted its motion, by way of the side chain of the histidine residue to which it was bonded, to the stretch of alpha helix of which the histidine was part.

The movement of the iron atom set off the sequence of change. "But the main difference between the two structures, oxy and deoxy, is those salt bridges that link together the two halves of the molecule in the deoxy form—and are missing in the oxy structure." In the fortnight since Perutz had spotted the trigger, he had worked out, from the Fourier maps of the difference between the two structures, the general sequence of events whereby, as the hemes were oxygenated one at a time, the salt bridges were broken a pair at a time, and the structure was thus released to take its oxygenated form.

"Hemoglobin is the prototype of a class of proteins—the others are enzymes—that are self-regulating," Perutz said. "These are the allosteric en-

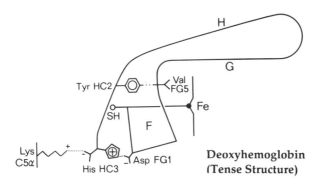

Deoxyhemoglobin (Tense Structure)

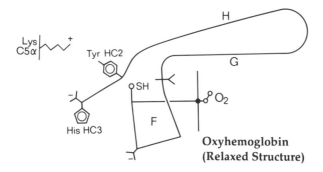

Oxyhemoglobin (Relaxed Structure)

Last stretches of beta chain: extra bonds in deoxy structure are formed by the two end residues, tyrosine and histidine.

zymes. Monod and his colleagues first produced a general theory of allosteric enzymes in 1963—and that was based in large part on my preliminary results with hemoglobin. But the mechanism I have now found has the essential features that a self-regulating, allosteric protein *will* have, according to Monod's scheme." Combination with oxygen at the hemes on the alpha chains had an effect that was transmitted throughout the structure of the molecule to alter the local configuration, and so the oxygen affinity, of the hemes on the beta chains. Breaking of the salt bridges provided the means of the change, called for by the theory of allostery, from the tense state to the relaxed state—and, indeed, met exactly a prediction, in the paper that Monod published in 1965 with Wyman and Changeux, that deoxyhemoglobin would turn out to be the tense structure. To be sure, hemoglobin was an unusual allosteric protein in one way: the theory called for separate reactive sites and regulatory sites, a requirement hemoglobin met by having each of its four sites serve the two distinct functions—reacting locally with oxygen and regulating at a distance the reactivity of the other sites.

Obviously a lot of detail was still unknown about what went on to transmit the strain through the molecule from hemes to salt bridges, Perutz said. "It will take another year or two of consolidation. But the most remarkable feature of the whole thing is that evolution itself should have used the varying diameters of the iron atom to produce this respiratory carrier!"

❖ ❖ ❖

Consolidation took seven years. Monod, when Perutz sent him a copy of the draft of his first paper announcing the mechanism, wrote back at once and generously: "Of course this must be right!" Bragg told him, "I am sure that this is how it works." Perutz was astonished at the opposition that sprung up in some other quarters. "Disbelief and indignation—I had expected naïvely that people would be delighted to see this problem solved. In fact, everybody was angry!"

Opposition forced the pace and variety of experiment. Many details of Perutz's mechanism changed. More refined calculation of Fourier maps found, for example, that on oxygenation the iron atom moved not three quarters of an angstrom but six tenths. Perutz relaxed somewhat the rigid mechanical progression of his scheme, realizing—in part on Crick's urging—that it made no difference to the mechanism, and much better chemical theory, to suppose that oxygenation or deoxygenation might begin at any of the four hemes and progress through the molecule. Thus it was possible that an oxygen molecule, by getting into the heme pocket on the beta chain, prising it open and combining with the iron, thereby transmitted an effect through the structure to change the affinity at the alpha hemes.

The most important modification restated—in fact, revised several times—that vivid, simple idea that the iron atom changes its diameter upon oxygenation. In its final form, in 1977, Perutz put the idea in terms of the modern theory of electron orbitals. "The reason why the bonds get longer is actually very sophisticated," Perutz said that fall. "When I used to talk about the hole between the four nitrogens being too small for the iron to fit in, that isn't really a good explanation. The real explanation is quantum mechanical. And is concerned with the symmetry of the orbitals into which the unpaired electrons go"—the four unpaired electrons of the iron atom in the deoxy, paramagnetic state. "Which is different from the symmetry of the orbitals of the porphyrin ring that surrounds the iron. And it is really the repulsion between the orbitals of the iron and the orbitals of the nitrogen—which is something the theoretical chemists nowadays call 'antibonding'—which produces the movement of the iron out of the plane of the porphyrin. That was brought home to me by Robin"—Perutz's son, himself a physical chemist. The two had been discussing a paper recently published by a group "who claimed to have disproved my mechanism of heme-heme interaction with a new method of measuring iron-nitrogen bond distances. They thought they had disproved me by showing that the hole between the four nitrogens is not really too small for the iron atom to fit in. And Robin then pointed out to me that this whole way of looking at the problem is much too naïve, and that the displacement of the iron from the plane of the four nitrogens really arises, as I said, from the repulsion between orbitals of different symmetry."

As the details were filled in and refined, a beautiful economy of structure and function in hemoglobin emerged. The salt bridges between the subunits, the iron atoms in the hemes—these are the two ends of the molecular piston. Perutz found two more salt bridges, for a total of eight. They provide the bonds that hold the molecule in the tense, deoxy structure. Their breaking or reformation mark the allosteric transition. And at the same time, the salt bridges explain the influence on the oxygen-affinity curve of all the various chemicals and mutations that had been puzzling for so long. "All agents which lower the oxygen affinity do so either by strengthening existing salt bridges in the T [tense] structure or by adding new ones," Perutz wrote in 1978. "Not all those extra bonds are between the subunits; some occur within the subunits and oppose the subtle structural changes which the subunits undergo on combination with oxygen." Finally, the salt bridges make the Bohr effect—they account both for the lowering of the oxygen affinity of hemoglobin by the presence of hydrogen ions in the surrounding water, and the converse uptake of hydrogen ions by the molecule as oxygen is released.

Thus, the physical phenomena that add up to the Bohr effect are the stimulus at the other end of the cycle. The salt bridges in hemoglobin connect strongly acid side groups, which are negatively charged, to weakly alkaline ones, which are positively charged. Hydrogen ions, when present in solution, add positive charge to bases, and thus strengthen salt bridges and make easier the formation of new ones. And thus, where the transition from the deoxy to the oxy state at the first heme in the molecule, in passage through the lungs, sets off the allosteric change from tense to relaxed that makes further oxygenation more likely, the presence of hydrogen ions in the blood of tissues and veins, due to carbon dioxide, initiates the allosteric transition in the reverse direction from relaxed to tense structure. That transition itself brings acidic groups into the proximity of weakly basic ones and thereby reduces the work that must be done to make more salt bridges. That transition itself transmits the strain to the heme-linked histidines of hemes still oxygenated, to pull the iron atom out of the plane of the porphyrin ring and so make unloading the oxygen more likely. The hemoglobin molecule is an oxygen pump. With every breath and every beat of heart and pulse, the cycle goes on.

The unity of structure and function in hemoglobin is dynamic—whereas that unity in deoxyribonucleic acid is static, passive. The unity of hemoglobin is not, perhaps, so instantaneously obvious as that of DNA. It is similarly ineluctable.

CONCLUSION, 1978

"Always the same impasse"

The rise of molecular biology is one of the important intellectual and scientific movements of the age. It has naturally attracted comment and interpretation from several of the scientists who took part in it and have since glanced back, and from outside observers as well. These accounts lay stress now on one, now on another aspect of that rise; they are mostly brief, and are no more than loosely coordinated in any deliberate way. Yet from them has emerged what may be taken as the usual view, the standard view—to put no fancier a name to it—of the origins and early development of the science. We can now examine the adequacy of that view and then attempt to say what sort of revolution in science the rise of molecular biology was.

That standard view has several elements. For one thing, new people came into biology, and most famously the physicists: Max Delbrück, Leo Szilard, Francis Crick, Maurice Wilkins, on an eccentric orbit George Gamow; Linus Pauling the physical chemist; from the older generation Sir Lawrence Bragg and Desmond Bernal; allied to them the straight chemists Max Perutz and John Kendrew; of course others. With the new men came new programs. Among these, the most highly articulated was the program of Delbrück, Salvador Luria, and the group that formed around them, the American phage group. With their fastidious rigor, their insistence on the simplest biological systems, their distrust of biochemists and earlier microbiologists, their self-conscious marking off of the group from others—their snobbery—the phage group has attracted attention even to a disproportionate degree. Another program, that of Pauling and some of his many colleagues and students in the United States, and the related enterprise of the crystallographers in England, was to investigate the structures of the large biological molecules—of proteins, in the first place—with the explicit idea that molecular structure should eventually illuminate physiological functioning. A more general urge—widely felt in the mid-thirties and voiced as a program by Warren Weaver and his associates at the Rockefeller Foundation—was to put physical chemistry to work to clear up some of the mysteries of biology. Weaver's interest produced the name "molecular biology" and, without having to go into much detail beyond that name, found the seed money for the work of, among others, Delbrück, Pauling, and Perutz. Then, after the second world war, another group of new men formed, this time native to microbiology, around André Lwoff at the Institut Pasteur in Paris.

The new men and their programs were in touch with one another, particularly and most continually at the California Institute of Technology, where Pauling was, once Delbrück settled there in 1947. Within a few years, the phage group and Lwoff's unit at the Institut Pasteur in Paris were exchanging younger people so regularly that in effect they interpenetrated. One reason, obviously, was that the programs were fundamentally similar in using genetic manipulation of micro-organisms to isolate elementary biological events for scrutiny—and this despite the distinctions that some of Delbrück's young enthusiasts felt between those who worked with phage and the others who worked with bacteria, and despite Delbrück's refusal to believe, for several years, in the phenomenon of lysogeny. But the great merger, as we have seen, was that precipitated by James Watson's moving to the Cavendish Laboratory, in Cambridge, to learn crystallography, and finding Crick there.

These elements, the most obvious in the growth of molecular biology, have suggested several attractive ways to state what happened simply yet more formally. For instance, the sciences have sometimes been thought to form ladders: by one convention, mathematics as most fundamental, mathematical physics, physics, physical chemistry, chemistry, organic chemistry and bio-chemistry, cellular physiology, and so on to ecology. (What such ladders represent, though, is unclear. Is a step upward a decrease in abstraction? An increase in complexity? Perhaps merely an increase in the size of the things treated? A decrease in that factor of academic prestige which one might label "difficult ultimateness"?) In such terms, molecular biology is taken to be the conquest of biochemistry by what has been called its "antidiscipline"—the science more basic in the hierarchy, physics or physical chemistry. This simple dialectic is made more plausible by the very name "molecular" biology, by the fact that physicists moved into the science—and by the attacks of some from older disciplines, most noisily the biochemist Erwin Chargaff, who decry molecular biology as reductionism. A moment's inspection shows this diagram to be too simple. Consider the physicists and the phage group, for instance. Not only did their program and personnel overlap with those of the microbiologists around Lwoff but, even before the discovery of the structure of DNA, the work of both groups was merging with that of Joshua Lederberg and William Hayes. In what way was the contribution of the physicists, to all that, uniquely physical?

More general and more useful is John Kendrew's version of the standard view. Kendrew proposed, first in 1965, that molecular biology was the confluence of two currents, which he labelled "information and conformation." Kendrew's play with words, a memorable one, introduces an unfortunate stroke of hindsight. Information theory was a program of the nineteen-forties and after, by which one would expect mathematicians and physicists to have influenced molecular biology; yet, in point of fact, it was almost entirely ineffectual. Except, perhaps, for Szilard's intervention in the development of the repressor model for control of enzyme biosynthesis—an intervention, assuredly dramatic, whose practical importance is easy to exaggerate—information theory has been useful to molecular biology mostly as a way to talk, later, about discoveries and models that were reached without appeal to its principles. Kendrew used "information" in this retrospective way. He meant by it, first, the moiety of molecular biology rooted in genetics, particularly of micro-organisms and including the genetical elucidation of biochemical pathways and controls, as by Jacques Monod and François Jacob; second, the moiety springing from studies of the structures of the large biological molecules, as by Pauling and by the British crystallographers, culminating in the elucidations of the relations of structure to function in DNA and proteins. These currents in the early years of molecular biology can be more accurately labelled the one-dimensional (thus, the genetic and sequential) and the three-dimensional (thus, the stereochemical or structural). Put that way, the distinction moves beyond the limits of the standard view well towards the crux of the matter. The interplay between the one-dimensional and the three-dimensional has been implicit through this book.

But the central element in the standard view of the early history of molecular biology is the ten- or fifteen-year struggle of opinion over the identity of the substance of the gene: the change from the conviction at the end of the thirties that genes were made of protein to the suspicion, and then the confirmation, that genes are made of nucleic acid. The change in characterization of the genetic substance, and the discoveries and arguments that made the change persuasive, have been generally supposed to be the transformation that was most fundamental to the revolution in modern biology.

All these elements were present, to be sure. None the less, I think that the standard view of the origin of molecular biology is incomplete and misleading. The synthesis from which molecular biology grew was far more complex than the standard view acknowledges. The central transformation was different. To replace the standard view, I offer a couple of observations. They seem to me nakedly elementary.

Molecular biology arose in the synthesis of particular lines from five distinct disciplines. The synthesis began in the mid-thirties; the merger of genetics and x-ray crystallography, with Watson and Crick and the structure of DNA, was a triumph of that synthesis approaching its maturity. Each of the disciplines in the synthesis was developing vigorously even as their coming together began. I detect no useful hierarchy of abstraction or reduction, or even any consistent pattern of conflict, among these disciplines; strains *within* several of them were undoubtedly important. Physical chemistry was but one of the five. As advanced so dramatically by Pauling, for example, it was closely allied to—and got much of its power and detail from—the crystallography of the Braggs and their school in England. The third discipline was genetics— itself becoming biochemical. The fourth was microbiology—emerging in the forties from a transformation of its own. The fifth was biochemistry—itself in revolution in the same years.

But beyond that, the change that occurred, from the late thirties, in the identification of the substance of the gene—not protein but nucleic acid—has obscured a more fundamental change that took place at the same time in the way that the large molecules of the cell—nucleic acids and proteins both—were understood. And then when participants began to look back at the changes, molecular biologists had become absorbed with the details of how the plan in the DNA is moved out to be expressed as protein. This preoccupation surely reinforced—illogically but psychologically—the tacit assumption that the change from protein to DNA as the carrier of the plan, a change as it were in the anti-parallel direction, was the significant movement in the history.

The more fundamental transformation in biology from the thirties to the sixties was the working out of the idea of specificity. Biological specificity in its present sense had a number of forerunners. Today, three considerations subsume all the rest. The first is linearity: the long-chain biological molecules, both proteins and nucleic acids, are specific in sequence. The second is structure: the large biological molecules owe their functional specificity to specific configurations. This pairing, obviously, restates the interplay of the one-dimensional with the three-dimensional. The third consideration is that linear specificity determines three-dimensional specificity. This present sense of biological specificity was first clearly formulated by Crick in the mid-fifties when he put forward the sequence hypothesis.

By the late thirties, several problems that had bothered previous generations of biologists were settled. In particular, proteins and nucleic acids were understood to be very large molecules, made of long chains of subunits. Enzymes were understood to be proteins. Biologists certainly spoke of specificity. The phenomena they were most interested in—genes (whatever they were in substance), enzymes and antibodies (known to be protein)—were highly specific in action. Yet specificity, as a ruling way of thinking about biochemical events, was lacking that entire set of meanings—those distinct dimensions—it had acquired by 1970. In particular, though the action of enzymes (and genes) had to be in some sense specific, when biochemists of the thirties turned to their substances, their compositions, their chemical makeup, they thought of them in general, nonspecific ways. And they thought about nucleic acids and proteins, as substances, in strictly comparable ways—the chief difference being that nucleic acids, supposed to be less interesting, were thought about less elaborately.

Nucleic acids were believed, following Phoebus Aaron Levene's tetra-nucleotide hypothesis, to be composed of a monotonous rotation of the four nucleotides—repeating sets containing adenine, cytosine, guanine, thymine (or uracil). Proteins, though built of assortments of more than a score of different kinds of amino acids, were similarly believed to be multiples of fundamental units, many times repeated. Theodor Svedberg, for example, was convinced that he had discerned, with the aid of the ultracentrifuge, that proteins were made of multiples of units of 17,000 molecular weight. In 1937, Max Bergmann and Carl Niemann, at the Rockefeller Institute, claimed that they had found a general chemical law ruling the total number of amino acids in a given protein and the proportionate number of each kind of amino acid in that total. They thought this law was the result of rules governing the uniform spacing of each kind of amino acid along the polypeptide chain. Thus, they said, in fibroin, the protein of silk, "Each alanine residue is separated from the adjacent alanine residues by 3 other residues. . . . Each tyrosine residue is separated from the adjacent tyrosine residues by 15 other residues." And from "the superposition of many different individual frequencies" of perfectly spaced amino acids arose, they thought, the "law which determines the total number of residues" in the molecule—usually, they said, "288 units or a whole number multiple thereof." Specificity of a protein's function was due to differences in amino-acid compositions: "The physico-chemical and biological properties of a protein are based, in the last analysis, on the frequencies with which its constituent amino acid residues recur within its peptide chain."

The tetranucleotide hypothesis was generally accepted; the Bergmann-Niemann formulas could hardly have been regarded by many biochemists as conclusive. In either case, however, these ideas represented the traditional direction in which explanations of the chemical structures of these substances were to be sought. Neither for nucleic acids nor for proteins had any presentable alternative been suggested.

Explanations of the biochemical behavior of proteins sound a similar discord, to present-day ears, between specific functioning and nonspecific chemistry. Proteins were held to exchange amino acids with their neighbors in "dynamic equilibrium." They were thought to be synthesized in reversible chemical reactions catalyzed by the same enzymes that digested them. Again, in 1940 Pauling introduced a "theory of the structure and process of formation of antibodies" in which the extreme specificity of an antibody to its antigen was added in a folding-to-fit that took place, as a separate step, after the protein had been synthesized. Once more, specificity of action had to be obtained from a substance thought not to be fully specific in its original makeup.

In the five disciplines that eventually contributed to the formation of molecular biology, ideas of specificity had widely different standing and character. In microbiology, the relevance of specificity in the present-day sense awaited resolution of the long, piecemeal shift of opinion about what sort of creatures micro-organisms were. Alternatives were not clearly set against each other across the entire field while the shift was going on. One sees now that the question kept coming back to whether micro-organisms were comparable to higher organisms in the most fundamental respects, and in particular whether their life processes were determined by genes. The principal opposing idea, in various guises, was that micro-organisms were switched, modified, adapted in response to their immediate environment. Oswald Avery demonstrated in 1923 that individual types of pneumococci have different and specific polysaccharides making up their protective exterior capsules. He established that bacteria are of distinct, true-breeding varieties. If the working out of the idea of specificity was the fundamental transformation I suggest, then Chargaff was right to say in eulogy that Avery made two discoveries that ought to have claimed the attention of the Nobel selectors. Frederick Griffith's discovery of bacterial transformation followed in 1928, as we saw, and was pursued in Avery's laboratory; the dénouement in the mid-forties overshadowed Avery's earlier discovery. In the nineteen-thirties, meanwhile, Lwoff showed that micro-organisms have nutritional requirements like those of multi-celled creatures. Not quite single-handedly, Lwoff established micro-organisms as biochemical entities—a crucial step both conceptually and technically. In 1943, Luria and Delbrück demonstrated that bacteria acquire resistance, in particular to attack by phage, not by adaptation but by genetic mutation; nine months later, independently in German-occupied Paris, Jacques Monod and Alice Audureau demonstrated that bacteria acquire so-called adaptive enzymes by genetic mutation, as a precondition for what was later renamed the induction of the synthesis of the enzyme. With that, bacteria were brought at last into Mendel's realm. It remained for Joshua Lederberg, by his announcement in 1946 that bacteria mate, to set off the exploitation of classical genetics in bacteria.

Mendelian genetics itself had always been highly specific, of course, and from very early the genetic specificity—the map—was understood to form a strictly linear array. That the line of genes has no branches was known decades before DNA or proteins were shown to be unbranched. From Archibald Garrod's elucidation of inborn errors of metabolism, in the first decade of the century, Mendelian genetics had always had a biochemical line of research; in the nineteen-thirties, genetics was coming down from Thomas Hunt Morgan's insistence on the abstractness of the gene and the linear sequences. It was turned decisively biochemical by the experiments of Boris Ephrussi and George Beadle on enzymes that determine eye color in flies; these soon set Beadle working with Edward Tatum on the demonstration with molds that what a gene does is specify an enzyme.

In physical chemistry, specificity was not abstract and not linear, but concretely physical and three-dimensional. Some of the most powerful strategies of molecular biology are rooted in Pauling's rules for close packing of atoms and the local balancing of electrical charges in the structures of molecules; in the precision with which the lengths and angles of chemical bonds can be known; in the importance of hydrogen bonds in the structures of large biological molecules. Such ideas were apparent in principle and understood to matter to biology by a few people at the end of the thirties. They were seen in steadily sharper detail and brought home to many in the late forties and early fifties.

Crystallography, in its way, was also permeated with specificity. Hemoglobin had been crystallized in the first half of the nineteenth century. In 1909, Edward Tyson Reichert and Amos Peaslee Brown published their atlas of photomicrographs of hemoglobin crystals from more than a hundred different species—and the crystals took many forms, including, in single species, different forms when oxygenated or not. But crystals could grow only from identical molecules laid down in repeated arrays: this became incontrovertible even at the beginning of x-ray crystallography, and clearly applicable to proteins when Bernal, in 1934, showed how to get diffraction patterns from protein crystals, proving them to be regular to atomic dimensions. Then in 1938 Felix Haurowitz saw hemoglobin, before him on the stage of his microscope, transform its crystal structure as it changed from the deoxy to the oxy form. Specificity in the present-day sense of unique molecular structures was never a surprise to the crystallographers.

Having recapitulated the growth of ideas of specificity in four of the five disciplines that came together in molecular biology, we can return for a moment to Delbrück, Luria, the new men and the phage group. By no means were they so isolated as some of them have made out. In the decisive case, Delbrück and Luria were both vividly aware in the early forties that Avery was finding that the transforming factor was DNA. What was lacking was not the detailed knowledge of Avery's work, nor an appreciation that it could be crucial, but the possibility that DNA could be specific—and not, Delbrück said to me, "a stupid substance." It was hardly more clear just how proteins could be specific if genes were proteins.

The fifth discipline in the synthesis that formed molecular biology was biochemistry. The events in the cell that molecular biology has been concerned to understand are inescapably biochemical, as nobody disputes. Yet the standard view of the rise of molecular biology has somewhat taken for granted, for example, the radical sharpening of ideas of specificity represented by Fritz Lipmann's elucidation of the way energy is supplied to the steps of cellular reactions. And it grievously undervalues the work of the two biochemists who proved decisive in changing the way people thought about specificity.

The man who released the present-day understanding of molecular specificity in living processes was Frederick Sanger. His determination, beginning in the mid-forties, of the amino-acid sequences of bovine insulin proved that they have no general periodicities. His methods and this surprising result had many consequences, of course: the most general and profound was that proteins are entirely and uniquely specified. Sanger's conclusion was forceful by 1949, when he published his first definitive results in a paper appearing in June and went that same month to the yearly symposium in Cold Spring Harbor. Even before that, the news was spreading. As we saw, Crick, in Cambridge, knew of Sanger's progress peptide by peptide, month by month. The result was definitive by 1951, when Sanger published the complete sequence of one of insulin's two chains. Sydney Brenner remembered going to a talk Sanger gave at that time—the excitement, especially among the younger scientists, as they emerged. "At last, we knew what proteins were." With that vanished any possibility of a general law, a physical or chemical rule, for their assembly. "With that, you absolutely needed a code," Monod said.

A similar transformation, at just the same time, was overtaking nucleic acids. The precision of this parallel is an elementary fact lost in the usual accounts of the rise of molecular biology. The role that Sanger played for proteins was played for nucleic acids, but in a minor key, by Chargaff. That is, although he never learned to sequence DNA, Chargaff's demolition of the tetranucleotide hypothesis by 1949 and 1950 released the possibility, and made it highly likely, that nucleic acids were specific in sequence, too.

The structure of DNA then made specificity comprehensible. "Nobody, absolutely nobody, until the day of the Watson-Crick structure, had thought that the specificity might be carried in this exceedingly simple way, by a sequence, by a code," Delbrück said. "This was the greatest surprise for everyone." The rest followed. In September of 1957, before the Society for Experimental Biology, Crick gave the paper "On Protein Synthesis," in which he formulated the sequence hypothesis and the central dogma. By then, the notion of specificity in biology was recognizably that of the present day. Once again, nucleic acids and proteins were thought about in strictly parallel terms.

A subject of broadest and deepest interest, a grand synthesis of several lines of inquiry, a transformation of the ruling ways that scientists formulated their ideas in that subject—of course there's no question that the rise of molecular biology meets the obvious criteria and is one of the great revolutions in the history of science. But what sort of revolution, more exactly? Historians of science, in trying to account for revolutionary change, have relied upon the history of physics almost exclusively, and in physics have appealed to certain great set-piece battles. The important and fascinating cases, since the beginning of the history of science, have been those associated with the eras of Copernicus, Newton, Einstein, and the quantum. In each of these, it can be asserted that one cluster of ideas, closely interrelated and fully worked out, was overturned and replaced by another cluster of ideas, closely interrelated and fully worked out. Other examples, of slightly less magnitude, and still typically from the history of physics, can then be fitted summarily to that model of scientific revolution. I put it in rudimentary form, but something like this is the dominant idea of the pattern of a great scientific revolution.

The rise of molecular biology asks for a different model. Copernican astronomy, Newtonian physics, relativity, quantum mechanics—but biology has no such towering, overarching theory save the theory of evolution by mutation and natural selection. (And—obviously, but the false analogy is sometimes met—the predecessor that the theory of evolution overthrew was not a scientific theory.) Biology has proceeded not by great set-piece battles but by multiple small-scale encounters—guerrilla actions—across the landscape. In biology, no large-scale, closely interlocking, fully worked out, ruling set of ideas has ever been overthrown. In the normal way of growth of the science, variant local states of knowledge and understanding may persist for considerable periods. In the tent of our understanding of the phenomena of life, among the panels covered with brilliant pictures that seem to tell a continuous tale, suddenly a new panel begins to appear and grow. Revolution in biology, from the beginnings of biochemistry and the study of cells, and surely in the rise of molecular biology and on to the present day, has taken place not by overturnings but by openings-up.

Is it not so, further, that physics too, even at classical periods, has sometimes gone like that? The discovery of radioactivity in the last decade of the last century was an absolute surprise. It opened an enormous new area— which was exploited much in the manner, fast-breaking and at first localized, then spreading and generating new strains and consequential discoveries, that is characteristic of the guerrilla combat of biology. Be that as it may, physics in the present era, which is to say since the second world war, from theories of cosmology to fundamental particles, seems breathtakingly agile, promiscuously receptive to new notions, adaptable, and free of doctrinal rigidity. There is a revolution proceeding now in physics, I am told: it seems to be following the model that has been characteristic of the permanent revolution in biology.

❖ ❖ ❖

"My impression is that other people have gone into molecular biology with the same *general motives* that I had, but sometimes with a difference in point of view," Francis Crick said. "*I* went into it, for example, to try to show that you can explain all these phenomena—the term molecular biology wasn't common then, certainly I didn't know it, but the phrase I had in my mind was 'the borderline between the living and the dead'—that you could explain these phenomena just by the laws of ordinary physics and chemistry. But then you ought to consider Max Delbrück. He went into it because he hoped that by looking at biological things you would find *new* laws of physics and chemistry. And yet, you see, it's very remarkable that these two approaches have in the end amounted to much the same thing."

The Nobel prize in physiology or medicine in 1969 was awarded to the three founders of the American phage group, Salvador Luria, Alfred Hershey, and Max Delbrück. The Nobel prize in literature that year was awarded to Samuel Beckett. Beckett did not refuse the prize, but would not come to Stockholm to receive it. He did not say why. But his absence was of a piece with what was known of the man and what was evident from his writing, the growing and now reclusive shyness, the creative austerity that had turned an Irish abundance of language into a conviction of the terrifying difficulty of saying anything at all. Delbrück, in the lecture he gave on receiving the prize, compared the scientist and the artist. Speaking, with grave courtesy, in Swedish, he said:

> The medium in which [the scientist] works does not lend itself to the delight of the listener's ear. When he designs his experiments or executes them with devoted attention to the details he may say to himself, "This is my composition; the pipette is my clarinet." And the orchestra may include instruments of the most subtle design. To others, however, his music is as silent as the music of the spheres. He may say to himself, "My story is an everlasting possession, not a prize composition which is heard and forgotten," but he fools only himself. The books of the great scientists are gathering dust on the shelves of learned libraries. And rightly so. The scientist addresses an infinitesimal audience of fellow composers. His message is not devoid of universality but its universality is disembodied and anonymous. While the artist's communication is linked forever with its original form, that of the scientist is modified, amplified, fused with the ideas and results of others, and melts into the stream of knowledge and ideas which forms our culture. The scientist has in common with the artist only this: that he can find no better retreat from the world and also no stronger link with the world than his work.

When he had first learned of the Nobel prizes, Delbrück said, he had looked forward with a thrill of anticipation to hearing Beckett's lecture. He pointed out the confrontation implicit in the Nobel ceremonies, where "scientists are brought together with a writer." He went on:

> Again the scientists can look back on a life during which their work addressed a diminutive audience, while the writer, in the present instance Samuel Beckett, has had the deepest impact on men in all walks of life. We find, however, a strange inversion when we come to talking about our work. While the scientists seem elated to the point of garrulousness at the chance of talking about themselves and their work, Samuel Beckett, for good and valid reasons, finds it necessary to maintain a total silence with respect to himself, his work, and his critics.

Almost, one supposes, Delbrück regretted not having stayed away himself.

One day in July of 1972, in the course of the conversation with Delbrück sitting in the shade of a tree at the top of the long lawn at the Cold Spring Harbor Laboratory, I asked him about his interest in Beckett. In certain passages, Beckett had conveyed as well as anyone the nature of scientific intuition, and with that the desperately, irreducibly obsessive quality—

beyond all measure of reason—of the scientist's preoccupation with his problem. That much I knew Delbrück thought, because he had told me so once before. But now his answer was hesitant and tangential.

"I am very— This is difficult to talk about," he said. "I am very poor at it. It seems to me there is a very fundamental conflict between science and—and existence. In science we pretend that we are immortal. I mean, we ignore the fact that we—that any one of us will be non-existent. And not only in science; in daily life we always pretend that we are immortal." He laughed lightly. "Our whole mode of thinking in life, even more so in science, I think, is acting as if we were immortal, as if death didn't exist—and certainly as if the death of the scientist didn't exist." Children's voices drifted up the lawn. I went back to Beckett and the obsessiveness of scientific work. "As part of anxiety," Delbrück said in response to that. "If you look at the father of existentialism, Heidegger, he indeed characterizes existence, human existence, as a—first of all, as a being toward death, a kind of being which is toward death. And a being whose essence is anxiety; and a kind of being whose essence is care—*Sorge*." He tasted the German word a second time, slowly. "And whose existence— You are *there* before you know that you *are*.

"A scientist of course tries to do more than just relieve his anxiety," Delbrück said, more quickly, and grinned. "He also tries to make a fairly complete picture of the phenomena he is trying to encompass. And some people try to encompass our own existence. Existentialists, you might say. And— well, I don't know. I have nothing sensible to say about it. Except to express my disquiet—my disquietude at the way science pretends that the scientist is immortal." I asked what difference that made. "I suspect that it implies a loss of areas of reality you just can't express, that way," he said. All science was inevitably reductionist at that level? "Not reductionist," he said. "I think science, as it runs now, leads a deprived existence. This comes out very clearly when you try to do psychobiology. D'you want to look at anxiety as a hormonal phenomenon or do you want to look at it as an existential phenomenon? And if you look at it as an existential phenomenon then can you incorporate it into your picture of science?" Very quietly and simply, he said, "I'm not expressing myself clearly. And I can't because I'm not clear in my mind." Then, strongly, he said, "I have the feeling that the kind of psychology that could come out from a point of view of the immortality of the scientist can only be a very dull psychology. Which would miss many aspects which you would like to understand."

Several years later, in London, I told Beckett of what Delbrück had said about him. We were coming out of the Royal Court Theatre, after a rehearsal he had directed, into the sun of a late spring afternoon, blinking a little and stiff. "I read that talk of Delbrück's you gave me," Beckett said. "He does me too much honor." His voice was deep, quiet, melodious; his manner inward, thoughtful—in fact startlingly like Delbrück's. We went into the pub next to the theatre for a whisky. I described Delbrück's work briefly, as an effort to apprehend life by stripping it to its most elementary processes. Beckett was listening attentively and warily. Now he stirred as if in recognition.

"When one looks at one's work coldly—I can hardly bear to look at a work again once it is finished—one sees that one reaches *the same impasse. Always the same impasse*," Beckett said. "Each time one thinks one starts fresh, new—yet each time one reaches the same impasse. There are many ways to begin, many roads to it, but always the same impasse at the end." He asked about Delbrück's work: "Have these discoveries any therapeutic use?" And, "Could they be visualized? Seen?" He said, "I'm not trying to find answers. Rather, shapes."

Outside again, Beckett said, "I am ignorant of science. My work consists of little attempts to make shapes with words—attempts that always break down in the same way. It's curious: one knows that they are going to break down but still one persists."

On 4 December 1975, Monod and I had lunch at his apartment in Paris. The day was cool, sunny, hazy. When I arrived, the housekeeper answered the door. Monod came home a few minutes later. For the first time, I noticed him limp. We talked about Ephrussi, and Morgan's interest in embryology at Caltech when Monod was there in 1936. Monod told an ironical story about how Watson had got him to come to Cold Spring Harbor to give a talk about some of the ideas in *Chance and Necessity*—and how he discovered only at the last minute that it was not a serious occasion but "a pep talk to the millionaires who have these big, beautiful houses around there. In fact, it was a trap." Lunch was perfect for the season, the city, the occasion—and curiously distancing, in consequence. The table was set formally for two. The napkins were heavy. The wineglasses were slightly chipped. The housekeeper brought avocado pears with a vinaigrette. Then she came around with two small steaks, medium, and the excellent crisp French string beans and small yellow-fleshed potatoes. Monod served a decent claret from a bottle by his place, the label turned away. He drank almost none himself. The sun streamed in, golden and autumnal. The room seemed entirely of the nineteenth century. The afternoon was plucked from time. We talked again about the Lysenko controversy, and about the work he had done at the Institut Pasteur during the war. The housekeeper brought a salad of lettuce, the dressing *fines herbes*. She brought a plate of five cheeses. Monod poured a little more wine.

"I have always thought of myself as an amateur in science," he said. "I don't feel bound to keep on doing science. Why should I?" The housekeeper brought a bowl of fruit. "Why do you want to oblige me to go on working at the bench if I prefer to write a book?" I asked if he had the book in mind. "Yes, I do. To some extent. I have the title. That's the great secret, of course. *L'Homme et Temps—Man and Time*. It's to be a series of essays. Where the central theme is the nature of time, and the relation of man to time, as an individual, as a biological object, and in history, as societies in history—what concepts men have had of time, depending on the cultures in which they were living."

A month after that, Monod learned that he had leukemia. He continued as Director of the Institut Pasteur, though his medical regimen grew increasingly onerous, including repeated transfusions. In May of 1976, he went for a few days' visit to his family home in Cannes. While there, he suffered a massive hemorrhage of the liver, and died in the hospital in Cannes on May 31. Some time later, his elder brother told me that Monod's last intelligible words had been "Je cherche á comprendre." I am trying to understand.

EPILOGUE

"We can put duck and orange DNA together—with a probability of one.*"*

ON THE TRANSFORMATION
OF MOLECULAR BIOLOGY, 1970–1995

In 1968, 1969, 1970, the leading molecular biologists, the founders, viewed what they had wrought and were confident of the future. They had assembled an overarching and coherent account, a beautiful account, of the fundamental processes of life: namely, the transmission of heritable characters down the generations and their expression in the building of each cell. These processes work, of course, through the genes, long known in principle, but now understood to be specific sequences of subunits in nucleic acids—the nucleotides, four kinds distinguished by four types of bases. The double-helical structure of DNA, held together by pairs of bases between the strands, together with the unique base-pairing rules, explain at a glance how the genes are reproduced to be handed down. The specific sequences determine, according to what had come to be called the genetic code, the construction by mechanisms in the cell—dependent, again, on base pairing—of corresponding sequences of amino acids, constituting protein molecules, which are the active agents, tens of thousands of different specific agents, that grow and maintain the organism. Feedback loops among the proteins and between proteins and nucleic acids control the biochemistry of the cell, and, indeed, turn on and off the activity of the genes themselves. The understanding of nucleic-acid sequences went far, also, to explain how genes mutate yet maintain their remarkable stability. Molecular biologists had recast the sciences they drew on, overthrowing established preconceptions, rephrasing the very modes of understanding in terms of the lumps and bumps—the shapes and the distribution of charges—of molecules and their aggregates, which determine the interplay in the living cell. Thus they had answered questions that had troubled profound observant intellects for twenty-five hundred years. They had reached a culmination of a program of research and discovery that had been gathering momentum since Gregor Mendel and Charles Darwin, for more than a century. They had explained heredity.

Or, at least, they had done these things in outline, with certain bacteria and a few other of those minimal creatures, single-celled and without cell nuclei, called prokaryotes, together with some viruses that live in them. Those who had put that coherent outline together, particularly the leaders, the founders, thought it surely could be expanded to encompass higher organisms, eukaryotes. Classical molecular biology of course must be consolidated, "filling in all of the biochemistry so that what we know in outline we also know in detail." Francis Crick put it this way in 1970. The effort would require much drudgery. Now, though, it was time to ask the fundamental questions of classical molecular biology all over again for higher organisms. Jacques Monod had asserted that what was true for *E. coli* would be true for the elephant. Indeed, "The basic biological mechanisms," Crick said then, "turn out, with minor variations, to be the same throughout nature—just as we had assumed." Nucleic acids, the genetic code, the fact of control by feedback loops—these are universal. Yet how to ask the fundamental questions about genes and their expression all over again for higher organisms was far from evident.

Here and in what immediately follows I shall pull together selectively points that molecular biologists made to me at the end of the nineteen-sixties and early in the seventies. Though I presented these at various places throughout the book (particularly in chapter 4), the aim of this broad recapitulation is to establish what can be called the strategic position of molecular biology then—necessary if we are to understand the development of the science in the ensuing decades and the concomitant deep changes in the structure of the community and in the aims of many of the scientists.

Bacteria and bacterial viruses had been the main experimental subjects. Bacteria lead simple lives. They are far smaller than animal cells—proportionately as a honey-bee to a cat. Their cells contain no nuclei, they make far fewer substances, they possess far less genetic material—not multiple chromosomes made up of DNA and proteins interwoven but one long loop of naked DNA plus a few more genes on small extra rings called plasmids. Though the results made a revolution, the methods had been as simple as the creatures. They had mostly sprung from the discovery early in the nineteen-fifties of bacterial sex. In 1951, Joshua Lederberg and Norton Zinder had demonstrated that bacteria sometimes exchange genes when a virus incorporates bits of genetic information from one host cell and transports it to others—the maneuver they called transduction. By 1952, experiments had shown that sometimes one bacterial cell (promptly termed the male) will transmit a copy of its genes to another (the female). In 1955, Élie Wollman and François Jacob, in Paris, had found that, by agitating the culture in a Waring blender, bacterial mating could be stopped when only part of the genes had been piped across. Exploiting such means, molecular biologists were able to move selected portions of genetic material from one strain of bacteria into another, and then determine how the inserted genes express themselves and how they are regulated. The general pattern of gene action had been brought out in large part by manipulating a few genes at a time in the reproduction of bacteria.

Complex, multi-cellular organisms present the problems of development, or how the fertilized ovum, a single cell, becomes the adult with cells differentiated into many kinds of tissues and organs, with the capacity to heal wounds, even to regenerate lost tissues. "At this point, one moves beyond classical molecular biology," Crick said. And development, Sydney Brenner said, is at the center. "Development ultimately includes *all* of biology: and it will all have to be put on a molecular basis." Brenner himself had chosen a simple worm, a nematode, *Caenorhabditis elegans*, barely visible to the naked eye, the adult male possessing exactly 1,031 somatic cells and a rudimentary nervous system, and to begin was asking how the worm's nerve cells, from the fertilized egg and proceeding through cell division after cell division, grow to be in the right places.

In sum, molecular biology in the strict sense was turning to development and differentiation, embryology at the level of molecular mechanisms. Yet in the broader sense molecular biology was making an imperial, expansionist claim, to all the rest of biology. The drive was strong. The founders were expecting new problems, entire fields, "to go molecular." Candidates were of many kinds.

Perhaps oddly (or so it struck me then), at the time muscle seemed a prime example of the new set of problems: how chemical energy is converted to mechanical force. "We want the detailed molecular interactions," Brenner said. Two years earlier, indeed, Max Perutz had told me that Hugh Huxley, a colleague, and others elsewhere were taking on muscle by adapting x-ray diffraction to larger molecular assemblages. Perutz himself was staying with the relation of structure to function in the hemoglobin molecule, yet just then, as he was at last getting the structure out in fine detail, his emphasis had begun shifting ever more to the functional side. James Watson, at Harvard, took a different slant: what probably needed to be attacked next, he thought, was the cell membrane, its composition, structure, and active functioning. Brenner called for this, too. For instance, "We know that cells can pump selected sorts of molecules into themselves through their membranes against tremendous electrochemical gradients," he said. "Yet we don't know anything about these pumps; we don't know their molecular structure." But the membrane, Watson said gloomily, had for too long been a vague conceptual catch-all to which baffling phenomena could be relegated, in pseudo-explanation; to solve the membrane would be very hard, like trying to understand the commerce and politics of a large city by observing it from a mile up.

Immunology—now *there* was a field that in the previous fifteen years had already been radically reshaped by basic theoretical considerations arising from a molecular approach. Though the phenomena were tricky and experiments often damnably imprecise, immunology appeared to be yielding to assault by an alliance of molecular biologists with immunologists. The work was going on at many places, an emergent network of scientists and centers, chief among them the Walter and Eliza Hall Institute, in Melbourne; the Basel Institute of Immunology; the Rockefeller University, in New York; Stanford University and the Salk Institute, in California.

Then neurobiology—ah, this was the problem alluring beyond others. Even a quarter-century earlier, Crick had hesitated between neurobiology and the nascent molecular biology. Now, many were moving to the nervous system. Seymour Benzer, for instance, had turned to the behavioral genetics of fruit flies, trusty *Drosophila,* and was accumulating behavioral mutations, flies that danced, flies that ran from light instead of towards it, and one bizarre mutation he called "drop dead." Benzer had from a friend a funny cartoon, a family tree of molecular biologists switching to neurobiology; photographs of six or eight familiar faces were perched among its branches. Molecular biology of the nervous system was poised to grow fast, Crick said; it was "attractively mysterious, one of the last real secrets in biology." Furthermore, the nervous system might profitably be attacked as a problem in developmental biology, as Brenner was doing with his worm.

Meanwhile, in Cambridge, Frederick Sanger, quietly, persistently, and with technical genius was working away at methods for sequencing nucleic acids as he had previously taught the world to sequence proteins; where proteins, though, offered twenty different amino acids, 'nucleic acids had only four types of nucleotides. In the late nineteen-sixties, Sanger was coaxing out the secrets of RNA.

The molecular biology that distinguishes higher organisms from bacteria defines what is required to explain how the single fertilized egg grows to be the adult creature of specialized cells and tissues. To solve development and differentiation, to understand embryology, was conceptually and technically prior to all else on the consensus list. Embryology had stood for nearly a century as the most intractable problem in biology. For those working at the bench, the difficulties were daunting. Some could not have been foreseen. The next two decades also held some great, unpredicted discoveries: lots that's true of Monod and his pet elephant is irrelevant to *E. coli.* Life was full of surprises.

❖ ❖ ❖

Science comprises many subjects, many approaches. Scientists observe no single, unitary scientific method: as one said, "You get there how you can." The sciences as practiced display two characteristics that illuminate the era in molecular biology that began at the end of the nineteen-sixties. Most sciences are dependent on their own technology, on development of instruments for observation and techniques for manipulation. Much new science works just beyond the full stretch of its technology. Biologists dislike so-called targetted research, where funding sources specify goals (as in the "war on cancer" of the early nineteen-seventies) in part because the prerequisite technological innovations cannot usually be ordered ahead. Secondly, scientists like anyone confronting a new problem will start with what they already know. Neither in theoretical speculation nor at the bench do they often sail beyond sight of the shore. (This is simple and obvious, but not banal.) They habitually proceed by tame analogy, by small modifications of ideas and methods that have been successful. Baffled by higher organisms, the next generation of molecular biologists looked to adapt as directly as possible the methods that had worked with bacteria and bacteriophage. In particular, they reinvented sex—bacterial sex, applied to higher organisms.

Normal sexual reproduction of higher organisms shuffles and redeals the enormous gene sets of the two parents. The results are all but impossible to interpret at the fine scale of molecular genetics. Yet the problem of development and differentiation can be restated as one of finding the controls that turn some sets of genes on, others off. Molecular biologists saw that they urgently needed methods analogous to what had worked so brilliantly with bacteria and phage, ways of moving short, known groups of genes from one animal cell to another, to detect what they do. Thus, an early requirement was to learn to grow cells from higher organisms not as coherent tissues but as bulk cultures of individual cells, like bacteria. By 1970, also, a variety of enzymes had been discovered that bacteria make, in order to protect themselves from viruses by chopping foreign DNA at particular sites; these are called restriction enzymes, because they restrict the range of viruses. Still other enzymes had been found that cells use to repair their own DNA. Once biologists got such enzymes, they realized that they could use them to cut and recombine segments of genetic material at will. (Hence the technical term, recombinant DNA, for molecular biology of this sort.) Another idea, again by analogy with what had worked with bacteria, was to paste selected genes into animal viruses in order to carry them into animal cells. Especially interesting here were tumor viruses that don't kill the cell but commandeer its biochemical machinery. And biologists learned that they could keep right on using bacteria, putting segments of genes from higher organisms into them, where they would multiply in step.

"The most important discovery last year was the one by Baltimore and Temin," Max Perutz said in February of 1971. "You must pay close attention to that." The previous June, *Nature* had published back-to-back a pair of papers, one by David Baltimore, the other by Howard Temin and Satoshi Mizutani, announcing a phenomenon of great promise, one that to many seemed also to shake the conceptual foundations of molecular biology. Years earlier, when André Lwoff had proved that lysogenic viruses insert their DNA into the chromosome of their host bacteria, where it lies dormant as what he called the prophage, reproducing along with each cell cycle, he had speculated whether animal-cell viruses could function in the same way. The suggestion had been taken up by Howard Temin, at the University of Wisconsin. He worked with Rous sarcoma virus, which gives chickens warts and has as its genetic material not DNA but RNA. Other RNA viruses were known, for example tobacco-mosaic virus, but these forced their host cells to make more viral RNA by way of an RNA intermediate. Temin became convinced that Rous sarcoma virus goes through an intermediate phase in which its genome exists as DNA inserted into the DNA of infected chicken cells. By the mid-sixties he was publishing paper after paper about the DNA provirus hypothesis. He had little success convincing others: his experiments seemed overly elaborate, his data equivocal. Nor had he sought direct evidence of the enzyme that would have to be present to translate RNA into DNA.

In the winter of 1969–'70, a postdoctoral fellow, Satoshi Mizutani, urged that they do so and started the experiments. That March, at the Massachusetts Institute of Technology, David Baltimore began the biochemistry with another RNA virus, one that gives mice leukemia, to see what enzyme it used to carry out synthesis of its genome. Thursday morning, 28 May 1970, at the Tenth International Cancer Congress, in Houston, Temin gave a brief talk announcing proof that his virus indeed had the necessary enzyme for RNA-directed DNA synthesis. Baltimore's paper with the same claim arrived at *Nature* the following Tuesday, June 2. Baltimore went to the annual Cold Spring Harbor Symposium, where Watson, who was running it, stuck him in as the last speaker, on the eleventh. Temin and Mizutani's paper got to *Nature* on the fifteenth. John Maddox, the editor, hardly paused for peer review: the papers led the issue of June 27. Many molecular biologists saw immediately that reverse transcription, once the enzymes were actually isolated and made available, would provide a unique tool. One who heard Baltimore at Cold Spring Harbor remembered the excitement: "We realized we could make DNA copies of active genes"—that is, of messenger RNA—"and this would be a way to follow gene expression with the tools of DNA manipulation that were just coming out. I think this point was not obvious to all, but to those of us who saw it, it meant that embryology was now a branch of molecular biology."

The papers were accompanied by uproar about the central dogma. Watson's formula, in 1952 even before he and Crick got the structure, had been "DNA makes RNA makes protein," and that had become the standard statement: it appeared to rule out movement from RNA to DNA. The lead editorial in that June 27 *Nature* carried the headline CENTRAL DOGMA REVERSED, and others excitedly joined in: Crick had been shown wrong, or at least oversimple. So in *Nature* for 8 August 1970 Crick responded with a two-and-a-half page article asserting waspishly that from the beginning in 1957 he had analyzed all possible transfers of detailed, residue-by-residue, sequential information among DNA, RNA, and protein, and had concluded that DNA-to-DNA, DNA-to-RNA, and RNA-to-protein were known, while all transfers from protein, either to nucleic acids or to protein, were unknown and postulated never to occur. A chief reason was that the machinery of translation is complex and none exists, no trace, for back-translation. About the three remaining transfers, DNA-to-protein, RNA-to-RNA or RNA-to-DNA, although no evidence or necessity for them were known in 1957, they could not be ruled out. Thus, Crick's original statement in 1957 had been precise and spare: "Once (sequential) information has passed into protein it cannot get out again." (This was the article for which he drew the triangular diagram of the central dogma reproduced on page 335.)

Anyway, the first attempts to put the new methods together for the insertion of known genes from one organism into cells of another took place in 1971, independently and apparently at the same time, in two laboratories at Stanford University Medical School. One scientist was Peter Lobban, a graduate student working alone. Another was Paul Berg, chairman of the Biochemistry Department, with a group around him. In the course of a long conversation in Cambridge, England, in September of 1976, Berg explained how his work five years earlier had been designed to develop the full selection of tricks that exploitation of bacterial sex and transduction had provided to classical molecular biology. "The question we wanted to ask was, first, Are bacterial genes expressed in an animal cell?" Berg said. "And if they're not, at what point in the read-out mechanism are they blocked?"

Far more than these initial steps was of course required. The array of restriction enzymes for cutting DNA at chosen places grew sophisticated;

much of this early work was done by Herbert Boyer, at Stanford. Another discovery was that plasmids, those small rings of DNA that bacteria carry, can be extracted, have a few genes inserted, and then be put into another bacterial cell; the bacteria then multiply normally, plasmids, foreign DNA and all. The first efficient ways to do this selectively were invented by Stanley Cohen, yet another scientist at Stanford. Then Boyer and Cohen collaborated in the use of plasmids to carry animal genes into bacterial cells. Early in 1972, two of Berg's colleagues, Janet Mertz and Ronald Davis, showed that certain restriction enzymes cut DNA into fragments with ends that rejoin readily; sticky ends made hybrid DNAs far easier to construct—even hybrids between species. Bacteria can reduplicate explosively, thirty to fifty times a day, limited only by the food supply: thirty doublings would give a billion-fold increase, fifty replications the same increase in cell number as from a fertilized egg to an adult human. At once, molecular biologists had a way to grow large quantities of one or a few known, identical genes—the process called cloning.

❖ ❖ ❖

Early in the nineteen-seventies, the thought had occurred to several scientists, notably Berg, that with some of these methods it might prove possible to cure genetic diseases, of the kind where an infant is missing a gene or bears a defective gene, and as a result fails to make an essential protein and suffers the consequences. Familiar examples include sickle-cell anemia and cystic fibrosis; many such single-gene-defect disorders are inborn errors of metabolism, and most in fact are individually rare, some very rare indeed. It was evident from the first that to correct such disorders genetically would be complicated. Those who pondered and discussed the problem, sophisticated as they were in the ways of bacterial genetics, saw clearly that the would-be gene therapist must isolate the normal gene and grow it in sufficient quantity; identify the specific cell type or tissue where it is needed; discover the regulatory mechanisms governing the expression of this particular gene, and possibly supply those, too; figure out how to deliver the gene to its target cell, or even to its particular location on a particular chromosome, and find a suitable vector, or carrier, to do this—animal-cell virus? bacterial plasmid? the patient's own cells, bone-marrow cells perhaps, extracted, treated, and returned? From disorder to disorder, these requirements may vary widely: that a general procedure could be found has always seemed unlikely. (So it has proved.)

I must emphasize, though, that in those days before even the term "recombinant DNA" was coined, hardly anyone was thinking much of what we now call bio- or genetic technology. To be sure, the imaginable medical applications were attractive, on the one hand because they offered interesting technical difficulties and sharp research focus, on the other because they suggested ways to get more funding. None the less, the technology of recombinant DNA was invented and is still today being elaborated not primarily to make money, nor even to cure disease, but to carry on the main scientific work of molecular biology—to find means to crack open a profound scientific problem. The point is important. Other commentators, reconstructing the history later and in the light of commercial applications and the performance of new biotechnology companies on the stock market, have been eager to see the first steps in genetic technology as driven by entrepreneurial lust and the scientists as venal from the start. I was interviewing them at the time. They spoke only of the problems and the excitement of finding the means to solve them. Yet by the mid-nineteen-seventies molecular biologists had assembled an unprecedented tool-kit and were beginning to recognize its power.

Brenner put the matter pungently in conversation in 1975. "I would expect this technique to be comparable in impact to the use of radioactive isotopes as tracers in biology—and that, as you know, is extremely important. It's going to be very widespread. Simply, it's going to allow us to tackle for the first time problems of the molecular genetics of higher organisms—of anything. Elephants. Sea urchins. It'll make a lot of things obsolete, the possible easier, the impossible possible." He thought recombinant DNA would be "probably more profound in the questions it can reach" than any previous methods. "Why? *Because these things self-replicate.* You have a way of enhancing yield. You can detect things. And you can put together combinations— You may say, Look, even biological systems since the dawn of time have not had an opportunity to explore the complete range of combinations now suddenly possible. And, in principle, for any new combination we can make the probability—*one.* One is a very large number in these things. We can put duck and orange DNA together—with a probability of one."

Unprecedented powers might carry unprecedented risks. As early as the summer of 1972, several biologists surmised that some combinations unlikely in nature might be dangerous. Some worried that bacteria could be constructed containing genes for tumor viruses or for toxins, and that they might escape from the lab, a man-made plague that could evade the body's defenses. Others asked whether hybrids between organisms widely separated by evolution might violate some natural species barrier. A chain of circumstances led to the formation of a committee, with Berg as chairman, that met in April of 1974 and called for an international conference the following spring and, in the meantime, for a worldwide voluntary moratorium on the kinds of work that looked most dangerous, like fooling with genes for bacterial toxins. Late in February of 1975, a hundred and forty-six molecular biologists, mostly from the United States but including delegations from Japan, Australia, Israel, the Soviet Union, and ten other European countries, met at Asilomar, a conference center on the California coast near Monterey, to discuss what the dangers might be. In the light of what was known and what was uncertain then, the moratorium and the meeting were no doubt prudent steps. At Asilomar, none the less, a taint of self-congratulation hung in the air as biologists compared their public spirit with the nuclear physicists' thirty years earlier. Several people there, though, had begun to think that the worst danger had already been allowed to escape—a plague of public hysteria and governmental regulation.

By the end of 1975, the putative dangers of recombinant DNA had been picked up not only by the press but as a political cause by a ragged assortment of individuals and groups. Regulations ("guidelines") for research were established by the National Institutes of Health. Senator Edward Kennedy held hearings. State legislatures entertained bills. University towns saw agitation, including Princeton, Ann Arbor, and most controversially Cambridge, Massachusetts, where the mayor used the controversy as a stick to beat Harvard and MIT with. Yet scientists' fears were receding fast. All the biologists who had signed the original appeal for the moratorium and the conference now agreed that the dangers had been exaggerated and the practical effects pernicious. (Over two decades, the guidelines have been progressively relaxed. A Recombinant Advisory Committee still sits at the National Institutes of Health, with its approval required, for example, of protocols for trials of gene therapy.)

In England, meanwhile, Sanger had switched from RNA and was trying to determine how to read directly the messages encoded on strands of DNA. In 1975, he published a method by which a gene of average length could be sequenced by any graduate student in a month or two of work—today it is done in hours—and the result read off as easily as consulting a railway timetable. (In 1980, as we have noted, Sanger became the third scientist in history to receive a second Nobel prize in science.)

❖ ❖ ❖

At the same period, the structure, the organization of the science was changing radically. Matthew Meselson said to me not long ago that molecular biology, classical molecular biology, had had a linearity: a main-line sequence of discoveries, the canonical papers appearing as though in just the order one would want to teach the topics to beginning undergraduates. But after about 1970, he said, "the field was like a river hitting sand." He called this "the great dispersal." Laboratories and groups were multiplying. They were working on the next generation of problems, yet by comparison with the past, Meselson said, most problems seemed unambitious. Roger Kornberg, a younger biologist, had pointed out similar things to me: in those days from the nineteen-seventies on, the very processes of discovery were smeared out, a small step taken here, capitalized upon there, adapted over yonder, and each small step published as a small paper.

The founding scientists and their first recruits were not superseded, but they were joined by ever more young scientists. The annual assemblies of molecular biologists at the Cold Spring Harbor Symposia suggest the pattern. As a reference point, the symposium in 1941, when the subject was genes and chromosomes, had 85 participants who heard 33 papers strung out over two weeks. At the Cold Spring Harbor Symposium in the second week in June of 1953, organized by Max Delbrück but where Watson presented the double helix, 281 scientists attended and 41 papers were read. By 1970, at the symposium on transcription of genetic material, where Baltimore presented reverse transcription, 325 attended and 97 papers were read. In 1980, when the subject was movable genetic elements, 302 came and 117 papers were read. In 1989, devoted to immunological recognition—itself a sub-subdivision of immunology—446 people packed themselves in for a week to listen to 114 speakers; the papers had to be published as two volumes.

The newcomers might be capable, thoroughly informed, sometimes brilliant, but were likely to be narrowly focussed, for these cohorts were the first to have been trained in molecular biology as such rather than coming into the field from elsewhere. The times were rich: the lure of medical applications and the intrinsic interest of the science attracted better funding—in rough proportion to the size of the enterprise and the real costs of equipping labs and doing experiments—than the science was ever to see again. With growth came dispersion; with dispersion emerged new networks of communication (culminating today in electronic communication). With growth came increased and more ruthless competition and a new alacrity in publication. No longer could a scientist brood over a problem for years and then come up with a magisterial summation as in the paper by Avery, MacCleod, and McCarty identifying DNA as the transforming principle in pneumococcus. Never again would a laboratory chief impose a self-denying ban on competition as when Sir Lawrence Bragg told Watson and Crick in the fall of 1951 that the structure of DNA was a problem belonging to Wilkins and Franklin at King's College London, on which they were not to trespass.

In classical molecular biology, Meselson also said, the field had had great teachers, great mentors, great schools and styles: Linus Pauling, Bragg, Delbrück, Lwoff and Monod and Jacob, Perutz, Crick. After the early nineteen-seventies, this was no longer the characteristic structure of the field. "The new generations, by and large, don't have towering figures as we did, to look up to—who set the standards of scientific evidence, of conduct, of ambition." With growth, the passing-on of distinctive styles in science declined. With growth, true mentoring, mentoring by unspoken example reinforced by admiration, respect, emulation, a touch of fear—this, too, was attenuated. That, he thought, was the most grievous loss.

Discoveries changed character. Much of the most interesting work done from the nineteen-seventies forward has been in development of technical methods: recombinant DNA began and defines this shift. To analyze the new science and characterize its makers in further detail will be the subject of another book; here I can do no more than tick off several leading innovations.

Monoclonal antibodies are one of these. Mammals are capable of making antibodies specific to any one of hundreds of thousands of different substances. The antibody response is, of course, crucial in infectious illness. The cells that make antibodies are central to the immune system and are called B-lymphocytes; B-cells in culture go through several generations but soon die out. Each B-cell produces just one type of antibody, but in nature the immune system produces mixtures of many different B-cells. Antibodies being exquisitely sensitive to minute quantities of the substance to which they are specific, such mixtures have long been uniquely useful in biological research. To isolate cells that only make antibodies of one known specificity and that can be cultured for many generations would provide an unprecedented tool for laboratory as well as clinic. In 1975, Cesar Milstein and Georges Köhler devised a method for making hybrid cells with just these properties. B-lymphocytes may suffer a type of cancer called myeloma; myeloma cells—having lost controls on growth and proliferation, which is why they are cancers—survive in culture. Properly maintained, they multiply indefinitely: they are termed immortal. So Milstein and Köhler figured out how to fuse myeloma cells with normal B-cells, in bulk, and then to grow just the hybrids that make the desired antibody. Such cells are called hybridomas. They produce monoclonal antibodies. Milstein and Köhler had made not a discovery but an invention: in an astonishing bureaucratic failure, the British neglected to patent it.

In 1977, Sanger simplified his method of sequencing DNA into a version that has since been automated. Early in the nineteen-eighties, Marvin Carruthers, at the University of Colorado, devised a way to synthesize strands of DNA of any desired base sequence. He and Leroy Hood, then at the California Institute of Technology, went on to invent instruments that make such strands automatically. Sanger's work and Carruthers's, complementary to each other, are primary to almost everything later.

Among scores of other technical advances, the most generally useful has been the polymerase chain reaction, an ingenious invention of the late nineteen-eighties by Kary Mullis and others at Cetus Corporation, in Berkeley, California. The polymerase chain reaction multiplies chosen DNA sequences *in vitro,* that is, with no need to put the sequence into a cell: cloning without clones. The procedure is direct and fast. The polymerase chain reaction has been called the most revolutionary new technique in molecular biology in the nineteen-eighties. Cetus patented the process. In the summer of 1991, in a complex financial dance, Cetus merged into a smaller California biotechnology company, called Chiron, while selling the rights to the process to Hoffman-La Roche, a big Swiss pharmaceutical company, for three hundred million dollars. The polymerase chain reaction has enhanced many-fold the effectiveness of other genetic technology, particularly the use of reverse transcriptase. Minute quantities of messenger RNAs can be multiplied into many DNA copies, which are then used for close study of gene expression: the combination has realized the promise of reverse transcriptase to open up embryology.

Over all, the technology of genetic experiment and analysis has done more than facilitate research and theory. It has driven research and theory. Sometimes the new technology *is* the science.

To be sure, some discoveries went well beyond the technical. In the mid-nineteen-forties, Barbara McClintock, working with corn plants, had begun to get glimpses of unusual genetic phenomena. Some of these revealed the action of what were evidently control elements: in 1960 when Monod and Jacob published their great review of the controls on the rates at which bacteria express genes, McClintock saw at once the parallels with her own observations, and wrote papers to explicate these. She provided the first evidence that genetic regulation in higher organisms might be similar to regulation in bacteria. But some of her observations went deeper. These, and the reception of them, require an historical digression.

Extending the methods of classical Mendelian genetics, and with unequalled sense for what lay behind the apparently unruly behavior of the data, McClintock had found evidence that some genes move from place to place and will have different effects—often affecting nearby genes—depending on where they sit. She presented her observations in papers in 1950 and '51. She was a distinguished geneticist, president of the Genetics Society of America in 1945, the third woman to be elected member of the National Academy of Sciences. Evidence of genes that move was not altogether unprecedented.

However, she drove some of her interpretations deeper still. She believed that movable genetic elements with different effects depending on location might explain development and differentiation, and that they must have evolutionary significance—that they signified directed mutation, Lamarckian inheritance. Such speculations were met with skepticism, bafflement, incredulity. An historical account of McClintock has become widely accepted: that her papers of 1950 and '51 were ignored and that she was isolated within the community of geneticists and emergent molecular biologists. She certainly believed so. She told me of receiving few requests for reprints, of a meeting at which a long-time colleague turned his back on her. Others—not she herself—have held her an example of a woman, working in a feminist mode, her results running against the grain of a male-dominated style of genetic thought.

Recently, however, a young historian of science, Nathaniel Comfort, has looked again at the reception of McClintock's ideas about movable genetic elements. As part of a larger biographical study of McClintock, he has shown that ideas of mutable genes, movable genes, and related phenomena were familiar to geneticists in the early nineteen-fifties, that other geneticists were working on transposable elements, and that her two papers on these were known and cited, in general favorably. The paper from 1950 was even included in an anthology, *Classic Papers in Genetics,* published before the end of the decade. What was not accepted was her belief that transposable elements explain development and demonstrate directed mutation.

Then in the mid-seventies geneticists found jumping genes in many other organisms, isolated them by standard molecular methods, and named them transposons. Suddenly McClintock's observations of twenty years earlier were comprehensible to all. Her vindication was the Cold Spring Harbor Symposium in June of 1980, devoted entirely to movable genetic elements. She was awarded the Nobel prize in physiology or medicine in 1983, at the age of 81. Her speculations about the developmental and evolutionary roles of movable elements are still shrugged off—though not by all.

Another most unsettling discovery was that in higher organisms many genes are interrupted, in the sequence on the DNA, by stretches of bases that bear no relation to the gene, that appear to have no function at all. Gilbert christened these intervening sequences, or introns; the stretches of DNA that are part of the structural gene are called exons. The phenomenon complicates transcription. After the gene is transcribed, the RNA must be edited to snip out the introns. At the same time, nucleotides are added to the transcript at each end, a cap and a tail. Only then does the messenger RNA pass through the nuclear membrane into the cytoplasm, where ribosomes and the rest of the translation machinery wait. The controls on eukaryotic transcription and editing are by no means fully understood. The explanation for the existence of introns and the editing process may lie buried deep in the evolutionary past.

Enzymes are proteins. John Howard Northrop and his colleagues had proved this beyond doubt in the early nineteen-thirties. Sidney Altman, who at the Medical Research Council laboratory in Cambridge early in 1971 demonstrated to me how one extracts DNA, was working at that time on mutants of a particular transfer RNA, to see whether altered structures of the molecules would change, and so reveal, the way they function. This work had led him to a new enzyme that snips off bits of the transfer RNA to tailor it to its final form. In the fall of that year he moved to Yale. The enzyme, christened ribonuclease P, turned out to be made of a polypeptide chain, protein, intertwined with a strand of RNA. Such a combination was not in itself a surprise. Nucleic acids are associated with proteins, after all, in several of the complex structures that carry out central processes of cells: notably, chromosomes are constructed of DNA and proteins, while each of the two subunits of ribosomes is built of a number of different proteins and a strand of RNA. But then Altman and his colleagues found that in ribonuclease P the RNA by itself will perform the enzyme's catalytic function—sluggishly, to be sure, but unmistakably. Unbelievable, against all expectations: Altman devised ingenious experiments to convince himself, then more to convince others. In the clinching experiment, he transcribed the RNA strand directly from the gene, precluding all possibility of contamination with protein. Yet it took months to get the unbelievable result published. Meanwhile, at the Howard Hughes Medical Institute laboratory at the University of Colorado, a group headed by Thomas Cech found a different RNA molecule that has catalytic activity. Enzymologists at first were outraged by the notion that these could be thought of as proper enzymes; yet since then still more of them have been found. Cech and Altman took the Nobel prize in chemistry in 1989. The discovery has revolutionized at least one field, the origin of life: if RNA molecules can be catalysts as well as information carriers, then we can speculate that the present DNA, RNA, protein world was preceded by what has come to be called the RNA world.

More generally, regulatory mechanisms in higher organisms are turning out to be unexpectedly different from those in bacteria, more diverse and more complex. In contrast to the negative regulation of Monod and Jacob's repressor model—not even universal in bacteria—regulation in higher organisms is typically positive: genes must be turned on, usually by a combination of steps. Yet we also know from recent work that DNA carries certain controls on development and differentiation so fundamental that they appear all across the animal kingdom, primates to insects. Morgan, who in 1933 had published *Embryology and Genetics* but had told Boris Ephrussi a year later that he was unable to connect the two halves of his title, would have been astonished to see his fruit flies become a favored tool of developmental biologists.

The question that comes first in developmental biology, simplest to state yet most baffling of all: How does the fertilized egg, that single cell, take that initial step, differentiating between front and back, top and bottom, of the organism? How does the egg *know*—and that's the way biologists typically put it—which way is up, or forward? *Drosophila* provided the experimental system with which Christiane Nüsslein-Volhard and several colleagues, in western Germany in the nineteen-eighties, found elements of the explanation and followed them out in considerable molecular detail, a pursuit still going on. We can but sketch the work and its import. Her esssential decision was to search systematically for mutations that affect not adult flies but the very earliest embryos. Others had noticed mutated embryos from time to time, but had not recognized what they offered. To find such mutations is complicated by several factors. Many genes take part in the first hours of the embryo's life, and a mutation in one may be compensated by others that have overlapping functions. Some crucial mutations are not of the new organism but of the mother, affecting whether and how she generates functional eggs. Nüsslein-Volhard had to screen embryos on a vast scale, eventually working with more that twenty thousand fly families. She set the screening up as a production line; the work required some simple new gadgetry, but no high-technolgical breakthroughs.

The results proved intricate in detail, yet in principle simple. It turns out that the egg of the fruit fly is already marked front and rear, top and bottom, before it is fertilized. The mother in the very process of egg formation deposits at one location strands of messenger RNA—not a gene, but the gene's transcript—for a protein called bicoid. The site of those strands of messenger is thenceforth the head end. Similarly though in more complicated ways she marks the opposite end of the cell, and top and bottom. Thus in a contribution entirely from the mother and independent of the chromosomes of the egg nucleus, the egg is potentially differentiated even at the moment of fertilization; and from that moment those messenger strands begin to function, their biochemical products diffusing through the cytoplasm of the cell. The result is a gradient of chemical concentration, indeed overlapping gradients as the products of front and rear, top and bottom, interact variously, depending on concentration, to form the initial patterns that establish the fates of the cell lines emerging as division starts up. Flies are highly segmental: these patterns determine the differences along the sequence of segments. Molecular biologists at first responded with incredulity, even scorn, to notions of diffusion and gradients of determining biochemical factors. Yet these themes would have seemed entirely familiar to embryologists of the nineteen-twenties and thirties, although the differences are in fact considerable, above all Nüsslein-Volhard's plenitude of genetical and biochemical detail. In December of 1995, she shared the Nobel prize in physiology or medicine. Some now believe that Nüsslein-Volhard ranks just below Mendel or Watson and Crick in the pantheon. The question still open is just how far these processes in fruit flies can be generalized to vertebrates.

❖ ❖ ❖

In the mid-eighties, Robert Sinsheimer, a molecular biologist from Caltech who had recently become Chancellor of the University of California branch at Santa Cruz, together with other scientists, declared that the time and technology had arrived to map and sequence the entire human genome, that is, to locate and identify every gene and control element and to obtain the full nucleotide sequence of every chromosome, from the long arm of chromosome 1 through the short arm of Y. Sooner or later this would have to be done: nothing less, proponents said, would allow the necessary analysis of the development of the human embryo to adulthood, nothing less would establish the evolutionary relationship of man to other animals, while on the practical side nothing less would bring adequate understanding, diagnosis, prevention, or cure of genetically based diseases. In 1986, at one of the early meetings on the subject, in a phrase he has regretted, Walter Gilbert called the complete genome sequence "the holy grail" of biology. Even from the first steps, advocates proclaimed, benefits would begin to flow.

Initially, many biologists opposed the enterprise. The technology was by no means adequate, they said. (Proponents: The project will drive the technology, sequencing will be automated.) The data will be all but useless, kilometre-long strings of As, Ts, Cs, Gs, in unpredictable order, few segments identifiable, hard to store, impossible to analyze. (Proponents: Computer memories and speeds double every eighteen months; programs will be sophisticated.) The costs will be prohibitive; biology will become like physics, "big science"; in this era of tightening research budgets the project will divert resources from more valuable research, starve the rest of us. (Proponents: As always happens, economies of scale will drive costs down. We'll automate. We'll not tithe the rest of biology but secure new money.) The grunt work will be mindlessly repetitious, will drain personnel, drive young biologists away. (Proponents: Automation! And the project with its results will be as exciting and important as anything in molecular biology can ever be again.) The kinds of knowledge the project will make available about individuals pose grave dangers—legal, social, ethical. (Proponents: True.)

From the start, the project was promoted enthusiastically by the Department of Energy, which has an institutional presence in human genetics that goes back to the first research on the effects of radiation from atomic weapons on survivors of Hiroshima and Nagasaki. The National Institutes of Health was slow to support the project, but could not let it pass by. The politics were tortuous. The Congress bought the human genome project and began to pay for it. Scores of books, perhaps a hundred, have been written about the politics, the science, and the social and legal implications; several of these are good, and we need not précis them here.*

In ten years from the first discussions, the human-genome project has become the central and by now the defining enterprise in human genetics. Its goals and results are indeed of inestimable value. While the project is far from being the totality of molecular biology, it impinges on all else; it will indeed enable all else. It has brought genetics and its supposed applications to the fascinated attention of the public—and has generated expectations that are in some cases wildly premature at best. It raises concerns. Although by the turn of the millennium these may seem overblown, at present they appear grave. I find myself obliged to say why I think the human-genome project has been misrepresented and oversold.

Throughout, the project—indeed, the public perception of molecular genetics in total, today—is beset by semantic problems, problems of systematic imprecision of language, often betraying imprecision of thought. These begin with the primary level, mapping and sequencing. For T. H. Morgan and

---

*Best for the history and politics is *The Gene Wars: Science, Politics, and the Human Genome,* by Robert Cook-Deegan (Norton, 1994). For crucial years in the mid-nineteen-eighties, Cook-Deegan worked at the Office of Technology Assessment, an information-gathering creature of the United States Congress, where he ran projects including one on gene therapy and another, the big one, on the genome project. He got to observe most of the players and much of the action; he mastered the documents and conducted scores of interviews.

Best on the science for the non-specialist reader is *Signs of Life: The Language and Meanings of DNA,* by Robert Pollack (Houghton Mifflin/Viking, 1994). Pollack is of the rare breed whose science itself is permeated and in part shaped by an acute historical, literary, and critical sensibility. His is the most distinguished book about science I have seen so far this decade; it is a sensuous delight to read.

Alfred H. Sturtevant, map and sequence were interchangeable—a linear array of factors that determine known characters, spaced in proportion to the frequencies with which these characters are inherited independently. But the gene seemed unitary then. Today, the conceptual difference between genetic sequences and maps is important yet often neglected even by geneticists. The sequence is now defined and directly accessible. It is the order of the base-pairs, the nucleotides. The sequence has often been called "the ultimate map," but this is a considerable confusion—for the sequence is, rather, the ultimate territory of the genome project, and its one-to-one representation by an immensely long string of As Ts Gs Cs, is by itself, without analysis, useless. Maps of whatever kind never reproduce the territory mapped. They are at once less than the territory and more. They are abstractions from the territory, leaving much out; and they are labelled, the features of the terrain identified. Every natural territory generates many different maps, and the differences spring from the uses to which the maps are to be put. The political map differs from the road map which differs from the map of natural resources as detected from land-mapping satellites, and so on. All maps share one general characteristic: the abstracting and labelling establish a set of relationships. In short, by omitting much of the territory maps gain explanatory power. Maps of the genome will locate and distinguish all the different components, labelling their functions and interactions. The sequence by itself is dry and uninformative. The map is going to be more complex than we can yet imagine—more than we shall ever be able to take in. (Computer programs will feed us the sets of relations we need to know.) The contrast is rooted in a single cause: the human genome was not designed, it evolved.

However you consider it, that label *The Human-Genome Project* is of course multiply misleading. Take it apart. "*The* human genome." As we all know, there is no one human genome. For gene mapping, the reference DNAs come from individuals in 70 large families, 53 in Utah and 13 in France; they are held in Paris. They are used for establishing linkages—that is, for relating a gene newly isolated to others on the same chromosome. At the level of the mapping of genes the differences among individuals may be unimportant. The aim here—the philosopher Philip Kitcher points out—"is to do for our genetics the kind of thing that Vesalius and the great anatomists that followed him did for our phenotype."

That is a straightforward sensible goal for gene mapping, at an immediate level, anyway, although mapping has other levels and purposes, too. Yet when we come to sequencing, on a scale at least a thousand times finer than gene mapping, we will be interested in precisely those stretches that we know to contain genes and control elements and where variation is high. Here we uncover clues to disease; here may be buried our talents. Here, the preconception that there is one human genome, implicit but unexamined in the name of the project and built into the way the project is being carried out, is both false and pernicious.

The human population is characterized by enormous genetic variation, comparable not to any wild species but only to domesticated animals such as cattle or dogs: this is so because for a hundred thousand years we have lived in many ways as domesticated animals. To sequence stretches of DNA from one or a few individuals and so build up a single composite total sequence—as in fact the project is doing—altogether fails to capture this variation. Instead, it perpetuates the deep error called essentialism, which has persisted in zoology since Aristotle or before, and which says or assumes that every species has a standard, characteristic type, from which any variation is a falling away. That's the way zoologists classified species, fitted new ones into the system. Darwin replaced essentialism with the understanding that a species is a population, and that variation is what characterizes a population. This realization underlies the theory of evolution and is the foundation of the success of the Darwinian method. From Darwin to the present day, the sources and nature of variation have been among the central questions of evolution.

To study human genetic variation only when it is thought to have clinical significance, as the project is in danger of doing, reinforces the essentialist error. A person has by recent estimate perhaps two hundred thousand genes; the human population counts millions of alleles. Those that relate to genetic diseases are a tiny proportion of the total. To understand the human population at the genetic level, to exploit what the genome project ideally should be telling us, we require a large number of human genomes.

Now take "the *human* genome." Many genomes other than the human are necessarily part of the project, from disease viruses through bacteria and yeast, nematode and fruit fly, to mouse and chimpanzee. As I noted, comparisons among species are even now yielding provocative clues for immediate problems of developmental biology, as for ultimate questions of evolution. In the summer of 1995, when the first complete nucleotide sequence of a bacterium was published, suddenly we knew the creature had some hundreds of genes whose function had never been detected; and we found ourselves astonished at the degree of similarity, on an evolutionary scale, between many bacterial genes and some genes of higher animals.

Move on to "*the* project." Despite the designation, which marks it a single entity, and the assurances that have been made to the money agencies and ultimately the Congress that the thing can be wrapped up not long past the turn of the millennium, in fact and for excellent reasons the research has no foreseeable terminus. Extending and exploiting it will occupy scientists and soak up funds well past the middle of the next century. Yet the label promotes the project. Try titling it accurately—for example, "the genomes business." Falls flat; won't sell.

The selling of the genome project, more generally the way new discoveries in human genetics are announced, has led the public to dangerously over-simplified beliefs about what genes do—and to inflated hopes about what scientists and physicians will be able to do with them. Again the problem is semantic. In its most extreme form we see this in the press almost weekly: the fallacious phrase "the gene for" some disease or defect. It is announced that molecular biologists have discovered the gene for schizophrenia. The gene for diabetes. The gene for alcoholism. The gene for Huntington's chorea. The gene for homosexuality. The gene for Alzheimer's disease. Another gene for schizophrenia. The gene for obesity. Another gene for pancreatic cancer (did you miss the first one?). The gene for vivacity. With honorable exceptions, specialists have failed clearly and insistently to convey certain of the elements of genetics to the media and to the public through them—and cannot pass the blame to the sensationalism of journalists or the ignorance of that public.

We must bring certain standard terms into common use. Genetic diseases? But we know that what gets discovered is not "the gene for." Indeed, it's not quite correct even to use the phrase "the mutation for." What we mean, and what biologists must say and get the media to say and the public to realize, is that one discovers "an allele that—." Of course there's a legitimate meaning for what a gene is for, namely, whatever it is that "the allele that" disables. The undamaged gene at that location does one thing—usually, in our present ignorance, something we don't yet know—and that is what it is for; but we have found a different form ("allelomorph," from the Greek, in William Bateson's term from 1902) that leads to defect in the organism. To write of an allele that results in a defect as "the gene for defect D" is a total displacement, a reversal of meaning. This is not a trivial matter. Consider how much confusion is caused by the notion that we discover "a gene for breast cancer"—and the next month we find another. The incorrect usage is so widespread, so entrenched, that we cannot rescue it, any more than we can restore the word "gay" to mean debonair, joyous, again. For biologists consistently to speak and think in terms of various alleles that do A or B or C—this would clear up much of the confusion at once. That is just the most obvious instance of the necessity of clear terminology to straighten thinking and deflate expectations. There is a precept here. Less is not more. Less is sometimes disastrously less. Oversimplification creates confusion that no reader can comprehend: fuller explanation will bring enlightenment. More can simplify.

Here, perhaps the most fundamental principle in genetics is that genes act in concert—with other genes, and under the influence of environmental factors—to grow the adult organism. This principle goes back to the earliest days. In the first two decades of this century, a bitter scientific controversy raged between the Mendelians and the proponents of an alternative approach. The anti-Mendelians protested especially against the theory that all inheritance is particulate, due to discrete, independent, stable factors, when, as anyone can plainly see, most variation is continuous. As one of many corollaries, they objected—although the term "holism" was coined only in 1926 and "reductionism" entered the language much later still, in the nineteen-forties—that an organism must be treated as a whole, its organization not to be understood by analyzing it to bits. Here is what T. H. Morgan, the leading Mendelian of the day, had to say to the anti-Mendelians in 1915.

His first precept was that each gene has "manifold effects." He illustrated the point with the white-eyed mutant of fruit flies (a mutant much studied by the Morgan group; the discovery of it, five years earlier, had led to the understanding of sex-linked characters). "We find that the white eye is only one of the characteristics that such a mutant race [i.e., strain] shows," Morgan wrote, and went on,

> The solubility of the yellow pigment of the body is also affected; the productivity [fecundity] of the individual also; and the viability is lower than in the wild fly. All of these peculiarities are found whenever the white eye emerges from a cross, and are not separable from the white eye condition. It follows that whatever is in the germ plasm that produces white eyes, it also produces these other modifications as well. . . . Many examples of this manifold effect are known to students of heredity.

It is perhaps not going too far to say that any change in the germ plasm may produce many kinds of effects on the body.

Morgan next wrote of the variability of characters, which had seemed to opponents of the idea of genes to be a crushing disproof. His second precept: "All characters are variable, but there is at present abundant evidence to show that much of this variability is due to the external conditions that the embryo encounters during its development. Such differences as these are not transmitted in kind—they remain only so long as the environment that produces them remains." In short, as the organism develops, any character in the organism is the result not of genes alone but of interactions of genetic processes with various and variable environmental factors.

Morgan's third precept was that, contrariwise, "Characters that are indistinguishable may be the product of different genes. We find, in experience, that we can not safely infer from the appearance of the character what gene is producing it. There are at least three white races [strains, varieties] of fowls produced by different genes," and he gave other instances. This led directly to his fourth precept: "Each character is the product of many genes." Indeed, "It might perhaps not be a very great exaggeration to say that every gene in the germ plasm affects every part of the body, or, in other words, that the whole germ plasm is instrumental in producing each and every part of the body."

Fourscore years have passed since Morgan wrote that paper, about the time that separated Gettysburg from Valley Forge. But have we come far? Gettysburg and Valley Forge are but ninety miles apart. Despite the rich tapestry of understanding woven in molecular genetics since the nineteen-forties and in the molecular biology of development since the early seventies, what Morgan wrote remains central to understanding, to explaining, all we have learned. Genes act in concert. In the innumerable interactions of the developmental process, any single gene may affect several characters. The technical term is pleiotropy. And any character will be influenced by many genes. The term is polygeny. Development cannot take place without its environment, and the outcome will be pervasively affected by the environment.

Morgan's precepts establish a crucial and absolute distinction. While only a single allele, or very few, are involved in any given one of a high proportion of genetically related diseases, general heritable qualities of body or mind, our longevity, our intelligence, our particular talents, are the product of many genes, in exquisite balance among themselves and with the environment. This is why positive eugenics, breeding people for the enhancement of general qualities, cannot work, and why the extreme of positive eugenics, direct intervention in the germ line to improve such qualities, is a forbidden experiment: almost certainly you would upset the exquisite balance and engender a new human who would be seriously defective. All genes work in concert.

Has the practical payout of molecular genetics been exaggerated? From the first stirrings of the methods of recombinant DNA, scientists dwelt upon their potential for the cure of genetically based disease. Twenty years elapsed: in September of 1990 the first clinical trial of gene therapy, approved by the Recombinant Advisory Committee of the National Institutes of Health, began. The lapse of time reflected both the supreme difficulty of the task and the exemplary caution of the committee. (At least two unsanctioned attempts at gene therapy had been made before then, one in Germany the other in Israel, both instigated by American scientists—both of whose careers were severely damaged in consequence.) The patient was a girl lacking the gene for the protein adenosine deaminase, and therefore suffering severe combined immune deficiency, effectively a total lack of the immune response. The disorder dooms any child born with the defect to live in isolation and to die early, but it is extremely rare.

Five years later, by June 1995, the Recombinant Advisory Committee had approved 106 clinical protocols, and 597 patients had already undergone experiments in gene transfer. The rush reflected the enormous pressure to show results. By that time, the NIH was spending some two hundred million dollars a year on research into gene therapy, and the interest of biotechnology companies had grown so great that they were thought to be spending as much again, or more. Early in 1995, Harold Varmus, director of the NIH, himself a Nobel-prize retrovirologist and a shrewd, skeptical realist, appointed a committee to look into gene therapy, its results so far, its promise, and the research the NIH ought to pursue. The committee met repeatedly, heard many views including those of the enthusiasts, and reported in the first week in December of 1995. Among their conclusions:

> While the expectations and the promise of gene therapy are great, clinical efficacy has not been definitively demonstrated at this time in any gene therapy protocol, despite anecdotal claims of successful therapy and the initiation of more than 100 . . . approved protocols. Significant problems remain in all basic aspects of gene therapy. Major difficulties at the basic level include shortcomings in all current gene transfer vectors and an inadequate understanding of the biological interaction of these vectors with the host. In the enthusiasm to proceed to clinical trials, basic studies of disease pathophysiology, which are likely to be critical to the eventual success of gene therapy, have not been given adequate attention. . . . Overselling of the results of laboratory and clinical studies by investigators and their sponsors—be they academic, federal, or industrial—has led to the mistaken and widespread perception that gene therapy is further developed and more successful than it actually is. Such inaccurate portrayals threaten confidence in the integrity of the field.

The committee reprimanded the would-be gene therapists themselves in language for this sort of document startlingly blunt: "Investigators in the field and their supporters need to be more restrained in their public discussion of findings, publications, and immediate prospects for the successful implementation of gene therapy approaches."

Cures are not coming soon and even if possible will for decades be inordinately expensive. More modestly, though, prevention is paying out even today. For a large and rapidly increasing number of disorders, we are now able to test adults to see whether they carry a recessive allele that, if they were to choose another carrier as mate, would risk begetting an afflicted child. A couple where both carry the allele can choose not to have children with each other. Jews of eastern European ancestry are subject to several disorders of genetic origin, of which the best publicized is Tay-Sachs disease: testing for the relevant alleles before marriage has become widespread among Jews of that descent. Many American blacks are doing the same to avoid the risk of sickle-cell anemia. Alternatively, early in pregnancy cells can be drawn from amniotic fluid and alleles that carry the risk of defects detected. At that stage, prevention of course requires abortion. Some ethicists fret that to indulge the wish for a perfect baby seems the perverse extreme of consumerism; the reality is that a great deal of misery could be avoided. The choices must be free, of course; yet in a speech at one of the meetings in 1993 marking the fortieth anniversary of the discovery of the structure of DNA, Watson observed that Huntington's disease—caused by a dominant allele that often doesn't manifest itself until the person with it has had children—could be all but eliminated in a generation if everyone who has relations afflicted by the disease were to take the test before a child was conceived.

Despite the promises, even the rewards, the most valuable product of the genome project is still and always not for the clinic but for the science. Look again at maps and sequences. Even at the first level, the locations of structural genes, mapping has progressed far enough to establish interesting scientific complications. For reasons rooted in the evolutionary history of mammals, genes for one organ or one system or one function may not confine themselves to some one chromosome. Genes for the enzymes for the several steps in a biochemical pathway are not to be found lying together in an operon as in the bacterial model; they may be on one chromosome or on different chromosomes. Even in the case where a protein is made up of two different chains, all too often the genes for the different chains are not on the same chromosome. Thus, the gene for the alpha chain of hemoglobin is located on chromosome 16, that for the beta chain on chromosome 11.

At the next level, the maps must account for regulation, for genetic sequences that are not structural but control elements. This will surely include whole classes of controls we have hardly glimpsed. The maps must also account for the immense amount of DNA that has no function we know, or that may be functionless.

The penultimate level of mapping is the developmental: we want the map to show us what gets turned on and off in temporal order during the human creature's lifetime—what Brenner and his followers, now multitudinous, have succeeded in doing for his worm, what Nüsslein-Volhard and her school have well begun for the fruit fly. Special versions of the developmental maps will be clinical, dealing, say, with inborn errors of metabolism or with the cancers and their prevention.

The ultimate map will extract from the sequences and from all the preceding levels of mapping, like layers from an archeological dig, whatever evidence about the evolution of man the steps of that evolution have not obliterated.

❖ ❖ ❖

I have recently been re-reading Jim Watson's *The Double Helix*. I found it nostalgic and immeasurably sad. How remote from us, how different from what we know now, that time was. Indeed, though Watson wrote it just fifteen years after the events, already the harsh immediacy of his prose could not conceal the distancing that that prose accomplishes. If ever there was a golden age in science, those early days of molecular biology are surely a candidate.

Perhaps there have been others. Morgan's group at Columbia in the decade from 1908, say? Göttingen for quantum physicists in the late nineteen-twenties? Gowland Hopkins' biochemistry laboratory at Cambridge at much that same time? For molecular biology, we can specify the period and places. From the early nineteen-forties through the early fifties, the small, slowly expanding group around Max Delbrück, who worked with the genetics of bacteriophage and in the summers gathered at Cold Spring Harbor. From the early nineteen-fifties and the rest of the decade, the Medical Research Council unit in the Cavendish Laboratory at Cambridge, and André Lwoff's garret at the Institut Pasteur in Paris.

What is a golden age like, in a science? Watson and Crick pointed out to me as we rode in an elevator in Dallas not long ago that the obvious first requirement is that the science be just a-borning, nascent, small. Certainly that small size at the start of a new field is crucial. An important aspect of this is that the people in the new field have arrived there from elsewhere. The small size maximizes the collision frequency, the intensity of intellectual interactions; the variety of starting points maximizes the interplay, the scope and angle of intersection of ideas. In short—and we see this in the arts, too—new intellectual movements spring up between the paving stones of more rigid, established approaches and disciplines.

Again, at the beginning of a new field both concepts and experiments can have a freshness and simplicity that, among much else, may mean that the actual costs of research are low, in absolute terms but especially in relation to the payout. And that payout! Great principles, often radically simplifying theories clearing away overgrown, encrusted complexities. Watson remarks elsewhere that he got his doctorate quickly, in part because there was a lot of biology, chiefly biochemistry, that he didn't bother to learn because he saw that in five years it would no longer be relevant, probably no longer defended as true.

Yet a golden age in science is more than small size, modest needs, and big ideas. A true golden age is also an age of innocence. It thrives, for a while, in the competitive harsh ocean of the twentieth century, as an island of idealism, and of play, and of, at the same time, an austere devotion to intellectual enthusiasm and openness. The phage group has been called "one of the little communities of intellectual purpose and excitement that constitute the only genuine utopias of the twentieth century." Competition surely there will be, but competition that's more unifying of the tight small community than divisive—competition for the approval, respect, and lively immediate intellectual response of your colleagues and mentors.

Listen to Watson, this time from the essay he wrote a couple of years before *The Double Helix* and published in the volume prepared to honor Max Delbrück on his sixtieth birthday, *Phage and the Origins of Molecular Biology*. He's describing his induction into the phage group:

> As the summer passed on I liked Cold Spring Harbor more and more, both for its intrinsic beauty and for the honest ways in which good and bad science got sorted out. . . . Most evenings we would stand in front of Blackford Hall or Hooper House, hoping for some excitement, sometimes joking whether we would see Demerec [the director then] going into an unused room to turn off an unnecessary light. . . . On other evenings, we played baseball next to Barbara McClintock's corn-

field, into which the ball all too often went. . . . When August began the Lurias went home to Bloomington because Zella would soon have a child. This left Dulbecco and me even more free to swim at the sand spit or to canoe out into the harbor, often in search of clams or mussels.

Surely, Max Delbrück created the ethos of the phage group in the nineteen-forties—and this was something he brought from the Bohr circle in Copenhagen of nearly fifteen years earlier. Surely, André Lwoff created something similar at the Pasteur: one has only to look at the photographs of the team eating lunch together in the attic in the early nineteen-fifties to experience an almost physical awareness of the rush of excitement that filled so much of those days there. We know from *The Double Helix* that the unit at the Cavendish, increasingly biological in the midst of a physics laboratory, was another such island of intellectual play and enthusiasm and dedication to getting the answers.

Delbrück was one of the most seductive intellects of our time: the charm of the man and his ideas and style are palpable in everything he touched. But style is surely essential to a golden age. And such style is an integral component of leadership—which is, of course, another essential. Morgan, Rutherford, Bohr, Bragg. Pauling. Delbrück. Lwoff. Crick. These men had styles of leadership that attracted younger scientists of great individuality and ambition—scientists like Monod or Jacob, Watson or Brenner, who would never have fit comfortably or creatively into more traditional laboratories. André Lwoff, in particular, was a leader and scientific stylist who has been unfortunately neglected in Anglo-American accounts of the origins of molecular biology. His review of *The Double Helix* reveals a man of immense rectitude, compassion, and, indeed, wisdom. His work was wildly original in its day—and, with all that, he remains the one scientist I know of who could be intentionally funny in a serious scientific paper.

One more example—small but telling—of the high style in molecular biology. Sydney Brenner once remarked to me, speaking of the time in the early nineteen-sixties when the primary question was the nature of the genetic code, that one attempted to create experiments, to invent experiments, where the absolute minimum of data would produce the maximum insight.

The golden age, the age of innocence, has been followed by our age of brass. Brenner himself, for a quarter-century now, has been driving a project almost without style: his brute-force attack on the total developmental biology of the nematode *Caenorhabditis elegans*. Watson has been for two decades himself the director of Cold Spring Harbor Laboratory: consider how he has transmuted it into a powerhouse of big biology, a place where the young Jim Watson could hardly have flourished. Yet one cannot simply be sentimental about such changes: Brenner's worm has after all come to throw brilliant light on aspects of development that could not have been got out in other ways. Watson's massive enterprise, led by his invaluable intuition, has produced a lot of good science.

Within five years from Asilomar, molecular biology was generating an entirely opposite kind of excitement. By 1980, the headlines were about the creation of wealth: biologists and venture capitalists convinced each other to the extent of hundreds of millions of dollars that genetic technology would rival computers and silicon chips. Some observers claim that such commercial pressures corrupt the basic science, in at least two ways. They fear that the free communication among laboratories that was a glory of the nineteen-fifties and sixties is at risk, or even destroyed, when discoveries become trade secrets and new techniques, even new life forms and new genetic sequences, cannot be published until the lawyers have filed for patents. Beyond that, the attraction of enormous profit is said to divert scientists from basic research and to distort the allocation of funds.

Yet such conclusions were not self-evident. At the time, perhaps naïvely, I had hoped for a different outcome. Communication, first: good scientists have historically been close-mouthed until their results are sure and written up; a few, indeed, are pathologically secretive. On the other side, molecular biology in general moves faster than almost any other science: what you hold back today your rivals will have in print next week. Even the patent process need not be crippling: scientists fortunate enough to work at the university or company whose lawyers are responsive have not found publication delayed.

None the less, during the nineteen-eighties an alternative view of the science grew up—one that competes today for the allegiance of graduate students and postdocs, and that bears an unpleasant resemblance to the attitudes and practices that at the same period became so conspicuous in the law and especially in high finance. Speaking a while ago of a notorious case of scientific fraud, Watson said, "It's a story of the eighties." Exactly so. Those remarkable small self-governing scientific democracies of the classical era, represented by the phage group, the garret at the Institut Pasteur, the RNA Tie Club, the Medical Research Council unit at the Cavendish—few such collaborations are now to be found. (One is the far-flung network of those working on *C. elegans,* Brenner's worm: but Brenner was long Crick's closest collaborator at that Medical Research Council laboratory.) They are steadily being superseded by a different sort of laboratory life and ethos: laboratories that are large and rigidly hierarchical, and an ethos driven by careerism, in which doing science is in large part a way to secure more grants, to get promotions, to gain power. Perhaps to get rich.

❖ ❖ ❖

A further, far deeper structural transformation is now affecting science in general, molecular biology in particular. It will continue to operate and with increasing and irresistible force. It relates to the phenomenal growth of molecular biology—of sciences at large. But the problem that now envelops the sciences, conditioning everything else, is not one of growth in the simple sense. Rather, it is the transition of the sciences from exponential growth to the steady state. In 1963, Derek de Solla Price, who was an historian of science, published a graph showing that scientific activity had been expanding at an exponential rate for three hundred years or more—indeed, that the output of scientific papers was doubling every fifteen years. At that rate, Price said, by the turn of the millennium every man, woman, and child in the United States would be spending every working minute doing research and writing it up. Derek Price was a peppery, amusing man, and everyone laughed and said, What a droll idea! Like most assertions of Malthusian limits, this one was taken seriously by few. Yet we have been creeping up on those limits. The transition to a steady state is producing enormous systemic strains, some obvious, some subtle.

For most scientists, the obvious sign of the transition is the shortage of funds and the much increased competition for what funds are available. Along with this, they feel an ever-increasing pressure from politicians and government agencies for directed or targeted research, with slogans like "the war on X" or "national needs." The practice of science in the mode that has proved so successful since the second world war has had definitive features: good funding on the one hand, and on the other self-governance within the scientific community, in the choice of the next problems to be attacked and in the evaluation of results. Parsimony and political interference: these seem to most scientists to be the characteristics of an antithetical situation. One longs for a return to normality. Of course, we are not going to get any return to such a nostalgic normality.

The intercourse between academic laboratory and industry is changing the career structures of molecular genetics. At least from the early nineteen-seventies onwards, the line between so-called pure and applied research has grown ever more tenuous. It is now vanishingly thin. The trick one laboratory devises to take a tiny step forward in the puzzle of cell development has often turned out to allow manipulations of some other sort of cell for the production of something commercially profitable; indeed, we can no longer predict which kind of advance will come first and in what sort of laboratory, academic or biotechnological. Often today you'll find that the laboratories in the biotechnology company are better equipped and funded than those in the university department across town; these days, much of the work done in them is just as publishable in the journals, just as prize-worthy.

The transition to the steady state brings with it a change in the patterns of work and advancement in molecular biology. The young biologist today sees not one but two career ladders, one academic the other high-tech industrial— and the best of them soon realize that as they ascend they will be able to step from one to the other, either way. These developments, too, though they may seem encouraging to the more talented young scientists, are changing their goals and the overall nature of the enterprise.

Along with these changes, I have recently awakened to another development, one that must be profoundly demoralizing. Peter Medawar, in his book *Advice to a Young Scientist,* pointed out that one of the most important decisions a beginner will make is the choice of research problem. Pick one that's important, he urged, one that looks ripe but that will call for sustained hard work and talent. In recent years, attending conferences, listening to papers, wandering through poster sessions, I have seen great care, hard work, earnest dedicated thought—devoted to tiny problems. Molecular biology has become fractal.

In the midst of a great structural transformation, its longer-term consequences are not predictable in detail—at the microlevel, so to speak. But keep in mind such things as the changes in the career structures of the biological sciences, the changes in the aims and conditions and organization of biological research, and the changes in the ways research is evaluated. The golden age, or even its successor the grand imperium of science—if these ever existed they are forever gone. I have taken to crying out to scientists: The barbarians are in the city: civilize them if you can, but work with them as you must.

❖ ❖ ❖

A quarter-century of following the development of molecular biology, acquaintance with hundreds of molecular biologists and friendship with many, perhaps gives me license to ask a last question: Are molecular biologists doomed forever to be arguing the practical benefits of their research? Must you forever be selling cures? We must never let go of the principle that has served the sciences so well from the second world war. Yes, science brings immense practical rewards: but we must teach and teach again that often the delay between discovery and application is long, decades, perhaps, and rarely can we tell which results will turn out to be useful.

Beyond that, science offers the highest sort of human satisfaction. It provides a unique and sublime pleasure; it fills a need nothing else can touch. Molecular biologists, geneticists, ecologists, immunologists, neurobiologists are forgetting what is for ourselves and for the lay public the greatest gift we bring. Consider the physicists, the cosmologists, or the archeologists, or the paleontologists. They understand better than biologists do what scientists have to offer that, after all, is exciting, memorable, treasured. When Arthur Eddington led a team of astronomers, in 1919, to measure the bending of starlight by the gravitational attraction of the sun, and so provided the first confirmation of Einstein's general theory of relativity, the news made headlines in *The Times* of London. When Tsung-Dao Lee and Chen Ning Yang overthrew the quantum principle of the parity of weak interactions, their recondite but revolutionary discovery that the universe displays handedness appeared on the front page of *The New York Times*. The big bang and black holes have been taken into common speech. The meteorite-impact explanation of mass extinctions, the excavation of a vast imperial tomb in China defended by an army of porcelain soldiers, the achievement of a young archeologist deciphering the impenetrable Mayan inscriptions and so overturning received notions of a thousand years of that history, the collision of fragments of a comet with the planet Jupiter, the detection of a possible planet circling a star relatively near yet inconceivably remote: these fill no bellies, conquer no diseases, process no peace—yet we all know of them. Who would pay for the space shuttle or the Hubble telescope for their practical return on investment? When Hubble looks deep into space and time and finds fifty times more galaxies than we knew, it tells us about the nature and beginnings of the universe. Who would support for reasons of utility the expedition that brought home to us that woman of mystery we all know—Lucy, she of the Ethiopian desert? She whispers to us about where and what we came from.

Ernst Mayr, last of the great founding scientists of neo-Darwinism, wrote some years ago that biology has two strands that run through all else —genetics, which asks proximate questions, questions that begin with *How?*, and evolution, which asks ultimate questions, questions that begin with *Why?* In molecular biology, molecular genetics, in for example the long-term promise of the genome project at its fullest, including all those other genomes and their evolutionary resemblances and distances, the proximate questions now promise to reach ultimate answers.

We must always remember that what we offer beyond all else is the understanding of origins—the origins of the human race, the origin and nature of life itself. This is what the public most deeply wants to know. This is after all what validates the choice of science as a vocation.

Starting in the mid-nineteen-seventies and continuing to the present day, various facts and opinions have been published about the life and career of Rosalind Franklin which are demonstrably wrong. In response to these, and in consultation with editors at an American monthly magazine of general interest, I wrote the essay that follows. I thought of it, and still do, as a defense of Franklin against a coarse misappropriation of her experience as a scientist. The editors bought it, paid me the fee agreed upon, but then decided it was too technical and too polemical for their audience.

Although in parts the essay necessarily overlaps some of the first three chapters of The Eighth Day of Creation, *it tells the story of the discovery of the structure of DNA from a new and different point of view—Rosalind Franklin's point of view, or at least over her shoulder exclusively. Nobody had attempted this before. The structure, I suppose, is what the editors feared was too technical. The essay puts that story to a new and different use—the evaluation of Franklin's true position in the laboratory at King's College London, and of her career as a scientist who happened to be a woman and who died at the age of 37. This, I suppose, is what was too polemical.*

AFTERWORD I

❖ ──

In defense of Rosalind Franklin:
The myth of the wronged heroine

An immense basement corridor—high, dimly lit, flagged in stone, lined between massive arches with dark and featureless oaken lockers rooted there since the nineteenth century—leads for hundreds of feet beneath one of the principal older buildings of King's College of the University of London. Northwards, the corridor almost cuts under the Strand, one of the busiest thoroughfares of the Metropolis, but here no sound of bus or lorry penetrates, nor can one see from here the Thames flowing silently past the Victoria Embankment, to the south. At intervals along the corridor, doors open to laboratories and offices, some windowless, some with windows through which daylight can filter down from an areaway. The corridor whispers to our imaginations about one of the pivotal episodes in all of modern science, the discovery in the early nineteen-fifties of the three-dimensional molecular structure of DNA—deoxyribonucleic acid, the stuff that genes are made of—and about the discovery's melodramatic and still controversial circumstances. The corridor hints of the situation of Rosalind Franklin, a scientist of ability, determination, and integrity who, at King's College London in grim professional isolation, brought out certain features that are essential to the structure; who none the less was beaten to the discovery because others, unknown to her, had access to her data; and who did not live to share the Nobel prize.

This drama has everything—exploration of the unknown; low comedy and urgent seriousness; savage competition, vaulting intelligence, abrupt changes of fortune, sudden understandings; eccentric and brilliant people, men of honor and of less than honor; a heroine, perhaps wronged; and a treasure to be achieved that was unique and transcendent. This is one of the great stories, and is so, to begin with, because the explanatory power of the discovery itself cannot be overstated. It was made at the Cavendish Laboratory, of Cambridge University, at the end of February of 1953, worked out in the next weeks, and published in a brief report in the British weekly journal *Nature* for April 25, signed by J. D. Watson and F. H. C. Crick.

There the scientific community at large first encountered the simple barber-pole model of what soon was called the double helix. DNA is two complementary chains, and when the two chains separate each can assemble on it-self, taking molecular components from the surrounding cellular soup, a duplicate of its former complementary partner—producing two identical double helices. The unique structure of DNA instantly illuminated what had been the most central yet most baffling question in biology, namely, how heredity works in physical, chemical terms—both in the transmission of hereditary characters with great fidelity down the generations and in the expression of those characters as the individual organism grows. Those dual functions are built in, physically, and with surpassing parsimony and elegance. The structure does more than explain. It is beautiful.

To be sure, the details of gene reproduction and gene expression turned out, over the next three decades, to be stupefyingly intricate. Yet they do yield. Knowing the structure and pursuing its manifold consequences, we understand the chemical mechanisms of life with a richness and to a fine scale undreamed of in 1953, and we may be confident that far more is to come. Then, the scientists interested in the problems numbered no more than a few dozen; today, the roster is uncountable but in the tens of thousands. James Watson and Francis Crick, each in his way, have become powerful, famous, and rich. The principles of the workings of the double helix are taught to every highschool pupil, are part of the thoughtful layman's intellectual tool box. Now the practical applications are emerging. They bid fair to transform the world.

❖❖❖

As the consequences of the discovery have grown, so too has the unquiet legend of Rosalind Franklin—that she could have got the structure out first, that the passage of her data to Watson and Crick was a breach of faith, that she has been deprived of full credit for the crucial elements she did contribute, that she was ill-used by the research group she was part of at King's College London. The particular claims have been wrapped up repeatedly in the charge that Franklin's work at King's and her entire career suffered because she was a woman, a "lonely" woman, trying to make her way in science, in the "man's world" of science. The Franklin of 1953 has been adopted as an emblem of the obstacles women encounter who pursue professional careers nowadays.

Watson himself provoked the legend, though hardly intending it. Early in 1968, he published a memoir of the discovery of the structure of DNA, *The Double Helix*. This is a great book, but strange—awkward, indirect, a style vivid but lumpy, and straining with barely repressed resentments. In the previous two years, Watson had circulated copies of the first draft—under the curious title *Honest Jim*—to many of those who appeared in its pages, and almost to a man they were enraged. The draft was scandalous about how the discovery was made and about the behaviour of almost everyone involved: the scientist who ran off with another's wife, the graduate student who had to leave Cambridge because he got some other scientist's *au pair* girl pregnant, and much more of the same—with, perhaps as bad, what scientists said behind each other's backs about each other's work and personalities. Crick led a thundering attack against publication. Watson made a lot of cuts, fuzzed a lot of other details. Over continued angry protests, the book came out. Biologists seized upon it with fascination. The lay public did, too.

One person in the book who did not get heavily censored was Rosalind Franklin. Watson described her as unattractive, badly dressed and groomed, opinionated, stubborn, hot tempered—a caricature of a woman who has given up her sexuality for science. He also made clear that her science was wrongheaded. *The Double Helix* appeared a few months short of ten years after Franklin's death.

The Franklin of legend is not Watson's caricature but a creature of the response that it provoked. The memoir touched off extraordinary reviews—by no means unequivocally favorable, but coruscating with imaginative response and surely an index of the book's power. All mentioned the treatment of Franklin. In a magisterial, witty, and scathing review in *Scientific American*, the French microbiologist André Lwoff—a scientist of encompassing humanity—analyzed Watson's psychopathology shrewdly, and said, "His portrait of Rosalind Franklin is cruel. His remarks concerning the way she dresses and her lack of charm are quite unacceptable. At the very least the fact that all the work of Watson and Crick starts with Rosalind Franklin's x-ray pictures and that Jim has exploited Rosalind's results should have inclined him to indulgence."

The feminist appropriation of Franklin began soon after. In 1971, Elizabeth Janeway, in her book *Man's World, Woman's Place,* used Watson's treatment of Franklin as an instance of a widespread attitude of men towards women in science—a point that in itself is no doubt correct but that tells us of Watson not of Franklin. In 1975, a biography of Franklin appeared, by Anne Sayre, a writer of short fiction, who had known Franklin and whose husband is an x-ray crystallographer, Franklin's specialty. Sayre's motive was a generous sense of outrage at Watson; her book, unfortunately, was shallow. She asserted that science at King's College London was a man's domain that excluded women socially and scientifically. She stated as fact that "At the time there was one other woman scientist on the laboratory staff besides Rosalind."

The legend has taken on independent life. In an otherwise perceptive study, *Women in Science*, Vivian Gornick says that she was led to look into the problems of women scientists in the nineteen-eighties by those of Rosalind Franklin thirty years earlier. The portrait of Franklin that piqued her, though, was not Watson's but, she says to my astonishment, my own—in *The Eighth Day of Creation*. Gornick is embarrassingly generous to my book, yet then she adds, about Franklin at King's:

> As I read on in *The Eighth Day* the conviction began to grow in me that woman *qua* woman Franklin was something of a permanent freak in their midst, and that it was this unalterable fact that fed so disastrously into an alienation that might have been ameliorated had she been a man. As a woman, she was a historic stranger in the King's College Laboratory: They had never really had them, never wanted them, never known what to do with them; women in the lab just didn't seem right.

In a score of conversations in recent years, men as well as women in the sciences or professions have confidently, sometimes almost unshakably asserted to me a similar view of Franklin.

That view is false. Franklin had a formidably unpleasant time at King's. But these were not the reasons. Nor does my book say they were—yet the facts of her predicament and career have not been told fully nor the errors of other accounts firmly set right.

❖ ❖ ❖

In January of 1953, Franklin was at one corner of a competition the full heat of which she could not have felt. Scientists at three laboratories were going after the structure of DNA. Linus Pauling, at the California Institute of Technology, was the foremost physical chemist of his day, and his day had already shone brilliantly for a quarter century. By Christmas of 1952, Pauling had a structure to propose for DNA and had completed a precise table-top model of it. At the Cavendish, in Cambridge, Watson was but a visitor, a young American postdoctoral fellow with a burning conviction that the structure of DNA would reveal the secrets of the gene. Crick, though eleven years older, had had his education interrupted by naval research during the war and was still completing his doctorate, for which he was studying the structure of hemoglobin. A year earlier, the two had essayed a structure of DNA that to their embarrassment had proved to be a piece of chemical nonsense. They had then been ordered by the director of the laboratory, Sir Lawrence Bragg, the Cavendish Professor, to leave DNA alone; the problem belonged to King's. But they had learned of Pauling's attempt, as Franklin had not, and they were restless under the ban.

At King's College London, two scientists took DNA as their problem, Franklin and Maurice Wilkins. They were not on speaking terms. That fact is the poisoned root both of the history of the discovery and of the legend of Rosalind Franklin.

Franklin was born in 1920 into a rich, philanthropic, upper-middle-class Jewish family; her ancestors had come to London in the eighteenth century. She went to St. Paul's Girls' School, in London, an academically excellent private school whose graduates were prepared to do well in the professions they took up. At Cambridge University she read chemistry; on her final examinations she achieved not a first—that is, not the highest honors—but a top second. In 1947, she was hired by a French government laboratory in Paris. There she worked with carbon compounds, coal and graphite. She made herself an able and even an original physical chemist. Certain of her papers are among those that founded the science and technology of high-strength carbon fibres.

Franklin became adept with the technique called x-ray diffraction. The science of x-ray crystallography had been started by Lawrence Bragg when he was what we would now call a graduate student, before the first world war; x-ray diffraction was born, as it remains to this day, the chief tool for determining the three-dimensional arrangements of the atoms in molecules that form crystals. But while in Paris Franklin did not work with large molecules or with those of biological interest.

She gave herself with zest and dedication to research, to laboratory life, to Paris in the late nineteen-forties. A colleague, slightly younger, was the crystallographer Vittorio Luzatti; Franklin, Luzatti, and his wife became close friends. A photograph he took when the three were on a walking trip in the south of France in 1950 or '51 shows Franklin at a table of an evening, in discussion, completing some bit of mending: she had glossy black hair, a high forehead, brilliant eyes and a soft mouth—a face attentive, attractive, feminine, alive with sense and intelligence. "She was devoted to her work, extremely devoted," Luzatti said in a conversation some years ago. She was animated and stubborn in argument in or out of science, he added. At times, anyway, "She was not very easy to get along with . . . to some extent. She was a very warm personality; she had good friends, many people liked her very much; some disliked her just as much—hmm?" And, he said, she enjoyed Paris wonderfully.

A while ago, Luzatti and I spent a day exploring the Paris Franklin had known. The laboratory where they had worked had small rooms, but light and airy, now converted to offices, strung along a corridor under the eaves of a building off the Quai Henri IV. One apartment where she had lived was up a dim stairway across the grit and pigeons of an unmistakably Parisian courtyard. A lively bistro where members of the laboratory had sometimes lunched and argued was perched deliciously on a bluff at the back of the Latin Quarter. Franklin lived and worked four years in Paris. She had her farewell dinner at the Luzattis'. I asked him why she had left. "Well, she said she was English, and if she wanted a scientific career in England, it was time to go home."

❖ ❖ ❖

The laboratory Franklin joined at King's College London was an ambitious venture. Immediately after the war, a Scottish physicist, John Randall, had proposed the idea of a team of physicists, chemists, and assorted biologists who would cooperatively study "cells, especially living cells, their components and products"—which, if you think about it, is a large wedge of the entire biological cheese. His multidisciplinary multiproblematical enterprise Randall later modestly called "an experiment in biophysics." He sold the scheme to the Medical Research Council, which is the British government agency that funds basic as well as applied research in biology. He enlisted Maurice Wilkins, a young physicist who had been a graduate student under him before the war, who had then worked on radar and the atomic bomb, and who now wanted to apply physics to problems of life rather than of death. Randall was appointed Wheatstone Professor of experimental physics at King's College London. They established the Medical Research Council Biophysics Unit there in 1946 and began to recruit. Randall's office was upstairs; everyone else was billeted along the basement corridor. Wilkins appears to have been, informally, second in command.

Wilkins became interested in DNA. He approached it with a suitably large battery of techniques. X-ray crystallography he mostly had to teach himself. Early in 1950, using some excellent new extracts of DNA, he and a graduate student, Raymond Gosling, got x-ray patterns that were tantalizingly richer in information than anyone else's so far. A structure was surely implicit in them, but the mathematical transformations necessary to get from pattern to molecular structure were daunting. He put the x-ray pictures aside and suggested to Randall that they needed to hire a specialist.

❖ ❖ ❖

Zest, gaiety, friendship, light and air, good science vigorously argued—in January of 1951, Franklin left these in Paris for the descent into the basement corridor at King's. In the courtyard by the building, a large bomb crater was still unfilled. (A later colleague of Wilkins's wrote, several years ago, that Franklin suffered from claustrophobia which the underground laboratories exacerbated, but I know of no evidence and the speculation is unnecessary to explain the difficulties that followed.) Randall had helped her secure a three-year research fellowship. From correspondence with him, she knew that she was to apply her skill at x-ray crystallography to DNA. At a meeting with Franklin and others—but Wilkins was on holiday—Randall handed the problem over to her, together with prints of the best x-ray patterns and the graduate student Gosling.

In the next days, Franklin met the others in the laboratory—nearly three dozen scientists and technicians. They saw a young woman of middle height, slim, dark, well and neatly dressed, determined, brisk, sometimes brusque. Among the others, at some point she met Wilkins. He was tall, lean, with a long neck and thin, sharp features—the kind of good looks that sit well on a clergyman with a left-wing social conscience. His voice was deep and resonant. To put the matter gently, Wilkins was a romantic of science, meditative, speculative, markedly indecisive, dreaming of an ideal broad collaborative social process of doing science—a pacifist, preferring to negotiate with a problem rather than to conquer it. With people he was, as he remains, shy, passive, indirect to the point of deviousness. Some who worked with him remark that he could respond to vigorous disagreement only by turning aside. Franklin had research from Paris to finish writing up. Wilkins turned over to her the supply of DNA that yielded sharp patterns. She began ordering new equipment and modifying what was already there. She thought, with reason, that she had been given the structure of DNA as an independent project. Only by summer was she well enough begun to realize that Wilkins expected to remain involved in it. Yet when she confronted him, he backed away. Their temperaments mixed explosively.

Early in the fall, Franklin and Gosling began getting a new and different x-ray-diffraction pattern from DNA. She was a superb experimentalist, patient, dextrous, untiring. The equipment shone a needle-thin beam of low-intensity x rays through a fine fibre of DNA. The repeating atomic arrangements, made as the molecules stacked in crystalline form, deflected the beam systematically, and thus gave an abstract pattern of spots on the film. Each picture took hours of exposure. The fibre had to be kept at constant high humidity; DNA in live cells is wet, after all, and as it dries its molecular structure alters. At medium humidity, she got patterns like Wilkins' best, which they called the A form. At high humidity, the pattern changed, becoming bolder and simpler—cruciform, with clearly marked bars on the four arms and a strong smeared arc at top and bottom. This became the B form. It was Franklin's great experimental advance.

Wilkins told her that the strong cross suggested the molecule was a helix. She told him that that was jumping to conclusions. One Friday evening in November, she gave a colloquium to tell interested members of the lab about her results. Her notes for this talk are detailed. They show, for example, that she was comparing her results to work Pauling had done with helical structures of protein chains. In her audience was a skinny, awkward young man with short-cropped hair, who stared at her pop-eyed and wrote down nothing. This was her first encounter with Watson. He had arrived at the Cavendish and had met Crick about the time Franklin first discerned the B form. He was at the colloquium on Wilkins's invitation. A fortnight later, Wilkins told her that

Watson and Crick had built a model of a structure of DNA. A group from King's travelled up by train to see it. She recognized instantly and said bluntly that the model was incorrect. Watson, taking no notes, had misremembered the values she had reported. This was her first meeting with Crick.

In public that winter, she rebuffed as premature the suggestion that DNA was helical, and along with that any idea of trying to build a model of the molecule. No: the problem next was to get better pictures and to read back by mathematical analysis from the flat A and B patterns to the structure in the round. Relations with Wilkins grew steadily colder. Her notes and unpublished reports from that winter show beyond doubt that she understood the helical argument well. However, she chose to concentrate analysis on the A form. Gosling was a useful pair of hands. She thought alone. Wilkins, shut out, told his frustrations to an old friend, Crick, and increasingly to a new one, Watson.

Franklin, in the spring of 1952, got two superb new pictures of the B form. But for other reasons—spurious ones, in the event—she became convinced the structure of the A form, at least, was not helical. That summer, she also began to look for a new lab, back in Paris or elsewhere in London. Her notebooks for the period are copious, and through their patient record of experiment and theoretical explorations shine fortitude and intelligence. She needed a collaborator.

The situation at the Cavendish was exactly the reverse: volatile collaborative enthusiasm and no data.

❖ ❖ ❖

Doubts had been growing for some time about the productivity of Randall's biophysics unit. Towards the end of 1952, the Medical Research Council formed a special committee, including heads of other of the council's research units, to look into the work there. Randall asked his staff for reports of progress on each current project, and assembled them in twelve mimeographed foolscap pages, "Notes on Current Research." Franklin and Gosling took five paragraphs to report their x-ray studies in technical detail. When the committee visited the lab, on December 15, the document was distributed. It was not confidential. But Franklin did not realize that winter and may never have learned that a copy made its way to the Cavendish and eventually to Crick and Watson.

The first page listed the scientific staff from Professor J. T. Randall, F.R.S., down to research students—31 scientists in all. Of those 31, eight were women. In the laconic English style, initials hid their sex. Dr. H. B. Fell, now Dame Honor Fell, was the unit's Senior Biological Advisor; she was Director of the Strangeways Laboratory, in Cambridge, but came in every week to give an experienced ear and counsel to each research team. Dr. E. J. (Jean) Hanson was the unit's senior full-time biologist and an authority on muscle. Dr. A. V. W. Martin is now Dr. Angela Martin Brown and is still at King's. Dr. M. (Marjorie) B. M'Ewen has retired to her native Scotland; Miss M. I. Pratt, then a tutorial student, is now Dr. Margaret Pratt North, living in Leeds. Dr. R. E. Franklin was a Turner and Newall Fellow. Miss P. Cowan, a Nuffield Research Fellow, is now Dr. Pauline Cowan Harrison, Reader in Biochemistry at the University of Sheffield. Miss J. Towers was listed as holder of a Carnegie

Scholarship. Not listed were Dr. Mary Fraser, the unit's first chemist, who had left the laboratory earlier in the year to have a baby and who now lives in Australia, and Sylvia Fitton Jackson, who was one of the lab technicians—but an unusual technician, for she had already published research at the laboratory, and soon afterwards, at Randall's urging, went to Cambridge, where with no previous degrees she took a Ph.D. and is now a fellow of Darwin College. Women, in other words, worked as scientists at every level from top to bottom and held between a third and a quarter of the professional posts.

Of these ten women, two, Franklin and Hanson, had died by the time I inquired. Of the eight living, I talked with five and corresponded with several of those and two others; the eighth, Miss Towers, I was unable to trace. The seven agreed that women at their laboratory were treated equitably. (Sylvia Jackson's career is striking evidence.) They suggested several differences between their situation and much of the rest of British science of the time. The biophysics unit was not an establishment crusted with history. Distinctions of rank or seniority had not grown up. The scientists were young, fresh, new at the work, and their convictions were typical of a liberated post-war generation.

Of Franklin's position, Dame Honor Fell said in conversation, "I *do* know a great deal about that unit. And I never saw a sign of sex discrimination. Of course, Franklin was a rather difficult character—and she disliked Wilkins with whom unfortunately she was working. They were *both* I think rather difficult people. But I don't believe she was discriminated against because she was a woman. I never saw a sign of it—and I'm sure I would have done." She added, "I heard a *lot*—including a lot that I've kept in a separate part of my mind and did not discuss with the director—and I never saw a sign of that."

Mary Fraser wrote to me at length from Australia. She said that in the biophysics unit "the atmosphere was very casual and you were accepted as a scientist irrespective of your sex." She went on:

> Remember that Britain was still suffering from wartime shortages and food was still rationed in 1952. . . . There was little social entertaining outside of lab hours. Thus informal coffee get-togethers were the accepted form of social gathering and after lunch someone's lab was usually selected and Sylvia Jackson usually took charge and made gallons of coffee in a huge beaker. . . .
>
> Now Rosalind Franklin arrived and I suppose we assumed she would fit into the casual role of relaxing amidst the beakers, balances, centrifuges and petri dishes—but she didn't. Rosalind didn't seem to want to mix. . . . Her manner and speech was rather brusque and everyone automatically switched-off, clammed-up and obviously never got to know her. . . .
>
> Dedicated people like Rosalind Franklin (the sex is irrelevant) great artists, scientists, writers, mountaineers, sportsmen or the Florence Nightingales have an obsession and their fellow human beings are secondary to their overriding passion.

Sylvia Jackson, in a conversation in the gardens by the river at Darwin College, said, "Randall's unit was totally non-biassed. Jean Hanson, Angela Brown, and myself were all of much the same age." She added, "Every Saturday, we used to go along to the Strand Palace Hotel"—several blocks from the lab—"for lunch. Maurice, Jean, Angela, myself," and two other King's scientists she named. "Or we would go to a pub in Epping, by car. But not Franklin. She did not basically take part in that sort of thing. Too intent."

Though I had not asked directly, Jackson offered, "My impression of Rosalind Franklin was twofold. She was absolutely dedicated, a tremendously hard worker; she strode along rather quickly; she was enormously friendly if you gave her half a chance. But I found her *formidable*."

In short, those of Franklin's colleagues at King's who were women unanimously reject the view that her troubles there arose because she was shut out as a woman. Responding decades after the events, these women know that professional women often do face barriers, systematic barriers that men do not meet. None the less, they reject as unhistoric and anachronistic the use of Rosalind Franklin as an emblem for the condition of women in science.

❖ ❖ ❖

Late on the afternoon of Friday, January 30, Franklin was in her lab off the basement corridor, examining an x-ray picture over a light box. The door opened. She looked up and saw Watson. He advanced, and handed her a copy of a typewritten manuscript, which had come to the Cavendish from Linus Pauling and which proposed a structure for DNA. Franklin looked at it. The structure had three strands coiled with the backbones inside, very tight, and the bases pointing outward. Watson said that Pauling had made an elementary error. The structure could not be correct. She, too, saw at once that it could not be right, for a reason Watson did not yet know—because the x-ray pictures Pauling had used were prewar and did not separate A and B patterns. Watson began to lecture her about helices. She found all this intensely irritating—with reason. She stepped from behind the lab bench, moved towards Watson—a vigorous, compact young woman. Watson, over six feet and gawky, snatched the typescript and ducked back to the door just as Wilkins put his head in. The ludicrous encounter ended. Franklin did not realize, that winter, and may never have learned that in the next few moments as the two men walked down the corridor Wilkins stepped into another room, brought out a photograph, and showed it to Watson. It was a print of one of the excellent x-ray patterns in the B form that Franklin had obtained nine months earlier.

Franklin had arranged to leave King's for the laboratory of the crystallographer Desmond Bernal, at Birkbeck College of the University of London, by mid-March. On February 10, her notebooks show that she took up at last the B form of DNA and the evidence that the structure was helical. She reviewed the mathematics of the patterns helices make. But then she turned away once more. She did not come back to the B structure until February 23. That day, she began with the photograph Wilkins had shown Watson three weeks earlier. She determined the diameter of the helix; she confirmed a previous conclusion that the backbones were coiled at the outside of that diameter. That afternoon and all the next day, she argued back and forth from the B form to the A, and so finally drove herself to the understanding that the A form must be helical, too. She concluded that the helix was made not of three but of two chains. In hours, she confuted errors that for nine months had held her spellbound.

Her notes on structure B stop with February 24. Several days later, she and Gosling began to draft a paper to state their conclusions about the B structure thus far. About March 12, Wilkins learned by telephone that Watson and Crick had a structure and a completed model and were writing a paper. By then, Franklin was unpacking at Birkbeck. She first heard of the structure on the 18th, and at once asked that she and Gosling publish their material on the B form simultaneously.

Once more she took the train to Cambridge. The room at the Cavendish was crowded with four desks in the corners and a table in the middle. On the table stood the model, nearly six feet tall, more than a yard wide, and mostly empty space. Its various subunits were made of thin brass rods and thinner sheet metal, welded together so that each atom was represented at the correct angle and distance from its neighbors. Such components Watson had then assembled by means of brass sleeves fitting over the rods and held by set screws—a fiddly business with a spidery, skeletal result.

Franklin immediately recognized the essential facts of the structure. They are few. The two backbones are on the outside, coiled up and down around a common axis. The double helix is 20 angstrom units in diameter (79 billionths of an inch). The backbones hook themselves together again and again, across the middle, by joining molecular subunits called bases. The pairs of bases are 3.4 angstroms apart, and the helix makes a complete turn in ten pairs, 34 angstroms. All this, Franklin saw, agreed with what she had calculated. (It could not do otherwise. Watson had learned the crucial dimensions on his brief inspection of the photograph Wilkins had shown him seven weeks earlier. His scientific imagination is intensely visual. He had understood instantly facts that Franklin had only since figured out.)

She looked more closely. The backbones of the double helix are long chains, chemically identical. But they have one peculiarity. The way their subunits string together, the chains point in opposite directions—which is to say, one runs up, the other down, and the molecule is symmetrical, looking the same when turned end for end. This fact was new to Franklin. Again, she saw it was not in conflict with her analysis. (It hardly could be. Crick had discovered that the chains ran in opposite directions when, in mid-February, he read Franklin and Gosling's section of the Medical Research Council committee's report on Randall's biophysics unit. By the coincidence of a technical similarity between crystallographic details of the B form and of the hemoglobin Crick was already working on, this had been his turn to find in Franklin's data what she had failed to see.)

She looked more closely still. The bases in DNA are of just four kinds, and each kind pairs readily with just one of the other kinds. The unique pairing of the bases is the central feature of the structure. It means that each strand is complementary to the other and that when the two unzip each can assemble on itself a duplicate of its former partner. She will have seen the genetical consequences: the sequence of the bases along a strand enciphers, as with a four-letter alphabet, the specifications the organism needs to build itself. A gene is such a sequence, a sentence in that alphabet. The geometry of the only pairs of bases that would fit within the double helix was Watson's discovery.

❖ ❖ ❖

Were Watson and Crick pirates? Say, rather, that they were uncannily lucky. They were lucky first in each other—in their complementary combination of knowledge, talents, and temperaments. They were lucky second in the extraordinary series of things that came their way. But they were more than lucky: hungrily they made their luck, with speed, wit, and guile they exploited it. Though Franklin's data was essential, Watson and then Crick had understood it as she had not. Then the final insight, fitting the bases together within the double helix: this was not forced by Franklin's material. The last sentence but one in their letter to *Nature* reads: "We have also been stimulated by a knowledge of the general nature of the unpublished experimental results and ideas of Dr. M. H. F. Wilkins, Dr. R. E. Franklin and their co-workers at King's College, London." That ambiguous shimmer perhaps sums it all.

Franklin was poignantly unlucky. She had no collaborator. It's been said that Watson was her collaborator. She was stubborn—a virtue in science but with limitations, for she was too unwilling to speculate early about the helical evidence, too set on analyzing the A form by classical mathematical means, and far too rigidly opposed to building models. She was doubly unlucky in Wilkins. Their preclusive scientific incompatibility stiffened her approach. He, shut out, had no understanding scientific auditors but Watson and Crick.

Could she have got it first? She had not perceived that the backbones run in opposite directions. She had not started modelling the B form as a double helix and so had yet even to encounter the problem of fitting the bases inside. Furthermore, she was moving. Randall, mean-spiritedly, no doubt set on by Wilkins, made her agree to wind up and publish what she had on DNA, then leave the problem behind. And yet, and still, she had been so close, two half-steps away, that she saw at once that the Watson and Crick structure was essentially correct. Watson was surprised at her gracious assent.

At Birkbeck in her new laboratory, Franklin worked with the structure of tobacco-mosaic virus, and the work went brilliantly. In Cambridge, Bragg retired from the Cavendish; the unit studying the structures of biological molecules grew fast. Discussions began about establishing the unit as a separate entity, with its own building. Watson was in the States. Crick and Franklin became friends professionally and personally. Franklin was invited to be a senior member of the Cambridge unit's scientific staff. She had cancer. She spent her last recuperation with the Cricks. She died in April of 1958 at the age of 37. The Medical Research Council Laboratory of Molecular Biology was built several years later. It has been, for a quarter century, perhaps the greatest biological laboratory in the world.

A Nobel prize is never given posthumously. None has ever been divided among more than three scientists. In 1962, the Nobel prize in physiology or medicine was awarded to Watson, Crick, and Wilkins. Had Franklin lived would she have got it? The Nobel committees have sometimes made quirky awards, omissions, and downright mistakes; but we cannot doubt that the value of her work was known. In any event, Nobel prizes are a decoration. Rosalind Franklin had a distinguished career as a scientist, cut short by early death. To say otherwise, as those must do who claim her, despite the facts, as an example of a woman blocked in her scientific career because she was a woman, diminishes what she accomplished.

Erwin Chargaff, biochemist, has been known chiefly for two things: his obser-
vation of the one-to-one ratios of the bases adenine to thymine and of guanine
to cytosine, and his bitter hatred of molecular biology. Just as the original edi-
tion of The Eighth Day of Creation *was going to press, Chargaff published a*
memoir: Heraclitean Fire: Sketches from a Life before Nature *(New York:*
Rockefeller University Press, 1978). The bitterness was there, intensified,
coarsened, yoked to a conviction that he had been denied recognition for dis-
covering the base ratios, an essential element of the structure of DNA. At the
invitation of Edward Shils, editor of the scholarly quarterly Minerva, *I*
prepared a critical survey of other historical accounts of molecular biology—
papers, several books—that had appeared by the time mine was published. The
survey included a review of Chargaff's memoir, which I have adapted for the
essay that follows.

AFTERWORD II

What did Erwin Chargaff contribute?

"Both proteins and nucleic acids were hopelessly inadequately characterized, in those days." So Max Delbrück told me, to explain the tentative response of biologists in the nineteen-forties to the paper by Oswald Avery, Colin Macleod, and Maclyn McCarty, on the identification of the transforming principle in pneumococcus as DNA. The man who responded to Avery's paper and began the correct characterization of DNA was Erwin Chargaff.

The memoirs of the biochemist Chargaff—*Heraclitean Fire*—plunge the commentator who knows the facts into an unhappy conflict of pieties. Erwin Chargaff has been, above all, a man of honor who has refused to let himself become inured to the times' decay. He has protested in countless short pieces addressed to his peers in science or on occasion to a wider audience—and in the past has done so with a fine and witty rage. We must cherish him for that. Chargaff is a man of classical education, which he has kept sharp and shining. He is a master of many languages (reading, by his count, in 15) and a master of language—a mordant stylist and a scathing polemicist. We can admire all this. Chargaff trained as a chemist in Vienna after the first world war, and earned his living at biochemistry for fifty years, forty of them at Columbia University's College of Physicians and Surgeons, eventually becoming chairman of the department there; yet he likes to say that he strayed into science, that biochemistry was but an avocation, that he has always been an outsider at the inside. In the late nineteen-forties, Chargaff made certain simple discoveries about the chemical composition of DNA. His discoveries were central—enabling—in the rise of molecular biology. He has been honoured for those discoveries—though with the parsimonious justice of a ten-per-cent tip. Yet at the time he made the discoveries he did not fully understand their consequences, and to this day he does not properly value the work he did. This compound failure of judgement—predictive, retrospective—has poisoned his relation to the world of science for a quarter of a century. Now he is old. He has published his apology for his life. It is intermittently fascinating, sometimes beautiful, but, in the end, a treason to himself. Chargaff has demeaned his gift: where he used to be funny and nasty, here the humour has left him. That lapse one might pass in silence. But he is also trying to revise history.

Chargaff, on encountering Avery's paper of 1944, was decisively moved to take up research into DNA. Methods for separating and accurately measuring tiny quantities of complex and closely similar biological substances, such as the different nucleotides, were only then becoming available. Adapting the methods to DNA took Chargaff several years. By 1949, he and his colleagues had analysed the DNAs of a variety of different organisms. They found the four different bases appearing in DNA in proportions that, although constant in all tissues of a given species, varied widely from one species to another, only in a few species approaching equal representation of the four. "The results serve to disprove the tetranucleotide hypothesis," Chargaff wrote in 1949 in a review paper published the following year. Then he added—he now informs us, only when the paper stood in proof—a brief observation about a curious possible regularity that he noticed, after all, in the proportions of the bases:

> It is, however, noteworthy—whether this is more than accidental, cannot yet be said—that in all desoxypentose nucleic acids [DNAs] examined thus far the molar [that is, molecule-to-molecule] ratios of total purines to total pyrimidines, and also of adenine to thymine and of guanine to cytosine, were not far from 1.

A young biochemist of exceptional talent said to me a while ago that Chargaff has found the beauty of science to lie in the mystery of its problems, even to the point where he could not bear to see them solved. "That's one of the most remarkable characteristics of most scientists," I was told. "At some level, eventually, they fondle their problems."

The dominant mood of Chargaff's memoir is nostalgia. The passings he laments include "the last glow of a calm, sunlit period . . ., the dying years of the Austrian monarchy," into which he was born in 1905; his mother, in images out of his childhood, seen "floating behind a screen of tears," who was "deported into nothingness from Vienna in 1943"; the distinct second-person-singular pronoun in English, "victim to a grammatical egalitarianism that has corroded the poetic core of the language." Almost above all, however, Chargaff regrets the disappearance of science and in particular chemistry "of the old observance"—that is, small-scale, inexpensive, reflective, and, indeed, reverently non-reductionist before the mysteries of life. Chargaff cuts his nostalgia with contemptuous hatreds. He excoriates the totality of those American institutions whose gigantism, competitiveness, faceless hypocrisy, and ethical insensitivity have led, among other things, to the monstrous misuse of science in atomic weapons and, he thinks, in genetic engineering. Detesting most things he cares to mention about the country that took him in, he secretes a special sac of venom for the place with which he was mostly closely associated: Columbia University treated him shabbily at the time of his retirement, refusing to endorse new grant applications, changing the locks on his old laboratory, stranding him with a pension of thirty per cent of his salary. But worst of all he scorns the science he unintentionally helped to found—molecular biology. And where his other hatreds, arresting enough in detail and in phrasing, do not these days seem so original, here lie the pathetic self-betrayals of his book.

The central episode in Chargaff's memoir, by all evidence a psychological pivot of his life—the place where wit now fails him and where he is conjuring the past to be different—was his meeting, in Cambridge in the last week of May of 1952, two younger scientists interested in DNA: Francis Crick and James Watson. The two were then virtually unknown. Chargaff, on the other hand, appeared in the ascendant. Columbia made him a full professor that year; he was invited to lecture that summer at the Weizmann Institute, in Israel, and in several European cities, and was to read a paper about DNA at the Second International Congress of Biochemistry, meeting in Paris; he had hopes of escaping Columbia and the United States to a chair in Switzerland (though this came to naught). Crick had never heard of Chargaff before his colleague Kendrew arranged their meeting; Watson has written that he had previously read of Chargaff's work. Chargaff knew more about the chemistry of DNA than either Watson or Crick did then, and he told them of his findings. Nine months later, Watson and Crick solved the structure of the DNA molecule. The double helix turned out to have immense significance in explaining how genes function—and to have Chargaff's one-to-one ratios built in. The two strands of the structure, their backbones curling up the outside, barber-pole fashion, have their bases projecting inwards—and the whole thing is held together by chemical bonds between the bases, in which adenine on one strand always pairs with thymine on the other, and guanine always pairs with cytosine. Thus, Watson and Crick's structure of DNA immediately solved the mystery of how the gene makes identical copies of itself: the two strands separate, and each one forms a new, complementary second strand on itself, the nucleotides of the new strand taking position according to the pairing rules, A to T or T to A, and G to C or C to G.

In 1972, in the course of a conversation, I asked Chargaff whether he had perceived the consequences of his discovery of the one-to-one ratios of adenine to thymine and guanine to cytosine. "Yes and no," he answered then. "No, I did not construct a double helix." Priority is the only form of property in intellectual discoveries (as Sir Peter Medawar has pointed out), so it is silly to suppose that scientists ought not be concerned with questions of priority. Now Chargaff describes his encounter with Crick and Watson in May of 1952 in three pages placed at the exact middle of his memoir. "The first impression was indeed far from favorable," he writes, and goes on:

> I seem to have missed the shiver of recognition of a historical moment: a change in the rhythm of the heartbeats of biology. . . . The impression: one [Crick], thirty-five years old; the looks of a fading racing tout, something out of Hogarth ("The Rake's Progress"); Cruikshank, Daumier; an incessant falsetto, with occasional nuggets glittering in the turbid stream of prattle. The other [Watson], quite undeveloped at twenty-three, a grin, more sly than sheepish; saying little, nothing of consequence.

But then Chargaff writes:

> I told them all I knew. If they had heard before about the pairing rules, they concealed it. But as they did not seem to know much about anything, I was not unduly surprised. I mentioned our early attempts to explain the complementarity relationships by the assumption that, in the nucleic acid chain, adenylic was always next to thymidylic acid and cytidylic next to guanylic acid [these are the names of the nucleotides carrying the various bases]. . . . I believe that the double-stranded model of DNA came about as a consequence of our conversation.

And he goes on:

> When, in 1953, Watson and Crick published their first note on the double helix, they did not acknowledge my help and cited only a short paper of ours which had appeared in 1952 shortly before theirs, but not, as would have been natural, my 1950 or 1951 reviews.

Indeed, though Chargaff does not say so in his memoir, some months after Watson and Crick's paper he wrote a letter chiding Crick for not having cited his work adequately. Discovery of the structure of DNA touched off a widespread search for the means by which the genetic information is read off the DNA and translated to specify the building of the organism—that is, the search for the genetic code and for the biochemistry by which the cell makes proteins. Chargaff, though he does not say this, either, in his memoir, at first tried to take part in the elucidation of the genetic code. By the mid-nineteen-fifties, though, Crick was to a great degree dominating the direction and style of this effort—an *arbiter* and an *elegantia* that Chargaff found intolerable. By 1958, Chargaff was denouncing molecular biology and its practitioners for arrogance, ignorance, reductionism, and self-serving sensationalism. In 1962, the Nobel prize in physiology or medicine was awarded to Watson and Crick, together with Maurice Wilkins.

Whatever the basis and justification, item by item, of Chargaff's dislikes in present-day science and among its practitioners and institutions, it is the case that for a decade at least his attacks, at their most amusing and their most telling—and even in dead earnest on such a possibly important issue as the regulation of experiments that modify inheritance by moving fragments of DNA from one organism to another—have been discounted by many of his peers as impelled by enmity and disappointment. Enmity? At one time, Chargaff could write of molecular biologists, "That in our day such pygmies throw such giant shadows only shows how late in the day it has become" —an oblique personal reference and a nice inversion of the established conceit that has scientists seeing farther because perched on the shoulders of their giant predecessors. Now he gives us "the looks of a fading racing tout"—and the reader must wince with embarrassment. Disappointment? Chargaff's claim for credit is now broader and more insistent than before. He thus invites scrutiny. "Such things are only susceptible of a later judgement: *Quando judex est venturus / Cuncta stricte discussurus!*", he writes; but we need not postpone scrutiny so long, for several things can be said with assurance about how Chargaff understood his discoveries at the time he met Watson and Crick, and about the use they made of them.

Chargaff now writes, as we just saw, that he told Watson and Crick in 1952 "about the pairing rules" and "the complementarity relationships." Eight pages earlier in his memoir, writing of the experiments of 1948 and 1949 and the paper of 1950 based on them, he speaks of "the regularities that I then used to call the complementarity relationships and that are now known as base-pairing." However, in that review of 1950 (I quoted the full relevant passage above), Chargaff wrote, "The molar ratios" —molecule to molecule— "of adenine to thymine and of guanine to cytosine, were not far from 1." In a review Chargaff published in 1951, he gave more evidence but said again, "The ratios of adenine to thymine and of guanine to cytosine were near 1." That observation, in slightly varying language, in fact became his standard.

Once, in a paper of 1951 about the DNA of salmon sperm, he speculated about the reason, then turned away:

> Not only the ratio of purines to pyrimidines but also that of adenine to thymine and of guanine to cytosine equals 1. As the number of examples of such regularity increases, the question will become pertinent whether it is merely accidental or whether it is an expression of certain structural principles that are shared by many desoxypentose nucleic acids, despite far reaching differences in their individual composition and the absence of a recognisable periodicity in their nucleotide sequence. It is believed that the time has not yet come to attempt an answer.

And once, at about the same time, he faced the problem of the structural significance of the base ratios straight on—and came to a conclusion that now seems astonishing. In a major review that he published in 1951 but that he did not choose to reprint, 20 years later, in his collected major papers, he wrote:

> If one postulated an ideal case in which a desoxypentose nucleic acid exhibited ratios of adenine to guanine and of thymine to cytosine that were both 1·4 and ratios of adenine to thymine, of guanine to cytosine, and of purines to pyrimidines, all equalling one, a simple construction could, for instance, assume that a subunit consisting of 24 nucleotides contained 7 dinucleotides, in which adenylic acid was linked to thymidylic acid, and 5 dinucleotides, in which guanylic and cytidylic acids were united, all these distributed in a certain pattern. The experimental results have, however, disproved this simplified assumption.

The trouble was that Chargaff considered DNA only as single-stranded, never seriously wondering whether the molecules could be duplex.

Indeed, nowhere in any paper that Chargaff published before Watson and Crick announced their structure does he refer to the one-to-one ratios as "complementary" or exhibiting "complementarity," or as "base-pairing" or as being "pairing rules." Chargaff's correspondence for that period is deposited at the Survey of Sources for the History of Biochemistry and Molecular Biology, at the American Philosophical Society Library, in Philadelphia. Though the correspondence itself is still closed, in 1977 the survey issued a thorough analysis and index of the Chargaff papers. The analysis describes Chargaff's letters to Crick, after the DNA structure, as "confined . . . to a scolding for Crick's failure to cite Chargaff's base-ratio principle in relation to the specific base pairing stereochemical interpretation advanced by Watson and Crick."

Surely, that catches the point: base-pairing and complementarity are notions that express the structural, three-dimensional conclusion that Chargaff never reached. In particular, they are terms that apply naturally to a two-stranded structure, with chemical bonds between paired bases and a method of replication, of the general sort that Watson and Crick proposed. Watson and Crick got there by model-building, the power of which they learned from Linus Pauling.

Fortunately, we are able to determine more closely what Chargaff actually had in mind about the structure of DNA when he first met Watson and Crick. By then, Chargaff's laboratory had turned still further away from any conception of the structure of DNA to which the notion of complementarity could be applied, or that could have suggested a physical, local, "base-pairing" consequence to the one-to-one ratios. At the Congress of Biochemistry in Paris that July, Chargaff read a paper that reviewed the published work and mentioned the base ratios in the same language as before. But Chargaff's name was on another paper presented in Paris, as co-author with a young colleague at Columbia, Christoff Tamm. An abstract of this was published in the proceedings of the congress, and the full results of the work were put into a longer paper by Chargaff, Tamm, and two others, and sent off to the *Journal of Biological Chemistry* that December. Chargaff and his associates were taking tentative first steps in a problem of heroic difficulty—to find a way to read out the specific sequence of nucleotides, the genetic message, in DNA. They had begun by controlled chemical degradation of DNA, using a mild acid that stripped the purines (adenine and guanine) off the strands, leaving the backbones with the pyrimidine bases (thymine and cytosine) still attached. This substance they then treated in further ways that told them whether pyrimidines were significantly bunched. Tamm said in Paris, and the group's paper repeated, "The structure of DNA that emerges from these experiments is that of a chain in which tracts of pyrimidine nucleotides alternate with stretches in which purine nucleotides predominate." Thus, the one-to-one ratios did not apply at each location along the molecule: Chargaff had abandoned any literal, local pairing of bases, and any meaning for his base ratios except a general, statistical one. If he told Watson and Crick all he knew, he misled them.

Of course, the resolution emerges automatically from the Watson-Crick model, for *that* DNA had two strands—which Chargaff's chemical treatments were disrupting. And yet he now writes, "I believe that the double-stranded model of DNA came about as a consequence of our conversation." Fortunately, we have authoritative, independent evidence of what Watson and Crick had on their minds in the five or six months before they got the structure out. In the autumn of 1952, Jerry Donohue, a physical chemist and crystallographer trained at the California Institute of Technology under Linus Pauling, came to the Cavendish Laboratory for a postdoctoral year. Donohue was assigned a desk in the same room as Watson and Crick. Donohue is no particular partisan of theirs. He is a scientist of prickly individuality and an almost belligerent insistence on accuracy of detail. At a critical moment in February of 1953, Donohue corrected Watson and Crick about the structure of individual bases, which they had got wrong because much of the published literature showed them wrong. Donohue's correction enabled Watson and Crick to fit the bases into the double helix. Later in the nineteen-fifties, Donohue raised the point that other structures of DNA might also be compatible with the crystallographic evidence, implying that Watson and Crick's canvassing of possible structures had been hasty and incomplete; Donohue and Gunther Stent even proposed an alternative model. Donohue had a running argument with Crick about that, which ended only in 1970 with Crick's challenging him, in a letter in *Science*, to "build such a model and publish the coordinates." In 1978, Donohue wrote a review of Chargaff's memoir for *Nature*. He found the memoir, on reflection and overall, "delightfully enthralling." But about Chargaff's belief that the double-stranded model of DNA came about as a result of the conversation with Watson and Crick in May of 1952, Donohue wrote this:

> Now, there are only a very few people who are in a position to know whether or not this belief of his is true. I am one of them. I categorically state that it simply *is not* true. Shortly afterward, when I came on the scene, at such times as they were

thinking about DNA, they worried more about the density, the pitch of the helix; pairing, if it came up, was "like with like", not complementarity. They were not, in fact, even using the correct chemical structures of the bases. When the final model of DNA was discovered—more or less by accident—it wasn't Chargaff's rules that made the model, but the model that made the rules.

Just so: a model may be a kind of theory, and Watson and Crick's model of DNA, with the physical pairing of the bases that they discovered, transformed what had been an unexplained anomaly, the base ratios, into one of the crucial structural facts about the gene. (The celebrated parallel is the anomalous perihelion of Mercury. It had been certain since 1845 that the orbit of Mercury was walking around the sun so that its perihelion, or point of closest approach to the sun, was advancing more than Newtonian dynamics could in any way predict, by the tiny amount of 43 seconds of arc per century. But in 1916, when Einstein put forward the general theory of relativity, the physicist Karl Schwarzschild fitted the orbit of Mercury into Einstein's equations—and found that the 43 seconds of arc were exactly accounted for. An unexplained observational anomaly was dramatically elevated into a necessary consequence.)

Chargaff's brooding on his bitterness has so fixated his attention that he has failed to understand the true magnitude of what he did accomplish. He, along with most other participants and observers, has been misled by the standard view of molecular biology—by its attention to the change that occurred from the conviction, in the late nineteen-thirties, that genetic information was carried by protein to the knowledge that it is carried by DNA. That change did take place, as we have amply reviewed, but consideration of it has obscured the more fundamental transformation that was going on at that same time in the way that the large molecules of the cell—nucleic acids and proteins alike—were understood.

Where Frederick Sanger's sequencing of insulin in the late nineteen-forties and early fifties established that the sequence of amino acids in a polypeptide chain is unique and yet predictable by no chemical rule, Chargaff at the same time, though he never sequenced any DNA, demonstrated by destroying the tetranucleotide hypothesis that DNA is free to be specific in sequence in the same way that proteins are. That is the discovery that made molecular biology possible and for which we must honor Erwin Chargaff.

Notes

The list of interviews and conversations is followed by a list of short titles of books and articles referred to frequently and of journals whose titles are abbreviated in the notes; for the ease of the general reader, and of the scholar who is not an historian of science, the titles of most journals are given in full in the notes. The notes themselves then follow; they are keyed to page numbers and to tag phrases.

Interviews and Conversations

Sidney Altman, December 1970, Medical Research Council Laboratory of Molecular Biology, Cambridge, England (hereafter MRC Laboratory, Cambridge)

Bruce Ames, 10 August 1978, Department of Biochemistry, University of California, Berkeley; 7 May 1978, by telephone to his house in Berkeley

Alice Audureau, 22 September 1976, by telephone to her house in La Baule, France

David Baltimore, 12 February 1971, Massachusetts Institute of Technology

Leslie Barnett, January 1971, Cavendish Laboratory, Cambridge

Samuel Beckett, 18 May 1976, Royal Court Theatre and adjoining pub, London

Seymour Benzer, 25 February 1971, California Institute of Technology; 30 April and 1 May 1978, by telephone

Paul Berg, 6 September 1974, by the Cam, Cambridge; 18 September 1976, Peterhouse, Cambridge; 9 August 1978, Department of Biochemistry, Stanford University

David Blow, January 1971, MRC Laboratory, Cambridge

Sir Lawrence Bragg, 28 January 1971, his apartment, The Boltons, London

Sydney Brenner, 8 May 1968, MRC Laboratory, Cambridge; with Crick, September 1970, MRC Laboratory, Cambridge; 26 January 1971, King's College, Cambridge; talk, 8 June 1972, Institute for Advanced Study, Princeton; 13 March 1975, King's College, Cambridge; with Crick, 19 November 1975, MRC Laboratory, Cambridge; 7 February 1976, 28 February 1976, March 1977, 14 May 1977 and 21 March 1978, MRC Laboratory, Cambridge

Angela Martin Brown, 2 September 1976, King's College, London

John Cairns, April 1974, Imperial Cancer Research Fund Laboratory, Mill Hill, London

Torbjörn Caspersson, December 1976, by telephone to Stockholm

Erwin Chargaff, 15 February 1972, College of Physicians and Surgeons, Columbia University, New York

Georges Cohen, 13 November 1976, Institut Pasteur, Paris

Melvin Cohn, March 1971, cafeteria breakfast, New York City

Pauline Cowan: see Pauline Cowan Harrison

Heinrich Cramer, 22 April 1976, Max-Planck-Institut für Experimentelle Medizin, Göttingen, West Germany

Francis Crick, September 1970 and 1 September 1971, MRC Laboratory, Cambridge; 3 July 1975, Middlesex Hospital, London; 10 September 1975 and 19 and 20 November 1975, MRC Laboratory, Cambridge; informal conversation, 10 July 1976, Swaffham Bulbeck, Cambridgeshire; interview, 28 June 1977, Cold Spring Harbor Laboratory, Cold Spring Harbor, Long Island

Odile Crick, informal conversation, 10 July 1976, Swaffham Bulbeck; interview, 24 August 1976, the Cricks' house, Portugal Place, Cambridge

Cedric Davern, conversation, 31 May 1978, Cold Spring Harbor Laboratory

Max Delbrück, 25 February 1971, his office at the California Institute of Technology, Pasadena, California; 9 July 1972, Cold Spring Harbor Laboratory

Jerry Donohue, summer 1974, his office, University of Pennsylvania

René Dubos, 24 April 1978, by telephone to Rockefeller University, New York City

Renato Dulbecco, 23 February 1971, Salk Institute, La Jolla, California (with assistant); 19 March 1976, Imperial Cancer Research Fund Laboratory, Lincoln's Inn Fields, London

Manfred Eigen, 12 November 1975, lecture at Cambridge; 21 and 24 April 1976, Max-Planck-Institut für Biophysikalische Chemie, Göttingen

Vladimir Alexandrovich Engelhardt, 16 February 1977, Fellows' Room of the Royal Society, London

Boris Ephrussi, 5 December 1975 and 13 November 1976, Centre National de la Recherche Scientifique, Gif-sur-Yvette

Claudine Escoffier-Lambiotte, 14 November 1976, her apartment, rue Murillo, Paris

Dame Honor Fell, 28 January 1977, Addenbrooks Hospital, Cambridge

Bertrand Fourcade, April 1971, Travellers' Club, Paris

Heinz Fraenkel-Conrat, 10 August 1978, Virus Laboratory, University of California, Berkeley

Alan Garen, 23 January 1977, Algonquin Hotel lobby, New York City

Walter Gilbert, 13 February 1971, Biological Laboratories, Harvard; 14 February 1971, his house, Upland Road, Cambridge, Massachusetts; October 1973, Biological Laboratories, Harvard

Ray Gosling, 21 July 1975, Guy's Hospital, London

François Gros, 12 October 1971, Institut de la Biologie Moléculaire, Faculté des Sciences, University of Paris; 15 November 1976, Institut Pasteur, Paris

Marianne Grunberg-Manago, 16 November 1976, Institut de Biologie Physico-Chimique, Paris

Roger Guillemin, 22 February 1971, Salk Institute, La Jolla, California

Pauline Cowan Harrison, 22 January 1977, by telephone to the University of Sheffield, England

Felix Haurowitz, conversation, 1 June 1978, New York Academy of Sciences symposium at the Barbizon Plaza Hotel; 19 January 1972, by telephone to Bloomington, Indiana

William Hayes, 5 October 1976, Royal Society, London

Leon Heppel, 26 July 1977, by telephone to Cornell University, Ithaca, New York

Mahlon Hoagland, 9 August 1977, Rockland, Maine

Dorothy Crowfoot Hodgkin, 8 January 1971, her laboratory at the University of Oxford

David Hogness, 16 February 1977, Royal Society, London

Robert Holley, 22 February 1971, Salk Institute, La Jolla, California

Bernard Horecker, 27 February 1978, his apartment on East 64th Street, New York City

Rollin Hotchkiss, 12 June 1978 and 18 July 1978, Rockefeller University, New York City

Hugh Huxley, 19 February 1971, MRC Laboratory, Cambridge

Vernon Ingram, March 1971 and 11 August 1977, Massachusetts Institute of Technology

François Jacob, September 1970, 9 December 1970, and 3 December 1975, Institut Pasteur, Paris; 28 and 29 May 1977, his apartment, rue Guyamer, Paris; 5 May 1978, Dana Palmer House, Harvard

Sylvia Fitton Jackson, 30 June 1976, Darwin College, Cambridge

Eugène Jungelson, 17 November 1976 and 27 May 1977, his office, Boulevard St. Germain, Paris

Herman Kalckar, September 1973, Harvard Medical School

John Kendrew, 13 and 14 January 1971, MRC Laboratory, Cambridge; 11 November 1975, office of *The Journal of Molecular Biology,* Cambridge

Richard D. Keynes, 9 February 1976, Physiology Laboratories and Graduate Student Center, Cambridge

Aaron Klug, August 1972, 6 June 1975, 27 June 1975, October 1975, and 19 January 1976, MRC Laboratory, Cambridge

Arthur Kornberg, 11 August 1978, Department of Biochemistry, Stanford University

Roger Kornberg, conversation, 19 November 1974; interview, 12 September 1975, Cosin Court, Peterhouse, Cambridge; 11 August 1978, Stanford University Medical School

Joshua Lederberg, 1 August 1977, by telephone to Stanford University; interview, 9 August 1978, Stanford University

Edwin Lennox, second week in February 1971, Christ Cella's, New York City; 7 February 1978, MRC Laboratory, Cambridge

Leonard Lerman, 4 August 1977, by telephone to the Department of Biological Sciences, State University of New York, Albany, New York

Anne Levine, 21 September 1978, New York City

Fritz Lipmann, 19 March 1971 and 7 July 1977, Rockefeller University, New York City

William Lipscomb, 20 June 1978, his office, Harvard University

Salvador Luria, 17 March 1971, Massachusetts Institute of Technology; conversation, 3 October 1973, Lexington, Massachusetts; interview, 4 October 1973, Massachusetts Institute of Technology

Vittorio Luzzati, 2 December 1975, Centre National de la Recherche Scientifique, Gif-sur-Yvette

André Lwoff, 3 December 1975, Institut Pasteur, Paris; 15 November 1976, lunch at Café de l'Institut, Paris; 13 June 1977, by telephone

Werner Maas, 1 March 1978, by telephone to his office at the New York University Medical Center

Ole Maaløe, 1 April 1977, by telephone to Copenhagen

John Maddox, December 1970, editorial office of *Nature*, Little Essex Street, London

Angela Martin: see Angela Martin Brown

Heinrich Matthaei, 22 April 1976, Max-Planck-Institut für Experimentelle Medizin, Göttingen

Barbara McClintock, June 1973 and 30 July 1978, Cold Spring Harbor Laboratory, Long Island

Zhores A. Medvedev, 16 February 1976, National Institute for Medical Research, Mill Hill, London

Janet Mertz, 25 November 1976, MRC Laboratory, Cambridge

Matthew Meselson, March 1971 and 3 October 1971, Lexington, Massachusetts; 28 August 1975, Boston Airport; 8 August 1977, his house in Kirkland Place, Cambridge, Massachusetts

Avrian Mitchison, 30 December 1970, London

Jacques Monod, September 1970, 9 December 1970, and 5 October 1971, Institut Pasteur, Paris; 1 December 1975 and 4 December 1975, his apartment, Avenue de la Bourdonnais, Paris

Olivier Monod, 13 November 1976, his apartment, rue de la Grande Chaumière, Paris

Phillipe Monod, 15 November 1976, and by telephone 16 November 1976, his apartment, rue de la Tombe Issoir, Paris

Marshall Nirenberg, 11 March 1971 and 25 July 1977, National Institutes of Health, Bethesda, Maryland

Margaret Pratt North, 21 January 1977, by telephone to Leeds; letter, 2 February 1977

Geneviève Noufflard, 16 November 1976, her apartment, 61 rue de Varennes, Paris

Severo Ochoa, 21 April 1978, by telephone to his apartment in New York City; interview, 14 August 1978, New York University Club

Arthur Pardee, 23 February 1978, his house in Brookline, Massachusetts

Linus Pauling, 1 March 1971, Stanford University; 23 December 1975, Peter Pauling's London apartment

Peter Pauling, 1 February 1970, laboratory at University College, London

David Perrin, 15 November 1976, Institut Pasteur, Paris

Max Perutz, 8 May 1968, MRC Laboratory, Cambridge; 29 and 30 May 1968, Royal Society, London; 3 April 1970, train from London to Cambridge; August 1970, his house, Sedley Taylor Road, Cambridge; September 1970, MRC Laboratory, Cambridge; 4, 5, and 6 December 1970, his house; 5 January 1971 and 23 January 1971, MRC Laboratory, Cambridge; 2 October 1971, with Adrienne Weil, Paris; April 1972, seminar series and interview at Cornell University; 18 December 1974, conversation, MRC Laboratory, Cambridge; interview, 15 February 1975, MRC Laboratory, Cambridge; 23 May 1975, conversation, Peterhouse and old Cavendish Laboratory, Cambridge; 19 January 1976, lunch, MRC Laboratory, Cambridge; 23 January 1976, lunch with Klug, MRC Laboratory, Cambridge; 31 January 1976, 7 February 1976, 11 November 1977, 19 November 1977, and 3 February 1978, MRC Laboratory, Cambridge

David Phillips, 8 January 1971, his laboratory at the University of Oxford

Robert Pollack, 3 June 1973 and 27 June 1973, Cold Spring Harbor Laboratory; 30 July 1975, editorial office of *The New Yorker*, New York City

Margaret Pratt: see Margaret Pratt North

Mark Ptashne, 10 August 1977, and checking interview, 6 May 1978, Biological Laboratories, Harvard

Alexander Rich, 12 February 1971, 17 March 1971, office at Massachusetts Institute of Technology;

❖

August 1975, by telephone; 12 February 1977 and 9 August 1977, office at Massachusetts Institute of Technology

Stanfield Rogers, 6 September 1971, Paris

Warren Ruderman, 27 April 1978, by telephone to Interactive Radiation, Inc., New Jersey

Andrei Sakharov, 16 March 1978, his apartment in Moscow

Frederick Sanger, 6 January 1971, 22 January 1971, and 2 February 1976, MRC Laboratory, Cambridge

Anne Sayre, June 1972, her house, St. James, Long Island

Howard Schachman, conversation, 4 April 1978, MRC Laboratory, Cambridge

Maxine Singer, 25 July 1977, National Institutes of Health, Bethesda, Maryland

S. Jonathan Singer, 23 February 1971, University of California, San Diego

John Smith, January 1977, MRC Laboratory, Cambridge: with Brenner, 28 February 1976, MRC Laboratory, Cambridge

Franklin W. Stahl, 4 June 1973, Cold Spring Harbor Laboratory

Roger Stanier, 26 May 1977, Institut Pasteur, Paris

Gunther Stent, 24 September 1977, at his brother's apartment, Maida Vale, London

Michael Stoker, 28 January 1971, Imperial Cancer Research Fund Laboratory, Lincoln's Inn Fields, London

Alexander Stokes, 11 August 1976, his house at Welwyn Garden City

Hewson Swift, October 1973, his house in Hyde Park, Chicago

Andrew Gabriel Szent-Gyorgi, 12 May 1978, by telephone to Brandeis University, Waltham, Massachusetts

Heinz Georg Terheggen, 23 October 1971 and 2 February 1972, by telephone to Cologne

Charles Thomas, 6 October 1973, Harvard Medical School

Alfred Tissières, 16 February 1977, Royal Society, London

Alexander Lord Todd, 19 July 1976, Christ's College, Cambridge

Andrew Travers, January 1971, MRC Laboratory, Cambridge

Agnès Ullman, 16 November 1976, Institut Pasteur, Paris

James Watson, 1 December 1970, Imperial Cancer Research Fund, Lincoln's Inn Fields, London; 16 March 1971 and 24 and 25 September 1973, Biological Laboratories, Harvard; 30 September 1973, on plane to Washington; three lectures, fall 1973, Science Center, Harvard University; 5 October 1973, Biological Laboratories, Harvard

Klaus Weber, 22 April 1976, Max-Planck-Institut für Biologische Physik, Göttingen

Adrienne Weil, conversation with Perutz, 2 October 1971, Paris

Maurice Wilkins, 26 June 1971, Versailles and Paris; 15 September 1975 and 12 March 1976, King's College London; September 1976, Lansdowne Club, London; 24 November 1978, by telephone to his office

Élie Wollman, 15 November 1976, Institut Pasteur, Paris; 17 June 1977 and 5 July 1978, by telephone to Institut Pasteur, Paris

Women at King's College London: see Angela Martin Brown, Dame Honor Fell, Pauline Cowan Harrison, Sylvia Fitton Jackson, and Margaret Pratt North; also correspondence with Mary Fraser and Marjorie M'Ewen

Gerard Wyatt, 26 April and 29 April 1978, by telephone to his house in Kingston, Ontario

Charles Yanofsky, 1 March 1971, Stanford University

Paul Zamecnik, 5 August 1977, Massachusetts General Hospital, Boston; 5 May 1978, by telephone to Massachusetts General Hospital; conversation, 1 June 1978, at New York Academy of Sciences symposium, Barbizon Plaza Hotel, New York

Norton Zinder, 1 August 1977, Rockefeller University, New York

Abbreviations for Published Works Frequently Cited in the Notes

Barcroft Memorial, Haemoglobin—F. J. W. Roughton and J. C. Kendrew, eds., *Haemoglobin: A Symposium Based on a Conference Held at Cambridge in June 1948 in Memory of Sir Joseph Barcroft* (London: Butterworths, 1949).

Bragg, *Development of X-ray Analysis*—Sir Lawrence Bragg, *The Development of X-ray Analysis*, ed. David C. Phillips and Henry Lipson (London: G. Bell and Sons, 1975).

Bragg, "History"—Sir Lawrence Bragg, "The History of X-ray Analysis," *Contemporary Physics* 6 (February 1965): 161–171.

Bragg, "How Proteins Were Not Solved"—Sir Lawrence Bragg, "First Stages in the Analysis of Proteins," *Reports of Progress in Physics* 28 (1965): 1–16; the short title was Bragg's own when he delivered the paper before the X-ray Analysis Group on 15 November 1963.

Brookhaven Symposia 12, 1959—*Structure and Function of Genetic Elements: Report of Symposium held June 1–3, 1959 (Upton, N.Y.: Brookhaven National Laboratory, 1959).*

C. R. Acad. Sci.—*Comptes rendus Hebdomadaires des séances de l'Académie des Sciences.*

C.S.H. Symp. Q. Biol.—*Cold Spring Harbor Symposia on Quantitative Biology*

Chargaff, *Essays*—Erwin Chargaff, *Essays on Nucleic Acids* (Amsterdam: Elsevier Publishing, 1963).

Chargaff, "Grammar"—Erwin Chargaff, "Preface to a Grammar of Biology," *Science* 172 (14 May 1971).

Chargaff, "What Really Is DNA?"—Erwin Chargaff, "What Really Is DNA? Remarks on the Changing Aspects of a Scientific Concept," *Progress in Nucleic Acid Research* 8 (1968).

Ciba Symposium on Viruses, 1956—G. E. W. Wolstenholme and Elaine C. P. Millar, eds., *Ciba Foundation Symposium on the Nature of Viruses* (held in London, 26–28 March 1956) (London: Churchill, 1957).

Daedalus, Fall 1970—Gerald Holton and Stephen R. Graubard, eds., *The Making of Modern Science: Biographical Studies, Daedalus,* Fall 1970.

The Double Helix—James D. Watson, *The Double Helix: A Personal Account of the Discovery of the Structure of DNA* (New York: Atheneum, 1968).

Dubos, *The Professor*—René J. Dubos, *The Professor, the Institute, and DNA* (New York: The Rockefeller University Press, 1976).

Edsall, "Blood and Hemoglobin"—John T. Edsall, "Blood and Hemoglobin: The Evolution of Knowledge of Functional Adaptation in a Biochemical System," *Journal of the History of Biology* 5 (Fall 1972): 205–257.

Ford Symposium on Enzymes, 1955—Oliver H. Gaebler, ed., *Enzymes: Units of Biological Structure and Function,* Henry Ford Hospital International Symposium (held in Detroit, Mich., 1–3 November 1955) (New York: Academic Press, 1956).

McCollum-Pratt Symposium 1956—William D. McElroy and Bentley Glass, eds., *A Symposium on the Chemical Basis of Heredity,* sponsored by the McCollum-Pratt Institute of the Johns Hopkins University . . .(held 19–22 June 1956) (Baltimore: The Johns Hopkins Press, 1957).

Oak Ridge Symposium, April 1955—Oak Ridge National Laboratory, *Symposium on Structure of Enzymes and Proteins* . . . Gatlinburg, Tenn., 4–6 April 1955, published as *Journal of Cellular and Comparative Physiology* 47, supplement 1 (1956).

Of Microbes and Life–Jacques Monod and Ernest Borek, eds., *Les Microbes et la vie: Of Microbes and Life* (New York: Columbia University Press, 1971).

Olby, *Path*—Robert Olby, *The Path to the Double Helix* (London: Macmillan, 1974).

Perutz, *Proteins and Nucleic Acids*—Max F. Perutz, *Proteins and Nucleic Acids: Structure and Function,* Eighth Weizmann Memorial Lecture Series, April 1961 (Amsterdam: Elsevier Publishing, 1962).

Phage and . . . Molecular Biology—John Cairns, Gunther S. Stent, and James D. Watson, eds., *Phage and the Origins of Molecular Biology* (Cold Spring Harbor, Long Island: Cold Spring Harbor Laboratory of Molecular Biology, 1966).

Proc. N. A. S.—*Proceedings of the National Academy of Sciences* of the U.S.

Les Prix Nobel en . . .—*Les Prix Nobel en* 1962, etc. (Stockholm: Imprimerie Royale P. Norstedt & Söner); 1995 (Stockholm: Almqvist & Wiksell International, in press).

Royaumont Colloquium 1952—André Lwoff, ed., *Le Bactériophage: Premier Colloque International,* Royaumont 1952. *Annales de l'Institut Pasteur,* Tirages special, 84 (January 1953): 1–318.

Schrödinger, *What Is Life?*—Erwin Schrödinger, *What Is Life? The Physical Aspect of the Living Cell,* based on lectures delivered . . . at Trinity College, Dublin, in February 1943 (Cambridge: At the University Press, 1944).

Stent, *Bacterial Viruses*—Gunther S. Stent, *Molecular Biology of Bacterial Viruses* (San Francisco: W. H. Freeman, 1963).

Stent, *Molecular Genetics*—Gunther S. Stent, *Molecular Genetics: An Introductory Narrative* (San Francisco: W. H. Freeman, 1971).

Structural Chemistry and Molecular Biology—Alexander Rich and Norman Davidson, eds., *Structural Chemistry and Molecular Biology: A Volume Dedicated to Linus Pauling by His Friends* (San Francisco and London: Freeman, 1968).

Symp. Soc. Exp. Biol. 1, held 1946—Symposia of the Society for Experimental Biology 1, *Nucleic Acid* (held in Cambridge, England, July 1946) (Cambridge: At the University Press, 1947).

◆

Symp. Soc. Exp. Biol. 12, held 1957—Symposia of the Society for Experimental Biology 12, *The Biological Replication of Macromolecules* (held in September 1957) (Cambridge: At the University Press, 1958).

Watson, *Molecular Biology*—James D. Watson, *Molecular Biology of the Gene*, 2nd ed. (New York: W.A. Benjamin, 1970). Where I quote the first edition, the reference is given in full.

Zamecnik, Harvey Lecture—Paul C. Zamecnik, "Historical and Current Aspects of the Problem of Protein Synthesis" (lecture delivered 21 May 1959), *Harvey Lectures* 54 (1960): 256–281.

Zamecnik, "Historical Account—Paul C. Zamecnik, "An Historical Account of Protein Synthesis, with Current Overtones—A Personalized View," *C. S. H. Symp. Q. Biol.* 34 (1969): 1–16.

1 *"He was a very remarkable fellow. Even more odd then, than later."*

PAGE

3   Perutz—The conversation took place on 29 May 1968.

3   "modest mood"—*The Double Helix*, p. 7.

4   "underemployed"—Interviews, Watson, 1 December 1970 and 16 March 1971.

4   Nobel photo—*Svenska Dagbladet*, 11 December 1962, p. 1.

5   Watson, "all the metabolic"—Watson, *Molecular Biology*, p. 99.

5   Crick—Interview, Crick, September 1970.

5   Rich, "origin of life"—Interview, Rich, 17 March 1971.

6   Benzer—Interview, Benzer, 25 February 1971.

6   Brenner—Interview, Brenner, 26 January 1971.

6   Monod—Interview, Monod, September 1970.

6   "most famous event"—*The Double Helix*, p. 222.

7   Perutz, difficulty of structures—Interview, Perutz, 8 March 1968.

7   Perutz—Interview, 8 March 1968.

8   Chargaff, "outsider"—Erwin Chargaff, "A Fever of Reason," *Annual Review of Biochemistry* 44 (1975).

9   Wilkins, "Midas' gold"—Interview, Wilkins, 26 June 1971.

10   Crick on Watson's book—Francis Crick, "Foreword" to Olby, *Path*, p. v; see also Francis Crick, "The Double Helix: A Personal View," *Nature* 248 (26 April 1974): 768.

11   Miescher—Miescher's biographer was his uncle, Wilhelm His, and the biography appears in Miescher's collected works, *Die histochemischen und physiologischen Arbeiten von Friedrich Miescher, gesammelt und herausgegeben von seinen Freunden*, 2 vols. (Leipzig: F. C. W. Vogel, 1897), vol. 1, pp. 5–32. Accounts based on that biography, and going on to the later history of nucleic acids, include Alfred Mirsky, "The Discovery of DNA," *Scientific American*, June 1968, pp. 78–84; J. N. Davidson, "Introduction" to *The Nucleic Acids*, vol. 1 (New York: Academic Press, 1955), pp. 1–2; Chargaff, "Grammar," pp. 637, 638; Chargaff, "What Really Is DNA?" pp. 300–303.

11   a new, unexpected compound—Letter from Friedrich Miescher to Wilhelm His, 26 February 1869, in Miescher, *Die histochemischen und physiologischen Arbeiten*, vol. 1, pp. 33–38; Friedrich Miescher, "Ueber die chemische Zusammensetzung der Eiterzellen," *Hoppe-Seyler's medicinisch-chemischen Untersuchungen* 4 (1871), reprinted in Miescher, *Die histochemischen und physiologischen Arbeiten*, vol. 2, pp. 3–23.

11   Hoppe-Seyler repeated the work—Wilhelm His, in Miescher, *Die histochemischen und physiologischen Arbeiten*, vol. 1, pp. 9, 10n.

11   Chargaff, "Miescher . . . realized"—Chargaff, "What Really Is DNA?" p. 302.

11   "sleepwalking"—Interview, Chargaff; letter, Chargaff to Judson, 25 September 1976.

12   molecular alphabet—Letter, Friedrich Miescher to Wilhelm His, 17 December 1892, in Miescher, *Die histochemischen und physiologischen Arbeiten*, vol. 1, pp. 116–117, *trad. auct.* My attention was drawn to the letter by Robert Olby and Erich Posner, "An Early Reference to Genetic Coding," *Nature* 215 (1967): 556, though my sense of its significance differs from theirs.

12   Chargaff, "vanishing cap"—Chargaff, "Grammar," pp. 637–638.

12   "reticent and intense"—Chargaff, "What Really Is DNA?" p. 300.

13   Crick, "twins"—F. H. C. Crick, "The Structure of the Hereditary Material," *Scientific American*, October 1954, p. 58.

13   Chargaff's scorn—Erwin Chargaff, "Deoxypentose Nucleoproteins and Their Prosthetic Groups," *Symposia of the Society for Experimental Biology* 9 (1955); reprinted in Chargaff, *Essays*, p. 53n.

13   tetranucleotide hypothesis—P. A. Levene and L. W. Bass, *Nucleic Acids*, American Chemical Society Monograph Series (New York: Chemical Catalog Company, 1931); see also M. R. Pollock, "The Discovery of DNA" (Third Griffith Memorial Lecture, delivered at the

General Meeting of the Society for General Microbiology, 6 April 1970), *Journal of General Microbiology* 63 (1970): 5–6.

13 belief persisting that genes are protein—See, for example, the extended, painfully puzzled discussion that follows Rollin D. Hotchkiss, "Bacterial Transformation," in *Symposium on Genetic Recombination*, held at Oak Ridge National Laboratory, Oak Ridge, Tennessee, 19–21 April 1954, particularly the exchange between the cytologist Kenneth Cooper and Hotchkiss, *Journal of Cellular and Comparative Physiology* 45 (1955): 1–22, esp. pp. 18–21; also Rollin Hotchkiss, "Gene, Transforming Principle, and DNA," in *Phage and . . . Molecular Biology*, pp. 180–200.

13 Avery et al.—Oswald T. Avery, Colin M. MacLeod, and Maclyn McCarty, "Induction of Transformation by a Desoxyribonucleic Acid Fraction Isolated from Pneumococcus Type III," *Journal of Experimental Medicine* 79 (1 February 1944): 137–158; "Received for publication 1 November, 1943."

14 Altman—Interview, Sidney Altman.

16 Medawar, anti-historical—P. B. Medawar, *The Art of the Soluble* (London: Methuen, 1967), p. 151.

17 thoroughly understood—Watson, *Molecular Biology*, p. 74.

17 Rough and Smooth—Avery, MacLeod, and McCarty, "Induction of Transformation," p. 140.

17 Griffith, R. and S.—Frederick Griffith, "The Influence of Immune Serum on the Biological Properties of Pneumococci," *Reports on Public Health and Medical Subjects*, no. 18 (London: His Majesty's Stationery Office, 1923), pp. 1–13. Avery's biographer, René Dubos, points out that the original discovery of a difference between Smooth virulent and Rough nonvirulent colonies was made a year earlier, with a different bacterial species, the Shigella dysentery bacilli, by J. A. Arkwright: Dubos, *The Professor*, p. 129.

18 antibodies to polysaccharide—Dubos, *The Professor*, pp. 105–106; Michael Heidelberger and Oswald T. Avery, "Immunological Relationships of Cell Constituents of Pneumococcus," *Proceedings of the Society for Experimental Biology and Medicine* 20 (1923): 435–436. Dubos, p. 201, lists later papers of Avery's that developed the discovery.

18 Griffith, startling discovery—Frederick Griffith, "The Significance of Pneumococcal Types," *Journal of Hygiene* 27 (January 1928): 141–144. Griffith also achieved transformation in other combinations of R and S strains, but that of RII to SIII is the one picked by Avery, MacLeod, and McCarty, "Induction of Transformation," p. 137.

18 Avery's disbelief—This passage was written before publication of Dubos's biography of Avery; but the reference that supersedes others is therein, Dubos, *The Professor*, p. 136.

18 confirmed in Berlin—F. Neufeld and W. Levinthal, "Beiträge zur Variabilität der Pneumokokken," *Zeitschrift für Immunitätsforschung* 55 (1928): 324–340.

18 repeated at Rockefeller—Martin H. Dawson, "The Transformation of Pneumococcal Types," *Journal of Experimental Medicine* 51 (1930): 99–122 and 123–147.

18 leaving out the mice—M. H. Dawson and R. Sia, "*In vitro* Transformation of Pneumococcal Types," *Journal of Experimental Medicine* 54 (1931): 681–699 and 701–710; "Received for publication July 10, 1931."

18 Alloway—J. Alloway, "The Transformation *in vitro* of R Pneumococci into S Forms of Different Specific Types by the Use of Filtered Pneumococcus Extracts," *Journal of Experimental Medicine* 55 (1932): 91–99; "Received for publication October 17, 1931."

18 "substance responsible"—Rollin D. Hotchkiss, "Oswald T. Avery: 1877–1955," *Genetics* 51 (1965): 2. See also Dubos, *The Professor*, p. 105.

18 "out the window"—Hotchkiss, "Gene, Transforming Principle, and DNA," p. 184.

19 "mucoid colonies"—Avery, MacLeod, and McCarty, "Induction of Transformation," pp. 140, 142.

19 friends remember—e.g., interview, Luria, October 1973; interview, Kalckar; interview, Delbrück, 9 July 1971; Dubos, *The Professor*, passim but particularly pp. 161–164; Hotchkiss, "Oswald T. Avery" and "Gene, Transforming Principle, and DNA," passim.

19 "might be a nucleic acid"—Hotchkiss, "Oswald T. Avery: 1877–1955." *Genetics* 51 (1965): 5.

19 methods—Avery, MacLeod, and McCarty, "Induction of Transformation," pp. 142–143.

19 tests—Avery, MacLeod, and McCarty, "Induction of Transformation," pp. 144–151.

20 conclusion—Avery, MacLeod, and McCarty, "Induction of Transformation," pp. 152–155.

21 letter to Roy Avery—Handwritten letter from Oswald Avery to Roy Avery, 26 May 1943. No transcript of this letter so far printed is altogether accurate; the best appears in Dubos, *The Professor*, pp. 217–220. I have excerpted directly from a machine copy of the original, kindly provided by the Manuscript Unit, Archives and Records Library Service, Tennessee State Library and Archives, Nashville, Tennessee.

22 opposition—Telephone conversation with René J. Dubos, 24 April 1978, New York, N.Y.: "I happen to be one of the few who was active at the time who is still alive," Dubos said; he also told me that in his biography of Avery (*The Professor*), he had understated

Mirsky's opposition because Mirsky had been his colleague at the Rockefeller Institute. Mirsky's "implacable" opposition is also confirmed by Hotchkiss, interview, 12 June 1978.

22   Willstätter—Richard Willstätter, "Probleme und Methoden der Enzymforschung," *Die Naturwissenschaften* 15 (1927): 585–596, summarizes his developed view. See also Hotchkiss, "Gene, Transforming Principle, and DNA," p. 190.

22   Northrop, Sumner—John H. Northrop, "Crystalline Pepsin: I. Isolation and Tests of Purity; II. General Properties and Experimental Methods," *Journal of General Physiology* 13 (20 July 1930): 739–766 and 767–780; James B. Sumner, "The Isolation and Crystallization of the Enzyme Urease: Preliminary Paper," *Journal of Biological Chemistry* 69 (August 1926): 435–441; see also John H. Northrop, "Biochemists, Biologists, and William of Occam," *Annual Review of Biochemistry* 30 (1961): 1–9.

22   many visitors—Visitors, in the early forties, included Max Delbrück (interview, 9 July 1972), Salvador Luria (interview, October 1973), Herman Kalckar (interview); Theodosius Dobzhansky, quoted in Olby, *Path* p. 189; MacFarlane Burnet, for whom see his *Changing Patterns: An Atypical Autobiography* (London: Heinemann, 1968), p. 81. An extended list of the contacts with Avery's laboratory and citations and appreciations of his work, before and after the great paper of 1944, was given by Rollin Hotchkiss, "The Identification of Nucleic Acids as Genetic Determinants" (paper delivered at the Symposium on the Origins of Modern Biochemistry, New York, N.Y., 2 June 1978), *Annals of the New York Academy of Sciences*, in press.

23   Avery's effect on Chargaff, Lederberg—Chargaff, interview, and "Grammar," p. 639; Lederberg, interview (he supplied, e.g., the word "shocked"), and Joshua Lederberg and Harriet A. Zuckerman. "From Schizomycetes to Bacterial Sexuality: A Case Study of Discontinuity of Science" (unpublished), pp. 2·3 and 2·4.

23   Nobel Prize—Arne Tiselius, of the Nobel Foundation, was quoted in *Scientific Research*, October 1967, p. 53, as saying, "That Avery never received the prize is lamentable, and had he not died when he did I think he would almost certainly have gotten it." See also Göran Liljestrand, "The Prize in Physiology or Medicine," in *Nobel: The Man and His Prizes*, ed. Nobel Foundation (Amsterdam: Elsevier, 1962), p. 281; Dubos, *The Professor*, p. 159.

23   Chargaff, "rarer instance"—Chargaff, "Grammar," p. 639.

23   Crick, boldness—Interview, 1 September 1971.

24   Watson, lectures and conversations—The lectures took place the last week in September of 1973; the interviews were on 24 and 25 September 1973, at the Biological Laboratories and the Faculty Club, Harvard University; the airplane ride to Washington was the evening of 30 September 1973.

29   advanced ornithology—Letter to Judson from the Office of the Registrar, University of Michigan, 8 October 1976.

29   "doing it yourself"—Interview, aboard plane, 30 September 1973.

29   Wright on Avery—Letter, Sewall Wright to Joshua Lederberg, 19 September 1973, cited and quoted in Olby, *Path*, bibliography, p. 494.

29   Watson on Schrödinger, graduate applications, etc.—J. D. Watson, "Growing Up in the Phage Group," in *Phage and . . . Molecular Biology*, pp. 239–40.

29   Muller's work—See Hermann Joseph Muller, *Studies in Genetics: The Selected Papers of H. J. Muller* (Bloomington: Indiana University Press, 1962); his bibliography is on pp. 591–610.

30   tobacco-mosaic virus—Marinus Willem Beijerinck, "Ueber ein contagium vivum fluidum als Ursache der Fleckenkrankheit der Tabaksblätter," *Verhandelingen der Koninklyke akademie van Wettenschappen te Amsterdam* 65 (December 1898): 3–21. Translation by James Johnson appears in *Phytopathological Classics* 7 (1942): 33–52.

30   Twort—F. W. Twort, "An Investigation on the Nature of Ultra-Microscopic Viruses," *The Lancet*, 4 December 1915, pp. 1241–1243.

30   d'Hérelle—Felix H. d'Hérelle, "Sur un microbe invisible antagoniste des bacilles dysentériques," *C. R. Acad. Sci.* 165 (1917): 373–374.

30   diarrhoea of locusts—Felix d'Hérelle, "The Bacteriophage," in *Science News*, 14 (Harmondsworth: Penguin, 1949), p. 44, quoted in Stent, *Bacterial Viruses*, pp. 4–5.

30   Muller on phage—Hermann J. Muller, "Variation Due to Change in the Individual Gene" (Paper delivered at Symposium on the Origin of Variations at the Thirty-ninth Annual Meeting of the American Society of Naturalists, Toronto, 29 December 1921), *The American Naturalist* 56 (January–February 1922): 48–49. The reprinting of this passage in Muller's selected papers, *Studies in Genetics*, contains five misprints.

31   Muller's leap—H. J. Muller, "The Gene" (Pilgrim Trust Lecture, delivered before the Royal Society of London, 1 November 1945), *Proceedings of the Royal Society of London*, B, 134 (1947): 1–37; reprinted in part in Muller, *Studies in Genetics*; the matter quoted in the text

appears in the "Foreword" by Joshua Lederberg to that volume, p. 5. For a sketch of Muller's role in the origins of molecular biology, see Elof Axel Carlson, "An Unacknowledged Founding of Molecular Biology: H. J. Muller's Contributions to Gene Theory," *Journal of the History of Biology* 4 (1971): 149–170.

31   Lederberg—"Foreword" to Muller, *Studies in Genetics*, p. 5.

31   Luria—S. E. Luria, "Mutations of Bacteria and of Bacteriophage," in *Phage and . . . Molecular Biology*, p. 177.

31   intellectual emigration—The classic study is, of course, Donald Fleming, "Emigré Physicists and the Biological Revolution," in *The Intellectual Migration*, ed. Donald Fleming and Bernard Bailyn (Cambridge, Mass.: The Belknap Press of Harvard University Press, 1969), pp. 152–189; it bears saying, however, that Fleming's article is at heart an imaginative and vividly inflected reading of the collection of pieces by various members of the American phage group in *Phage and . . . Molecular Biology*, together with Watson's *The Double Helix*. For Perutz: interview, Perutz, 4 December 1970. For Szilard: Leo Szilard, "Reminiscences," ed. Gertrud Weiss Szilard and Kathleen R. Winsor, in *The Intellectual Migration*, ed. Fleming and Bailyn, and Edward Shils, "Leo Szilard: A Memoir," *Encounter*, December 1964, pp. 35–41.

32   Delbrück—Gunther S. Stent, "Introduction: Waiting for the Paradox," in *Phage and . . . Molecular Biology*, pp. 3–6; Max Delbrück, "A Physicist Looks at Biology," *Transactions of the Connecticut Academy of Arts and Sciences* 38 (December 1949): 173–190, reprinted in *Phage and . . . Molecular Biology*, pp. 9–22; Max Delbrück, "Personal Records Questionnaire of the Royal Society," 6 pp., copy on file in the Delbrück Papers, California Institute of Technology Archives; Max Delbrück, "Homo Scientificus According to Beckett" (Talk delivered to Chemistry and Society Lecture Series, California Institute of Technology, 24 February 1971), privately circulated, pp. 4–5; Max Delbrück, "A Physicist's Renewed Look at Biology: Twenty Years Later" (Nobel Prize lecture, Stockholm, 10 December 1969), *Science* 168 (12 June 1970): 1312–1315.

32   Ellis and Delbrück—Emory L. Ellis, "Bacteriophage: One-Step Growth," in *Phage and . . . Molecular Biology*, pp. 58–62.

32   d'Hérelle's steps—Felix d'Hérelle, *The Bacteriophage and Its Behavior* (Baltimore: Williams and Wilkins, 1926), cited and quoted in Ellis, "Bacteriophage," pp. 56, 57.

32   "I don't believe it"—Ellis, "Bacteriophage," p. 58.

32   Delbrück on the first phage experiments—Max and Mary Bruce Delbrück, "Bacterial Viruses and Sex," *Scientific American*, November 1948, reprinted in *The Molecular Basis of Life*, ed. Robert H. Haynes and Philip C. Hanawalt (San Francisco: W. H. Freeman, 1968), where matter quoted is at p. 113.

33   one-step growth experiment—Emory L. Ellis and Max Delbrück, "The Growth of Bacteriophage," *Journal of General Physiology* 22 (1939): 365–384; "Accepted for publication, September 7, 1938."

33   Delbrück to the Harvey Society—Max Delbrück, "Experiments with Bacterial Viruses (Bacteriophages)" (Lecture delivered on 17 January 1946), *Harvey Lectures* 41 (1946): 161–162.

34   Luria meets Delbrück—Luria, "Mutations of Bacteria," p. 173; interview, Luria, October 1973.

34   Delbrück on Cold Spring Harbor and on Hershey—Luria, "Mutations of Bacteria," pp. 174, 175.

34   Luria on 1941 Symposium—Salvador E. Luria, "Early Days in the Molecular Biology of Bacteriophage," in *Proceedings of the Conference on the History of Biochemistry . . . May 21–23, 1970* (Brookline, Mass.: American Academy of Arts and Sciences, n.d.), p. 24.

35   misfit in society—Delbrück, "*Homo Scientificus . . .* p. 4.

35   one-step growth—Ellis and Delbrück, "The Growth of Bacteriophage," p. 365.

35   Anderson, Luria, electron microscopes—Thomas F. Anderson, "Electron Microscopy of Phages," in *Phage and . . . Molecular Biology*, pp. 63–65; S. E. Luria and T. F. Anderson, "Identification and Characterization of Bacteriophages with the Electron Microscope," *Proc. N. A. S.* 28 (1942): 127–130.

36   phage treaty—Anderson, "Electron Microscopy," p. 73.

36   "last stronghold of Lamarckism"—Quoted in Stent, *Molecular Genetics*, p. 151.

37   fluctuation test—Luria, "Mutations of Bacteria," pp. 174–175; S. E. Luria and M. Delbrück, "Mutations of Bacteria from Virus Sensitivity to Virus Resistance," *Genetics* 28 (November 1943): 491–511; "Received May 29, 1943."

37   "intelligent and simpleminded"—Interview, Meselson, 16 March 1971.

37   fluctuation test compared to Mendel—Stent, *Molecular Genetics*, p. 148.

37   morphology and chemical composition of phage—Summarized in Stent, *Molecular Genetics*, pp. 303–306.

38   few phage enter—Anderson, "Electron Microscopy of Phages," in *Phage and . . . Molecular Biology*, p. 72.

38   osmotic shock—Anderson, "Electron Microscopy," pp. 75–76; Thomas F. Anderson, "Destruction of Bacterial Viruses by Osmotic Shock," *Journal of Applied Physics* 21 (1950): 70; Roger M. Herriott, "Nucleic-acid-free T2 Virus 'Ghosts' with Specific Biological Action," *Journal of Bacteriology* 61 (1951): 252–254.

38   "little hypodermic"—Letter, Roger Herriott to Alfred Hershey, 16 November 1951, quoted in Alfred Hershey, "The Injection of DNA into Cells by Phage," in *Phage and . . . Molecular Biology*, p. 102.

38   impact of Avery's paper—For counting and mapping of distribution of *The Journal of Experimental Medicine*, and of citations, see H. V. Wyatt, "When Does Information Become Knowledge?" *Nature* 235 (14 January 1972): 86–89. For simplistic sociology, see, e.g., Nicholas C. Mullins, "The Development of a Scientific Specialty: The Phage Group and the Origins of Molecular Biology," *Minerva* 10 (1972): 51–82. But the chief focus of the controversy over the influence of Avery's discovery is the paper by Gunther S. Stent, "Prematurity and Uniqueness in Scientific Discovery," *Scientific American*, December 1972, pp. 84–93. Stent, a clever man, a strong, vivid writer, and himself a prominent member of the phage group, asserted that Avery's work was ignored not simply because Avery was not a member of the phage group, but because the discovery was in some way "premature"—and he drew the comparison with the long neglect of Mendel's discoveries. Stent's assertion has been contradicted by several other scientists who were active in the late forties and early fifties, most notably and effectively by Rollin Hotchkiss, "The Identification of Nucleic Acids as Genetic Determinants," and by Joshua Lederberg and Harriet A. Zuckerman, "From Schizomycetes to Bacterial Sexuality"; see also Dubos, *The Professor*, pp. 155–159.

39   Delbrück at Cold Spring Harbor—Interview, Delbrück, 9 July 1972.

42   paradigms—Thomas S. Kuhn, *The Structure of Scientific Revolutions*, 2nd ed., enl. (Chicago: The University of Chicago Press, 1970).

43   Gilbert, "big house on the hill"—Interview, Gilbert, October 1973.

43   Luria—Conversation, Salvador Luria and Matthew Meselson, 3 October 1973, at Meselson's house in Lexington, Massachusetts; interview with Luria the next day.

43   Dobzhansky—Theodosius Dobzhansky, *Genetics and the Origin of Species*, 2nd ed. (New York: Columbia University Press, 1941), pp. 47–50.

44   Boivin—André Boivin, A. Delauney, Roger Vendrely, and Y. Lehoult, "L'acide thymonucléique polymérisé, principe paraissant susceptible de déterminer la spécificité sérologique et l'équipement enzymatique des bactéries: signification pour la biochimie de l'hérédité," *Experientia* 1 (1945): 334–335. See also André Boivin and Roger Vendrely, "Rôle de l'acide désoxy-ribonucléique hautement polymérisé dans le déterminisme des caractéres héréditaires des bactéries. Signification pour la biochimie générale de l'hérédité" *Helvetica Chimica Acta* 29 (1946): 1338–1344.

45   Watson and Luria—Watson, "Growing Up in the Phage Group," 239–240.

46   multiplicity reactivation—S. E. Luria, "Reactivation of Irradiated Bacteriophage by Transfer of Self-Reproducing Units," *Proc. N. A. S.* 33 (1947): 253ff. J. D. Watson, "The Properties of X-ray-inactivated Bacteriophage: I. Inactivation by Direct Effect," *Journal of Bacteriology* 60 (December 1950): 697–718; J. D. Watson, "The Properties of X-ray-inactivated Bacteriophage: II. Inactivation by Indirect Effects," *Journal of Bacteriology* 63 (April 1951): 474–485. For discussion and later developments, see Stent, *Bacterial Viruses*, pp. 289–300.

47   Watson at Cold Spring Harbor—Watson, "Growing Up in the Phage Group," pp. 241–242.

47   Novick and Szilard—Aaron Novick, "Phenotypic Mixing," in *Phage and . . . Molecular Biology*, pp. 134–135.

48   Delbrück and Bohr—Interview, Delbrück, 9 July 1972; for Bohr and his circle, Stefan Rozental, ed., *Niels Bohr: His Life and Work as Seen by His Friends and Colleagues* (Amsterdam: North-Holland Publishing, 1967).

48   Watson and Szilard—Watson, "Growing Up in the Phage Group," p. 242.

48   Haurowitz—Conversation, Haurowitz, 2 June 1978, at Symposium on the Origins of Modern Biochemistry, sponsored by the New York Academy of Sciences; Felix Haurowitz, *Chemistry and Biology of Proteins* (New York: Academic Press, 1950), p. 345; Felix Haurowitz, "Important Discoveries in Biochemistry Between 1920 and 1940." *Proceedings of the Conference on the History of Biochemistry and Molecular Biology . . . May 21–23, 1970* (Brookline, Mass.: American Academy of Arts and Sciences, n.d.), p. 15.

48   Watson in Pasadena—Watson, "Growing Up in the Phage Group," pp. 242–243.

48   Watson to Europe—"Growing Up in the Phage Group," pp. 243–244.

48   Merck fellowship—*The Double Helix* p. 46.

49   Watson's interest attracted—Lloyd M. Kozloff and Frank W. Putnam, "Biochemical Studies of Virus Reproduction: III. The Origin of Virus Phosphorus in the *Escherichia coli* T6 Bacteriophage System," *Journal of Biological Chemistry* 182 (1950): 229–242.

49   "first serious student"—*The Double Helix*, p. 23.

2 *"DNA, you know, is Midas' gold. Everybody who touches it goes mad."*

PAGE

51   "humility . . . not conducive"—P. B. Medawar, *The Art of the Soluble* (London: Methuen, 1967), p. 110.

51   Pauling's books—Linus Pauling, *The Nature of the Chemical Bond*, 1st ed. (Ithaca: Cornell University Press, 1939); *General Chemistry*, 1st ed. (San Francisco: W. H. Freeman, 1947).

52   tribute to Pauling—The point has been acknowledged freely, and most recently by Watson in his "Welcoming Remarks" to the 1978 Cold Spring Harbor Symposium on Quantitative Biology, 31 May 1978, in which he observed that this was the twenty-fifth anniversary of his presentation of the structure of DNA at Cold Spring Harbor, and said, "The original structure was a monument to the way of thinking of Linus Pauling."

52   Crick like Pauling—Letter, James Watson to Max Delbrück and his wife, 9 December 1951, p. 2; original in Delbrück Papers, California Institute of Technology Archives.

52   Pauling interview—Interview, Pauling, 1 March 1978; supplemented and checked in interview, Pauling, 23 December 1975; see also Linus Pauling, "Fifty Years of Progress in Structural Chemistry and Molecular Biology," in *Daedalus*, Fall 1970, pp. 988–1014; "Fifty Years of Physical Chemistry in the California Institute of Technology," *Annual Review of Physical Chemistry* 16 (1965): 1–14.

53   Weaver and "molecular biology"—Warren Weaver, "Molecular Biology: Origin of the Term" (letter), *Science* 170 (1970): 581–582; "The Natural Sciences," in *Annual Report of the Rockefeller Foundation*, 1938, p. 203.

53   Pauling's hemoglobin—Linus Pauling, "The Oxygen Equilibrium of Hemoglobin and Its Structural Determination," *Proc. N. A. S.* 21 (1935): 186–191. The paper was not an important contribution to the study of hemoglobin, though two papers Pauling published the next year, with Charles Coryell, were of great value: see chapter 9, below.

53   autobiographical note—Pauling, "Fifty Years of Progress," pp. 988–989.

54   newest X-ray-diffraction pictures—The claim is expressed, for example, in the Editor's Note to Pauling, "Fifty Years of Progress," p. 1009. In a conversation, Wilkins said, "I would summarize my general reaction as follows, that if Pauling had been able to get his passport I think he would have wrapped the whole DNA thing up." Interview, Wilkins, 15 September 1975.

55   Pauling and quantum mechanics—Pauling, "Fifty Years of Progress," pp. 992–997.

56   "But it was a short note"—Letter, Linus Pauling to Judson, 8 October 1976, p. 4; Linus Pauling, "The Shared-Electron Chemical Bond," *Proc. N. A. S.* 14 (1928): 359.

56   Bragg's crystal structures—Bragg, "History," pp. 166–168; Bragg, *Development of X-ray Analysis*, pp. 148–158.

57   Pauling met the elder Bragg—Letter, Linus Pauling to Judson, 8 October 1976, p. 2.

57   Pauling's rules—Linus Pauling, "The Principles Determining the Structure of Complex Ionic Crystals," *Journal of the American Chemical Society* 51 (5 April 1929): 1010–1026; "Received September 5, 1928."

57   225 substances—Pauling, "Fifty Years of Progress," p. 997.

57   Bragg's chagrin—Interview, Perutz, 15 February 1975, later checked and confirmed. Aaron Klug, at that laboratory, on reading this passage said that he had learned independently of Bragg's chagrin and sense of rivalry with Pauling from R. W. James, who had been a colleague of Bragg's before the second world war and was Professor of Crystallography in South Africa when Klug was an undergraduate there in the late forties; interview, Klug, 19 January 1976.

57   great essay—Linus Pauling, "The Nature of the Chemical Bond: Application of Results Obtained from the Quantum Mechanics and from a Theory of Paramagnetic Susceptibility to the Structure of Molecules," *Journal of the American Chemical Society* 53 (6 April 1931): 1367–1400; "Received February 17, 1931."

57   Bernal on Pauling—J. D. Bernal, "The Pattern of Linus Pauling's Work in Relation to Molecular Biology," in *Structural Chemistry and Molecular Biology*, p. 370.

60   Bernal's discovery—J. D. Bernal and D. Crowfoot, "X-ray Photographs of Crystalline Pepsin," *Nature* 133 (26 May 1934): 794–795; the letter is dated May 17. Dorothy Crowfoot Hodgkin and Dennis Parker Riley, "Some Ancient History of Protein X-ray Analysis," in *Structural Chemistry and Molecular Biology*, pp. 15–16.

60   cyclols—The cyclol theory was the creation of Dorothy Wrinch; its latest expressions were I. Langmuir and D. M. Wrinch, "Vector Maps and Crystal Analysis," *Nature* 142 (1938): 581–583; D. M. Wrinch, "The Structure of the Insulin Molecule," *Science* 88 (1938): 148–149.

60   cages demolished—Linus Pauling and Carl Niemann, *Journal of the American Chemical Society* 61 (1939): 1860–1867.

61   Astbury at Leeds—Olby, *Path*, pp. 38, 41–53, is detailed and amusing about Astbury's early career.

61   wool and keratin—W. T. Astbury and A. Street, "X-ray Studies of the Structure of Hair, Wool, and Related Fibres," *Philosophical Transactions of the Royal Society* of London, A, 230 (1932): 75–101; Sir Lawrence Bragg, J. C. Kendrew, and M. F. Perutz, "Polypeptide Chain Configurations in Crystalline Proteins," *Proceedings of the Royal Society* of London, A, 203 (1950): 321–357, discusses Astbury's work at pp. 321–324; see also W. L. (Sir Lawrence) Bragg, "X-ray Analysis of Proteins" (36th Guthrie Lecture, delivered to the Physical Society, London, 12 March 1952), *Proceedings of the Physical Society*, B 55 (1952): 833–837.

62   Astbury on DNA—William T. Astbury and Florence Bell, "Some Recent Developments in the X-ray Study of Proteins and Related Structures," *C. S. H. Symp. Q. Biol.* 6 (1938): 109–118; W. T. Astbury and F. O. Bell, "X-ray Study of Thymonucleic Acid," *Nature* 141 (1938): 747–748; W. T. Astbury, "X-ray Studies of Nucleic Acids," *Symp. Soc. Exp. Biol.* 1, pp. 67–76.

62   lecture to the Swedish Academy—Linus Pauling, "Modern Structural Chemistry" (Nobel Lecture, 11 December 1954), in *Les Prix Nobel en 1954*, p. 98.

63   the paper model—Pauling told me the story in the second interview, 23 December 1975, in London.

64   "Pauling was very interested"—Interview, Perutz, 31 January 1976.

65   Bragg, Kendrew, and Perutz—"Polypeptide Chain Configuration in Crystalline Proteins"; Bragg, "How Proteins Were Not Solved," pp. 6–7.

65   Singer—Interview, S. J. Singer.

66   Courtaulds—C. H. Bamford, W. E. Hanby, and F. Happey, *Proceedings of the Royal Society* of London, A, 205 (1951): 30–46; conversation with Perutz, 23 May 1975.

66   Pauling's first announcement—Linus Pauling and Robert B. Corey, "Two Hydrogen-Bonded Spiral Configurations of the Polypeptide Chain," *Journal of the American Chemical Society* 72 (1950): 5349.

67   Watson in Copenhagen, Naples; reaction to alpha helix—*The Double Helix*, pp. 24–28, 31–33, 35–38; J. D. Watson and O. Maaloe, "The Transfer of Radioactive Phosphorus from Parental to Progeny Phage," *Proc. N. A. S.* 37 (August 1951): 507–513.

68   Pauling's structures—Linus Pauling, Robert B. Corey, and H. R. Branson, "The Structure of Proteins: Two Hydrogen Bonded Helical Configurations of the Polypeptide Chain," *Proc. N. A. S.* 37 (April 1951): 205–211; the bonanza in May was seven papers in that issue of the same *Proceedings*, pp. 235–285.

69   Todd to Bragg—Interview Todd.

69   Perutz's confirmation—Conversation Perutz, May 1975, later checked and confirmed; M. F. Perutz, "New X-ray Evidence on the Configuration of Polypeptide Chains; Polypeptide Chains in Poly-γ-benzyl-L-glutamate, Keratin and Haemoglobin," and H. E. Huxley and M. F. Perutz, "Polypeptide Chains in Frog Sartorius Muscle," both *Nature* 167 (30 June 1951): 1053 et seq. Perutz amplified the significance of these findings in "The 1.5-A. Reflexion from Proteins and Polypeptides," *Nature* 168 (13 October 1951): 653

70   "Wilkins's experimental work"—Pauling overlooked Rosalind Franklin's role, for which see chapter 3.

70   schizophrenia—Linus Pauling, "Orthomolecular Somatic and Psychiatric Medicine" (Communication to the Thirteenth International Convention on Vital Substances, Nutrition and the Diseases of Civilization, at Luxembourg, 18–24 December 1967). *Zeitschrift Vitalstoffe–Zivilisationskrankheiten* 1 (1968): unpaged; "Orthomolecular Psychiatry," *Science* 160 (19 April 1968): 265–271.

71   Vivonex 100—Vivonex Corporation, Mountain View, California.

72   data confirmed by theory—Sir Arthur Eddington, *New Pathways in Science* (Cambridge: At the University Press, 1935), p. 211.

72   Kalckar—Interview, Kalckar, October 1973.

72   resistance to penicillin—Rollin D. Hotchkiss, "Transfer of Penicillin Resistance in Pneumococci by the Deoxyribonucleate Derived from Resistant Cultures," *C. S. H. Symp. Q. Biol.* 16 (1951): 457–461.

72   remained cautious—Hotchkiss. "Gene, Transforming Principle, and DNA," in *Phage and . . . Molecular Biology*, p. 194.

73   Boivin—André Boivin, Roger Vendrely, et Colette Vendrely, "L'acide désoxyribonucléique du noyau cellulaire, dépositaires des caractères héréditaires; arguments d'ordre analytique," *C. R. Acad. Sci.* 226 (March 1948): 1061–1063; A. Boivin, R. Vendrely, et R. Tulasne, "La spécificité des acides nucléiques chez les êtres vivants, spécialement chez les bactéries," *Colloques internationaux du Centre National de la Recherche Scientifique* 8: *Unités Biologiques Douées de Continuité Génétique* (colloquium held in Paris, June-July 1948, published 1949): 67–78.

73   corollary characteristics—For turnover, Einar Hammarsten and George de Hevesy, "Rate of Renewal of Ribo- and Desoxyribonucleic Acids," *Acta Physiologica Scandinavica* 11

(1946): 225–243. For constancy, even starving, J. N. Davidson, "The Distribution of Nucleic Acids in Tissues," *Symp. Soc. Exp. Biol.* 1, held 1946, pp. 77–85, particularly p. 83, where he cites earlier studies; constancy even after prolonged starvation of proteins, Paul Mandel, Lila Mandel, and Monique Jacob, "Sur le comportement comparé, au cours du jeûne protéique prolongé, des deux acides nucléiques des tissus animaux et sur sa signification," *C. R. Acad. Sci.* 226 (1948): 2019–2021; constancy in plants as in animals, Hewson Swift, "The Constancy of Desoxyribose Nucleic Acid in Plant Nuclei," *Proc. N. A. S.* 36 (November 1950): 643–654.

73 workers less guarded—Mandel, Mandel, and Jacob, "Sur le comportement comparé," p. 2021.

73 Chargaff's conversion—Chargaff. "Grammar," p. 639.

73 Chargaff's paper—Edwin Chargaff, "Chemical Specificity of Nucleic Acids and Mechanism of Their Enzymic Degradation," *Experientia* 6 (1950): 201–209; reprinted in Chargaff, *Essays*, pp. 1–24.

74 "enormous" specificity—Chargaff, *Essays*, p. 21.

75 Chargaff ratios—Chargaff, *Essays*, p. 13.

75 "degradation of . . . science"—Erwin Chargaff, "A Quick Climb Up Mount Olympus," *Science* 159 (29 March 1968): 1449.

75 "such pygmies"—Chargaff, "Grammar," p. 641.

75 Wilkins's talk—The meeting on theoretical biology was held by the Institut de la Vie, *Troisième Conférence Internationale: De La Physique Théorique à la Biologie*, Versailles, 21–26 June 1971; Wilkins's talk was "Social Implications of Biology," delivered Saturday morning, June 26, and the quotations are directly from my notes, and differ slightly from the text published later, in *From Theoretical Physics to Biology*, ed. M. Marois (Basel: Karger, 1973), pp. 451–467.

76 cloning—John B. Gurdon, "Transplanted Nuclei and Cell Differentiation," *Scientific American*, December 1968, pp. 24–35; James D. Watson, "Moving Towards Clonal Man: An Example of Scientific Inevitability?" *The Atlantic Monthly*, May, 1971; J. D. Watson, "Potential Consequences of Experimentation with Human Eggs" (Testimony to the Twelfth Meeting of the Panel on Science and Technology, Committee on Science and Astronautics, U. S. House of Representatives, 28 January 1971).

76 Musil—Robert Musil, *The Man without Qualities*, tr. by Eithne Wilkins and Ernst Kaiser (London: Secker and Warburg, 1953), vol. 1 p. 254.

77 Wilkins's career—Interview, Wilkins 26 June 1971, in Versailles and Paris; additional biographical details from *Les Prix Nobel en 1962*, pp. 74–75, 126–128.

77 Randall's biophysics unit—J. T. Randall, "An Experiment in Biophysics," *Proceedings of the Royal Society* of London, A, 208 (1951): 1–24.

78 "I had spun a very thin fibre"—Wilkins, interview 26 June 1971; see also Wilkins, *Les Prix Nobel en 1962*, p. 128.

78 Wilkins's and Gosling's procedures—Wilkins, pp. 128–130. Interviews, Wilkins, 26 June 1971, 15 September 1975, and 12 March 1976. Interview, Gosling. Wilkins has also kindly let me have a copy of "X-ray Diffraction Studies of DNA in King's College from 1950 Onwards," a draft of a historical account, prepared in the spring of 1976.

79 X-ray equipment—Interview, Gosling; letter, Walter E. Spear to Judson, 16 February 1976.

79 couple of notes—M. H. F. Wilkins, "Some Aspects of Microspectrography," *Discussions of the Faraday Society* 9 (1950): 363–369; the meeting took place in Cambridge, September 25–28, and in this paper Wilkins was primarily concerned with instruments that had been built in the Biophysics Unit at King's College London, but he discussed the dichroism of nucleic acid, pp. 368–369; W. E. Seeds and M. H. F. Wilkins, "Ultra-violet Microspectrographic Studies of Nucleoproteins and Crystals of Biological Interest," ibid., pp. 417–423; this paper, they said, was "preliminary" (p. 417), and indeed for nucleic acids went little further than the work of Torbjörn Caspersson ten years earlier (p. 422); for Caspersson, see chapter 5, below; M. H. F. Wilkins, R. G. Gosling, and W. E. Seeds, "Nucleic Acid: An Extensible Molecule?" *Nature* 167 (12 May 1951): 759–760.

79 Naples meeting—M. H. F. Wilkins, "Ultraviolet Dichroism and Molecular Structure in Living Cells," *Publications of the Zoological Station of Naples* 23 (1951): supplement, pp. 104–114.

80 if Pauling had seen Wilkins's photos—See first note to p. 74, above.

80 Wrote about Wilkins—*The Double Helix*, pp. 225–226.

81 several articles—L. D. Hamilton, "DNA: Models and Reality," *Nature* 218 (18 May 1968): 633–637; Aaron Klug, "Rosalind Franklin and the Discovery of the Structure of DNA," *Nature* 219 (24 August 1968): 808–810 and 843–844, "Corrigenda," p. 879, "Addendum," p. 880.

81 Franklin's early life—Muriel Franklin, *Rosalind* (London: privately printed), pp. 7, 8, 13, 15. This is a memorial by Franklin's mother.

81   coals and chars—Klug, "Rosalind Franklin"; interviews, Klug, 27 June 1975 and 19 January 1976; interview, Gosling; J. D. Bernal, "Rosalind Franklin," obituary notice in *Nature*, reprinted in Muriel Franklin, *Rosalind*, pp. 25--27; letter, Rosalind Franklin to John Randall, dated "8/7/50," original in Klug's file of Franklin's papers; letter, Charles Coulson (Professor of Theoretical Physics, King's College London) to Rosalind Franklin, 17 July 1950, original in Klug's file; and other correspondence in that file, passim.

81   exploring Franklin's Paris, Luzatti on why she left—22 December 1982.

82   "change in programme—Letter, John Randall to R. Franklin, dated "4th December, 1950"; original in Klug's file of Franklin's papers.

82   meeting at turn of the year—Interview, Gosling; Alexander Stokes, interview, 11 August 1976, was not able to recall the meeting in any detail.

84   Bragg like Mozart—M. F. Perutz, "Bragg, Protein Crystallography and the Cavendish Laboratory," *Acta Crystallographica* A26 (March 1970): 184.

84   Perutz on Bragg—Interview, Perutz, 6 December 1970.

84   Bragg fostered at the Cavendish—Perutz, "Bragg . . . and the Cavendish Laboratory," p. 185; interview, Bragg.

84   Perutz's first meeting with Bragg—Perutz, "Bragg . . . and the Cavendish Laboratory," p. 184.

84   Bragg's last paper—W. L. Bragg, "The Determination of the Coordinates of Heavy Atoms in Protein Crystals," *Acta Crystallographica* 11 (1958): 70–75.

85   "tears of emotion"—M. F. Perutz, "Sir Lawrence Bragg," *Nature* 233 (3 September 1971): 75.

85   conversation with Bragg—Interview, 28 January 1971; further details of the history of X-ray crystallography from Sir Lawrence Bragg, "History," pp. 161–165; Bragg, *Development of X-ray Analysis*, pp. 14–23.

85   "obscure German journal"—W. Friedrich, P. Knipping, and M. Laue, "Interferenz-Erscheinungen bei Röntgenstrahlen," *Sitzungsberichte der mathematisch-physikalischen Klasse der Königlich Bayerischen Akademie der Wissenschaften*, 8 June 1912, pp. 303–322.

85   the courts of St. John's—This detail from J. G. Crowther, *The Cavendish Laboratory 1874–1974* (London: Macmillan, 1974), p. 272.

86   Bragg's spatial visualization—For testimony besides Perutz's, see Crowther, *The Cavendish Laboratory*, p. 313. Bragg was in fact an amateur draughtsman of some skill.

86   lunch at the Athenaeum—Perutz, "Bragg . . . and the Cavendish Laboratory," p. 185.

87   Bragg on Crick—Interview, 28 January 1971.

87   Crick's early life—Biographical details from "Dr. Francis Crick, FRS: Notes on His Early Career," summary prepared from Crick's personal files at the Medical Research Council in March 1970, by the council's Information Group: copy supplied to me 8 March 1977; supplemented from Robert Olby, "Francis Crick, DNA, and the Central Dogma," in *Daedalus*, Fall 1970, pp. 938–941. Crick, however, has read the passage containing these details: checking interview, Crick, 10 September 1975.

87   Monod on Crick—Interview, Monod, September 1970.

87   Crick's motive in entering biology—Interview, Crick, 1 September 1971.

88   "division between the living and the non-living"—The passage from Crick's application to the Medical Research Council is quoted in "Dr. Francis Crick, FRS: Notes on His Early Career."

88   Sanger biography—Interview, Sanger (videotaped for archive of the Biochemical Society), Cambridge, 13 November 1987; Nobel lecture, *Les Prix Nobel en 1958* (Stockholm: Nobel-stiftung, 1959), pp. 134–146, reprinted in *Science* 129 (1959): 1340–1345; Frederick Sanger, "Sequences, Sequences, and Sequences," draft manuscript prepared for *Annual Review of Biochemistry* (57).

89   Sanger's papers on insulin—F. Sanger, "Fractionation of Oxidized Insulin," *Biochemical Journal* 44 (1949): 126–128; F. Sanger, "The Terminal Peptides of Insulin," *Biochemical Journal* 45 (1949): 563–574; F. Sanger and H. Tuppy, "The Amino-acid Sequence in the Phenylalanyl Chain of Insulin. 1 & 2," *Biochemical Journal* 49 (1951): 463–480 and 481–490; F. Sanger and E. O. P. Thompson, "The Amino-acid Sequence of the Glycyl Chain of Insulin. 1 & 2," *Biochemical Journal* 53 (1953): 353–365 and 366–374; F. Sanger, E. O. P. Thompson, and Ruth Kitai, "The Amide Groups of Insulin," *Biochemical Journal* 59 (1955): 509ff.

89   Askonas on Sanger, Crick—Conversations with Askonas, 24 July 1995, San Francisco, and 14 March 1996, London.

89   Crick's approach to the Cavendish—Interview, Perutz, 6 December 1970; interview, Crick, 10 September 1975; letter, Georg Kreisel to Judson, 2 September 1976.

90   disastrous paper—Bragg, "How Proteins Were Not Solved," p. 7.

90   Crick's autobiography—Francis Crick, *What Mad Pursuit: A Personal View of Scientific Discovery* (New York: Basic Books, 1988).

90   "hatbox" and other models—See chapter 9.

90 "They were . . . really thinking about proteins"—Interview, Crick, 1 September 1971.

90 Crick on meeting Watson—Interview, Crick, 1 September 1971.

91 details of Cavendish quarters—Interview and tour of the building with Perutz, 23 May 1975.

91 Bragg liked Watson's enthusiasm—Interview, Bragg.

93 Crick's "fewest possible assumptions"—Interview, Crick, 10 September 1975.

93 Astbury's 1947 picture and interpretation—William T. Astbury, "X-ray Studies of Nucleic Acids," in *Symp. Soc. Exp. Biol.* 1, held 1946, plate 1, figs. 2a and b, pp. 66–68, 71, 76.

94 Todd's nucleic-acid backbone—Interview, Todd; Alexander Todd and D. M. Brown, "Nucleotides: Part X. Some Observations on the Structure and Chemical Behaviour of the Nucleic Acids," *Journal of the Chemical Society* (1952): 52–58; "Received August 15, 1951."

94 Furberg at Birkbeck—Letter, Furberg to Judson, 16 March 1976, establishes the biographical and scientific details.

94 Furberg's nucleotide "absolutely essential"—Conversation. Crick, 10 July 1976.

94 Furberg's models—Sven Furberg, "Crystal Structure of Cytidine," *Nature* 164 (2 July 1949): 22; "An X-ray Study of Some Nucleosides and Nucleotides" (Ph.D. diss., University of London Birkbeck College, August 1949); idem, "The Crystal Structure of Cytidine," *Acta Crystallographica* 3 (September 1950): 325–333; idem, "An X-ray Study of the Stereochemistry of the Nucleosides," *Acta Chemica Scandinavica* 4 (1950): 751–761; idem, "X-ray Studies on the Decomposition Products of the Nucleic Acids" (abstract of paper read on Furberg's behalf by C. H. Carlisle), *Transactions of the Faraday Society* 46 (September 1950), reprint unpaged; idem, "On the Structure of Nucleic Acids," *Acta Chemica Scandinavia* 6 (1952): 634–640—this last being the one that finally got Furberg's models into print. A photographic copy of pages 88–96 of Furberg's dissertation, and reprints of all of these papers, signed by Rosalind Franklin, are in Aaron Klug's file of her papers, in most cases accompanied by loose sheets of notes on the papers, in her hand; the notes are terse and undated but the details she singled out suggest in most cases that she read the papers while actively working on the structure, even though she may sometimes have acquired the reprints later.

94 Furberg's model at King's College London—Randall, "An Experiment in Biophysics," plate 2, fig. 19.

94 Bernal, "the key"—J. D. Bernal, "The Material Theory of Life" (review of James D. Watson, *The Double Helix*), *Labour Monthly*, July 1968, pp. 324, 325.

94 Furberg and Franklin did not meet—Letter, Sven Furberg to Judson, 16 March 1976, and again 3 August 1976.

95 Watson's report of Delbrück's comment on alpha helix—*The Double Helix*, p. 40.

95 Wilkins at Hardy Club—Minute book of the Hardy Club, p. 14; the minute book is in the possession of the club's then secretary, Richard D. Keynes, Professor of Physiology in the University of Cambridge. For each meeting, Keynes listed the place, the speaker and his subject, and those present; Watson was not then a member and was not present that evening as a guest; for his conversation with Crick and Wilkins afterwards, see *The Double Helix*, pp. 54, 56.

95 Wilkins, Franklin, the July 1951 meeting—Interviews, Wilkins, 15 September 1975 and 12 March 1976; Wilkins, letters to Judson, 12 July 1976, 22 July 1976, 28 September 1976.

95 Wilkins in Naples—Interviews, 15 September 1975 and 12 March 1976; Wilkins, "X-ray Diffraction Studies of DNA in King's College from 1950 Onwards"; the squid sperm resulted in the paper, M. H. F. Wilkins and J. T. Randall, "Crystallinity in Sperm Heads: Molecular Structure of Nucleoprotein *in Vivo*," *Biochimica et Biophysica Acta* 10 (1953): 192–193; M. H. F. Wilkins and B. Battaglia, "Note on the Preparation of Specimens of Oriented Sperm Heads for X-ray Diffraction and Infrared Absorption Studies and on Some Pseudo-Molecular Behaviour of Sperm," ibid., 11 (1953): 412–415.

96 Wilkins meeting Chargaff in U.S.; Franklin's refusal of cooperation—Interview, 15 September 1975; Wilkins, "X-Ray Diffraction Studies of DNA in King's College from 1950 Onwards"; Wilkins, letters to Judson, 12 July 1976, 28 September 1976.

96 Perutz, Vand, and the development of helical-diffraction theory—Interview, Perutz, 15 February 1975, and further conversation that spring; interviews, Crick, 1 September 1971 and 10 September 1975; interview, Stokes. W. Cochran and F. H. C. Crick, "Evidence for the Pauling-Corey α-Helix in Synthetic Polypeptides," *Nature* 169 (9 February 1952): 234, letter dated "Dec. 14"; W. Cochran, F. H. C. Crick, and V. Vand, "The Structure of Synthetic Polypeptides. I. The Transform of Atoms on a Helix," *Acta Crystallographica* 5 (September 1952): 581–586, "Received 16 February 1952"; *The Double Helix*, pp. 62–67; Matthew & Son Ltd., "Wine Tasting" (program dated "October 31st, 1951"), original in Crick's files.

97 Watson's fellowship—*The Double Helix*, pp. 43–47; letter, Roy Markham to Judson, 15 September 1976. Olby, *Path*, pp. 306, 308–309, gives further details and quotes correspondence.

97   Watson's version—*The Double Helix*, pp. 68–70.

97   colloquium—Rosalind Franklin, manuscript headed "Colloquium Nov. 1951"; original in Aaron Klug's file of Franklin's papers. Olby, *Path*, pp. 349–350. published what he described as "an edited version of her notes"; the editing amounted to filling out her contractions, abbreviations, and grammatically incomplete sentences, but in the process Olby made two significant errors and an interesting omission. In her original, at one point under her heading "Evidence for spiral structure," Franklin actually wrote that "near-hexagonal packing suggests that there is only one helix (containing possibly more than 1 chain) per lattice point Density measurements (24 residues/27 A) suggest more than 1 chain"—except that for the words "more than," in both appearances, she used the mathematical sign >. Olby omitted the sign and failed to supply the words, in both instances—thus reversing Franklin's meaning. Olby also omitted the "cf Pauling" quoted in my text.

98   Wilkins acknowledged—Interview, Wilkins, 26 June 1971.

98   "complete change in picture"—Franklin's notes used a horizontal arrow where I supplied the word "towards."

98   "did . . . show helical features"—Letter, Wilkins to Judson, 12 July 1976.

100   Franklin told Klug—Interview, Klug, 19 January 1976.

100   Wilkins already knew—Interview, Gosling.

100   Perutz on X-ray diffraction—Interview, Perutz, 15 February 1975.

101   crystallites in DNA fibres—The fact was stressed to me by Klug, interview, 27 June 1975.

103   "Francis . . . visibly annoyed"—*The Double Helix*, pp. 75–77.

103   draft paper—F. H. C. Crick and J. D. Watson, "The Structure of Sodium thymonucleate: a possible approach," unpublished, undated manuscript in Crick's hand; p. 8. Crick has the original. We discussed it in the interview of 10 September 1975.

105   diagrams in enol forms in nineteen-fifties—The authoritative, available reference at the Cavendish was J. N. Davidson, *The Biochemistry of the Nucleic Acids* (London: Methuen), of which the first edition was published 1950, the second, revised, 1953; for diagrams of thymine, uracil, and guanine only in enol forms, see 2nd ed., rev., pp. 7, 8.

106   model building; the confrontation with the King's group—*The Double Helix*, pp. 83–96; interview, Crick, 10 September 1975; interview, Gosling; for comment on Franklin's approach, interview, Klug, 19 January 1976.

106   "not a shred of evidence"—*The Double Helix*, p. 94—yet on p. 122, Watson reports that Wilkins told him on May 1 of 1952 that "now" Franklin "was insisting that her data told her DNA was *not* a helix" (Watson's italics). The shift of emphasis, between November and May, from the lack of evidence *for* a helical structure to positive evidence *against* such a structure is interesting (hardly conclusive) because Wilkins has claimed that Franklin was unequivocally against a helical structure as early as the fall of 1951. For Franklin's colloquium notes, see pp. 118–121; for her opinion in February 1952 that DNA in both forms, crystalline and wet, was unmistakably helical, Rosalind E. Franklin, "Interim Annual Report: January 1st 1951—January 1st 1952" (unpublished report to Turner and Newall fellowship committee, dated 7 February 1952), pp. 2, 3, 4–5; carbon copy in Klug's file of Franklin's papers. When the flat contradiction between those documents and Wilkins's claim was pointed out to him, he suggested that she had been "a little bit two-faced," saying one thing to him at King's College in the winter of 1951–'52 while putting a different view in her report: interview, Wilkins, 15 September 1975. Gosling, who was present at the Cambridge confrontation, recalled that Franklin was against model building at that stage in solution of a structure as complicated as DNA appeared to be, but not that she was then against a helical structure: interview, Gosling. Stokes doubted that Franklin was anti-helical before March 1952—which fits with the time of the first X-ray-diffraction photo that first gave her evidence, as she thought, that DNA could not be helical: interview, Stokes. For a discussion of that photo and her interpretation, see pp. 134, 141 et seq.

106   "what Jim said wasn't reasonable"—Interview, Crick, 10 September 1975.

106   Bragg's ban—*The Double Helix*, pp. 97–99.

107   Crick's coiled coil—F. H. C. Crick, "Is α-Keratin a Coiled Coil?" *Nature* 170 (1952): 882–883.

107   Crick's Christmas present—*The Double Helix*, p. 101.

107   Watson's fellowship moved to England—Letter, Watson to Delbrück, 20 May 1952; original in Delbrück Papers, California Institute of Technology archive.

108   Luria denied passport—Interview, Luria, 17 March 1971; *The Double Helix*, pp. 118–119.

108   Luria's paper—S. E. Luria, "An Analysis of Bacteriophage Multiplication." in *The Nature of Virus Multiplication: Second Symposium of the Society for General Microbiology*, held at Oxford University, 16 and 17 April 1952 (Cambridge: At the University Press, 1953), pp. 99–113.

108   Oxford meeting—Interview, Hayes; interview, Gunther Stent; Watson's contribution to the

discussion, after Luria, "An Analysis of Bacteriophage Multiplication." pp. 113–116; *The Double Helix*, pp. 118–121.

108 Hershey on the Waring Blender experiment—A. D. Hershey, "The Injection of DNA into Cells by Phage,"in *Phage and . . . Molecular Biology*, pp. 100–108; quotes from p. 104.

108 Anderson's wry wonder—Anderson, "Electron Microscopy of Phages," in *Phage and . . . Molecular Biology*, p. 76.

108 the tedious paper—A. D. Hershey and Martha Chase, "Independent Functions of Viral Protein and Nucleic Acid in Growth of Bacteriophage," *Journal of General Physiology* 36 (1952): 39–56; "Received for publication, April 9, 1952."

108 spread word by letters—Not only to Watson; see, e.g., letter, Ole Maaløe to Alfred Hershey, 2 April 1952; original in archives of California Institute of Technology.

109 Delbrück on Hershey and Chase—Interview, 9 July 1972.

109 zipper—L. F. Hewitt, in discussion of Luria, "An Analysis of Bacteriophage Multiplication," p. 116; Watson's "hat and hatbox," ibid.

109 Watson on Wilkins on diameter of nucleic acid—Watson in discussion in *Second Symposium of the Society for General Microbiology*, p. 171.

109 Royal Society meeting—Program, with chief participants, in file of The Royal Society, London.

109 Pauling's passport—*The New York Times*, 12 May 1952, p. 8; *ibid.*, 13 May, p. 10, cites letters to *The Times*, London; interview, Pauling, 23 December 1975.

109 Crick, "scandal"—Interview, Crick, 3 July 1975.

109 Watson's reaction—*The Double Helix*, p. 117.

110 If Pauling had visited—Wilkins said he would have shown Pauling the pictures: interview, 12 March 1976.

110 Wilkins's research plans for 1952—Letter, Wilkins to Crick, "On train Innsbruck-Zurich," undated, late winter 1952.

111 Franklin, "helix in the wet state"—Franklin, "Interim Annual Report: . . . January 1st 1952," p. 3.

112 Franklin to Gosling—Interview, Gosling.

112 "double orientation"—Franklin, notebook entry for 18 April 1952; in Klug's file of Franklin's papers.

112 in Franklin's defense—Klug, "Rosalind Franklin and the Discovery of the Structure of DNA," *Nature* 219 (24 August 1968): 843–844.

112 Franklin in audience May 1—The only person to place her at the meeting was Dr. Pauline Cowan Harrison, who had known Franklin since the summer of 1951, and who became her colleague at King's College London in the fall of 1952; Dr. Harrison was at the Royal Society meeting and vividly remembered a conversation during which Franklin warned her against coming to King's. Telephone interview, Harrison.

112 Wilkins to Watson—*The Double Helix*, p. 122.

112 Franklin's techniques—Interviews, Gosling; Klug, 19 January 1976; Klug, "Rosalind Franklin and the Discovery of the Structure of DNA," *Nature* 219 (24 August 1968): 810, 844; Franklin's notebooks for April and May 1952, in Klug's file; Rosalind E. Franklin and R. G. Gosling, "The Structure of Sodium Thymonucleate Fibres: I. The Influence of Water Content," *Acta Crystallographica* 6 (1953): 673–677, in particular 674.

114 Wilkins led by Franklin's assertions—Wilkins, "X-ray Diffraction Studies of DNA in King's College," pp. 33–35; interview, Wilkins, 26 June 1971.

114 Franklin's restlessness—For idea of returning to Paris, letter, Franklin to Anne Sayre, 1 March 1952, cited and quoted in Sayre, *Rosalind Franklin and DNA* (New York: Norton, 1975), pp. 137–138 and 208n. For joining Bernal, letter, Franklin to J. D. Bernal, 19 June 1952; carbon copy in Klug's file of Franklin's papers.

114 crystallization of TMV—Wendell M. Stanley, "Isolation of a Crystalline Protein Possessing the Properties of Tobacco-Mosaic Virus," *Science* 81 (1935): 644–645.

115 Bernal's striking TMV result—J. D. Bernal and I. Fankuchen, "X-ray and Crystallographic Studies of Plant Virus Preparations," *Journal of General Physiology* 25 (1941): 111–165.

115 Watson's TMV photo—*The Double Helix*, pp. 124–125.

115 "the Jim Watson type of science"—Interview, Kendrew, 13 January 1971.

115 Pauling's interest, spring 1952—Interview, Pauling, 1 March 1971; interview, Rich, 9 August 1977.

116 Crick, "any pairing at all?"—Interview, 3 July 1975.

116 Crick in full spate—*The Double Helix*, p. 126.

116 Griffith already speculating—Olby, "Francis Crick, DNA, and the Central Dogma," in *Daedalus*, Fall 1970, pp. 956–957. Olby interviewed Griffith in May 1968; Griffith died in 1972. By Olby's account, Griffith had been thinking about bases paired like dominoes on the flat rather than like playing cards interleaved, but said nothing to Crick about his own scheme; Crick, however, was surprised and skeptical to learn that: interview, 10

September 1955. Dr. Olby has kindly allowed me to listen to the tape of his interview with Griffith, but unfortunately it does not settle what Griffith had in mind.

117   in line for tea—Quotations are from an interview, Francis Crick to Robert Olby, 8 March 1968; Dr. Olby has kindly allowed me to hear the tape, which I transcribed afresh.

117   Delbrück on complementary replication—Interview, 9 July 1972; Linus Pauling and Max Delbrück, "The Nature of the Intermolecular Forces Operative in Biological Processes," *Science* 92 (26 July 1940): 77–79; the paper by Pascual Jordan is cited therein. Watson's "well worn": *The Double Helix*, p. 127.

118   Crick's meeting Franklin—Interview, Crick, 3 July 1975.

118   Crick on Franklin's claimed double orientation—Interviews, 3 July 1975 and 10 September 1975.

119   encounter with Chargaff—*The Double Helix*, p. 130. Chargaff, with the aid of his diary for the year, dated his visit to Cambridge as 24 to 27 May 1952: Erwin Chargaff, *Heraclitean Fire: Sketches from a Life before Nature* (New York: Rockefeller University Press, 1978), pp. 100–103, and checking query, by telephone, from Sara Spencer at *The New Yorker*.

119   Chargaff—Interview, Chargaff, 15 February 1972; letter. Chargaff to Judson, 25 September 1976.

120   Crick's account of conversation with Chargaff—Interview conducted with Crick by Robert Olby, 8 March 1968.

120   "nowadays . . . published it straight away"—Crick, interview, 1 September 1971; he returned to the idea in our conversation 3 July 1975; his week of experiments that failed to find base pairing was discussed in those interviews.

121   "death . . . of helix"—A photocopy of the card was supplied to me by Alexander Stokes; interview, Stokes, 11 August 1976.

121   Watson at Paris congress—*The Double Helix*, p. 132; Erwin Chargaff, "The Nucleic Acids of Microorganisms," in *Symposium sur le Métabolisme Microbien*, IIe *Congrès International de Biochimie* (Paris: Société d'Édition d'Enseignment Supérieur, 1952), pp. 41–46; Christoph Tamm and Erwin Chargaff, "Observations on the Distribution Density of Individual Nucleotides Within a Desoxyribonucleic Acid Chain" (abstract), in *Résumés des Communications, IIe Congrès International de Biochimie* (Paris: Masson, 1952) p. 206; Tamm did not, obviously, address the question whether the DNA molecule was made of more than one chain.

122   Pauling on short notice—His passport had been issued after he had signed an affidavit saying he was not a Communist; *The New York Times*, 16 July 1952, p. 2.

122   Royaumont—*The Double Helix*. pp. 133–138; the talk on TMV is not included in the proceedings of *Colloquium Royaumont* 1952, supplement to *Annales de l'Institut Pasteur* 84 (1953); André Lwoff, "Truth, Truth, What Is Truth?" (review of *The Double Helix*), *Scientific American*, July 1968, p. 134.

122   boyish, fit, vital—photo in *The Double Helix*, p. 139.

122   bacterial conjugation—Joshua Lederberg and Edward L. Tatum, "Novel Genotypes in Mixed Cultures of Biochemical Mutants of Bacteria," *C. S. H. Symp. Q. Biol.* 11 (1946): 113–114; *idem*, "Gene Recombination in *Escherichia coli*," *Nature* 158 (1946): 558.

122   linear chromosome in bacteria—Joshua Lederberg, "Gene Recombination and Linked Segregations in *Escherichia coli*," *Genetics* 32 (1947): 505–525.

122   "rabbinical complexity"—*The Double Helix*, p. 142.

123   Watson met Hayes—*The Double Helix*, p. 141; interview, Hayes.

123   bacterial gender—J. Lederberg, L. L. Cavalli, and E. M. Lederberg, "Sex Compatibility in *Escherichia coli*," *Genetics* 37 (1952): 720–730; William Hayes, "Recombination in *Bact. coli* K12: Unidirectional Transfer of Genetic Material," *Nature* 169 (1952): 118–119.

123   "seemed a good one at the time"—William Hayes, "Sexual Differentiation in Bacteria," in *Phage and . . . Molecular Biology*, p. 210.

123   Watson on beating Lederberg—*The Double Helix*, p. 143.

123   wrote paper together—James D. Watson and William Hayes, "Genetic Exchange in *Escherichia coli* K12: Evidence for Three Linkage Groups," *Proc. N. A. S.* 39 (1953): 416–426.

123   relationship of Watson to Hayes—Interview, Hayes.

123   Franklin's notebook—In Klug's file of her papers.

3 *"Then they ask you, 'What is the significance of DNA for mankind, Dr. Watson?'"*

PAGE

125   Crick on politeness in science—British Broadcasting Company, "The Prizewinners" (television program), broadcast Tuesday, 11 December 1962, 9:25 to 10:25 P.M.—the day after the award of Nobel Prizes to Perutz and Kendrew and to Crick, Watson, and Wilkins.

125   impressions of Franklin—Interviews, Klug, Gosling, Wilkins, Luzatti, Stokes, Francis Crick, Odile Crick; conversation with Adrienne Weil; interviews with Angela Martin Brown,

Dame Honor Fell, Sylvia Fitton Jackson, Anne Sayre, and by telephone with Pauline Cowan Harrison and Margaret Pratt North.

126 the nickname "Rosy"—Letter, Wilkins to Judson, 12 July 1976.

126 Franklin and the standing of women in Randall's laboratory—Interviews with Angela Martin Brown, Dame Honor Fell, Sylvia Fitton Jackson; by telephone with Pauline Cowan Harrison and Margaret Pratt North; letter, Marjorie M'Ewen to Judson, 15 September 1976; letter, Margaret Pratt North to Judson, 2 February 1977.

126 Klug, Crick, Gosling on Franklin—Interviews: Klug, 27 June 1975; Crick, 3 July 1975; Gosling, 21 July 1975.

127 Patterson synthesis—J. C. Kendrew and M. F. Perutz, "The Application of X-ray Crystallography to the Study of Biological Macromolecules," in *Barcroft Memorial, Haemoglobin*, p. 171.

127 Its defects—e.g., Bragg, "How Proteins Were Not Solved," p. 4.

127 Poincaré—Henri Poincaré "L'invention mathématique," *Bulletin de L'Institut Général Psychologique* 8 (Mai-Juin 1908): 179–180; *trad. auct.*

128 Franklin and Gosling's procedure—Rosalind E. Franklin and R. G. Gosling, "The Structure of Sodium Thymonucleate Fibres. II. The Cylindrically Symmetrical Patterson Function," *Acta Crystallographica* 6 (1935): 680. Interview, Gosling.

128 Franklin's notebook entries—Notebook in Klug's file of Franklin's papers.

129 Watson's autumn—*The Double Helix*, pp. 147, 149, 151–153; interview, Hayes.

129 Wilkins's autumn—M. H. F. Wilkins, "X-ray Diffraction Studies of DNA in King's College from 1950 Onwards" (unpublished manuscript, March 1976), p. 35; H. G. Davies, M. H. F. Wilkins, J. Chayen, and L. F. La Cour, "The Use of the Interference Microscope to Determine Dry Mass in Living Cells and as a Quantitative Cytochemical Method," *Quarterly Journal of Microscopical Science* 95 (September 1954): 271–304.

129 Pauling's autumn—Linus Pauling and Robert B. Corey, "Compound Helical Configurations of Polypeptide Chains: Structure of Proteins of the α-Keratin Type," *Nature* 171 (1953): 59; F. H. C. Crick, "Is α-Keratin a Coiled Coil?" *Nature* 170 (1952): 882–883.

129 Pauling's reading list—Provided by the citations at the end of the paper offering a structure for DNA, Linus Pauling and Robert B. Corey, "A Proposed Structure for the Nucleic Acids," *Proc. N. A. S.* 39 (1953): 96–97.

130 Pauling on Todd, Chargaff—Interview, Pauling, 23 December 1975.

130 Robley Williams's seminar, Pauling's triple helix—Linus Pauling, "Molecular Basis of Biological Specificity" (contribution to special section, "Molecular Biology Comes of Age"), *Nature* 248 (26 April 1974): 771. Interview, Rich, 9 August 1977.

130 preparing two papers—R. E. Franklin and R. G. Gosling, "The Structure of Sodium Thymonucleate Fibers. I. The Influence of Water Content," and . . . II. "The Cylindrically Symmetrical Patterson Function," *Acta Crystallographica* 6 (1953): 673–677 and 678–685. For the timing, and the fact that they had no model in mind yet: Franklin's notebooks, in Klug's file; interview, Gosling; and sources in note following.

131 the report for the Biophysics Research Committee—Medical Research Council, Biophysics Research Unit, Wheatstone Physics Laboratory, King's College London, "Notes on Current Research Prepared for the Visit of the Biophysics Research Committee 15th December, 1952," mimeographed, 12 pp., 5 December 1952. The first page is a staff list. "Nucleic Acid Research," described pp. 5–8, comprised "Desoxyribose Nucleic Acid and Nucleoprotein Structure (M. H. F. Wilkins)," p. 6, and "X-ray Studies of Calf Thymus D.N.A. (R. E. Franklin and R. G. Gosling)," pp. 6–8. For the provenance of the visiting committee and its makeup, see M. F. Perutz, "DNA Helix" (letter to the editor), *Science* 164 (27 June 1969): 1537–1538, and letters from M. H. F. Wilkins and James D. Watson, ibid., 1538–1539.

131 news of Pauling's structure—*The Double Helix*, pp. 156–158.

131 Pauling's Christmas—Conversation with Leonard Lerman, start canteen of Medical Research Council Laboratory of Molecular Biology, autumn 1974, confirmed by telephone 4 August 1977; interview, Pauling, 23 December 1975; Linus Pauling and Robert B. Corey "A Proposed Structure for the Nucleic Acids," pp. 84–97; letter, Linus Pauling to Peter Pauling, 31 December 1952, cited and partly quoted in Olby, *Path*, p. 394; Linus Pauling and Robert B. Corey, "Structure of the Nucleic Acids," *Nature* 171 (21 February 1953), p. 346.

132 Frankin's letter to Corey—I have not been able to trace her letter, but Corey's reply, which dates and describes hers, is in Klug's file of Franklin's papers: letter, 13 April 1953, Robert B. Corey to Rosalind Franklin.

132 Pauling's press conference—*The New York Times*, 20 January 1953, p. 27 col. 3; 25 January 1953, section IV, p. 9 col. 6.

132 Franklin's letter—Letter, R. E. Franklin to J. D. Bernal, dated "January 14th" with no year; but the year is of course confirmed by the date and day, "Wednesday January 28th," in text of letter. Klug obtained a copy of the letter from Bernal's files.

132   her new notebook—Like the old, in Klug's file of Franklin's papers.

133   wintry January day—Yes, the weather was chilly: see the reports that week in *The Times*.

133   Pauling mailed ms.—Letter, Linus Pauling to Peter Pauling, 21 January 1953, cited in Olby, *Path*, p. 394; *The Double Helix*, p. 159.

133   Watson's reaction to manuscript—*The Double Helix*, p. 247.

134   the erroneous paper—Linus Pauling and Robert B. Corey, "A Proposed Structure for the Nucleic Acids," pp. 84, 85, 87, 93, 96; interview, Pauling, 1 March 1971.

135   "not an acid at all"—*The Double Helix*, pp. 160–161.

136   the encounter with Franklin—*The Double Helix*, pp. 164–167; interview, Wilkins, 15 September 1975, though he suggested Watson's account may have been slightly exaggerated.

136   the "B" photo—*The Double Helix*, pp. 167–171; interviews; Wilkins, 15 September 1975 and 12 March 1976; interview, Klug, 19 January 1976; interview, Watson, 5 October 1973.

138   Bragg's reaction to Watson—Interview, Crick, 1 September 1971; conversation, Perutz, 23 January 1975, as well as earlier conversation, Paris, 2 October 1971.

139   Pauling's triple too tight—Letter, Linus Pauling to Peter Pauling, 4 February 1953, cited and in part quoted in Olby, *Path*, pp. 382–383; letter, Robert B. Corey to Rosalind Franklin, 13 April 1953, original in Klug's file of Franklin's papers.

140   bases on the inside—Crick in interview with Robert Olby, 10 July 1968, quoted in Olby, *Daedalus*, Fall 1970, pp. 961 and 984n.; *The Double Helix*, pp. 177–178.

140   Wilkins to Sunday lunch—Letter, Wilkins to Crick, dated "Thursday," and written after Franklin's symposium of January 28, to which it refers; Peter Pauling, "DNA: The Race That Never Was?" *New Scientist* 58 (31 May 1973): 559, interview, Crick, 1 September 1971; *The Double Helix*, p. 179.

140   Franklin's notes—Notebook, and loose sheet noting reaction to the Pauling-Corey structure, in Klug's file of Franklin's papers.

141   Crick gets MRC committee report—*The Double Helix*, pp. 181, 182; André Lwoff, "Truth, Truth, What Is Truth?" *Scientific American*, July 1968, p. 134; Erwin Chargaff, "A Quick Climb Up Mount Olympus," *Science* 159 (29 March 1968): 1449; M. F. Perutz, "DNA Helix," pp. 1537–1538, and letters from M. H. F. Wilkins and James D. Watson, ibid., 1538–1539; interview, Crick, 10 September 1975.

141   Franklin's paragraphs—R. E. Franklin and R. G. Gosling, "X-ray Studies of Calf Thymus D.N.A.," in Medical Research Council, Biophysics Research Unit . . ., pp. 6–8; the dimensions they gave for the unit cell were: $a = 22.0$ A; $b = 39.8$ A; $c = 28.1$ A; $\beta = 96.5°$.

142   the crucial fact of the dyad—This passage and the ensuing one, about Watson's first trying to build the model with the sugars too close, was specifically checked with Crick and the wording of the quotes revised, interview, 10 September 1975; see also Olby, *Daedalus*, Fall 1970, p. 961.

143   "ludicrous to say [Jim] didn't understand"—Interview, Crick, 3 July 1975.

143   "biological objects came in pairs"—*The Double Helix*, p. 171.

143   "a very good reason"—Interview, Crick, 10 September 1975; the argument was first published in F. H. C. Crick and J. D. Watson, "The Complementary Structure of Deoxyribonucleic Acid," *Proceedings of the Royal Society* of London, A, 223 (1954): 82–83.

144   standard monographs—James Norman Davidson, *The Biochemistry of the Nucleic Acids* (London: Methuen, 1950). The diagrams are in chap. 2. J. Masson Gulland, "The Structures of Nucleic acids," in *Symp. Soc. Exp. Biol.* 1, held 1946, pp. 1–14, and in particular J. Masson Gulland and D. O. Jordan, "The Macromolecular Behaviour of Nucleic Acids," in ibid., pp. 56–65; these contain references to their earlier work.

144   Pauling to his son—Letter, Linus Pauling to Peter Pauling, 18 February 1953, cited and quoted in Olby, *Path*, pp. 383, 409.

145   Broomhead's papers—June M. Broomhead, "The Structures of Pyrimidines and Purines: II. A Determination of the Structure of Adenine Hydrochloride by X-ray Methods," *Acta Crystallographica* 1 (1948): 324–329, and "The Structures of Pyrimidines and Purines: IV. The Crystal Structure of Guanine Hydrochloride and Its Relation to That of Adenine Hydrochloride," ibid. 4 (1951): 92–100. Copies of both of these were in Franklin's file, each with a sheet of her notes on the paper.

145   Watson's like-with-like—*The Double Helix*, pp. 182–188; letter, Watson to Delbrück, 20 February 1953, original now in the Delbrück Papers in the archive of the California Institute of Technology.

146   "torn to shreds"—*The Double Helix*, p. 189; interview, Donohue; Jerry Donohue, "The Hydrogen Bond in Organic Crystals," *Journal of Physical Chemistry* 56 (1952): 502–510.

147   how much more Donohue might have said—Interview, Watson, 5 October 1973; interview, Donohue.

147   Franklin's failure—The notebook with Klug's annotations is in his file of her papers; the paper was Rosalind E. Franklin and R. G. Gosling, "A Note on Molecular Configuration in Sodium Thymonucleate—Rough Draft" (typescript, 5 pp.), in Klug's file; interviews, Klug, 19 January 1976, and 6 June 1975.

148 Crick's insight—Interview, Crick, 1 September 1971.

149 Watson's insight—Interview, Watson, 5 October 1973; *The Double Helix*, pp. 194, 196.

149 a poem about science—Edna St. Vincent Millay, "Euclid alone has looked on Beauty bare," in her *Collected Poems*, ed. Norma Millay (New York: Harper, 1956), p. 605; copyright 1923, 1951 by Edna St. Vincent Millay and Norma Millay Ellis.

151 "our dark lady"—Letter, Wilkins to Crick, dated "Saturday"; Franklin began at Birkbeck College London on Monday, March 16, so that the week following Wilkins's letter was her last at King's College—see her "Annual Report: 1st January, 1953–1st January, 1954" to the Turner and Newall fellowship committee, copy in Klug's file of Franklin's papers; interview, Crick, 10 September 1975.

151 building, viewing the model—Interviews, Crick, 3 July and 10 September 1975; *The Double Helix*, pp. 198–205, 214; interview, Todd; interview, Bragg; the detail of G. F. S. Searle's reaction to the model is from J. G. Crowther, *The Cavendish Laboratory 1874–1974* (London: Macmillan, 1974), pp. 288–289, though Crowther puts Searle's age too high by a year.

151 Watson's correspondence—Letter, Pauling to Watson, 5 March 1953, cited and quoted in Olby, *Path*, p. 409; letter, Watson to Delbrück, 12 March 1953, original now in the Delbrück Papers at the California Institute of Technology.

152 sense of dénouement—Interview, Delbrück, 9 July 1972.

153 Wilkins, Franklin learn of the model—Rosalind E. Franklin and R. G. Gosling, "A Note on Molecular Configuration"; letter, Wilkins to Crick, 18 March 1953. Wilkins has told me that he believes he went to Cambridge to see the model *before* writing his letter of the 18th, but this is doubtful: the letter, though not clear on the point, implies otherwise; *The Double Helix*, pp. 208–209, has Watson present when Wilkins saw the model, and it is certain that Watson was in Paris for most if not all of the few days between Kendrew's telephoning Wilkins and the 18th, for Gerard Wyatt has found an entry in his diary for 1953 that places Watson at the Institut Pasteur when Wyatt visited there Monday March 16—telephone conversation, Wyatt, Saturday 29 April 1978; neither Kendrew nor Crick, in interviews, were clear about this timing.

153 "Jim Watson and I . . ."—Letter, Crick to Michael Crick, 19 March 1953.

153 Wyatt's contribution—Telephone conversation, Wyatt, 26 April 1978; interview, Crick, 3 July 1975; Gerald R. Wyatt, "The Nucleic Acids of Some Insect Viruses," *Journal of General Physiology* 36 (1952): 201–205.

154 Pauling's visit—*The Double Helix*, pp. 222–223. The Solvay meeting took place from 6 to 14 April 1953, and the announcement of the structure of DNA was introduced after Bragg's paper, as a "note complémentaire": J. D. Watson and F. H. C. Crick, "The Stereochemical Structure of DNA," in Institut International de Chimie Solvay, Neuvième Conseil de Chimie . . . *Les Protéines: Rapports et Discussions* (Brussels: R. Stoops, 1953), pp. 110–112; Pauling's remarks in the discussion, p. 113; list of participants, pp. 9–10.

154 the papers in *Nature*—J. D. Watson and F. H. C. Crick, "A Structure for Deoxyribose Nucleic Acid," *Nature* 171 (25 April 1953): 737–738; M. H. F. Wilkins, A. R. Stokes, and H. R. Wilson, "Molecular Structure of Deoxypentose Nucleic Acids" ibid., pp. 738–740; Rosalind E. Franklin and R. G. Gosling, "Molecular Configuration in Sodium Thymonucleate," ibid., pp. 740–741.

155 coy statement—Stent, *Molecular Genetics*, p. 204; Crick says the statement was a claim to priority: see next reference, p. 766.

155 twenty-first anniversary—Francis Crick, "*The Double Helix*: A Personal View," *Nature* 248 (26 April 1974): 768.

155 discovery hard to pin down—Interview, Crick, 10 September 1975; interview, Watson, 5 October 1973.

157 "I'm Jim, I'm smart"—Interview, Wilkins, 26 June 1971.

157 Watson on his book—Interview, Watson, 5 October 1973.

157 incandescently angry letter—Letter, Crick to Watson, 13 April 1967.

157 *The Loose Screw*—Francis Crick, "*The Double Helix*," p. 768.

157 lust for Nobel Prize—Interview, Crick, 1 September 1971.

158 Bragg kind to Watson—Sir Lawrence Bragg, "Foreword" to *The Double Helix*, pp. vii–ix; interview, Bragg.

158 "are you doing science?"—Interview, Watson, 24 September 1973; also 1 December 1970.

159 residue of nostalgia—*The Double Helix*, p. 223; Watson, *Molecular Biology*, pp. 66–67.

159 the later papers—J. D. Watson and F. H. C. Crick, "Genetical Implications of the Structure of Deoxyribonucleic Acid," *Nature* 171 (30 May 1953): 964–967; J. D. Watson and F. H. C. Crick, "The Structure of DNA," *C. S. H. Symp. Q. Biol.* 18 (1953): 123–131; F. H. C. Crick and J. D. Watson, "The Complementary Structure of Deoxyribonucleic Acid," *Proceedings of the Royal Society* of London A223 (1954): pp. 80–96.

160 Watson objects to BBC—Letter, Watson to Crick, 9 October 1953.

161 Franklin ordered off DNA—Conversation, Crick, 10 July 1976; letter, Randall to Franklin, 17 April 1953; Rosalind E. Franklin and R G. Gosling, "Evidence for 2-Chain Helix in Crystalline Structure of Sodium Deoxyribonucleate," *Nature* 172 (25 July 1953): 156–157; Rosalind E. Franklin and R. G. Gosling, "The Structure of Sodium Thymonucleate Fibres: III. The Three-Dimensional Patterson Function," *Acta Crystallographica* 8 (1955): 151–156; interview, Klug, 19 January 1976.

162 semi-conservative replication—Max Delbrück and Gunther S. Stent, "On the Mechanism of DNA Replication," in *McCollum-Pratt Symposium*, 1956, pp. 699–736; interview, Stent; Cyrus Levinthal, The Mechanism of DNA Replication and Genetic Recombination in Phage," *Proc. N. A. S.* 42 (1956): 394; Gunther S. Stent and Niels K. Jerne, "The Distribution of Parental Phosphorus Atoms Among Bacteriophage Progeny," ibid. 41 (1955): 704; interview, Meselson, October 1971.

163 the beauty of the experiment—Watson, *Molecular Biology*, p. 298; interview, Cairns.

163 Haldane's "astonishing words"—J. B. S. Haldane, *New Paths in Genetics* (London: George Allen & Unwin, 1941), p. 44—and yet, in the next sentence, he takes it for granted that genes are made of protein; letter, Meselson to Judson, 8 May 1976.

164 the Meselson-Stahl experiment—Interviews, Meselson, October 1971 and 8 August 1977; interview, Stahl; Matthew Meselson and Franklin W. Stahl, "Demonstration of the Semiconservative Mode of DNA Replication," in *Phage and . . . Molecular Biology*, pp. 246–251; Matthew Meselson, Franklin W. Stahl, and Jerome Vinograd, "Equilibrium Sedimentation of Macromolecules in Density Gradients," *Proc. N. A. S.* 43 (1957): 581–583; Matthew Meselson and Franklin W. Stahl, "The Replication of DNA in *Escherichia Coli*," ibid. 44 (1958): 671–682.

165 heavy nitrogen—Telephone conversation, Dr. Warren Ruderman, 27 April 1978; Ruderman in the mid fifties was president of Isomet Corporation, the firm that supplied the heavy nitrogen; the manufacturing process was developed by Dr. Thomas Ivan Taylor, at Columbia University. Linus Pauling once told me—letter, 8 October 1976—that *he* had obtained the ^{15}N for Meselson: as a Foreign Member of the Soviet Academy of Sciences, he wrote to the Academy's president to obtain the isotope. But Pauling misremembered: his request was made later, and for another isotope, when Meselson needed some heavy carbon, ^{13}C, not then made at sufficient purity in the United States, for a different experiment using density-gradient centrifugation.

166 "clean as a whistle!"—Letter, Meselson to Watson, 8 November 1957.

167 Franklin's later work—See, e.g., Rosalind E. Franklin, A. Klug, and K. C. Holmes, "X-ray Diffraction Studies of the Structure and Morphology of Tobacco Mosaic Virus," in *Ciba Symposium on Viruses* 1956, pp. 39–52.

167 Wilkins's later work on DNA—Wilkins's Nobel lecture, 11 December 1962, explains the work and gives full references; M. H. F. Wilkins, "The Molecular Configuration of Nucleic Acids," in *Les Prix Nobel en 1962*, pp. 126–154.

167 "mustn't give the impression"—Interview, Crick, 10 September 1975.

168 "*the dominant motive*"—Interview, Watson, 16 March 1971; also 1 December 1970.

169 "that young Watson"—Interview, Bragg.

4 *On T. H. Morgan's Deviation and the Secret of Life*

PAGE

177 "anything that interests molecular biologists"—Francis Crick, "Molecular Biology in the Year 2000," *Nature* 228 (14 November 1970): 613n; interview, Perutz, 8 May 1968.

177 Crick's role—Interview, Monod, September 1970.

178 "I do my thinking both ways"—Interview, Crick, 1 September 1971.

178 "molecular biology is an ambiguous term"—Interview, Crick and Brenner, 15 September 1970, supplemented and clarified in the case of Brenner's contribution by the interview with him, 8 May 1968, which covered almost identical ground; for Crick's contribution, see also F. H. C. Crick, "Molecular Biology and Medical Research," *Journal of the Mount Sinai Hospital* 36 (May-June 1969): 178–188, and F. H. C. Crick, "Recent Research in Molecular Biology: Introduction," in *British Medical Bulletin* 21 (September 1965): 183–186, which articles present several of Crick's favorite points of that period in expanded and somewhat more formal terms. The text of the interview was read and commented on by Crick, and more briefly by Brenner, in November 1975.

179 Watson in his textbook—J. D. Watson, *Molecular Biology of the Gene*, 1st ed. (New York: W. A. Benjamin, 1965), pp. 99–100.

181 Mendel—Gregor Mendel, *Experiments in Plant-Hybridisation*, translation of the Royal Horticultural Society, corrected and annotated by W. Bateson (Cambridge, Mass.: Harvard University Press, 1946), pp. 1–2.

181 Morgan—Thomas Hunt Morgan, "The Relation of Genetics to Physiology and Medicine,"

Nobel prize lecture delivered 4 June 1934, in Nobel Foundation, *Nobel Lectures . . .: Physiology or Medicine, 1922–1941* (Amsterdam: Elsevier, 1965), pp. 315–316; Thomas Hunt Morgan, *Embryology and Genetics* (New York: Columbia University Press, 1934), p. 9.

184   "outside the magic circle"—Crick has made the point repeatedly; for this exact wording, see his interview with Robert Olby, 8 March 1968, as quoted and cited in Olby, "Francis Crick, DNA, and the Central Dogma," in *Daedalus*, Fall 1970, pp. 970, 986n.

184   "It's quite ludicrous"—Interview, Crick, 1 September 1971.

185   Molecular biology's origins—See, e.g., John F. Kendrew, "Information and Conformation in Biology," in *Structural Chemistry and Molecular Biology*, pp. 187–197; Gunther S. Stent, "That Was the Molecular Biology That Was," *Science* 168 (1968): 390–395.

185   Loeb—Jacques Loeb, *The Mechanistic Conception of Life*, ed. Donald Fleming (Cambridge, Mass.: The Belknap Press of Harvard University Press, 1964), p. 5.

185   "the aim of molecular biology"—Interviews, Monod, September and December 1970.

186   doubts of reality of large molecules—See, for history, John H. Northrop, "Biochemists, Biologists, and William of Occam," *Annual Review of Biochemistry* 30 (1961): 1–10; John T. Edsall, "Proteins as Macromolecules: An Essay on the Development of the Macromolecule Concept and Some of Its Vicissitudes," in *Perspectives in the Biochemistry of Large Molecules*, (*Archives of Biochemistry and Biophysics* Supplement 1, 1962), pp. 12–20; N. W. Pirie, "Patterns of Assumption About Large Molecules," ibid., pp. 21–29.

188   Medawar—Peter B. Medawar, *The Art of the Soluble* (London: Methuen, 1967), p. 105.

188   Perutz on hemoglobin—Max F. Perutz, "Haemoglobin: The Molecular Lung," *New Scientist*, 17 June 1971, pp. 676–679; he described hemoglobin the same way in his Croonian Lecture, as delivered 30 May 1968, though the phrases did not survive in the printed version (M. F. Perutz, "The Haemoglobin Molecule," *Proceedings of the Royal Society* of London, B, 173 [1969]: 113–140).

189   Garrod; Beadle and Tatum—Sir Archibald Garrod, *Inborn Errors of Metabolism*, 2nd ed. (London: Oxford University Press, 1923, first ed. 1909); George W. Beadle and Edward L. Tatum, "Genetic Control of Biochemical Reactions in *Neurospora*," *Proc. N. A. S.* 27 (1941): 499–506, was the first paper, but the explicit statement of the one-to one relation between gene and enzyme was George Beadle, "Genes and the Chemistry of the Organism," *American Scientist* 34 (1946): 31–53, 76; see also George W. Beadle, "Biochemical Genetics: Some Recollections,"in *Phage and . . . Molecular Biology*, pp. 23–32.

191   Crick sat back—F. H. C. Crick, "The Genetic Code: III," *Scientific American*, October 1966, p. 62.

192   wine night at King's—Interview, Sydney Brenner, 26 January 1971.

196   "the barbarism of the 20th century"—Interview, Chargaff; letter, Chargaff to Judson, 25 September 1976; Chargaff, "What Really Is DNA?" pp. 298, 327, 330; Chargaff, "Grammar," p. 641.

5 *"The number of the beast"*

PAGE

234   Medawar—Peter B. Medawar, *Induction and Intuition in Scientific Thought* (Philadelphia: American Philosophical Society, 1969), p. 48.

234   essential to include errors—Interview, Monod, 1 December 1975; interview, Crick, 20 November 1975.

236   Sanger's latest results—F. Sanger and E. O. P. Thompson, "The Amino-Acid Sequence in the Glycyl Chain of Insulin," 1 and 2 *Biochemical Journal* 53 (1953): 353–374.

236   Gamow's scheme—Interview, Crick with Brenner, 19 November 1975; F. H. C. Crick, "The Genetic Code—Yesterday, Today, and Tomorrow," *C. S. H. Symp. Q. Biol.* 31 (1966): 4.

237   Watson, "strange feeling"—Letter, Watson to Delbrück, 22 March 1953; original in Delbrück Papers, California Institute of Technology.

237   Delbrück's reaction—Letter, Delbrück to Watson, 14 April 1953; letter, Delbrück to Bohr, 14 April 1953; originals in Delbrück Papers. Letter, Delbrück to Judson, 26 June 1978.

238   a group came to Cambridge—Interviews, Brenner, 7 February 1976; Orgel; letter, Jack D. Dunitz to Judson, 22 March 1976.

238   Brenner's background—Interview, 7 February 1976, with some biographical detail from the transcript of an interview with Brenner conducted on 21 May 1975, in Cambridge, by Charles Weiner of the Massachusetts Institute of Technology, of which Brenner allowed me to make a copy; further detail from school and university records and correspondence in Brenner's early papers.

238   "some work on phage"—Letter, Sir Cyril Hinshelwood to Brenner. 29 May 1952.

238   labelling DNA with dyes—Orgel, interview, also recalled the idea.

238   Brenner meeting Crick and Watson—Interview, Crick with Brenner, 19 November 1975; interview, Brenner, 7 February 1976.

239   coiling problem—Letter, Watson to Delbrück, 21 May 1953, original in Delbrück Papers, California Institute of Technology; interview, Crick, 28 June 1977.

239   Cold Spring Harbor paper—J. D. Watson and Francis Crick, "The Structure of DNA," *C. S. H. Symp. Q Biol.* 18 (1953): 127, 130.

240   Hershey's reaction—A. D. Hershey, "Functional Differentiation within Particles of Bacteriophage T2," *C. S. H. Symp. Q. Biol.* 18 (1953): 138.

240   "people didn't *believe* in the code"—Interview, Crick, 20 November 1975.

241   Zamecnik's formula—Zamecnik, Harvey Lecture, p. 256.

241   RNA hard to work with—Interview, John Smith.

242   Caspersson and Schultz—Letter, Torbjörn Caspersson to Judson, 22 December 1976; interview, Monod, 1 December 1975; for Caspersson's methods, T. Caspersson and Jack Schultz, "Cytochemical Measurements in the Study of the Gene," in *Genetics in the 20th Century*, ed. L. C. Dunn (New York: Macmillan, 1951), pp. 155–171; T. Caspersson and Jack Schultz, "Nucleic Acid Metabolism of the Chromosomes in Relation to Gene Reproduction," *Nature* 142 (13 August 1938): 294–295; idem, "Ribonucleic Acids in Both Nucleus and Cytoplasm, and the Function of the Nucleolus," *Proc. N. A. S.* 26 (1940): 507–515; Jack Schultz, T. Caspersson. and L. Aquilonius, "The Genetic Control of Nucleolar Composition," ibid., pp. 515–523.

242   "the first statement of a correlation"—T. Caspersson and Jack Schultz, "Pentose Nucleotides in the Cytoplasm of Growing Tissues," *Nature* 143 (8 April 1939): 602–603.

243   more evidence after Schultz—Torbjörn Caspersson, "Nukleinsäureketten und Genvermehrung," *Chromosoma* 1 (1940): 605–619; idem, "Studien über die Eiweissumsatz der Zelle," *Die Naturwissenschaften* 29 (17 January 1941): 33–43; letter, Caspersson to Judson, 22 December 1976.

243   Brachet's account—Jean Brachet, "La localisation des acides pentosenucléiques dans les tissus animaux et les oeufs d'amphibiens en voie de développement." *Archives de Biologie* 53 (1942): 207–257, discussion of Caspersson, pp. 252–255.

243   small particles—Torbjörn Caspersson, "The Relations between Nucleic Acid and Protein Synthesis," in *Symp. Soc. Exp. Biol.* 1, held 1946, p. 147; Jean Brachet, "Nucleic Acids in the Cell and the Embryo," in ibid., p. 213–215, 222.

243   fragmentary and elusive—Caspersson and Schultz, "Nucleic Acid Metabolism," p. 295; Caspersson, "Studien über den Eiweissumsatz," p. 43, where the wording in German is almost identical to that quoted, which is from Caspersson, "The Relations," p. 136; Brachet, "Nucleic Acids in the Cell," p. 217; interview, Crick, 19 November 1975.

243   Claude's particles—Albert Claude, "Concentration and Purification of Chicken Tumor I Agent," *Science* 87 (20 May 1938): 467–468; idem, "Chemical Composition of the Tumor-Producing Fraction of Chicken Tumor I," *Science* 90 (1 September 1939): 213–214; idem, "Particulate Components of Normal and Tumor Cells," *Science* 91 (19 January 1940): 77–78.

244   Caspersson's scheme—Caspersson, "The Relations," p. 137.

244   Davidson saw clearly—James N. Davidson, "The Distribution of Nucleic Acids in Tissues," *Symp. Soc. Exp. Biol.* 1, held 1946, pp. 80, 82.

244   DNA makes RNA makes protein—André Boivin and Roger Vendrely, "Sur le role possible des deux acides nucléiques dans la cellule vivante," *Experientia* 3 (15 January 1947): 32–34, submitted 15 November 1946.

245   Brachet's intuitive flash—Brachet, "Nucleic Acids in the Cell," pp. 214–215.

245   Astbury's exhortation—William T. Astbury, "X-ray Studies of Nucleic Acids," in *Symp. Soc. Exp. Biol.* 1, held 1946, p. 70.

245   Lipmann and energy—Interview, Lipmann, 19 March 1971, corrected and supplemented on the basis of the interview 7 July 1977; Fritz Lipmann, "Development of the Acetylation Problem: A Personal Account," in *Les Prix Nobel en 1953*, pp. 151–177, where, however, the chief interest is Lipmann's related discovery of Coenzyme A.

246   the accidental clue—Fritz Lipmann, "Die Dehydrierung der Brenztraubsäuren," *Enzymologia* 4 (1937): 65–72.

247   ATP in muscle—The key reference is Otto Meyerhof, E. Lundsgaard, and H. Blaschko, "Über die Energetik der Muskelkontraktion bei aufgehobener Milchsäurebildung," *Die Naturwissenschaften* 18 (1930): 787, cited in Dorothy M. Needham, *Machina Carnis: The Biochemistry of Muscular Contraction in Its Historical Development* (Cambridge: At the University Press, 1971); for recognition of the role of ATP in yielding muscle energy see, e.g., p. 122.

247   imaginative leap—Fritz Lipmann, "Metabolic Generation and Utilization of Phosphate Bond Energy," *Advances in Enzymology* 1 (1941): 99–162.

248   the squiggle—ibid., p. 101 and passim; Herman M. Kalckar, "Lipmann and the 'Squiggle,'" in *Current Aspects of Biochemical Energetics* (New York: Academic Press, 1966), pp. 1–8.

248   Nobel lecture—Lipmann, "Development of the Acetylation Problem," pp. 155, 176, 175.

248 information theory—Claude E. Shannon, "A Mathematical Theory of Communication," *The Bell System Technical Journal* 27 (1948): 379–423 and 623–656—note his credit to Wiener, p. 652; idem, "Communication Theory of Secrecy Systems," ibid., 28 (1949): 656–715, the history of the paper given p. 656n; Claude E. Shannon and W. Wiener, *The Mathematical Theory of Communication* (Urbana: University of Illinois Press, 1949); John von Neumann, *Theory of Self-Reproducing Automata*, ed. and completed by Arthur W. Burks (Urbana: University of Illinois Press, 1966)—for the remarks on the relation between a complex machine and its description, p. 47 et seq., and for building such a machine, Lecture 5; see also John Kemeny, "Man Viewed as a Machine," *Scientific American*, April 1955, pp. 58–67, based on a talk von Neumann gave in Princeton in March 1953.

250 comically little result—See, e.g., Henry Quastler, ed., *Essays on the Use of Information Theory in Biology* (Urbana: University of Illinois Press, 1953), which was compiled from lectures and discussions at the University of Illinois the previous summer. Sir Peter Medawar is fond of citing, as a forerunner, H. Kalmus, "A Cybernetical Aspect of Genetics," *Journal of Heredity* 41 (1950): 19–22, but although Kalmus sets out to discuss genes as messages, he then proceeds to get the point exactly wrong, saying that the message conveyed by a gene is not to be compared "to a sequence of signals" but rather is chemical in nature, like a hormone, which betrays a fatal lack of understanding of the requisite specificity.

250 earliest mention of coding—Erwin Schrödinger, *What Is Life?* (Cambridge: At the University Press, 1944), "based on lectures delivered . . . at Trinity College, Dublin, in February 1943," pp. 60–62; Schrödinger made an error in calculating one of the numerical examples, in the first edition here cited, which I have corrected from the 1969 edition.

251 Crick's sequence hypothesis—Interview, 19 November 1975.

251 One that Crick missed—Kurt G. Stern, "Nucleoproteins and Gene Structure," *Yale Journal of Biology and Medicine* 19 (1947): 944, 945.

252 Hinshelwood's theoretical speculation—P. C. Caldwell and Sir Cyril Hinshelwood, "Some Considerations on Autosynthesis in Bacteria," *Journal of the Chemical Society* (1950): 3156–3159: "Received July 26, 1950"; quotes from pp. 3156, 3157. For Pauling on antibodies, the essential papers are Linus Pauling, "A Theory of the Structure and Process of Formation of Antibodies," *Journal of the American Chemical Society* 62 (130): 2643; idem, "Molecular Structure and Intermolecular Forces," in *The Specificity of Serological Reactions*, ed. Karl Landsteiner, rev. ed. (Cambridge: Harvard University Press 1945), pp. 275–293; and the one cited by Caldwell and Hinshelwood; which is Pauling, "Antibodies and Specific Biological Forces," *Endeavour* 7 (1948): 43ff. For Chargaff on nucleic acids, his article in *Experientia* 6 (1950), reprinted in Chargaff, *Essays*, p. 21. It must be said, however, that Chargaff has sometimes remembered his ideas of that time as having been somewhat more specific than the original text justifies: for example, in 1975 he wrote (*The Sciences*, August–September, p. 23) that his paper of 1950 suggested for the first time "that the order in which the nucleotides were arranged could form a text"—a word, and a clarity of concept, not present in the original.

252 Crick on Hinshelwood—F. H. C. Crick, "The Genetic Code—Yesterday, Today, and Tomorrow," *C. S. H. Symp. Q. Biol.* 31 (1966): 3–4.

253 Dounce's reminiscence—A. L. Dounce, opening remarks in "Nucleoproteins: Round Table Discussion," in *Oak Ridge Symposium*, April 1955, p. 103.

253 Dounce's paper—Alexander L. Dounce, "Duplicating Mechanism for Peptide Chain and Nucleic Acid Synthesis," *Enzymologia* 15 (1952): 251, 253; letter, Dounce to Judson, 25 July 1978.

254 Campbell and Work—P. N. Campbell and T. S. Work, "Biosynthesis of Proteins," *Nature* 171 (6 June 1953): 997–1001.

254 Dalgliesh—C. E. Dalgliesh, "The Template Theory and the Role of Transpeptidation in Protein Biosynthesis," *Nature* 171 (6 June 1953): 1027–1028.

256 Gamow's idea—Crick, "Genetic Code," p. 4. George Gamow, "Protein Synthesis by DNA-Molecules" (mimeographed 10 pp.): apparently the earliest surviving version of the paper later published, though the acknowledgments at the end, an unnumbered last sheet, include thanks to Crick for "helpful discussion"; the paper bears the name C. G. H. Tompkins, Gamow's fictional amateur of science, as coauthor. The original copy is in Crick's file.

256 the twenty standard amino acids—Interview, Crick, 19 November 1975; Gamow, "Protein Synthesis," p. 2; Crick discussed the list of standard amino acids in F. H. C. Crick, "On Degenerate Templates and the Adaptor Hypothesis: A note for the RNA Tie Club" (mimeographed, 17 pp., n.d. but sent out January 1975), and more publicly in F. H. C. Crick, "On Protein Synthesis," in *Symp. Soc. Exp. Biol.* 12, held 1957, pp. 139–140.

258 "audacious and . . . unsupported"—Stent, *Molecular Genetics*, p. 35.

262 28 "uncommon amino acids"—E. Bricas and C. Fromageot, "Naturally Occurring Peptides," in *Advances in Protein Chemistry* 8 (1953): 6–7 (table).

◆

263   Watson's help to Brenner—Letter, Maaløe to Brenner, undated, but filed with Brenner's correspondence in order received, July 1953; letter, Watson to Brenner, 17 July 1953. "Wednesday is all right for Bill Hayes. He . . . will have lunch with us"; letter, J. M. Mitchison to Brenner, dated "4 Aug" and suggesting he come to Cambridge to discuss possible work in Edinburgh.

263   Caltech conference—For contemporary summary, John C. Kendrew, "Structure of Proteins," *Nature* 173 (9 January 1954): 57–58; interview, Huxley; interview, Crick, 19 November 1975; interview, Perutz, 4–5 December 1970.

264   Crick met Gamow—Interview, Luzatti; interview, Crick, 19 November 1975.

265   rules applied to insulin—The draft—Gamow, "Protein Synthesis,"—in Crick's possession does *not* contain the ensuing discussion of the consequences of the Glu-Glu-Cys-Cys sequence; it was added in 1954 and appears in George Gamow, "Possible Mathematical Relation between Deoxyribonucleic Acid and Proteins," *Det Kongelige Danske Videnskabernes Selskab: Biologiske Meddelelser* 22, no. 3 (1954): 1–13, at pp. 10–11.

265   "sufficiently precise to admit disproof"—F. H. C. Crick, "The Present Position of the Coding Problem," in *Structure and Function of Genetic Elements: Brookhaven Symposia in Biology* 12 (1959): 35; George Gamow, "Protein Synthesis," "Possible Mathematical Relation between Deoxyribonucleic Acid and Proteins," "Possible Relation between Deoxyribonucleic Acid and Protein Structures," *Nature* 173 (13 February 1954): 318.

266   Dunitz from Caltech—Letter, Jack Dunitz to Sydney Brenner, 5 November 1953; original in Brenner's papers.

266   Watson to Crick—Letters, James Watson to Francis Crick, 12 November 1953. 16 December 1953, 13 February 1954; spelling and punctuation as in originals, which are in Crick's files.

267   Gamow to Crick—Letter, George Gamow to Francis Crick, 8 March 1954; spelling and punctuation as in original, which is in Crick's files.

268   Franklin to Crick—Her letter does not survive; he replied, letter, Francis Crick to Rosalind Franklin, 21 March 1954—original in Klug's file of Franklin's papers.

268   Dunitz to Brenner—Letter, Jack Dunitz to Sydney Brenner, 30 March 1954; original in Brenner papers.

268   Delbrück's results with *Phycomyces*—Interview, Delbrück, 9 July 1972.

269   RNA Tie Club—"Organization Charter," undated, listing aims, emblems, and sixteen members; circular letter, Gamow to members, on RNA Tie Club letter paper, 4 July 1955, and listing twenty members; note, dated 16 October 1972, Martynas Yčas to Librarian of Congress, with details of history of RNA Tie Club for the Gamow papers there; all kindly provided to me by Dr. Yčas, with letter, 11 February 1976, bearing yet further details.

269   National Academy symposium on nucleic acids—Linus Pauling, "Symposium on the Structure and Function of Nucleic Acids," *Proc. N. A. S.* 40 (1954): 747–748, which dates the meeting as the afternoon of 26 April 1954; the papers by Todd, Crick, and Watson are published, ibid., pp. 748–764.

269   Gamow to Crick—Letter, George Gamow to Francis Crick, 27 May 1954.

270   electron microscopy of interiors of cells—Claude, "Particulate Components . . . *Science* 91 (1940), pp. 77–78. Keith R. Porter, *Journal of Experimental Medicine* 97 (1953): 727, cited in George E. Palade, "Microsomes and Ribonucleoprotein Particles," in *Microsomal Particles and Protein Synthesis*, ed. Richard B. Roberts (London, etc.: Pergamon Press), pp. 36–49, of which pp. 36–40 are an historical account; George E. Palade, "A Small Particulate Component of the Cytoplasm, "*Journal of Biophysical and Biochemical Cytology* 1 (1955): 59–68, where the footnote to p. 59 says: "The substance of this article was presented in 1953 at the eleventh annual meeting of the Electron Microscope Society in Pocono Manor, Pennsylvania"; George E. Palade and Philip Siekovitz, "Liver Microsomes," ibid. (1956): 171–200; Alfred Tissières, "Ribosome Research: Historical Background," in *Ribosomes*, ed. M. Nomura, A. Tissières, and P. Lengyel (Cold Spring Harbor, N.Y.: Cold Spring Harbor Laboratory, 1974), pp. 3–12.

270   particles in bacteria—See previous note; also, S. E. Luria, M. Delbrück, and T. F. Anderson, "Electron Microscope Studies of Bacterial Viruses," *Journal of Bacteriology* 46 (1943): 60; H. C. Schachman, A. B. Pardee, and R. Y. Stanier, "Studies on the Macromolecular Organization of Microbial Cells," *Archives of Biochemistry and Biophysics* 38 (1952): 245.

271   particles in strings—Interview, Crick, 20 November 1975.

271   prevailing doctrines about proteins—Interview, Zamecnik; interview, Lipmann, 7 July 1977; interview, Hoagland; for Zamecnik's own account of the state of the science as he entered it, see Zamecnik, "Historical Account," pp. 1–3, and Zamecnik, Harvey Lecture, pp. 256–258; for an independent statement of some of the doctrines Zamecnik and others overthrew, see, e.g., Henry Borsook, "The Biosynthesis of Peptides and Proteins," *Proceedings of the Third International Congress of Biochemistry, Brussels 1955* (New York: Academic Press, 1956), pp. 92–104.

271   Zamecnik's experiments—The crucial papers on the role of microsomes were two: Elizabeth

B. Keller, Paul C. Zamecnik, and Robert B. Loftfield, "The Role of Microsomes in the In-corporation of Amino Acids into Proteins," *Journal of Histochemistry and Cytochemistry* 2 (September 1954): 378–386, which describes the experiments with live rats (pp. 378–383), and cites previous work in Zamecnik's lab and elsewhere; John W. Littlefield, Elizabeth B. Keller, Jerome Gross, and Paul C. Zamecnik, "Studies on Cytoplasmic Ribonucleoprotein Particles from the Liver of the Rat," *Journal of Biological Chemistry* 217 (November 1955): 111–123, "Received for publication, February 24, 1955," which was the definitive demonstration using the cell-free system.

272   cell-free system—Early attempts at a cell-free system using *E. coli*, interview, Zamecnik; first reports of rat-liver cell-free system, Philip Siekevitz and Paul C. Zamecnik, "In vitro In-corporation of 1-C^{14}-DL-alanine into Proteins of Rat-liver Granular Fractions," *Federation Proceedings* 10 (March 1951): 246; Paul C. Zamecnik, "Incorporation of Radioactivity from DL-leucine-1-C^{14} into Proteins of Rat Liver Homogenate," *Federation Proceedings* 12 (March 1953): 295; Paul C. Zamecnik and Elizabeth B. Keller, "Relation between Phosphate Energy Donors and Incorporation of Labeled Amino Acids into Proteins," *Journal of Biological Chemistry* 209 (1954): 337–354.

272   four essential components—Zamecnik, "Historical Account," p. 4.

273   meeting Watson—Zamecnik, "Historical Account," pp. 4–5.

274   Benzer—Interview, Benzer; Seymour Benzer, "Adventures in the rII Region," in *Phage and . . . Molecular Biology*, pp. 157, 159–161.

275   Lwoff's unit at the Pasteur—See *Of Microbes and Life*, passim, and, in particular, François Jacob, "La belle époque," pp. 98–104.

276   the size of the gene—Guido Pontecorvo, "Genetic Formulation of Gene Structure and Gene Action," *Advances in Enzymology* 13 (1952): 121–149.

277   Benzer back in Purdue—Benzer, "Adventures," pp, 160–163, and interview; Seymour Ben-zer, "Genetic Fine Structure" (lecture delivered 15 September 1960), *Harvey Lectures* 56 (1961): 1–21. Benzer's first experiments were with phage T2; he soon shifted to T4. Seymour Benzer, "Fine Structure of a Genetic Region in Bacteriophage," *Proc. N. A. S.* 41 (1955): 344–354; idem, "The Elementary Units of Heredity," in *McCollum-Pratt Symposium* 1956, pp. 70–93.

278   meeting Brenner—Interview, Benzer; interview, Brenner, 7 February 1976; Benzer, "Adventures in the rII Region," in *Phage and . . . Molecular Biology*, p. 162.

279   Brenner invited to Woods Hole—Letter, James Watson to Sydney Brenner, 26 July 1954; original in Brenner's papers.

280   Crick and the rest at Woods Hole—Interviews, Crick, 19 and 20 November 1975; interviews, Brenner, 7 February and 28 February 1976. Details of the points Crick and the others discussed were first put on paper the following January, F. H. C. Crick, "On Degenerate Templates and the Adaptor Hypothesis"; the paper is dated by letters, Crick to Brenner, 3 and 12 January 1955.

282   the hoax party—Crick's account checked by telephone with Andrew Gabriel Szent-Gyorgi.

282   encounter with Ephrussi—Interviews, Crick, 19 November 1975 and 28 June 1977; inter-view, Ephrussi, 13 November 1976; George W. Beadle, "Biochemical Genetics: Some Recollections," in *Phage and . . . Molecular Biology*, pp. 26–27.

283   new piece of anatomy—Interview, Crick, 20 November 1975; telephone conversation, Werner Maas.

284   Franklin's urgent letter—Letter, Franklin to Bernal, 2 September 1954; original in Klug's file of Franklin's papers.

284   "clusters within clusters"—Letter, Brenner to Benzer, 11 October 1954.

284   overlapping codes—Interview, Brenner, 7 February 1976; Brenner found his original work-sheet, and kindly allowed me to make a photographic copy.

285   Brenner stopped in Washington—Interview, Brenner, 7 February 1976; interview, Rich, 9 August 1977.

285   origin of term "adaptor"—Interview, Crick, 20 November 1975, borne out by letter, Crick to Brenner, 4 November 1955; original in Brenner's papers.

285   "worse . . . than . . . nightmares" Letter, Brenner to Alan Garen, 3 January 1955, carbon copy in Brenner's papers.

286   "Dearest Sydneyboy"—Letter, Stent to Brenner, 6 December 1954; letters, Crick to Brenner, 3 and 12 January 1955; originals all in Brenner's papers.

6 *"My mind was, that a dogma was an idea for which there was no reasonable evidence. You see?!"*

PAGE

287   Crick's family background, relation to Cambridge—Interviews, Crick, 1 September 1971 and 28 June 1977; details of family from Robert Olby, "Francis Crick, DNA, and the Central Dogma," *Daedalus*, Fall 1970, pp. 938–942.

288   Watson describes the Crick circle—Letter, Watson to Delbrück, 9 December 1951; original in the Delbrück Papers in the archives of the California Institute of Technology.

288   the Mastership of Caius—I first learned of the offer from others at Gonville and Caius College, not from Crick.

288   the adaptor note—F. H. C. Crick, "On Degenerate Templates and the Adaptor Hypothesis: A Note for the RNA Tie Club" (undated, mimeographed, 17 pp. plus cover sheet; on evidence of Crick's letters to Brenner, completed mid January 1955—see previous chapter).

289   epigraph—Kai Kā'ūs Ibn Iskandar, *A Mirror for Princes: the Qābūs Nāma . . .*, tr. Reuben Levy (London: Creṡset Press, 1951).

290   Kendrew's computer program—J. M. Bennett and J. C. Kendrew, "The Computation of Fourier Syntheses with a Digital Electronic Calculating Machine," *Acta Crystallographica* 5 (January 1952): 109–136; "Received 28 July 1951."

290   quotes from, details of, adaptor paper—Discussed with Crick, interview 28 June 1978, with manuscript.

293   Crick, Tie-Club note "negative"—Crick to Brenner, 12 January 1955; original of this and all other letters to or from Brenner are in his file, unless noted otherwise.

294   Mirsky and Allfrey—For a late statement of their work and view, with citations, V. Allfrey, A. E. Mirsky, and S. Osawa, "The Nucleus and Protein Synthesis," *McCollum-Pratt Symposium* 1956, pp. 200–231.

294   energy for forming peptide bonds—Interview, Zamecnik; interview, Hoagland; Mahlon B. Hoagland, "An Enzymic Mechanism for Amino Acid Activation in Animal Tissues," *Biochimica et Biophysica Acta* 16 (1955): 288–289, "Received December 4, 1954"; Zamecnik, "Historical Account," p. 4. The first full paper on amino-acid activation was Mahlon B. Hoagland, Elizabeth B. Keller, and Paul C. Zamecnik, "Enzymatic Carboxyl Activation of Amino Acids," *Journal of Biological Chemistry* 218 (January 1956): 345–358; "Received . . . June 3, 1955."

295   Gamow, "loose tringles" code—Letter, Gamow to Brenner, 22 January 1955; punctuation, orthography as in original.

295   Gamow's review of the coding problem—George Gamow, Alexander Rich, and Martynas Yčas, "The Problem of Information Transfer from the Nucleic Acids to Proteins," *Advances in Biological and Medical Physics* 4 (1956): 23–68; interview, Brenner, 7 February 1956.

296   Watson wrote from Caltech—Watson to Brenner, 11 March 1955.

296   symposium at Oak Ridge—Published as *Symposium on the Structure of Enzymes and Proteins . . . Gatlinburg, Tenn. April 4–6, 1955, Journal of Cellular and Comparative Physiology* 47, supp. 1 (1956). Borsook's remark, Henry Borsook, "The Biosynthesis of Peptides and Proteins," ibid., p. 57. Zamecnik's report, Paul Charles Zamecnik, Elizabeth Beach Keller, John Walley Littlefield, Mahlon Bush Hoagland, and Robert Berner Loftfield, "The Mechanism of Incorporation of Labeled Amino Acids into Protein," ibid., pp. 81–102. Dounce's proposals, Alexander Dounce, leading discussion on "Nucleoproteins," ibid., pp. 105–106. Borsook's "mechanism for transporting," ibid., pp. 51, 53. Chargaff's "prophetic question," Erwin Chargaff, contribution to general discussion following Zamecnik's paper, ibid., p. 92—and Zamecnik's description of the question was made on the cover of a reprint of the paper and the discussion he sent me in 1977.

297   Vincent's astonishing suggestion—Walter Sampson Vincent, contribution to discussion following Zamecnik's paper, ibid., pp. 97, 98. Crick pointed out to me, interview 28 June 1977, that Vincent, using starfish, probably had spotted a kind of RNA called heteronuclear RNA, which is found in higher organisms but not in prokaryotes; Vincent's suggestion of a function for a metabollically very active RNA fraction is still remarkable.

298   announcement of enzyme synthesizing RNA—Interview, Grunberg-Manago; interview, Ochoa; Marianne Grunberg-Manago and Severo Ochoa, "Enzymatic Synthesis and Breakdown of Polynucleotides; Polynucleotide Phosphorylase," *Journal of the American Chemical Society* 77 (1955): 3165–3166; Marianne Grunberg-Manago, Priscilla J. Ortiz, and Severo Ochoa, "Enzymatic Synthesis of Nucleic Acid like Polynucleotides," *Science* 122 (11 November 1955): 907–910.

299   Crick to Paris—Interview, Crick, 28 June 1977; interview, Ephrussi, 13 November 1976.

299   Benzer, "crossing mutants"—Benzer to Brenner, 26 May 1955.

300   Stent's three items of news—Stent to Brenner, 12 June 1955; interview, Stent; interview, Fraenkel-Conrat, with manuscript; H. Fraenkel-Conrat and Robley C. Williams, "Reconstitution of Active Tobacco Mosaic Virus from Its Inactive Protein and Nucleic Acid Components," *Proc. N. A. S.* 41 (1955): 690–698; H. Fraenkel-Conrat, "The Role of the Nucleic Acid in the Reconstitution of Active Tobacco Mosaic Virus," *Journal of the American Chemical Society* 78 (20 February 1956): 882–883; A. Gierer and G. Schramm, "Infectivity of Ribonucleic Acid from Tobacco Mosaic Virus," *Nature* 177 (1956): 702.

301   Crick's footnote—Letter, Crick to Brenner, 12 January 1955.

301   Ingram responds to Ephrussi's challenge—Interviews, Ingram; interviews, Crick, 19 and 20 November 1975 and 28 June 1977.

302   Pauling's account of discovery of nature of sickle-cell anemia—Linus Pauling, "Fifty Years of Progress in Structural Chemistry and Molecular Biology," *Daedalus*, Fall 1970, p. 1011; interview, Pauling, 1 March 1971.

303   the paper—Linus Pauling et al., "Sickle Cell Anemia, a Molecular Disease," *Science* (25 November 1949): 543–548, quotation at p. 547; "Based on a paper presented . . . in April, 1949."

303   Neel's note—J. V. Neel, "The Inheritance of Sickle Cell Anemia," *Science* 110 (1949): 64–66.

304   Perutz on sickling—M. F. Perutz and J. M. Mitchison, "State of Haemoglobin in Sickle-Cell Anaemia," *Nature* 166 (21 October 1950): 677–678; see chapter 9, below.

307   Ingram's development of "finger prints"—Interviews, Ingram, the second one with draft; interviews, Crick, 19 and 20 November 1975 and 28 June 1977, last of these with draft; V. M. Ingram, "A Specific Chemical Difference Between the Globins of Normal Human and Sickle-Cell Anaemia Haemoglobin," *Nature* 178 (13 October 1956): 792–794; diagrams from that paper.

307   Crick's nine-page letter—Crick to Brenner, 6 July 1955.

308   Crick's reference to DNA structure, "Wilkins & Co"—M. Feughelman, R. Langridge, W. E. Seeds, A. R. Stokes, H. R. Wilson, C. W. Hooper, M. H. F. Wilkins, R. K. Barclay, L. D. Hamilton, "Molecular Structure of Deoxyribose Nucleic Acid and Nucleoprotein," *Nature* 175 (1955): 834–836

308   Crick's reference to Benzer—Seymour Benzer, "Fine Structure of a Genetic Region in Bacteriophage," *Proc. N.A.S.* 41 (1955): 344–354.

309   "first official" Tie-Club letter—Gamow to Brenner, 4 July 1955.

309   long letter from Stent—Stent to Brenner, dated 1 and 5 August 1955.

309   combination code—George Gamow and Martynas Yčas, "Statistical Correlation of Protein and Ribonucleic Acid Composition," *Proc. N. A. S.* 41 (December 1955): 1011–1019, "Appendix I: Random Division of a Unit Length into *n* Parts (After J. von Neumann)," pp. 1016–1017.

310   air-letter to Brenner—Crick to Brenner, 30 August 1955; Brenner's reply does not survive but is described in Crick's, next note.

310   Crick's gloomy letter—Crick to Brenner, 7 October 1955.

310   position overturned—Crick to Brenner, 11 October 1955.

310   succeeding letters—Crick to Brenner, 20 October, 21 October, and 4 November 1955.

312   Perutz to London to ask for Brenner's appointment—Interview, Perutz, 6 December 1970.

312   Crick offers job—Crick to Brenner, 30 December 1955.

312   Zamecnik and Hoagland, fall 1955—Interview, Zamecnik; interview, Hoagland; Paul C. Zamecnik, "Historical Account," pp. 5, 6; Zamecnik has sent me machine copies of the relevant pages of his laboratory notebooks, showing the first experiment on 31 October 1955, and a note that day calling for greater care in washing; a repeat of the experiment on 3 November, with the same result; a note on 10 November that the radioleucine is "tightly bonded to the RNA," and a first diagram of an explanation, on 18 November, showing bonding of amino acid to RNA as a first, independent step.

313   Zamecnik, "anxious to remove all doubts"—Zamecnik, "Historical Account," pp. 6–7.

313   problem put aside for six months—Interview, Zamecnik, and conversation with him, 1 June 1978; interview, Hoagland.

313   Crick's brief note—Crick to Brenner, 17 January 1956.

313   Orgel talks on code—Crick to Brenner, 13 February 1956; interview, Orgel; interview, Crick, 28 June 1977.

314   Wilkins, Crick, at symposium on nucleic acids—M. H. F. Wilkins, "Molecular Structure of Deoxyribose Nucleic Acid and Nucleoprotein and Possible Implications in Protein Synthesis," *Biochemical Society Symposium* 14 (1956), quoted remark at p. 24; F. H. C. Crick, contribution to discussion, ibid., p. 26.

315   Ciba Symposium, March 1956—F. H. C. Crick and J. D. Watson, "Virus Structure: General Principles," in *Ciba Foundation Symposium on the Nature of Viruses*, ed. G. E. W. Wolstenholme and Elaine C. P. Millar (London, J. & A. Churchill, 1957), pp. 5–13, quoted matter at pp. 8, 12–13; letter, Crick to Monod, 31 December 1961, original in Monod correspondence at Institut Pasteur. A companion paper to the one delivered at the Ciba Symposium was F. H. C. Crick and J. D. Watson, "Structure of Small Viruses," *Nature* 177 (10 March 1956): 473–475. Interviews, Crick, 20 November 1975, 28 June 1977.

316   crystallizing microsomes for X-ray diffraction—Letter, Crick to Brenner, 6 June 1956, written from Alexander Rich's lab at the National Institutes of Health, Bethesda, Maryland, where Watson was also visiting: Crick, after replying to practical and scientific matters raised in a letter (not surviving) from Brenner, goes on to say, "We should certainly try to crystallize microsomes if Alex and Jim haven't done it before we start. As to other things . . ."

316   Crick's science bulletins—Crick to Brenner, 30 April 1956.

316   comma-free code—Interviews, Crick, 20 November 1975 and 28 June 1977; interview, Orgel; F. H. C. Crick, J. S. Griffith, and L. E. Orgel, "Codes Without Commas," *Proc. N. A. S.* 43 (1957): 416–421.

318   Dalgliesh's point—C. E. Dalgliesh, "The Template Theory and the Role of Transpeptidation in Protein Biosynthesis," *Nature* 171 (6 June 1953): 1027–1028.

319   Delbrück's warning to Stent—Letter, Stent to Brenner, 1 and 5 August 1955.

319   work in Rich's lab on RNA—Alexander Rich, "The Structure of Synthetic Polyribonucleotides and the Spontaneous Formation of a New Two-Stranded Helical Molecule," *McCollum-Pratt Symposium* 1956, pp. 557–561. The paper was inserted into the program late, and illustrates the way conferences were accelerating communication that summer; see also Watson's paper at that symposium, "X-Ray Studies on RNA and the Synthetic Polyribonucleotides," pp. 552–556, which reports other aspects of the work that spring.

319   Crick's long letter about the conferences—Crick to Brenner, dated Tuesday, 17 July 1956 at Ann Arbor, and the following Friday, from Bethesda.

320   Stent's idea—Not presented in Stent's paper at McCollum-Pratt Symposium; interview, Stent.

320   Kornberg's "most exciting story"—Arthur Kornberg, "Pathways of Enzymatic Synthesis of Nucleotides and Polynucleotides," McCollum-Pratt Symposium 1956, pp. 579–608.

320   Benzer's report—Seymour Benzer, "The Elementary Units of Heredity," *McCollum-Pratt Symposium* 1956, pp. 70–93—a paper usually taken to be his definitive review of the work in the fine structure of the gene; Lederberg's observation, reported by Crick, is not preserved in the published discussion.

321   Jacob's report—François Jacob and Élie L. Wollman, "Genetic Aspects of Lysogeny," *McCollum-Pratt Symposium* 1956, pp. 468–499.

321   Zamecnik's work—Zamecnik did not give a paper at the McCollum-Pratt Symposium, but at the Gordon Research Conference on Proteins and Nucleic Acids, 25–29 June, that summer, he gave a talk, "Studies on the Role of RNA and Nucleotides in Protein Synthesis," on 27 June. All papers and discussions at Gordon Conferences are unpublished.

321   Volkin's short, puzzling paper—Elliot Volkin and L. Astrachan, "RNA Metabolism in T2-Infected *Escherichia coli*," *McCollum-Pratt Symposium* 1956, pp. 686–695; quoted matter at 686 and 694, question and reply at 695. The first publication of the work was Elliot Volkin and L. Astrachan, "Intracellular Distribution of Labeled Ribonucleic Acid After Phage Infection of *Escherichia coli*," *Virology* 2 (August 1956): 433–437; "Accepted April 2, 1956."

323   Holley by way of the energy—Interview, Holley; Robert W. Holley, "Alanine Transfer RNA" (Nobel Lecture, 12 December 1968), *Les Prix Nobel en 1968*, pp. 1–2; Robert W. Holley, "An Alanine-dependent, Ribonuclease-inhibited Conversion of AMP to ATP, and Its Possible Relationship to Protein Synthesis," *Journal of the American Chemical Society* 79 (5 February 1957): 658–662, "Received August 3, 1956." As there was a disagreement between Holley and Zamecnik, later, about priority for discovery of transfer RNA, I want to state explicitly that the judgment I express about Holley's first paper is my own, written before my interviews with Zamecnik and Hoagland.

323   Berg in by way of enzymes—Interviews, Berg; letter, Berg to Judson, 25 July 1978, with comments on this passage in draft; Paul Berg, "Acyl Adenylates: The Interaction of Adenosine Triphosphate and L-Methionine," *Journal of Biological Chemistry* 222 (1956): 1025–1034, the preliminary isolation; Paul Berg and E. J. Ofengand, "An Enzymatic Mechanism for Linking Amino Acids to RNA," *Proc. N.A.S.* 44 (1958): 78–86.

324   Zamecnik and Hoagland—Zamecnik's notebooks indicate that he returned to the experiments with radioactively labeled leucine, to trace its binding to soluble RNA, on 12 July 1956, getting a clear positive result.

324   the paper—Mahlon B. Hoagland, Paul C. Zamecnik, and Mary L. Stephenson, "Intermediate Reactions in Protein Biosynthesis," *Biochimica et Biophysica Acta* 24 (1957): 215–216; "Received January 16th, 1957."

324   Watson's visit—Interview, Hoagland.

324   biochemists converged—A few weeks later, independently, a team of biochemists in Japan reported similar results: Kikuo Ogata and Hiroyoshi Nohara, "The Possible Role of the Ribonucleic Acid (RNA) of the pH 5 Enzyme in Amino Acid Activation," *Biochimica et Biophysica Acta* 25 (1957): 659–660; "These results were presented at . . . the 29th annual meeting of the Japanese Biochemical Society . . . on October 31, 1956."

324   "soul-satisfying harmony"—Mahlon B. Hoagland, "Nucleic Acids and Proteins," *Scientific American*, December 1959, p. 59.

325   "trinucleotide or a bit bigger"—Interview, Crick, 20 November 1975; F. H. C. Crick, "On Protein Synthesis," *Symp. Soc. Exp. Biol.* 12, held 1957, p. 156.

325 "nothing against its being large"—The definitive paper on role of soluble RNAs: Mahlon B. Hoagland, Mary L. Stephenson, Jesse F. Scott, Liselotte I. Hecht, and Paul C. Zamecnik, "A Soluble Ribonucleic Acid Intermediate in Protein Synthesis," *Journal of Biological Chemistry* 231 (March 1958): 241–257.

325 Crick back in Cambridge—Crick to Brenner, 21 August 1956 and 30 August 1956.

326 Ingram's finger-print technique—V. M. Ingram, "A Specific Chemical Difference Between the Globins of Normal Human and Sickle-Cell Anaemia Haemoglobin," *Nature* 178 (1956): 792–794; V. M. Ingram, "Gene Mutations in Human Haemoglobin: The Chemical Difference Between Normal and Sickle Cell Haemoglobin," *Nature* 180 (17 August 1957): 326–328.

326 microsomal particles proved recalcitrant—Interview, Tissières; Alfred Tissières, "Ribosome Research: Historical Background," in *Ribosomes*, ed. M. Nomura, A. Tissières, and Peter Lengyel (Cold Spring Harbor: Cold Spring Harbor Laboratory, 1974): 6–7. For earliest work on microsomes, see chapter 5, above. For status of problem in early 1955, J. W. Littlefield, E. B. Keller, J. Gross, and P. C. Zamecnik, "Studies on Cytoplasmic Ribonucleoprotein Particles from the Liver of the Rat," *Journal of Biological Chemistry* 217 (1955): 111–123; for slightly later period, see the collection of papers in Richard B. Roberts, ed., *Microsomal Particles and Protein Synthesis* (London: Pergamon, 1958): the papers were presented at the First Symposium of the Biophysical Society, Cambridge, Mass., 5, 6, and 8 February 1958.

326 microsomal particles in yeast, pea seedlings—F. C. Chao and H. K. Schachman, "The Isolation and Characterization of a Macromolecular Ribonucleoprotein from Yeast," *Archives of Biochemistry and Biophysics* 61 (1956): 220ff.; P. O. P. Tso, J. Bonner, and J. Vinograd, "Microsomal nucleoprotein particles from pea seedlings," *Journal of Biophysical and Biochemical Cytology* 2 (1956): 451ff; Benjamin D. Hall and Paul Doty, "The Configurational Properties of Ribonucleic Acid Isolated from Microsomal Particles of Calf Liver," ibid., pp. 27–35. For Watson's own account of his part in the work, see his Nobel Lecture, "Involvement of RNA in the Synthesis of Proteins," *Les Prix Nobel en 1962*.

326 Fu-Chuan Chao at Atlantic City—Fu-Chuan Chao, "Dissociation of Macromolecular Ribonucleoprotein of Yeast," *Archives of Biochemistry and Biophysics* 70 (1957): 426–431; "A preliminary report of this work was presented before the 130th Meeting of the American Chemical Society at Atlantic City, September, 1956."

327 Brenner's note to the Tie Club—S. Brenner, "On the Impossibility of All Overlapping Triplet Codes: A Note for the RNA Tie Club" (Mimeographed, title page plus three pages of text plus seven pages of tables and references; September 1956), quoted line of conclusion, p. 3.

327 Crick's brief note—Crick to Brenner, 1 October 1956, again 22 October 1956.

328 Crick, "all about phage tails"—Crick to Brenner, 25 October 1956.

328 Gamow's letter—Gamow to Brenner, 24 November 1956.

329 Brenner left "theoretician no real problem"—F. H. C. Crick, "The Genetic Code—Yesterday, Today, and Tomorrow," *C. S. H. Symp. Q. Biol.* 31 (1966): 5.

329 Gamow, "completely lost the belief"—Letter to Brenner, 13 April 1957; then again, 4 June 1957.

329 Brenner arrives Cambridge—Interviews, Brenner, March 1977 and 14 May 1977; the notes he took on Perutz's lectures are among his correspondence.

329 not beakers but vats—Letter, Streisinger to Brenner, 19 March 1957; Brenner to Hoover, Ltd., 2 April 1957, and reply.

329 Delbrück, "What are your plans?"—Letter, Delbrück to Brenner, 2 April 1957.

329 competition at Gordon Conference—Benzer to Brenner, 13 July 1957; Cyrus Levinthal to Brenner, 9 July 1957.

330 sugar and starches at symposium—M. Stacey, "The Biosynthesis of Oligo- and Polysaccharides," *Symp Soc. Exp. Biol.* 12, held 1957, quoted matter at p. 185.

330 Crick's great paper at symposium—F. H. C. Crick, "On Protein Synthesis," *Symp. Soc. Exp. Biol.* 12, held 1957, pp. 138–163.

330 more than 1500 papers in 1956, 2,000 in 1957—See the review, Daniel Steinberg and Elemer Mihalyi, "The Chemistry of Proteins," *Annual Review of Biochemistry* 26 (1957): 373–418, their count on p. 373.

330 quotations from Crick's paper—"In the protein molecule," p. 139; "at first sight paradoxical," pp. 138–139; "what can we guess," pp. 153–155; "it may surprise the reader," p. 143; "instructive to compare," p. 196; the five indispensable paragraphs, pp. 152–153.

333 Crick on the central dogma—Interview, Crick, 20 November 1975.

334 his piece clarifying the dogma—Francis Crick, "Central Dogma of Molecular Biology," *Nature* 227 (8 August 1970): 461–563, quoted matter at 562.

336 specialty in their own right—CCA sequence at terminus, Liselotte I. Hecht, Mary L. Stephenson, and Paul C. Zamecnik, "Dependence of Amino Acid Binding to Soluble Ribonucleic Acid on Cytidine Triphosphate," *Biochimica et Biophysica Acta* 29 (1958):

460–461; Liselotte I. Hecht et al., "Nucleoside Triphosphates as Precursors of Ribonucleic Acid End Groups in a Mammalian System," *Journal of Biological Chemistry* 233 (October 1958): 954–963.

336   Crick in the lab—Interview, Crick, 20 November 1975; interview, Hoagland.

336   Benzer's and Lipmann's pretty, witty experiment—F. Chapeville, F. Lipmann, G. von Ehrenstein, B. Weisblum, W. J. Ray, and S. Benzer, "On the Role of Soluble Ribonucleic Acid in Coding for Nucleic Acids," *Proc. N.A. S.* 48 (1962): 1086–1092.

336   Holley's seven years—Interview, Holley; work described in Robert W. Holley, "Alanine Transfer RNA" (Nobel Lecture, 12 December 1968), *Les Prix Nobel en 1968.*

337   structures of transfer RNA—J. E. Ladner, A. Jack, J. D. Robertus, R. S. Brown, D. Rhodes, B. F. C. Clark, A. Klug, "Structure of Yeast Phenylalanine Transfer RNA at 2.5 A. Resolution," *Proc. N. A. S.* 72 (November 1975): 4414–4418; G. J. Quigley, A. H. J. Wang, N. C. Seeman, F. L. Suddath, A. Rich, J. L. Sussman, and S. H. Kim, "Hydrogen Bonding in Yeast Phenylalanine Transfer RNA," *Proc. N. A. S.* 72 (December 1975): 4866–4870.

338   transfer RNAs and origin of life—F. H. C. Crick, S. Brenner, A. Klug, and G. Pieczenik, "A Speculation on the Origin of Protein Biosynthesis," *Origins of Life* 7 (1976).

338   Tissières to Harvard—Interview, Tissières; A. Tissières and J. D. Watson, "Ribonucleoprotein Particles from *Escherichia coli,*" *Nature* 182 (1958): 778–780.

338   "ribosome"—Richard B. Roberts, "Foreword," to R. B. Roberts, ed., *Microsomal Particles and Protein Synthesis,* p. viii.

338   "capped sphere or acorn shape—A. Tissières et al., "Ribonucleoprotein Particles from *Echerichia coli,*" *Journal of Molecular Biology* 1 (1959): 221–233; see also Benjamin D. Hall and Paul Doty, "The Preparation and Physical Chemical Properties of Ribonucleic Acid from Microsomal Particles," ibid., pp. 111–126.

338   colinearity intractable—Interview, Benzer.

339   diversion to proflavine mutations—S. Brenner, S. Benzer, and L. Barnett. "Distribution of Proflavin-Induced Mutations in the Genetic Fine Structure," *Nature* 182 (11 October 1958): 983–985.

339   "lectures on elementary genetics"—Interview, Crick, 20 November 1975.

339   two biochemists in Moscow—A. N. Belozersky and A. S. Spirin, "A Correlation between the Compositions of Deoxyribonucleic and Ribonucleic Acids," *Nature* 182 (12 July 1958): 111–112, containing reference to *Biokhimia;* the French group, Ki Yong Lee, R. Wahl, and E. Barbu, "Contenu en bases puriques et pyrimidiques des acides désoxyribonucléiques des bactéries," *Annales de l'Institut Pasteur* 91 (1956): 212–224. A third, influential, but later paper was Noboru Sueoka, Julius Marmur, and Paul Doty, "Dependence of the Density of Deoxyribonucleic Acids on Guanine-Cytosine Content," *Nature* 183 (23 May 1959): 1429–1431.

340   Delbrück's fascination with comma-free code—S. W. Golomb, L. R. Welch, and M. Delbrück, "Construction and Properties of Comma-Free Codes," *Kongelige Danske Videnskabernes Selskab, Biologiske Meddelelser* 23 (No. 9, 1958): 1–34, quoted matter pp. 9, 10, 11.

341   alternate coding schemes—V. V. Chavchanidze, "The Primary 'Alphabet' of Deoxyribonucleic Acid," *Biofizika* 3 (1958): 391–395, cited and described in Martynas Yčas, *The Biological Code* (Amsterdam and London: North-Holland, 1969), p. 33n; Yčas's own scheme described loc. cit.

341   Sinsheimer's two-letter code—Robert L. Sinsheimer, "Is the Nucleic Acid Message in a Two-symbol Code?" *Journal of Molecular Biology* 1 (1959): 218–220; Crick's characterization of the idea, "The Genetic Code—Yesterday, Today, and Tomorrow," *C. S. H. Symp. Q. Biol.* 31 (1966): 5.

341   encouraging news—Cited in Alfred Tissières, "Ribosome Research: Historical Background," in *Ribosomes,* pp. ix–x, 3–12.

341   Crick at Brookhaven—Interviews, Crick, 3 July 1975, 20 November 1975, 28 June 1977; F. H. C. Crick, "The Present Position of the Coding Problem," *Brookhaven Symposia* 12, 1959, pp. 35–39.

7   *"The gene was something in the minds of people as inaccessible as the material of the galaxies."*

PAGE

343   modest symposium—*Microbial Genetics: Tenth Symposium of the Society for General Microbiology* (held at the Royal Institution, London, April 1960) (Cambridge: At the University Press, 1960).

343   Cambridge weather—The Times, 16 April 1960.

344   discussion at King's—Interviews, Crick, Brenner, Jacob, Garen; telephone conversation and correspondence with Maaloe; details in chapter 8.

344 Jacob's vocation—Interview, Jacob, 3 December 1975; interview, Monod, 1 December 1975; interview, Ephrussi, 5 December 1975.

345 Wollman on Pasteur—Interview, Wollman.

345 Lwoff's attic—Interview, Georges Cohen; he took me on a tour of the space; interview, Jacob, 28 May 1977; see also the various reminiscences in *Of Microbes and Life*, passim.

345 Jacob on Lwoff and Monod—Interview, Jacob, 28 May 1977.

347 Lwoff's early work—Interview, Jacob, 28 and 29 May 1977; interview, Lwoff, 3 December 1975; Paul Fildes, "André Lwoff: an Appreciation," in *Of Microbes and Life*, pp. 13–15; Jacques Monod, "Du microbe à l'homme," ibid., pp. 1–9; B. C. J. G. Knight, "On the Origins of 'Growth Factors,'" ibid., pp. 16–18; R. Y. Stanier, "*L'évolution physiologique*: A Retrospective Appreciation," ibid., pp. 70–76.

347 Monod's biography—Interviews, Monod, 1 and 4 December 1975; interview, Philippe Monod; letter, Philippe Monod to Judson, 23 June 1977, in response to a series of specific questions: Philippe was Jacques' elder brother; Jacques Monod, autobiographical statement in *Les Prix Nobel en 1965*, pp. 135–137. The account of Monod's life that André Lwoff published in *Biographical Memoirs of Fellows of the Royal Society* 23 (1977): 385–412, is inaccurate in many small details.

348 "scientific Puritan"—quoted in Gordon Rattray Taylor, "Jacques Monod," *Science Year* 1971, p. 382.

350 Ephrussi—Interviews, 5 December 1975 and 13 November 1976.

350 wreck of the *Pourquoi-Pas?*—*New York Times*, 17 September 1936, p. 1, col. 7 and p.3, col. 1; 18 September, p. 3, col. 1. The ship had been built by Charcot for Antarctic exploration in the early years of the century; the expedition was to have been Charcot's last; four scientists besides Charcot were killed in the wreck; the hurricane that sank the ship also tore up the eastern seaboard of the United States, killing hundreds of people.

351 Monod's conducting—Interview, Philippe Monod: he kindly showed me a collection of playbills and reviews for the seasons that La Cantate produced in 1938 and 1939, including, for example, laudatory reviews in *Figaro* and *La République*, both of 14 May 1939, of the sixth concert of the group that season. The offer of a job conducting in Pasadena appears, however, to have been somewhat inflated as a family legend. In a letter to Jacques Monod, September 19, 1936, from Edward C. Barrett, Controller, California Institute of Technology, Monod was offered the job of director of the Institute's Glee Club, at a salary of $500 for the year. Philippe Monod has the original.

352 Lwoff's advice to Monod to switch to bacteria—André Lwoff, "Jacques Monod," *Biographical Memoirs of Fellows of the Royal Society*, 23 (1977): 387.

353 Monod's first encounter with adaptation to lactose among *E. coli*—Interviews, Monod, 1 and 4 December 1975; Jacques Monod, "De l'adaptation enzymatique aux transitions allostérique" (Nobel prize lecture delivered 11 December 1965), in *Les Prix Nobel en 1965*, pp. 1 and 2 of Monod's printed text (my copy is a reprint, with pages renumbered).

353 "no interest . . . to the Sorbonne"—André Lwoff, "Jacques Monod (1910–1976)," *La Nouvelle Presse Médicale*, 25 September 1976, p. 2002.

353 Monod's activities in the Resistance—Interviews, Monod, 1 and 4 December 1975; interview and correspondence, Philippe Monod; interview, Geneviève Noufflard; interviews, Eugène Jungelson; interview, Anne Levine; interview, André Lwoff, 15 November 1976, and telephone conversation 13 June 1977; letter with comments on draft of this chapter, 1 June 1977; letter, 31 May 1978, to Judson from Claude Lévy, secretary-general, Comité d'Histoire de la 2 Guerre Mondiale.

354 "dynamic equilibrium"—J. Yudkin, "Enzyme Variation in Micro-organisms," *Biological Reviews* 13 (1938): 93ff. Rudolf Schoenheimer, *The Dynamic State of Body Constituents* (Cambridge, Mass.: Harvard University Press, 1942).

354 Monod's papers on *diauxie*—Jacques Monod, "Sur la nature de la phénomène de diauxie," *Annales de l'Institut Pasteur* 71 (1945): 37–40, and the concept of a "surface" is at page 40; a tangential paper is Jacques Monod, "Inhibition de l'adaptation enzymatique chez *B. coli* en présence de 2-4 dinitrophénol," *Annales de l'Institut Pasteur* 70 (1944): 381–384. Note that both papers are dated as having been received by 2 December 1943; the late publication date is due to the wartime restrictions on publication.

355 "stupid kind of molecule"—Interview, Delbrück, 9 July 1972.

355 the October meeting in Geneva—Henri Noguères, *Histoire de la Résistance en France de 1940 à 1945*, vol. 4 (Paris: Laffont, 1976), pp. 34–39, where, however, although Philippe Monod's presence is mentioned Jacques' is not; for Jacques' presence, interview with Philippe Monod.

355 Réseau Vélites and arrest of Croland—Letter to Judson from Claude Lévy, secretary-general, Comité d'Histoiré de la 2 Guerre Mondiale, 16 May 1978, accompanied by archival material including an interview with Croland's sister taken on 21 March 1947.

355 Monod's work at Institut Pasteur—Interview, 1 December 1975; telephone conversation,

Audureau; J. Monod and A. Audureau, "Mutation et adaptation enzymatique chez *Escherichia coli-mutabile*," *Annales de l'Institut Pasteur* 72 (1946): 868–878.

356   fluctuation test: S. E. Luria and M. Delbrück, "Mutations of Bacteria from Virus Sensitivity to Virus Resistance," *Genetics* 28 (November 1943): 491–511; "Received May 29, 1943."

356   Noufflard—Interview, Geneviève Noufflard; she also allowed me to make a copy of an autobiographical memoir she had written, in English, in 1946 or 1947.

357   Pierremain—Interviews, Eugéne Jungelson.

358   Stendhal's Malivert—André Lwoff, comment on draft of this chapter, with letter to Judson, 1 June 1977; interview, Claudine Escoffier-Lambiotte.

358   Monod's report, 14 July 1944—"Rapport Hebdomadaire No. 1—E[tat] M[ajeur] N[ational] 3e Bureau," dated 14 July 1944, p. 3; original in possession of Mlle Noufflard.

359   Jacob's war—Interviews, Jacob, 3 December 1975, 28 and 29 May 1977.

359   order to the citizens of Paris—"E.M.N. 3e B. à Region P. 1," dated "20/8/44" and signed "Malivert"; original hand-written copy and original carbon copy in possession of Mlle Noufflard.

360   liberation of Ministry of War—Interview, Noufflard; see also "The Talk of the Town: Mademoiselle Vernier," *The New Yorker*, 14 July 1945, p. 16, for a contemporary interview with Mlle Noufflard.

360   Monod encounters Luria-Delbrück and Avery papers—Jacques Monod, "De l'adaptation enzymatique . . .," *Les Prix Nobel en 1965*, p. 3 (in reprint's numbering).

361   Monod at Cold Spring Harbor, 1946—Jacques Monod, "Du microbe à l'homme," in *Of Microbes and Life*, p. 7.

361   Delbrück's comment—Max Delbrück, in discussion following David Bonner, "Biochemical Mutations in Neurospora," *C. S. H. Symp. Q. Biol.* 11 (1946): 14–23; Delbrück at pp. 22–23.

361   Delbrück's and other papers—M. Delbrück and W. T. Bailey, Jr., "Induced Mutations in Bacterial Viruses," *C. S. H. Symp. Q. Biol.* 11 (1946): 33–37; Alfred Hershey, "Spontaneous Mutations in Bacterial Viruses," ibid., pp. 67–76; S. E. Luria, "Spontaneous Bacterial Mutations to Resistance to Antibacterial Agents," ibid., pp. 130–137; André Lwoff, "Some Problems Connected with Spontaneous Biochemical Mutations in Bacteria," ibid., pp. 139–155; J. Monod, contribution to discussion after S. Spiegelman, "Factors Controlling Enzymatic Constitution," ibid., Monod at pp. 274–275.

362   Lederberg's announcement—Joshua Lederberg and E. L. Tatum, "Novel Genotypes in Mixed Cultures of Biochemical Mutants of Bacteria," ibid., pp. 113–114. Interview, Lederberg, 9 August 1978; for circumstances of the discovery, Joshua Lederberg and Harriet Zuckerman, "From Schizomycetes to Bacterial Sexuality: A Case Study of Discontinuity in Science" (unpublished ms. dated 31 January 1977), p. 2:4.

363   Lederberg's dissertation—Results published as Joshua Lederberg, "Gene Recombination and Linked Segregations in *E. coli*," *Genetics* 32 (1947): 505–525.

363   Monod's decision to stick to enzymes—Jacques Monod, "De l'adaptation enzymatique . . .," *Les Prix Nobel en 1965*, p. 3 (in reprint's numbering); Jacques Monod, "The Phenomenon of Enzymatic Adaptation," *Growth* 11 (1947): 223–289.

363   a lactase extracted—Jacques Monod, Anne-Marie Torriani, and Joël Gribetz, "Sur une lactase extràite d'une souche d'*Escherichia coli mutabile*," *C. R. Acad. Sci.* 227 (1948): 315–316; Jacques Monod, "Facteurs génétiques et facteurs chimiques spécifiques dans la synthèse des enzymes bactériens," in *Unités Biologiques Douées de Continuité Génétique: Colloques Internationaux du Centre National de la Recherche Scientifique* 8 (Paris, Juin—Juillet 1948) (Paris: C.N.R.S., 1949): 181–199, in particular 189–194.

363   Lysenko—For background, the standard sources are Zhores A. Medvedev, *The Rise and Fall of T. D. Lysenko*, tr. I. Michael Lerner (New York and London: Columbia University Press, 1969) and David Joravsky, *The Lysenko Affair* (Cambridge, Mass.: Harvard University Press, 1970). For Monod's reaction, see his preface to Medvedev's book, in the French edition, *Grandeur et chute de Lysenko* (Paris: Gallimard, 1970), pp. 7–15.

364   articles in *Combat*—Marcel Prenant, *Combat*, 14 September 1948, pp. 6, 4; Jacques Monod, "'La victoire de Lyssenko n'a aucun caractère scientifique' estime le Dr Jacques Monod," *Combat*, 15 September 1948, pp. 1 and 6.

365   "very important to me"—Interviews, Monod, 1 and 4 December 1975.

366   lysogeny—Lwoff, "The Prophage and I," in *Phage and . . . Molecular Biology*, pp. 88–99; interviews, Lwoff, 3 December 1975 and 15 November 1976; Eugène Wollman and Elisabeth Wollman, "Régénération des bactériophages chez le *B. megatherium* lysogène," *Comptes rendus societé biologique* 122 (1936): 190–192, and later papers cited by François Jacob and Élie L. Wollman, *Sexuality and the Genetics of Bacteria* (New York: Academic Press, 1961), p. 365; F. M. Burnet and M. McKie, "Observations on a Permanent Lysogenic Strain of *B. enteritidis gaertner*," *Australian Journal of Experimental Biology and Medicine* 6 (1929): 277–284; Niels Ole Kjeldgaard, "The Unmasking of the Unseen," in *Of Microbes and Life*, pp. 89–93.

366   Delbrück's disbelief—Interview, Wollman; interview, Stent; Élie Wollman, "Bacterial Conjugation,"in *Phage and . . . Molecular Biology*, pp. 216–225.

366   the Wollmans seized by Gestapo—Pierre Nicolle, "Eugène Wollman," *Annales de l'Institut Pasteur* 72 (1946): 855–858; interview, Élie Wollman.

367   Lwoff on his methods—"The Prophage and I," *in Phage and . . . Molecular Biology*, pp. 89, 91; the discovery of induction, ibid., pp. 92–93; André Lwoff, L. Siminovitch, and N. Kjeldgaard, "Induction de la production de bactèriophages chez une bactérie lysogéne," *Annales de l'Institut Pasteur* 79 (1950): 815–859; André Lwoff, "Lysogeny," *Bacteriological Reviews* 17 (1953): 269–337.

367   the implications of lysogeny—Lwoff, "Lysogeny," *Bacteriological Reviews* 17 (1953): 287; interview, Lwoff, 3 December 1975.

368   Zinder and Lederberg—Norton Zinder and Joshua Lederberg, "Genetic Exchange in Salmonella," *Journal of Bacteriology* 64 (1952): 679–699; B. D. Davis, "Non-filtrability of the Agents of Genetic Recombination in *E. coli*," *Journal of Bacteriology* 60 (1950): 507–508.

369   discovery of phage *lambda*—Élie Wollman, "Bacterial Conjugation," in *Phage and . . . Molecular Biology*, pp. 216–225; Esther M. Lederberg, "Lysogenicity in *E. coli* K12," *Genetics* 36 (1951): 560.

370   Cohn's arrival—Interview, Cohn; Alvin M. Pappenheimer, Jr., "The Evolution of an Infectious Disease of Man," in *Of Microbes and Life*, p. 174.

371   first work with β-galactosidase—M. Cohn and J. Monod, "Purification et propriétés de la β-galactosidase (lactase) d'*Escherichia coli*," *Biochimica et Biophysica Acta* 7 (1951): 153–174; interview, Cohn; Monod, "De l'adaptation enzymatique . . ., *Les Prix Nobel en 1965*, p. 5 (in reprint).

371   Cohn's antiserum test—M. Cohn and Anne-Marie Torriani, "Immunochemical Studies with the β-galactosidase and Structurally Related Proteins of *Escherichi coli*," *Journal of Immunology* 69 (1952): 471–491.

372   discovery of the galactosides—Interview, Cohn; interview, Monod, 1 December 1975; Jacques Monod, Germaine Cohen-Bazire, and Melvin Cohn, "Sur la biosynthèse de la β-galactosidase (lactase) chez *Escherichia coli*: la spécificité de l'induction," *Biochimica et Biophysica Acta* 7 (1951): 585–599.

374   Jacob's entry into science—Interviews, 3 December 1975 and 28 May 1977.

375   first meeting Lwoff—François Jacob, "La belle époque," in *Of Microbes and Life*, pp. 98–104; quotations from p. 99, *trad. auct.*

375   Wollman's return—Interview, Wollman; Wollman, "Bacterial Conjugation," in *Phage and . . . Molecular Biology*, pp. 216–225.

376   Lwoff, Stent, meeting Hayes—Interview, Hayes; interview, Stent.

376   Lwoff carried the news back to Wollman—Telephone conversation, Wollman, 5 July 1978.

376   Hayes's discovery—William Hayes, "Recombination in *Bact. coli* K12: Unidirectional Transfer of Genetic Material," *Nature* 169 (1952): 118–119; "Observations on a Transmissible Agent Determining Sexual Differentiation in *Bact. coli*," *Journal of General Microbiology* 8 (1953): 72–88.

377   the events behind the curves—J. Monod, A.M. Pappenheimer, Jr., and G. Cohen-Bazire, "La cinétique de la biosynthèse de la β-galactosidase chez *E. coli* considérée comme fonction de la croissance," *Biochimica et Biophysica Acta* 9 (1952): 648–660, quotation at 650, *trad. auct.*

378   Hogness's arrival—Interview, Hogness.

378   "a remarkable intuition"—Lwoff, "The Prophage and I," in *Phage and . . . Molecular Biology*, p. 95.

379   Monod's search for universal mechanism—Quotation is from Monod, "De l'adaptation enzymatique . . .," *Les Prix Nobel en 1965*, p. 6 (in reprint); Jacques Monod and G. Cohen-Bazire, "L'effet d'inhibition spécifique dans la biosynthèse de la tryptophanedesmase chez *Aerobacter aerogenes*," *C.R. Acad. Sci.* 236 (1953): 530–532; Melvin Cohn and Jacques Monod, "Specific Inhibition and Induction of Enzyme Biosynthèsis," *Third Symposium of the Society for General Microbiology* 1953 (Cambridge: At the University Press, 1953), pp. 132–149.

379   Szilard picked it up—Aaron Novick and Leo Szilard, "Experiments with the Chemostat on the Rates of Amino Acid Synthesis in Bacteria," in *Dynamics of Growth Processes* (Princeton: Princeton University Press, 1954), pp. 21–32; Aaron Novick, "Phenotypic Mixing," in *Phage and . . . Molecular Biology*, pp. 137, 140.

379   change of terminology to "induction"—M. Cohn, et al., "Terminology of Enzyme Formation," *Nature* 172 (12 December 1953): 1096 (dated "Oct. 8").

379   "enzyme repression—Henry J. Vogel, "Repression and Induction as Control Mechanisms of Enzyme Biogenesis: The 'Adaptive' Formation of Acetylornithinase," in *McCollum-Pratt Symposium* 1956, pp. 276–289.

380   double bluff—Monod, "De l'adaptation enzymatique . . .," *Les Prix Nobel en 1965*, p. 11 (in

reprint); letter, Jacques Monod to Melvin Cohn, 28 February 1958, p. 4 (carbon copy in Monod correspondence files at Institut Pasteur, Paris).

380   experiment of characteristic conceptual simplicity—Interview, Monod, 1 December 1975; interview, Hogness; interview, Cohn; David S. Hogness, Melvin Cohn, and Jacques Monod, "Studies on the Induced Synthesis of β-galactosidase in *Escherichia coli*: the Kinetics and Mechanism of Sulfur Incorporation," *Biochimica et Biophysica Acta* 16 (1955): 99–116; quotation is at p. 107; quotation Monod read to me is at p. 114.

381   Meselson heard Monod—Interview, Meselson, 8 August 1977.

382   Jacob, Wollman, Hayes at Cold Spring Harbor 1953—François Jacob and Élie L. Wollman, "Induction of Phage Development in Lysogenic Bacteria," *C. S. H. Symp. Q. Biol.* 18 (1953): 101–121; W. Hayes, "The Mechanism of Genetic Recombination in *Escherichia coli*," ibid., pp. 75–93.

382   Wollman and Jacob on bacterial conjugation—Interview, Wollman; interviews, Jacob, 3 December 1975 and 28 May 1977.

385   their experiment—Élie L. Wollman and François Jacob, "Sur le mécanisme du transfert de matériel génétique au cours de la recombinaison chez *Escherichia coli K12*," *C. R. Acad. Sci.* 240 (séance du 20 Juin 1955): 2449–2451; François Jacob and Élie L. Wollman, "Étapes de la recombinaison génétique chez *Escherichia coli K12*," ibid. (séance du 27 Juin 1955): 2566–2568.

385   Georges Cohen—Interview, Cohen; Georges Cohen, "Découverte d'une perméase dans un grenier," in *Of Microbes and Life*, pp. 94–97; letter, Jacques Monod to Dr. Burckhardt Helferich, Chemisches Institut, Universität Bonn, 21 October 1954, introducing Cohen and Howard Rickenberg, who want to learn to make TMG radioactive; reply, Helferich to Monod, 29 October 1954, saying that Cohen and Rickenberg will be welcome (carbon copy of the former and original of latter in Monod correspondence, Institut Pasteur).

386   discovery of the permease—Howard V. Rickenberg et al., "La galactoside-perméase d'*Escherichia coli*," *Annales de l'Institut Pasteur* 91 (1956): 829–857; George N. Cohen and Jacques Monod, "Bacterial Permeases," *Bacteriological Reviews* 21 (1957): 169–194; interview, Cohen; Cohen, "Découverte d'une perméase dans un grenier," in *Of Microbes and Life*, pp. 94–97.

386   Crick's visit to Paris—Letter, Francis Crick to Jacques Monod, 4 April 1955, original in Monod correspondence at Institut Pasteur; Linus Pauling, "A Theory of the Structure and Process of Formation of Antibodies," *Journal of the American Chemical Society* 62 (1940): 2643–2657; Linus Pauling and Dan H. Campbell, "The Manufacture of Antibodies in Vitro," *Journal of Experimental Medicine* 76 (1942): 211–220; interview, Crick, 28 June 1977; interview, Jacob, 3 December 1975.

387   permease and inducibility—Interview, Monod, 1 December 1975; H. V. Rickenberg et al., "La galactoside-perméase d'*Escherichia coli*," *Annales de l'Institut Pasteur* 91 (1956): 847–849, 855. As late as the winter of 1957–58, Monod still thought that the gene I was probably located between the two structural genes, for on 28 February 1958 he wrote, in a letter to Melvin Cohn, "It seems that the constitutive mutation is between the galactosidase and the permease" (carbon of original in Monod correspondence at Institut Pasteur).

388   first genetic maps of *E. coli*—The best contemporary summary of the early work is François Jacob and Élie L. Wollman, "Genetic and Physical Determinations of Chromosomal Segments in *Escherichia coli*," in *Symp. Soc. Exp. Biol.* 12, held 1957, pp. 75–92; the first timed map, described in my text, is reproduced p. 77. Relevant early papers include François Jacob and Élie L. Wollman, "Sur les processus de conjugaison et de recombinaison génétique chez *E. coli*. I. L'induction par conjugaison ou induction zygotique," *Annales de l'Institut Pasteur* 91 (1956): 486–510; F. Jacob and É. L. Wollman, "Recombinaison génétique et mutants de fertilité chez *E. coli K12*," *C. R. Acad. Sci.* 242 (1956): 303–306. Interview, Wollman; interview, Jacob, 3 December 1975. For the full treatment of the topic, François Jacob and Élie L. Wollman, *Sexuality and the Genetics of Bacteria* (New York and London: Academic Press, 1961).

388   Jacob in U.S. in 1956—É. L. Wollman, F. Jacob, and W. Hayes, "Conjugation and genetic recombination in *Escherichia coli*," *C. S. H. Symp. Q. Biol.* 21 (1956): 141–162, which includes the electron micrograph taken by Anderson; François Jacob and Élie L. Wollman, "Genetic Aspects of Lysogeny," in *McCollum-Pratt Symposium* 1956, pp. 468–498, the paper that Crick's letter referred to; letter, Crick to Brenner, 17 July 1956, original in Brenner's papers.

390   collaboration between Jacob and Monod—Interview, Monod, 1 December 1975; interviews, Jacob, 3 December 1975, 28 and 29 May 1977.

390   "first glimpse of the operon"—Interviews, Monod, 5 October 1971 and 1 December 1975.

392   design of experiment in spring 1957—Interviews just cited; letter, Monod to Arthur Pardee, 16 May 1957, carbon copy in Monod correspondence at Institut Pasteur.

393   microsomal particles in *E. coli*—Howard K. Schachman, Arthur B. Pardee, and Roger Y.

Stanier, "Studies on the Macromolecular Organization of Microbial Cells," *Archives of Biochemistry and Biophysics* 38 (1952): 245–260; "Received November 5, 1951."

393 Pardee—Interview, Pardee; conversation, Schachman; interview, Stent.

393 Pardee's meeting Monod in 1952—Arthur B. Pardee, "The PaJaMo Experiment" (draft, photocopied, of contribution to memorial volume for Jacques Monod, ed. André Lwoff; draft prepared early 1978), p. 2.

393 Pardee's permease—A. B. Pardee, "An Inducible Mechanism for Accumulation of Melibiose in *Escherichia coli*," *Journal of Bacteriology* 73 (1957): 376–385.

393 striking new example of feedback—Richard A. Yates and Arthur B. Pardee, "Control of Pyrimidine Biosynthesis in *Escherichia coli* by a Feed-back Mechanism," *Journal of Biological Chemistry* 221 (1956): 757–770. Two notes on the timing of the discovery appear in this paper: "Presented in part at the Pacific Slope Biochemical Conference . . . December 30, 1954" and "Received for publication, November 21, 1955."

393 Umbarger's feedback mechanism—H. Edwin Umbarger, "Evidence for a Negative-Feedback Mechanism in the Biosynthesis of Isoleucine," *Science* 123 (1956): 848; "Received 21 October 1955."

393 radiophosphorus to disrupt DNA—Interview, Pardee; interview, Stent; Elizabeth McFall, Arthur B. Pardee, and Gunther S. Stent, "Effects of Radiophosphorus Decay on Some Synthetic Capacities of Bacteria," *Biochimica et Biophysica Acta* 27 (1958): 282–297; "Received July 31st, 1957."

393 correspondence with Monod—Letter, Monod to Pardee, 16 May 1957; letter, Pardee to Monod, discussing among other things the work with ^{32}P to disrupt DNA, 22 May 1957; carbon copy of the former and original of the latter in Monod correspondence, Institut Pasteur.

394 Pardee's work in fall 1957—Interview, Pardee; interview, Monod; Pardee, "The PaJaMo Experiment" (draft) pp. 2–4.

394 an opponent more imaginative than they—Interview, Pardee; interview, Monod; interview, Jacob, 3 December 1975; Pardee, "The PaJaMo Experiment" (draft), p. 5; letter, Monod to Cohn, 12 December 1957.

396 Szilard's intervention at Pasteur—Interviews, Jacob, 9 December 1970, 3 December 1975, 28 and 29 May 1977; interview, Monod, 1 December 1975; interview, Pardee; telephone conversation with Werner Maas; Luigi Gorini and Werner K. Maas, "The Potential for the Formation of a Biosynthetic Enzyme in *Escherichia coli*," *Biochimica et Biophysica Acta* 25 (1957)—on a copy of this paper sent me by Maas, he wrote, "This experiment was conceived in the Fall of 1956 by Szilard, Novick, Gorini and myself"; interview, Horecker; letter, Aaron Novick to Judson, 22 November 1977.

396 Maas at American Physiological Society—Werner K. Maas and Luigi Gorini, "Negative Feed-Back Control of the Formation of Biosynthetic Enzymes," in *Physiological Adaptation* (Washington, D.C.: American Physiological Society, 1958), pp. 151–158; Maas's quoted remark, p. 157.

397 Szilard in Berlin—No text of Szilard's talk appears to survive, so the one source is his reference to the sequence of the discussions, and to his Berlin talk, in Leo Szilard, "The Control of the Formation of Specific Protein in Bacteria and in Animal Cells," *Proc. N. A. S.* 46 (March 1960): 277–292, but in particular pp. 277–278.

397 the effect of Szilard's intervention—Interview, Pardee; interviews, Jacob, 3 December 1975, 28 and 29 May 1977; interview, Horecker. See also Kenneth Schaffner, "Logic of Discovery and Justification in Regulatory Genetics," *Studies in the History and Philosophy of Science* 4 (1974): 349–385: this paper represents the one previous attempt to unravel the strands of discovery that went into the development of the idea of the repressor, and it mentions Szilard's intervention; however, it came to my attention only after my own research was completed—and, with apologies to Dr. Schaffner, I must say that I found it hard to follow. My account is independent of his.

398 Monod's report to Cohn—Letter, Monod to Cohn, 28 February 1958.

399 Hoagland's visit—Interview, Hoagland.

399 first report of PaJaMo experiment—Arthur B. Pardee, François Jacob, and Jacques Monod, "Sur l'expression et le rôle des allèles 'inductible' et 'constitutif' dans la synthèse de la β-galactosidase chez des zygotes d'*Escherichia coli*," *C. R. Acad. Sci.* 246 (28 May 1958): 3125–3128; *trad. auct.*

400 symposium in Brussels—F. Jacob, "Transfer and Expression of Genetic Information in *Escherichia coli* K12," *Experimental Cell Research*, Supplement 6 (Proceedings of the Symposium held June 9–13, 1958, at l'Université Libre de Bruxelles, Belgium) (1959): 51–68, matter quoted at p. 67; A. B. Pardee, "Experiments on the Transfer of Information from DNA to Enzymes," ibid., pp. 142–151, matter quoted at p. 150; interview, Pardee; interview, Jacob, 28 May 1977.

400 implausible alternatives—Interview, Monod, 1 December 1975.

400 Jacob's realization of the parallel—Interviews, Jacob, 3 December 1975, 28 and 29 May 1977;

François Jacob, "The Switch" (draft of contribution to memorial volume for Jacques Monod, ed. André Lwoff; draft prepared early 1978), pp. 1, 9–12; François Jacob, "Genetic Control of Viral Functions" (Harvey Lecture delivered September 18, 1958), in *Harvey Lectures 1958–1959* (New York: Academic Press, 1960), pp. 1–39; material quoted at p. 24.

403 Monod's reception of Jacob's idea—Interview, Jacob, 3 December 1975; interview, Monod, 1 December 1975.

405 Pardee and Riley destroying the gene—Letter, Pardee to Monod, 8 December 1958.

406 professorship at Sorbonne—Letters, Monod to Cohn, 10 December 1958 and 21 January 1959, carbon copies in Monod correspondence, Institut Pasteur.

406 definitive PaJaMo paper—Letters, Pardee to Monod, 29 December 1958; Pardee to Monod, 9 January 1959; Monod to Pardee, 20 January 1959; Monod to Pardee, 27 January 1959, with revised draft; Pardee to Monod, 9 February 1959, commenting on revision and with new data; Pardee to Monod, 20 February 1959, with more comment; Monod to Pardee. 4 March 1959; Pardee to Monod, 31 March 1959, with final corrections; originals or carbons all in Monod correspondence, Institut Pasteur. Arthur B. Pardee, François Jacob, and Jacques Monod, "The Genetic Control and Cytoplasmic Expression of 'Inducibility' in the Synthesis of β-galactosidase by *E. coli,*" *Journal of Molecular Biology* 1 (1959): 165–178; "Received 16 March 1959."

406 Cohen's experiment with Jacob—Georges Cohen and François Jacob. "Sur la répression de la synthése des enzymes intervenant dans la formation du tryptophane chez *Escherichia coli,*" *C. R. Acad. Sci.* 248 (15 June 1959): 3490–3492.

406 Monod's effort to extract repressor—Letters, Monod to Pardee, 20 January 1959, 4 March 1959, 10 April 1959.

407 visit to Cambridge—Interview, Crick, 19 November 1975.

407 "ribosomes are crap"—Interview, David Perrin.

407 the conversation that developed the operon—Interview, Monod, 1 December 1975; interview, Cohen; interviews, Jacob, 3 December 1975, 28 and 29 May 1977; letter, Jacob to Judson, 28 June 1977, with further details of finding of operator-constitutive mutants.

409 conference in Copenhagen—Letter, Ole Maaløe to Judson, 17 May 1977. with date (13–19 September 1959), subject, and list of participants; interviews, Jacob, 3 December 1975, 28 and 29 May 1977.

410 theoretical note in October—François Jacob and Jacques Monod, "Gènes de structure et gènes de régulation dans la biosynthése des protéines," *C. R. Acad. Sci.* 249 (5 October 1959): 1282–1284; matter quoted at p. 1283; *trad. auct.* Letter, Monod to Cohn, 27 October 1959.

410 Gros's work implicating RNA—Interviews, Gros; A. Bussard et al., "Effets d'un analogue de l'uracil sur les propriétés d'une protèine enzymatique synthètisée en sa présence," *C. R. Acad. Sci.* 250 (1960): 4049–4051.

411 Monod's report to Cohn—Letter, Monod to Cohn, 28 January 1960, carbon copy in Monod correspondence, Institut Pasteur.

411 operator-constitutive mutants published—François Jacob et al., "L'opéron: groupe de gènes à expression coordonée par un opérateur," *C. R. Acad. Sci.* 250 (29 February 1960): 1727–1729; matter quoted is at p. 1729.

411 Crick on the French results—Interviews, Crick, 19 and 20 November 1975, 28 June 1977.

412 Note to RNA Tie Club—F. H. C. Crick and S. Brenner, "Some Footnotes on Protein Synthesis: A Note for the RNA Tie Club" (unpublished, mimeographed paper, 8 pages), December 1959; matter quoted at pp. 6–8; figure redrawn in exact copy.

413 Meselson's visit to Brenner, his current work—Interview, Meselson, 8 August 1977; letters, Brenner to Meselson, 15 January 1960 and 26 February 1960, originals in Meselson's correspondence files; letter, Jacob to Meselson, 19 April 1960, original in Meselson's files; C. I. Davern and Matthew Meselson, "The Molecular Conservation of Ribonucleic Acid During Bacterial Growth," *Journal of Molecular Biology* 2 (1960): 153–160, matter quoted at p. 153; conversation, Davern; letter, Monod to Pardee, 16 February 1960.

414 Hershey's fast RNA in 1953—A. D. Hershey, "Nucleic Acid Economy in Bacteria Infected with Bacteriophage T2: II. Phage Precursor Nucleic Acid," *Journal of General Physiology* 37 (1953): 1–23; matter quoted at pp. 18, 19; see also A. D. Hershey, "Conservation of Nucleic Acids During Bacterial Growth," ibid., 38 (1954): 145–148.

414 Volkin and Astrachan—Elliot Volkin and L. Astrachan, "Phosphorus Incorporation in *Escherichia coli* Ribonucleic Acid after Injection with Bacteriophage T2," *Virology* 2 (1956): 149–161; Elliot Volkin and L. Astrachan, "RNA Metabolism in T2-infected *Escherichia coli,*" *McCollum-Pratt Symposium* 1956, pp. 686–695, the apologetic portion at pp. 694–695; L. Astrachan and E. Volkin, "Properties of Ribonucleic Acid Turnover in T2-infected *Escherichia coli,*" *Biochimica et Biophysica Acta* 29 (1958): 536–544, quoted matter at pp. 543–544.

414 Jacob's awareness—Interview, Jacob, 28 May 1977.

415   Monod on the possibility that DNA made protein—Interview, Brenner, 14 May 1977; interview, Jacob, 28 May 1977; interview, Gros, 15 November 1976; interview, Monod, 1 December 1975.

415   Brenner spots Spiegelman note—Interviews, Brenner, 28 February 1976, March 1977, and 14 May 1977; M. Nomura, B. D. Hall, and S. Spiegelman. "Characterization of Ribonucleic Acid (RNA) Synthesized in Phage Infected *E. coli*," *Federation Proceedings* 19 (March 1960): 315.

415   Good Friday, 1960, meeting in King's—Interview, Brenner, 28 February 1976; interviews, Jacob, 3 December 1975, 28 and 29 May 1977; interviews, Crick, 20 November 1975 and 28 June 1977; interview, Garen; telephone conversation, Maaløe; letter, Maaløe to Judson, 17 May 1977; telephone conversation, Lerman; S. Brenner and F. H. C. Crick, "What Are the Properties of Genetic RNA? A Note for the RNA Tie Club" (Uncirculated, unpublished pair of draft manuscripts in Crick's hand, later of the two dated "May 1960"; 6 and 14 pages; original in Brenner's files).

417   Jacob's report of two lines of work—Monica Riley et al., "On the Expression of a Structural Gene," *Journal of Molecular Biology* 2 (1960): 216–225.

420   Crick to Robert Olby—Quoted in Robert Olby, "Francis Crick, DNA, and the Central Dogma," in *Daedalus*, Fall 1970, p. 970.

420   Crick's impetuous draft—S. Brenner and F. H. C. Crick, "What Are the Properties of Genetic RNA?" cited just above, earlier of the two versions.

421   Jacob back in Paris—Letter, Jacob to Meselson, 19 April 1960.

421   Brenner writes in more detail—Letter, Brenner to Meselson, 7 May 1960.

422   Crick's revised memorandum—"What Are the Properties of Genetic RNA?" longer version.

422   Spiegelman's full paper—Masayasu Nomura, Benjamin D. Hall, and S. Spiegelman, "Characterization of RNA Synthesized in *Escherichia coli* after Bacteriophage T2 Infection," *Journal of Molecular Biology* 2 (1960): 306–326; "Received 12 May 1960 and, in revised form, 27 July 1960."

422   Yčas's paper on yeast—Martynas Yčas and Walter Vincent, "A Ribonucleic Acid Fraction from Yeast Related in Composition to Deoxyribonucleic Acid," *Proc. N. A. S.* 46 (1960): 804–811.

428   Gros, Watson, Gilbert, looking for bacterial messenger—Interviews, Gros; interview, Gilbert, 13 February 1971; interviews, Meselson, March 1971 and 8 August 1977.

429   the messenger experiments—Interviews, Meselson, March 1971 and 8 August 1977; interviews, Brenner, 28 February 1976 and 14 May 1977; interviews, Jacob, 3 December 1975 and 28 May 1977; interview, Dulbecco, 19 March 1976; interview, Stent; unpublished laboratory notebooks, in the handwriting of Meselson, Brenner, and Jacob, for the entire period, originals in Brenner's files.

430   Brenner's controls—Interview, Brenner, 14 May 1977, with his laboratory notebooks; letter, Brenner to Meselson, 14 October 1960, original in Meselson's correspondence.

430   "messenger" christened that fall—In a letter to Meselson dated 28 July 1960—and on the eve of a fortnight away from Paris—Jacob wrote of work Meselson was planning on "the physico-chemistry of X R.N.A.," which establishes that the stuff was not named by then.

430   the messenger papers—S. Brenner, F. Jacob, and M. Meselson, "An Unstable Intermediate Carrying Information from Genes to Ribosomes for Protein Synthesis," *Nature* 190 (13 May 1961): 576–581; this paper credits the messenger hypothesis to Jacob and Monod, p. 576; F. Gros et al., "Unstable Ribonucleic Acid Revealed by Pulse Labeling of *Escherichia coli*," *Nature* 190 (13 May 1961): 581–585.

430   the theoretical summation—François Jacob and Jacques Monod, "Genetic Regulatory Mechanisms in the Synthesis of Proteins," *Journal of Molecular Biology* 3 (1961): 318–356, quoted passages at pp. 352–353 and 354. One element in the genetic mechanism of control in microorganisms was not discovered until three years later: the "promoter," which is a distinct genetic site, on the DNA, before the start of a gene (or an operon), at which molecules of the enzymes that make messenger RNA must attach in order to start transcribing from the DNA into RNA. Jacques Monod, François Jacob, and Agnes Ullman, "Le promoteur, élément génétique nécessaire a l'expression d'un opéron," *C. R. Acad. Sci.* 258 (1964): 3125–3128.

430   Jacob to Cambridge—Interview, Brenner, 14 May 1977; interview, Jacob, 28 May 1977.

431   "I think we *did* have a new idea"—Interview, Crick, 28 June 1977.

8 *"He wasn't a member of the club."*

PAGE

433   mutation according to Watson and Crick—J. D. Watson and F. H. C. Crick, "Genetical Implications of the Structure of Deoxyribonucleic Acid," *Nature* 171 (1953): 964–967.

434   Pardee on 5-bromouracil—R. M. Litman and A. B. Pardee, "Production of Bacteriophage Mutants by a Disturbance of Deoxyribonucleic Acid Metabolism," *Nature* 178 (1956): 529–532.

435   Benzer's curiosity provoked—Seymour Benzer and Ernst Freese, "Induction of Specific Mutations with 5-Bromouracil," *Proc. N. A. S.* 44 (1958): 112–119; "Communicated by M. Delbrück, December 6, 1957."

435   Meselson, Stahl, 5-bromouracil—Interview, Stahl; interviews, Meselson, 2 October 1971 and 8 August 1977; letter, Stahl to Brenner, with note by Meselson, 24 May 1957; letter, Brenner to Stahl, 7 June 1957; letter, Stahl to Brenner, 19 June 1957; originals or carbon copies in Brenner's correspondence files.

435   Brenner and acridine dyes—S. Brenner, S. Benzer, and L. Barnett, "Distribution of Proflavin-Induced Mutations in the Genetic Fine Structure," *Nature* 182 (11 October 1958): 983–985; interview, Benzer; Benzer, "Adventures in the *r*II Region," in *Phage and . . . Molecular Biology*, pp. 163–164; interviews, Crick, 1 September 1971 and 28 June 1977; interview, Brenner, 28 February 1976.

436   first coherent theoretical frame—Ernst Freese, "The Specific Mutagenic Effect of Base Analogues on Phage T4," *Journal of Molecular Biology* 1 (1959): 87–105, "Received 3 December 1958"; Freese, "The Difference Between Spontaneous and Base-Analogue Induced Mutations of Phage T4," *Proc. N. A. S.* 45 (1959): 622–633, "Communicated . . . February 24, 1959"; Freese, "On the Molecular Explanation of Spontaneous and Induced Mutations," in *Brookhaven Symposia* 12 (1959): 63–75.

437   Freese's first argument—Freese, "The Difference . . . ," p. 628.

437   Freese, "Some of the new bases . . ."—Freese, in *Brookhaven Symposia* 12, p. 69.

437   discussion after Freese's talk—*Brookhaven Symposia* 12, pp. 73–75.

437   Brenner's new evidence—*Brookhaven Symposia* 12, p. 74.

437   Lerman's explanation of acridines' action on DNA—L. Lerman, "Structural Considerations in the Interaction of DNA and Acridines," *Journal of Molecular Biology* 3 (1961): 18–30; "Received 15 August 1960."

438   messenger RNA turned coding upside down—Interview, Brenner, 28 February 1976.

438   Crick back at Cavendish November 14—Letter, Crick to Alexander Rich, Tuesday, 15 November 1960, "Back at work yesterday . . . ," original in Rich's correspondence files.

438   over lunch at the Eagle—Interview, Brenner, 28 February 1976.

439   into print without hesitation—S. Brenner et al., "The Theory of Mutagenesis," *Journal of Molecular Biology* 3 (1961): 121–124, quoted matter at pp. 122, 123, 123–124.

440   completely new support for messenger hypothesis—Benjamin D. Hall and S. Spiegelman, "Sequence Complementarity of T2-DNA and T2-Specific RNA," *Proc. N. A. S.* 47 (February 1961): 137–146; J. Marmur and D. Lane, "Strand Separation and Specific Recombination in Deoxyribonucleic Acids: Biological Studies," ibid. 46 (15 April 1960): 453–461; P. Doty et al., "Strand Separation and Specific Recombination in Deoxyribonucleic Acids: Physical Chemical Studies," ibid., 46 (15 April 1960), 461–476; J. Marmur, C. L. Schildkraut, and P. Doty, "Biological and Physical Chemical Aspects of Reversible Denaturation of Deoxyribonucleic Acids," in *The Molecular Basis of Neoplasia* (Austin: University of Texas Press, 1962), pp. 9–43.

441   loopy codes—Jacques R. Fresco and Bruce M. Alberts, "The Accommodation of Noncomplementary Bases in Helical Polyribonucleotides and Deoxyribonucleic Acid," *Proc. N. A. S.* 46 (1960): 311–321, "Communicated by Paul Doty, December 28, 1959"; Jacques R. Fresco, Bruce M. Alberts, and Paul Doty, "Some Molecular Details of the Secondary Structure of Ribonucleic Acid," *Nature* 188 (8 October 1960): 98–101; F. H. C. Crick and L. E. Orgel, "On Loopy Codes" (mimeographed, undated, but, on evidence of papers cited in text, written in winter 1960–'61; 5 pp.).

441   Crick in the laboratory—Interview, Leslie Barnett; interviews, Crick, 1 September 1971, 20 November 1975, 28 June 1977—this last with his lab notebooks in front of us; interviews, Brenner, 28 February 1976 and 21 March 1978; interview, Ames; interview, Stent.

442   notebook entries, 8 May 1961 and after—Laboratory notebooks, in Crick's hand, and in his files.

442   messenger papers—Cited end of chapter 7, note to p. 444; interview, Brenner, 28 February 1976.

443   Brenner at NIH—Interview, Brenner, 21 March 1978; interview, Matthaei.

443   Cold Spring Harbor meeting—participants, organization, in *C. S. H. Symp. Q. Biol.* 26 (1961): ix–xv; Brenner's talk, S. Brenner, "RNA, Ribosomes, and Protein Synthesis," ibid., pp. 101–110, matter quoted at pp. 101, 108.

444   Gilbert on messenger—Interview, Gilbert, 13 February 1971; for alternate views of RNA, see, e.g., H. Harris, "Turnover of Nuclear and Cytoplasmic Ribonucleic Acid in Two Types of Animal Cell, with some Further Observations on the Nucleolus," *Biochemical Journal* 73 (1959): 362–369; H. Harris, "The Relationship between the Synthesis of

Protein and the Synthesis of Ribonucleic Acid in the Connective-Tissue Cell," ibid., 74 (1960): 276–279; Henry Harris, *Nucleus and Cytoplasm* (Oxford: Clarendon Press, 1968), pp. 40–67.

444   Gilbert on Jacob and Monod—François Jacob and Jacques Monod, "Genetic Regulatory Mechanisms in the Synthesis of Proteins," *Journal of Molecular Biology* 3 (1961): 318–356; interview, Gilbert, 13 February 1971.

445   McClintock on control mechanisms—Interviews, McClintock; B. McClintock, Bibliography (mimeographed; through 1971); Barbara McClintock, "The Origin and Behavior of Mutable Loci in Maize," *Proc. N. A. S.* 36 (June 1950): 344–355; idem, "Chromosome Organization and Genic Expression," *C. S. H. Symp. Q. Biol.* 16 (1951): 13–47; idem, "Induction of Instability of Selected Loci in Maize," *Genetics* 38 (1953): 579–599; idem, "Controlling Elements and the Gene," *C. S. H. Symp. Q. Biol.* 21 (1956): 197–216.

445   McClintock's staff seminar—Interview, McClintock, 30 July 1978; McClintock, "Staff Meeting, November 28, 1960" (mimeographed outline, four pages), original in McClintock's files.

445   McClintock's paper—Barbara McClintock, "Some Parallels Between Gene Control Systems in Maize and in Bacteria," *The American Naturalist* 95 (September-October 1961): 265–277.

445   Monod's concluding summary—J. Monod and F. Jacob, "Teleonomic Mechanisms in Cellular Metabolism, Growth, and Differentiation," *C. S. H. Symp. Q. Biol.* 26 (1961): 389–401, matter quoted at pp. 394–395.

446   Crick's suppressors—His laboratory notebooks, in his files; interview, Crick, 28 June 1977; interviews, Brenner, 28 February 1976 and 21 March 1978; F. H. C. Crick et al., "General Nature of the Genetic Code for Proteins," *Nature* 192 (30 December 1961): 1227–1232, where the multiple levels of suppression are displayed in Fig. 2.

447   Crick at Chamonix—F. Crick, contribution to discussion after George Streisinger et al., "Genetic Studies Concerning the Lysozyme of Phage T4," *Proceeding of the 11th Annual Reunion of the Société de Chimie Physique*, June 1961 (Oxford: Pergamon, 1962), p. 188.

447   organization of Moscow meeting—Interview, Engelhardt; interview, Medvedev; interview, Perutz.

448   arrival in Moscow—Interview, Meselson, 2 October 1971.

448   Lysenko's career, standing in 1961—Interview, Medvedev; interview, Engelhardt; interview, Sakharov; Zhores A. Medvedev, *The Rise and Fall of T. D. Lysenko*, tr. I. Michael Lerner (New York: Columbia University Press, 1969), pp. 131–150.

449   Prezent on genes as fantasy—I. I. Prezent, *Biologiya v Shkole* 6 (December 1961), quoted in Medvedev, *The Rise and Fall of T. D. Lysenko*, p. 146.

449   Crick, Meselson, Perutz on Lysenko's absence in 1961—Interview, Crick, 28 June 1977; interview, Meselson, 3 October 1971; conversation, Perutz, fall 1976 (wording checked).

450   Medvedev on Russian science in 1961—Interview, Medvedev.

452   Engelhardt on Lysenko—Interview, Engelhardt.

453   Kendrew, Perutz opening talk—Published as John Kendrew, "The Structure of Globular Proteins," in *Proceedings of the Fifth International Congress of Biochemistry, Moscow, 10–16 August 1961, Volume I: Biological Structure and Function at the Molecular Level*, ed. V. A. Engelhardt (New York: Macmillan, 1963), models in Figs. 3 and 4, quoted matter at p. 4.

453   Nirenberg's first talk—Interview, Meselson, 3 October 1971; conversation, Gilbert, at Cold Spring Harbor, June 1973; interview, Tissières; interviews, Nirenberg; text of Nirenberg's paper, M. W. Nirenberg and H. Matthaei, "The Dependence of Cell-Free Protein Synthesis in *E. coli* upon Naturally Occurring or Synthetic Template RNA," in *Proceedings of the Fifth International Congress . . .*, pp. 184–189.

453   Nirenberg, Matthaei biographies, details of collaboration—Interviews, Nirenberg; interview, Matthaei; letter, with extended commentary on first draft of entire relevant passage of manuscript, Matthaei to Judson, 28 April 1978; supplementary details from interview, Maxine Singer, telephone conversation, Heppel.

455   Heppel, Singer, Ochoa work—Interviews, Singer and Ochoa; telephone conversation, Heppel; interview, John Smith.

455   Zamecnik's cell-free *E. coli* system—Interview, Zamecnik; Marvin R. Lamborg and Paul C. Zamecnik, "Amino Acid Incorporation into Protein by Extracts of *E. coli*," *Biochimica et Biophysica Acta* 42 (1960): 206–211. An earlier cell-free system from bacteria was described by D. Schachtschabel and W. Zillig, "Untersuchungen zur Biosynthese der Proteine. I. Ueber ein Einbau ^{14}C-markierte Aminosäuren ins Protein zellfreier Nucleoproteid-Enzym-Systeme aus *Escherichia coli*," *Hoppe-Seyler's Zeitung für physiologische Chemie* 314 (1959): 262, cited in Alfred Tissières, "Ribosome Research: Historical Background," in *Ribosomes*, ed. M. Nomura, A. Tissières, and P. Lengyel (Cold Spring Harbor: Cold Spring Harbor Laboratory, 1974), p. 8, 12—but both Tissières, in that citation, and Zamecnik, in interview and in a note to me on the face of a reprint of

◆

his paper with Lamborg, point out that the German paper was unknown to the American workers at the time; Zamecnik adds that "I can't evaluate how reliable this paper was," and as he said to me in the interview quoted in text, the problem of reliability had been troublesome with earlier attempts at bacterial cell-free systems.

456   Tissières's cell-free system—Interview, Tissières; interview, Zamecnik; Tissières, "Ribosome Research"; A. Tissières, D. Schlessinger, and Françoise Gros, "Amino Acid Incorporation into Proteins by *Escherichia coli* Ribosomes," *Proc. N. A. S.* 46 (15 November 1960): 1450–1463; interviews, Nirenberg.

456   Nirenberg's notes—Read in his office, 12 March 1971; machine copies of excepts furnished with letter, Nirenberg to Judson, 23 June 1978.

457   Nirenberg, Matthaei, preparing their cell-free system—Interviews, Nirenberg; interview, Matthaei; interview, Singer.

458   their first publication of system—Heinrich Matthaei and Marshall W. Nirenberg, "The Dependence of Cell-free Protein Synthesis in *E. coli* upon RNA Prepared from Ribosomes," *Biochemical and Biophysical Research Communications* 4 (1961), "Received March 22, 1961"; the term "messenger" is at p. 407.

459   complete sequence of tobacco-mosaic virus—A. Tsugita et al., "The Complete Amino Acid Sequence of the Protein of Tobacco-Mosaic Virus," *Proc. N. A. S.* 46 (1960): 1463–1469; interview, Fraenkel-Conrat.

459   Nirenberg's departure to Berkeley—Date from Matthaei's laboratory notebooks.

459   description of Matthaei's experimental work—Interview, Matthaei, with notebooks.

461   the more invidious myth—Zamecnik, "Historical Account," p. 9.

462   Monod on messenger "proof"—J. Monod and F. Jacob, "Teleonomic Mechanism . . .," p. 395.

462   Nirenberg learned of Matthaei's success—Interviews, Ninenberg; interview, Fraenkel-Conrat; interview, Matthaei.

462   the erroneous paper on tobacco-mosaic protein—A. Tsugita et al., "Demonstration of the Messenger Role of Viral RNA," *Proc. N. A. S.* 48 (1962): 846–853.

462   present view of how tobacco-mosaic-virus protein is specified—Tony R. Hunter et al., "Messenger RNA for the Coat Protein of Tobacco Mosaic Virus," *Nature* 260 (1976): 759–764.

463   slow spread of news—Interview, Maxine Singer; interview, Matthaei; interviews, Nirenberg; interview, Rich, 9 August 1977; interview, Tissières.

463   Nirenberg's reception in Moscow—Interview, Tissières; conversation, Gilbert, June 1973; interview, Meselson, 3 October 1971; interviews, Crick, 1 September 1971 and 28 June 1977; interviews, Nirenberg; interview, Perutz.

463   the paper—M. W. Nirenberg and H. Matthaei, "The Dependence of Cell-Free Protein Synthesis," pp. 184–189, quoted sentence at p. 189.

464   Nirenberg's and Matthaei's full results—J. Heinrich Matthaei and Marshall W. Nirenberg, "Characteristics and Stabilization of DNAase-Sensitive Protein Synthesis in *E. coli* Extracts," *Proc. N. A. S.* 47 (1961): 1580–1588; Marshall W. Nirenberg and J. Heinrich Matthaei, "The Dependence of Cell-Free Protein Synthesis in *E. coli* upon Naturally Occurring or Synthetic Polyribonucleotides," ibid., pp. 1588–1602.

464   chagrin of those who failed to try artificial messengers—Interview, Jacob, 28 May 1977; interview, Tissières; interview, Brenner, 28 February 1976; interview, Ochoa.

465   Crick on "uncles and aunts"—Interviews, Crick, 28 June 1977 and 20 November 1975; interviews, Brenner, 28 February 1976 and 21 March 1978; interview, Leslie Barnett; interview, Ames; interview, Stent.

467   Brenner to Paris for experiment with Jacob—Interview, Brenner, 21 March 1978; F. Jacob and S. Brenner, "Génétique physiologique: sur la régulation de la synthèse du DNA chez les bactéries: l'hypothèse du replicon," *C. R. Acad. Sci.* 256 (1963): 298–300.

467   the paper—F. H. C. Crick et al., "General Nature of the Genetic Code for Proteins," *Nature* 192 (30 December 1961): 1227–1232.

468   Crick on the test historians should apply—Interview, Crick, 20 November 1975.

468   mutagenesis and carcinogenesis, the Ames test—Interview, Ames; interview, Crick, 20 November 1975; Harvey J. Whitfield, Jr., Robert G. Martin, and Bruce N. Ames, "Classification of Aminotransferase (*C* Gene) Mutants in the Histidine Operon," *Journal of Molecular Biology* 21 (1966): 335–355; Bruce N. Ames and Harvey J. Whitfield, Jr., "Frameshift Mutations in Salmonella," *C. S. H. Symp. Q. Biol.* 31 (1966): 221–225; B. N. Ames, J. McCann, and E. Yamasaki, "Methods for Detecting Carcinogens and Mutagens with the *Salmonella*/mammalian Microsome Mutagenicity Test," *Mutation Research* 31 (December 1975): 347–364; M. Meselson and K. Russell, "Comparisons of Carcinogenic and Mutagenic Potency," in *Origins of Human Cancer*, ed. H. H. Hiatt, J. D. Watson, and J. A. Winsten (Cold Spring Harbor: Cold Spring Harbor Laboratory, 1977), vol. 4, book C, pp. 1473–1481; J. McCann and B. N. Ames, "The *Salmonella*/microsome Mutagenicity Test: Predictive Values for Animal Carcinogenicity," in ibid, pp. 1431–1450, which also

contains further references. For most recent discussion, see Bruce N. Ames and Kim Hooper, "Does Carcinogenic Potency Correlate with Mutagenic Potency in the Ames Assay?" *Nature* 274 (6 July 1978): 19–20, and references.

469   back from Moscow, a small, neat experiment—Marshall W. Nirenberg, J. Heinrich Matthaei, and Oliver W. Jones, "An Intermediate in the Biosynthesis of Polyphenylalanine Directed by Synthetic Template RNA," *Proc. N. A. S.* 48 (January 1962): 104–109.

469   competition to get the code—Interviews, Nirenberg; interview, Ochoa; interview, Crick, 28 June 1977; interview, Matthaei; interview, Maxine Singer; telephone conversation, Heppel; for work from Ochoa's lab, J. F. Speyer et al., "Synthetic Polynucleotides and the Amino Acid Code, IV," *Proc. N. A. S.* 48 (March 1962): 441–448, and previous papers in the series as cited therein; for further papers from Nirenberg's lab, see Marshall Nirenberg, "Protein Synthesis and the RNA Code" (Harvey Lecture delivered 20 February 1964), *Harvey Lectures . . . 1963–1964*, Series 59 (New York: Academic Press 1965), pp. 155–185.

469   Brenner coins "codon"—In Brenner's files is preserved a preliminary draft, mimeographed, undated, of a paper, Seymour Benzer and Sewell F. Champe, "Hereditary Alteration of the Genetic Code in *Escherichia coli*," with a note to Francis Crick on the first page, in Benzer's hand, saying, in part," 'codon' was coined by Sydney. Do you think it is useful?"

469   sloppy work and hasty publication—F. H. C. Crick, "The Recent Excitement in the Coding Problem," in *Progress in Nucleic Acid Research* 1 (New York: Academic Press, 1963), pp. 163–217, quoted matter at pp. 180, 215.

469   Nirenberg's second discovery—Marshall Nirenberg and Philip Leder, "RNA Codewords and Protein Synthesis: The Effect of Trinucleotides upon the Binding of sRNA to Ribosomes," *Science* 145 (1964): 1399–1407.

470   Khorana's work—H. G. Khorana, "Polynucleotide Synthesis and the Genetic Code" (Harvey Lecture delivered 17 November 1966), *Harvey Lectures . . . 1966-1967* Series 62 (New York: Academic Press, 1968), pp. 79–105.

470   Crick's table—F. H. C. Crick, "The Genetic Code—Yesterday, Today, and Tomorrow," *C. S. H. Symp. Q. Biol.* 31 (1966): 3–9, table at p. 1.

470   further history of elucidation of genetic code—See Crick, "The Genetic Code"; Carl R. Woese, "The Present Status of the Genetic Code," *Progress in Nucleic Acid Research and Molecular Biology* 7 (1969): 107–172; F. H. C. Crick, "The Genetic Code" (The Croonian Lecture, 1966), *Proceedings of the Royal Society* B 167 (1967): 331–347.

470   nonsense codons—The elucidation of these was intricate, and the collaborations and citations many: for accounts with references, see A. O. W. Stretton, S. Kaplan, and S. Brenner,"Nonsense Codons," *C. S. H. Symp. Q. Biol.* 31 (1966): 173–179; S. Brenner, "Nonsense Mutants and the Genetic Code: A Small Piece of Molecular Genetics" (Paper read at Rockefeller University, 11 November 1971, on receipt, with Seymour Benzer, of the 1971 Albert Lasker Award; 13 pp., mimeographed); Martynas Yčas, *The Biological Code* (Amsterdam and London: North-Holland Publishing, 1969), pp. 208–211. The final paper referred to in my text is F. H. C. Crick and S. Brenner, "The Absolute Sign of Certain Phase–Shift Mutants in Bacteriophage T4," *Journal of Molecular Biology* 26 (1967): 361–363.

470   "the machinery without . . . the biochemistry"—Interview, Brenner, 28 February 1976.

470   "a club of his own"—Interview, Meselson, 3 October 1971.

9 *"As always, I was driven on by wild expectations."*

PAGE

475   scientists on continuing in research or not—Interview, Perutz, 4–5 December 1970; interview, Monod, 4 December 1975; interviews, Kornberg, 12 September 1975, 11 August 1978; notes, Perutz, on copy of draft of this passage; interview, Kendrew, 13 January 1971.

479   Perutz on proteins, DNA, hemoglobin, models, etc.—Interview, Perutz, 8 May 1968.

480   Lipscomb's enzyme—Interview, Lipscomb; William N. Lipscomb et al., "The Structure of Carboxypeptidase A. VII. The 2.0-Å Resolution Studies of the Enzyme and of Its Complex with Glycyltyrosine, and Mechanistic Deductions," *Brookhaven Symposia in Biology* 21 (1968): 24–90, contains earlier references.

485   Perutz at Royal Society—Interview, 29 May 1968; M. F. Perutz, "The Haemoglobin Molecule" (The Croonian Lecture, 1968, delivered 30 May), *Proceedings of the Royal Society* B (1969): 113–140.

485   Perutz's biography—Interviews, Perutz, 4, 5, and 6 December 1970; "Max Ferdinand Perutz," autobiographical sketch in *Les Prix Nobel en 1962*, pp. 67–68.

◆

486   Bernal's discovery—J. D. Bernal and D. Crowfoot, "X-ray Photographs of Crystalline Pepsin," *Nature* (26 May 1934): 794–795; see also the letter immediately following, no separate title, from W. T. Astbury and R. Lomax, ibid., p. 795.

487   Perutz on Bernal—M. F. Perutz, "Bragg, Protein Crystallography and the Cavendish Laboratory," *Acta Crystallographica* A26 (1970): 183–185, quoted matter at p. 183.

487   Perutz's decision to try hemoglobin—Interviews, Perutz, 4 December 1970, 31 January and 7 February 1976, 19 November 1977; checking interview, Perutz, 3 February 1978, with draft of this chapter; conversation, Haurowitz.

488   choice between hemoglobin and chymotrypsin—J. D. Bernal, I. Fankuchen, and Max Perutz, "An X-ray Study of Chymotrypsin and Haemoglobin," *Nature* 141 (19 March 1938): 523–524.

488   history of research into hemoglobin—My chief and indispensable guide in the pages that follow has been the splendid monograph by John T. Edsall, "Blood and Hemoglobin"; I supplemented Edsall's monograph, for a finding list of early papers, with the opening, historical chapters of Edward Tyson Reichert and Amos Peaslee Brown, *The Crystallography of Hemoglobins* (Washington: The Carnegie Institution of Washington, 1909).

489   Magnus—Gustav Magnus, "Ueber die im Blute enthaltenen Gase, Sauerstoff, Stickstoff und Kohlensäure," *Annalen der Physik und Chemie* ed. J. C. Poggendorf 40 (1837): 583–606; the crucial conclusions that arterial as well as venous blood contains carbon dioxide and that there is a shift in proportion of oxygen to carbon dioxide between arterial and venous blood are at pp. 599–600.

489   earliest crystals of hemoglobin—F. L. Hünefeld, *Der Chemismus in der thierischen Organisation* (Leipzig: F. A. Brockhaus, 1840), pp. 158–163; curiously, Reichert and Brown describe the source of the crystals wrongly; a machine copy of the relevant pages of Hünefeld's book was kindly obtained for me by Jürgen Rimpau, at the University of Göttingen.

489   Reichert's observations—K. B. Reichert, "Beobachtungen über eine eiweissartige Substanz in Krystallform," *Archiv für Anatomie, Physiologie und wissenschaftliche Medicin . . .*, ed. Johannes Müller (1849): 197–251, the guinea pig at pp. 198–199.

489   Funke—Otto Funke, "Ueber das Milzvenenblut," *Zeitschrift für rationelle Medicin*, Heidelberg, N. F. v. 1 (1851): 172–218, details of technique at pp. 185–188.

489   Teichmann—L. Teichmann, "Ueber die Krystallisation der organischen Bestandtheile des Bluts," *Zeitschrift für rationelle Medicin*, N.F. 3 (1853): 375–388, the name "Hämin introduced, with description of the crystals, at pp. 384–385.

490   Hoppe-Seyler names hemoglobin—Felix Hoppe-Seyler, "Ueber die chemischen und optischen Eigenschaften des Blutfarbstoffs," *Archiv für pathologische Anatomie und Physiologie (Virchow's Archiv . . .)* 29 (1864): 233–235.

490   Hoppe's spectrographic work—Felix Hoppe, "Ueber das Verhalten des Blutfarbstoffes im Spectrum des Sonnenlichtes," *Archiv für pathologische Anatomie*, 23 (1862): 446–499.

490   Stokes's discovery—G. G. Stokes, "On the Reduction and Oxidation of the Colouring Matter of the Blood," *Proceedings of the Royal Society* of London 13 (1864): 355–364.

492   Reichert and Brown—Reichert and Brown, *The Crystallography of Hemoglobins*.

493   Conant's surprising discovery—J. B. Conant, "An Electrochemical Study of Hemoglobin," *Journal of Biological Chemistry* 57 (1923): 401–414; see also James B. Conant, "The Oxidation of Hemoglobin and Other Respiratory Pigments" (Harvey Lecture delivered 16 February 1933), *Harvey Lectures . . .* 28 (1933): 159–183, a discussion of the work of 1923 at pp. 162–164.

494   Zinoffsky's analysis of hemoglobin—O. Zinoffsky, "Ueber die Grösse des Hämoglobinmoleküls," *Hoppe-Seyler's Zeitschrift für physiologische Chemie* 10 (1886): 16–34—reference taken from Edsall, "Blood and Hemoglobin," p. 208 and n.

494   Reid; Hüfner—E. W. Reid, "Osmotic Pressure of Solutions of Hemoglobin," *Journal of Physiology* 33 (1905): 12–19; G. Hüfner and E. Gansser, "Ueber das Molekulargewicht des Oxyhämoglobins," *Archiv für Physiologie und Anatomie (Virchow's Archiv . . .)* (1907): 209–216—papers cited and the work analyzed by Edsall, "Blood and Hemoglobin," pp. 237–238.

494   Adair overturned all this—Adair's own brief historical account, G. S. Adair, "A Rapid and Accurate Method for the Measurement of the Osmotic Pressure of Haemoglobin," in *Barcroft Memorial, Haemoglobin*, pp. 191–195; his early results, G. S. Adair, *Thesis* (Cambridge: King's College, 1922–1923), cited in Adair, "A Rapid and Accurate Method . . .," pp. 191, 195n; the decisive paper, G. S. Adair, "A Critical Study of the Direct Method of Measuring the Osmotic Pressure of Haemoglobin," *Proceedings of the Royal Society* of London 108A (1925): 627–637, "Received April 1, 1924"; Edsall's comment, "Blood and Hemoglobin," p. 242; Adair took the method further in Gilbert S. Adair, "The Osmotic Pressure of Haemoglobin in the Absence of Salts," *Proceedings of the Royal Society* of London 109A (1925): 292–300.

495 Svedberg and the ultracentrifuge—T. Svedberg and Robin Fåhraeus, "A New Method for the Determination of the Molecular Weight of the Proteins," *Journal of the American Chemical Society* 48 (February 1926): 430–438, "Received August 4, 1925," quoted matter at pp. 433, 430, 438; T. Svedberg and J. B. Nichols, "The Application of the Oil Turbine Type of Ultracentrifuge to the Study of the Stability Region of Carbon Monoxide-Hemoglobin," *Journal of the American Chemical Society* 49 (November 1927): 2920–2934, quoted matter at pp. 2920 and 2928.

496 oxygen-dissociation of myoglobin, hemoglobin—See, e.g., Joyce M. Baldwin, "Structure and Function of Haemoglobin," *Progress in Biophysics and Molecular Biology* 29 (1975): 228–230.

496 Christian Bohr—A brief biographical account of the elder Bohr appears in Stefan Rozental, ed., *Niels Bohr: His Life and Work as Seen by His Friends and Colleagues* (Amsterdam: North-Holland Publishing, 1967), pp. 11–15—a source relied on also, in part, by Edsall.

496 Hüfner supposed hemoglobin behaved simply—His and other early work is discussed, once again, by Edsall, "Blood and Hemoglobin," pp. 213–214, 219–224; the citation, therein, p. 213n, is G. Hüfner, "Ueber das Gesetz der Dissociation des Oxyhämoglobins und über einige daran sich knüpfenden wichtigen Fragen aus der Biologie," *Archiv für Anatomie und Physiologie (Physiologisches Abtheilung)* (1890): 1–27.

497 Bohr's study, with colleagues, of oxygen-dissociation of hemoglobin—The apparatus and techniques, August Krogh, "Apparate und Methoden zur Bestimmung der Aufnahme von Gasen im Blute bei Verschiedenen Spannungen der Gase," *Skandinavisches Archiv für Physiologie* 16 (26 October 1904): 390–401; the classic paper, Christian Bohr, K. Hasselbalch, and August Krogh, "Ueber einen in biologischer Beziehung wichtigen Einfluss, den die Kohlensäurespannung des Blutes auf dessen Sauerstoffbindung übt," ibid., pp. 402–412. Edsall makes a discussion of Bohr's work and its consequences the centerpiece of his monograph: "Blood and Hemoglobin," pp. 219–248.

497 Haldane, "die of asphyxia"—J. S. Haldane and J. G. Priestley, *Respiration*, 2nd ed. (New Haven: Yale University Press, 1935).

498 carbon dioxide drove the oxygen affinity down further—the central point is made in Bohr, Hasselbalch, and Krogh, "Ueber einen in biologischer . . .," p. 403, and followed immediately by the observation that previous investigators had missed it because they worked at high partial pressures of oxygen; the three discuss the physiological importance of the phenomenon in their concluding paragraphs, pp. 411–412.

498 multiple dissociation curves—Chart redrawn from Bohr, Hasselbalch, and Krogh, "Ueber einen in biologischer Beziehung wichtigen Einfluss . . .," p. 409.

499 the effect subtler than Bohr knew—Interview, Perutz, 19 November 1977, and written comments on draft; Perutz's instruction on the significance of the Bohr effect has considerably shaped my account of the paper.

499 Krogh's claim that he, not Bohr, found the effect—Krogh made the claim to Professor F. J. W. Roughton, of Cambridge, at the time of the Barcroft Memorial Symposium in 1948 (*Barcroft Memorial, Haemoglobin*), and Roughton told Edsall of the conversation: Edsall, "Blood and Hemoglobin," pp. 226–227.

499 Barcroft's work, Hill's, the Haldanes'—The investigations summarized with such reckless speed here are discussed in careful, sensitive detail by Edsall, "Blood and Hemoglobin," pp. 229–242, with citations. The crucial paper was A. V. Hill, "The Possible Effects of the Aggregation of the Molecules of Haemoglobin on Its Dissociation Curve," *Journal of Physiology* 40 (1910): iv–vii, cited and discussed by Edsall, pp. 239–240.

499 Adair's equation—G. S. Adair, "The Hemoglobin System. VI. The Oxygen Dissociation Curve of Hemoglobin," *Journal of Biological Chemistry* 63 (1925): 529–545; this is the last of a series of six papers, some written with colleagues, that Adair published as a bloc under the general title "The Hemoglobin System," ibid., pp. 493–545.

499 Perutz on his taking-in of Adair's work and Bohr's—Interview, Perutz, 19 November 1977.

500 Porphyrin ring—Diagram from Joyce M. Baldwin, "Structure and Function of Haemoglobin," *Progress in Biophysics and Molecular Biology* 29 (1975): p. 247, slightly simplified.

501 Pauling began to think about hemoglobin—Interview, Pauling, 1 March 1971.

501 Pauling told Perutz his aim—Interview, Perutz, 11 November 1977.

502 Pauling's papers on hemoglobin—The first was Linus Pauling, "The Oxygen Equilibrium of Hemoglobin and Its Structural Interpretation," *Proc. N. A. S.* 21 (1935): 186–191, but that was not a successful or important contribution. The experimental techniques were described in Linus Pauling and Charles D. Coryell, "The Magnetic Properties and Structure of the Hemochromogens and Related Substances," *Proc. N. A. S.* 22 (1936): 159–163, the method at p. 161; the crucial paper, though, was submitted a month later; Linus Pauling and Charles D. Coryell, "The Magnetic Properties and Structure of Hemo-

globin, Oxyhemoglobin and Carbonmonoxy-hemoglobin," *Proc. N. A. S.* 22 (1936): 210–216, "Communicated March 19, 1936," quoted matter at pp. 212, 213–214.

502 the significance of Pauling's early work on hemoglobin—Interview, Perutz, 31 January 1976.

503 "didn't think it was significant"—Interview, Perutz, 11 November 1977.

503 in the crystal room—Interview, Perutz, 5 January 1971; comments by Perutz on two drafts.

506 Hodgkin's memoir—Dorothy Crowfoot Hodgkin and Dennis Parker Riley, "Some Ancient History of Protein X-Ray Analysis," in *Structural Chemistry and Molecular Biology,* ed. A. Rich and N. Davidson (San Francisco: Freeman, 1968), pp. 15–28, passage quoted at p.15.

506 Bernal's paper—J. D. Bernal and D. Crowfoot, "X-ray Photographs of Crystalline Pepsin," *Nature* 133 (26 May 1934): 794–795, together with the following note, no separate title, by W. T. Astbury and R. Lomax.

506 pepsin's unit cell in the 1935 paper—Perutz later found that Bernal and Crowfoot's dimensions for the unit cell had been incorrect.

507 discovery of x rays—The paper appeared in English promptly: W. C. Röntgen, "On a New Kind of Rays," *Nature* (23 January 1896): 274–276, "translated by Arthur Stanton from the *Sitzungsberichte der Würzburger Physik-medic. Gesellschaft, 1895.*"

507 early history of x-ray analysis—Bragg, *Development of X-ray Analysis,* pp. 1–6.

507 Thomson's "whip-crack"—Bragg, "History," p. 162.

508 Laue's paper—W. Friedrich, P. Knipping, and M. Laue, "Interferenz-Erscheinungen bei Röntgenstrahlen," *Sitzungsberichte der mathematisch physikalischen Klasse der Königlich Bayerischen Akademie der Wissenschaften,* 8 June 1912, pp. 303–322

508 Bragg on "the parental hypothesis"—Bragg, "History," p. 162.

509 Bragg, "the object"—"History," p. 164.

509 Bragg, "I was led to think"—Bragg, *Development of X-ray Analysis,* pp. 22–23.

511 Bragg's first paper—W. L. Bragg, "The Diffraction of Short Electromagnetic Waves by a Crystal," *Proceedings of the Cambridge Philosophical Society* 17 (1913): 43–57, "Read 11 November 1912"; his first structures and the scale for measurement of x-ray wavelengths, Bragg, *Development of X-ray Analysis,* pp. 29–30.

513 first results with hemoglobin and chymotrypsin—Interviews, Perutz, 4–6 December 1970, 31 January 1976, 19 November 1977; J. D. Bernal, I. Fankuchen, and Max Perutz, "An X-ray Study of Chymotrypsin and Haemoglobin," pp. 523–524.

513 Perutz a refugee—Interviews, Perutz, 4–6 December 1970, 3 February 1978.

513 Bragg, "cut and try"—Bragg, "How Proteins Were Not Solved," p. 834.

513 Bernal's talk with Pauling—J. D. Bernal, "The Pattern of Linus Pauling's Work in Relation to Molecular Biology," in *Structural Chemistry and Molecular Biology,* p. 372.

513 obscure paper—Hodgkin and Riley, "Some Ancient History . . .," in *Structural Chemistry and Molecular Biology,* p. 16.

513 Crowfoot's shrinkage—Dorothy Crowfoot and Dennis Riley, "An X-ray Study of Palmer's Lactoglobulin," *Nature* 141 (19 March 1938): 521–522; Hodgkin and Riley, "Some Ancient History . . .," pp. 16–18.

513 Bernal on "states of hydration"—Bernal, Fankuchen, and Perutz. "An X-ray Study of Chymotrypsin . . .," p. 524.

514 Haurowitz's discovery—Conversation, Haurowitz, with portion of manuscript; letter, Haurowitz to Judson, 4 February 1972, saying, in part, "I looked up my protocols of the year 1938 and found on p. 63 a description of the experiment which Max Perutz reported to you. It was performed between April 1 and 5, 1938." Interviews, Perutz, 4–6 December 1970, 19 November 1977; Felix Haurowitz, "Das Gleichgewicht zwischen Hämoglobin und Sauerstoff," *Hoppe-Seyler's Zeitschrift für physiologische Chemie* 254 (1938): 266–274.

514 Perutz's family at partition of Czechoslovakia—Interviews, Perutz, 4–6 December 1970, 3 February 1978; conversation, Haurowitz.

515 Bernal's politics—Well-known, but see, e.g., the energetic biographical study of British left-wing scientists in the nineteen-thirties, Gary Werskey, *The Visible College* (New York: Holt, Rinehart and Winston, 1979).

515 Bragg's support for Perutz—Interview, Bragg; letter, Bragg to Tisdale, 28 November 1938, copy in W. L. Bragg Papers at the Royal Institution.

516 diagrams of Fourier analysis—redrawn from Bragg, *The Development of X-ray Analysis.*

517 the elder Bragg noted the consequence—W. H. Bragg, the Bakerian Lecture to the Royal Society, 1915, quoted in Bragg, *The Development of X-ray Analysis,* p. 104.

518 Bragg, "condensations of density"—Bragg, "History," p. 265.

518 the phases must be found—Bragg, *The Development of X-ray Analysis,* p. 108.

518 Bragg, "obliging way"—Sir Lawrence Bragg, "X-ray Crystallography," *Scientific American,* July 1968, p. 69.

518   Bernal's Friday Evening—J. D. Bernal, "The Structure of Proteins" (Discourse delivered Friday 27 January 1939), *Proceedings of the Royal Institution of Great Britain* 30: 541–557.

519   Svedberg's multiples—Theodor Svedberg, "Opening Address," in "A Discussion on the Protein Molecule," (held 17 November 1938), *Proceedings of the Royal Society* of London 170A (1939): 41–42.

519   Perutz's internment—Interview, 31 January 1976.

520   drying and shrinking—M. F. Perutz, "X-ray Analysis of Haemoglobin," *Nature* 149 (2 May 1942): 491; "Crystal Structure of Oxyhaemoglobin," *Nature* 150 (12 September 1942): 324; Joy Boyes-Watson and M. F. Perutz, "X-ray Analysis of Haemoglobin," *Nature* 152 (1943): 714–715.

520   project Habbakuk:—Interview, Perutz, 31 January 1976. "I had that childish faith"—Interview, Perutz, 31 January 1976. Kendrew's biography—Interviews, Kendrew; interview, Perutz, 4–6 December 1970; autobiographical sketch, "John C. Kendrew," *Les Prix Nobel en 1962*, pp. 65–66.

523   Barcroft suggested fetal hemoglobin—Interview, Perutz, 4 December 1970.

523   results with hemoglobin by 1947—Joy Boyes-Watson, Edna Davidson, and M. F. Perutz, "An X-ray Study of Horse Methaemoglobin. I," *Proceedings of the Royal Society* of London A 191 (1947): 83–132, matter quoted at p. 122.

524   Bragg turned to Medical Research Council—Bragg, "How Proteins Were Not Solved," p. 3; M. F. Perutz, "Bragg, Protein Crystallography and the Cavendish Laboratory," *Acta Crystallographica* A 26 (March 1970): 185.

524   Pauling's visit to Perutz—Interview, Pauling, 23 December 1975; interview, Perutz, 31 January 1976; the complete account is, of course, in chapter 2, above.

524   Perutz and Kendrew on the phase problem and Patterson methods—J. C. Kendrew and M. F. Perutz, "The Application of X-ray Crystallography to the Study of Biological Macromolecules," in *Barcroft Memorial, Haemoglobin*, pp. 161–179, matter quoted at pp. 165–166.

525   principle of substituting atoms familiar in Bernal's lab—Interview, Hodgkin; Hodgkin and Riley, "Some Ancient History . . .," in *Structural Chemistry and Molecular Biology*, pp. 24–25.

525   Kendrew and Perutz warned—Kendrew and Perutz, "The Application of X-ray Crystallography . . .," p. 170.

525   Perutz's Patterson synthesis—M. F. Perutz, "An X-ray Study of Horse Methaemoglobin. II," *Proceedings of the Royal Society* of London A 195 (1949): 474–499, "Received 8 June 1948," matter quoted at pp. 475, 489, 474–475.

526   Kendrew looking for protein of his own—Interview, Kendrew, 13 and 14 January 1971; John C. Kendrew, "Myoglobin and the Structure of Proteins" (Nobel Lecture, 11 December 1962), *Les Prix Nobel en 1962*, p. 104.

526   Bragg, Patterson synthesis "a false star"—Bragg, "How Proteins Were Not Solved," pp. 4, 6.

526   "keep them at it—Interview," Bragg.

527   the disastrous paper on protein structure—Sir Lawrence Bragg, J. C. Kendrew, and M. F. Perutz, "Polypeptide Chain Configurations in Crystalline Proteins," *Proceedings of the Royal Society* of London A 203 (1950): 324–357.

527   Pauling, sickle-cell—Linus Pauling et al., "Sickle Cell Anemia, A Molecular Disease," *Science* 110 (1949): 543–548; Perutz's reaction, interview, Perutz, 31 January 1976.

527   Perutz, sickle-cell—M. F. Perutz and J. M. Mitchison, "State of Haemoglobin in Sickle-Cell Anaemia," *Nature* 166 (21 October 1950): 677–678; M. F. Perutz, A. M. Liquori, and F. Eirich, "X-ray and Solubility Studies of the Haemoglobin in Sickle-Cell Anaemia Patients," *Nature* 167 (9 June 1951): 929–931.

528   "What Mad Pursuit"—Published as F. H. C. Crick, "The Height of the Vector Rods in the Three-dimensional Patterson of Haemoglobin," *Acta Crystallographica* 5 (1952): 381–386.

528   Kendrew, computers—J. M. Bennett and J. C. Kendrew, "The Computation of Fourier Syntheses with a Digital Electronic Calculating Machine," *Acta Crystallographica* 5 (January 1952): 109–135.

528   Bragg's and Perutz's next three—W. L. Bragg, E. R. Howells, and M. F. Perutz, "Arrangement of Polypeptide Chains in Horse Methaemoglobin," *Acta Crystallographica* 5 (January 1952): 136–141; W. L. Bragg and M. F. Perutz, "The External Form of the Haemoglobin Molecule. I," *Acta Crystallographica* 5 (March 1952): 277–283, and "The External Form of the Haemoglobin Molecule. II," ibid., 5 (May 1952): 323–328.

528   Bragg, "rather modest result"—Bragg, "How Proteins Were Not Solved," p. 8.

528   Riggs, sickle-cell—Interviews, Perutz 4–6 December 1970, 31 January 1976, and 7 February 1976; Austen F. Riggs, "Sulfhydryl Groups and the Interaction Between the Hemes in Hemoglobin," *Journal of General Physiology* 36 (1952): 1–16, quotation at p. 15.

529   absolute intensity—Sir Lawrence Bragg and M. F. Perutz, "The Structure of Haemoglobin,"

Proceedings of the Royal Society of London A 213 (1952): 425–435; Bragg, "How Proteins Were Not Solved," p. 10.

529   early glimpses of isomorphous replacement—Interview, Perutz, 4 December 1970.

529   Bijvoet's strychnine—C. Bokhoven, J. C. Schoone, and J. M. Bijvoet, "The Fourier Synthesis of the Crystal Structure of Strychnine Sulphate Pentahydrate," *Acta Crystallographica* 4 (1951): 275–280; interview, Hodgkin.

530   first experiments with isomorphous replacement—Interviews, Ingram; interview, Perutz, 4 December 1970.

531   symmetry in proteins—Interview, Perutz, 4 December 1970.

532   first results—D. W. Green, V. M. Ingram, and M. F. Perutz, "The Structure of Haemoglobin IV. Sign Determination by the Isomorphous Replacement Method," *Proceedings of the Royal Society* of London A 225 (1954): 287–307; "Received 5 March 1954."

533   The Fourier projection—Sir Lawrence Bragg and M. F. Perutz, "The Structure of Haemoglobin VI. Fourier Projections on the 010 Plane," *Proceedings of the Royal Society* of London A 225 (1954): 315–329, matter quoted at p. 315, diagram reproduced from p. 320; interview, Bragg.

534   Hodgkin's suggestion—Interview, Perutz, 4 December 1970; interview, Hodgkin.

535   Kendrew switched to whale myoglobin—Interviews, Kendrew; interview, Perutz, 4–6 December 1970; J. C. Kendrew, "Preliminary X-ray Data for Horse and Whale Myoglobins," *Acta Crystallographica* 1 (December 1948): 336; Kendrew, "Myoglobin and the Structure of Proteins" (Nobel Lecture, 11 December 1962), *Les Prix Nobel en 1962*, p. 104.

536   heavy atoms onto myoglobin—Interviews Perutz, 4–6 December 1970 and 7 February 1976; interviews, Kendrew; interview, Phillips.

536   locating the heavy atoms—Interviews, Perutz, 4–6 December 1970; M. F. Perutz, "X-ray Analysis of Haemoglobin" (Nobel Lecture, 11 December 1962), *Les Prix Nobel en 1962*, pp. 86–90; Harker's method, David Harker, *Acta Crystallographica* 9 (1954): 1, cited in Perutz, Nobel Lecture; Bragg's graphic method, W. L. Bragg, "The Determination of the Coordinates of Heavy Atoms in Protein Crystals," *Acta Crystallographica* 11 (1958): 70–75.

536   solution of myoglobin inevitable—Interview, Kendrew, 13 and 14 January 1971; Kendrew, Nobel Lecture, pp. 106–108.

537   problems of hemoglobin yielding—Interviews, Perutz, 4–6 December 1970, 7 February 1976.

537   structure of sperm-whale myoglobin—J. C. Kendrew, G. Bodo, H. M. Dintzis, R. G. Parrish, and H. Wyckoff, "A Three-Dimensional Model of the Myoglobin Molecule Obtained by X-ray Analysis," *Nature* 181 (8 March 1958): 662–666; interview, Kendrew, 13 and 14 January 1971.

540   hemoglobin to low resolution, myoglobin to high—M. F. Perutz et al., "Structure of Haemoglobin: A Three-Dimensional Fourier Synthesis at R.5 Å. Resolution, Obtained by X-ray Analysis," *Nature* 185 (13 February 1960): 416–422; J. C. Kendrew et al., "Structure of Myoglobin: A Three-Dimensional Fourier Synthesis at 2 Å. Resolution," ibid., pp. 422–427.

540   computers and diffractometers—Kendrew, Nobel Lecture, pp. 107–108, 111–113; interviews, Kendrew.

541   hemoglobin's surprises—Perutz, Rossman, et al., "Structure of Haemoglobin," quotation at p. 421; heme groups widely separate, p. 417, quoted matter at p. 421.

542   Bragg's Nobel campaigning—Letter, Bragg to Nobel Committee for Physics, 9 December 1932; letter, Bragg to Hulthen, 9 January 1960; letter, Bragg to Westgren, "January 1960" (no day on carbon copy); letter, Perutz to Bragg, 21 June 1960; letters, Bragg to Pauling, 9 December 1959 and 4 January 1960; letter, Pauling to Bragg, 15 March 1969, enclosing carbon copies of Pauling's letters to the Nobel Committees for chemistry and physics; originals or carbon copies, as appropriate, in W. L. Bragg Papers at the Royal Institution.

10 *"I have discovered the second secret of life."*

PAGE

545   Muirhead's discovery, Monod's interest—Interview, Perutz, 4–6 December 1970; Monod later cited the conversation in a paper: see below.

545   Monod's concerns in fall 1961—Interview, Monod, 1 December 1975.

546   Jacob at Cold Spring Harbor—François Jacob and Jacques Monod, "On the Regulation of Gene Activity," *C. S. H. Symp. Q. Biol.* 26 (1961): 193–211, quoted matter at pp. 199–200.

546   Jacob on "stupidity" of RNA repressor—Interview, Jacob, 5 May 1978; François Jacob, Raquel Sussman, and Jacques Monod, "Sur la nature du répresseur assurant l'immunité des bactéries lysogènes," *C. R. Acad. Sci.* 254 (13 June 1962): 4214–4216.

547 Jacob and Monod at Cold Spring Harbor—"On the Regulation of Gene Activity," p. 200.

547 tryptophan inhibition, other examples of end-product feedback—Jacques Monod and Germaine Cohen-Bazire, "L'effet d'inhibition spécifique dans la biosynthèse de la tryptophane-desmase chez *Aerobacter aerogenes*," *C. R. Acad. Sci.* 236 (1953): 530–532; Aaron Novick and Leo Szilard, "Experiments with the Chemostat on the Rates of Amino Acid Synthesis in Bacteria," *Dynamics of Growth Processes* (Princeton: Princeton University Press, 1954), pp. 21–32; R. A. Yates and Arthur B. Pardee, "Control of Pyrimidine Biosynthesis in *Escherichia coli* by a Feed-Back Mechanism," *Journal of Biological Chemistry* 221 (1956): 757–770, "Received for publication, November 21, 1955"; H. Edwin Umbarger, "Evidence for a Negative-Feedback Mechanism in the Biosynthesis of Isoleucine," *Science* 123 (1956): 848, "21 October 1955"; E. R. Stadtman et al., "Feed-back Inhibition and Repression of Aspartokinase Activity in *Escherichia coli* and *Saccharomyces cerevisiae*," *Journal of Biological Chemistry* 236 (1961): 2033–2038.

548 Monod, "didn't know where it would lead—Interview, Monod, 1 December 1975.

548 Changeux's first note—Letter, Monod to Pardee, 7 January 1963, in which Monod also says he has just personally looked at Changeux's notebooks and found the first observation at 29 November 1959.

549 Pardee on Monod's response to his work, 1960—Interview, Pardee; further, Gerhart and Pardee had a paper at the Federation of Societies for Experimental Biology meetings in Atlantic City, March 1961; abstract, John C. Gerhart and Arthur B. Pardee, "Separation of Feedback Inhibition from Activity of Aspartate Transcarbamylase," *Federation Proceedings* 20 (1961): 224, the abstract suggesting clearly that the enzyme may have separate inhibition and active sites.

549 Changeux's tongue-tied paper—Jean-Pierre Changeux, "The Feedback Control Mechanism of Biosynthetic L-Threonine Deaminase by L-Isoleucine," *C. S. H. Symp. Q. Biol.* 26 (1061): 313–318; two distinct sites claimed at p. 317.

549 Umbarger's paper—H. Edwin Umbarger, "Feedback Control by Endproduct Inhibition," *C. S. H. Symp. Q. Biol.* 26 (1961): 301–318, quoted matter at p. 306.

549 closing summary—Jacques Monod and François Jacob, "Teleonomic Mechanisms in Cellular Metabolism, Growth, and Differentiation," *C. S. H. Symp. Q. Biol.* 26 (1961): 389–401, quoted matter at p. 391.

550 Pardee and Gerhart on mechanics—John C. Gerhart and Arthur B. Pardee, "The Enzymology of Control by Feedback Inhibition," *Journal of Biological Chemistry* 237 (March 1962): 891–896, quoted matter at p. 891.

551 Monod on the repressor—Interview, Monod, December 1970.

551 ways of detecting the enzyme—The fluorescent tag, letter, Monod to Dr. Gregorio Weber, Department of Biochemistry, University of Sheffield, England, 29 December 1961, in which Monod writes of trying to show that the inducer is "a dissociable activator of an enzyme responsible for the inactivation of the repressor substance," and then writes of "what we call the 'induction enzyme.'" Edward Lennox was at the Institut Pasteur that winter, and has confirmed to me in conversation, spring 1978, that Monod was then pursuing such a model of repressor action.

551 Monod, "let me give two examples"—Interview, Monod, 1 December 1975.

552 "second secret of life"—Interview, Agnès Ullma.

552 the first paper—Jacques Monod, Jean-Pierre Changeux, and François Jacob, "Allosteric Proteins and Cellular Control Systems," *Journal of Molecular Biology* 6 (1963): 306–309, "Received 19 December 1962"; the second paper, Jacques Monod, Jeffries Wyman, and Jean Pierre Changeux, "On the Nature of Allosteric Transitions: A Plausible Model," ibid. 12 (1965): 88–118.

552 Monod, first paper more important—Interview, Monod, December 1970.

552 quotations from first paper—Monod, Changeux, Jacob, "Allosteric Proteins": "general model," "proteins assumed to possess two," p. 307; Pardee's, Changeux's, and four other enzymes, pp. 308–310; "Effet des analogues de la L-thréonine et de la L-isoleucine sur la L-thréonine désaminase," *Journal of Molecular Biology* 4 (1962): 220–225; Perutz's hemoglobin, pp. 314, 316, 319–321, with the work of Perutz and Muirhead cited as "personal communication," though by then the paper was published; in the same sentence, Monod and his colleagues cite Haurowitz's observation, from 1938, that hemoglobin crystals change form in transition from the deoxy to the oxy state; "synthesis of messenger RNA . . . mediated by an allosteric transition of the repressor," p. 328; "the most serious objection," p. 325; "not necessarily a conformational alteration," p. 322; "a living system constantly fighting against," pp. 324–325.

554 Monod acknowledged debt to Perutz—Monod to Perutz, 3 September 1964.

555 Monod, "I would tend to say so"—Interview, Monod, 1 December 1975.

555 Crick's judgement of allostery—Interview, Crick, 1 September 1971.

555 deoxyhemoglobin model—Interview, Perutz, 4–6 December 1970.

555 the Nobel prizes—Perutz wrote an account of the entire journey to Stockholm, for distribu-

tion to his family and friends, "Stockholm 1962" (8 pages, mimeographed), and the account is based on that, with additional details from conversations and from the materials printed in *Les Prix Nobel en 1962*, passim.

557   Perutz addressed relation of hemoglobin's structure to function—M. F. Perutz, "X-ray Analysis of Haemoglobin" (Nobel Lecture, 11 December 1962), *Les Prix Nobel en 1962*, pp. 82–102, quoted matter at pp. 92, 99.

557   Pardee's exchange about Priority with Monod—Pardee to Monod, 29 December 1962; Monod to Pardee, 7 January 1963; Pardee to Monod, 10 January (misdated 1962); Monod to Pardee, 16 January 1963.

557   Jacob on Pardee's parallel work—Interview, Jacob, 5 May 1978.

557   Phillips' work with lysozyme—Interview, Phillips; interview, Bragg; interview, Kendrew, 13 and 14 January 1971.

557   Bragg's disgust—Letter, Bragg to Sir Harold Himsworth, Secretary to the Medical Research Council, 18 January 1968, reporting on a visit Bragg had made to Perutz's Medical Research Council Laboratory of Molecular Biology, in Cambridge; in the letter, Bragg observes that Kendrew has not even yet published the myoglobin structure, which is unfair to Kendrew's many colleagues in the work.

558   Phillips's pursuit of details of lysozyme—Interview, Phillips, supplemented with detail from two papers by Phillips and his colleagues, presented at the Royal Society discussion meeting: C. C. F. Blake et al., "On the Conformation of the Hen Egg-White Lysozyme Molecule," in "A Discussion on the Structure and Function of Lysozyme" (Discussion held . . . 3 February 1966), *Proceedings of the Royal Society* B 167 (18 April 1967): 365–377, and C. C. F. Blake et al., "Crystallographic Studies of the Activity of Hen Egg-White Lysozyme," ibid., pp. 378–388.

559   Bragg opened discussion—Sir Lawrence Bragg, "Introduction," *Proceedings of the Royal Society* of London B 167:349.

560   Perutz closed—M. F. Perutz, "Concluding Remarks," *Proceedings of the Royal Society* of London B 167: 448; audience reaction: comment by Perutz on draft of this passage.

560   Gilbert on repressor—Interview, Gilbert, 13 February 1971.

560   Monod's attempts to isolate repressor—Interview, Ullman, who showed me the work sheets.

561   Stent and Brenner's doubts—G. S. Stent, "The Operon on Its Third Anniversary," *Science* 144 (1964): 816; S. Brenner, "Theories of Gene Regulation," *British Medical Bulletin* 21 (1965): 244–248.

561   Ptashne, "the *real* question"—Interview, Ptashne.

561   Polyribosomes—Interview, Gilbert, 13 February 1971; interviews, Rich, 12 February and 17 March 1971; Alexander Rich, "Polyribosomes," *Scientific American*, December 1963 (my reprint without original paging), with references.

561   isolation of *lac* and *lambda* repressors—Interview, Gilbert, 13 February 1971; interview, Ptashne, and checking interview with manuscript.

561   "means that would not depend on . . . model"—Walter Gilbert and Benno Müller-Hill, "Isolation of the Lac Repressor," *Proc. N. A. S.* 56 (1966): 1891–1898; "Communicated . . . October 24, 1966," quote on p. 1891.

564   Ptashne's paper—Mark Ptashne, "Isolation of the Lambda Phage Repressor," *Proc. N, A. S.* 57 (1967): 306–313; "Communicated . . . December 27, 1966."

565   proofs that the repressors bound DNA—M. Ptashne, "Specific Binding of the *Lambda* Phage Repressor to *Lambda* DNA," *Nature* 214 (1967): 232–234; Walter Gilbert and Benno Müller-Hill, "The *lac* Operator is DNA," *Proc. N.A.S.* 58 (1967): 2415–2421.

566   Monod's book—Jacques Monod, *Le Hasard et la nécessité* (Paris: Editions du Seuil, 1970).

566   Monod's lecture to Collège de France—Jacques Monod, "De la biologie à l'ethique" (Leçon inaugurale, La Collège de France), reported and reprinted, *Le Monde*, 30 November 1967, pp. 10–11; matter quoted, p. 11, *trad. auct.*

567   Jacob's book—François Jacob, *La Logique du vivant: une histoire de l'hérédité* (Paris: Editions Gallimard, 1970).

567   Monod's public activities—Interview, Ullman; interview, Lwoff, 15 November 1976; interview, Claudine Escoffier-Lambiotte; interview, Olivier Monod; interviews, Jacob, 28 and 29 May 1977 and 5 May 1978.

568   oxyhaemoglobin of horse at high resolution—Bragg to Sir Harold Himsworth, 18 January 1968.

568   Perutz's Croonian Lecture—M. F. Perutz, "The Haemoglobin Molecule" (The Croonian Lecture, 1968, delivered 30 May), *Proceedings of the Royal Society* of London B 173 (1969): 113–140, quotations at pp. 117, 131, 134—and from my own notes, for Perutz's revision between delivery and publication removed some expressive images, including the final quotation.

571   unravelling the Bohr effect—Interviews, Perutz, 4–6 December 1970, 3 April 1970; com-

ments on draft of this passage; for summary of the complete work and citations of individual papers, John V. Kilmartin, "Interaction of Haemoglobin with Protons, CO_2 and 2,3, Diphosphoglycerate," *British Medical Bulletin* 32 (September 1976): 209–212.

573 paper on insect hemoglobin—Robert Huber, "The Atomic Structure of an Insect Haemoglobin" (Invited paper, read at Bragg Symposium 1970, on 3 April); interview, Perutz, 3 April 1970.

573 the trigger found—Interviews, Perutz, August and September 1970, 4–6 December 1970, 7 February 1976; interview, Kendrew, 14 January 1971.

577 modifications of mechanism—Perutz, interviews, 11 and 19 November 1977; M. F Perutz, "Introduction," and "Structure and Mechanism of Haemoglobin," *British Medical Bulletin* 32 (September 1976): 193–194 and 195–208; Joyce M. Baldwin, "A Model of Cooperative Oxygen Binding to Haemoglobin," ibid., pp. 213–218; Max F. Perutz, "Hemoglobin Structure and Respiratory Transport," *Scientific American,* December 1978, pp. 92–105, 198.

Conclusion *"Always the same impasse"*

PAGE

579 creation of the standard view—Accounts that emphasize the importance of physicists in creating molecular biology include Donald Fleming's excellent brisk essay, "Emigré Physicists and the Biological Revolution," in Donald Fleming and Bernard Bailyn, *The Intellectual Migration: Europe and America, 1930–1960* (Cambridge, Mass.: The Belknap Press of Harvard University Press, 1969), pp. 152–189; Gunther S. Stent, "Introduction: Waiting for the Paradox," in *Phage and . . . Molecular Biology,* pp. 3–8—a volume, one recalls, dedicated to Max Delbrück on his sixtieth birthday; several of the essays in Alexander Rich and Norman Davidson, eds., *Structural Chemistry and Molecular Biology;* Eugene L. Hess, "Origins of Molecular Biology," *Science* 168 (8 May 1970): 664–669; Gunther S. Stent, "That Was the Molecular Biology That Was," *Science* 160 (26 April 1968): 390–395; Pierre Thuillier, "Comment est née la biologie moléculaire," *La Recherche,* May 1972, pp. 439–448, a clever synthesis of English-language accounts that illustrates the dominance of those accounts and provides, in fact, a good summary of the standard view. An account that accepts the standard view in order to correct and extend it is Elof Axel Carson, "An Unacknowledged Founding of Molecular Biology: H. J. Muller's Contributions to Gene Theory, 1910–1936." How crystallography, specifically, and genetics first collided in Europe is told by a participant, C. H. Waddington, in "Some European Contributions to the Prehistory of Molecular Biology," *Nature* 221 (25 January 1969): 318–321. The phage group dominates the accounts by Stent, Fleming, and the derivative Thuillier. The dates of all these suggest the obvious, that they were stimulated at least partly by the appearance of James Watson's *The Double Helix* (New York: Atheneum, 1968—and serialized in *The Atlantic Monthly,* January and February 1968).

580 the ladder of sciences and their "anti-disciplines"—A telling illustration of the wide acceptance of the standard view of the origins of molecular biology is provided by a recent passage in the controversy over the claims of a different discipline, sociobiology. Chief proponent of sociobiology has been Edward O. Wilson, at Harvard, who has outlined a program leading to what he calls "the new synthesis" of population genetics and behavioral science. Wilson has used the standard view of the history of molecular biology as a familiar example to explain the process of synthesis between disciplines and "anti-disciplines" he had in mind for sociobiology. In an introductory essay to an anthology of critiques of sociobiology, the Harvard historian of science Gerald Holton accepts that use of the standard view of molecular biology, saying of Wilson, in part: "Going far beyond the so-called Modern Synthesis of Mendelian genetics and biochemistry, he envisages a 'juncture' of neurobiology and sociobiology with social science . . . In the evolution of molecular biology, 'progress over a large part of biology was fueled by a competition among the various attitudes and themata derived from biology and chemistry—the discipline and its anti-discipline.' Wilson feels that a similar process will eventually occur for sociobiology." Gerald Holton, "The New Synthesis?," *Transaction: Social Science and Modern Society* 15 (September-October 1978): 18, and Holton's own quotations are from Edward O. Wilson, "Biology and the Social Sciences," *Daedalus* 106 (Fall 1977): 127–140.

580 Kendrew's version of the standard view—John C. Kendrew, "Information and Conformation in Biology," in *Structural Chemistry and Molecular Biology,* ed. A. Rich and N. Davidson (San Francisco: Freeman, 1968), pp. 187–197, "Based on the Herbert Spencer Lecture delivered at Oxford University, 1965."

581   central element of the standard view—Especially pronounced in *The Double Helix*, and in Gunther S. Stent, "Prematurity and Uniqueness in Scientific Discovery," *Scientific American*, December 1972, pp. 84–93; M. R. Pollock, "The Discovery of DNA: An Ironic Tale of Chance, Prejudice, and Insight," *Journal of General Microbiology* 63 (1970): 1–20; H. V. Wyatt, "When Does Information Become Knowledge?," *Nature* 235 (14 January 1972): 86–89. Robert Olby's history, *The Path to the Double Helix* (London: Macmillan, 1974) is constructed around what he sees as a "transformation of paradigms" between the nineteen-thirties and the late nineteen-forties and early nineteen-fifties, from the protein theory of the gene to the nucleoprotein theory to the nucleic-acid theory of the gene.

582   Svedberg's multiples—See, for a late formulation of his ideas of specificity and the multiples that he thought made up proteins, T. Svedberg, "A Discussion on the Protein Molecule: Opening Address," *Proceedings of the Royal Society* of London 170A (1939): 40–56.

582   Bergmann and Niemann—Max Bergmann and Carl Niemann, "Newer Biological Aspects of Protein Chemistry," *Science* 86 (27 August 1937): 187–190.

583   Bergmann—Niemann hardly regarded as conclusive—For an illustration of the influence of the hypothesis upon a singularly thoughtful young protein biochemist, see Paul C. Zamecnik, "An Historical Account of Protein Synthesis, with Current Overtones—A Personalized View," *C. S. H. Symp. Q. Biol.* 34 (1969): 1–2.

583   explanations of biochemical behavior of proteins, nineteen-thirties—Rudolf Schoenheimer, *The Dynamic State of Body Constituents* (Cambridge, Mass.: Harvard University Press, 1942); Max Bergmann and Heinz Fraenkel-Conrat, "The Role of Specificity in the Enzymatic Synthesis of Proteins," *Journal of Biological Chemistry* 119 (1937): 707–720; Max Bergmann. "A Classification of Proteolytic Enzymes," *Advances in Enzymology* 2 (1942): 49—also discussed by Zamecnik, "An Historical Account"; Linus Pauling, "A Theory of the Structure and Process of Formation of Antibodies," *Journal of the American Chemical Society* 62 (1940): 2643–2657; Linus Pauling and Dan H. Campbell, "The Manufacture of Antibodies in Vitro," *Journal of Experimental Medicine* 76 (1942): 211–220.

583   the five disciplines that eventually contributed—The passage in large part summarizes early work discussed in the various chapters of the book: citations to individual works will be found there.

586   Crick, " . . . same *general motives* . . ."—Interview, Crick, 1 September 1971.

587   Delbrück on Beckett—M. Delbrück, "A Physicist's Renewed Look at Biology: Twenty Years Later" (Nobel Lecture, 10 December 1969), text reprinted in *Science* 168 (12 June 1970): 1312–1315. For Delbrück's views on "new laws of physics and chemistry," see Max Delbrück, "A Physicist Looks at Biology," *Transactions of the Connecticut Academy of Arts and Sciences* 38 (December 1949): 173–190, reprinted in *Phage and . . . Molecular Biology*, pp. 9–22; Stent, "Waiting for the Paradox," ibid., pp. 3–8; Fleming, "Emigré Physicists . . . ."

587   the conversation with Delbrück—Interview, Delbrück, 9 July 1972: he had told me something of his opinion of Beckett in the previous interview, 25 February 1971. See also M. Delbrück, "*Homo Scientificus* According to Beckett" (Talk in Chemistry and Society Lecture Series, California Institute of Technology, 24 February 1971, transcript from tape, ed. Delbrück; photocopied, cover and 23 pp.); therein, Delbrück quoted an extended passage from Beckett's novel *Molloy*—for those familiar with the novel, it is the "sucking stones" passage—which, he said, described the scientist's intuition and the obsessive quality of his work.

588   conversation with Beckett—Interview, Beckett.

589   conversation with Monod—Interview, 4 December 1975.

589   Monod's death—Interview, Philippe Monod.

Epilogue *"We can put duck and orange DNA together—with a probability of* one.*"*

PAGE

592   done these things in outline—Here and in what follows, for sources refer to Chapter 4, pp. 177–197.

593   Perutz on muscle, Huxley—Interview, 8 May 1968, MRC Laboratory, Cambridge.

593   Watson on membranes—Interview, 16 March 1971, Biological Laboratories, Harvard University.

594   Benzer's behavioral mutants—Interview, 25 February 1971, California Institute of Technology.

594   "You get there how you can"—Roger Kornberg, in conversation, Stanford, spring of 1981.

595   Temin at Cancer Congress, Houston, 1970—Program, Tenth International Cancer Congress, for Thursday 28 May 1970, where Temin is scheduled to give a paper at 9:15 a.m. until

9:30 with the title "Role of DNA in the Replication of RNA Viruses." Howard M. Temin, "The DNA Provirus of RNA Sarcoma Viruses," in R. Lee Clark et al., eds., *Oncology 1970, Being the Proceedings of the Tenth International Cancer Congress* (Chicago: Year Book Medical Publishers, 1971), vol. I, pp. 776–780.

596   papers on reverse transcription—David Baltimore, "RNA-dependent DNA Polymerase in Virions of RNA Tumour Viruses," and Howard M. Temin and Satoshi Mizutani, "RNA-dependent DNA Polymerase in Virions of Rous Sarcoma Virus," *Nature* 226 (27 June 1970): 209–213. David Baltimore, "RNA-Dependent Synthesis of DNA by Virions of Mouse Leukemia Virus," *C. S. H. Symp. Q. Biol.* 35 (1970): 843–846.

596   remembered the excitement—Robert Pollack, in conversation, January 1996.

596   *Nature* on Central Dogma—"Central Dogma Reversed," *Nature* 226 (27 June 1970): 1198–1199.

596   Berg on origins of recombinant DNA—Interview, Paul Berg, 6 September 1974, Cambridge.

598   Brenner on recombinant DNA—Interview, 13 March 1975, King's College, Cambridge.

598   Asilomar meeting—see Horace Judson, "Fearful of Science," *Harper's*, June 1975.

598   Asilomar and response—James D. Watson and John Tooze, *The DNA Story: A Documentary History of Gene Cloning* (San Francisco: W. H. Freeman, 1981). Clifford Grobstein, *A Double Image of the Double Helix: The Recombinant-DNA Debate* (San Francisco: W. H. Freeman, 1979).

599   Sanger sequences DNA—Frederick Sanger, "The Croonian Lecture, 1975: Nucleotide Sequences in DNA," *Proceedings of the Royal Society* of London, B, 191 (1975): 317–333.

599   "a river hitting sand"—Interview, Matthew Meselson, spring 1995, Harvard University.

600   monoclonal antibodies—Interviews, Cesar Milstein, 9 and 10 August 1984, MRC Laboratory, Cambridge; interview, Edwin Lennox, 10 August 1984, MRC Laboratory, Cambridge. G. Köhler and C. Milstein, "Continuous cultures of fused cells secreting antibody of predefined specificity," *Nature* 256 (7 August 1975): 495–497.

601   Sanger simplified—F. Sanger, S. Nicklen, and A. R. Coulson, "DNA Sequencing with Chain Terminating Inhibitors," *Proc. N. A. S.* 74 (1977): 5463–5467.

601   polymerase chain reaction—K. Mullis, et al., "Specific Enzymatic Amplification of DNA In Vitro: The Polymerase Chain Reaction," *C. S. H. Symp. Q. Biol.* 51 (1986): 263–273.

601   sale of patent—Ann T. Thayer, "New Biotech Company Formed by Chiron, Cetus, Has Broader Focus," *Chemical and Engineering News*, 13 January 1992, pp. 9–10. David Dickson, "Patent on PCR Enzymes May Re-ignite Old Controversy," *Nature* 372 (17 November 1994): 212.

602   Comfort on McClintock—Nathaniel C. Comfort, "Two Genes, No Enzyme: A Second Look at Barbara McClintock and the 1951 Cold Spring Harbor Symposium," *Genetics* 140 (August 1995): 1161–1166. Nathaniel Comfort, "Jumping Genes Revisited: Data versus Interpretation in the Reception of Barbara McClintock's Transposable Elements," paper presented to the Joint Atlantic Seminar on the History of Biology, 13 April 1996, Harvard University.

603   RNA enzymes—Sidney Altman, "Enzymatic Cleavage of RNA by RNA," in *Les Prix Nobel en 1989*, pp. 137–160. Thomas R. Cech, "Self-splicing and Enzymatic Activity of an Intervening Sequence RNA from *Tetrahymena*," in *Les Prix Nobel en 1989*, pp. 165–188. These cite all relevant papers.

603   early development in *Drosophila*—Interview, Nancy Hopkins, 15 April 1996, her laboratory at MIT. Christiane Nüsslein-Volhard, "The Identification of Genes Controlling Development in Flies and Fishes," in *Les Prix Nobel en 1995*. Peter A. Lawrence, *The Making of a Fly: The Genetics of Animal Design* (Oxford: Blackwell Scientific Publications, 1992).

604   history of the human genome project—for details and citations, see Robert Cook-Deegan, *The Gene Wars: Science, Politics, and the Human Genome* (New York: W. W. Norton, 1994); Daniel J. Kevles and Leroy Hood, eds., *The Code of Codes* (Cambridge, Massachusetts: Harvard University Press, 1992).

606   Kitcher—Quotation is from a conversation in April 1996.

607   human variation and essentialism—This passage is profoundly influenced by conversations in March and April of 1996 with Robert Pollack.

608   Morgan against the anti-Mendelians—T. H. Morgan, "The Theory of the Gene," *The American Naturalist* 51 (September 1917): 513–544, quotations at 513, pp. 517–520.

610   report on gene therapy—*Recombinant DNA Research Volume 20: Documents Relating to "NIH Guidelines for Research Involving Recombinant DNA Molecules" August 1994—December 1994*, NIH Publication No. 95-3993, U.S. Department of Health and Human Services, Public Health Service, National Institutes of Health Report of Recombinant Advisory Committee, December 1995.

612   Watson at Cold Spring Harbor—James D. Watson, "Growing up in the Phage Group," in *Phage and . . . Molecular Biology*, pp. 239–245.

614   "a story of the eighties"—Watson, personal communication.

614   exponential growth of science—Derek J. de Solla Price, *Little Science, Big Science* (New York: Columbia University Press, 1963).

614   transition to the steady state—See Susan E. Cozzens, et al., eds., *The Research System in Transition* (Dordrecht: Kluwer Academic Publishers, 1990), passim, but especially John Ziman, "What is Happening To Science?" pp. 23–34. John Ziman, *Prometheus Bound: Science in a Dynamic Steady State* (Cambridge: Cambridge University Press, 1994). Horace Freeland Judson, "The World We Have Lost," in Donald A. Chambers, ed., *DNA: The Double Helix*, Annals of the New York Academy of Sciences vol. 758 (1995): 427–440.

615   Medawar's advice—Peter Medawar, *Advice to a Young Scientist* (London: Oxford University Press, 1979).

616   proximate and ultimate questions—Ernst Mayr, *The Growth of Biological Thought: Diversity, Evolution, and Inheritance* (Cambridge, Massachusetts: Harvard University Press, 1982), chapters 1 and 2.

Afterword I *In Defense of Rosalind Franklin: The Myth of the Wronged Heroine*

PAGE

621   Lwoff's review of *The Double Helix*—André Lwoff, "Truth, truth, what is truth (about how the structure of DNA was discovered)?", *Scientific American*, 219, July 1968, pp. 133–138. All the principal reviews of the book except for Irwin Chargaff's are reprinted in James Watson, *The Double Helix*, ed. Gunther S. Stent, Norton Classics edition (New York: W. W. Norton, 1980).

621   Janeway—Elizabeth Janeway, *Man's World, Women's Place* (New York: Morrow, 1971).

621   Sayre's biography—Anne Sayre, *Rosalind Franklin and DNA* (New York: W. W. Norton, 1975), p. 99.

621   Gornick on Franklin—Vivian Gornick, *Women in Science* (New York: Simon & Schuster, 1983).

622   Franklin's family, education, etc.—See notes to Chapter 2, pp. 649–656.

622   Luzatti on Franklin in Paris—Interviews, 2 and 3 December 1975; conversation, while exploring Franklin's Paris, 22 December 1982.

626   Dame Honor Fell—Interview, 28 January 1977, Addenbrooks Hospital, Immunology Department, Cambridge.

626   Mary Fraser—Letter, Mary Fraser, February 1977.

626   Sylvia Jackson—Interview, Sylvia Fitton Jackson, 30 June 1976, Darwin College, Cambridge.

626   other women at King's College London—Angela Martin Brown, 2 September 1976, King's College London; Pauline Cowan Harrison, telephone interview, 22 January 1977; Marjorie M'Ewen, letter, 15 September 1976; Margaret Pratt North, interview, 21 January 1977, and letter, 2 February 1977.

Afterword II *What did Erwin Chargaff Contribute?*

PAGE

630   base ratios—Chargaff, Erwin, "Chemical Specificity of Nucleic Acids and Mechanism of their Enzymatic Degradation," *Experientia*, VI (1950) pp. 201–209, passage quoted at p. 206; reprinted in Chargaff, Erwin, *Essays on Nucleic Acids* (Amsterdam: Elsevier, 1964), pp. 1–24.

634   1950–'51 reviews—Chargaff, Erwin, *Heraclitean Fire*, pp. 101–103; earlier phrases quoted are at pp. 8, 9, 13, and 19.

634   pygmies—Chargaff, Erwin, "Grammar," p. 641.

634   not far from 1—Chargaff, Erwin, "Structure and Function of Nucleic Acids as Cell Constituents," *Federation Proceedings*, X (1951), p. 655; reprinted in Chargaff, E., *Essays*, pp. 25–38, quotation at p. 27.

634   his standard—For example, Chargaff, Erwin, Zamenhof, Steven, Brawerman, George and Kerin, Leonard, "Bacterial Desoxypentose Nucleic Acids of Unusual Composition," *Journal of the American Chemical Society*, LXXII, 8 (15 August, 1950), p 3,825; Chargaff, Erwin, Lipshitz, Rakoma, and Green, Charlotte, "Composition of the Desoxypentose Nucleic Acids of Four Genera of Sea-Urchin," *Journal of Biological Chemistry*, CXCV, 1 (March 1952), p. 156; Chargaff, Erwin and Lipshitz, Rakoma, "Composition of Mammalian Desoxyribonucleic Acids," *Journal of the American Chemical Society*, LXXV, 15 (5 August, 1953), p. 3,659, "received March 12, 1953," or within a few days of the discovery by Watson and Crick.

635   time has not yet come—Chargaff, Erwin, Lipshitz, Rakoma, Green, Charlotte, and Hodes, M.E., "The Composition of the Desoxyribonucleic Acid of Salmon Sperm," *Journal of Biological Chemistry*, CXCII, 1 (September 1951), pp. 223–230, quotation at p. 229, with notes to his references.

635   simplified assumption—Chargaff, Erwin, "Some Recent Studies on the Composition and Structure of Nucleic Acids," *Journal of Cellular and Comparative Physiology*, XXXVIII, Supplement I (1951), pp. 41–59, quotation at pp. 46–47.

635   stereochemical interpretation—Abir-Am, Pnina, "The Erwin Chargaff Papers: Archival Source Material Report No. 7" (Philadelphia: Survey of Sources for the History of Biochemistry and Molecular Biology, American Philosophical Society Library, 1977), a machine-copied report, irregularly paged, but the reference is to p. 7 of the analytical introduction.

636   purine nucleotides predominate—Tamm, Christoff and Chargaff, Erwin, "Observations on the Distribution Density of Individual Nucleotides within a Desoxyribonucleic Acid Chain" (abstract), in *Résumés des Communications, II*ᵉ *Congrès International de Biochimie*, (Paris: Masson, 1951), p. 205. Chargaff does not list the paper in his lifetime bibliography in *Heraclitean Fire*, pp. 229–252. Tamm did not address the question whether the DNA molecule was made of more than one chain. The later paper is Tamm, Christoff, Shapiro, Hermann S., Lipshitz, Rakoma and Chargaff, Erwin, "Distribution Density of Nucleotides within a Desoxyribonucleic Acid Chain," *Journal of Biological Chemistry*, CCIII, 2 (August 1953), pp. 673–688, the quotation at p. 685; the base ratios are restated, in Chargaff's usual language, at p. 683.

636   publish coordinates—Crick, F. H. C., "DNA: Test of Structure?", *Science*, CLXVII, 3,926 (27 March, 1970), p. 1,694.

637   model that made the rules—Donohue, Jerry, "Fragments of Chargaff," *Nature*, CCLXXVI, 5,684 (9 November, 1978), pp. 133–135, p. 135.

Index